PROCEEDINGS OF THE
5TH INTERNATIONAL CONFERENCE ON

ELECTRO-RHEOLOGICAL FLUIDS, MAGNETO-RHEOLOGICAL SUSPENSIONS AND ASSOCIATED TECHNOLOGY

PROCEEDINGS OF THE
5TH INTERNATIONAL CONFERENCE ON

ELECTRO-RHEOLOGICAL FLUIDS, MAGNETO-RHEOLOGICAL SUSPENSIONS AND ASSOCIATED TECHNOLOGY

Sheffield, UK 10–14 July 1995

Editor

W. A. Bullough

Thermodynamics and Fluid Mechanics Groups
Department of Mechanical and Process Engineering
The University of Sheffield, SMMART, Sheffield, U.K.

World Scientific
Singapore • New Jersey • London • Hong Kong

Published by

World Scientific Publishing Co Pte Ltd
P O Box 128, Farrer Road, Singapore 912805
USA office: Suite 1B, 1060 Main Street, River Edge, NJ 07661
UK office: 57 Shelton Street, Covent Garden, London WC2H 9HE

ELECTRO-RHEOLOGICAL FLUIDS, MAGNETO-RHEOLOGICAL SYSPENSIONS AND ASSOCIATED TECHNOLOGY
Proceedings of the 5th International Conference

ISBN 981-02-2676-4

Printed in Singapore.

PREFACE

The twin scientific fields of electrorheological fluids and magnetorheological suspensions and their associated technologies are intensely multi-disciplinary. Publication of research papers generally takes place in literally dozens of different speciality journals. It is therefore important that a unifying conference is held periodically and that the proceedings from the conferences are published in a regular fashion. The present volume is the third published in sequence by World Scientific Publishing Company on behalf of international colleagues in the field.

The theme of the 5th International Conference was SYNERGY - generated by the interaction of different disciplines relevant to ERF and MRS investigations. To stimulate this theme, all lecture sessions included a mixture of papers - one session may have included applications, methodology, particle dynamics, structure characteristics and whatever is germane to the objective of furthering the standing of the subject. 'Lead-in' lectures were given by experts who had not recently been able to explain their work to colleagues in their own discipline. They were also charged with justifying the relevance of their area of work to the ERF/MRS field as a whole. For the same reason (to encourage synergy) the papers are distributed in the proceedings as per the lecture schedule.

All is not known about the mechanisms of ERF and MRS, particularly in respect of the unsteady and high shear rate zones of operation - at least. Hence the editor encouraged referees to be tolerant towards papers containing some controversy.

In their entirety, the proceedings contain probably the widest yet most up to date and best guide to what is going on in the ERF/MRS field at the present time. They provide a firm base from which the third wave of (flexibly operated) machines will be launched, an invaluable handbook for researchers and a guide to potential appliers of the technology.

Conference hosts SMMART made a conscious effort to make it easy for delegates to attend the meeting. This resulted in the presence in Sheffield in July 1995 of the great majority of top researchers in this field and a fine collection of papers. Without the 'up-front' generosity of the donors this result would not have been achieved. The organiser/editor wishes to place on record his gratitude to the sponsors and to Mrs Norma Parkes who undertook many of the organisational tasks in a thoroughly professional manner. Thanks are also due to Professor H. Block who acted as Conference Chairman and to Professor F. E. Filisko as Rapporteur - his summing up is given at the end of the papers. Session chairmen and referees are to be congratulated for their diligence and promptness without which these proceedings would not have appeared so quickly after the conference.

W. A. Bullough, Chairman
Smart Machines, Materials and Related Technologies Unit
Sheffield, October, 1995

CONTENTS

ER V: The Present and the Future

Rapporteur's Comments - Frank E Filisko

The renaissance of ER started in the late 1980's with the discovery of "intrinsic" ER fluids which worked without the need for any other "extrinsic" component such as adsorbed water or other surfactant. However this was just the beginning, the fluids were in their embryonic stage. The immediate activity however was not in the materials but in the applications. Inventors however quickly discovered that the fluids were almost impossible to obtain because there were no fluids being developed which were readily available to the public. More of concern however was that virtually all of the inventors concluded that although the fluids worked, they did not work to the degree that was expected. They were "too weak" to be practical. This evaluation clearly evolved from preconceived ideas of what the materials were expected to do, and the disappointment led to overlooking properties of the fluids which are truly unique. This oversight still exists mostly because inventors do not know as yet how to design devices making use of their unique properties. This was followed by a period of waiting for the "super fluid" which was substantially lengthened by numerous fallacious reports of super strong fluids. All of these were false but it was becoming clearer that a trial-and-error approach to improving ER fluids was not the way to go. The phenomenon leading to ER was more complex than imagined. This misconception, that the cause of ER is intuitively obvious, still pervades many in the field. These reports however slowed down the evolution of ER technology because it prevented people from accepting the current properties of the materials and beginning to design based upon these, as opposed to properties which were unrealistic.

For the first time at this meeting I sensed that people are accepting the levels of 4-7 kPa yield stress and actually turn away when someone reports something much stronger. Although I have no doubt that ER fluids will get tremendously stronger, it will not be easy. But it is important that the current levels are accepted and ER device designers work with these as realistic parameters.

A second factor which was presented at this meeting and is extremely important to the development of ER are the announcements of at least two commercial ER fluids by Bridgestone in Japan and Bayer in Germany. As I understand, both companies are going into a marketing stage with their fluids on a trial basis. Making these fluids readily available, in quantity, and with consistent property levels is a major step in the evolution of ER.

A third factor which was more evident at this meeting was a greater acceptance and appreciation of the unique properties and potential of ER materials instead of judging them based upon preconceived ideas of what they should do. Such advantages are very low power, high reliability and simplicity of devices, lower cost devices, much smaller packaging, and high speed operation.

It was tremendously refreshing at this meeting to notice that the number of papers talking about chain models was dramatically reduced. The obsession with chains seems to be over and hopefully will allow researchers to progress beyond this "intuitively obvious" consequence of putting dielectric particles in an electric field. That this occurs is a fact, that it is essential to ER has not been shown. What has been shown is that virtually all appropriate suspensions form chains, but of these the majority do not show ER activity. What is important is that researchers are looking beyond chaining for mechanisms associated with ER activity.

One of the more important axioms resulting from the discovery of "intrinsic" ER materials, i.e. those that do not need adsorbed water to work, is that ER activity could be associated with the basic chemistry and physical properties of materials and didn't need an "extrinsic" adsorbed surfactant. Many companies have recognized this and have approached the development of ER fluids from fundamental chemical principles. The fluids of Bayer, Bridgestone and Nippon-Shokubai are all very different but have similar property levels. Other companies and researchers in ER are also taking fundamental approaches to the development of very different ER fluids. I suspect, as there are many different plastics with very different properties, that there will be as well be many different ER fluid formulations.

One of the surest signs of the ultimate frustration and loss of interest in a field is the dwindling attendance at meetings and the lack of influx of new researchers. Not only has this meeting been very well attended but it is very obvious that many new and young researchers are getting involved in ER and the number is increasing. This is a sure sign of continuing and expanding viability of the field and I'm sure if this continues, ER technology will be a reality in the early 21st century.

We have also seen at this meeting a growing interest in properties of magnetorheological (MR) fluids. These materials have been around longer than ER and have been around continuously, commercialised by Ferrofluidics Corp. for decades and studied intensely by the Russians. Although they are "stronger" than ER at current values by about 7X, the simplicity of ER is lacking. Despite MR fluids being around and commercially available for so long, they have progressed only into very limited and specialised markets. Will their use now expand? Historically its not likely, at least not in large quantities. I suspect that those who begin to work with MR fluids because only of the added strength will discover the limitations which are well known and have prevented the substantial use of these materials. But maybe ER and MR together will result in a synergism of properties and significantly better ERMR fluids. Its something which will have to be evaluated at the next meeting in Japan. However I suspect that within a few years, ER fluids alone will reach strengths competitive or better than MR.

What about the immediate future of electrorheology? I believe that the situation is appropriate for the field to "take off". What it needs however is a catalyst, a Sputnik. The first commercial or more probably operational ER device could do that. I see such a device coming out in the near future and I hope it is catalytic. There will also be an increase in the number of commercial ER fluids. Many companies have been working in the area and many feel that they have fluids which are good enough to be used in devices. I hope so. But the importance of this is that developed fluids will be available to more and more inventors and increase the chances of the "Sputnik". Finally, as mentioned, I see ER fluids getting better and better, in many ways other than just strengths. The property levels will go up, not only as a result of fluids getting stronger, but as a result of better design of devices, i.e. devices designed to take advantage of the unique characteristics of ER fluids. But the process will be slow. It will involve better understanding of the properties, better understanding of the molecular mechanisms involved, and the recognition of applications especially appropriate to ER fluids, i.e. something that couldn't be done any other way.

MAGNETORHEOLOGICAL FINISHING

W. I. KORDONSKI
Byelocorp Scientific, Inc., 70 Pine Str., New York, NY 10270, USA
and
S. D. JACOBS
University of Rochester, 250 East River Road, Rochester, NY 12623, USA

Abstract

The technology of finishing for optics, ceramics, and semiconductors is one of the most promising uses of the magnetorheological effect. It perfectly coupled with computer control, allowing in quantity production the unique accuracy and quality of a polished surface to be achieved. The polishing process may appear as follows. A part rotating on the spindle is brought into contact with an magnetorheological polishing (MRP) fluid which is set in motion by the moving wall. In the region where the part and the MRP fluid are brought into contact, the applied magnetic field creates the conditions necessary for the material removal from the part surface. The material removal takes place in a certain region contacting the surface of the part which can be called the polishing spot or zone. The polishing process comes to the program-simulated movement of the polishing spot over the part surface. The mechanism of the material removal in the contact zone is considered as a process governed by the particularities of the Bingham flow in the contact zone. The problem like the hydrodynamic theory of lubrication is treated for plastic film. As this takes place the shear stresses distribution in the film is obtained from the experimental measurements of the pressure distribution in the contact spot. Reasonable correlation between calculated and experimental magnitudes of the material removal rate for glass polishing lends support to the validity of the approach.

1. Introduction

Traditional methods of finishing optics, ceramics and wafers involve one or more of the following techniques. The most commonly used technique is abrasive polishing [1]. In this technique, the part is lapped in a slurry containing small abrasive particles (~1-3mm in dia), which remove material from the surface to be polished. Limitations of this method is that the ultimate size of the abrasive particles dictate the quality of the final finish, and a high local pressure (on the abrasive particle size scale) can create subsurface damage that will remain after the finishing. This technique also tends to generate subsurface damage by generating mechanical stresses as a result of physical contact between the tool and the surface to be polished. Subsurface damage is significantly removed by the chemical-mechanical polishing [2].

To create "smart" loading and to overcome some problems with these traditional finishing methods, magnetic field assisted finishing processes were developed. In the method called "Magnetic Fluid Grinding", a driven part is submerged in a colloidal magnetic fluid that contains nonmagnetic abrasive particles. The operation is performed above permanent magnets. The magnets and fluid create buoyancy forces that levitate abrasive particles in the fluid and provide the load for the process [3]. The magnetic buoyant force of abrasive particles is too low to give efficient removal rates and shape accuracy is poorly controlled. Special elements called "floats" are introduced to overcome these problems [4,5]. This method is conceptually similar to traditional lapping based on the physical contact between the tool and the surface to be treated. Another type of the magnetic field assisted method, "Magnetic Abrasive Finishing", uses a flexible magnetic abrasive brushes for finishing. The grinding characteristics can be altered by varying the magnetic field [6].

In order to obtain both high precision in surface shape and damage free surfaces, a *non-contact* polishing methods are most preferable. One noncontact method is called "Hydroplane Polishing". It is based on hydrodynamic principles or more specifically the phenomenon of hydrodynamic lubrication [7]. In the process the following hydrodynamic phenomenon and abrasive particles action are used. When a circular disk plate, on which plural inclining surfaces are formed in circumferential direction, rotates in fluid, hydrodynamic pressure generates through the fluid wedge and the part floats from the disk plate surface. The part is polished by abrasive particles which pass through the floating gap with the fluid, little by little. A hydrodynamic principle is used to provide noncontact polishing in the "Elastic Emission Machining" method [8]. In this technique the part is polished into an arbitrary form by scanning an elastic polyurethane ball over the part surface. This ball is rotated in polishing fluid and, due to hydrodynamic forces, floats above the part surface. The floating gap which is created by an elasto-hydrodynamic lubrication state is much larger than the diameter of the abrasive particles but is still very small.

In this paper, we introduce a new magnetic field assisted noncontact hydrodynamic polishing method [9]. The principle of this technique lies in the fact that conditions for successful noncontact polishing are created and controlled by utilizing the state of *flow of a magnetorheological polishing fluid* in a magnetic field

2. Experimental

Let us consider the process of magnetorheological finishing (MRF). A convex part is installed at some distance from a moving flat wall. An electromagnet, placed below the transparent to the magnetic field moving wall, generates a non uniform magnetic field in the neighborhood of the part. The magnetic field gradient is normal to the wall and magnetic field force lines penetrate the gap between the wall and the part surface in the direction that is perpendicular to the direction of the wall motion. The MRP fluid is delivered to the wall in the vicinity of the electromagnet pole pieces. It is pressed by the magnetic field gradient against the wall, acquires the wall velocity and becomes a plastic

Bingham medium [10] before it enters the gap. Thereafter a shear flow of plastic MRP fluid occurs through the gap resulting in material removal from the surface. A typical interferogram of the polishing spot at the surface of the glass spherical part, a profile of material removal along the central axis of the flow, a removal rate vs the moving wall velocity and the magnetic field intensity are shown in Figure 1a,b,c,d respectively. The removal rate increases with the velocity of the moving wall (see Fig.1c) but indicates a tendency to the saturation with the electromagnet current as it is shown in Fig.1d. Data have been obtained on a Zygo Mark IV Laser Interferometer with processing the interferogram of the part surface after test polishing of the stationary and nonrotating part over a given time interval. The polishing process employs a computer generated programm to move the part with through the polishing spot. For spherical surfaces the part is rotated about a pivot point that is located at the part center of curvature.

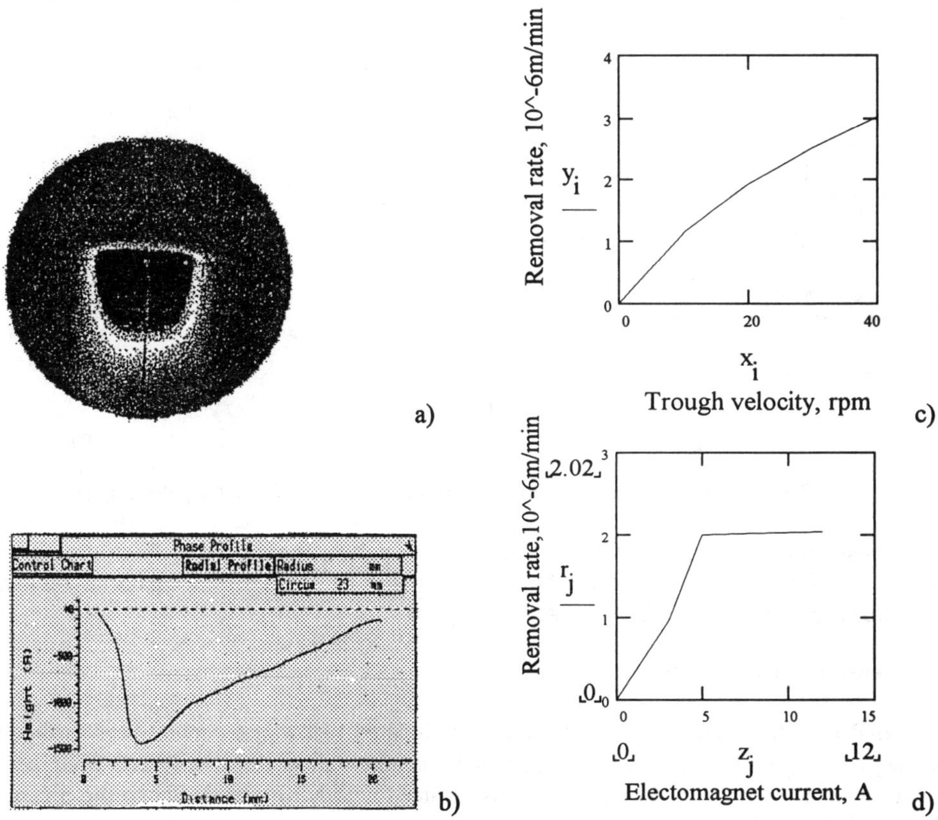

Figure 1. Material removal in the polishing spot

At the same time the part is rotated on its spindle to provide for uniform circumferential material removal from the part surface.

Another experimental parameter that is necessary for understanding the MR finishing process is the pressure distribution in the contact zone. Direct pressure measurements in the contact zone were conducted with a piezoelectric pressure probe fitted into a dummy part so that its small pressure receptive zone was located on the part axis, flush with its surface. The scheme of the experiment closely simulated the real polishing process, as shown in Figure 2a. In order to obtain the pressure distribution over the contact spot in the flow direction, the part was rotated about its radius of curvature. All pressure measurements were followed by polishing the spot into a glass part under the same conditions.

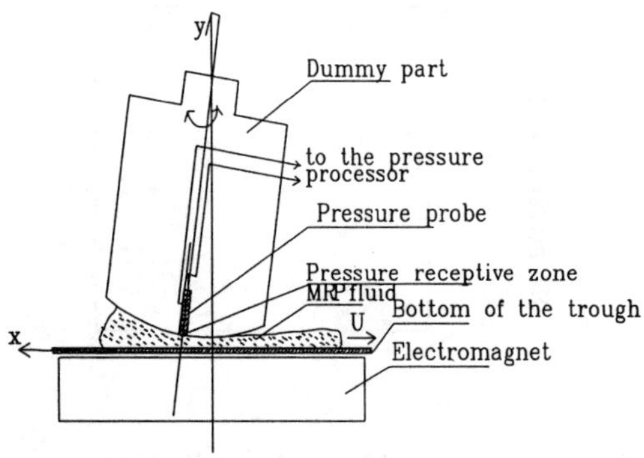

a)

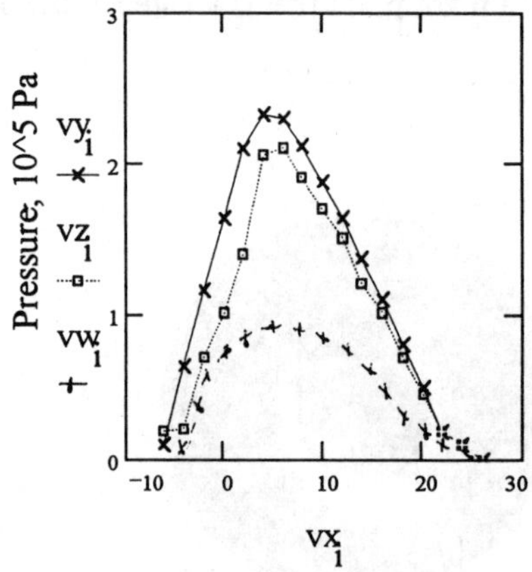

b)

Figure 2. Details of the experiment for pressure measurements in the polishing spot

The pressure distributions in the contact zone depend on the magnetic field intensity as shown in Figure 2b and follows roughly the shape of the spot for material removal (electromagnet current: vy-8A, vz-5A, vw-3A).

3. Theoretical consideration

MRP fluid flow caused by the moving wall located below the convex part surface is analagous to the classical problem commonly encountered in the theory and practice of hydrodynamic lubrication. Pressure profiles obtained with MRF are similar to the pressure distribution seen for journal bearings that operate in the hydrodynamic lubrication regime

[11]. So, it may be safely suggested that in the case of MR finishing the hydrodynamic situation is similar to the flow in the journal bearing.

Let us consider the simple two-dimensional MRP fluid flow in the gap between a spherical surface and the moving wall using the same approach as in the theory of lubrication. It is also assuming that MRP fluid rheological properties are governed by the Bingham equation [10]

$$\tau = \eta_0 \dot{\gamma} + \tau_o sign(\dot{\gamma}) \qquad (1)$$

where τ_0 is a yield shear parameter defined by the magnetic field intencity and γ is the shear rate. For shear stresses magnitude τ greater than τ_0 the material flows as a Newtonian fluid with viscosity η_0, otherwise the material remains rigid.

Refer to Figure 3. Two likely gap shapes for polishing with circular boundaries are illustrated. In each case the minimum gap thickness occurs at β=0 and has the magnitude h_0. The gap extends over the region $\beta_1<\beta<\beta_2$ ($x_1 <x<x_2$). Since appreciable pressure is developed only in the region of small angles, arc lengths along curved surfaces may be represented by the appropriate chords. Thus the gap is approximated by

$$h = h_0 (1+\alpha^2 x^2) \qquad (2)$$

where

$$\alpha^2 = \begin{cases} 1/(2h_0 R) & a) \\ [(R_2 - R_1)/2h_0 R_1 R_2] & b) \end{cases} \qquad (3)$$

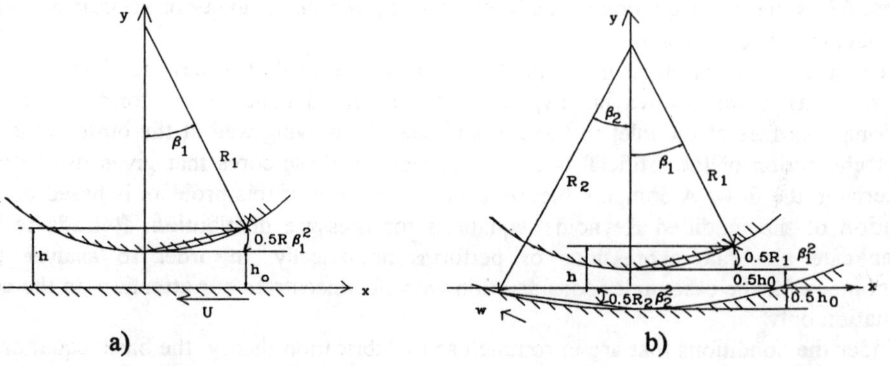

Figure 3. Parabolic geometry of the gap between moving wall and the part surface

At first, we will consider the case when the shear flow, initiated by the wall the moving with velocity U takes place in the gap at a given distance x along the gap and the shear

stress magnitude $|\tau|$ is greater than the yield stress τ_0 for all y. For two-dimensional flow, the Bingham model in this case reduces to the conventional lubrication theory result (symmetrical film) [11,12],

$$
0 \le y \le h: \qquad
\begin{aligned}
&P = P_a + \frac{2\eta_0 Ux}{h^2}, && a)\\
&u = U\left[1 - 3\left(1 - \frac{h^*}{h}\right)\frac{y}{h}\right]\left(1 - \frac{y}{h}\right) && b)\\
&\frac{dP}{dx} = \frac{6\eta_0 U}{\alpha h_0{}^2}\left(\frac{h - h^*}{h^3}\right). && c)
\end{aligned}
\qquad (4)
$$

Here h^* is the gap thickness where the pressure gradient becomes zero. The expressions for pressure, velocity and pressure gradient distribution along the gap are respectively presented. It is very important to notice that there is no effect on pressure due to the yield stress, but the magnitude of the shear stress increases directly with τ_0 [12].

In the case under consideration $P_a = p_{mag}$, and it means the magnetic component of total pressure. It does not depend on the flow and is determined by the expression

$$
p_{mag} = \mu_0 \int_0^H M_f(H)dH. \qquad (5)
$$

where M_f is the magnetization of the MRP fluid, μ_0 is a permeability of vacuum and H is the magnetic field intensity.

Generally, the shear stress in the fluid film may not exceed in places the yield value. In this case, as it was shown in [12], cores of unsheared material are formed near the stationary surface at the inlet to the gap and near the moving wall at the outlet. It is the throttling action of the artificial boundary surfaces of these cores that gives rise to new patterns in the flow. A complite theoretical consideration of this problem is based on the solution of the modified Reynolds' equations for pressure distribution. It leads to the complicate algebraic expressions or performs numerically. In order to analyze this phenomena for the case under consideration we will concentrate our attention on the core formation only.

Under the conditions that are in routine use by lubrication theory, the basic equation of flow may be shown to be

$$
\frac{\partial P}{\partial x} = \frac{\partial \tau_x}{\partial y} \qquad (6)
$$

By integrating Eq.6,

$$\tau_x = \frac{\partial P}{\partial x} y + \tau_{x1} \tag{7}$$

where τ_{x1} is the integration constant.

It is easily supposed that the pressure gradient is closely related to the core formation, because the shear stress distribution across the gap is defined by the pressure gradient magnitude. In the case of flow as illustrated in Fig.3a, Eq. 7 can be reduced to [12]

$$\tau_x = \frac{dP}{dx}\left(y - \frac{h}{2}\right) - \frac{\eta_0 U}{h} - \tau_0 \tag{8}$$

When the pressure gradient is equal to zero,

$$\tau_x = -\eta_0 \frac{U}{h} - \tau_0 \tag{9}$$

Because in this case shear stress is constant and then $|\tau_x| > \tau_0$, no core is formed. When the pressure gradient is present, the shear stress varies in y direction about a point $y=h/2$ and the slope of its distribution is equal to dP/dx. Consider the case of $dP/dx>0$. When the resultant shear stress on the stationary wall is less than the yield stress or $dP/dx>2hU/h^2$, a core is formed and it adheres to the upper surface (see below Fig.5). The thickness of sheared fluid in this case

$$h_s = \sqrt{\frac{2\eta_0 U}{|dP/dx|}} \tag{10}$$

and the range of core formation and the shear stress distribution, respectively, take the form

$$h - h_s = h - \sqrt{\frac{2\eta_0 U}{|dP/dx|}}, \quad \tau_x = \frac{dP}{dx}(h-y) - \sqrt{2\frac{dP}{dx}\eta_0 U} - \tau_0 \tag{11}$$

The maximum value of h-h_s corresponds to the condition when the shear stress on the stationary surface is equal to the yield stress. It is interesting that the extent of the core does not depend on τ_0.

In the case of $dP/dx<0$, the core formation occurs in the same manner as considered above, only the slope of the shear stress distribution is negative. The core adheres to the lower moving wall (see Fig. 5). When the pressure gradient magnitude increases further the shear stress on the stationary surface may exceed the yield point. Then, the core is

detached from the surface and floats in the film. The range for core formation in this case is

$$h_{s1} - h_s = \frac{2\tau_0}{dP/dx} \tag{12}$$

where h_{s1} is the upper boundary of the core. Similar considerations are also valid for the case of a core attached to the moving wall.

Using these basic statement for Bingham flow and the experimental data for the pressure distribution in the polishing zone, let us examine the flow that occurs under real polishing conditions. The configuration of the gap is assumed to be the same as in Fig. 3a.

Typical experimental pressure distribution and relevant distribution of the pressure gradient in the contact zone are presented in Figure 4a,b.

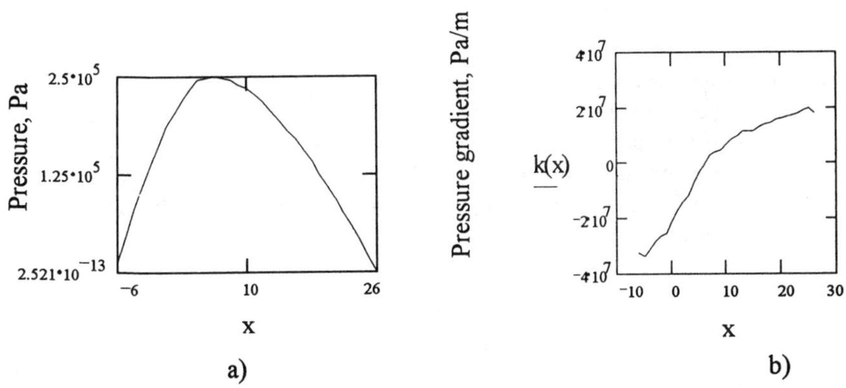

Figure 4. Pressure a) and the pressure gradient b) distribution in the contact spot

With the pressure gradient distribution in the hand and using Eq.8 it becomes possible to calculate the distribution of the shear stresses in the fluid film. Contour plot of the shear stresses is shown in Fig 5a. Here the vertical axis corresponds to *y*, mm, and the horizontal one to *x*, mm, (dashed line indicates the location of the part surface). As is seen from Fig.5a, zones of the shear stress that are lower than the yield point (τ_0=50kPa) are located at the upper right corner of the plot (the inlet to the contact region) and at the lower left corner (the outlet from the contact region). The cores of an unsheared fluid are formed in these locations in the film. Shear stresses on the moving wall (dashed line) and on the stationary surface (solid line) are shown in Figures 5b. It is easy to verify that the zones of unsheared fluid (stresses less than 50kPa) are attached to the moving wall at the outlet of the gap and to the stationary part surface at the inlet.

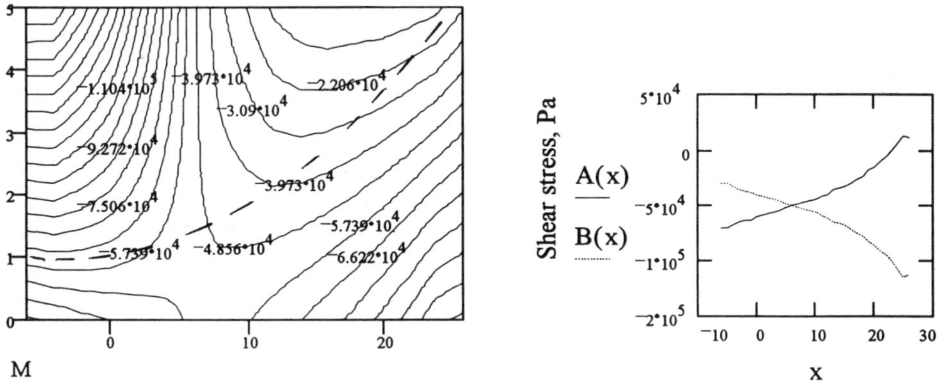

a) b)

Figure 5. Shear stresses distribution in the MR fluid film

The range of the sheared flow is defined by Eq.10 and is presented in Figure 6 whereas the range of the core formation at the outlet zone of the gap is defined by Eq. 11 and is shown in Fig. 7.

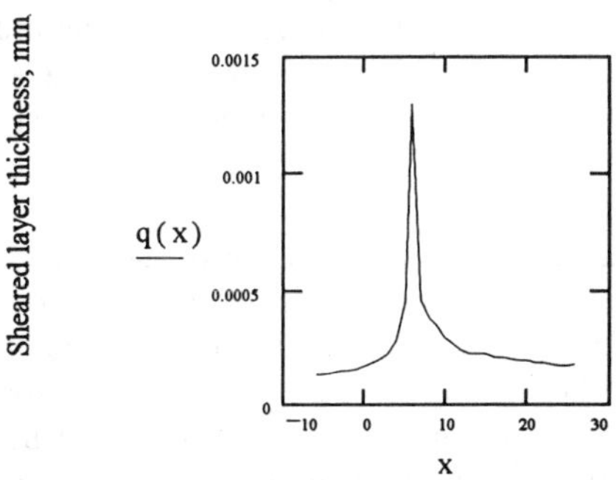

Figure 6. Thickness of the layer of the sheared fluid

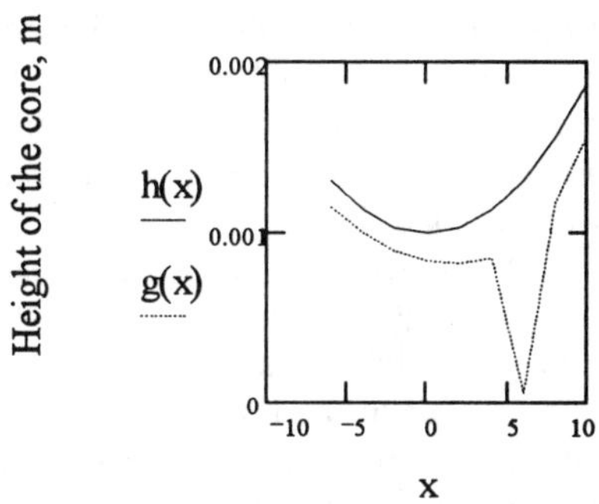

Figure 7. Gap geometry formed by the core

In Fig. 7 the dashed line is the boundary of the core and the solid line shows the boundary of the part surface. As may be inferred from the graph, the rigid core attached to the moving wall forms an artificial surface with a shape that is close to the shape of the part. The boundary of this surface is located at some distance from the part surface and the shear flow takes place in this gap. Assuming that the geometry of this new gap falls into the category b) in Fig.3, the minimum of the gap thickness h_0 is determined to be ~0.1mm for the experimental conditions (initial gap is 1mm). It is felt that the Newtonian flow occurs in this new artificial gap. In this case according to Eq.6a, the calculated pressure distribution is as presented in Fig. 8 (solid line). For comparison, the experimental pressure distribution is also plotted in Fig. 8 (dashed line).

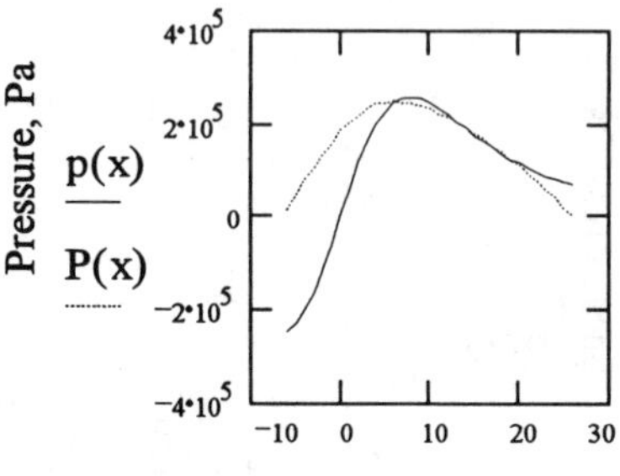

Figure 8. Pressure distribution in the contact zone

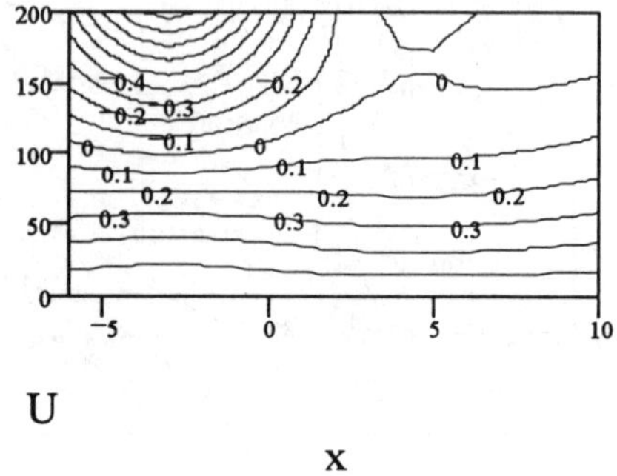

Figure 9. Fluid flow streamlines in the gap

As noted above, the theoretical approach used here is a two- dimensional one and much simplifies the real situation, particularly at the inlet of the contact zone. Therefore, in the following we will assess only the region in the neighborhood of the outlet zone. The fluid flow streamlines for this case can be defined by Eq.4b and are presented in Fig. 9 (the vertical axis corresponds to the y coordinate in μm and moving wall velocity is 0.6m/s). The main conclusion which can be done from these consideration is that generated by magnetic field MR fluid plastic (Bingham) properties provide a dramatic change of the patern of the flow due to the formation of the rigid core attached to the moving wall. As a result we have much higher fluid velocity near the part surface than in the case of Newtonian fluid flow.

Now, the pressure of the abrasive particles against the surface to be polished, or to be more specific, the interaction between the particle and the surface is the major factor to be considered. In this connection we have to define the forces that act on the particle when this one moves in the fluid flow generated by the moving rigid boundary with respect to the stationary surface. Here we shall not consider this problem in detail. It can be only pointed out that the low Reynolds number particle motion analyses shows the dominant role of the particle/surface collision. As this takes place, the kinetic energy of the particle converts to elastic strain energy over a small volume of the part material. If the corresponding stress becomes high, it can activate preexisting defects or generate new defects and cause plastic deformation leading to fracture. When a particle of radius r, velocity u_p, density r_p,Young's modules E_p, Poisson's ratio ν_p arrives with angle α on the surface of the part of Young's modules E_w and Poisson's ratio ν_w, and if kinetic energy component of the particle in the direction perpendicular to the part surface converts completely to the strain energy of the part, than the normal load, L_N, and pressure P are given by the following expressions, respectively [14].

$$L_N = \left(\frac{10}{3}\pi\right)^{\frac{3}{5}} \rho_p^{\frac{3}{5}} K_0^{-\frac{6}{5}} r^2 \left(u_p \sin\alpha\right)^{\frac{6}{5}}, \quad P = \frac{L_N k}{\pi r_p^2}, \quad K_0 = \left[\frac{3}{4}\left(\frac{1-\nu_p^2}{E_p} + \frac{1-\nu_w^2}{E_w}\right)\right]^{\frac{1}{3}}$$

Using standard for MR polishing parameters and taking the fill factor k=0.5, the magnitude of the mean particle (fluid) velocity near the surface u_p=0.1m/s and, for example, α=30° we obtain L_N=8.6x10^{-6}N for magnetic particle and the magnitude of 1.7x10^{-6}N for abrasive particle. In this case the "effective" working pressure is ~1MPa. It is significant that the magnitude of these forces correlates well to the magnitude of the load per grain (~10^{-7}N) that is convential to the pitch glass polishing [1].

4. Conclusions

In the context of the undertaken analyses the process of the material removal under magnetorheological finishing can be presented as follows. A layer of the magnetorheological polishing fluid is set into motion by a solid wall and enters the moderately large gap that is formed by the convex surface of the part to be polished and the moving wall. The magnetic field penetrates the surface/fluid contact zone, and converts initially low viscosity fluid into a plastic (Bingham) medium. The Bingham flow in the gap is similar to the flow of the lubricant in the journal bearing results in the formation of an unsheared fluid (core) that is attached to the moving wall. As this takes place an artificial moving boundary is formed very close to the surface of the part and the shear flow of low viscous MRP fluid occurs in this new gap. The geometry of this gap is defined by the distribution of the shear stresses in the fluid film and is controlled by the magnetic field intensity. The thickness of the artificial gap is calculated to be about 50-100μm. This provides high fluid velocity in the vicinity of the surface to be polished. Carried by the moving fluid solid particles bombard the surface of the part over the contact zone causing Hertzian penetration of the surface by the particles and converting their kinetic energy to strain energy resulting in the material removal. Both magnetic and classical abrasive particles execute a beneficial mechanical job in weakening the binding forces at the part surface. Abrasive particles, such as CeO_2, may provide an additional contribution to the polishing kinetics due to chemical effects.

In the light of the above, the advantages of magnetorheological finishing can be summarized as follows.

1. Magnetorheological finishing is a noncontact method of treatment that creates no surface and subsurface damages during processing.

2. Sufficiently thin working gap between the surface to be polished and the moving wall is created by the magnetic field resulting in the high velocity of the abrasive particles near the surface and, so, high efficiency of finishing. It is significant that this quality is provided by a simple mechanics. There is no limitation on the size of the particles.
3. There are no wearing parts or system elements to be periodically restored and continuous circulation of the working fluid via the polishing zone provides stable and reproducible conditions for finishing.
4. The mechanism of the magnetorheological finishing provides predictable distribution of the material removal over the polishing spot. Programmable scanning of the polishing spot over the surface provides high accurate deterministic finishing of the part.

5. Acknowledgements

The authors acknowledges the expertise, dedication and teamwork of D. Golini, B. Puchebner, D. Strafford, E. Fess, D. Pietrowski, Y. Hsu and G. Gorodkin. Support for this effort is provided by Byelocorp Scientific, Inc and the Center for Optics Manufacturing University of Rochester

6. References

1. L. M. Cook. *Journal of Non-Crystalline Solids,* **120,** 152-171, (1990)
2. I. Ali, S. R. Roy and G. Shinn. *Solid State technilogy.* **October 1994.**
3. N. Umehara, *Annals of the CIRP,* **43**, 185-188, (1994)
4. Y. Saito, h. Nikura, T. Oshio and T. Hanaoka. *Proceedings of the 6th International Conference on Production Engineering,* Osaka, 335-340, (1987)
5. T. Childs, S. Mahmood and H. Yoon. *Proc. Instn. Mech. Engrs.* **208,** 47-59, (1994)
6. M. Fox, K. Shimura and R. Komanduri. *Annals of the CIRP,* **43,** 181- 184, (1994)
7. J. Gormley, M. Manfra and A.Calawa. *Rev. Sci. Instrum.* **52(8),** 1256-1259, (1991)
8. Y. Mori, K. Yamauchi and K. Endo. *Journal of Japan Society of Precision Engineering,* **10(1),** 24-28, (1988)
9. W. Kordonski, I. Prokhorov, B. Kashevsky, S. Jacobs, B. Puchenber, Y. Hsu, D. Pietrowski, D. Strafford and E. Fess. *OSA OF&T Workshop Digest,* **13**,104-109,(1994)
10. Z. P. Shulman, W. I. Kordonsky, E. A. Zaltsgendler, I. V. Prokhorov, B. M. Khusid and S. A. Demchuk. *Int. J. Multiphase Flow,* **12(6),** 935-955, (1986)
11. W. A. Gross, L. A. Matsch, V. Castelli, A. Eshel, J. H. Vohr and M. Wildmann. *Fluid Film Lubrication. A Wiley- Interscience Publication,* 774, (1980)
12. J. A. Tichy. *J. Rheol.* **35(4),** 477-496, (1991)
13. S. Wada, H. Hayashi and K. Haga.. *Bulletin of the JSME,* **16(92),** 422-431, (1973)
14. Y. Mori, N. Ikawa, K. Sugiyama, T. Okuda and K. Yamauchi. *Journal of Japan Society of Precision Engineering,* **51(6)**, 1187-1194, (1985)

EFFECT OF DIELECTRIC PROPERTY OF HYDROUS DISPERSOID ON ELECTRORHEOLOGY

Akiko KAWAI,
Chemical Systems Department , National Institute of Materials and Chemical Research
1-1 Higashi, Tsukuba Ibaraki 305, Japan

and

Kunio UCHIDA[1*], Kunio KAMIYA[1*], Akihiro GOTOH[1*],
Satoshi YODA[1*], Kei URABE[2*] and Fumikazu IKAZAKI[1*]
[*1] *Chemical Systems Department,* [*2] *Composite Materials Department*
National Institute of Materials and Chemical Research
1-1 Higashi, Tsukuba Ibaraki 305, Japan

ABSTRACT

The effect of the dielectric property was investigated of titania and microcrystalline cellulose particles dispersed in a silicon oil on Electrorheology(ER). A commom understanding of the mechanism of ER is the electrical interaction by the interfacial polarization of dispersed particles. Block et al. reported the relaxation frequency of ER fluids, i. e. the rate of polarization, was very important to have an appreciable ER effect.

We found that the free water on the cellulose and the titania particles dispersed in the silicon oil, which is respectively dehydrated below the temperature of ca. 110 and 300 degree centigrade, was necessary for an appreciable ER effect. The relation was measured between the water content and the dielectric properties such as dielectric constant ε' and dielectric loss factor ε'', as the water adsorbed on the particles was estimated to alter the dielectric properties of the ER fluid. Both the dielectric constant and the relaxation frequency, which is defined by the applied frequency for a peak of the dielectric loss factor, increased with the increase of the water content. The relation was discussed between the ER effect and the dielectric properties.

1. Introduction

Fluids whose viscosity changes reversibly by an external electric field are known as electrorheological fluids. A common understanding of the mechanism of electrorheology(ER) is the electric interaction by the interfacial polarization of dispersed particles[1). In the case of hydrous systems, Uejima[2)] studied the ER effect in the suspension of microcrystalline cellulose. He found the ER effect increased with increasing water content until it leveled off. Sugimto et al.[3)] reported the apparent viscosity of ion exchange resins in an silicon oil increased with the increase of water content under the application of electric field. Uejima and Sugimoto measured the dielectric property, but no relation was investigated between the dielectric property and the ER effect. Block et al.[4] studied the dielectric property of semi-conducting polymers as ER fluids. The relaxation frequency of the dielectric constant was very important to have an appreciable ER effect. The relaxation frequency is defined by the applied frequency for a peak of the dielectric los factor ε''. He also measured[5] the dielectric spectra of dispersions of various polymethacrylate salts containing 15% water. The relaxation frequencies depended on the salts utilized.

We have been paying special attention to the interface of a dispersoid and a dispersion solvent, especially the adsorbed water on the dispersoid. In our experiment silica and titania partcles dispersed in an silicon oil were used. The free water on the silica particles was necessary and there was an appropriate quantity of the free water to get higher ER effect[6]. The free water was defined as the adsorbed water which was dehydrated below the temperature of 200℃. We also found that the amount of the bound water which was dehydrated above 200℃ affected the ER effect[7]. In the case of titania the free water on the surface was necessary and there was also an appropriate quantity of the free water to get higher ER effect[8].

In this study, the ER effect of titania and microcrystalline cellulose dispersed in an silicon oil, the amount of the adsorbed water and the dielectric properties, i.e. dielectric constant ε' and dielectric loss factor ε'' of the electrorheological fluids were measured by a rotational viscometer. The relation was discussed between the dielectric properties and the ER effect.

2. Experimental apparatus and procedure

2.1 Sample Preparation and characterization

Two kinds of particles, i.e. titania and microcrystalline cellulose were used as dispersoids. Ultra-fine TiO_2 was supplied by Idemitsu Co. Ltd.. Titania was prepared by hydrolyzed method in gas phase. The particle was amorphous and spherical by the analysis of XRD and SEM observation. The primary particle size was about 17nm under the observation by TEM. The microcrystalline cellulose was supplied by Merck. The average diameter(D_{50}) was 7 μm(2-15 μm). The water content of these two kinds of particles was changed by heat treatment or exposing them to wet atmosphere. A vacuum oven(VM-25, Irie Seisaku Sho) was used for the heat treatment of the dispersoids. The temperature of the heat treatment was 50, 100, 150 and 200℃. The samples were held for 1hr at the temperature. TG(thermogravimetry) was used for the measurement of the water content after the control of the water content. The sample was dispersed into dimethyl silicon oil(Toshiba silicone Co. Ltd.). The viscosity of the oil is 100mPas at 20 ℃. The dispersion concentration was 10wt.%. The dielectric property of the ER fluid was measured with LCR meter(HP 4284A, Yokokawa Hewlett Packard Co. Ltd.) in the range from 30Hz to 1MHz. The dielectric consatant ε' and the dielectric loss factor ε'' are respectively the real and imaginary part of the complex dielectric constant.

2.2 Rheological Measurements

The ER effect of the fluids was measured with a rotational viscometer (VT550, HAAKE). Flow curve was measured. The ER fluid was held in a gap between the inner rod and the outer cell. The inner rod was rotated. The gap was 1mm. The DC Voltage was kept constant at 2kv. Measurement temperature was maintained at 20℃ by the circulation of water from a constant temperature bath. Shear rate was varied from 0 to $700s^{-1}$. The ER effect was expressed by $\eta_{E=E} - \eta_{E=0}$ (η means viscosity). It means the net increase of the viscosity at an applied electric field.

3. Results and discussions

3.1 Dielectric Property

The water content analyzed by TG was shown in Table 1. TG was an established analysis method of water content and several researchers[9) 10) 11)] used TG analysis in order to measure the water content of dispersoid of ER fluids. The water content of cellulose by TG was compared with that measured by Uejima's method. Uejima measured the weight of a dispersoid before and after heat treatment at 110°C for an hour and defined water content M as the following equation.

$$M=(W_1-W_2)/W_1*100$$

,where W_1 and W_2 are respectively the weight of a dispersoid before and after the heat treatment. The weight loss of cellulose by TG was 2.9wt.% and the M value of the cellulose was 2.6wt.%.

Table 1 water content of titania and microcrystalline cellulose

titania					
heat treatment temperature (°C)	–	50	100	150	200
water content (wt.%)	8.3	8.3	7.4	5.8	4.6
microcrystalline cellulose					
exposing time to wet atmosphere (day)	–	1	2	3 ···	5
water content (wt.%)	3.4	6.0	7.3	7.7 ···	10.5

The TG curves of the cellulose and the titania dispersoid were almost the same. They showed monotonous decrease. The difference of the two curves was only the end temperature of the curve. The weight loss of the adsorbed water contained in the cellulose and the titania was completed at 110 and 300°C respectively. Most of the adsorbed water was thought to be free water.

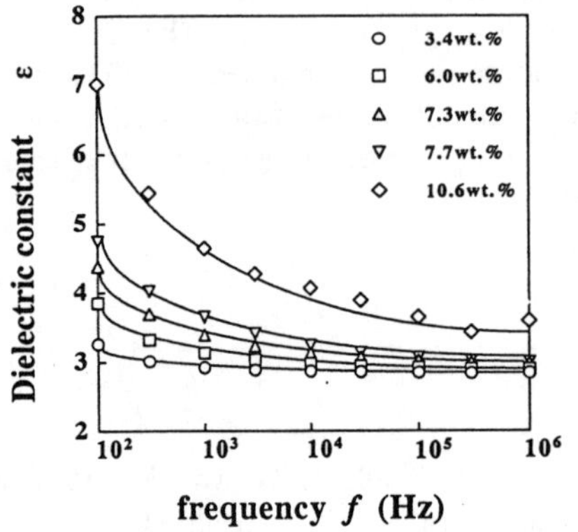

Fig.1(a) Dielectric constant of ER fluid which contain cellulose with different water content

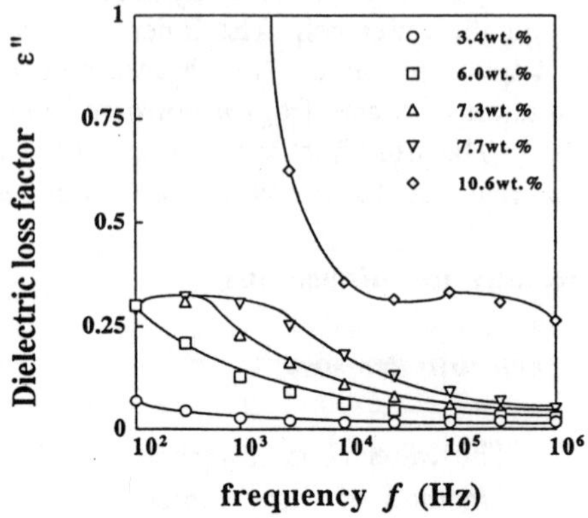

Fig.1(b) Dielectric loss factor of ER fluid which contain cellulose with different water content

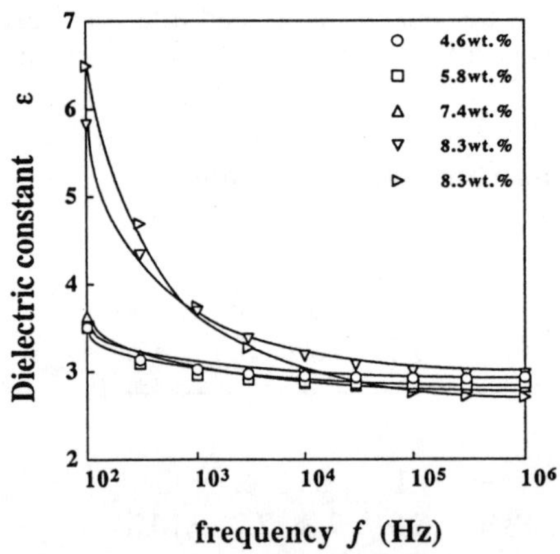

Fig.2(a) Dielectric constant of ER fluid which contain titania with different water content

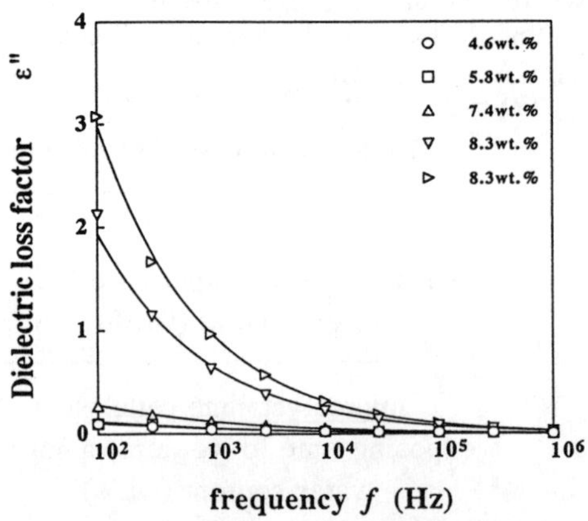

Fig.2(b) Dielectric loss factor of ER fluid which contain titania with different water content

The dielectric property with different water content was shown in Figs.1 and 2. There are four kinds of polarizations, i.e. interfacial polarization, dipole polarization, electronic polarization and atomic polarization. In our study, the measurement of the dilectric properties was in the range of interfacial polarization that was considered to contribute to an ER effect.

In Figs.1 and 2 the dielectric constant of the both ER fluids decreased with increasing frequency. The value of the dielectric constant at high water content was larger than that at low water content. This was because the dielectric constant of water itself was larger than the dispersoid. The more the water, the larger the dielectric constant. The effect of the adsorbed water content on the dielectric loss factor ε'' was, however, different in the case of cellulose and titania. In the case of the cellulose, there was a peak corresponding to an anomalous absorption in the range of the measurement frequency. The peak is the relaxation frequency of the adsorbed water on the cellulose. But in the case of the titania, the dielectric loss factor decreased monotonously. The dielectric loss factor of water itself has an anomalous absorption at the frequency of about 10^{10}Hz. The adsorbed water on the both particles are not the same as bulk water from the standpoint of the dielectric property. The relaxation frequency of the titania is etimated to be less than 100Hz. It means the rate of the interfacial polarization of the adsorbed water on the titania is lower than that on the cellulose, because the relaxation frequency corresponds to the rate of the polarization. In other words, the adsorbed water on the titania is bound to the surface more tightly than the adsorbed water on the cellulose. In the water content of 7.7 and 10.6 wt.% on the cellulose the relaxation frequency was respectively ca. 10^3 and 10^5Hz.

3.2 ER Effect Measurement

The ER effect of the both fluids increased with increasing water content as shown in Figs.3 and 4. The ER effect of the cellulose was several times higher than that of the

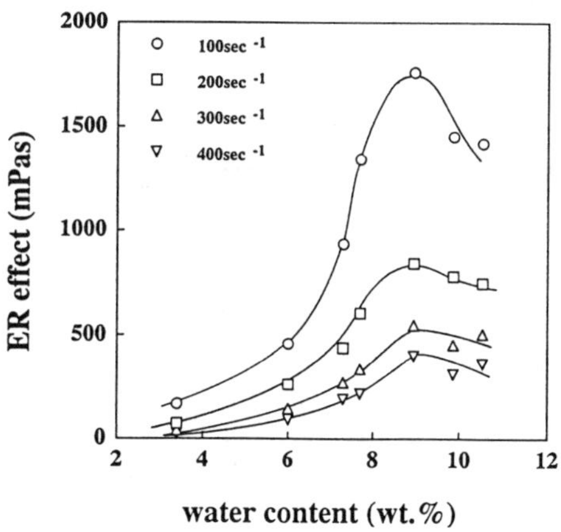

Fig.3 Relation between the ER effect of cellulose and water content

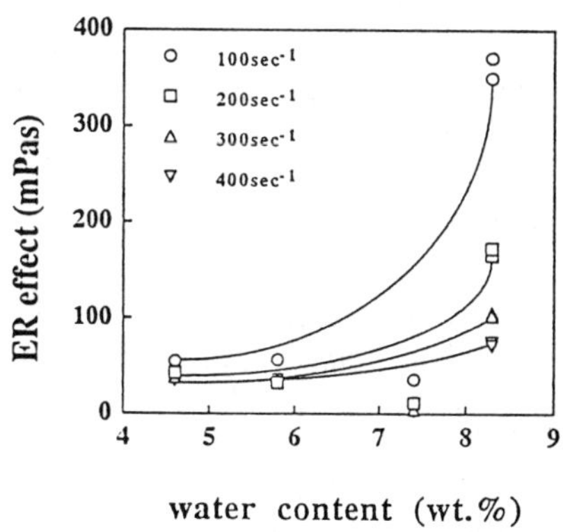

Fig.4 Relation between the ER effect of titania and water content

titania. This difference was considered to come from the difference of the dielecric property above-described. The dielectric constant of the titania was higher than that of the cellulose. In this study, the dielectric constant did not contribute to the ER effect. The dielectric loss factor of the both samples was however quite different, especially the relaxation frequency. In the titania with the slow rate of the polarization less than 100Hz, the ER effect was not so high as in the cellulose with the rate of the interfacial polarization of 10^3-10^5 Hz. The voltage of 2kV could not be stably applied to the fluid containing the cellulose with the adsorbed water of 10.6 wt.%. The relaxation frequency of 10^5 Hz was too high to have an appreciable ER effect in the case of cellulose.

Too high rate of polarization was found to be as bad as too low rate of polarization from the standpoint of the ER of hydrous systems. It is the same result as that of Block et al.[4) 5)] in the case of semi-conducting polymers and various polymethacrylate salts containing 15% water.

4. Conclusions

The relation was investigated between the ER effect and the dielectric properties of hydrous ER fluids containing titania and microcrystalline cellulose. The followings were induced from the results of the measurements on the cellulose and the titania which contained different water content.

1.The free water adsorbed on the cellulose and the titania gave the different effect on the dielectric property.

2.The dielectric loss factor of the electrorheological fluid containing the cellulose had an anomalous absorption and the frequency of the absorption was shifted to a higher frequency with the increase of the water content. The dielectric loss factor of the titania however did not have the anomalous absoption in the range of 100-10^6 Hz.

3.For getting higher ER effect, the fluid was expected to have an anomalous absorption in the frequency range of interfacial polarization.

5. Acknowledgements

We acknowledge EKO Instruments Trading Co. Ltd. in support of our work in the area of ER measurement.

6. References

1. A. Sugimoto et al., *RIIS-RR-22-3* 11 (1973) 1.
2. H. Uejima, *Jpn. J. Appl. Phys.* 11 (1972) 319.
3. M. Sugimoto, *Bull. JSME* 20 (1977) 1476.
4. H. Block, P. Rattray and T. Watson, *Proc. Int. Conf. on ER Fluids, Illinois USA* (1991) p93(World Scientific,1992).
5. H. Block et al., *Langmuir* 6 (1990) 6.

6. A. Kawai et al., *Adv. Powder Technol.* **5** (1994) 129.
7. A. Kawai et al., *J. Soc. Powder Tech. Japan* **31** (1994) 647.
8. A. Kawai et al., *Proc. Pacific Conf. on Rheology and Polymer Processing, Kyoto JAPAN* (1994) p58.
9. Y. Otsubo et al., *J. colloid Interf. Sci.* **150** (1992) 324.
10. W. Wong et al., *Proc. 2nd Int. Conf. on ER fluids, North Carolina, USA* (1989) p191
11. H. Conrad et al., *ibid*, p252

COMMERCIAL MAGNETO-RHEOLOGICAL FLUID DEVICES

J.D.Carlson, D.M.Catanzarite and K.A.St. Clair
Lord Corporation
Cary, NC 27511 USA
(jdcarlson@crd.lord.com)

Abstract

Controllable magnetorheological (MR) fluid devices have reached the stage where they are in commercial production. Such devices are finding application in a variety of real world situations ranging from active vibration control to aerobic exercise equipment. Examples of several, commercial MR fluid devices are described and the comparative ability of MR and ER fluids to meet the needs of practical devices is discussed.

1. Introduction

Controllable fluids are materials that respond to an applied electric or magnetic field with a change in their rheological behavior. Typically, this change is manifested when the fluids are sheared by the development of a yield stress that is more or less proportional to the magnitude of the applied field. These materials are commonly referred to as electrorheological (ER) or magnetorheological (MR) fluids. Interest in controllable fluids derives from their ability to provide simple, quiet, rapid-response interfaces between electronic controls and mechanical systems. That controllable fluids have the potential to radically change the way electromechanical devices are designed and operated has long been recognized.

Willis Winslow is generally credited as the first person to recognize the potential of controllable fluids in the 1940s with the first ER fluid patent[1] and paper[2] describing the ER effect. Since that time publications describing the properties, behavior and application of ER fluids have abounded. ER fluids have been the fodder for numerous theses and for many years the imminent, widespread application of ER fluids has been heralded. However, in spite glowing predictions and the expenditure of sizable sums of R&D money, the commercial success of ER fluids has remained elusive.

MR fluids are considerably less well known than their ER fluid analogs. Like ER fluids, MR are non-colloidal suspensions of polarizable particles having a size on the order of a few microns. MR fluids should not be confused with colloidal ferrofluids in which the particles are about 1000 times smaller than those found in typical MR fluids. Except for a flurry of interest after their initial discovery, there has been scant information published about MR fluids. Only recently has a resurgence in interest[3,4,5,6,7,8] in MR fluids been seen.

The initial discovery and development of MR fluids and devices can be credited to Jacob Rabinow at the US National Bureau of Standards[9,10,11] in the 1940s. Interestingly, this work was almost concurrent with Winslow's ER fluid work. The late 1940s and early 1950s actually saw more patents and publications relating to MR than to ER. While Rabinow's work is largely overlooked today, Winslow discussed the work on MR fluids going on at the National Bureau of Standards in his seminal paper on ER fluids.

The purpose of this paper is to examine some of the fundamental design principles for devices utilizing controllable fluids, to compare the ability of ER and MR fluids to meet the needs of practical devices and finally to show examples of several MR fluid devices that have recently reached large scale commercial application.

2. State-of-the-Art Controllable Fluids

State of the art ER fluids[5,12,13] routinely exhibit dynamic yield strengths in the range of 3 to 5 kPa for electric fields on the order of 3 to 5 kV/mm and off-state viscosity in the range of 0.20 to 0.30 Pa-s at 25°C. The maximum strength of a given ER fluid is generally limited by the electric field breakdown strength of the fluid. Operational temperature ranges for ER fluids depend on the type of polarization mechanism used to activate the fluid. Low conductivity, DC, ER fluids polarized by an ionic conduction mechanism are typically operational over the range of 10°C to 90°C. Such fluids have a room temperature conductivity such that they require on the order of 5 $\mu A/cm^2$ at an electric field of 3 kV/mm. Non-ionic ER fluids are generally usable over a broader temperature range, typically -25°C to +125°C. While they may exhibit a negligible DC conductivity, such non-ionic fluids are often used with high-frequency AC fields in order to minimize electrophoretic effects. In this case one must be able to provide sufficient non-dissipative, displacement current, perhaps as much as 1-2 mA/cm^2 at 3kV/mm(RMS), because of the relatively large device capacitance.

MR fluids[5,8,13] routinely exhibit dynamic yield strengths of 50 - 100 kPa for applied magnetic fields of 150-250 kA/m (~2 - 3 kOe) and off-state viscosity of 0.20 to 0.30 Pa-s at 25°C. The ultimate strength of MR fluids is limited by magnetic saturation. Operational temperatures for MR fluids easily span -40°C to +150°C. MR fluids are generally limited by the properties of the carrier fluid used in the formulation rather than the details of the polarization mechanism. Dissipative currents and joule energy loss in the MR fluid are not a concern. One is able to effectively use permanent magnets to energize MR fluids with no steady-state power requirement at all. Shear stress versus shear rate curves at various magnetic field strengths for a typical MR fluid are shown in Figure 1. A chart comparing state-of-the-art ER and MR fluids is given in Table 1.

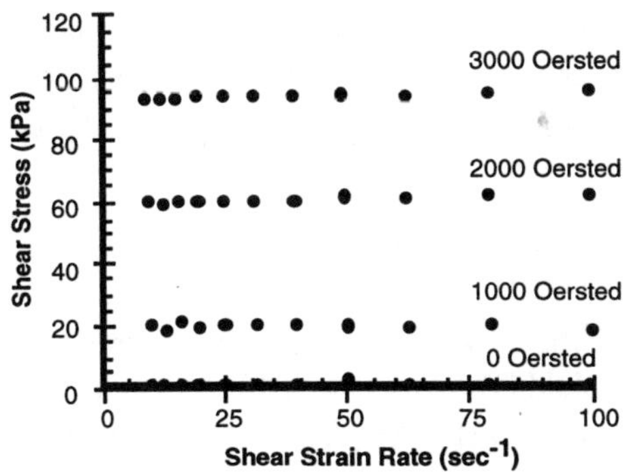

Figure 1 - Yield stress versus shear rate for a typical MR fluid (Lord VersaFlo™ MR-100)

Table 1 - Comparison of properties of typical ER and MR Fluids

Property	ER Fluid	MR Fluid
Yield Strength (Field)	2 - 5 kPa (3-5 kV/mm) Field limited by breakdown	50 - 100 kPa (150 - 250 kA/m) Field limited by saturation
Viscosity (no field)	0.2 - 0.3 Pa-s @ 25 °C	0.2 - 0.3 Pa-s @ 25 °C
Operating Temperature	+10 to +90 °C (ionic, DC) -25 to +125 °C (non-ionic, AC)	-40 to +150 °C (limited by carrier fluid)
Current Density	2 - 15 mA/cm^2 (4kV/mm, 25°C) (x10 -x100 @ 90°C)	can energize with permanent magnets
SpecificGravity	1 - 2.5	3 - 4
Ancillary Materials	Any (conductive surfaces)	Iron/Steel
Color	Any, Opaque or Transparent	Brown,Black,Gray/Opaque

3. Controllable Fluid Device Design

Most devices that use controllable fluids can be classified as having either fixed poles (valve mode) or relatively moveable poles (direct-shear mode). Diagrams of these two basic operational modes are shown in Figure 2. Examples of valve mode devices include servo-valves, dampers and shock absorbers. Examples of direct-shear mode devices include clutches, brakes, chucking and locking devices.

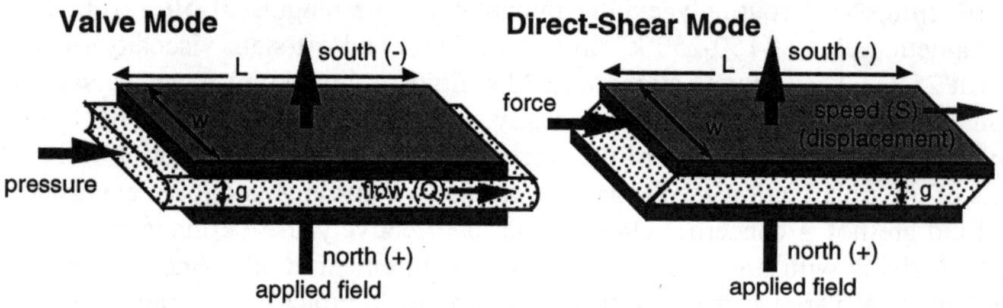

Figure 2 - Basic operational modes for controllable fluid devices.

The pressure drop developed in a valve mode device can be divided into a field independent viscous component ΔP_η and an applied field dependent induced yield stress component ΔP_τ. These pressures may be approximated by[14,15]:

$$\Delta P_\eta = \frac{12\,\eta\,Q\,L}{g^3\,w} \tag{1}$$

$$\Delta P_\tau = \frac{c\,\tau\,L}{g} \tag{2}$$

where η is the viscosity with no applied field and t is the yield stress developed in response to applied electric field E or magnetic field H. The parameter c has a value that ranges from a minimum value of 2 (for $\Delta P_\tau/\Delta P_\eta$ less than ~1) to a maximum value of 3 (for $\Delta P_\tau/\Delta P_\eta$ greater than ~100). The total pressure drop in a valve mode device is approximately equal to the sum of ΔP_η and ΔP_τ.

In a like manner, the force developed by a direct-shear device can be divided into viscous F_η and an induced yield stress F_τ components:

$$F_\eta = \frac{\eta\, S\, L\, w}{g} \tag{3}$$

$$F_\tau = \tau\, L\, w \tag{4}$$

The total force developed by a direct-shear device is the sum of F_η and F_τ.

While Eqs. 1-4 are certainly useful in the design of controllable fluid devices, they often do not provide the best insight into the significance of the various parameters. It is often useful to algebraically manipulate the above equations to provide a different derived design equation[16,17]:

$$V = \frac{12}{c^2}\left(\frac{\eta}{\tau^2}\right)\left(\frac{\Delta P_\tau}{\Delta P_\eta}\right) Q\, \Delta P_\tau \tag{5}$$

Equation 5 gives the <u>minimum active fluid volume,</u> V = Lwg, that is necessary in order to achieve the desired control ratio ($\Delta P_\tau/\Delta P_\eta$) at a given flow rate Q with a specified controlled pressure drop ΔP_τ. The key aspect to note in this equation is that it is the ratio of the fluid viscosity to the <u>square</u> of the yield stress (η/τ^2) that controls the active fluid volume .

Similarly, for direct-shear devices, an equation that give the necessary minimum active fluid volume is easily derived:

$$V = \left(\frac{\eta}{\tau^2}\right)\left(\frac{F_\tau}{F_\eta}\right) F_\tau\, S \tag{6}$$

Once again, for a given control ratio F_t/F_h, controlled force F_t and speed S, the minimum active fluid volume depends on the ratio of viscosity to yield stress <u>squared</u> (η/τ^2). Thus, for both valve mode and direct-shear mode devices, the <u>minimum</u> volume of fluid that must be acted upon by the applied field is governed by the ratio of fluid viscosity to the <u>square</u> of the induced yield stress (η/τ^2).

4. ER Versus MR Fluid Devices

4.1. Device Size

As we have shown above the minimum amount of controllable fluid in the working region of the device depends on the ratio (η/τ^2). One comparison of the relative effectiveness of ER and MR fluids can be obtained by considering the relative active fluid volumes necessary to achieve comparable device performance, i.e., the same control ratio, force and speed:

$$\frac{V_{ER}}{V_{MR}} = \frac{\left(\dfrac{\eta_{ER}}{\tau_{ER}^2}\right)}{\left(\dfrac{\eta_{MR}}{\tau_{MR}^2}\right)} \tag{7}$$

Since the off-state viscosity of ER and MR fluids are roughly comparable[5] this expression may be further simplified to yield:

$$\frac{V_{ER}}{V_{MR}} = \left(\frac{\tau_{MR}}{\tau_{ER}}\right)^2 \tag{8}$$

For typical, state-of-the-art, controllable fluids, i.e. ER fluids with $\tau_{ER} \sim$ 3-5 kPa and MR fluids with $t_{MR} \sim$ 50-100 kPa, a V_{ER}/V_{MR} ratio of 100-1000 results.

This is an obvious over simplification of the complete device situation. Naturally, the size of a device also includes such factors as the electrodes and high voltage insulation for ER fluids and the electromagnetic coil and steel pole pieces for the MR fluids. The large size of the above volume ratio does point out, however, that because of their very large yield strengths, MR fluid device may be substantially smaller than ER fluid devices capable of similar performance.

4.2 Electric Power

A direct comparison of the Joule losses in ER and MR fluid devices is difficult. In ER fluid devices such losses are due to electric conduction through the fluid while in MR fluid devices such losses are present in the coil used to produce the magnetic field. ER fluid losses can vary widely depending on the temperature of the fluid. Losses in the coil used to activate the MR fluid will vary widely depending on the specifics of how the coil is wound, i.e. wire material, gauge and coil geometry.

An instructive comparison can be made, however, by considering the energy needed to establish the appropriate electric or magnetic field throughout the volume of controllable fluid, i.e., how much energy must be provided by an external power supply to either charge the capacitance represented by V_{ER} or build the magnetic field throughout V_{MR}.

The energy W_{ER} needed for an ER fluid device equals the fluid volume times the energy density in the fluid:

$$W_{ER} = V_{ER}\,(1/2\,\kappa\,\varepsilon_o\,E^2) \tag{9}$$

where κ is the relative dielectric constant of the ER fluid, ε_o is the vacuum permittivity and E is the applied electric field. Similarly, the energy required for an MR fluid device is given by:

$$W_{MR} = V_{MR}\,(1/2\,B\,H) \tag{10}$$

The ratio of energy required for an ER device compared to that for a comparably performing MR device is thus given by:

$$\frac{W_{ER}}{W_{MR}} = \left(\frac{V_{ER}}{V_{MR}}\right)\frac{\kappa\,\varepsilon_o\,E^2}{B\,H} \tag{11}$$

Assuming a volume ratio of 100 from above and using typical values for the remaining parameters[5,6,12,13] (κ = 10, ε_o = 8.85x10^{-12}F/m, E = 4 MV/m, B = .8 T and H = 2x10^5 A/m) one finds that the field energy requirements for comparable ER and MR fluid device are approximately equal, i.e., $W_{ER}/W_{MR} \sim 1$. The means that power requirements for the external power supplies will be approximately equal for ER and MR fluid devices that provide comparable performance. The primary difference will be that the ER supply must provide high voltage at low current while the MR power supply provides higher current at only a low voltage.

4.3. Sensitivity to Temperature and Contaminants

Apart from the relatively low yield strength exhibited by most ER fluids, factors which have limited their use in many practical applications include temperature sensitivity and sensitivity to the presence of contaminants. The importance of both of these factors stems from the fact that all ER fluids ultimately depend on the movement of ions or electric charge to function. These factors are generally not a primary concern in the design of MR fluid devices.

5. Commercial MR Fluid Device Examples

While numerous examples of ER fluid devices have been described in the literature, few examples of MR devices are to be found. The MR fluid devices described below are examples of recently developed hardware currently being marketed by Lord Corporation under the Rheonetic™ name. These MR fluid devices are capable of meeting the requirements of real world applications and are presently being manufactured in significant volumes for original equipment manufacturers.

5.1. MR Fluid Linear Damper

The Rheonetic linear damper shown in Figure 3 is designed for use as a secondary suspension element in on- and off-highway vehicles[18]. Also shown is a family of measured force versus velocity curves for this damper.

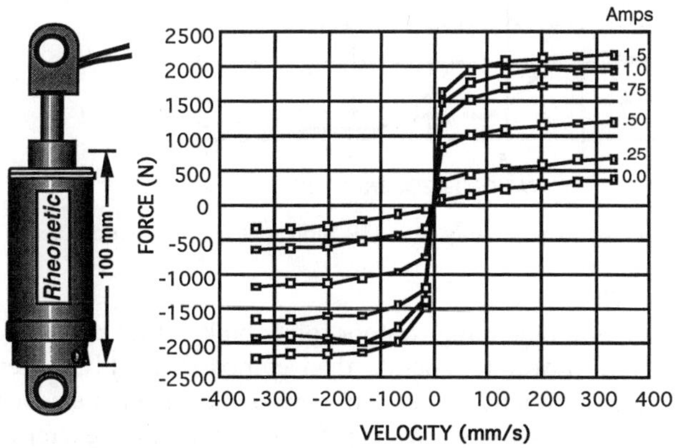

Figure 3 - Rheonetic™ MR fluid seat damper and measured force versus velocity curves.

This damper is a direct replacement for passive dampers used in seat suspensions. The damper has a diameter of 35mm. A magnetically controlled valve located in the internal piston regulates the flow of MR fluid within the damper. The damper is capable of generating enough force (>2000 N) to dissipate mechanical energy at the rate of over

600 watts and can accommodate a total stroke of about 50 mm. An input power of 4 watts is required to operate the damper at its nominal design current of 1 amp. Although the damper contains a total of about 50 ml of MR fluid, the actual amount of MR fluid that is activated at any given instant is only about 0.3 ml. A comparably performing ER fluid damper needs to activate about 30ml of ER fluid at any given instant.

The performance of a suspended seat equipped with the MR fluid damper is shown in Figure 4. A suspended seat acts largely like a single degree of freedom oscillator. If the seat is lightly damped it displays very desirable vibration isolation at high frequency, i.e., low transmissibility. However, with low damping the seat displays a troublesome large amplitude resonance at low frequency. Adding a large amount of fixed (passive) damping can control the resonance; however, one then degrades performance at high frequency. The power of the controllable MR fluid damper is shown by the curve labeled "controlled damping" in Figure 4. This is the transmissibility one obtains using a simple feedback controller that senses real-time seat position. Proper tuning of the parameters of the feedback system enable one to obtain the best of both worlds - elimination of the resonance and retention of good high frequency isolation.

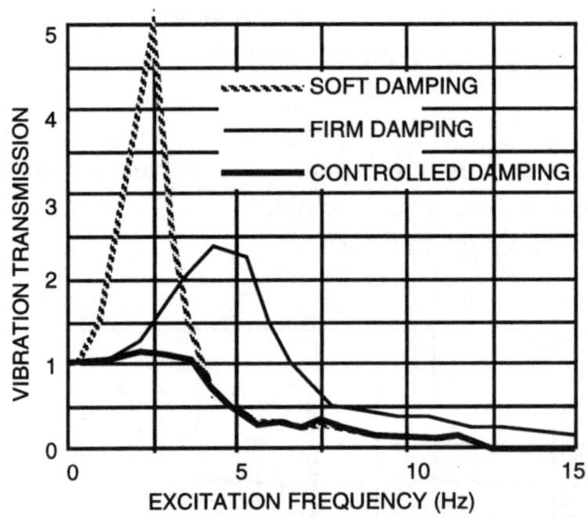

Figure 4 - Vibration transmissibility for a MR fluid controlled suspended seat.

5.2. MR Fluid Rotary Brake

MR brakes are smooth-acting, proportional brakes that are more compact and require substantially less power to operate than comparable eddy-current or magnetic-hysteresis brakes. Their simplicity and ease of control makes them a cost effective choice for a wide variety of applications ranging from controllable exercise equipment to precision tension control. MR fluid brakes are compact, smooth-acting and require low-power to operate. They are capable of high-torque at low-speed. An MR fluid brake that is currently being manufactured and sold as a controllable resistance element for programmable aerobic exercise equipment, e.g. stairclimbers and cyclers, is shown in Figure 5.

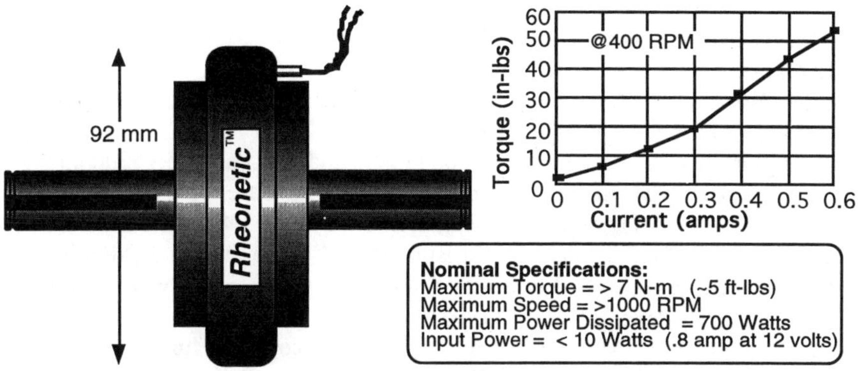

Figure 5 - Rheonetic™ MRB-2107 brake used in aerobic exercise equipment.

5.3. MR Fluid Vibration Damper

A small, MR fluid vibration damper that is being used for real-time, active-control of damping in numerous industrial application is shown in Figure 6. This damper may be also be used as a locking device. This damper does not require dynamic, sliding seals; the small amplitudes encountered in many vibratory situations allow bonded elastomeric elements to be used instead. Typically these dampers are used as controllable dashpots in parallel with separate spring elements. Alternatively, stiffness may be added to the elastomeric elements so that the damper functions as a controllable mount.

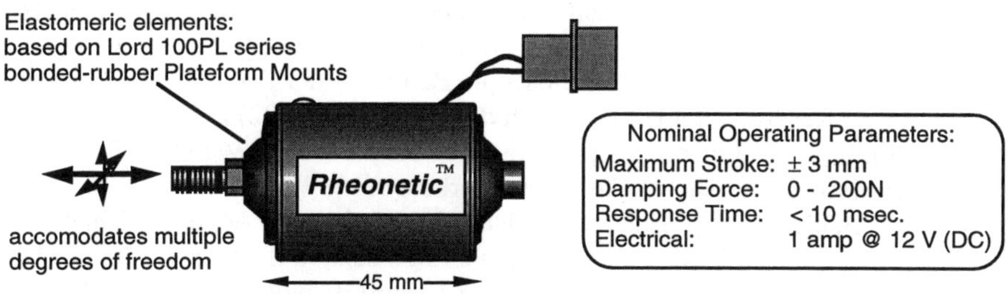

Figure 6 - Rheonetic™ MRD-1004 vibration damper.

6. Conclusion

Controllable fluid technology has seen a long infancy. The relatively low strength, temperature and contaminant sensitivity and need for high-voltage of most ER fluids has impeded their widespread commercialization. Unencumbered by these hurdles MR fluids are poised to enable controllable fluid technology to grow out of its infancy into a viable, commercially successful realm.

References

1. W.M. Winslow, "Induced Fibration of Suspensions", *J. Appl. Phys.*, **20** (1949) 1137-1140.
2. W. M. Winslow, "Method and Means For Translating Electrical Impulses Into Mechanical Force", U.S. Patent 2,417,850 (1947).
3. E.M. Shtarkman, U.S.Patent 4,992,360 (1991) and U.S.Patent 5,167,850 (1992).
4. W. Kordonsky, Magnetorheological Effect As a Base of New Devices and Technologies", J. Mag. and Mag. Mat, **122** (1993) 395-398.
5. K. D. Weiss, T. G. Duclos, J. D. Carlson, M. J. Chrzan and A. J. Margida, "High Strength Magneto- and Electro-rheological Fluids", Society of Automotive Engineers, SAE Paper # 932451 (1993).
6. J. D. Carlson and M. J. Chrzan, "Magnetorheological Fluid Dampers", U.S. Patent 5,277,282 (1994) and J. D. Carlson, M. J. Chrzan and F. O. James, "Magnetorheological Fluid Devices", U.S.Patent 5,284,330 (1994).
7. J. D. Carlson, "The Promise of Controllable Fluids", in *Actuator 94*, H. Borgmann and K. Lenz, eds., AXON Technologie, Bremen (1994) 266-270.
8. J. D. Carlson and Keith D. Weiss, "A Growing Attraction To Magnetic Fluids", Machine Design, Aug. 8 (1994) 61-66.
9. J. Rabinow, "The Magnetic Fluid Clutch", *AIEE Trans.*, **67** (1948)1308-1315.
10. "Magnetic Fluid Clutch", *National Bureau of Standards Technical News Bulletin*, **32 (#4)** (1948) 54-60.
11. J. Rabinow, "Magnetic Fluid Torque and Force Transmitting Device", U.S. Patent 2,575,360 (1951).
12. K. Weiss, J. D. Carlson and J. Coulter, "Review: Material Aspects of Electrorheological Systems", J. Intell. Mat. Sys. & Struct., **4** (1993) 13-34.
13. K. Weiss and J. D. Carlson, "Macroscopic Behavior of ER Fluids: Techniques for Measuring Response Time", Int. J. Mod. Phys.: B, **6** (1992) 2609-2623.
14. R. W. Phillips, "Engineering Applications of Fluids with a Variable Yield Stress", Ph.D. Thesis, University of California, Berkeley (1969).
15. T. G. Duclos, "An Externally Tunable Hydraulic Mount Which Uses Electrorheological Fluid", Soc.Automotive Engineers, SAE Paper # 870963 (1987).
16. T. G. Duclos, "Design of Devices Using Electro-rheological Fluids", Society of Automotive Engineers, SAE Paper # 881134 (1988).
17. J.D.Carlson and T.G.Duclos, "ER Fluid Clutches and Brakes - Fluid Property and Mechanical Design Considerations", in *Electrorheological Fluids*, J. D. Carlson, A. F. Sprecher and H. Conrad, eds., Technomic, Lancaster (1990) 353-367.
18. W. Leventon, "Fluid Damper May Make Truck Seats Safer", Design News, Oct. 4 (1993) 185-186.

BASIC OBSERVATIONS ON ELECTRO-RHEOLOGICAL FLUIDS

J.E.STANGROOM

ER Fluid Developments Ltd., Vincent Works, Brough,
Bradwell, Sheffield S30 2HG, U.K.

ABSTRACT

The early observations which led to the "Water Bridge" theory of ER fluids are described. The key features are, firstly that a very wide range of materials show some ER response, and, secondly, that although the electric field evokes the inter-particle force, it can be shown that the latter is not purely electrostatic.

Editor's note: This paper was invited by the editor on the grounds that in Proceedings of international conferences there should be a place for controversial articles. Though the referees have asked for changes that the author could not make, and the editor does not take sides in the scientific disputations, the editor considers that the paper contains descriptions of a substantial amount of experimental experience to warrant publication.

1. Introduction

In his last book[1], the Nobel prizewinner, Konrad Lorenz, wrote:- "We tend to forget that description is the foundation of all science. I do not mean to question the value of experiments, but observation must come first, in order to generate the questions for experimentation to answer." This Paper presents my early observations, which have never been published before and provide the foundation for all subsequent work.

The work started in 1967, when I became involved in attempts by the British Ministry of Defence to find better ER fluids for military applications:- the only fluid then available, silica gel in mineral oil, was abrasive, tended to settle out, and was not really active enough. Rather than trying to improve existing fluids, I decided to investigate the basic mechanism, hoping that this would eventually lead to better materials.

2. Microscope Observations

Being originally a biologist, I started by looking at ER fluids under a microscope. I glued two pieces of 0.1mm copper foil to a microscope slide with a 1mm gap between them for the ER fluid and connected them to a 200 V. d.c. power supply .

After a few false starts - the ER fluids had to be diluted to about 5 - 10% solids before I could see anything - I was able to see that the field made the particles form chains between the electrodes. Later, I cemented a coverslip over the foils so I could observe the action of an ER "valve", using a hypodermic syringe as a pump. When the

ER fluids were slightly disturbed, the chains dragged along the surface of the electrodes; at the same time, some particles became detached and re-attached themselves to existing chains, so the structure progressively changed from a more or less random initial lattice to definite fibres. More violent disturbance broke the chains completely, but short lengths of chain persisted, even though they were free of the electrodes, and re-attached wherever they had the opportunity. The electric field seemed to evoke a force of attraction between the particles, causing them to form into chains, and it was this which gave rise to the changes in flow properties observed in starch suspensions. Silica gel behaved similarly, although it tended to cling to one electrode (electrophoresis).

To learn more about the inter-particle force, I decided to observe as many solids and oils as I could find, and look for common features between those which behaved similarly to starch and silica gel. The behaviour of suspensions in an electric field under the microscope could be roughly divided into four types:-

a) The particles were apparently unaffected by the field. This was rare.

b) The particles migrated to one of the electrodes ("Electrophoresis"). This was probably the most common.

c) In one suspension, of powdered Nylon (Perlon) in isopropyl alcohol, the d.c. field caused the particles to be violently agitated, forming "vortices" which whirled round at high speed. The effect was so striking that it was seriously considered as a means of breaking up the boundary layer and improving heat transfer. This was finally identified as a form of electrostatic "pith ball" effect.

d) A wide range of materials formed chains, providing the base liquid was non-conducting. This included cellulose, "Sephadex"and other ion-exchange resins, vermiculite, various ceramics, carbon black and powdered aluminium.

Carbon black and aluminium suggested that the mechanism might be that shown in Fig 1. With no particles, the electric field between the electrodes would be linear, as shown by the dotted line. With conducting particles, on the other hand, charges would be induced to locally cancel the electric field (solid line) and these induced charges would attract each other electrostatically. Other solids, such as ceramics, would behave as conductors due to water adsorbed on their surfaces.

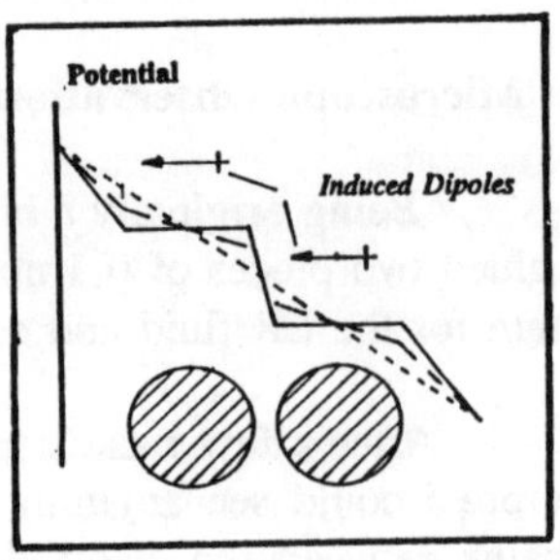

Figure 1 Basic Electrostatic Model

The currents passed by ER fluids under the microscope were too small to be easily measured, but an accident reminded me of their importance. The power supply had a low output impedance and no safety trip, so a flash-over while repeating the test with powdered aluminium in hydrocarbon oil copper-plated the microscope objective:- current through an ER fluid was clearly an important practical consideration. Very low resistance particles, like aluminium, could cause flash-over, but if the particles had a high resistance, the voltage drop in the particles themselves would reduce the induced dipoles (pecked line in Fig 1).

One possible compromise was to use highly conducting particles, but surround them with an insulating layer to prevent high currents. After some difficulty, I obtained a small sample of carbon particles coated with epoxy-resin. A small sample of this in mineral oil was tested under the microscope. When the field was switched on, the particles formed chains, just like starch or silica gel. However, after a few seconds with the field applied, the chains spontaneously broke in the middle and collapsed onto both electrodes.

Eventually I realised that drawing a straight line to represent the potential without added particles in Fig 1 assumes a uniform field. With 0.1mm foils separated by 1mm, the field will be anything but uniform; in fact, it will resemble the dotted line in Fig 2. Once this is realised, the "bi-directional electrophoresis" needs no further explanation: it is, in fact "Dielectrophoresis", a well-known phenomenon. However, if dielectrophoresis occurs when suspensions of conducting particles in insulating oils are tested under the microscope, then the behaviour of suspensions of starch, silica gel and many other materials is anomalous. The formation of persistent chains which re-form even when broken by external forces must be due to a force which is not purely electrostatic, although it is evoked by an electric field.

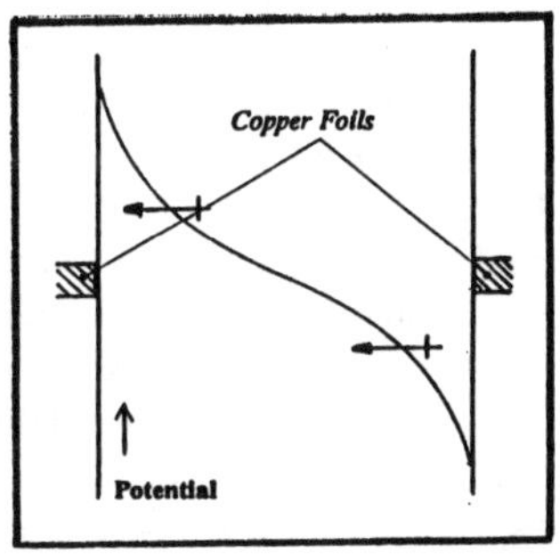

Figure 2 Modified Electrostatic Model

2. Quantitative Experiments

"Static Shear" measurements were used to compare different ER fluids:- we simply applied an increasing torque to a sample of ER fluid in a stationary Couette apparatus, and noted the stress at which shearing occurred. This was simple, easy to do, needed only a small amount of sample, and eliminated one parameter. With virtually no mechanical energy dissipated in the sample, temperature control was not a problem, and, by restricting ourselves to d.c., we reduced the number of variables to an absolute minimum. All these features allowed us to compare ER fluids very quickly.

In the first tests, a starch-based ER fluid in a constant field was sheared every ten minutes. The results are shown diagrammatically in Fig 3. The yield stress fell steadily, reaching zero in an hour or less. If the polarity was then reversed, the yield stress rose steadily, reached a peak which was more or less on the same level as the initial readings

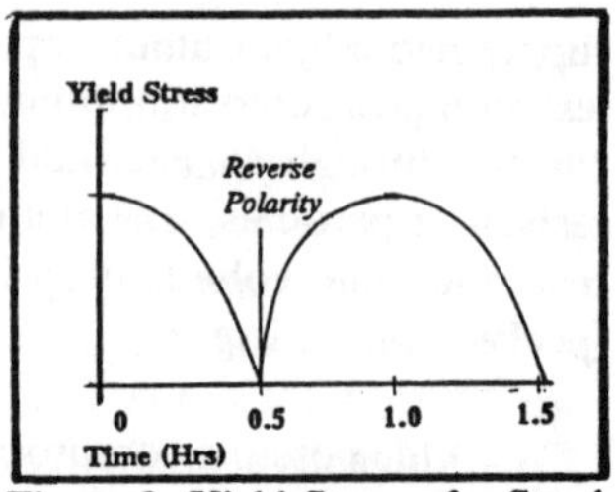

Figure 3 Yield Stress of a Starch ER Fluid

and then steadily declined to zero again. It was as if "something" was being propelled through the ER fluid from one electrode to the other, but it needed to extend throughout the whole ER fluid to give a good shear stress. In support of this, it was found that taking the field off and stirring the ER fluid vigorously "reset" the whole system back to the original state. These observations were made with maize starch; potato starch behaved rather differently. Eventually, it was found that washing with a continuous slow flow of distilled water for about a fortnight removed this effect, so that the yield stress remained constant. This sensitivity to trace contaminants was the first suggestion of a chemical mechanism. Although the response of a single sample of washed starch was now constant, it was still very variable between samples, and very poor by our present standards:- a yield stress of 1kPa was outstanding. When used at the same volume fraction, starch and silica gel were about equally active, but the former had a very considerable advantage in that its current density was an order of magnitude lower than that of silica gel.

The static yield stress measurement proved to be very rapid and convenient, and it was necessary to adopt some purely descriptive "Figure of Merit" to compare different ER fluids. Initially, we followed basic electrostatic theory and plotted the yield stress against $(\text{Field})^2$, but we found that, regardless of theory, a linear model with a "threshold" value fitted the results much better. The current, which was obviously not Ohmic, was initially fitted to a power law of the form (AE^n). However, there was no correlation of A and n values between related samples; furthermore, a variable value of n, implying variations in the conduction mechanism, was hard to accept. Our present quadratic representation was adopted:-

$$J = PE + QE^2$$

where P and Q are constants of the ER fluid, and J and E are respectively the current density and the field. This gives an excellent representation of the experimental results and the values of P and Q can be correlated between samples.

Many of the materials which looked promising under the microscope proved to have very poor response; this included the ceramics, carbon, vermiculite, and the polystyrene-based ion-exchange resins. Cellulose was also poor, although it, like starch, is a polymer of glucose, but "Sephadex", a further glucose polymer derived from dextran, behaved like starch. Clearly, although many materials could show traces of ER behaviour, the physico-chemical requirements for an effective ER solid were more complex than previously realised.

The essential clue was provided by an early (1943) paper by Kruyt and van Selms[2], who showed that starch, powdered coal and other hydrophilic solids dispersed in a hydrophobic oil solidified when very small amounts of water were added. The water formed "bridges" between the hydrophilic particles so that energy had to be expended against the surface energy at the oil-water interface in order to disrupt the resulting structure. The materials used by Kruyt and van Selms were clearly similar to those used in ER fluids, and the non-specific requirements for materials to show this effect - hydrophilic solid and hydrophobic liquid - would explain why so many materials that were apparently active under the microscope. Furthermore, such a water-based mechanism would lead to a yield point, rather than a change in viscosity. However, it was not clear how the application of an electric field could cause water to appear in a suspension.

In the manufacture of china clay, water is removed from "slip" by passing it between two mesh electrodes with high voltage between them:- electro-osmosis drives the water out through one electrode. To test whether starch could show electro-osmosis, I made a glass "U" tube, small enough to go into a centrifuge cup. This was filled with a suspension of starch in water, and spun down, so that the starch formed a packed plug at the bottom. Two electrodes were then placed in the arms of the "U" tube and a small voltage applied. The water rapidly rose in the arm containing the negative electrode; reversing the polarity reversed the effect.

Although electro-osmosis could clearly move water through starch, it was not clear how an electric field could affect water, which is uncharged. Electro-osmosis, is, in fact, the reverse of electrophoresis. In both, the solid particles acquire a charge with respect to the medium (water) by ionisation or selective adsorption of ions so they experience forces in opposite directions in an electric field. In electrophoresis, the medium is stationary and the particles move; in electro-osmosis the solid is held in place, so the medium moves through it. If internal electro-osmosis was, in fact, the basis of the mechanism of ER fluids, anything which increased the charge separation between the solid matrix and the water within it would enhance the ER effect.

In the "U" tube experiment, the water moved towards the negative electrode, suggesting that it had picked up positive ions. Starch is a polymer of glucose; the only active groups it contains are hydroxyl groups which are only weakly ionised. What was required was a more strongly ionisable group. Dr.(now Professor) E.Haslam, who had supervised my PhD., drew my attention to "Alginic Acid", a polymer of glucuronic acid which occurs in kelp. Glucose and glucuronic acid are shown in Fig 4: the only difference is the replacement of a hydroxyl group in glucose by carboxyl (-COOH). Carboxyl hydrogen is easily lost as a positive hydrogen ion which is taken up by the water. This would clearly increase the charge separation, and, hopefully, the ER effect.

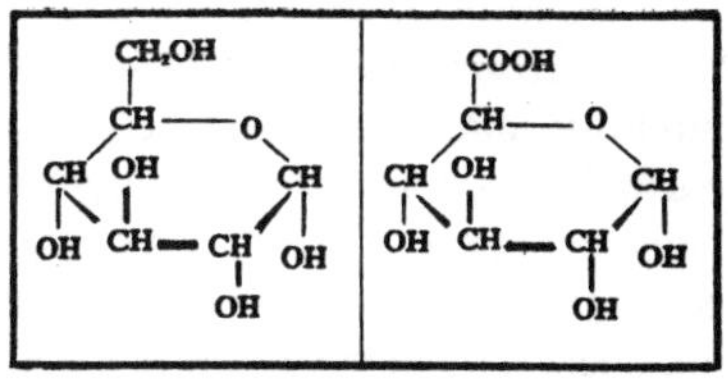

Figure 4 Glucose (left) and Glucuronic Acid

When finely-powdered alginic acid was suspended in an oil, and tested in the static rig, the yield stresses observed exceeded those of starch by a factor of about three. This was extremely encouraging, but the results were still very variable. Alginic acid, like starch, showed a time-dependent fall-off of yield stress when held continuously in a field for any length of time. This could be avoided by washing the alginic acid with 0.1N hydrochloric acid for several days, followed by distilled water. "Washed" alginic acid gave ER fluids which were reproducible and three times more active than any previously reported. This gave strong support to the basic hypothesis.

It was surmised that the water adsorbed in a solid particle could only form a bridge when electro-osmosis had concentrated it at one end; normally the water would be dispersed throughout the entire particle so its local concentration would not be high enough. This provided a rationale, for an earlier, and very puzzling observation, namely that cellulose fails to make a good ER fluid, even though it is chemically identical to starch. Cellulose is an inert, structural material so its molecules are close-packed in a pseudo-crystalline arrangement. Starch, being a labile food store, must have an open structure to make it readily available when required; this open structure enables it to adsorb much more water than cellulose and also allows the water to move through it when a field is applied.

If water is central to the ER response, then its concentration within the particle should be critical. Firstly, Kruyt and van Selms had noticed that if they added too much water to their suspensions, the yield stress fell:- the water now extended right around the particles and could act as a lubricant. The same effect was noticed in ER fluids in which the solid contained a great deal of adsorbed water:- the yield stress increased with the field only up to a certain point and sometimes actually fell as the field was increased further. Secondly, a completely dry solid should be inactive. A sample of alginic acid was dried in vacuo for a month over a large amount of fresh phosphorus pentoxide, removed from the desiccator as quickly as possible, suspended in oil and tested. Although its response was poor, it was still active. However, it was suspected that the sample had picked up water during the transfer, so the experiment was repeated but modified so that the sample was covered with oil before the air reached it. The resultant suspension showed no trace of ER activity. This extreme sensitivity of ER fluids to traces of water is worth bearing in mind when evaluating claims of "anhydrous" ER fluids.

In discussing water adsorbed onto solids, it is important to distinguish between mobility and availability. If the water-holding sites are close to each other, and only partially occupied, the energy barrier to be overcome when a water molecule transfers from one to another may be quite low, just as the energy barrier for electron transfer between atoms in a metal is low. On the other hand, completely removing a water molecule from a hydrophilic material may require a great deal of energy, equivalent to the electronic work function in a metal. Processes such as drying, freezing and evaporation involve the latter, so that it is very much more difficult to remove water molecules from partially hydrated materials than from liquid water. This allows "water based" ER fluids to function at temperatures well below 0°C and well above 100°C, just as strong brine has a much longer liquid temperature range than pure water.

Exact measurement of the amount of water adsorbed on a solid remains a significant problem. Water-generating side reactions with the reagents have been observed with Karl Fischer titration, and the water is so strongly held that the energetic methods needed to remove it will almost certainly modify the solid matrix irreversibly. Hydrogen exchange between the bound water and the active groups on the matrix make it difficult to apply spectroscopic measurements such as NMR or IR. Present methods, based on the conduction behaviour, are purely comparative:- they allow one sample to be compared with another but cannot specify an absolute amount.

The "water-bridge" approach involves too many unknowns for mathematical modelling to be reliable:- very little is known about the properties of thin films of water in a strong electric field between two hydrophilic surfaces. Eric Marson, in the Department of Applied Mathematics at Sheffield University, measured the "in-line" (i.e. parallel to the applied field) tension forces in a series of ER fluids with high solids content and then calculated the inter-particle forces from standard electrostatic theory. He showed that the forces he had observed were much greater than predicted by pure electrostatics, but agreed remarkably well with a "water-bridge" model. This must be regarded as indicative rather than firm evidence, since Marson had to make several simplifications.

3. Synthetic Solids

The variable results from acid-washed alginic acid made it desirable to find a purely synthetic alternative. Initially, vinyl acetate was co-polymerised with acrylic acid and the product hydrolysed to remove the acetate groups. Unfortunately, the solid product charred on drying, even at low temperature, almost certainly due to the elimination of water by internal lactones. However, it was realised that since the extent of ionisation was limited by the build-up of negative charge on the polymer chain, there would be no disadvantage in replacing all the hydroxyl groups by carboxyl, and using a homopolymer of methacrylic acid. Polymethacrylic acid, which is available commercially as "Versicol", was first used, but it was later found that cross-linked insoluble polymers were preferable, since the particles did not "cold-weld" together and become difficult to re-disperse. Other ionisable hydrophilic polymers, based on phenol/formaldehyde, have also been successful.

Synthetic materials allowed us to experiment with counter-ions other than hydrogen. This was not possible with alginic acid, which, having a very high molecular weight, became an intractable viscous mass when treated with alkali, whereas the monomers for the synthetic materials could be made into salts before being polymerised. As well as offering the possibility of improved ER fluids, salts offered another possible test of the "water bridge" theory.

Salts of polymeric acids are completely ionised, but the ions will tend to stay together as ion pairs. In water, counter-ions will be lost until prevented by the increasing negative charge on the polymer. Ion separation is driven by the hydration energy of the liberated ions, chiefly the positive ion. Thus, the activity of the ER fluids should follow the hydration energies of the counter ions.

Alkaline earth metal (e.g. Calcium) salts of polymethacrylic acid were very hard, and showed little ER activity:- presumably, the high charge density of the metal ion gave the ion pair bond significant covalent character. Trivalent ions, such as aluminium and chromium tended to cross-link the polymer, just as chromium salts are used for tanning leather; the complexes were labile, and the ER effects were not reproducible. The monovalent ions gave the most interesting results. Very bulky ions, such as guanidinium and tetramethyl ammonium gave deliquescent solids which were difficult to test - presumably the bond between the ion pairs was weak enough, even in media of low dielectic constant (e.g. air), to allow the ion to readily pick up water. The ER activity of the polymethacrylates of the alkali metals are shown in Fig 6., plotted against ionic hydration energy. The correlation is quite good.

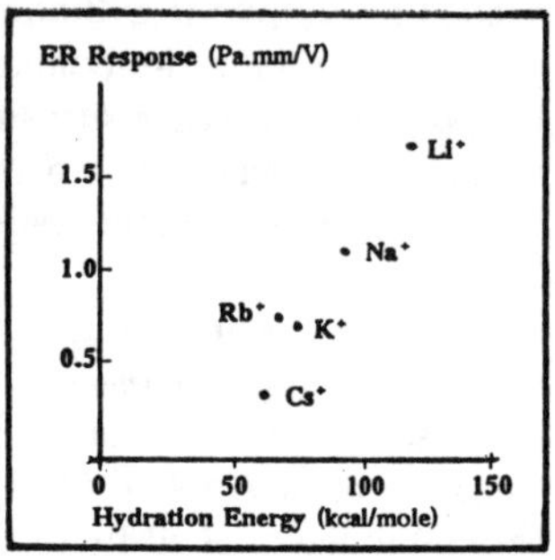

Figure 5 ER Activity and Hydration Energy

Conclusion

The "water-bridge" hypothesis brings together a whole range of disparate observations, including others published elsewhere[3] into a coherent structure. The effects described are seen in many ER fluids:- all the "anhydrous" ER fluids we have examined behave like as ours when their water content is varied. This approach is at least a useful guide for further investigation. On the other hand, because one mechanism may be partially correct without implying that all others must necessarily be incorrect. ER fluids are much more complex than they first appear, and, given their chemical diversity, it seems unlikely that they all work in exactly the same way.

References

1. Konrad Lorenz, *Here I am - where are you?* (Piper GmbH 1988; English translation Harper Collins, London, 1992)

2. H.R.Kruyt and F.G. van Selms. *Rec. trav. chim.* **62** (1943) 407, 415

3. L.F.Evans, I.Harness, P.R.Kermode and J.E.Stangroom. *Proc. Int. Conf. ER Fluids*. (1991) 154.

Closed Loop Controlled ER-Actuator

C. WOLFF

Institut für fluidtechnische Antriebe und Steuerungen (IFAS)
Rheinisch-Westfälische Technische Hochschule Aachen
Steinbachstraße 53
52074 Aachen, Germany

Abstract

The results of the investigation regarding the suitability of ERF when applied in hydraulics have shown so far that constructing electrorheological flow resistors for the control of pressure and volume flow is possible in principle. One of the main advantages when using the ER-technology in hydraulic systems can be seen in the high reaction rate of the ER-effect. The investigations presented in this article document the dynamic qualities of ER-fluids by means of a practical exploitation for the control of a cylinder actuator. Due to the particular possibilities for design of ER-control resistors a compact cylinder has resulted which differs considerably from traditional cylinder actuators in its construction and dynamic behaviour.

1 Introduction

Since the discovery of the ER-effect by the American W.Winslow more than 5O years ago a large number of ER-employment have been developed and examined. In the field of hydraulic applications the strong abrasive effect of the particles merged in the ER-fluids has interfered with a commercial utilisation. By the development of new types of ERF by Bayer AG an application in hydraulic installations has been enabled. It is exactly here that the advantages of the ER-technology, such as a high reaction rate of the effect and a simple construction of valves, can be fully exploited.

The elementary suitability of these ERF for use in hydraulics has already been proved in earlier studies[1]. For further investigation of the efficiency and of the performance limitations of the ERF an ER-actuator has been constructed which can be operated in a hydraulic circuit. By means of a position control system implemented on a micro-computer a freely position able, highly dynamical actuator has been developed, excelling by outstanding system dynamics. Additionally, by integrating the valve function into the cylinder case an extremely compact and simply constructed positioning unit could be developed that has no agitated elements except for the piston.

2 Choosing the hydraulic circuit concept

In general different concepts of circuit can be employed for activating hydraulic cylinders. Making the choice depends mainly on the expected system response and on dynamics as well as on circuit-technical requirements. Figure 2.1 shows two exemplary circuit concepts for the activation of a hydraulic cylinder. Here each flow resistor is to be considered as a separately controllable control edge. The left part represents a 4-edge-control unit as it can be employed for controlling a synchronising cylinder. In this case two resistors are simultaneously switched and each connected to the supply- or tank line by means of the bridge circuit represented. This kind of circuit is noted for a high reinforcement of velocity and power. The disadvantage results from the fact that four flow resistances are required which have to be alternatively activated by two power supply units.

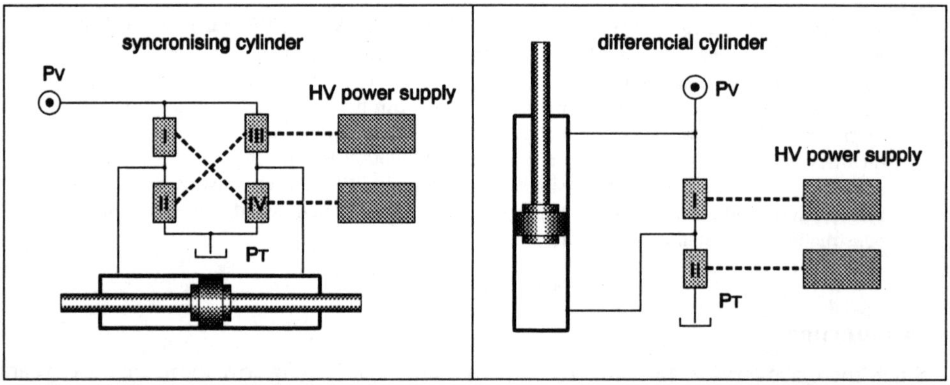

Fig. 2.1 Construction principle of a cylinder control unit with ER-flow resistors

In order to reduce the constructive requirements the number of resistors can be halved by using a differential cylinder. Regarding the circuit concept shown in the right part of the illustration the cylinder chamber in the large piston area is activated via two ER-resistors. Supply pressure is constantly effective on the smaller piston area. For extending the cylinder valve II is closed and valve I is opened. At this moment supply pressure takes effect on both piston areas and the cylinder is extended due to the difference of the piston areas. In the reverse case the lower tank pressure takes effect on the large piston area and the cylinder runs in because of the pressure difference adjacent to the piston.

The disadvantage of the two concepts presented here results from the fact that at least two flow resistances and two power supply units are required for control. In order to keep requirements for fabrication and construction as low as possible the actuator is to be operated by one control edge and one power supply. This can be realised by the circuit variants represented in figure 2.2.

Variant II and III enable the positioning of the driving unit by means of a constant and an adjustable control resistor. The running-in of the cylinder being realised mechanically by means of a spring with variant II and hydraulically via supply pressure with variant III. With both of these concepts supply pressure is constantly higher than load pressure. Supply pressure, however, conforms to load pressure with variant I and only one consumer at a time can be connected to supply. As this variant is the simplest solution for a one-edge control unit and only one driving unit is installed in the test stand, circuit variant I was taken up for realising the actuator.

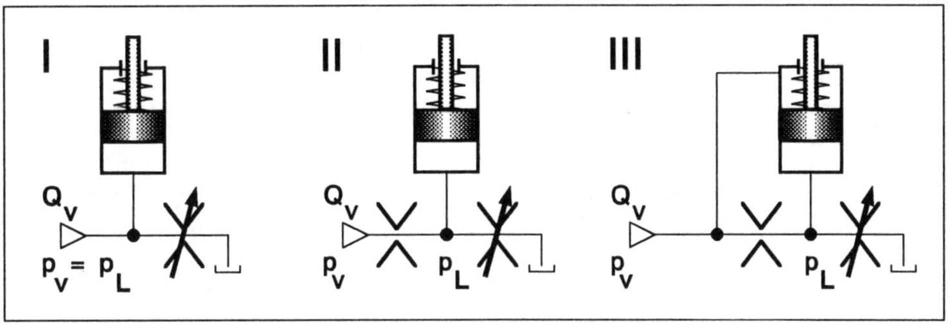

Fig. 2.2 Simple circuit concepts for cylinder driving units

3 Construction of the ER-actuator

3.1 Working principle

The basic diagram shown in figure 3.1 reveals the schematic construction of the actuator. Differing from variant I shown in figure 2.2 the small piston area is connected to the tank line. Thus it is not ambient pressure that is effective here, but the tank pressure increased by flow losses. Thereby the restoring forces are reduced.

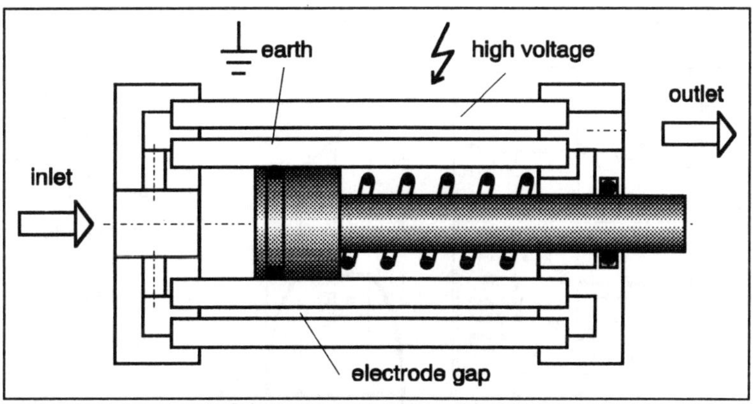

Fig. 3.1 Schematic construction of the ER-actuator

The piston is let in a cylinder liner, the periphery of which also forms the earthend part of the ER-annular gap. Another tube is arranged around the cylinder liner, forming with its internal face the external part of the electrode gap. This ring is insulated from the other component parts and connected to high voltage. Depending on the field intensity pressure is being built up by the electrode gap. As long as the pressure required for extending the piston has not been reached, the volume flow as a whole discharges via the resistor.

With pressure exceeding load pressure in dependence on the field strength, the piston is accelerated and the piston rod is extended. When the valve is completely closed, the driving unit reaches its maximum extending velocity. Exactly at the moment when the cylinder is totally extended, the whole volume flow discharges via the ER-resistor. This way a limitation of pressure within the system is automatically ensured rendering a separate pressure relieve valve unnecessary. As soon as the field strength within the gap is reduced, the system pressure decreases and the piston is pushed back by the power of the preloaded spring or an external load. The volume flow via the ER-valve here results from the total of the pump volume flow and of the chamber volume flow displaced by the piston. While the extending velocity is limited by the adjusted volume flow of the pump, the maximum running-in velocity depends on the flow resistance of the ER-gap. As in the Newtonian case pressure above the gap increases linearly with the volume flow, the piston accelerates as long as a balance of powers between the external power and the chamber pressure of the cylinder ensures.

3.2 *Theoretical dimensioning of the ER-resistor*

For laying out the geometry of the gap it was initially appointed that the actuator is to produce an external power of approx. 600 N. The pressure generated by the ER-resistor was fixed at 20 bar. Hereby results a theoretical piston diameter of 19,54 mm. In order to simplify the selection of the sealing a piston diameter of 20 mm is chosen and taken as a basis for the further calculations. The inner diameter of the gap was fixed at 30 mm, so that the cylinder liner has a still sufficient wall thickness of 5 mm in the range of the gap. The length of the gap was fixed at 100 mm and the gap height at 0,75 mm.

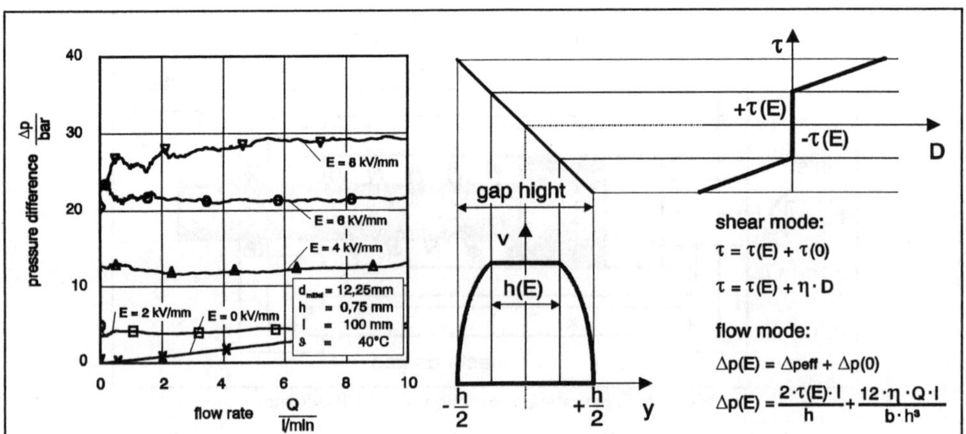

Fig. 3.2 Comparison measurement for rating the power limits of the ER-actuator

By means of the comparison measurement represented in figure 3.2 and of the simplified calculation formulas according to the Bingham - model it can now be estimated if the demanded difference of pressure is built up with the given measures[2]. As a calculation basis for the determination of the yield stress $\tau(E)$ there are comparison measurements carried out at an annular gap with a length of 100 mm, an average diameter of 12,25 mm and a height of 0,75 mm.

According to the Bingham-model the actual pressure difference only depends on the geometrical ratio l/h and the theoretical yield stress $\tau(E)$. With gap length and gap height being identical for both gaps it results:

$$\Delta p_{eff,\,actuator} = \Delta p_{eff,\,gap} = \Delta p(E) - \Delta p(0) = 26{,}5\ \text{bar} \qquad (1).$$

The values were acquired with a volume flow of 4l/min, since the actuator is to be operated for this pump volume flow. With a basic viscosity of 34 mPas the Newtonian part of the pressure difference of the actuator resistor can be calculated to 0,66 bar. Hence it follows that a total pressure difference for the actuator of 27,2 bar can theoretically be reached. So with a piston diameter of 20 mm a maximum control force of 854,5 N can be calculated.

The next step provided a dimensioning of the restoring spring for running-in the piston. In this case the starting point of the calculation is the pressure difference built up by the ER-resistor. With a volume flow of 4 l/min the maximum extending velocity amounts to 0,21 m/s when the ER-valve is closed. Provided that the piston velocities should be identical when extending or running in, the required spring tension can be calculated via the balance of powers at the piston. With a volume flow of 8 l/min when running in the piston there is hence a pressure difference of 1,34 bar .

3.3 *Technical design*

Regarding the design of the actuator, great importance was attached, as already mentioned, to a compact construction. That is why the advantages of the ER-technology, concerning the simple construction of valves, can be graphically documented. As shown in figure 3.3 the actuator consists in the main of several nested cylinder liners and two housing parts. These parts, fitted on the face of the liners, serve as guide and centring for the tubes and are braced with each other by four threaded rods. The inner cylinder liner fulfils a double function. On the one hand it serves as a cylinder liner and guide for the piston, on the other hand as an earthend inner electrode of the ER-flow resistor. Another tube is arranged around the cylinder liner, forming the external ring of the electrode gap. The outer electrode is let in a further tube which is coated inside by synthetic material for insulation. The connecting blocks are equipped with drill holes for the hydraulic connections as well as for temperature- and pressure transducers. In order to ensure a flexible formation of the high-voltage connection the voltage supply is let outside via a threaded rod. Because of this, however, the volume of the construction had to be slightly enlarged due to the required insulations.

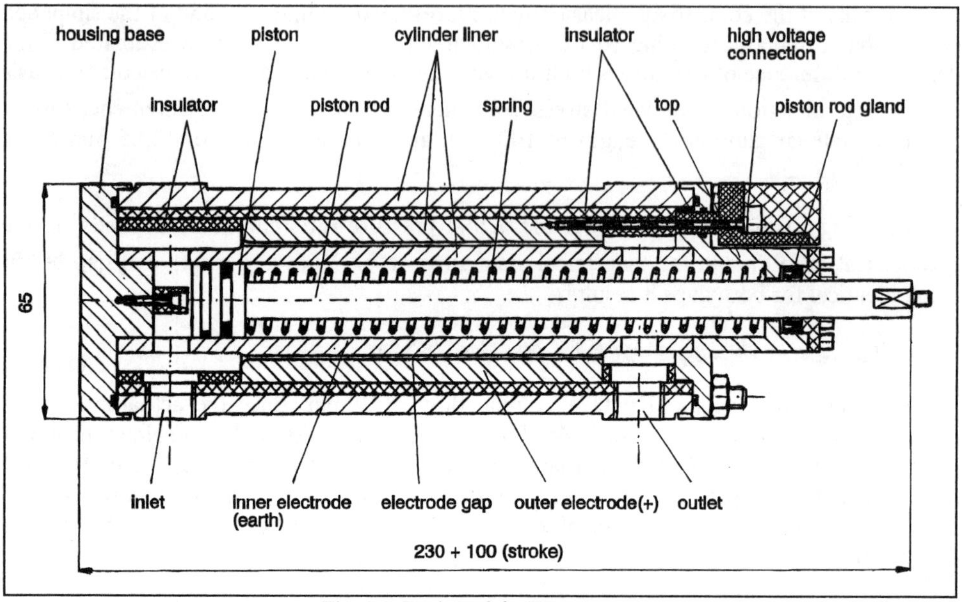

Fig. 3.3 Design of the ER-actuator

4 Description of the control system

4.1 Hydraulic test stand and measuring technique

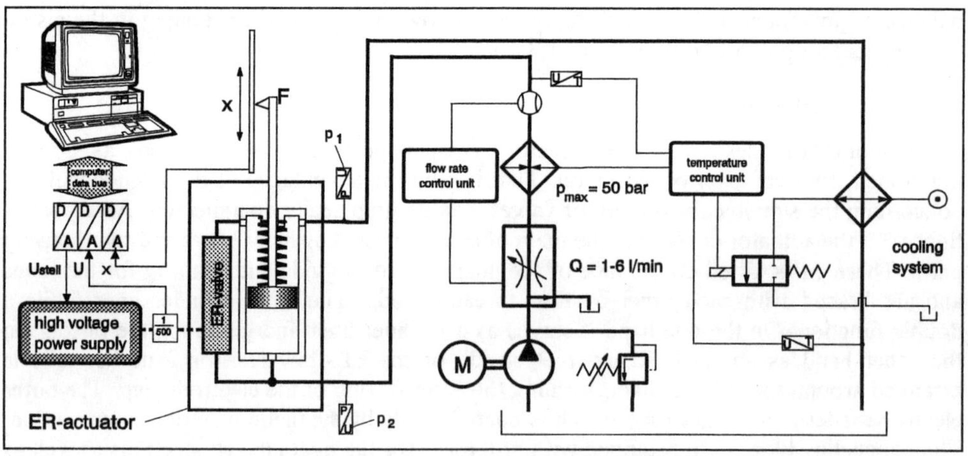

Fig. 4.1 Schematic construction of the test stand

The actuator was operated in the test stand represented in figure 4.1. The pressure supply is carried out by means of a gear pump delivering a constant volume flow of 6 l/min at a rotational speed of 1500 r.p.m. Via a three-way flow control valve the volume flow can be adjusted to the desired value. By means of transducers the load pressure difference, volume flow, temperature as well as the present position of the piston can be recorded for further processing in a micro computer.

4.2 Construction of the control unit

For examining the positioning behaviour of the actuator the driving unit is operated in a closed positioning loop. In order to adjust the control parameters and at the same time to enable a flexible reaction to the special demands of the fluid, the controller structure was implemented on a micro computer. With a positioning control the position of the piston is recorded as a controlled variable by a position transducer and transferred to the controller via an AD-converter. With the aid of the programme further state variables such as velocity and acceleration can be calculated by differentiation from the position signal. By means of a modified PID-controller the correcting variable is then determined and given to the supply terminals.

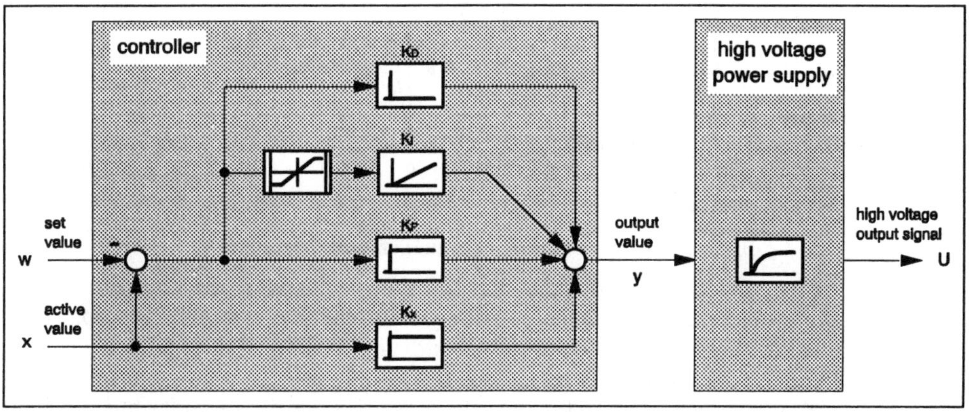

Fig. 4.2 Controller structure of the modified PID-controller

Figure 4.2 shows the structure of the controller. In addition to a conventional PID-controller, consisting of a proportional, integrating and differentiating part, this structure is implemented with an offset voltage proportional to position. By employing a switching integrator the drifting-off of the I-part can be avoided. For optimising the positioning response the controller can be parameterised differently in a rectangular activation for both set values. With the computer in use scanning times up to 0,5 ms are realisable this way. As shown in the graph, the dynamic systems behaviour of the high-voltage amplifier in the development of the controller structure must be taken into account when using ER-technology in closed loops, as it decisively influences the dynamic behaviour of the positioning system.

5 Results

5.1 Influence of voltage supply on step function response behaviour

In order to demonstrate the mentioned influence of the dynamics of the high-voltage supply source, figure 5.1 illustrates the step function response behaviour of the actuator with two different devices of power supply. The supply source were activated by a rectangular set value of voltage, the actual voltage and the piston position being measured.

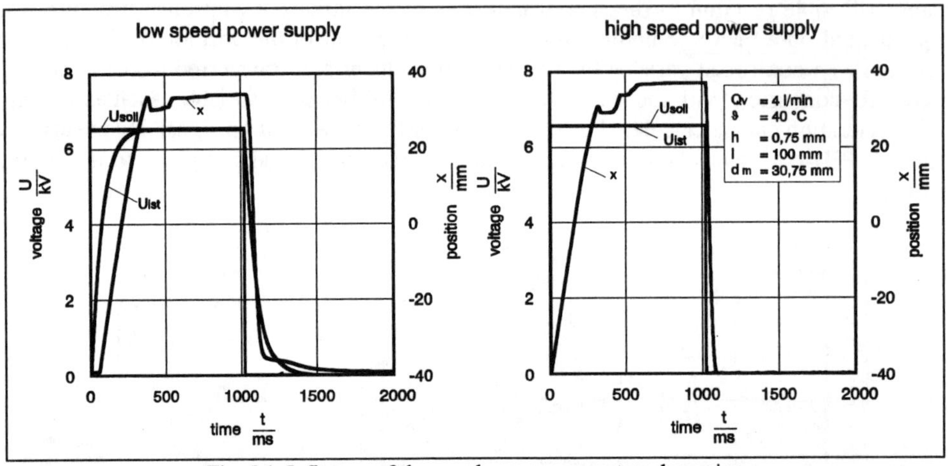

Fig. 5.1 Influence of the supply source on system dynamics

The left side of the graph shows the step function response of the actuator when a constant-current power unit is used, the output signal of which is smoothed by capacitors. Since in this case the high-voltage field is only retardingly established due to the integrating behaviour of the capacitors, the actuator, too, reacts only after a delay time of approx. 100 ms. As well when switching off the field the actuator goes back into its final position with considerable retardation. Here, too, first the capacitors have to be discharged. In the right part of the graph, however, a highly dynamical voltage amplifier was applied, building up and taking down the electric field immediately after the set value decrease. Delay time now amounts to only a few milliseconds and cannot be seen in this representation. This comparison impressively documents the fundamental influence of the dynamics of the high-voltage power supply on the complete system behaviour. In order to render the dynamics of the positioning system as high as possible for the following investigations, the highly dynamical voltage amplifier was employed for activation.

5.2 Positioning behaviour of the ER-actuator

Figure 5.2 demonstrates the positioning behaviour of the actuator with a rectangular set value signal and with the driving unit being loaded with a strong spring in addition to the restoring spring.

Here, too, the excellent dynamics of the driving unit becomes obvious. The actuator moves immediately after switching the field on resp. off. Moreover, the constant course of velocity when extending shows that the valve is completely closed and that the actuator reaches its maximum extending velocity of 0,2 m/s.

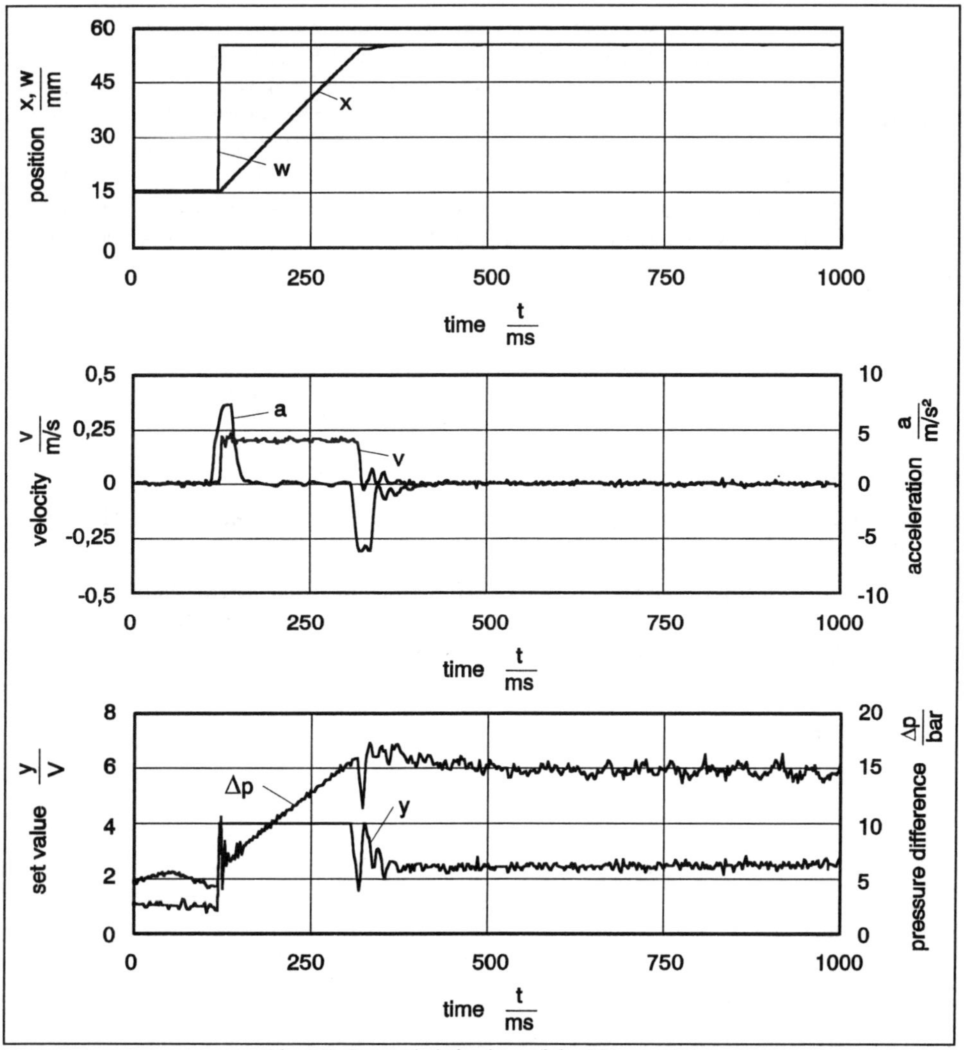

Fig. 5.2 Dynamic behaviour of the ER-actuator

Another indication for the fast operating ER-valve is the pressure peak immediately after switching on the field. Due to the hydraulic inductivity of the ERF the static pressure within the piston chamber rises with the sudden braking of the liquid column. By this increase in pressure the piston can be accelerated considerably faster and reaches a constant extending velocity already after approx. 3 ms.

6. Summary

This article summarises first results of investigations dealing with the construction of a freely position able cylinder driving unit by means of the ER-technology. As a result an extremely compact and simply constructed ER-actuator has been developed showing a very good dynamical behaviour. The positioning behaviour that was presented is based on a concept of control known in the art, that has not yet been adapted to the special demands of this technology. The application of most modern micro electronics has, however, enabled the development of new control strategies, which can cope with the extremely high dynamics of ER-valves. Thus there is the possibility of further improving the control behaviour here presented as well as the exactitude of position.

7 Acknowledgements

This study was supported by Bayer AG, Leverkusen and Schenck AG, Darmstadt.

8 References

1. Ch. Wolff, Elektrorheologische Flüssigkeiten - neue Möglichkeiten in der Hydraulik? **5**, Ölhydraulik und Pneumatik (1994), p290-296.

2. Z. Lou, R. D. Erwin and F. E. Filisko, *Behaviors of electrorheological valves and bridges,* Conference on Electrorheological (ER) Fluids, Southern Illinois University at Carbondale, Illinois, Oct. 15-16, 1991, p398-423.

Finite Element Analysis of Electrorheological Fluids

R. Tao, Qi Jiang, H. K. Sim*
Department of Physics, Southern Illinois University, Carbondale, IL 62901, USA

ABSTRACT

The Laplace equation for an infinite chain of dielectric particles in an electrorheological fluid is solved with the finite element analysis (FEA) method. The FEA results reveal that the dipole model gives a good approximation in calculating energy, even if neighbor particles are very close and the mismatch is moderate. However, the dipole model underestimates the pair attractive force when the particles are in or near contact. The conductor model is a good approximation when the two particles are near contact and the mismatch is very big. A formula to calculate the neighbor pair attractive force for the whole range is proposed. The FEA results agree with this formula reasonably well.

1. Introduction

An electrorheological (ER) fluid consists of fine dielectric particles suspended in a liquid of low dielectric constant.[1–5] In recent years, many attempts have been made to provide detailed theoretical descriptions for ER fluids.[6–8] A widely adapted approach is to treat each particle as a point-dipole associated with a hard core.[6,7] This dipole model has some success in explaining and predicting the ER effect. For example, a prediction based on the dipole model that the ground state of an ER fluid is a body-centered tetragonal (bct) lattice[7] was experimentally confirmed.[9]. Various computer simulations based on the dipole model also produced some good results.[10–12]

When the ER fluids have a small dielectric mismatch between the particles and the base liquid, the dipolar approximation may be sufficient. However higher-multipole contributions may be important when the dielectric mismatch is large and the particles are nearly touching.[13–17] Recently, a conduction model is proposed to explain the ER effect under dc or low ac fields.[18] Since this model is quite different from the dipole model, it is important to have knowledge about the following information: Under what condition is the dipolar approximation valid? Under what condition is the conduction model valid? To answer the above questions, we must have some exact numerical solutions to compare with.

When an ER fluid is in an electric field, particles form chains between the two electrodes. A physical chain and its infinite images are equivalent to an infinite chain. Therefore, in this paper we will solve the Laplace equation for an infinite chain system with the finite element analysis (FEA) method. Our ER system has spheric particles of dielectric constant ϵ_p and the liquid of dielectric constant ϵ_f. The results are clearly determined by two parameters, the dielectric mismatch,

⋆ On sabbatical leave from Department of Physics, National University of Singapore

$\kappa = \epsilon_p/\epsilon_f$, and the distance between two neighbor particles d in the unit of sphere radius a.

A similar problem was studied in Reference 13 with truncated multipole expansion for a chain of up to 40 particles. Their result indicates that the interaction force and energy between two touching spheres are divergent even if the spheres have a finite dielectric constant. Recently, Anderson also studied this problem and finds that for a large κ, the attractive force between two touching particles is proportional to κ^2, not divergent.[14]

Our FEA method enables us to solve the system exactly for the whole range of interest. We will report about the electric field, interaction energy, and attractive force as a function of κ and d. Based on our FEA data, we have proposed a formula to calculate the pair attractive force. This formula does not only agree with the conductor model and Anderson's result at a very large κ and $d \to 2a$, but also approaches the dipole model when κ is not very big.

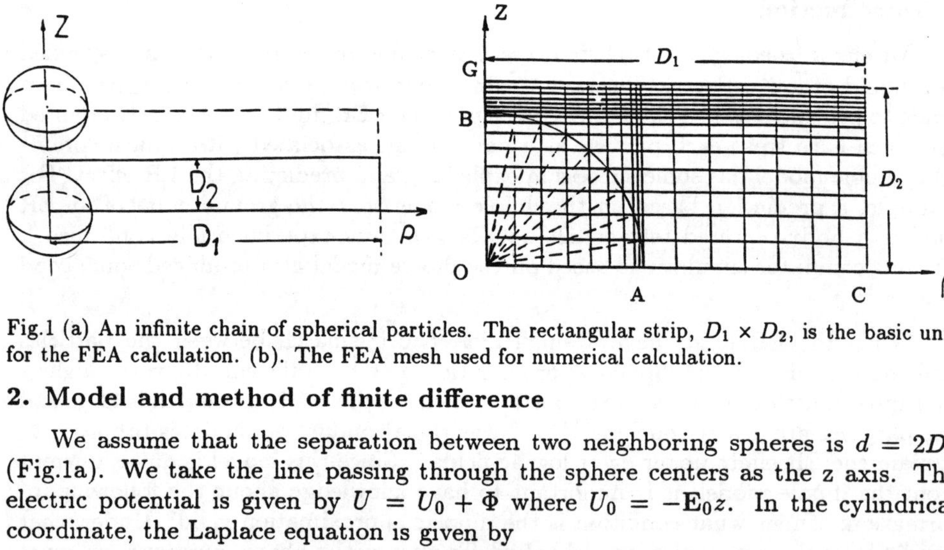

Fig.1 (a) An infinite chain of spherical particles. The rectangular strip, $D_1 \times D_2$, is the basic unit for the FEA calculation. (b). The FEA mesh used for numerical calculation.

2. Model and method of finite difference

We assume that the separation between two neighboring spheres is $d = 2D_2$ (Fig.1a). We take the line passing through the sphere centers as the z axis. The electric potential is given by $U = U_0 + \Phi$ where $U_0 = -\mathbf{E}_0 z$. In the cylindrical coordinate, the Laplace equation is given by

$$\frac{\partial^2 \Phi}{\partial \rho^2} + \frac{1}{\rho}\frac{\partial \Phi}{\partial \rho} + \frac{\partial^2 \Phi}{\partial z^2} = 0. \tag{2.1}$$

U and Φ are functions of ρ and z only. The symmetry provides $\Phi(\rho, z) = -\Phi(\rho, -z)$. We only need to solve the Laplace equation (2.1) for a two-dimensional strip in Fig.1. Theoretically, D_1 should extend to infinity. We have found in our calculation that as $D_1 \geq 10a$, the solution is almost independent of the values of D_1. In applying the FEA method, we generate a mesh in Fig.1b to discretize the Laplace equation. We first divide the angle $\angle AOB$ evenly into a large number of divisions. Then, at all cross-points of the circle and the radii, we draw lines parallel to the z axis and ρ axis. Outside the circle, we further divide $\overline{AC}$ evenly into a large number of divisions. We then divide $\overline{BG}$ evenly into many divisions. The boundary conditions are given by $\Phi(\rho, 0) = \Phi(\rho, D_2) = 0$, $\partial\Phi(0, z)/\partial\rho = 0$, and $\Phi(D_1, z) = 0$. On the sphere surface U is continuous and

$$\kappa\,(\rho\partial U/\partial\rho + z\partial U/\partial z)\,|_{r=a-} = (\rho\partial U/\partial\rho + z\partial U/\partial z)\,|_{r=a+} \tag{2.2}$$

where $r = \sqrt{\rho^2 + z^2}$. Our present results are only from the dielectric mismatch model which is valid for ER fluids made of perfectly insulating materials or for ER fluids in a high ac field. However, our method can be easily extended to include cases with non-vanishing conductivity.

Since $\mathbf{E}_0$ and hence the applied voltage is maintained as a constant, the following electrostatic energy W characterize the interaction,

$$W = \frac{1}{8\pi}\left(\int \mathbf{E}\cdot\mathbf{D}d\mathbf{r} - \int \mathbf{E}_0\cdot\mathbf{D}_0 d\mathbf{r}\right) \tag{2.3}$$

where $\mathbf{D}_0 = \epsilon_f\mathbf{E}_0$ and $\mathbf{D} = \epsilon(\mathbf{r})\mathbf{E}$ with $\epsilon(\mathbf{r}) = \epsilon_f$ or ϵ_p according to the region. After the thermal energy is ignored, W should be maximum in the ground state. The interaction energy per particle is given by

$$w = \frac{W}{N} = \frac{(\epsilon_p - \epsilon_f)}{8\pi}\int_{sphere} \mathbf{E}_0\cdot\mathbf{E}d\mathbf{r} = \frac{1}{2}\epsilon_f\mathbf{E}_0\mathbf{p} \tag{2.4}$$

where $\mathbf{p}$ is the induced dipole moment of the sphere. Therefore, in the preferred state each sphere has the strongest dipole moment. When we stretch the chain, the elastic force per particle is given by $f_e = \partial w/\partial d$. The neighbor pair force is given by

$$\mathbf{f}_{ij} = \int \frac{\sigma_i\sigma_j\mathbf{r}_{ij}}{\epsilon_f r_{ij}^3} ds_i ds_j. \tag{2.5}$$

Usually, f_e is stronger than the neighbor pair force from Eq.(2.5).

Under the dipolar approximation, the force between two dipoles in the chain is attractive in the z direction, $f_{ij}^d = -6p^2/(\epsilon_f d^4)$. With the self-consistent method, we can find the local field, $\mathbf{E}_{loc} = \mathbf{E}_0/(1 - 4\beta\mathbf{a}^3\zeta(3)/\mathbf{d}^3)$ where $\zeta(3) = 1 + 1/2^3 + 1/3^3 + \cdots$=1.20206 and $\beta = (\kappa - 1)/(\kappa + 2)$. The induced dipole $\mathbf{p} = \epsilon_f\beta a^3\mathbf{E}_{loc}$.

The attractive force between two neighbor dipoles is given by

$$f_{pair} = -\frac{6\beta^2 \epsilon_f E_0^2 a^6/d^4}{(1-4.8082\beta a^3/d^3)^2}. \qquad (2.6)$$

Under the dipolar approximation, $w = \epsilon_f \mathbf{E_0} \cdot \mathbf{p}/2 = \epsilon_f a^3 \beta E_0^2/[2(1-4.8082\beta a^3/d^3)]$. When we stretch the chain, the elastic force per particle is $f_e = \partial w/\partial d = \zeta(3) f_{pair}$, 20% stronger than f_{pair}.

Our conductor model has all dielectric particles in the chain replaced by metallic spheres. This model can be realized from the dielectric mismatch model by letting $\kappa = \epsilon_p/\epsilon_f \to \infty$. The electric field is concentrated between two neighbor metallic spheres. The potential difference between two neighbor spheres is $E_0 d$. The electrostatic energy can be estimated by

$$w = \frac{1}{8\pi}\int_0^{\pi/2} \frac{2\pi a^2 \sin\theta\cos\theta d\theta (E_0 d)^2}{d-2a\cos\theta} = \frac{\epsilon_f E_0^2 d^3}{16}[\ln(\frac{d}{d-2a}) - \frac{2a}{d}]. \qquad (2.7)$$

The attractive force between these neighbor spheres is approximated by

$$f = -\frac{\epsilon_f E_0^2 d^2}{16}[\frac{2a}{d-2a} + \frac{4a}{d} - 3\ln(\frac{d}{d-2a})]. \qquad (2.8)$$

The results from the above two models will be compared with our finite element calculations.

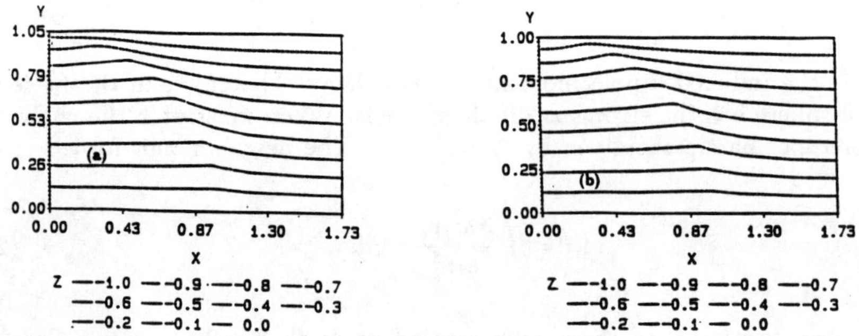

Fig.2 The electric potential U (in unit of $E_0 a$). (a) $\kappa = 2.5$ and $d = 2.1a$. (b) $\kappa = 2.5$ and $d = 2.0001a$.

3. FEA results and discussions

In our calculation, we vary κ from 2 to 50000 and make the distance d from $2.1a$ approaching to $2a$.

3.1. Electric Fields.

Fig.2a shows the electric potential U at $\kappa = 2.5$ and $d = 2.1a$, while Fig.2b shows U at $\kappa = 2.5$ and $d = 2.0001a$. In both cases, the electric field inside the sphere is quite uniform, indicating that the dipolar approximation is quite good at $\kappa = 2.5$ even if the two spheres are very close.

When $\kappa = 10$, the electric field at $d = 2.1a$ is still not too far from the dipolar approximation (Fig.3a). However, at $d = 2.0001a$ (Fig.3b), the field is different from the dipolar approximation because the field inside the sphere is no longer uniform.

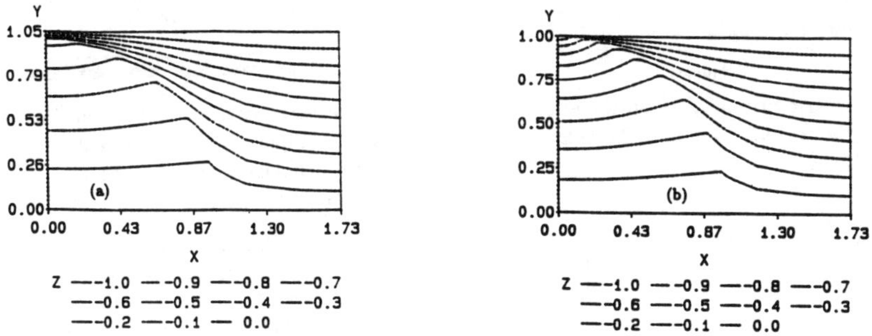

Fig.3 The electric potential U (in unit of E_0a). (a) $\kappa = 10$ and $d = 2.1a$. (b) $\kappa = 10$ and $d = 2.0001a$.

When $\kappa = 5000$, the electric field is quite different. A comparison of Fig.4a and Fig.4b shows that as the spheres get closer, the electric field is more concentrated between the two spheres. In addition, the electric potential inside the sphere is almost a constant which indicates that the conductor sphere model may be good for this situation.

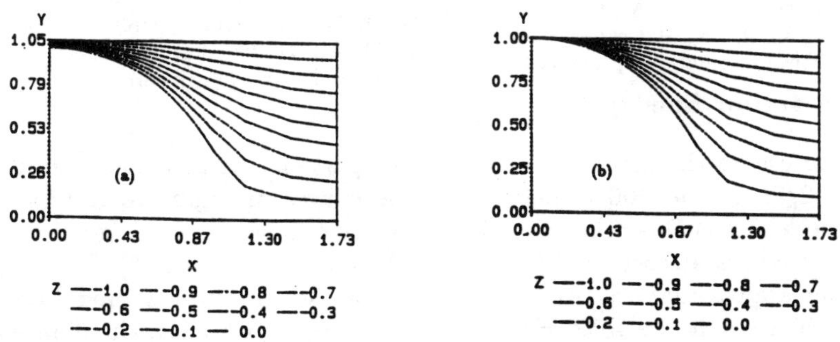

Fig.4 The electric potential U (in unit of E_0a). (a) $\kappa = 5000$ and $d = 2.1a$. (b) $\kappa = 5000$ and $d = 2.0001a$.

To examine the detail of the electric field, in Fig.5 we plot U right outside the sphere surface as a function of angle θ for $\kappa = 50000$, and $d = 2.0002a$, $d = 2.002a$, $d = 2.02a$, and $d = 2.2a$ respectively. It is clear that at $\kappa = 50000$, the dielectric particles are close to conductor particles. Indeed, U only deviates from zero near $\theta = 0$ or π where the surface charge is concentrated. In the case $d = 2.2a$, U is almost zero on the whole surface, $|U/(E_0 a)| \leq 5 \times 10^{-3}$, indicating very low surface charge density. As d is reducing, the electric field gets stronger in the polar region.

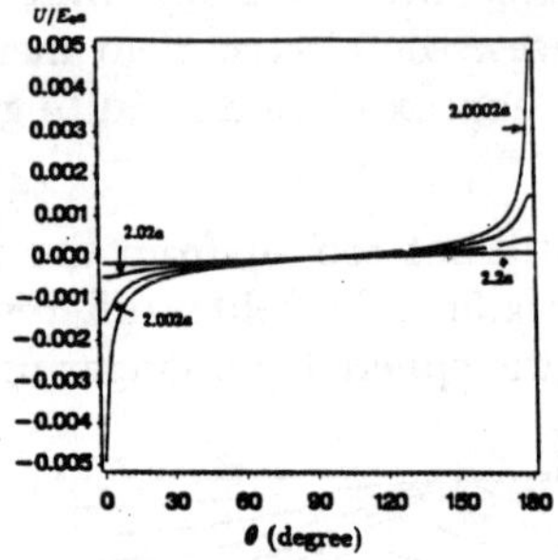

Fig.5 The electric potential U on the sphere surface vs θ at $\kappa = 50000$ for $d = 2.0002a$, $2.002a$, $2.02a$, and $2.2a$.

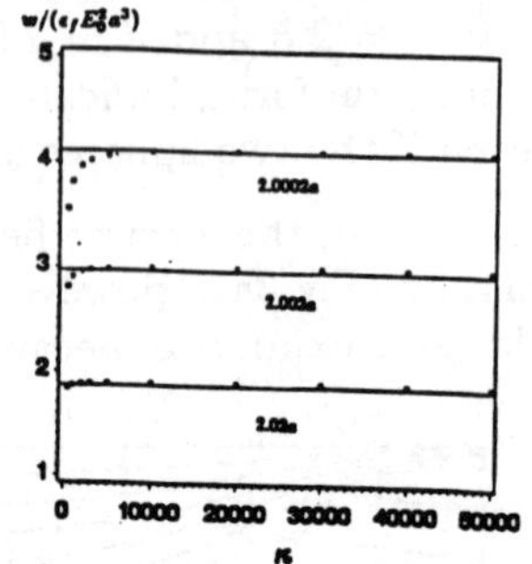

Fig.7 The electrostatic energy per particle w vs κ for $d = 2.0002a$, $2.002a$, and $2.02a$ respectively. The three horizontal lines are the results from the conductor model.

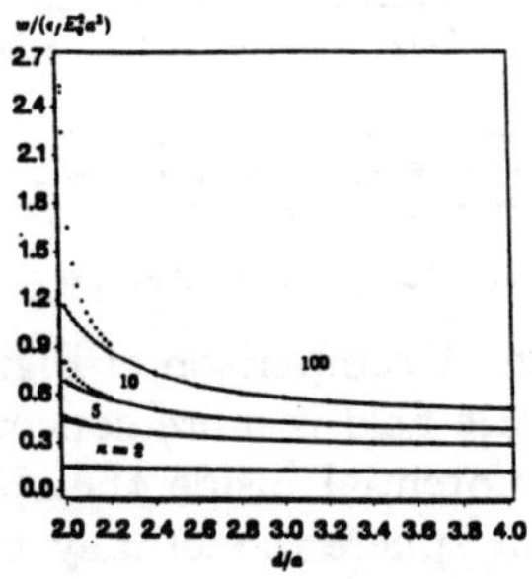

Fig.6 The electrostatic energy per particle w vs d for $\kappa = 2$, 5, 10, and 100 respectively. The curves are from the dipolar approximation.

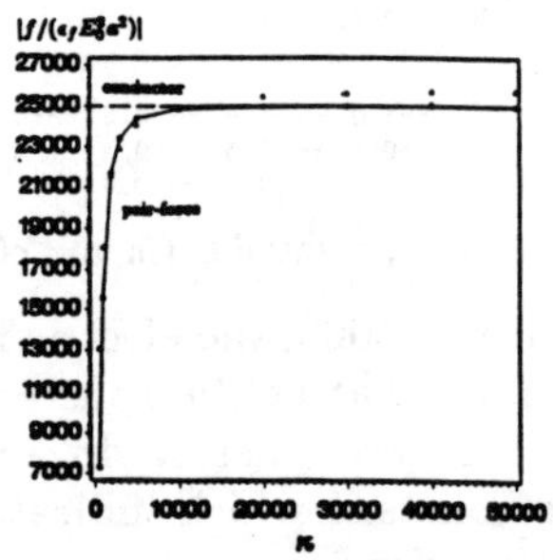

Fig.8 The strength of the attractive force between two neighbor spheres f vs κ at $d = 2.00002a$. Dashed line: conductor model; △: FEA results; □: the pair-force Eq.(3.1).

3.2. Interaction Energy

Fig.6 presents the interaction energy per particle, w, as a function of spacing d at $\kappa = 2, 5, 10$, and 100 respectively. The curves in Fig.6 are results from the dipolar model. It is clear that the dipolar approximation is quite good for $\kappa \leq 10$. At $\kappa = 2$, there is almost no difference in the energy between the FEA result and the dipolar approximation. At $\kappa = 10$ and $d = 2a$, the dipolar model only differs from the FEA results by 15%. However, at $\kappa = 100$ and $d = 2a$, the difference from the FEA calculation and the dipole model is quite big.

In Fig.7 we plot w vs κ at $d = 2.02a, 2.002a$ and $2.0002a$. The three horizontal lines are the results of the conductor model in Eq.(2.7). As κ increases, w approaches a plateau. However, it is clear that Eq.(2.7) underestimates w at $d = 2.02a$ and $d = 2.002a$. The situation is improved at $d = 2.0002a$. This is easy to understand because Eq.(2.7) only considers the electric field between two neighbor spheres and only as $d \to 2a$ this approximation is good.

3.3. Attracting Force

To have an empirical fit to the FEA results, we introduce the following formula which give a good approximation to the neighbor pair attractive force,

$$f_{pair} = -\epsilon_f E_0^2 a^2 \frac{a^4 \beta^2 (6d - 4a)}{d^4 (d - 2a\beta^{(1-\beta)/2})}. \tag{3.1}$$

This formula seems to cover the whole range of interest. If β is small or d is big, Eq.(3.1) recovers the dipolar result. If we take $\beta = 1$ for the conductor spheres, then Eq.(3.1) is identical to the leading term of the conductor model. Results from Eq.(3.1) have been plotted in Fig.8 and Fig.9.

In Fig.8 we plot this attractive force as a function of κ at $2.00002a$. The horizontal dashed line is the result of the conductor model. It is clear that the pair force formula (3.1) matches the FEA result very well.

Fig.9a, b, and c illustrate the neighboring attractive force vs distance d between two neighboring spheres at $\kappa = 2$, 5, and 10 respectively. The results from the dipole model are also plotted for comparison. The deviation from the dipole model comes when the spheres begin to contact. The FEA calculation shows that the attractive force between two touching dielectric spheres is finite but stronger than the result from the dipole model. The higher κ, the stronger the attractive force. For example at $\kappa = 10$, the FEA result is 6 times stronger than the dipole model. Figures 9a, b, and c also show that Eq.(3.1) fits the FEA data quite well.

When the spheres are in contact, $d = 2a$, Eq.(3.1) gives the neighbor attractive force,

$$f_{pair} = -\frac{\epsilon_f E_0^2 a^2 \beta^2}{4(1 - \beta^{(1-\beta)/2})} \tag{3.2}$$

If $\kappa \gg 1$, then $\beta \to 1$, we can approximate $\beta^{(1-\beta)/2} \approx 1 - (1-\beta)^2/2$. Thus we have the leading term

$$f_{pair} \approx -\frac{1}{18} \epsilon_f E_0^2 a^2 (\kappa - 1)^2. \tag{3.3}$$

Anderson's result[14] has this force $-0.52\epsilon_f E_0^2 a^2 \kappa^2/(4\pi) = -0.04138\epsilon_f E_0^2 a^2 \kappa^2$. Since $\kappa \gg 1$, both results are proportional to κ^2. However, our force is slightly stronger than Anderson's result in the above limit.

We also note that the pair force of our FEA result in Eq.(3.1) is proportional to E_0^2, while the the conduction model[18] has the leading term of the pair force proportional to E_0. The conduction model assumes that the conductivity of the base liquid is a function of the local field which, in turn, limits the upper value of the local field. Our present FEA calculation does not place any cap on the local field. Therefore, the above difference should be understandable.

Acknowledgement. This research is supported by Illinois Consortium for Advanced Radiation Sources and US Naval Surface Warfare Center.

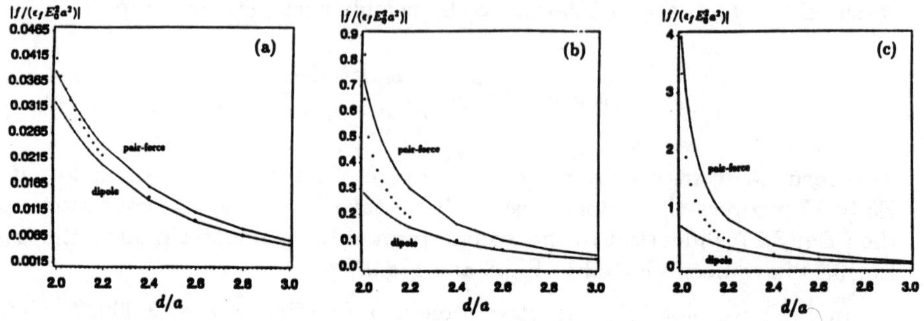

Fig.9 The strength of the attractive force between two neighbor spheres f vs d. (a) $\kappa = 2$, (b)$\kappa = 5$, and (c) $\kappa = 10$. Results from Eq.(3.1) and the dipole model are also included.

References

1. For example, see, *Electrorheological Fluids*, edited by R. Tao and G. D. Roy (World Scientific Publishing Comp., Singapore, 1994).
2. W. M. Winslow, *J. Appl. Phys.* **20**, 1137 (1949).
3. R. Tao, J.T. Woestman, and N. K. Jaggi, *Appl. Phys. Lett.* 55, 1844, (1989)
4. H. Block and J. P. Kelly, US Patent 4,687,589 (1987).
5. F. E. Filisko and W. E. Armstrong, US patent No. 4 744 914 (1988).
6. T. C. Halsey and W. Toor, *Phys. Rev. Lett.* **65**, 2820(1990).
7. R. Tao and J. M. Sun, *Phys. Rev. Lett.* **67**, 398(1991).
8. R. Tao, *Phys. Rev. E* **47**, 423 (1993).
9. T. J. Chen, R. N. Zitter, and R. Tao, *Phys. Rev. Lett.* **68**, 2555(1992).
10. R. Tao and Q. Jiang, *Phys. Rev. Lett* **73**, 205 (1994).
11. D. J. Klingenberg, F. van Swol, and C. F. Zukoski, *J. Chem. Phys.* **91**, 7888 (1989); **94**, 6160 (1991).
12. J. R. Melrose and D. M. Heyes, *J. Chem. Phys.* **98**, 5873 (1993).
13. Y. Chen, A. F. Sprecher, and H. Conrad, *J. Appl. Phys.* **70**, 6796 (1991).
14. R. A. Anderson, *Langmuir*, **10**, 2917 (1994).
15. L. C. Davis, *Appl. Phys. Lett.* **60**, 319(1992); . *J. Appl. Phys.* **72**, 1334 (1992); *Phys. Rev. A* **46**, R719 (1992).
16. H. J. H. Clercx and G. Bossis, *Phys. Rev. E* **48**, 2721 (1993).
17. R. Friedberg and Y. Yu, *Phys. Rev. B* **46**, 6582 (1992).
18. N. Felici, J. F. Foulc, and P. Atten, in Ref. 1, 139.

DIELECTRIC EVIDENCE FOR THE MATERIAL DESIGN OF ER FLUIDS

Tian Hao and Yuanze Xu
Institute of Chemistry, Academia Sinica, Beijing, 100080, China

ABSTRACT
The dielectric and electrorheological measurements of anhydrous ER fluids based upon oxidized polyacrylonitrile and zeolite under AC electric field are described. It was found that the yield strength was not only determined by dielectric constant of suspensions, but also by dielectric loss. The formulation based on Wagner model, which describes the polarization of heterogeneous dielectrics, was employed to predict the yield stress as a function of electric frequency. The good agreement between experimental and theoretical predictions indicated that conductivity of the dispersed particle and the dielectric constants of ER suspension are decisive factors in the design of ER fluids. The way of material design is discussed concerning working frequency, rate and temperature ranges.

1. Introduction

1.1. Interfacial Polarization

An important objective of the research on ER fluids is to be able to predict their strength in terms of structural and compositional properties of ER fluids. A generally accepted concept is that a deep polarization will occur when an electric field is applied to an ER fluid, and the polarized solid particles fibrillate between the two electrodes providing the mechanical strength[1,2]. It is known that dielectrics can be polarized by means of various mechanisms. The electron and atom processes are fast processes ranging 10^{-12}---10^{-14} second. However, considering the response time of most ER fluids scaled at milliseconds, the slow polarization should be regarded as dominant factors in ER activity. The Debye theory of dielectric behavior describes the polarization arising from the orientation of permanent dipoles, and Wagner theory[3,4,5] predicts the so-called interfacial polarization of heterogeneous dielectrics, one component of which has a considerably higher conductance than the other. In most cases, Debye polarization is dominated in the homogeneous system, and Wagner polarization in heterogeneous system. Since the ER fluids we study are suspensions of semi-conducting solid particles dispersed in dielectric oil, it is of great interest to study Wagner theory in some more detail.

1.2. Wagner Model

A good insulator having some particles of semi-conducting material embedded in it would show the peculiarities in dielectric behavior that can be predicted by Wagner's Model[3-5]. Wagner's analysis assumed that the conducting particles were spheres, sparsely distributed throughout an insulator of comparatively low dielectric loss. This assumption does reflect some situations in real ER suspensions, where semiconducting particles keep isotropic form and the media conductivity is much less than that of the particles, although the field distortions caused by the formation of chain, column or network structures is not taken into account.

The main results of the model are given in Eq. 1-5. The real component of the complex dielectric constant of the suspension is given by:

$$\varepsilon' = \varepsilon_\infty[1+K/(1+\omega^2 t_r^2)] \quad (1)$$

where

$$\varepsilon_\infty = \varepsilon_1[1+3\phi(\varepsilon_2-\varepsilon_1)/(2\varepsilon_1+\varepsilon_2)] \quad (2)$$

$$K = 9\phi\varepsilon_1/(2\varepsilon_1+\varepsilon_2) \quad (3)$$

$$t_r = (2\varepsilon_1+\varepsilon_2)/4\pi\sigma_2 \quad (4)$$

where ε_1 and ε_2 are the dielectric constants of the continuous (insulating) and dispersed (conducting) phases, σ_2 is the conductivity of the latter (in e.g.s. electrostatic units, 1 e.s.u.=(1/9)x10^{-11} S), ϕ is the volume fraction of conducting particles, ω is the angular frequency of alternating electric field, ε_∞ is the relative permittivity of suspension at very high frequency and t_r is the relaxation time of polarization. The dielectric loss tangent is given by:

$$\tan\delta = K\omega t/(1+K+\omega^2 t_r^2) \quad (5)$$

1.3. Percolation Concentration

A few studies on dielectric properties relating to ER effect have been reported in the literature[2,8-11]. Some inconsistent results may ascribe to the fact that the structure of the charged ER suspension changes along with particle concentration [2]. It was found that a percolation transition of viscosity along with concentration exists for both polymer-based ER suspensions and zeolite suspensions[7]. The critical volume fraction around 37% is independent of the electric field strength. In this concentration range there exists a 3-dimensional percolated network which changes its structure very little when charged. This structure is more stable and homogeneous than the chain or column structure formed through particle migration and rotation in lower concentration. Although lower concentrations, say 20%, has been widely employed in research papers, it has little practical significance due to the low yield stress, inhomogenity and two-phases flow causing long response time and instability in application.

In this work article concentration 35% was employed for two sets of typical dry ER fluids in order to establish the relationship between ER effect and dielectric properties of the suspended material. Wagner model would be verified and used to correlate ER strength to the dielectric properties and particle conductivity. Based on this, the ER design of response time, the working temperature and frequency range of ER fluids and other ER phenomena will also be discussed.

2. Experimental

2.1. Material

Four kinds of oxidized polyacrylonitrile(O-PAN) having conductivity of 1.1x10^{-11}, 2.3x10^{-9}, 1.5x10^{-7} and 3.1x10^{-5} S/m respectively were prepared in a special process[12], then dispersed in silicone oil(50cp) at a volume fraction of 35%, and named OP1, OP2, OP3 and OP4. Aluminosilicate particles of 1μm in size were treated at different temperatures and dispersed in silicone oil(50cp)[13] named as AS1, AS2 and AS3 respectively. The data of barium titanate and ceramics are also listed for comparison in Table1(all offered by manufacturer).

Table 1: The electric properties of the used materials(20°C)

Materials	Conductivity(S/m)	Dielectric Constant
Silicone oil	$1x10^{-14}$	2.0
Oxidized Polyacrylonitrile	$1.1x10^{-11}$	1.86
	$2.3x10^{-9}$	3.97
	$1.5x10^{-7}$	5.30
	$3.1x10^{-5}$	6.37
Aluminosilicate	$1x10^{-9}$~10^{-12}	5~11
$BaTiO_3$	$1x10^{-7}$	4500
Ceramics	$1x10^{-7}$	12000

2.2. Measurements

Dielectric measurements were carried out at ambient temperature using a dielectric instrument (Type Q16), made by Shanghai Electric Instrument Factory. Rheological measurements were made at steady state(γ= 0.3 s^{-1}) by using a modified parallel-plate Rheometrics Mechanic Spectrometer(Model 605) made by. Rheometrics Company, U.S.A. Considering ER fluids behave as a solid under electric field[17] , the stress obtained at any shear rate can be regarded as their yield stress once they are yielded. The alternating electric source was made by our lab., supplying 600v(frequency ranging 0.1--100 kHz).

3. Results and Discussion

3. 1. Dielectric and Electrorheological Behaviors

The fact that non-ER materials, barium titanate and ceramics, have very high dielectric constant as compared with good ER materials, O-PAN and aluminosilicate, suggests that there is no simple relation between ER effect and dielectric constant of the dispersed materials. ER effect was hardly observed in sample OP1. Samples OP2 and OP3 exhibit strong ER strength, i.e. several thousands times as high as the zero-field values. Samples AS 1 and AS2 also exhibit very strong ER effect.

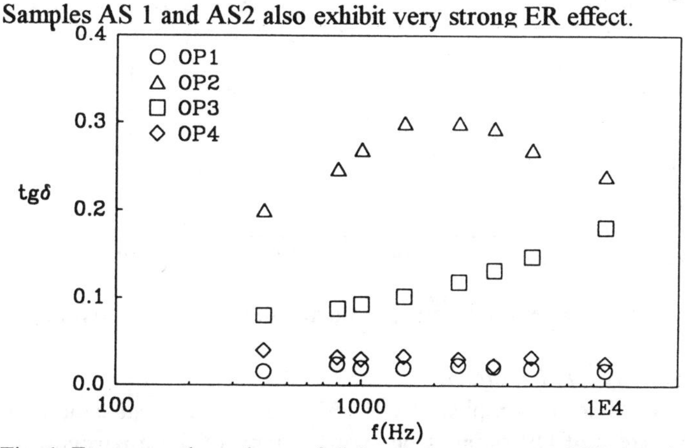

Fig. 1. Frequency dependence of dielectric loss tangent of OP samples.

The dielectric loss tangent of OP suspensions at ambient temperature as function of frequency are given in Figure 1. For samples OP2 and OP3 the relaxation peaks are located or nearly located in the frequency range of measurements. Sample OP1 and OP4 show very low values of loss tangent, and their relaxation peaks are supposed to be in the lower and higher frequency ranges respectively. Figure 2 shows dielectric constant of AS samples in the frequency range of 400-10000 Hz.

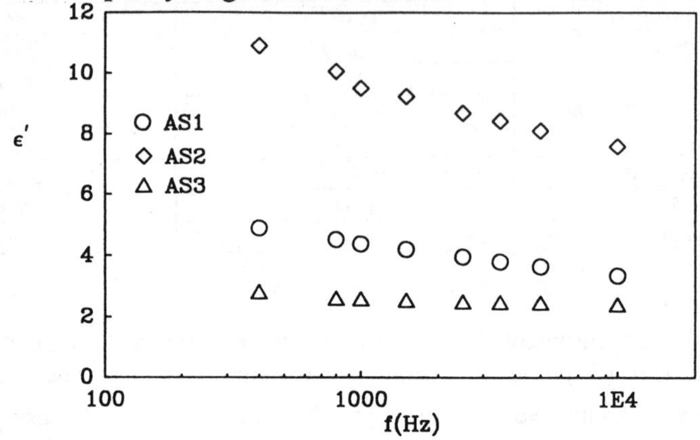

Fig. 2. Frequency dependence of dielectric constant of AS samples.

Yield stress τ of all OP and AS samples does not change monotonocally with increasing dielectric constant or loss tangent even for the same series samples. Fig.3 shows the dependence of τ to tgδ. What could be noticed here is that for samples of strong ER effect, like OP2, OP3 AS 1 and AS2, not only dielectric constant but also loss tangent have relatively high values around 0.1 in the frequency range around 1000 Hz. From material design point of view, material having large dielectric loss and dielectric constant is preferred for the high performance of ER fluids.

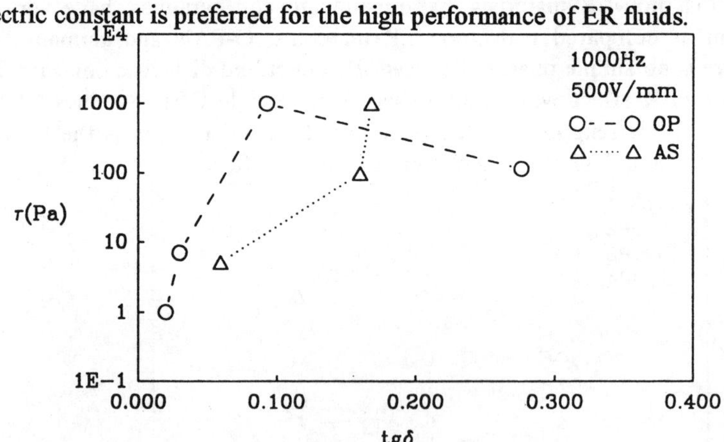

Fig. 3. Yield stress of OP and AS samples against their dielectric loss tangent.

3.2. Particle Conductivity and ER effect.

The O-PAN particles in samples of OP series are of different conductivity. The dependence of yield stress of OP samples on conductivity of these particles is given in Figure 4. The fact that a peak appears at a conductivity value around 10^{-7} S/m is in good agreement with the result obtained by Block[2]. Figure 5 is conductivity dependence of the current density for the OP samples. It is worth to emphasize that the current density of suspensions in high electric field is dominated by conductivity of the solid particles. In the OP series sample, OP3 has the strongest ER effect, but its current density is too large for the practical application.

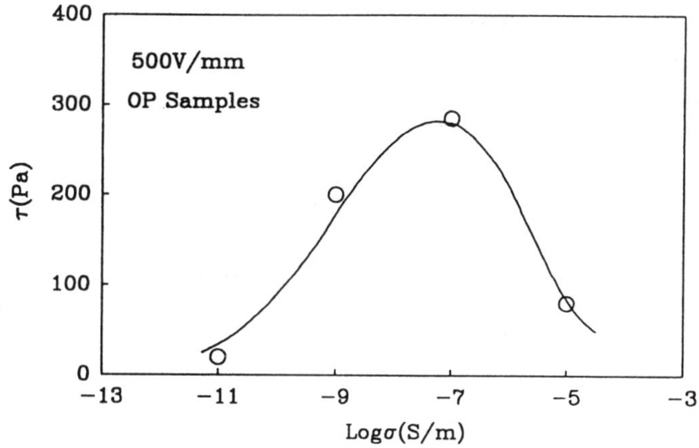

Fig.4. Conductivity dependence of yield stresses of OP series samples measured at 0.5kV/mm.

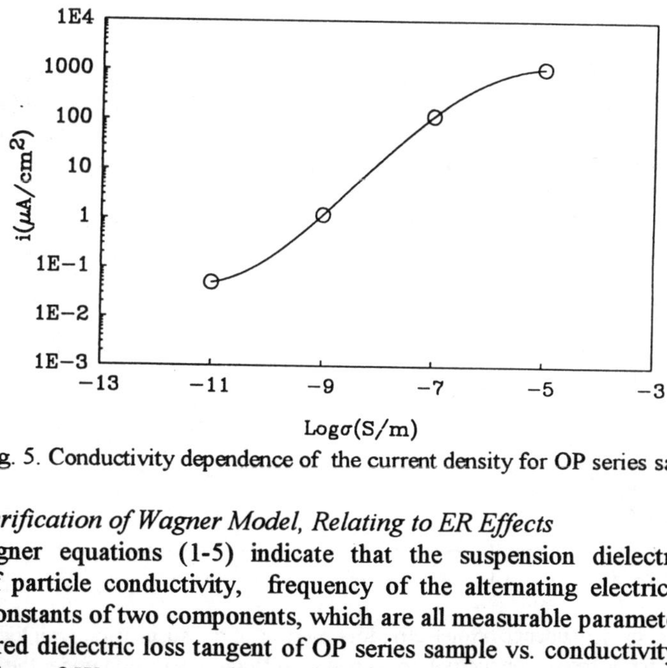

Fig. 5. Conductivity dependence of the current density for OP series samples

3.3. The Verification of Wagner Model, Relating to ER Effects

Wagner equations (1-5) indicate that the suspension dielectric constant is function of particle conductivity, frequency of the alternating electric field, and the dielectric constants of two components, which are all measurable parameters.

The measured dielectric loss tangent of OP series sample vs. conductivity together with the prediction of Wagner equation are shown in Figure 6. We use the frequency as 1000Hz, dielectric constant ε_2 and conductivity σ_2 data from Table 1. The theoretical prediction agrees well with the experimental curve. What is also worthy to mention here is that the dielectric loss tangent approaches maximum at particle conductivity 10^{-9} S/m coincides where the ER fluid exhibit the strongest ER response. It may suggest the close relation of ER activity and the Wagner polarization.

It is reasonable to assume that the ER response time is equal to the relaxation time of interfacial polarization described by Wagner equation (4). This is an inverse proportional relation between the relaxation time and particle conductivity. That is, the larger conductivity of the dispersed material corresponds to the faster response time of ER fluid. Also, the effect of conductivity is considerably large compared with the effect of dielectric constant of two components. The calculated relaxation time of suspension with the particle conductivity 10^{-9} S/m is about 1 millisecond, which provided a good explanation that the response time of many ER fluids is scaled at 1 millisecond and the conductivity of the dispersed material for most of ER fluids is around 10^{-9} S/m.

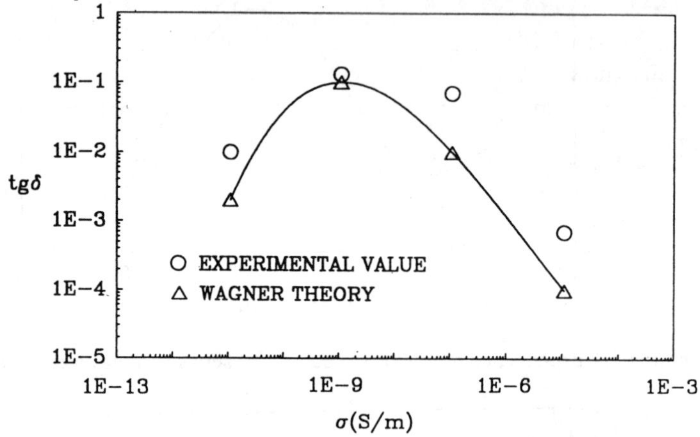

Fig. 6. Conductivity dependence of dielectric loss tangent of OP samples and the Wagner model prediction using of ε_1=2, Φ=0.35 and f=1000Hz

Another dominant factor in respect to the dielectric property of the suspension is the frequency of the applied external electric field. Figure 7 shows the dielectric loss

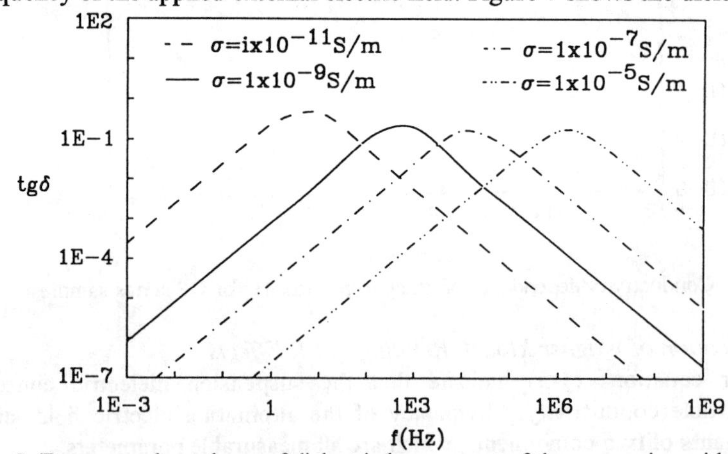

Fig. 7. Frequency dependence of dielectric loss tangent of the suspension with the dispersed materials having different conductivity obtained with Wagner equation under the assumption at ε_1=2, ε_2=10 and Φ=0.35.

tangent of suspension with the dispersed particles having different conductivity vs. frequency predicted by Wagner model under the assumption of ε_1=2, ε_2=10 and Φ=0.35. The peaks of dielectric loss tangent corresponding to different particle conductivity appear at 10Hz, 10^3Hz, 10^4Hz and 10^6Hz respectively. Comparing to Figure 4 we see tgδ maximum of OP2 really locates at 1000Hz and that of other three samples distribute in a reasonable way similar to the prediction of Wagner model.

A clear argument to justify the application of Wagner to ER fluids is the frequency dependence of yield stress together with Wagner prediction of dielectric constant against frequency for OP2 under 0.3kV/mm which is plotted in Figure 8. The curves fall in the same frequency range. In order to find the correlation between yield stress and dielectric constant, $(\tau-\tau_0)$ against $(\varepsilon'-\varepsilon_\infty)^2$ is plotted in Figure 9, where τ and τ_0 are yield stress of OP2 under 0.3kV/mm and zero electric field respectively, ε' is dielectric constant of system predicted with Wagner model, and ε_∞ is infinite frequency relative permittivity. Obviously, $(\tau-\tau_0)$ is the increment of yield stress resulted from the external electric field,

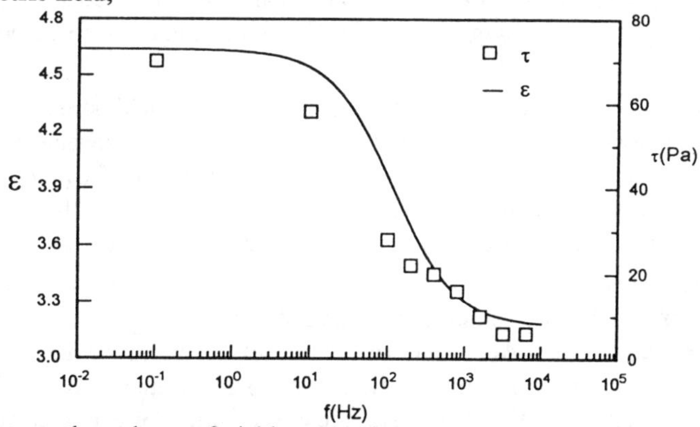

Fig, 8. Frequency dependence of yield stress of OP2 measured in experimental (E=0.3kV/mm) and dielectric constant dependence on frequency obtained with Wagner model under the assumption of $\varepsilon_1=2$, $\varepsilon_2=10$ and $\Phi=0.35$..

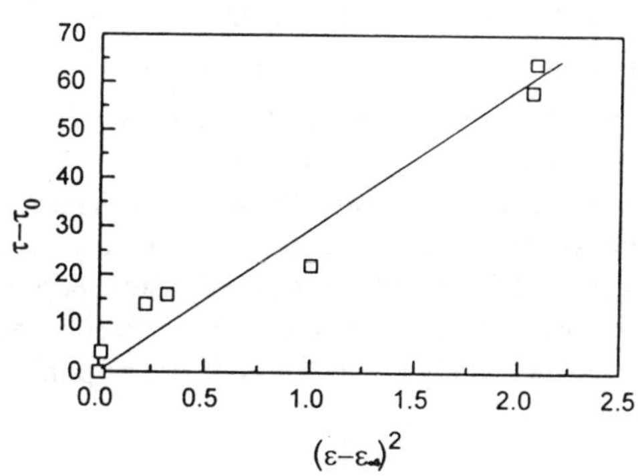

Fig,9. The increment of yield stress $(\tau-\tau_0)$ and against the increment of dielectric constant of system $(\varepsilon'-\varepsilon_\infty)^2$

whereas $(\varepsilon'-\varepsilon_\infty)$ is approximately scaled the magnitude of slow polarization. A linear relation between $(\tau-\tau_0)$ and $(\varepsilon'-\varepsilon_\infty)^2$ is found and can be expressed approximately as follows:

$$(\tau-\tau_0) = 30(\varepsilon'-\varepsilon_\infty)^2 \qquad (6)$$

The above equation obtained under E=0.3kV/mm, Φ=0.35. Experimentally, yield stress is directly proportional to the square of electric field strength E[10-14] and particle volume fraction Φ[7]. The model also happens to predict $(\varepsilon'-\varepsilon_\infty) \propto \Phi$. So the yield stress of concentrated ER fluid under electric field can be written as:

$$(\tau-\tau_0) = K'E^2(\varepsilon'-\varepsilon_\infty)^2 \qquad (7)$$

where constant K'=270 in our case, E in kV/mm. E is assumed as 3kV/mm, and $(\varepsilon'-\varepsilon_\infty)$ =1.5 under assumption of ε_1=2, ε_2=10 and Φ=0.35. The calculated $(\tau-\tau_0)$ value is near 7KPa, a reasonable value.

3.4. About the ER Fluid Design

Equation (7) summarizes how the mechanical strength relates to the material parameters. This provides a thinking line for the ER fluid design, more comprehensively than before[6].

Working Frequency As we mentioned before, the yield stress drops as AC frequency above dielectric transition range, which is mainly determined by particle conductivity. The higher the conductivity, the higher the working frequency as long as the conductivity of the suspension, which is no less than one tenth of the particle conductivity, is still acceptable. The mechanical working frequency or working shear rate should be somehow near the AC frequency[2], although the essential mechanism of the response time in various dynamic situations needs further study.

Working Temperature According to the model, temperature dependence of ER strength should rely on the variation of dielectric constant and particle conductivity against temperature, especially the latter. The ideal ER fluids are expected to work in a very wide temperature range 240~420K. These requirements, i.e., that the dielectric constant and particle conductivity change slightly with temperature, however, are still far from being met in practice. On the basis of time-temperature equivalency rule, if an ER fluid can not work in low temperature range, it may show obvious ER activity once the temperature gets higher or at lower frequency. This has been proved in OP fluids[7] as well as other fluids[16] with the limitation of too high current density at high temperature caused by high particle conductivity.

Function of Water: The ever generally accepted "Water" mechanism of ER can also be understood with Wagner model. The trace water absorbed in any material can make the apparent dielectric loss tangent and dielectric constant of system larger and the interfacial polarization make ER fluids showing much stronger ER effect. Water content is also limited by too large leakage current leading to the loss of ER response.

4. Conclusion

It was found experimentally that certain value of dielectric loss tangent of the dispersed material is necessarily required for high performance ER fluids. Wagner model, dealing with the interfacial polarization for systems of semiconducting spheres dispersed in an insulating medium, successfully predicts the dielectric properties of ER fluids as functions of frequency of the applied external A.C. field. Wagner model also predicts the right dependency of dielectric loss tangent on the conductivity of the dispersed material showing maximum tgδ around 0.10 at the particle conductivity around 10^{-9} S/m at frequency 1000Hz. The correlation of ER strength and dielectric constants leads to the relation of mechanical strength of ER fluid to material variables: dielectric constants, especially, the conductivity of dispersed particles.

Therefore, we see a relative simple scheme: not only the working frequency range but also the working temperature range of ER fluid are dominated by the relaxation time of Wagner polarization, which together with suspension conductivity are mainly determined *by the* conductivity of the dispersed material and dielectric constant of two components. It should be mentioned here that the correlation of Wagner polarization and ER responses is still empirical We do not know why this model of dilute dielectrics works well for concentrated ER suspension, where the electric field is distorted by dense packed polarized particles. Also, the micro-mechanism of Wagner polarization are worth much further study.

Acknowledgment: This work is supported by National Natural Science Foundation of China and Polymer Physics Laboratory, Academia Sinica. Many helps from Prof. M.Xu, Prof.. Z. Xu, Prof. Y.Chen, and Ms.Y.Li are highly acknowledged.

References:

1. W.M. Winslow. J. Appl. Phys. 20 (1949) 1137; Y.F.Deinega and G.V. Vinogradov. Rheol. Acta 23 (1984)636; J.C.Jordan and M.T.Shaw IEEE Trans. Elco. Inst. 24 (1989) 849; D.J.Klingenberg and C.F.Zukoski. Langmuir. 6 (1990) 15; M. Whitfle and W. A.Bullough. Nature 358 (1992) 373.
2. H.Block. J.P.Kelly. A. Qin and T. Watson. Lang. muir. 6 (1990) 6; K.M.Blackwood and H.Block. Trends in Polymer Science. 1 (1993) 98; K.M.Blackwood. H.Block. P.Rattray. G.Tsangaris and D.N.Vorobiev, in Electrorheological Fluids. ed. R. Tao & G.D.Roy (World Scientific. Singapore. (1994) *3.*
3. K.W. Wagner. Archiv Fuer Electrotechnik. 1914. 2. 371
4. S.O.Morgan. Trans. Am. Electrochem. Soc. 65 (1934) 109.
5. R.W. Sillars. J.I.E.E., 80 (1937) 378.
6. Y.Xu. R. Liang. T. Hao. Q. Zhang. Z. Xu. and Y. Chen. in Proc. Int. Confl. Electrorheological Fluids ed. R. Tao (World Sci. Press. Singapore. New Jersey. London. Hong Kong. 1992) p. 129.
7. T.Hao. Y.Chen. Z.Xu. YXu. and Y.Huang. Chinese J. of Polymer Sci., 12 (1994) 97.
8. D.L. Klass. D.L.;Martinek. T.W.J. Appl. Phys. 38 (1967) 75.
9. Uejima. H. Jpn. J. Appl. Phys. 11 (1972) 319.
10. H. Conrad. A.F. Sprecher. Jr., Y.Choi. and Y.Chen. J. Rheol. 35(7) (1991) 1393.
11. K.D.Weiss and J.D.Carlson. Proc. 3rd. Intn. Conf. on ER fluids. Ed. R. Tao. (World Sci. Press. Singapore. New Jersey. London. Hong Kong. 1992) 264
12. Yuanze Xu and Ruifeng Liang. J. Rheol., 35(7) (1991) 1335.
13. Tian Hao. M.S. Thesis. Institute of Chemistry. Academia Sinica. (1992).
14. L.Marshall. C.F.Zukoski and J.W. Goodwin. J. Chem. Soc. Faraday. Trans., 5(6).(1989)2785.
15. D.JKlingenberg. F. van Swol and C.F.Zukoski. J. Chem. Phys. 91(12) (1989) 7888.
16. U.Y.Treasurer. F.E.Filisko. and L.H.Radzilowski. J.Rheol 35(6) (1991) 1051.
17. R.Tao and J.M.Sun, Phys.Rev. Lett. 67, (1991)398; Tian-Jie Chen, R.N.Zitter and R.Tao, Proc. 3rd Intn.Conf. on ER fluids, ed.R.Tao World Scientific Press, (1992)17.

STEADY–STATE SOLUTION OF AN ER CLUTCH AT HARMONIC LOAD

VLADIMÍR ORAVSKÝ
Institute of Materials and Machine Mechanics,
Slovak Academy of Sciences
Račianska 75, P.O.Box 75, 830 08 Bratislava 38, Slovakia
and
VLADIMÍR KRČ–JEDINÝ
branch of Institute of Materials and Machine Mechanics,
Severná 14, P.O.Box 8, 036 80 Martin, Slovakia

ABSTRACT

A concentric electrorheological clutch (ERC) constituting the central part of a broader system: electro–hydro–aggregate (EHA) with an electrodrive on one side and a loading machine on the other side is considered. Harmonic loading moment is taken into account. Dynamic model of EHA is presented in a dimensionless form and linearized in the vicinity of mean values of the sought steady–state solution. Due to limited extent of the paper further course of solution is described only briefly whereby averaging of variable viscosity of ER fluid is performed. As solution special Bessel, i. e., Kelvin functions and matrix calculus are utilized. Methodology of 3-level analysis is briefly described. Some results of preliminary analysis are presented. Lastly some open problems are pointed out.

1. Introduction

This paper links up another similar paper [8] in this Proceedings. The conception, methodology and system in question are the same: a compound electro–hydro–aggregate (**EHA**) consisting of an electrodrive (induction motor) **ED**, a loading machine (brake) **B** and electrorheological clutch (**ERC**) between them. ERC itself consists of two concentric discs with electrorheological fluid (**ERF**) between them (Fig.1 in Ref. [8]). It is well known [1] ERF has variable viscosity that can easily be controlled by D. C. field in wide range and so advantageously utilized in many technical applications, ERC being one of them. There are two types, arrangements of EHA: EHAI when the inner disc of ERC is connected with ED and the outer one with B and conversely for EHAII (Fig. 2 in Ref. [8]).
The difference between the paper [8] and this one is that here a dynamic model of EHA is built up and investigated. This means that for ED a nonlinear dynamic moment characteristic (**dynMCH**) instead of linear quasistatic one is taken into account and for brake B a harmonically variable loading moment instead of a constant one is considered. As to model of ERF, it is again supposed to be a homogenous incompressible Bingham plastic, and flow in ERC to be laminar and only tangential. The aim of the paper is to get a theoretical foundation for experimental verification, if and to what extent the above quasistatic model of ERF is valid also in dynamic conditions or if a more complicated dynamic model in Ref. [3] must be used. Due to limited extent oh the paper, attention will be focused mostly on ERC and other data about all EHA will be only concise and informative. To be short also the model will be presented directly in dimensionless (nondimensional, **ndim**) form.

2. Description of dynamic model

A more detailed computation scheme for derivation of dyn model of EHA is given in Fig.1.

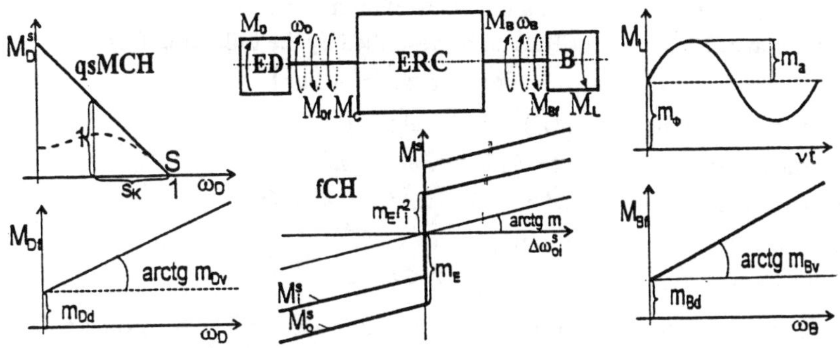

Fig. 1 Computational scheme of EHA

ED produces at angular velocity (**angV**) ω_D a dynamic driving moment M_D given by the following nonlinear dynMCH [5]:

$$dM_D/dt + s_K M_D = (1-\kappa_D)(1-\omega_D), \qquad d\kappa_D/dt + s_K \kappa_D = M_D(1-\omega_D) \tag{1}$$

where κ_D is an auxiliary quantity watching nonlinearity of dynMCH and s_K is the critical slip of ED. Corresponding quasistatic (**qs**) MCH one gets for constant M_D^s, κ_D^s(qs quantities are distinguished by the upper index s). Then after excluding κ_D^s it results from Eq.(1):

$$M_D^s = (1-\omega_D)s_K / \left[s_K^2 + (1-\omega_D)^2\right] \tag{2}$$

This qs MCH is plotted in Fig. 1 (left up, dashed line). Note that arbitrary dim quantities and parameters (**Pm**) are non-dimensionalized according to patterns:

$$M = \overline{M}/(2\overline{M}_K), \quad \omega = \overline{\omega}/\overline{\omega}_S, \quad J = \overline{J}/\overline{J}_D, \quad r = \overline{r}/\overline{r}_o, \tag{3}$$

where $\overline{M}_K$ is the maximum qs moment of ED, $\overline{\omega}_S$ synchronous speed of ED, $\overline{J}_D$ moment of inertia of ED and $\overline{r}_o$ radius of outer disc of ERC (dim quantities and Pm-s are distinguished by upper strips). Due to choose (3) one gets advantageously:

$$M_{\max}^s = 1/2, \quad \omega_S = 1, \quad J_D = 1, \quad r_o = \tag{4}$$

Brake B is supposed to produce harmonic loading moment: $M_L = m_\Phi + m_a \sin \nu t$ (5)
with adjustable mean value m_Φ, amplitude m_a and frequency ν (Fig. 1, right up). Moreover on both sides of ERC there act mechanical friction moments:

$$M_{Df} = m_{Dd} + m_{Dv}\omega_D, \quad M_{Bf} = m_{Bd} + m_{Bv}\omega_B \tag{6}$$

with components m_{Dd}, m_{Bd} of dry friction and m_{Dv}, m_{Bv} of viscous friction (Fig. 1, down left and right). Note, that ndim Pm-s m_Φ and m_{Bd} can be unified into one ndimPm of load [8]:

$$m_L = m_\Phi + m_{Bd} \tag{7}$$

Finally ERC acts on its input and output shafts with moments M_C and M_B which coincide according to arrangement (Fig. 2 in Ref. [8]) either with outer moment M_o or inner moment M_i (Fig. 1 in Ref. [8]) as follows:

for EHAI: $M_C = M_o$, $\omega_D = \omega_o$, $J_C = J_o$, $M_B = M_i$, $\omega_B = \omega_i$, $J_B = J_i$ (8)

for EHAII: $M_C = M_i$, $\omega_D = \omega_i$, $J_C = J_i$, $M_B = M_o$, $\omega_B = \omega_o$, $J_B = J_o$ (9)

whereby J_i, J_o, J_B, J_C are corresponding moments of inertia of discs.

Tangential motion of ERF within ERC can be described by the following Eqs [8]:

$$\rho\,\partial u/\partial t = \mu\left(\partial^2 u/\partial r^2 + \Omega/r\right) \quad (10)$$

with boundary conditions $u(r_i) = \omega_i r_i,\; u(1) = \omega_o$ (11)

where ndim shear rate $\Omega = \bar{\dot{\gamma}}/\bar{\omega}_S = \partial u/\partial r - u/r$ (12)

ndim friction moment $M(r) = \overline{M}/(2\,\overline{M}_K) = m_o\,\mu\,\Omega r^2$ (13)

ndim dyn viscosity $\mu = \bar{\mu}/\bar{\mu}_0 = 1 + \tau_E/|\Omega|$ (14)

circumferential V $u = \omega\, r \qquad r_i \le r \le 1$ (15)

Note, that ndim shear rate, friction moment and radius are normed according to Eq. (3). Analogously ndim dyn viscosity is normed regarding the basic dyn viscosity $\bar{\mu}_0$ without ER effect (so that $\mu_0 = 1$). Moreover ndim density ρ, ndim size of outer disc m_o, ndim shear due to ER effect τ_E and ndim inner radius of ERC r_i are normed as follows:

$$\rho = \bar{\rho}\,\bar{r}_o^2\,\bar{\omega}_S/\bar{\mu}_0,\quad m_o = \pi\bar{r}_o^2\,\bar{l}_o\,\bar{\mu}_0\,\bar{\omega}_S/\overline{M}_K,\quad \tau_E = \bar{\tau}_E/(\bar{\mu}_0\,\bar{\omega}_S),\quad r_i = \bar{r}_i/\bar{r}_o \quad (16)$$

where, along with known dim Pm-s, $\bar{l}_o$ is width of ERC.

Note also the mentioned moments M_i, M_o cohere with the moment (13) by couplings:

$$M_i = M(r_i), \qquad M_o = -M(1) \quad (17)$$

The corresponding qs moments M_i^s, M_o^s can be expressed by formulas [8]:

$$M_i^s = m\,\Delta\omega_{oi} + q_{oi}\,m_E\,r_i^2, \qquad M_o^s = -m\,\Delta\omega_{oi} - q_{oi}\,m_E \quad (18)$$

where $\Delta\omega_{oi} = \omega_o - \omega_i$, $\qquad q_{oi} = \Delta\omega_{oi}/|\Delta\omega_{oi}|$ (19)

m is relative ndim Pm of size of ERC with respect to size of ED,

m_E is ndim Pm of additional moment due to ER effect.

The last two ndimPm-s depend on primary ndimPm-s m_o, τ_E as follows [8]:

$$m_E = m_o\,\tau_E,\quad m = m_o/R_i,\quad \text{where } R_i = \left(1/r_i^2 - 1\right)/2 \quad (20)$$

Relations (18) are graphically expressed in Fig. 1 (down, amidst).

Finally ndim Eqs. of motion of both sides of ERC are:

$$(1 + J_D)\,d\omega_D/dt = 2m_K\left(M_D - M_{Df} - M_C\right),$$
$$(J_B + J_L)\,d\omega_B/dt = 2m_K\left(M_B + M_{Bf} - M_L\right) \quad (21)$$

where $m_K = \overline{M}_K/\left(J_D\,\bar{\omega}_S^2\right)$ is ndim size of ED and $J_D = 1$ or J_L are moments of inertia of ED or B. Eqs.(1÷21) constitute ndim dyn model of EHA with altogether 14 ndimPm-s from which: 4 ndimPm-s belong to ED (s_K, m_K, m_{Dd}, m_{Dv}), 6 ndimPm-s to ERC (r_i, m_o, τ_E, ρ, J_o, J_i) and 4 ndimPm-s to B (m_L, m_a, ν, J_L). (22)

The model is nonlinear (**nlin**) due to nlindyn MCH (1) and nlin dyn viscosity (14). The aim of investigation is to find the steady–state solution of the above model and analyze it regarding all ndim Pm-s (22). For solution an approximate analytic method will be chosen.

3. Steady–state solution in the frame of the 1st approximation

Before starting with solution, it is convenient to introduce a new ndim time: $\tau = \nu\, t$,(23) (derivatives with respect to τ will be denoted by dots) and consequently new auxiliary ndimPm-s:

$\nu_B = \nu\, J_{BL}/(2m_K)$, $\nu_D = \nu\, J_{DC}/(2m_K)$, where $J_{DL} = J_B + J_L$, $J_{DC} = 1 + J_C$ (24)

Then from Eqs. (1, 5÷7, 10, 21) one gets the following standard differential system of the 5th order:

$$\nu_D\,\dot{\omega}_D = M_D - m_{Dd} - m_{D\mathrm{v}}\,\omega_D - M_C \tag{25}$$
$$\nu_B\,\dot{\omega}_B = M_B - m_L - m_{B\mathrm{v}}\,\omega_B - m_a \sin\tau \tag{26}$$
$$\nu \dot{M}_D = (1-\kappa_D)(1-\omega_D) - s_K\, M_D \tag{27}$$
$$\nu\dot{\kappa}_D = M_D(1-\omega_D) - s_K\,\kappa_D \tag{28}$$
$$\nu\rho\dot{u} = \mu\left(\partial^2 u/\partial r^2 + \Omega/r\right) \tag{29}$$

which must be completed with Eqs. (11÷14, 17) and structural relations (8, 9). Note that from the differential (**dif**) system (25÷29) one easy turns into the qs system by setting:

$$\nu = 0 \quad\Rightarrow\quad \tau = 0,\ \ \nu_B = \nu_D = 0 \tag{30}$$

A steady–state solution can be generally sought in the form: $W(t) = W_\Phi + W_\Pi(\nu t)$, (31) where W_Φ is the mean value (defining so called working point Φ) and W_Π is a centered periodic component (with zero mean value). In the frame of the 1st approximation 2 important additional assumptions are adopted:

$$1.\ m_a = O(\varepsilon) \Rightarrow W_\Pi = O(\varepsilon), \qquad 2.\ h = 1 - r_i = O(\varepsilon), \tag{32}$$

i. e. amplitude of load (excitation), as well as all periodic components of response are assumed to be small of order of the small Pm ε and the same it holds for gap h of ERC. This ensures a possibility of lineartization of a system in vicinity of the working point Φ, i. e. neglecting of powers and mutual products of periodic components. In agreement with Eq. (31, 23) one seeks the steady–state solution of the system equations in the form:

$$(\tau) = W_\Phi + W_\Pi(\tau), \qquad \text{for } W = \omega_i, \omega_o, \omega_D, \omega_B, M_i, M_o, M_D, M_C, M_B, \kappa_D \tag{33}$$
$$W(r,\tau) = W_\Phi(r) + W_\Pi(r,\tau), \quad \text{for } W = u, \mu, M, \Omega \tag{34}$$

After substitution of Eqs.(33, 34) into Eqs. (25÷29, 11÷14), after the mentioned linearization and finally after separation of mean and periodic components (in agreement with the method of harmonic balance) also in Eqs. (8, 9), one gets the following systems

for mean values:	for periodic components:	
$M_{D\Phi} - m_{Dd} - m_{D\mathrm{v}}\,\omega_{D\Phi} - M_{C\Phi} = 0$	$\nu_D\,\dot{\omega}_{D\Pi} = M_{D\Pi} - m_{D\mathrm{v}}\,\omega_{D\Pi} - M_{C\Pi}$	(35)
$M_{B\Phi} - m_L - m_{B\mathrm{v}}\,\omega_{B\Phi} = 0$	$\nu_B\,\dot{\omega}_{B\Pi} = M_{B\Pi} - m_{B\mathrm{v}}\,\omega_{B\Pi} - m_a \sin\tau$	(36)
$(1-\kappa_{D\Phi})(1-\omega_{D\Phi}) - s_K\,M_{D\Phi} = 0$	$\nu\dot{M}_{D\Pi} = -(1-\kappa_{D\Phi})\omega_{D\Pi} - (1-\omega_{D\Phi})\omega_{D\Pi} - s_K\,M_{D\Pi}$	
$M_{D\Phi}(1-\omega_{D\Phi}) - s_K\,\kappa_{D\Phi} = 0$	$\nu\dot{\kappa}_{D\Pi} = M_{D\Pi}(1-\omega_{D\Phi}) - M_{D\Phi}\,\omega_{D\Pi} - s_K\,\kappa_{D\Pi}$	(38)
$\mu_\Phi\left(\partial^2 u_\Phi/\partial r^2 + \Omega_\Phi/r\right) = 0$	$\nu\rho\dot{u}_\Pi = \mu_\Phi\left(\partial^2 u_\Pi/\partial r^2 + \Omega_\Pi/r\right)$	(39)
where $\Omega_\Phi = \partial u_\Phi/\partial r - u_\Phi/r$	$\Omega_\Pi = \partial u_\Pi/\partial r - u_\Pi/r$	(40)
$\mu_\Phi = 1 + \tau_E/\lvert\Omega_\Phi\rvert$	$\mu_\Pi = -\tau_E\,\Omega_\Pi/\Omega_\Phi^2$	(41)
$M_\Phi(r) = m_o\,\mu_\Phi\,\Omega_\Phi\,r^2$	$M_\Pi(r) = m_o\left(\Omega_\Phi\mu_\Pi + \Omega_\Pi\mu_\Phi\right)r^2$	(42)
$M_{i\Phi} = M_\Phi(r_i),\ M_{o\Phi} = -M_\Phi(1)$	$M_{i\Pi} = M_\Pi(r_i),\ \ M_{o\Pi} = -M_\Pi(1)$	(43)

with boundary conditions:

$$u_\Phi(r_i)=\omega_{i\Phi}r_i,\ u_\Phi(1)=\omega_{o\Phi} \quad\Big|\quad u_\Pi(r_i)=\omega_{i\Pi}r_i,\ u_\Pi(1)=\omega_{o\Pi} \tag{44}$$

and with structural relations

for EHAI: $M_{C\Phi}=M_{o\Phi},\ M_{B\Phi}=M_{i\Phi},$ | $M_{C\Pi}=M_{o\Pi},\ M_{B\Pi}=M_{i\Pi},$ (45)

$\omega_{D\Phi}=\omega_{o\Phi},\ \omega_{B\Phi}=\omega_{i\Phi}$ | $\omega_{D\Pi}=\omega_{o\Pi},\ \omega_{B\Pi}=\omega_{i\Pi}$ (46)

for EHAII: $M_{C\Phi}=M_{i\Phi},\ M_{B\Phi}=M_{o\Phi},$ | $M_{C\Pi}=M_{i\Pi},\ M_{B\Pi}=M_{o\Pi},$ (47)

$\omega_{D\Phi}=\omega_{i\Phi},\ \omega_{B\Phi}=\omega_{o\Phi}$ | $\omega_{D\Pi}=\omega_{i\Pi},\ \omega_{B\Pi}=\omega_{o\Pi}$ (48)

First the mean system given by left Eqs. (35÷48) must be solved. Note that this system coincides with the qs system in Ref. [8]. The difference is that a nlinqsMCH (2) instead of linear one is considered. Then one obtains from the left Eqs. (37, 38):

$$M_{D\Phi}=x_\Phi\big/\left(1+x_\Phi^2\right),\ \ \kappa_{D\Phi}=M_{D\Phi}\,x_\Phi=x_\Phi^2\big/\left(1+x_\Phi^2\right) \tag{49}$$

where relative slip is introduced $x_\Phi=\left(1-\omega_{D\Phi}\right)/s_K$ (50)

Using the mentioned coincidence with Ref. [8], one gets readily for the "mean" boundary problem, defined by the left Eqs. (39÷48), the following solution:

$$M_{C\Phi}=m\,\Delta\omega_\Phi+m_C\,r_C,\ \ M_{B\Phi}=m\,\Delta\omega_\Phi+m_C\,r_B \tag{51}$$

where $\Delta\omega_\Phi=\omega_{D\Phi}-\omega_{B\Phi},\ \ m_C=q_\Phi\,m_E,\ \ q_\Phi=\Delta\omega_\Phi\big/\left|\Delta\omega_\Phi\right|$ (52)

while for EHAI: $r_C=1,\ \ r_B=r_i^2,$ for EHAII: $r_C=r_i^2,\ \ r_B=1.$ (53)

Then, after inverting Eqs. (49, 51, 52) into left Eqs. (35, 38) and after some manipulation, one gets the following cubic Eq.:

$$a_1\,x_\Phi^3-a_0\,x_\Phi^2+\left(1+a_1\right)x_\Phi-a_0=0 \tag{54}$$

where $a_0=m_D-m_B\,m_1,\ \ a_1=s_K\left(m_{D\mathrm{v}}+m_{B\mathrm{v}}\,m_1\right)$ (55)

and futher $m_D=m_{Dd}+m_{D\mathrm{v}}+m+m_C\,r_C,\ \ m_B=m+m_C\,r_B-m_L$ (56)

$$m_1=m/m_0,\ \ m_0=m+m_{B\mathrm{v}} \tag{57}$$

Eq. (54) can be solved by known methods. The sought solution x_Φ coincides either with one real root or with the minimum of 3 real roots. Then using Eqs. (50, 36, 49), the others mean values will be:

$$\omega_{D\Phi}=1-s_K\,x_\Phi,\ \ \omega_{B\Phi}=\left(m_B-s_K\,m\,x_\Phi\right)/m_0,\ \ \Delta\omega_\Phi=\omega_{D\Phi}-\omega_{B\Phi},\ \ \Omega_\Phi=\Delta\omega_\Phi\big/\left(R_i\,r^2\right) \tag{58}$$

At the end of this stage of calculation the Eq. (52) must be checked. The solution is inadmissible, if either: $\Delta\omega_\Phi\le 0$ or if $x_\Phi\ge 1$, (59)

i. e. if a frozen regime takes place or if the maximum of qsMCH gets over. Further the periodic system given by the right Eqs. (35÷48) will be solved.

4. Determination of centered periodic solution

First the most difficult "periodic" boundary problem given by right Eqs. (39÷48) will be solved, then periodic MCH from right Eqs. (37, 38) will be derived and solved and finally periodic solution of all EHA (35, 36) will be completed. With use of Eq. (41, 58) one gets from right Eqs. (39, 40, 42) the starting Eqs.:

$$\nu\rho\,\dot u_\Pi=\mu_\Phi\left[\partial^2u_\Pi/\partial r^2+\partial u/(r\,\partial r)-u_\Pi/r^2\right]\quad\text{with } u_\Pi(r_i)=\omega_{i\Pi}r_i,\ \ u_\Pi(1)=\omega_{o\Pi} \tag{60}$$

$$M_\Pi(r)=m_o\left(\partial u/\partial r-u/r\right)r^2\quad\text{with } M_{i\Pi}=M_\Pi(r_i),\ \ M_{o\Pi}=-M_\Pi(1) \tag{61}$$

whereby $\mu_\Phi(r)=1+\mu_{Ei}\,r^2$ with $\mu_{Ei}=\tau_E\,R_i\big/\left|\Delta\omega_\Phi\right|$ (62)

For purely viscous fluid, i. e. for $\tau_E = 0 \Rightarrow \mu_{Ei} = 0 \Rightarrow \mu_\Phi = 1$, the boundary problem (60) can redily be solved by means of Kelvin functions [4,6]. To solve it also for ERF, one must average the variable viscosity (62) within the gap of ERC, which is legitimate due to the 2nd assumption (32)[7]. In agreement with theory of dif. Eqs , requiring to isolate the highest derivative $\partial^2 u_\Pi/\partial r^2$, averaging of the reciprocal viscosity $\mu_\Phi^I = 1/\mu_\Phi$ must be performed:

$$\mu_\Phi^{I\Phi} = (1/\mu_\Phi)^\Phi = h_i^{-1}\int_{r_i}^{1}\frac{dr}{\mu_\Phi(r)} = \frac{1}{\mu_{E1}}\operatorname{arctg}\frac{\mu_{E1}}{1+r_i\,\mu_{Ei}}, \qquad \text{where } \mu_{E1} = (1-r_i)\sqrt{\mu_{Ei}} \tag{63}$$

Then it is convenient to denote: $\kappa^2 = \nu\rho\mu_\Phi^{I\Phi}$ (64)

and to introduce a new independent variable: $q = \kappa r$ (65)

Then boundary problem (60) obtains the form: $\boxed{\dot{u}_\Pi = u_\Pi'' + u_\Pi'/q - u_\Pi/q^2}$ (66)

whereby derivatives with respect to q are denoted by upper dashes. Eq. (66) which doesn't contain any ndimPm is the base for further solution. In view of harmonic load (5) in (36) and due to linearity of the system (35÷38, 66), one can seek centered periodic solution of all quantities in the following form: (67)

$$_\Pi(\tau) = W_c\cos\tau + W_s\sin\tau \quad \text{for } W = \omega_i, \omega_o, \omega_D, \omega_B, M_i, M_o, M_D, M_C, M_B, \kappa_D$$

$$_\Pi(q,\tau) = W_c(q)\cos\tau + W_s(q)\sin\tau \quad \text{for } W = u,\ M, \Omega \tag{68}$$

Then, after insertion of Eqs. (67, 68) into Eqs. (60, 61) and after comparison of equivalent terms at cos τ or sin τ , i. e. after application of the method of harmonic balance, one obtains the parallel systems:

$$u_s = u_c'' + u_c'/q - u_c/q^2 \qquad\Big|\qquad -u_c = u_s'' + u_s'/q - u_s/q^2 \tag{69}$$

$$M_c(r) = m_o\kappa(u_c' - u_c/q)r^2 \qquad\Big|\qquad M_s(r) = m_o\kappa(u_s' - u_s/q)r^2 \tag{70}$$

$$\text{where } u_c(\kappa r_i) = \omega_{ic},\ u_c(\kappa) = \omega_{oc} \qquad\Big|\qquad u_s(\kappa r_i) = \omega_{is},\ u_s(\kappa) = \omega_{os} \tag{71}$$

$$M_{ic} = M_c(\kappa r_i),\ M_{oc} = -M_c(\kappa) \qquad\Big|\qquad M_{is} = M_s(\kappa r_i),\ M_{os} = -M_s(\kappa) \tag{72}$$

Further somewhat lengthy solution can be presented here only concisely.
One can prove [4,6] that a general solution of the boundary problem (69, 71) is:

$$u_c(q) = b\cdot\operatorname{ber}_1 q + b^i\cdot\operatorname{bei}_1 q + k\cdot\operatorname{ker}_1 q + k^i\cdot\operatorname{kei}_1 q \tag{73}$$

$$u_s(q) = b^i\cdot\operatorname{ber}_1 q - b\cdot\operatorname{bei}_1 q + k^i\cdot\operatorname{ker}_1 q - k\cdot\operatorname{kei}_1 q \tag{74}$$

where $\operatorname{ber}_j q$, $\operatorname{bei}_j q$, $\operatorname{ker}_j q$, $\operatorname{kei}_j q$, $\quad j = 0, 1, \ldots$ (75)
are the special Bessel functions so called Kelvin functions [4] of the first order (j = 1) and b, b^i, k, k^i are unknown integration constants determinable from boundary conditions (71). After their application, one gets system of 4 lin algebraic Eqs., which can be written down and solved in a matrix form: $\mathbf{G}\mathbf{b} = \boldsymbol{\omega}_C \Rightarrow \boxed{\mathbf{b} = \mathbf{G}^{-1}\boldsymbol{\omega}_C}$ (76)
where $\mathbf{b} = col[b, b^i, k, k^i]$, $\boldsymbol{\omega}_C = col[\omega_{ic}, \omega_{is}, \omega_{oc}, \omega_{os}]$, $\mathbf{G} = [G_{jk}]$, $j,k = 1\div 4$ (77)
whereby elements $\mathbf{G}_{jk}$ are base functions (75) taken either for $q_i = \kappa\, r_i$ or for $q_o = \kappa$. (78)

Then it is possible to determine friction moments M_c, M_s (70) in the gap of ERC namely on its edges M_{ic}, M_{is}, M_{oc}, M_{os} according to (72). This again can concisely be written in matrix form as follows: (79)

$$\boxed{\mathbf{M}_C = m_o\mathbf{R}\mathbf{H}\mathbf{b} = m_o\mathbf{R}\mathbf{H}\mathbf{G}^{-1}\boldsymbol{\omega}_C}$$

where $\mathbf{M}_C = col[M_{ic}, M_{is}, M_{oc}, M_{os}], \mathbf{R} = diag[-r_i^2, -r_i^2, 1, 1], \mathbf{H} = [H_{jk}],\ j,k = 1 \div 4$ (80)
whereby H_{jk} are expressions of base functions (75) of the zero and first order (j = 0, 1) taken for q_i or q_o (78). Then a structural arrangement of EHA (Fig. 2 in Ref. [8]) according to relations (45÷48) can be considered after introduction of vectors:

$$\boldsymbol{\omega}_A = col\,[\omega_{Dc},\ \omega_{Ds},\ \omega_{Bc},\ \omega_{Bs}],\ \mathbf{M}_A = col\,[M_{Cc},\ M_{Cs},\ M_{Bc},\ M_{Bs}] \quad (81)$$

and trasformation matrix $\mathbf{T} = \begin{bmatrix} \tilde{\mathbf{0}} & \tilde{\mathbf{I}} \\ \tilde{\mathbf{I}} & \tilde{\mathbf{0}} \end{bmatrix}$, $\dim \mathbf{T} = 4 \times 4$ (82)
for which it holds $\mathbf{T}^{-1} = \mathbf{T}$ or $\mathbf{T}^2 = \mathbf{I}$, (83)
where $\tilde{\mathbf{0}}$ is nil matrix of size 2×2 and $\tilde{\mathbf{I}}$ or $\mathbf{I}$ is identity matrix of size 2×2 or 4×4.
Then one gets with use of(45 ÷ 48, 79 ÷ 83):

for EHAI: $\boldsymbol{\omega}_{AI} = \mathbf{T}\boldsymbol{\omega}_C,\ \mathbf{M}_{AI} = \mathbf{T}\overline{\mathbf{M}}_C = m_o \mathbf{TRHG}^{-1}\boldsymbol{\omega}_C = m_o \mathbf{TRHG}^{-1}\mathbf{T}\boldsymbol{\omega}_{AI}$ (84)

for EHAII: $\boldsymbol{\omega}_{AII} = \boldsymbol{\omega}_C,\quad \mathbf{M}_{AII} = \mathbf{M}_C = m_o\,\mathbf{RHG}^{-1}\boldsymbol{\omega}_C = m_o\,\mathbf{RHG}^{-1}\boldsymbol{\omega}_{AII}$ (85)

This can be unified by formula: $\boxed{\mathbf{M}_A = m_o \mathbf{D}_A \boldsymbol{\omega}_A}$ where $\mathbf{D}_A = \mathbf{R}_A \mathbf{H}\mathbf{G}_A^{-1}$ (86)
and for EHAI: $\mathbf{R}_A = \mathbf{TR},\ \mathbf{G}_A = \mathbf{TG}$, for EHAII: $\mathbf{R}_A = \mathbf{R},\ \mathbf{G}_A = \mathbf{G}$ (87)

Note that vector $\boldsymbol{\omega}_A$ (81) is unknown till now and will be determined after completization of solution of all EHA. Further from Eqs. (37, 38), after excluding κ_Π, one obtains "periodic" MCH of ED in the form:

$$\boxed{\nu^2 \ddot{M}_{D\Pi} + 2 s_K \nu \dot{M}_{D\Pi} + s_K x_0 M_{D\Pi} = -(\nu\dot{\omega}_{D\Pi} + s_K x_1 \omega_{D\Pi})/x_0} \quad (88)$$

where $x_0 = 1 + x_\Phi^2,\ x_1 = 1 - x_\Phi^2$ (89)
Using Eq. (68) for $W = \omega_D, M_D$, one gets solution of dif. Eq. (88) in the form

$$M_{Dc} = -M_{D1}\,\omega_{Dc} + M_{D2}\,\omega_{Ds},\qquad M_{Ds} = -M_{D2}\,\omega_{Dc} - M_{D1}\,\omega_{Ds} \quad (90)$$

$$\text{where}\quad M_{D1} = s_K \frac{\nu^2 + s_K^2 x_1}{(s_K^2 x_0 - \nu^2)^2 + 4 s_K^2 \nu^2},\quad M_{D2} = \frac{\nu}{x_0} \cdot \frac{\nu^2 + s_K^2 (1 - 3x_\Phi^2)}{(s_K^2 x_0 - \nu^2)^2 + 4 s_K^2 \nu^2} \quad (91)$$

Finally with use of Eq. (68) for $W = \omega_B, \omega_D, M_B, M_C, M_D$ as well as vectors (81) and Eq. (90), one obtains from dif Eqs. (35, 36) matrix form: $\quad \mathbf{C}_A \boldsymbol{\omega}_A = \mathbf{M}_A - \mathbf{f}$ (92)

$$\text{where } \mathbf{C}_A = \begin{bmatrix} \tilde{\mathbf{C}}_D & \tilde{\mathbf{0}} \\ \tilde{\mathbf{0}} & \tilde{\mathbf{C}}_B \end{bmatrix},\quad \tilde{\mathbf{C}}_D = \begin{bmatrix} c_{D1} & c_{D2} \\ -c_{D2} & c_{D1} \end{bmatrix},\quad \tilde{\mathbf{C}}_B = \begin{bmatrix} m_{B\nu} & \nu_B \\ -\nu_B & m_{B\nu} \end{bmatrix} \quad (93)$$

and $f = col\,[0,\ 0,\ 0,\ m_a],\ c_{D1} = m_{D\nu} + M_{D1},\qquad c_{D2} = \nu_D - M_{D2}$ (94)
Then using Eq. (86), one gets from Eq. (92):

$$(m_o \mathbf{D}_A - \mathbf{C}_A)\boldsymbol{\omega}_A = \mathbf{f} \quad \text{so that} \quad \boldsymbol{\omega}_A = \mathbf{A}^{-1}\mathbf{f}, \quad \text{where } \mathbf{A} = m_o \mathbf{D}_A - \mathbf{C}_A \quad (95)$$

As the vector **f** has only the last element, one can write instead of solution (95):

$$\boxed{\boldsymbol{\omega}_A = m_a\, col_4(\mathbf{A}^{-1})} \quad (96)$$

where the last column of inverse matrix $\mathbf{A}^{-1}$ is extracted. It results that harmonic components of angV $\boldsymbol{\omega}_A$ (96, 81) are proportional to the amplitude of excitation m_a while amplitudes of friction moment $\mathbf{M}_A$ (86, 81) are proportional to the product $m_o.m_a$, where size of outer disc m_o is given by Eq. (16). Now the steady–state solution is fully determined. There is no place here to excerpt and arrange a concise effective calculation algorithm including also an algorithm for calculation of Kelvin functions (75).

5. Analysis of some results

In a course of analysis of multiPm dyn systems, three levels are proved to be convenient and efficient:

1. preliminary analysis of influence of individual Pm-s upon a chosen quantity. This is based on a special sensitivity analysis and determination of a sequence of dominance [8]. This can help to reduce the volume of analysis significantly,
2. determination and analysis of resonant regions,
3. determination, analysis and representation of resonant peaks. This enables one to construct representative and expressive diagrams containing very compressed information.

Here only a short illustration of the first stage can be given. In the space of ndimPm-s (22), for 39366 combinations of values, one gets for driving angV ω_D of EHAI the following sequence of dominance: m_{Bv}, m_o, r_i, s_K, m_a, m_L, ν, m_{Dv}, m_{Dd}, τ_E, m_K, J_{DC}, ρ, J_{BL}.

So the three most influential Pm-s for dynamic response of ED seem to be: viscous friction of brake m_{Bv}, size m_o and radius r_i (16) of ERC. More detailed analysis will be presented in the conference lecture.

6. Conclusions

The presented analytical and numerical results should be verified experimentally to prove or disprove, if the quasistatic model of ERF (10÷14) can be used also in dynamic conditions (for harmonic load) or if more complicated model in Ref. [3] must be used. Also influence of temperature that has been neglected might be important and therefore incorporated into the model. This task might be contents for further investigation and activity.

Acknowledgements

This task has been partially supported by Grant Agency of the Slovak Academy of Sciences (Project No. 2/1135/94) which is hereby gratefully acknowledged.

References

1. *Electrorheological (ER) Fluids. A Research Need Assessment - Final Report.* (U. S. Department of Energy, Washington, D. C., May 1993).
2. R. Firoosian and al., *Dynamic torque response of an electrorheological clutch,* DE–Vol. 35 Rotating Machinery and Vehicle Dynamics ASME 1991, pp. 295÷301.
3. D. R. Gamota and F. E. Filisko, *Dynamic mechanical studies of ER materials: Moderate frequencies.* J. Rheol., **35** (3), April 1990, pp. 399÷425.
4. M. Abramowitz and I. A. Stegun: *Handbook of Math. Functions with Formulas, Graphs and Mathematical Tables* (Russian translation), (Nauka, Moscow, 1979).
5. V. Oravský, *Some new problems and results in dynamics of rotating machine aggregates.* In: Proc. of the 8th World Congress on the Theory of Machines and Mechanisms, Prague 1991, *Vol. 3, pp. 665÷668.*
6. V. Oravský, *Dynamic characteristic of hydroclutch with ER fluid and its solution for harmonic movement.* In: Proc. Dynamics of Fluid, Prague 1992, pp. 45÷46.
7. V. Oravský and V. Krč–Jediný, *Using of Kelvin functions for determination of steady–state motion of an electrorheological clutch* (in Slovak). In: Proc. of 10th Scientific Conf. VŠDS, Žilina 1994, pp. 146÷151.
8. V. Oravský, *Quasi–static efficiency of a concentric ER clutch.* In: This Proceedings.

ROTATION OF THE AXISIMMETRIC DIELECTRIC BODIES (DEB) IN ELECTRORHEOLOGICAL SUSPENSIONS (ERS)

ZINOVIY P. SHULMAN
A.V.Luikov Heat and Mass Transfer Institute, Belarussian Academy of Sciences

VALERIY NOSOV
Belarussian State Polytechnic Academy

ABSTRACT

Consideration is given to rotation of nonconducting bodies in electrorheological suspensions. Experimental data are presented describing the effect of an electric field on the rotation speed as well as on rheological and electrophysical properties of ERS. The dependence of speed on temperature and the rotor geometry is considered in some detail. Two examples of engineering applications are presented.

The effect is known since the previous century and was investigated by Quincke[1], Richardson[2] and Sumoto[3] independently of each other.

Different and sometimes contradictory opinions were stated about its physical nature. The first attempts to give a qualitative interpretation of the effect were presented in the works Lampa[4], Hideweiler[5] as well in the investigations by Vedy[6,7] and Sumoto[8]. Let's outline them briefly:

1. Any DEB rotates in a strong electric field (EF) till it exists.
2. The rotation is conditioned by electrostatic repulsion forces. The direction of rotation is ambiguous.
3. The conductivity of the surrounding the DEB semiinsulating medium must be in the range 10^{-6}-10^{-9} Ohm^{-1}.

Sumoto was the first to determine quantitative relationships. However, not the speed was measured but an indirect parameter, viz. the interelectrode current difference D_j at a rotating and braked rotor.

Dukchin, Deinega and Simonova[9] theoretically calculated the speed of rotation of DEB (RDEB) under the assumption of a very thin layer of electrically charged fluid in a Newtonian dielectric liquid. However, the calculations were not compared with known experimental results.

The problem of designing a dielectric motor did not receive recognition, development and propagation mainly because of very low power output.

In the laboratory of Rheophysics at Heat and Mass Transfer Institute, Belarussian Academy of Sciences, since 1966 a new special class of dielectric fluids was developed, viz. electrorheological suspensions (ERS). These suspensions were employed as working media in dielectric motors. The main feature of an ERS is a markedly pronounced electrorheological effect (ERE). By the way, this term was firstly introduced by Z.Shulman in 1966 before publication[10]. Heretofore a term "electroviscous effect" was used.

In this lecture we concentrate not on theoretical problems but on experimental facts only. The main aim is to extend our knowledge about ERSes and their properties. In experiments, typical ERSes were used[11]. For example, they consist of: SP, i.e. diatomite or aerosil particles (ε=7.2), CP (carrier media): transformer or vacuum oils (ε=2.1-2.4), activators: W, PEPA, DEA, OA, and a stabilizing surfactant: MOG. The ERSes were prepared prior to each experiment and were subjected to ultrasound dispersing to attain more uniform distribution of particle sizes. Furthermore, a sample of each ERS was taken for the use in experiments on RDEB and for measuring rheological and electrophysical characteristics.

Rheological characteristics. Measurements were made using a specially designed rotational rheoviscometer. Experiments were performed in the following ranges: viscosity $\eta_{eff}=2\cdot10^{-3}$-1.35 Pa·s; shear rate γ=20-210 s^{-1}; electric field strength E=0-3 kV/mm.

The dependencies of dielectric permittivity, ε, and dielectric losses, tgδ, on the electric field frequency, f, were obtained by Prof. Deinega's group (Ukrainian Academy of Sciences, Kiev). All the ERS flow curves were generalized by Shulman's model of nonlinear viscoplastic media[12] in its special three-parameter form (n=m)

$$\tau^{1/n} = \tau_0^{1/n} + \left(\mu_p \dot{\gamma}\right)^{1/n} \tag{1}$$

where τ, τ_0 and μ_p are shear stress, yield stress, and plasticity, respectively. The most popular Bingham model follows from Eq. (1) at n=1.

The flow curves shift markedly towards greater shear stresses with increasing EFS (Fig.1). The inclination to the γ axis varies only slightly in this case, especially for small activator contents. The yield stress changes markedly, it may be roughly described as $\tau_0 = KE^2$. Here the coefficient K is a function of the ERS composition and properties. The slopes of rheograms beyond the nonlinear section vary slightly, i.e. an EF weakly affects the plastic viscosity. Only for aerosil the rate of increase is noticeably higher. The curvature and extend of the nonlinear section of a rheogram increase with EFS at small and moderate shear rates. Hence, the ERS mechanical behaviour is characterized by nonlinear viscoplasticity of the suspension flow.

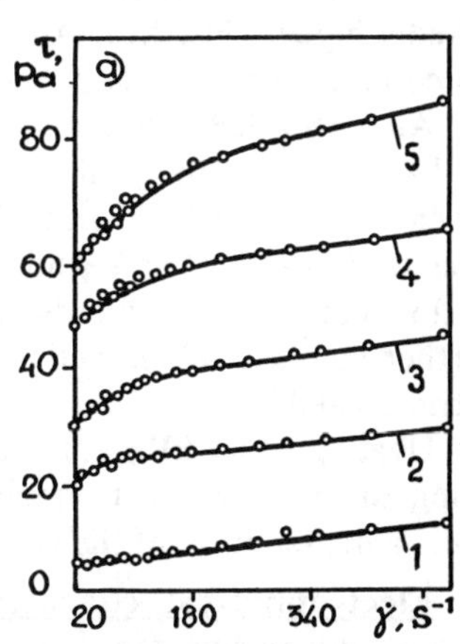

Fig.1 ERS flow curves (C_{diat}=7.5%, φ_w=6.5%): 1, E = 0; 2, E=1.0; 3, E=1.5; 4, E=2.0; 5, E=2.5 kV/mm

In the absence of particles the carrier media is less sensitive to external EF. The viscosity dependence on the particles content increases with

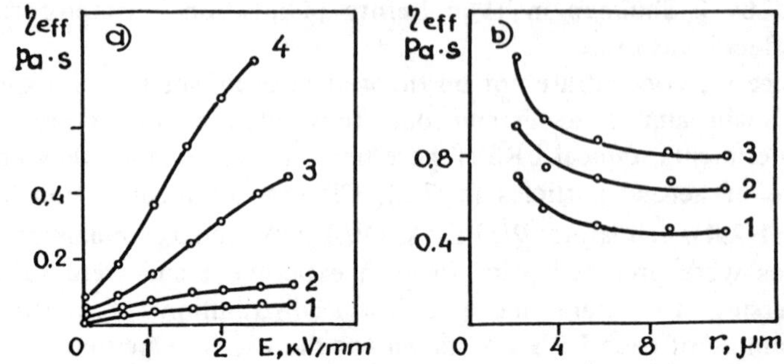

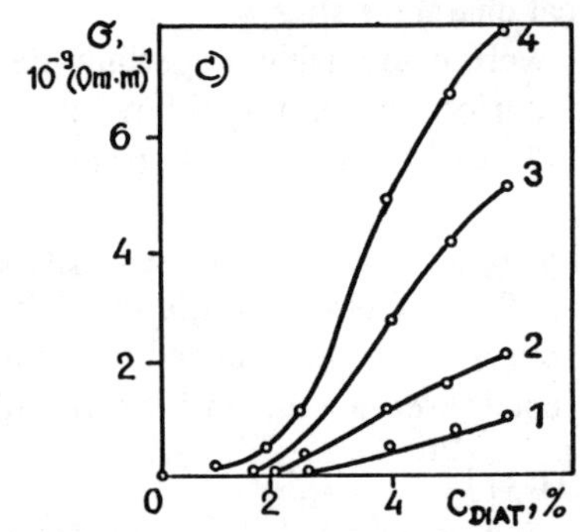

Fig.2 Characteristics of transformer oil-diatomite ERS (φ_w=5.5%): (a) γ=86 s^{-1} (1, C_{diat}=1%; 2, C_{diat}=2.5%; 3, C_{diat}=5%; 4, C_{diat}=10%); (b) C_{diat}=5%, E=3 kV/mm (1, γ=107.45 s^{-1}; 2, γ=71.6 s^{-1}; 3, γ=53.7 s^{-1}); (c): 1, E=0.5; 2, E=1.0; 3, E=2.0; 4, E=3.0 kV/mm

concentration C_{part} (Fig.2,a). For a very thinned ERS, this dependence is nearly linear. It become nonlinear for higher concentrations. The particle sizes do not change the character of the dependence $\eta(E,\gamma)$. But for smaller particles the ERS viscosity is markedly higher (Fig.2,b). An increase in the concentration of particles enhances the ERS conductivity roughly proportionally to the suspensions viscosity (Fig.2,c).

An ERS filled with diatomite dried to a constant weight or to a moisture content less than 2% is rheologically insensitive to an external EF (Fig.3). This behaviour is inherent in all ERSes with silica particles. A further increase in the activator concentration at first acts similary to an increase in EFS and the content of particles. The flow curves shift towards higher shear stresses without changing their slopes. But a further increase of the activator content enhances the conductivity to the point of electric breakdown.

Heating of an ERS enhances the shear stress to a certain degree. But at further heating the ERS viscosity begins to reduce and the flow curves become more flat. Hence, a degradation of the rheological factor occurs (Fig.4).

Electrophysical characteristics. Permittivity, ε, and dielectric losses, tgδ, increase with raising the diatomite concentration until they reach a plateau (Fig.5). Furthermore, the dielectric dispersion, which is a nonlinear dependence of ε on the field frequency, f, increases. Similar dependencies were obtained for ERSes with aerosil. This is in a good agreement with established general principles.

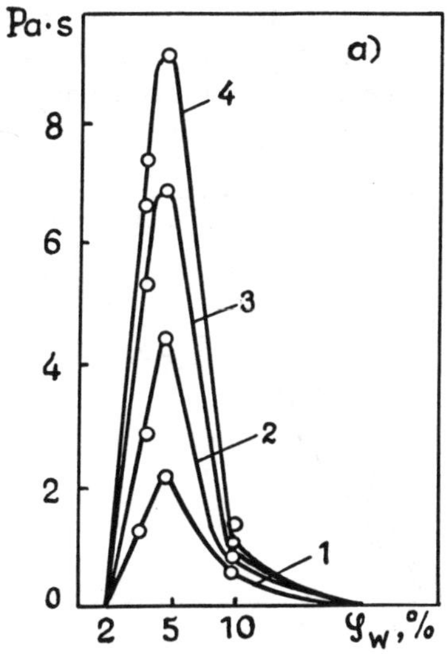

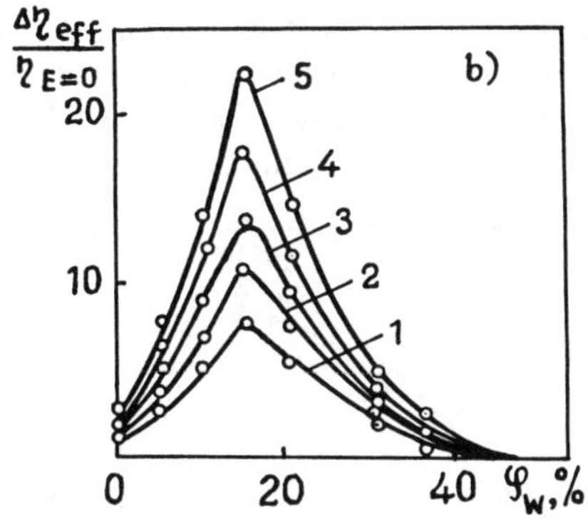

Fig.3 The influence of moisture (activator) content on electroviscosity:
(a) diatomite (C_{diat}=2.3%)-polyisobutilene solution ($C_{poyis.}$=1%) ERS, γ=72.5 s^{-1}; 1, E=3; 2, E=4.0; 3, E=5.0; 4, E=6.0 kV/mm;
(b) Polyvinyl alcohol (C_{pva}=2.3%)-vaseline oil ERS; γ=72.5 s^{-1}: 1, E=0.8; 2, E=1.2; 3, E=1.6; 4, E=2.0; 5, E=2.4 kV/mm (Deinega and Kolganich)

With increasing the activator (water) content the ε and tgδ values rise more

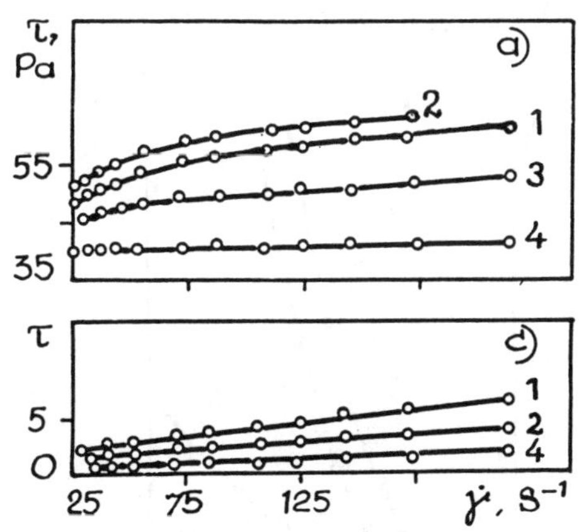

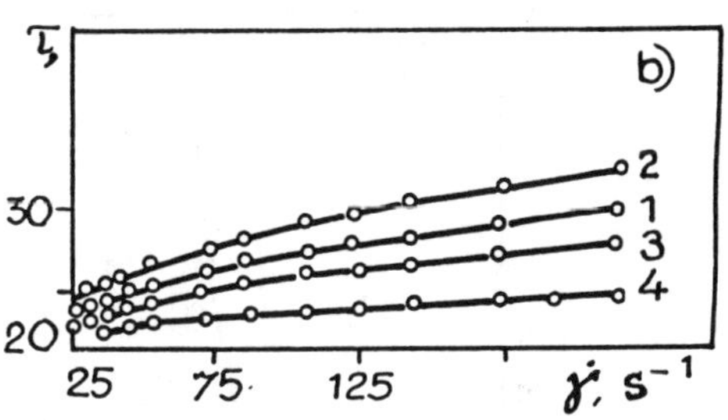

Fig.4 Flow curves of the transformer oil-diatomite ERS (C_{diat}=5 %; φ_w=6.5 %) (a) E=2.5 kV/mm; (b) E=1.5 kV/mm; (c) E=0; 1, T=293 K; 2, T=313 K; 3, T=333 K; 4, T=363 K.

intensively than with increasing the concentration of particles, no tendency to saturation at C_{act} is observed. After the attainment of a certain concentration C_{act}, the ε, tgδ and σ values increase avalanche-like. The dependencies of ε and δ on C_{aer} and C_{act} turn out to be similar (Fig.6).

The relationships between ε or tgδ and the electric field frequency for aerosil-containing ERSes at low frequencies (up to 1kHz) show a steep decline. In the region above 4-10 kHz the frequency has a weak influence on the electrophysical parameters. The frequency markedly affects ε and tgδ at higher C_{part} and C_{act}.

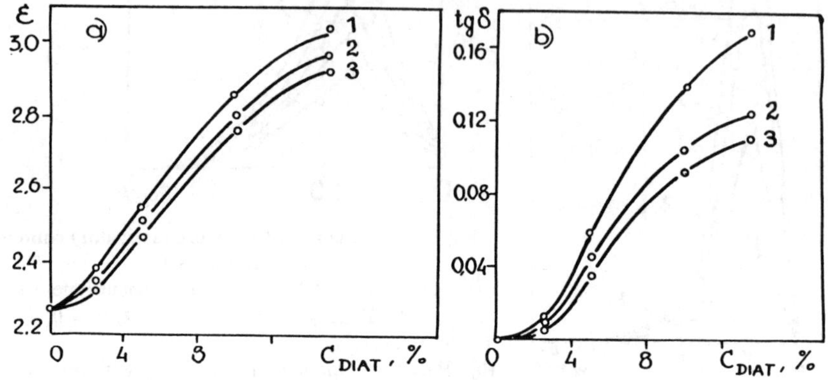

Fig.5 Electrophysical characteristics of the diatomite-transformer oil ERS. φ_w=5.5%: 1, f=400 Hz; 2, f=800 Hz; 3, $f=10^3$ Hz.

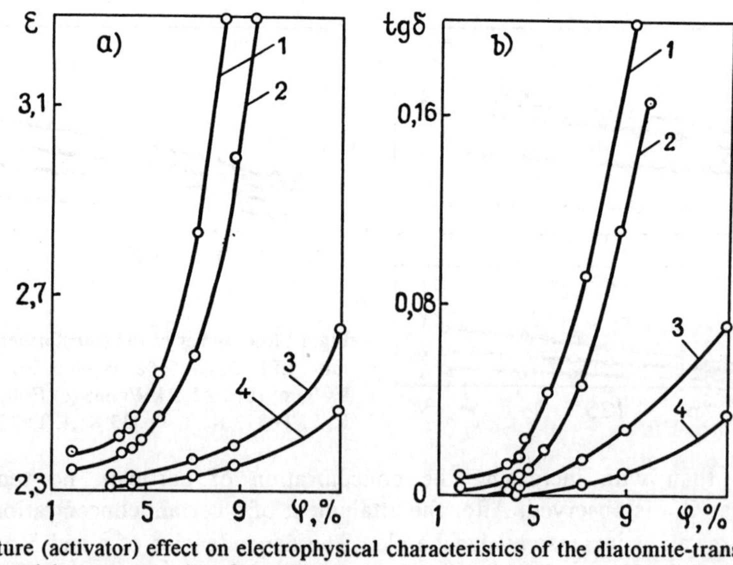

Fig.6 Moisture (activator) effect on electrophysical characteristics of the diatomite-transtormer oil ERS with particles concentration C_{diat}=5% (curves 1 and 2) and C_{diat}=2.5 % (curves 3 and 4). Frequencies: curves 1 and 3, f=400 Hz; curves 2 and 4, $f=5\cdot10^3$ Hz.

Characteristics of RDEB[13]. The experiments were conducted on two setups (Fig.7). The first one (I) is a two-electrode model on the basis of flat capacitor (a

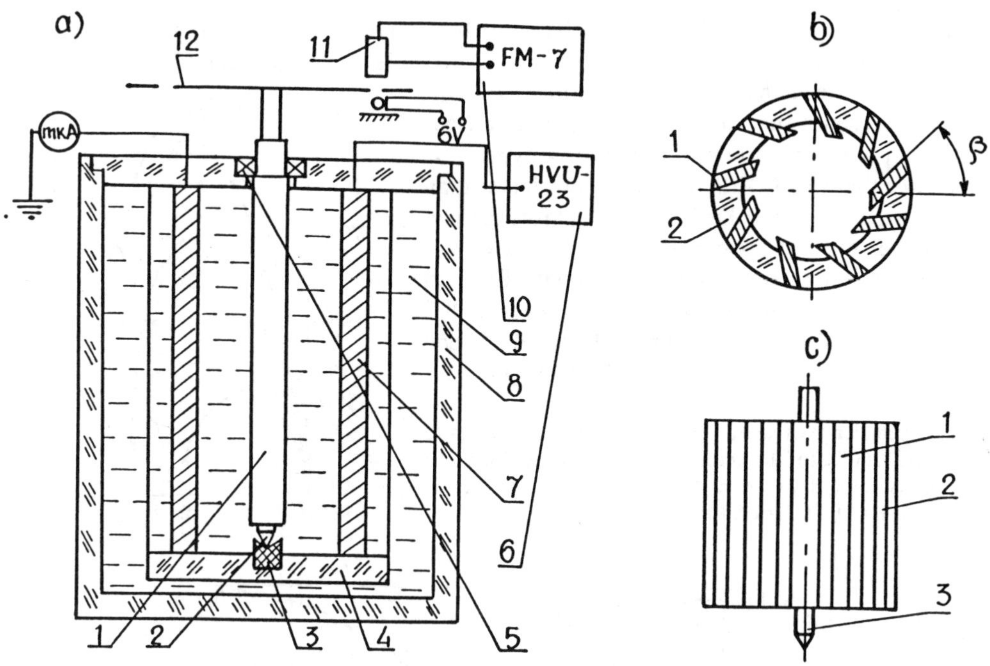

Fig.7 (a) Two-electrode high speed model with a uniform electric field: 1, dielectric rotor (d=4 mm); 2, hardened metallic needle; 3, rubby step bearing; 4, plexiglass box; 5, ball bearing; 6, stabilized high-voltage supply unit; 7, electrodes; 8, thermostat; 9, ERS (working medium); 10, frequency meter; 11, photodiode; 12, disc-interrupter (16 perforations);
(b) sketch of stator and rotor (c) of the multielectrode axisimmetric low-speed model. (b): 1, electrodes (2 to 9 pairs); 2, hollow plexiglass cylinder ; (c): 1, dielectric drum; 2, metallic strips (15 pieces); 3, shaft.

uniform field). In the model I, the DEB (rotor) diameter was 4 mm. The inter-electrode clearance could vary smoothly from 5 to 10 mm. In the model II the rotor diameter was 39 mm and stator diameter was 41 mm (Fig.7,b,c). The experiments with a uniform EF were performed in an effort to compare our results with analogous data obtained by other authors. On the other surface of the rotor (model II) there was a set of thin metallic strips (up to the optimal number of 15) for arranging a favorable EF. Their width, thickness and number were varied in experiments, the number of electrode pairs was varied from 2 to 9. Electrodes 1 had a form of flat metal plates inclined to the stator cylindrical surface at various angles (from 0° to 45°) (Fig.7,b).

At low aerosil concentrations (up to 1 wt%) the speed of RDEB is nearly a linear function of EFS throughout the region of measurements (Fig.8). With further ERS thickening this dependence become more and more curved due to an increase in electroviscosity η(E) and appropriate rise of the rotational resistance. At C_{part} >2.5% the speed reduces and the parameter $n=n(C_{part})$ shifts towards higher EFS, which is similar to the behaviour of diatomite ERSes. Fig.9,a presents the speed characteristics a glass rotor in pure transformer oil taken from work[14]. It indicates that the RDEB was slow (not faster than 16 s^{-1} at the voltage U=100 kV). Unfortunately, that paper did not specify the interelectrode gap, and hence it was impossible to estimate EFS. In our experiments the voltage did not exceed 8.5 kV, but even for low-concentrated ERS the speed of RDEB turned out to be tenfold higher (Fig.9).

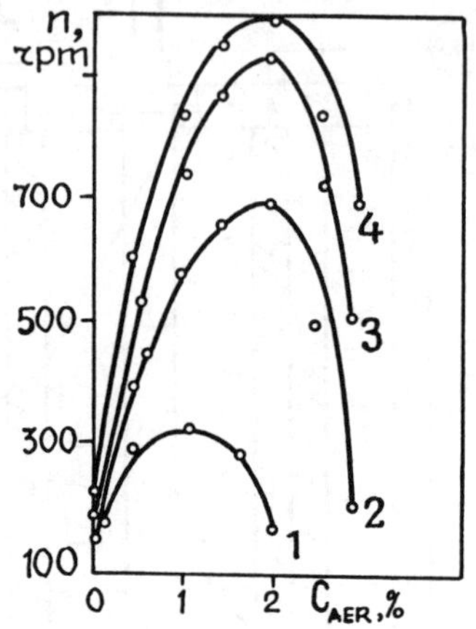

Fig.8 Dependence of speed on particles content; C_{PEPA}=20 %; MOG=200 % (of the aerosil mass). High speed model I (for glass rotor d_{rot}=4 mm; $H_{capac.}$=5 mm): 1, E=1.0; 2, E=1.4; 3, E=1.6; 4, E=1.8 kV/mm.

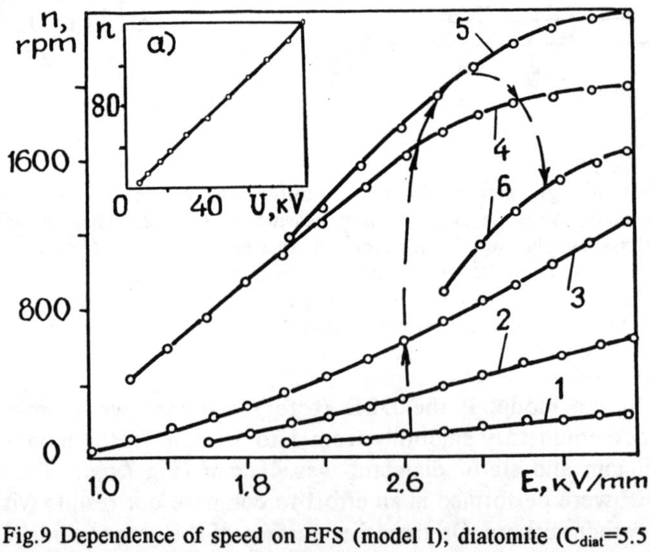

Fig.9 Dependence of speed on EFS (model I); diatomite (C_{diat}=5.5 %)-transformer oil ERS: 1, C_{diat}=0.1%; 2, C_{diat}=0.5%; 3, C_{diat}=1.0%; 4, C_{diat}= 5 %; 5, C_{diat}=10 %; 6, C_{diat}=15 %.

An increase in C_{part} markedly enhances the RDEB up to several thousands r.p.m., first linearly and then with lower rate of n(E). At C_{part}>10% the speed reduces and the curves shift towards higher EFS. Simultaneously, the electroconductivity sharply increases up to electric breakdown. For aerosil-containing ERS with PEPA activator and glass rotor (a uniform EF) the speed dependence on C_{part} is different-looking and has a shape of an upset parabola (Fig. 8). But this behaviour is not common to other kinds of particles (for example, diatomite with W-activator).

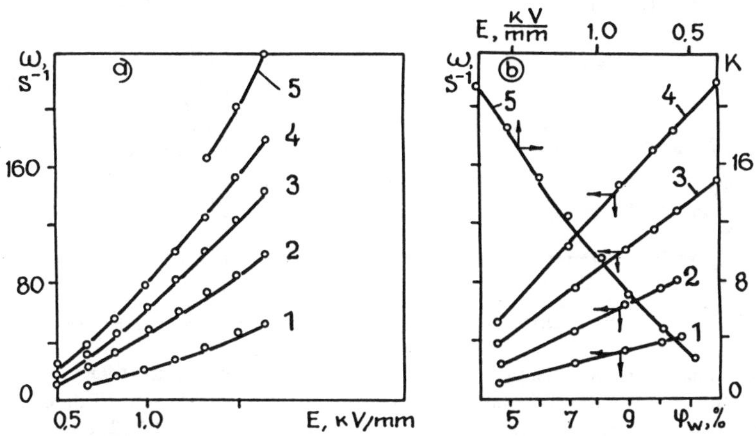

Fig. 10 Speed dependencies on EFS (a) and activator (water) content (b) for a diatomite (2.5 %)-transformer oil ERS:
(a): 1, φ_w=4.4%; 2, φ_w=7.0%; 3, φ_w=8.8%; 4, φ_w=10.5%; 5, φ_w=13%;
(b): 1, E=0.67; 2, E=1.0; 3, E=1.34; 4, E=1.6 kV/mm ; 5, K=K(E).

An ERS composed of diatomite, transformer oil and W-activator was investigated. In ERSes with low activator concentration the speed of RDEB increases linearly with EFS, as well as the slope of the parameter n=n(E) does. The speed of RDEB in low-concentrated ERS is a linear function of moisture content, φ_w, at E=const (Fig 10,b), the steepness of this line increases with E. All the experimental data were generalized by an empirical formula

$$\omega=K(\varphi_w-2) \quad \text{for } \varphi_w > 2\%, \qquad (\omega=\pi\cdot n/30) \qquad (2)$$

The coefficient K varies almost linearly with E (Fig.10,b) $K=9.75\cdot10^{-6}E-1.88$ (3)

Previously, mainly water-activated diatomite ERSes were investigated. Their principal drawback is gradual desorption of a W-activator from the particles surface, especially under heating. We invented a number of new ERSes used as working media with more stable activators.

The experiments were performed with ERSes based on four narrow fractions of diatomite particles (C_{part}=5%). The speed of RDEB and the interelectrode current depended nonmonotonously on the particles size (Fig. 11) whereas the electroviscosity increased monotonously with decreasing the particles size. Therefore for the finely ground particles, for example, of aerosil, the rotor speed is lower. Faster rotation for relatively coarse particles is an important advantage for engineering applications of ERSes. There is no longer any necessity for fine grinding of particles. Particles sized several microns provide a reasonable speed of RDEB. Addition of antisedimentators gives an effect to some extent similar to that of an activator. In the case of MOG as an admixture the falling branch of the speed curve rises and finally attains the saturation level (Fig.12).

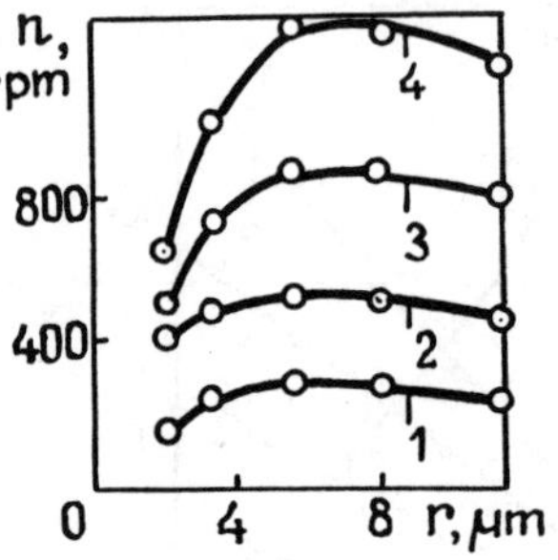

Fig.11 Speed versus particles size for a diatomite (2.5 %)-transformer oil ERS and glass rotor: 1, E=1.2; 2, E=2.0; 3, E=2.8; 4, E=3.6 kV/mm.

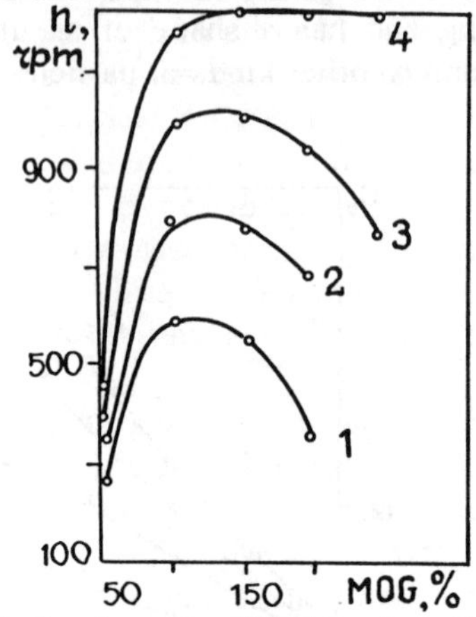

Fig.12 Speed versus MOG content for an aerosil (C_{aer}=1.5%)-oil ERS, φ_{PEPA}= 0.45%, ebonite rotor: 1, E=1.2; 2, E=1.4; 3, E=1.6; 4, E=1.8 kV/mm.

Control tests with a textolite DEB were performed. At initial moisture content φ_w=4.5% the rotor was motionless. But with raising the moisture content the relationship n(E) shifted towards higher EFS and became weaker, and finally the rotation came to a full stop.

The effect of ERS heating in the range 293-353 K (Fig.13) showed that (a) the speed of RDEB diminished; (b) the slope of the dependence n=n(E) lowered; (c) the dependence n=n(E) shifted towards higher EFS; (d) the linearity of the dependence n=n(E) retained throughout the temperature range; e) the speed of RDEB at E=const diminished. Taking into account the effect of temperature on electrophysical and rheological properties of suspensions, we can conclude that the temperature

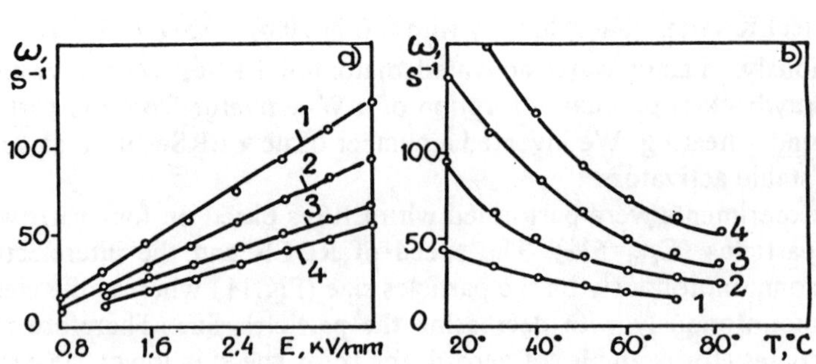

Fig.13 Speed characteristics for a diatomite (C_{diat}=2.5%)-transformer oil ERS and glass rotor: (a) φ_w=8.8%: 1, T=293 K; 2, T=303 K; 3, T=313 K; 4, T=323 K; (b) φ_w=9.9%: 1, E=1.2; 2, E=2.0; 3, E=2.8; 4, E=3.6 kV/mm.

dependence of the rotation speed is mainly determined by the activator content and properties. The rate of decrease of the rotation speed is greatest in the range 293-353 K. The dependence n=n(T) is distinct. Its steepness diminishes because of an increase in viscosity under ERS heating (Fig.13,a). For all activator concentrations, the higher the temperature, the more is the ERE depressed (Fig.13,b).

As indicated above, ERSes with high activator (water) content are most sensitive to heating because of the activator desorption from the particles surface. This hypothesis was verified in experiments on diatomite-containing ERS activated with DEA. In this case heating weakly affects the speed of RDEB. For highly thickened ERS under weak heating the speed of RDEB is either temperature independent or increases slightly because of the dilution of a carrier media (for example, transformer oil). However, subsequent heating enhances the activator desorption and thus the relationship n=n(E) becomes descending. The experiments were carried out with an ERS cooled to a room temperature after heating to 353K, all other conditions being the same. It appeared that the activator desorption was irreversible in nature (Fig.14).

The RDEB take place at non-cylindrical rotor geometry (square and rectangle rods), Fig.15. A non-axisymmetric DEB rotates in an ERS. The rotation speed of an non-axisymmetric DEB increases as its form approached to cylindrical.

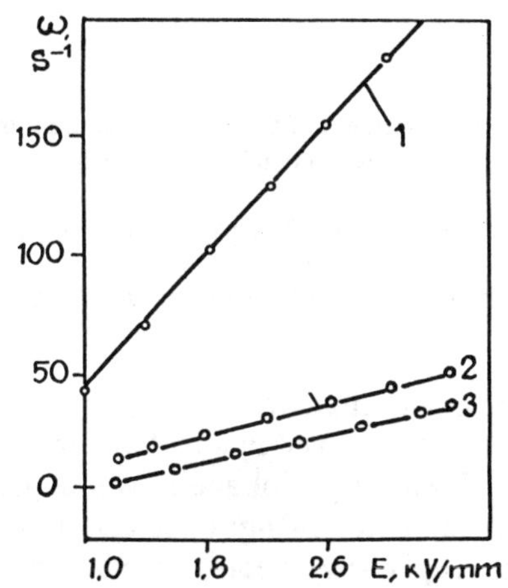

Fig.14 Effect of heating and cooling for a diatomite (2.5 %)-transformer oil ERS and glass rotor; φ_w=9.9%: 1, initial state (t=20°C); 2, t=80°C (after heating from 20°C); 3, t=20°C (cooling after preliminary heating to 80°C)

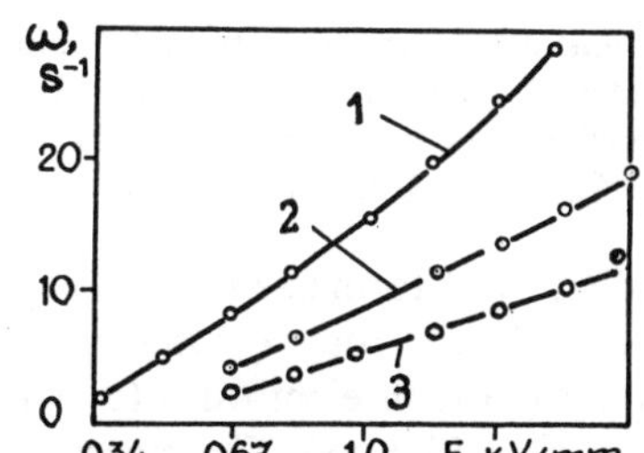

Fig.15 Speed - SEF correlations for various rotor geometries. All DEB made of organic glass. Diatomite (C=0.5 %) - transformer oil; φ_w=9.9 %; 1 - cylinder (d=4 mm); 2 - square (2.7×2.7 mm); 3 - rectangle (1.8×3.6 mm)

Some possible applications. (1) A method of measuring the moisture content of dielectric powders and solids. (Fig.16). A relationship n=n(E,φ_{act}) is used for testing a material immersed in an ERS.

Fig. 16 illustrates this for an ebonite rotor immersed into a diatomite ERS with C_{part}=2.5%. The relationship is linear in the range of moisture content of particles, which is most important for applications, viz. 3-13%. With increasing EFS both the speed and the slope of the $n=n(E,\varphi_{act})$ dependence increase.

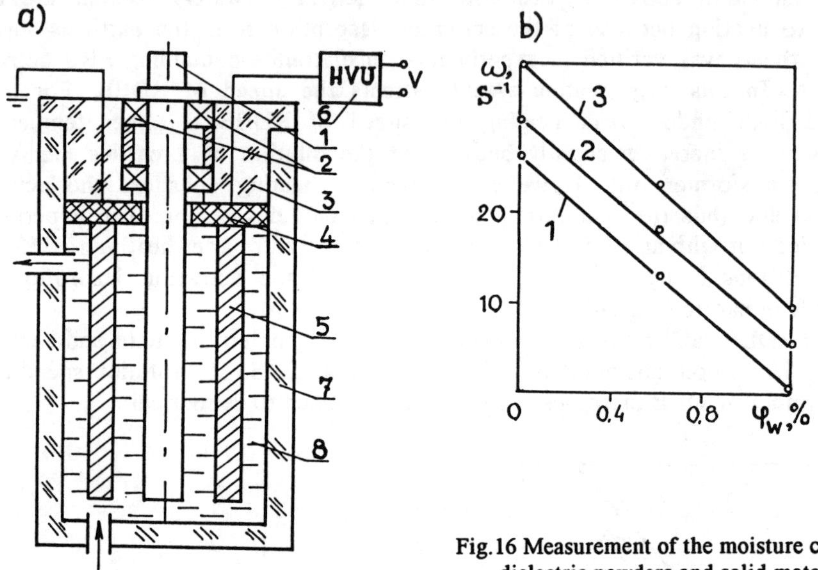

Fig. 16 Measurement of the moisture content of dielectric powders and solid materials

(a) design of the device: 1, rotor (ebonite, d=3-4 mm); 2, ball bearing; 3, cover; 4, insulator; 5, electrodes; 6, high voltage supply unit; 7, container; 8, ERS with wet particles;
(b) speed versus moisture content of textolite for a diatomite (2.5%)-transformer oil ERS: 1, E=1.34; 2, E=1.5; 3, E=1.67 kV/mm.

Empirical formulas (2), (3) are used for water as an activator.

A dielectric rotor (1) made, for example, of ebonite, is fixed in a ball bearing (2) of the lid (3) with electrodes (5) which is connected with high voltage power supply filled with an ERS (8). The measurement procedure is the following: (a) a certain volume of dielectric moistened powder is mixed with a specially selected standard non-conducting liquid; (b) this suspensions is placed into the device; (c) at a given voltage the n value is registered; (d) the moisture content is determined by relationship (2).

To determine the moisture content of a solid material, the following procedure is used: a) a rod (d=3-4 mm) of a material to be tested is inserted in the device filled with a known special ERS; b) the moisture content is determined by the n magnitude according to the graph (Fig. 16,b) of n versus φ_w at E=const.

(2) A generator of undamped mechanical oscillations (Fig.17,a). The generator operation is as follows: (a) after a voltage is applied to electrodes, the dielectric shaft begins rotating until it collides with the beam end with a stopper; (b) after the collision the shaft moves in the opposite direction until a new collision with the other stopper occurs, and so on. The collision frequency depends on E, the amplitude is controlled by the positions of stoppers. As is clear from Fig.17,b, the dependence f=f(E) is linear for any dielectric shaft.

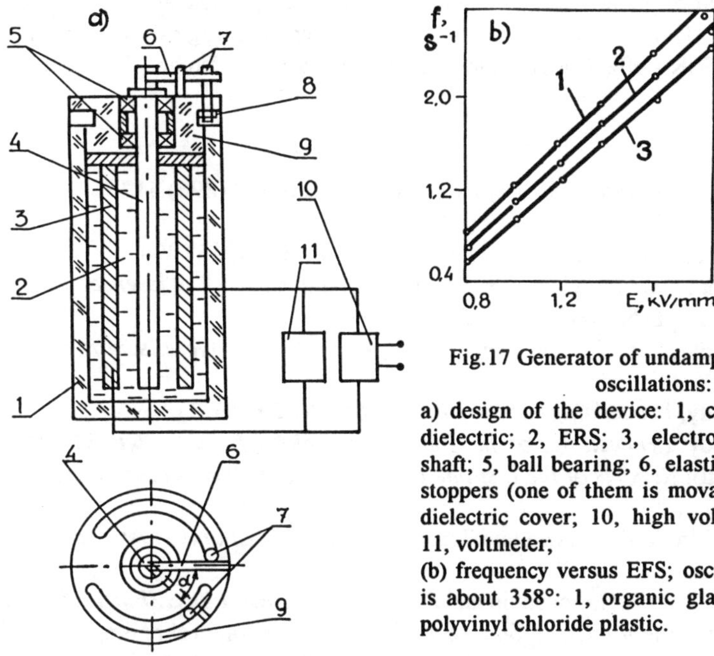

Fig.17 Generator of undamped mechanical oscillations:
a) design of the device: 1, container made of dielectric; 2, ERS; 3, electrodes; 4, dielectric shaft; 5, ball bearing; 6, elastic (steel) beam; 7, stoppers (one of them is movable); 8, screw; 9, dielectric cover; 10, high voltage supply unit; 11, voltmeter;
(b) frequency versus EFS; oscillation amplitude is about 358°: 1, organic glass; 2, ebonite; 3, polyvinyl chloride plastic.

Main advantages of an ER-motor (in brief).

1. No need in copper, commutators, windings, brushes, magnetic circuits, and many of expensive and deficient components of traditional electric motors.
2. Good hermetic features and a possibility to withstand high pressures.
3. Independence of its characteristics on spatial position (horizontal or inclined).
4. Durability of permanent operation (up to one year) and a large number of on/off cycles.

Nomenclature

τ, γ - shear stress and shear rate (Pa and s^{-1}, accordingly); τ_0 - yield stress (Pa); $\eta_{eff}=(\tau/\gamma)$ - effective viscosity (Pa·s); ω - angular velocity (s^{-1}); n - rotation speed (r.p.m.); T -

temperature (K or °C); U - voltage; E - Field strength (kV/mm; j - current (mA); σ - conductivity per unit volume $(Ohm \cdot m)^{-1}$, $(\Omega \cdot cm)^{-1}$; ε - dielectric permittivity; C - particles concentration (wt%); φ - activator concentration (% of C_{part}); φ_W for water as activator (% of C_{part}); φ_{PEPA}- for polyethylenepolyamine as activator (% of C_{part}); φ_{DEA} - for diethylamine as activator; C_{diat} - concentration of diatomite particles (wt%); C_{AER} - concentration of aerosil particles (wt%).

Abbreviations
MOG - monooleat of glicerol as surfactant (% of C_{part}); OA - oleic acid as a surfactant (% of C_{part}); ERE - electrorheological effect; ERS - electrorheological suspension; DEB - dielectric body; ESR - effect of spontaneous rotation; SA - surfactant; EFS - electric field strength.

References

1. G.Quincke, *Annalen der Physik und Chemie* **Bd.59** (1896) 417.
2. S.W.Richardson, *Nature* **119** (1927) 238.
3. J.Sumoto, *Journal of Physical Society, Japan* **10** (1955) 494.
4. A.Lampa, *Sitzungsberichte Königlich Akademie der Wissenschaften (Wien), Math. Naturwiss Kl.* **115**(2a) (1906) 1659.
5. A.Heydweiller, *Verhandlungen der physikalischen Gesellschaft zu Berlin* **16** (1897) 32.
6. L.G.Vedy, *Proceedings of the Physical Society* **39** (1927) 169.
7. L.G.Vedy, *Proceedings of the Cambridge Philosophical Society* **27** (1931) 91.
8. I.Sumoto, *Reports of the Institute of Physical and Chemical Researches (Tokyo)* **32** (1956) 41.
9. S.S.Dukhin, Y.F.Deinega and T.S.Simonova, in *Heat and Mass Transfer* (Minsk, 1972) p.157.
10. A.V.Luikov, R.G.Gorodkin and Z.P.Shulman, *Inzh. Phys. Zhurn.* **15** (1968) 589
11. In *Electrorheological Effect*, ed.A.V.Luikov (Nauka i Tekhnika, Minsk, 1972)
12. Z.P.Shulman, in *Heat and Mass Transfer* (Nauka i Tekhnika, Minsk, Vol.10, 1968) p.3.
13. Z.P.Shulman and V.M.Nosov, *Rotation of Non-Conductive Bodies in Electrorheological Suspensions*, (Nauka i Tekhnika, Minsk, 1985).
14. Y.S.Karpov, V.A.Krasnoperov, Y.T.Okunev and V.V.Pasuinkov, in *Physics of Dielectrics* (Moscow, 1960), p.124.

An Electrostatic Stepper Motor

E.C.Partington and Edward Wong Chun Kay
University of Hong Kong
Department of Industrial and Manufacturing Systems Engineering
&
W.A. Bullough
University of Sheffield
Department of Mechanical and Process Engineering
Mappin Street, Sheffield

ABSTRACT

This paper describes a new concept in pulse controlled motor and precision linear actuator techniques. Piezo translators [PZT] employed to provide reciprocating primary motion are connected to a load via a controllable electrorheological fluid [ERF] clutch to form a programmable speed and step-width drive. Ideal considerations are used to quantify the limiting potential of the drive and details are given of its development and progress.

1. Introduction

Electrostatic fields alter the physical properties of piezo crystals and electrorheological fluids leading to the possibility of direct electronic control over motion. PZT's afford very precise control over position. ER fluids effectively can be switched from liquid to "solid", providing lubrication in the liquid phase and adhesion when solid. This paper describes how these characteristics can be exploited in a controllable motor.

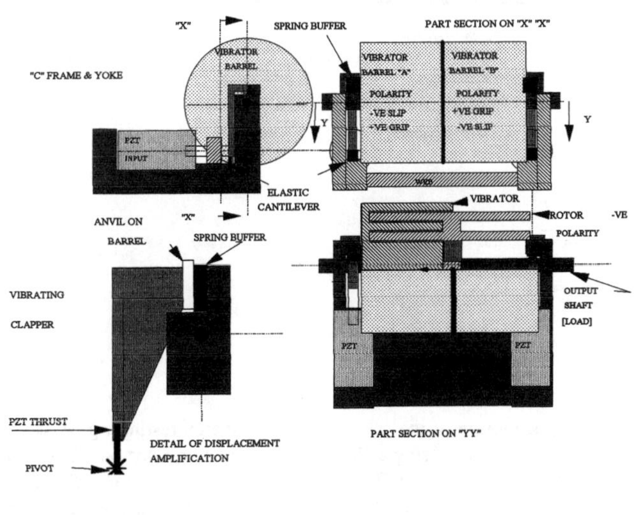

Figure 1 The reciprocating motion of paired PZT's is amplified by a lever system and mechanically rectified by twinned concentric ERF clutches acting in pulsed synchronism with the PZT's to give continuous output motion. ER fluid fills the interstices between the vibrating driver and the driven rotor. The buffer is a spring which stores the energy required for reverse motion.

2. System Design.

Piezo translators provide precise primary reciprocating motion proportional to charge when excited by suitable high voltage pulses. This reciprocating motion is electromechanically rectified by synchronised high voltage pulses across ERF filled clutches, providing forward or reversing uniform output motion. Reversed motion is simply obtained by the addition or subtraction of one output pulse to the alternating driver barrels. This is more fully explained in Wong et al [1].

2.1 Some Design Considerations

The design problems posed by the system are, the very small output motion of available PZT's, and the limiting static yield shear strength of available ERF [Typically around 10 kPa]. The first places heavy demands upon the stiffness and precision of the kinematic chain since any deformation, backlash or running tolerance occurring will absorb part of the minute displacement afforded by the PZT. This has been overcome by the addition of a cantilevered displacement magnifier pivoted about a flexural hinge [or a fulcrum bearing] [Fig 1]. Without displacement amplification the PZT's would have to be run at high frequencies to give even a modest output speed. High frequencies are problematical in that capacitive reactance varies inversely with frequency leading to excessive current demand. In our experiments values of frequency were kept below 200 Hz.

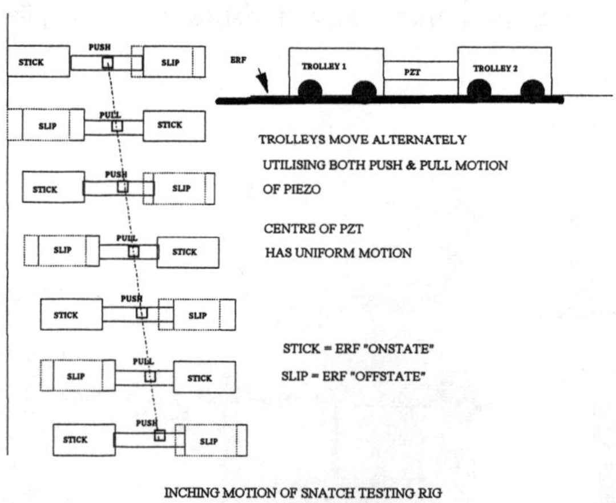

Figure 2 Showing the linear output motion of the "electronic slug" snatch testing rig. A sawtooth pulse train provides a reciprocating output from a single PZT. This is mechanically rectified by alternately activating the ERF layer in front and behind the PZT in synchronism with the sawtooth wave. Thus the trolleys move intermittently to inch the PZT along. It is interesting to note that a point at the centre of the PZT will move continuously with uniform motion.

Any inherent weakness in the yield strength of the activated ERF is compensated for by a design which maximises the shear area Thus the design, based on a nominal yield strength of 10 kPa, resulted in a shear area measured in 100's of cm^2 in order to provide for useful levels of torque output. Other problems have stemmed from a lack of quantification of the performance of ERF stiffness under the snatch loading

conditions of operation which are employed i.e. when subjected to very steep square waves at low shear rates.
An experimental system was designed to test these factors under real loading conditions and to prove the validity of the PZT/ERF compounded system. This was an inchworm type of linear actuator [labelled the "electronic slug" after its somewhat gastropodal progress over the ERF "slime"] using a single PZT mounted in between a pair of trolleys sliding over an ERF film [Fig 2]. Each trolley was switched alternately in synchronism with the expansion and contraction of the PZT to provide an intermittent motion. The system was used to measure both steady and intermittent loading behaviour.

The results of the "electronic slug" test series were promising enough to go on to the detailed design of the stepper motor which is described qualitatively below in section three.

3. The Prototype Rotary Motor

Consideration of the problems as well as the opportunities to be had from the system led to the present design [See Fig 1] which is intended eventually to prove the benefits of operation offered by electrostatic systems generally, such as; rapidity of response, precision, lightness and the ease with which the electromechanical system can be controlled via pulse techniques by real time computer systems. Using electrostatics we may approach more closely an analogue of the human brain-muscle system.

3.1. Ideal Performance

Since the long term objectives include the production of a light plastic arm joint for prosthetic use, the performance requirements were chosen to be those which would simulate the human arm. The desired performance was as follows: # a mass of 1 kg lifted at # an acceleration of 2g # at the radius of half a metre. Summing the deadload and the dynamic load, this translates to an approximate torque requirement of 15 Nm. Referred to the driving barrels of the motor, this torque is the product of the shear load transmissible by the ERF at the radius of the ERF layer. This, in turn, is defined by the product of the static yield shear strength of the ERF and the area of the ERF layer. This was established as 300 cm^2 at an aggregated mean rotor/driver barrel radius of 50mm. Taking the nominal static shear strength of the selected ERF as 10 kPa [i.e. 1 N/cm^2] , this translates to a total maximum shear load bearing strength of 300 N per barrel. Once more, this yields a maximum transmissible torque of 15 Nm as required. The PZT force/lever configuration can supply a torque of up to 20 Nm.

3.1.1. Speed Setting

This was set at the typical human arm throw of 1/4 revolution/sec which may be obtained variously by adjusting the peak voltage of the sawtooth pulse applied to the PZT's, and/or mechanically by adjustment of the cantilever displacement amplifier. The output speed in radians/sec is the product of the amplified PZT displacement per pulse at the tip of the clapper and the chosen pulse frequency divided by the radius at which the clapper strikes the load transmission block on the driver drums [Fig 3].

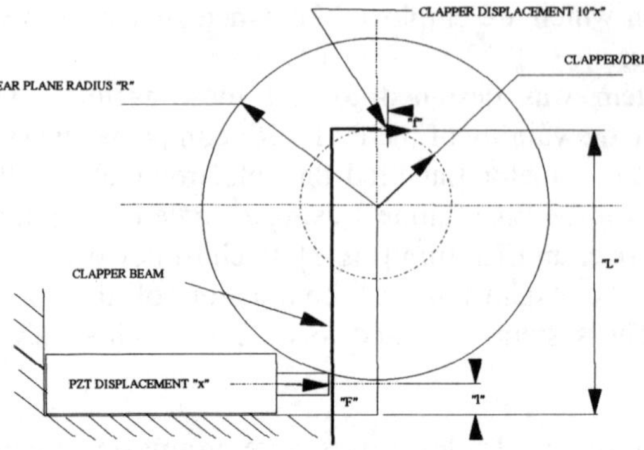

DISPLACEMENT AMPLIFICATION = L/l = 10

FORCE ATTENUATION l/L = 1/10

AVAILABLE TORQUE=[F]x[l/L]x[r]

TRANSMISSIBLE TORQUE=[ERF AREA]X[ERF YIELD STRENGTH]XR

OUTPUT SPEED = [FREQUENCY]X[10"x"]X[2 BARRELS]X[1/"r"] RAD/SEC

Figure 3 Showing how the PZT output is magnified to give the required values of displacement and force. A compounded lever system amplifies the PZT displacement by 10 times. The PZT output force is reduced by the same factor

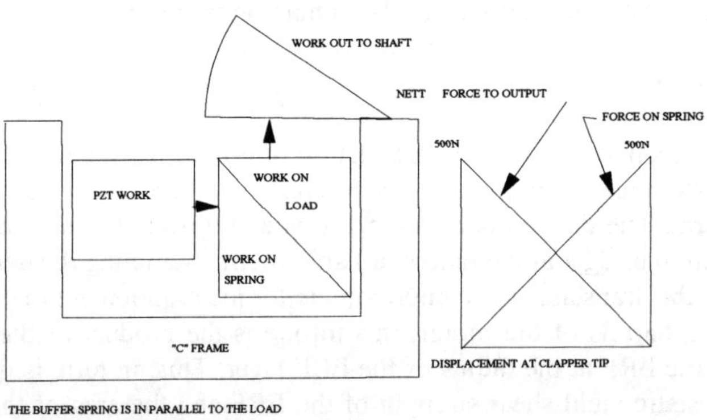

THE BUFFER SPRING IS IN PARALLEL TO THE LOAD
HALF OF THE PZT WORK IS STORED ELASTICALLY
TO PROVIDE REVERSING WORK
HALF OF THE PZT WORK IS OUTPUT TO THE SHAFT

NOTE: ALTHOUGH THE NETT FORCE TO THE BARREL GOES TO ZERO
THE ALTERNATE BARREL CATCHES THE LOAD AT MAXIMUM FORCE

EXAMPLE: TO LIFT A DEAD LOAD OF 25kg.
ABOUT 250N IS NEEDED TO HOLD THE LOAD
LEAVING FROM 250N TO ZERO TO ACCELERATE IT
AT ZERO NETT FORCE THE VELOCITY IS MAINTAINED

Figure 4 Showing the system energetics. PZT work is equally divided between work at the output shaft and the system elasticity as strain energy. The strain energy so stored is available for reversed motion.

Speed control at any cantilever displacement amplitude can be had by varying the pulsing frequency. For example at 50 Hz/barrel and 60 x 10^{-6} m with a displacement gain of 10, at a radius of 40mm, we have an output speed of 1.5 rad/sec, i.e. about a quarter turn a second, as required.

The limiting power capacity of the system is dictated by the maximum shear force transmissible by the ERF surface [which is the product of the shear area and the yield strength] and the frequency at which the ERF may be switched [which is limited by the time constant of the ERF]. The limiting speed is governed by the product of two variables # maximum effective frequency [in turn, a function of the ERF properties], # the amplified PZT output displacement. For any given PZT power supply the limiting torque/speed characteristic will be hyperbolic since power is defined as the product of torque and speed.

3.2 Piezo Drive Subsystem.

The PZT's are mounted within a "C" frame. The PZT output motion is amplified about 12.5 times [at the ERF shear plane] by a cantilever displacement amplifier. [See Figure 3] The motive force train is harnessed compressively within a stiff "C" frame. Piezo stacks are weak in tension, torsion and shear loading, so protection has been introduced by means of a side thrust eliminator and by designing the system so that the PZT is always in compression. The system kinematics are described below.

3.2.1 Kinematic Chains

3.2.1.1. Force and Displacement Chain

Figure 3 below shows the kinematic chain and the displacement amplifier. The extreme lower end of the lever is a flexural hinge cantilevered into the "C" frame whilst the rest is sufficiently rigid not to absorb the displacement. The PZT displacement is amplified by the lever some ten times. Referred to the ERF shear surface this is further magnified, giving a motion at the shear surface of 12.5 times the PZT motion. Thus, the available output force referred to the limiting shear surface is 1/12.5 of the PZT input of about 10,000 N, i.e 800 N [more than the maximum shear force of 300N transmissible by the ERF ratchets given an ERF yield strength of 10 kPa and an area of 300 cm^2 but ample to accommodate future, stronger, ER fluids]

3.2.1.2. Energy Chain

Figure 4 below shows the system work and strain energy relationships of the piezo drive sub-system. On the expansion of the PZT, the output work is shared equally by the applied load and the strain energy absorbed by a spring . The latter is used to provide reversing work energy. [See Figure 1 "Buffer"]. As the PZT contracts, the spring gives up its strain energy either as available reverse work if required or as compressive work on the contracting PZT. The PZT is all the time under compressive loading, which acts as a form of preload on the PZT. This is a favourable loading condition for piezo electric stacks.

3.3 Piezo Selection Criteria

The piezo translators were selected to give sufficient displacement and force to provide for the output work at the ERF shear surface. The output force was 10,000 N and maximum expansion was 60 micrometers.

3.4 ERF Clutch Subsystem

The design of the clutching [or ratcheting] system is a compromise between minimising size and maximising torque transmissibility. Shear area has been maximised at a low average radius by summing concentric cylinders. Figure 1 above shows the nesting driver and driven, [or, vibrator and rotor] concentric cylinders. The gap between the cylinders is a half millimetre. The applied voltage is around one and a half kilovolts d.c.

Since the design of the clutch involves ERF in the static shear mode, fluid selection was based upon the reasonable assumption that high torque transmissibility was to be had from high particle volume fractions which give a high shear modulus and a high static yield stress. It is also assumed that the speed of response [i.e. of fibre formation and the development of shear strength] increases with particle volume fraction[2]. On the other hand the effective dielectric constant also increases with volume fraction placing further constraints upon the maximum frequency operable. As the applied frequency rises the breakdown voltage may decrease resulting in reduced operating voltages, thus lowering the field strength and causing a severe loss in mechanical properties. Also, capacitive reactance varies inversely with frequency and hence increases current demand. A commercially available fluid of 45% particle fraction was selected. This mixture has a high viscosity but the designed slip speeds are low hence relatively little energy loss per stroke is anticipated.
Despite potential problems associated with high particle density, the gains to be had in terms of response and stiffness and yield strength dictated the use of the highest particle densities practicable. The size weight and inertia penalties for reducing the particle volume fraction are serious since shear area would have to be increased proportionately giving a bigger and heavier motor for a given torque output.

4. Discussion and Future Work

The potential advantages over conventional stepper motors and actuators which are being sought are:

1 The PZT/ERF system provides high torque slow motion as required for robotics, **gearlessly**.

2 Since the system is electrostatic rather than electromagnetic, much of the material from which a device is constructed can be of plastics

3 The system uses high voltage and low capacitance, and hence has low a time constant, affording a faster response than that of conventional stepper motors. Hence they are more suitable for high frequency pulse switching, and for mechatronic applications in general.

4 Square wave pulse control allows real time computer control over speed and position.

5 Speed control is achieved by varying either the pulse amplitude or its frequency.

6 Reversing is simply accomplished by rearranging the pulse sequence.

There is some expectation that square pulses may help to overcome the hysteretic tendencies in the torque transmission fluid. It is anticipated that new ER fluids may be more than twice as strong as that proposed here. This will lead to very significant reductions in weight and further enhance the attractiveness of the system to the aerospace industry.

Among the objectives for future work is the development of small plastic motors for artificial limbs. This will involve linear actuator systems as well as the rotary version described here.

Future designs may include magneto-strictors and MRS which may afford yet further improvements in performance.

We are looking into the concept of an electronic transmission for automobile use. This would replace the crankshaft, gearbox and conventional clutch with a pulse controlled ERF system.

The first and foremost objective of our work is to prove the feasibility of the prototype system here described.

5. Acknowledgements

We wish to acknowledge the contributions of the following people. Mr Leung Wai-keung, of the IMSE Department of HKU, for his continuous interest and technical advice, and Roger Leung, IMSE Department HKU, for his expert advice on real time computer control, Dr.P.Y.Foo of the HKU Department of Electrical and Electronic Engineering who advised us on the high voltage switching system, Dr Douglas Brookes of Advanced Fluid Systems Ltd for his advice concerning ER fluid, and Michael Donat of Phyziks Instrumente GmbH for advice on PZT usage.

6. References

1. Wong C.K., Partington E.C. and Bullough W.A. *"Design Characteristics Of a Piezoelectric/ER Fluid Motor", Proceedings of the 2nd International Conference on Mechatronics and Machine Vision in Practice,* 12th-14th September, 1995, The City University of Hong Kong. P 175-178

2. Bonnecaze R.T. and Brady J.F. *"Yield Stresses in Electrorheological Fluids" J. Rheol.* 36 [1] , January 1992 p73-115

STUDIES OF ER FLUIDS FEATURING RODLIKE PARTICLES

REX C. KANU
Polymer Science Program
Institute of Materials Science
University of Connecticut
Storrs, CT 06269-3136 USA

and

MONTGOMERY T. SHAW
Polymer Science Program and Department of Chemical Engineering
Institute of Materials Science
University of Connecticut
Storrs, CT 06269-3136 USA

ABSTRACT

Chaining of micron-sized polarizable particles in ER fluids is generally accepted to be responsible for the liquid-to-solid transitions on the application of an external electric field. It has been hypothesized that the strength of the particle-particle interactions solely determines the rheological properties of ER fluids. In our work, the particle's structure has been used to control interactions; for example, we have developed systems featuring rodlike particles. With such particles it should be possible to enhance the dielectric interaction of the particles as well as their mechanical interaction. The main goal of our effort has been to distinguish between these two mechanisms through measurements of the dielectric properties in conjunction with the rheological responses. Based on the experimental evidence thus far gathered, we can state that most, but not all, of the rheological effects are explainable in terms of the dielectric changes in the fluid.

1. Introduction

1.1 *Background*

Electrorheological (ER) fluids are two-phase materials that undergo a reversible liquid-to-solid transition upon application of an electric field transverse to the flow field. The force arising from the electrically induced polarization of the particles causes the formation of a columnar structure along the electric field and turns the suspension from a liquid to a solid.[1,2] The attractive features of ER fluids for practical application are the speed and reversibility of the liquid-to-solid transition. However, several well-known problems may restrict their application. One of the major problems is the low values of yield stress exhibited by many ER fluids.

To improve the rheological properties of the fluid, several approaches are possible, and many have been explored. The most direct is to substitute particles with higher polarizability.[3] Another is to reduce the relative permittivity of the matrix oil, or reduce its viscosity.[4] Particle size distribution can influence the performance of an ER fluid.[5] Shaped or skewed electrical fields have been shown arrange the particles more efficiently than uniform fields.[6] The fields could be brought to higher levels by increasing the dielectric strength of the matrix; however, this approach has apparently not been tried.

The variable of interest in this work is the particle shape; we have hypothesized that for some applications it may be possible to enhance the low-strain properties significantly by using rodlike particles. While this hypothesis has been advanced before,[7] there has been little attempt to explore the limits of the effects.

1.2 *Objective and Approach*

In this paper we describe the tailoring of the geometrical (and thereby electrical) anisotropy of the ER particles to enhance the performance of the resulting ER fluids. Of primary concern, and the main subject of this paper, is the effect of aspect ratio on the dielectric and rheological properties. We addressed this objective by formulating ER fluids using particles cut from highly polarizable fibers, thereby allowing precise control of the aspect ratio.

2. Experimental

The material use to make the ER particles was poly(p-phenylene-2,6-benzobisthiazole) [PBZT], a highly conjugated liquid crystalline polymer featuring a ladder-like structure (Figure 1). PBZT fibers were used because of their highly anisotropic dielectric properties; they are also chemically and thermally stable. Desired particle size and geometric aspect ratio were obtained by microtoming the fibers after embedding them in styrene monomer that was subsequently polymerized.[8]

Electrorheological measurements were performed on a Rheometrics System 4. Parallel-plate fixtures were used for measurements of the torque and normal force. The plates, with diameters up to 62.5 mm, were made of contoured stainless steel and were insulated from the rest of the system with spacers of machinable glass. The moving plate was electrically connected through a copper ring contacting a carbon brush. High voltage dc was supplied with a Bertan power supply Model 2 05A-05R connected to the bottom plate through a thin wire. For ac we used a current-limited transformer (60 Hz). The plate gap was set at about 1 mm for most measurements.

In the case of steady shear measurements, steady state was established before the high voltage was applied. Current was monitored, along with the torque and normal force. Dynamic measurements employed various strain levels to establish the effect of this variable. Samples were redispersed between each run.

Dielectric measurements of the suspensions were performed at 30 °C on an IMASS Time Domain spectrometer. By pre-aligning the particles we were able to measure roughly the dielectric properties of the ordered state. We also inferred these from the normal-stress response of the RMS.[9]

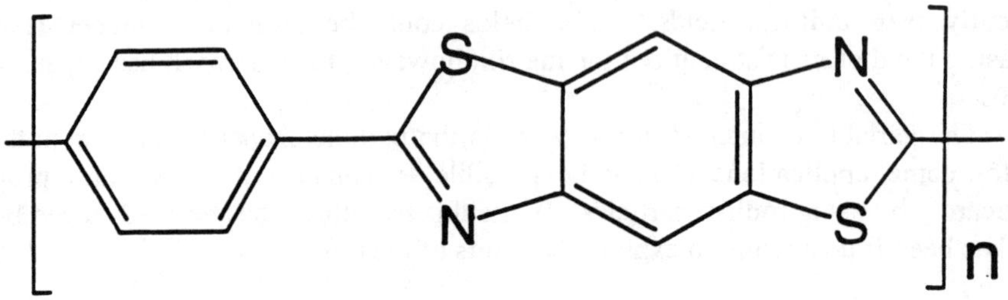

Figure 1. Chemical structure of polymer (PBZT) used to make particles for ER fluids.

3. Results and Discussion

Figure 2 suggests that the ER response of the fluid can be enhanced by increasing the aspect ratio of the particulate phase. The figure also shows that the real part of the complex shear modulus, G', scales as the square of the field, E, independent of aspect ratio. The dependence of modulus on aspect ratio is depicted in Figure 3, which also

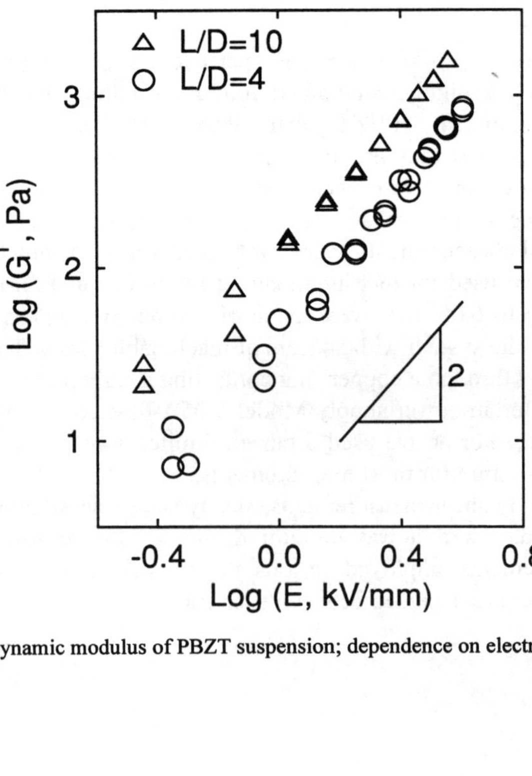

Figure 2. Dynamic modulus of PBZT suspension; dependence on electrical field.

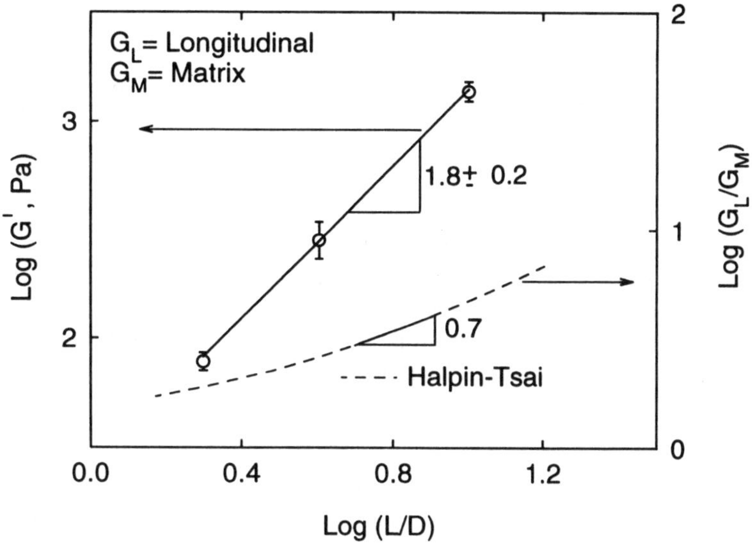

Figure 3. Dependence of the moduli of aligned ER fluids on aspect ratio of the PBZT particles. The dotted line is calculated from the Halpin-Tsai relationship for composites.

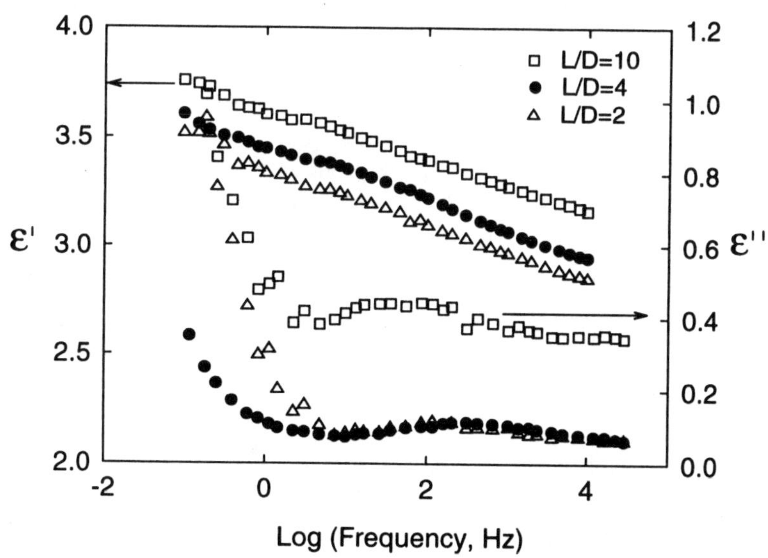

Figure 4. Dielectric data showing the movement of the Maxwell-Wagner-Sillars dispersion to lower frequencies with particle aspect ratio. Fibers were aligned before measurement.

shows what would be expected for a uniaxial composite at the same volume fraction reinforcement. Based on this result we hypothesized that the enhancement of modulus with aspect ratio is primarily a dielectric effect, as opposed to mechanical stress-transfer, the increased dipolar interactions being brought about by longer fibers. This led to investigation of the dielectric response of the suspensions. Increasing Maxwell-Wagner-Sillars polarization with L/D should be evidenced by decreasing frequency at the maximum of this dispersion. Figure 4 shows typical dielectric data for aligned suspensions for fibers of two aspect ratios. Clearly the dispersion is moving in the expected direction. We also expected, but did not find, an increase in dc conductivity of the aligned suspensions with L/D (see below).

Using the dielectric information, we calculated the anticipated stiffness of the aligned suspensions. Results are shown in Figure 5, where β is given by the expression

$$\beta = \frac{\varepsilon_p - \varepsilon_c}{\varepsilon_p + 2\varepsilon_c} \qquad (1)$$

and ε_p and ε_c are the relative permittivities of the particle and oil, respectively. While an improvement over the mechanical hypothesis, the suspensions still appear to be inadequately represented by the theory.

The conductivity of the aligned suspensions are shown in Figure 6. Although the variance for these measurements is high, the evidence suggests that the longer particles do not conduct more than the shorter particles. An explanation would be that the polarization of the particles is by internal channels in the fiber, and that these are effectively insulated from adjacent fibers.

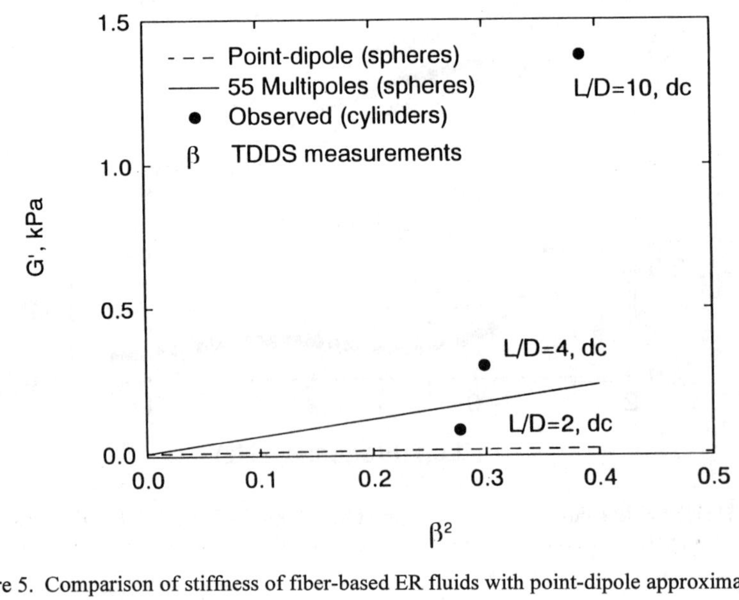

Figure 5. Comparison of stiffness of fiber-based ER fluids with point-dipole approximations.

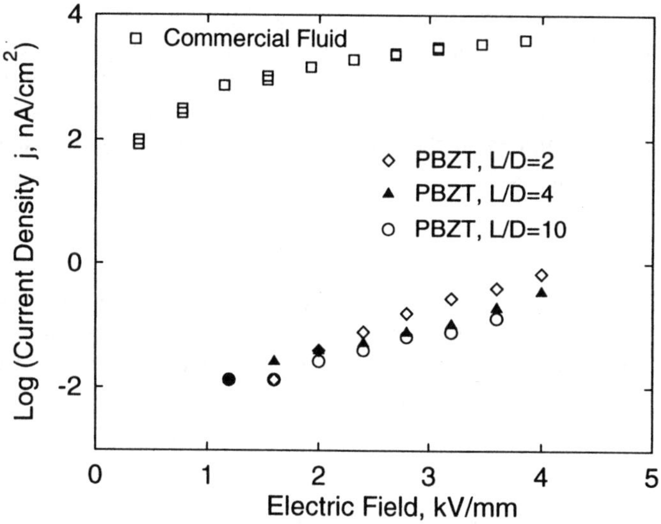

Figure 6. Conductivity of fiber-based ER fluids (15 vol %) showing that the conductivity is not likely to increase with aspect ratio. A commercial fluid is shown for comparison. T = 28 °C.

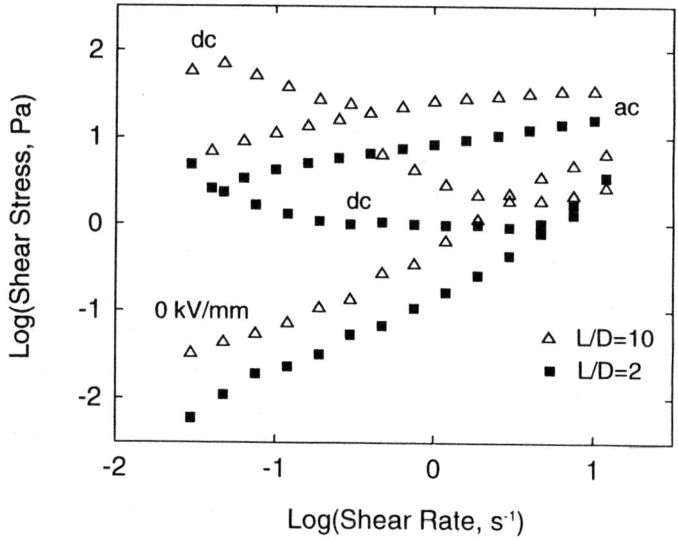

Figure 7. Response of PBZT-based ER fluid under steady shear flow with ac and dc fields.

The steady shear response of the PBZT-based fluid is shown in Figure 7. While dc, as opposed to ac, leads to higher stresses at low rates, the fluid loses its properties at higher shear rates. The origin of the more "normal" response with ac is unclear; however, we can hypothesize that the stress at high rates with the dc is abnormally low because the fluid is organized, but aligned in the flow direction. Light scattering studies may be helpful in testing this hypothesis.

4. Conclusions

Anhydrous ER fluids can be made by combining oil with particles made from chopped PBZT fibers. The conductivity of these fluids is very low. The static performance of the fluid can be enhanced by increasing the aspect ratio of the particles, which increases their polarizability. Steady-shear ER properties are less attractive.

5. Acknowledgments

The PBZT fibers were provided by through the courtesy of the Department of the Air Force, Wright-Patterson Air Force Base, Ohio, USA. Partial financial support of the project was through contract DAAL 03-89-G-0080 of the Army Research Office and by Texaco, Inc..

6. References

1. P. M. Adrian, and A. P. Gast, *Physics Fluids,* **31** (1988) 2757.
2. H. Block, J. P. Kelly, A. Qin, and T. Waton, *Langmuir*, **6** (1990) 6.
3. C. A. Randall, D. E. McCaulley, C. P. Bowen, T. R. Shrout and G. L. Messing, in Proc. 4th Int. Conf. on Electrorheological Fluids, *Electrorheological Fluids-Mechanisms, Properties Technology and Applications*, ed. R. Tao and G. D. Ray (World Scientific Publishing Co., Singapore, 1994) p. 60.
4. A. F. Sprecher, Y. Chen, Y. Choi and H. Conrad, in Proc. 4th Int. Conf. on Electrorheological Fluids, *Electrorheological Fluids-Mechanisms, Properties Technology and Applications*, eds. R. Tao and G. D. Ray (World Scientific Publishing Co., Singapore, 1994) p. 142.
5. Y.-H. Shih and H. Conrad, in Proc. 4th Int. Conf. on Electrorheological Fluids, *Electrorheological Fluids, Mechanisms, Properties Technology and Applications*, eds. R. Tao and G. D. Roy, (World Scientific Publishing Co., Singapore, 1994) p. 294.

6. P. V. Katsikopoulos and C. F. Zukoski, in Proc. 4th Int. Conf. on Electrorheological Fluids, *Electrorheological Fluids, Mechanisms, Properties Technology and Applications*, eds. R. Tao and G. D. Roy, (World Scientific Publishing Co., Singapore, 1994) p. 251.

7. C. E. Chaffey and S. G. Mason, *J. Colloid Sci.* **19** (1964) 525.

8. R. C. Kanu, and M. T. Shaw, in *Theoretical and Applied Rheology,* Vol. 2, eds. P. Moldenaers and R. Keunings, (Elsevier, Amsterdam, 1992) p. 766.

9. W.-C. Yu, M. T. Shaw, X.-Y. Huang and F. W. Harris, *J. Polym. Sci., Polym. Phys. Ed.* **32** (1994) 481.

Dynamics of an Electrorheological Valve.

M. Whittle[1]

Proctor Department of Food Science, The University of Leeds, Leeds LS2 9JT. United Kingdom.

R.J. Atkin

Applied Mathematics Section, School of Mathematics and Statistics, The Hicks Building, The University of Sheffield, Sheffield S3 7RH. United Kingdom.

W.A. Bullough

Department of Mechanical and Process Engineering, The University of Sheffield, Sheffield S1 3JD. United Kingdom.

Abstract

Time dependent solutions to the equations of motion for flow in an electro-rheological valve are developed and discussed in relation to the limiting speed of operation. The fluid is considered to be a homogeneous continuum obeying the Bingham plastic constitutive equation incorporating a yield stress and viscous component. The results show that the speed of response is determined principally by the plastic viscosity and electrode gap irrespective of the yield stress and corroborates earlier work.

1. Introduction

Electrorheological (ER) fluids are "smart" materials which undergo a rapid (~1ms) reversible change in rheological properties in response to electric fields of the order 10^6 V m^{-1}. The development of devices which

[1] Formerly at the Department of Mechanical and Process Engineering The University of Sheffield.

exploit this phenomenon promise rapid response times that outperform electro-magnetic mechanisms in some areas of operation[1,2]. The analysis of the overall response of these devices, identifying limitations and discovering optimum operating parameters is germane to this purpose. In an earlier publication[3] we have analysed the dynamic pressure response of an experimental valve and derived response times for the fluid itself. This involved a consideration of the coupling of the valve to the system and a numerical model for the response of the isolated valve. With recourse to methods developed in the study of ER clutches[4,5] we now examine the isolated valve response in more detail and present analytical solutions for the time dependent flow equation.

Under zero field conditions electrorheological fluids are generally characterised by a simple Newtonian viscosity η. When subjected to high electric fields ($E \sim 1\mathrm{MV\,m^{-1}}$) they develop a yield stress σ_o and the Bingham plastic constitutive equation

$$\sigma(E,\dot{\gamma}) = \pm\sigma_0(E) + \eta\dot{\gamma} \quad , \qquad (1.1.1)$$

for the shear stress σ, with $\dot{\gamma}$ denoting strain rate is often used to describe their behaviour [6]. The + (-) sign corresponds to positive (negative) strain rate. The difference $\sigma(E,\dot{\gamma}) - \sigma(0,\dot{\gamma})$ is often called the electro-stress. This change of rheology is the operational basis for ER devices and the yield strength attained for a given field is often used as a rough guide to the potential usefulness of a particular fluid.

The main effort in this work has been to derive time dependent mathematical solutions to the equation of motion in response to electrorheological changes represented by a step in the yield stress for the Bingham model. We make the assumption that this step is instantaneous and thus focus on the fluid dynamic consequences without concerning ourselves with the detailed response characteristics of the fluid. In real systems it will be necessary to take account of the small but finite time that real fluids take to respond[3,7]. For comparison with earlier work [3] we also examine the response to a pressure step for purely Newtonian flow. The results presented should also be relevant to devices employing magnetorheological (MR) fluids which operate in an analogous way using magnetic fields.

2. Theory and Analysis.

2.1 Basic Theory and Steady State Solutions.

The experimental valves and results which inspired this work have been described in several earlier publications[3,8]. The design comprises two parallel channels formed by a central cylindrical electrode of variable length and thickness which slides into an earthed annular cavity of fixed dimensions forming the valve body. The radius of the common electrode was the same for each valve (mean r = 28mm) and sufficiently large for the field in the gap h to be considered effectively uniform. The valve can therefore be modelled quite adequately as a pair of parallel plates of length l (Fig. 1) which is the case considered here. In the experimental set-up a nominally constant flow of ER fluid is maintained along the channels and is associated with a pressure difference ΔP between the ends. When an electric field E(t) is applied the fluid electro-stress appears and there is a consequent increase in the pressure drop across the valve. If the pressure is instead maintained constant then the flowrate must decrease. The constant pressure case is more amenable to analysis and is studied here.

The time dependent evolution of flow in the valve is determined by the balance of momentum flux and the force on the fluid determined by the pressure and stress gradients[9]. The full equation of motion is simplified by the symmetry of the valve to flow in a two dimensional channel which we discuss in the x-y plane (Fig. 1). For isothermal laminar flow of an incompressible Bingham plastic, assuming that the flow is in the x-direction, to satisfy the equation of continuity the x component of velocity u must be independent of x. In the absence of body forces it follows from the equations of motion that P depends only on x and t and

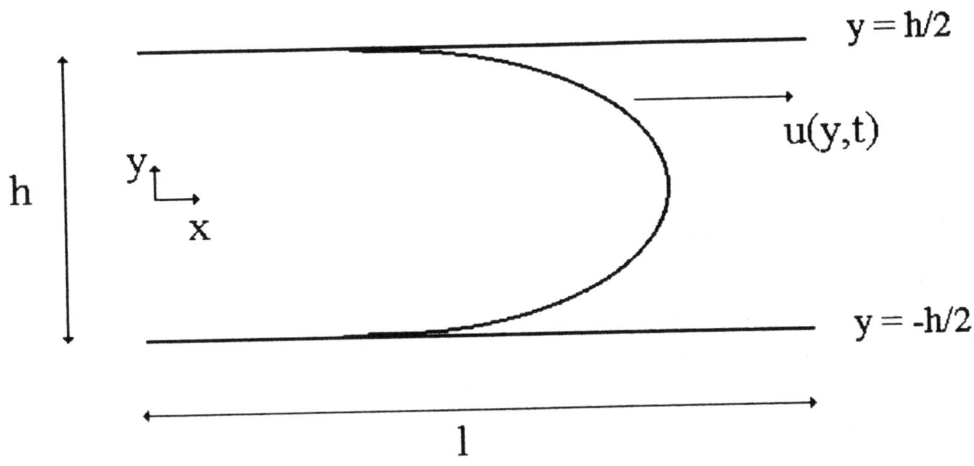

Fig. 1. Schematic diagram of flow $u(y,t)$ between parallel plates gap h.

$$\rho \frac{\partial u(y,t)}{\partial t} = \frac{\Delta P}{l} + \frac{\partial \sigma(y,t)}{\partial y} , \tag{2.1.1}$$

where ρ is the fluid density, $\partial P/\partial x = -\Delta P/l$ and $\sigma(y,t)$ is the shear stress, itself dependent on the velocity gradient $\dot{\gamma} = \partial u(y,t)/\partial y$. Substituting for the stress from Eq. (1.1.1) with the proviso that the field is constant so that $\sigma_o(E)$ is constant, for $y \neq 0$ we obtain

$$\eta \frac{\partial^2 u(y,t)}{\partial y^2} = -\frac{\Delta P}{l} + \rho \frac{\partial u(y,t)}{\partial t} , \tag{2.1.2}$$

assuming η to be constant. This is also the equation of motion for a Newtonian fluid. In the steady state $t \to \infty$ and for a Newtonian fluid the velocity profile $u(y)$ satisfying the no-slip boundary conditions on $y = \pm h/2$ is shown in many textbooks[9] to be parabolic and given by

$$u(y) = u_o\left(1 - \alpha y^2\right) , \tag{2.1.3}$$

where

$$u_o = \frac{\Delta P\, h^2}{8\eta l} \; ; \; \alpha = \frac{4}{h^2} \; . \tag{2.1.4}$$

Not all solutions of Eq. (2.1.2) correspond to a flow in a Bingham plastic since we also require $|\sigma| > \sigma_o$. For a Bingham plastic, plug flow occurs in the centre of the channel [10]. Fluid in this region moves as a solid body while the velocity profile between the plug and the walls is again parabolic and

$$u(y) = \frac{\Delta P}{8\eta l}\left[-4y^2 + 8\,y_o|y| + \left(h^2 - 4h\,y_o\right)\right] \; , \tag{2.1.5}$$

where

$$y_o = \frac{\sigma_o l}{\Delta P} \; . \tag{2.1.6}$$

The central region of this solution for $|y| < y_o$ is actually unphysical giving a velocity gradient that cannot be supported by the system and this region in fact corresponds to a plug moving with the velocity $u_p = u(y_o)$

$$u_p = u_o + \frac{\sigma_o}{2\eta}\left[y_o - h\right] \tag{2.1.7}$$

while for $y_o < |y| < h/2$

$$u(y) = u_o\left(1 - \alpha\, y^2\right) + \frac{\sigma_o}{\eta}\left[|y| - \frac{h}{2}\right] . \tag{2.1.8}$$

Eq. (2.1.8) is usually integrated over the gap to give the more familiar [10] steady-state relation between the pressure and the flow rate

$$\frac{h^3}{12}\left(\frac{\Delta P}{l}\right)^3 - \left(\frac{h^2\sigma_o}{4} + \frac{Q\eta}{b}\right)\left(\frac{\Delta P}{l}\right)^2 + \frac{\sigma_o^2}{3} = 0 \quad , \tag{2.1.9}$$

where Q is the flow rate and b the valve width.

In studying the time dependent solutions for electrorheology we are interested in the transition from parabolic profiles characterised by Eq. (2.1.3) to the plug flow solution Eq. (2.1.5). In section 2.3 we also examine solutions for a pressure jump in the Newtonian case so that comparison can be made with an earlier approximate solution [3].

2.2 Time Dependent Solutions and Response to an ER Change.

In this section we suppose that a steady flow profile, as described by Eq.(2.1.3), has been established in the absence of a field and that the rheological properties of the fluid then are instantaneously changed at time $t=0$ following the application of a field. This may be represented by

$$\sigma_o(t) = \sigma_o H(t) , \tag{2.2.1}$$

where $H(t)$ is the Heavyside step function, unity for $t>0$ and zero elsewhere.

Denoting the Laplace transform L of the velocity profile by

$$U(y,s) = L\{u(y,t)\} \tag{2.2.2}$$

Eq. (2.1.2) becomes

$$\lambda \frac{\partial^2 U(y,s)}{\partial y^2} = -\frac{1}{s}\frac{\Delta P}{\rho l} + sU(y,s) - u(y,0) , \tag{2.2.3}$$

where

$$\lambda = \frac{\eta}{\rho} \tag{2.2.4}$$

is the kinematic viscosity. The complementary function $U_{CF}(y,s)$ is obtained from

$$\lambda \frac{\partial^2 U(y,s)}{\partial y^2} - sU(y,s) = 0 \tag{2.2.5}$$

whose solution can be expressed in the form

$$U_{CF}(y,s) = A\cosh\left(y\sqrt{\frac{s}{\lambda}}\right) + B\sinh\left(y\sqrt{\frac{s}{\lambda}}\right), \tag{2.2.6}$$

where A and B are arbitrary constants. For our problem $u(y,0)$ is given by Eq. (2.1.3) and a particular integral for Eq. (2.2.3) is $u_o(1-\alpha y^2)/s$ and so the general solution can be written as

$$U(y,s) = A\cosh\left(y\sqrt{\frac{s}{\lambda}}\right) + B\sinh\left(y\sqrt{\frac{s}{\lambda}}\right) + \frac{1}{s}u_o(1-\alpha y^2). \tag{2.2.7}$$

We look for a symmetric solution $U(-y,s)=U(y,s); 0 \le y \le h/2$, which satisfies the boundary conditions

$$U\left(\pm\frac{h}{2}, s\right) = 0 . \tag{2.2.8}$$

Then at $y=h/2$ we obtain

$$A\cosh\left(\frac{h}{2}\sqrt{\frac{s}{\lambda}}\right) + B\sinh\left(\frac{h}{2}\sqrt{\frac{s}{\lambda}}\right) = 0 . \tag{2.2.9}$$

Substitution gives

$$U(y,s) = A\frac{\sinh\left(\Delta y\sqrt{\frac{s}{\lambda}}\right)}{\sinh\left(\frac{h}{2}\sqrt{\frac{s}{\lambda}}\right)} + \frac{1}{s}u_o\left(1-\alpha\,y^2\right), \qquad (2.2.10)$$

where $\Delta y = \frac{h}{2} - |y|$. We can evaluate the prefactor A by comparing Eq. (2.2.10) in the limit of small s (corresponding to large t) with the steady state solution Eqs. (2.1.7) and (2.1.8). Since

$$\underset{s\to 0}{Lim}\ \frac{\sinh\left(\Delta y\sqrt{\frac{s}{\lambda}}\right)}{\sinh\left(\frac{h}{2}\sqrt{\frac{s}{\lambda}}\right)} = \frac{\frac{h}{2}-|y|}{\frac{h}{2}} \qquad (2.2.11)$$

we obtain

$$\underset{s\to 0}{Lim}\ U(y,s) = A\left(\frac{\frac{h}{2}-|y|}{\frac{h}{2}}\right) + \frac{u_o}{s}\left(1-\alpha y^2\right) \qquad (2.2.12)$$

which leads to

$$\frac{1}{s}u(y) = A\left(\frac{\frac{h}{2}-|y|}{\frac{h}{2}}\right) + \frac{u_o}{s}\left(1-\alpha y^2\right). \qquad (2.2.13)$$

For agreement in this limit take $A = A'/s$, where comparing with Eq.(2.1.8) for $|y| > y_o$

$$A'\left(\frac{\frac{h}{2}-|y|}{\frac{h}{2}}\right) = -\frac{\sigma_o}{\eta}\left(\frac{h}{2} - |y|\right) \tag{2.2.14}$$

we can then identify

$$A' = -\frac{\sigma_o h}{2\eta} \tag{2.2.15}$$

so that

$$A = -\frac{1}{s}\frac{\sigma_o h}{2\eta} \quad . \tag{2.2.16}$$

With this substitution Eq. (2.2.10) can be inverted [11] to give

$$\begin{aligned} u(y,t) &= u_o\left(1-\alpha\, y^2\right) \\ &- \frac{\sigma_o}{\eta}\frac{h}{\pi}\left[\sum_{n=1}^{\infty}\frac{(-1)^{n+1}}{n}\left[1-e^{-t/\tau_n}\right]\sin\frac{2n\pi\Delta y}{h}\right] , \end{aligned} \tag{2.2.17}$$

where

$$\tau_n = \frac{h^2}{4n^2\pi^2\lambda} \quad . \tag{2.2.18}$$

Summed over 200 terms and evaluated at $t = 1$ ms, which is long compared to τ_1, this function is compared with the steady state solution in Fig. 2. Other parameters are given in Table 1. The sharp central spike arises because the sine function vanishes at this point. The central region, between maxima of the function, does not in any case represent a valid solution and is spanned by a plug as indicated in the figure. Away from the central singularity Eq.(2.2.17) is seen to be a close approximation to the steady state solution for a Bingham plastic. The profiles are plotted for a series of times in Fig. 3 to show the evolution of the profiles. Again in these profiles a plug must be imagined in the central region.

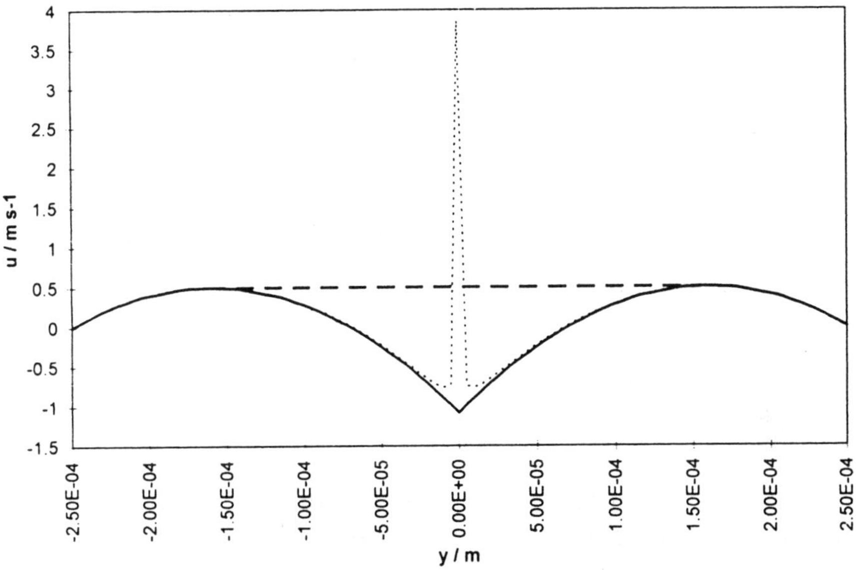

Fig. 2. Steady state velocity profiles. -------- from Eq. (2.1.5); .... from Eq.(2.2.17) with t = 1 ms. - - - - - - position of plug.

Gap	h / mm	0.5
Newtonian viscosity	η / Pa s	0.1012
Fluid density	ρ / kg m^{-3}	1445
Yield Stress	σ_0 / Pa	2000
Pressure	ΔP / bar	12.5

Table 1. Parameters used in this study.

n	τ_n / μs ER step	τ_n / μs Pressure step
1	90.4	361.7
2	22.6	40.2
3	10.0	14.5
4	5.6	7.4
5	3.6	4.5

Table 2. Values of τ_n for ER step Eq.(2.2.18) and pressure step Eq. (2.3.12) estimated with the parameters of Table 1.

For the parameters of Table 1 the first few time constants given by Eq. (2.2.18) are evaluated in Table 2.

The result can be generalised to incorporate the more complicated responses found in real ER fluids[3,7]. Writing

$$Z(y,s) = \frac{\sinh\left(\Delta y\sqrt{\frac{s}{\lambda}}\right)}{\sinh\left(\frac{h}{2}\sqrt{\frac{s}{\lambda}}\right)} \qquad (2.2.19)$$

the solution can be more generally expressed as

$$U(y,s) = F(s)Z(y,s) + u_o\left(1-\alpha\, y^2\right), \qquad (2.2.20)$$

where $F(s)$ is a function representing the input response. Similar expressions are obtained for the solutions to flow in radial and concentric clutches[4,5]. In the present case the input is a step function and $F(s)$ can be identified with A defined by Eq. (2.2.16)

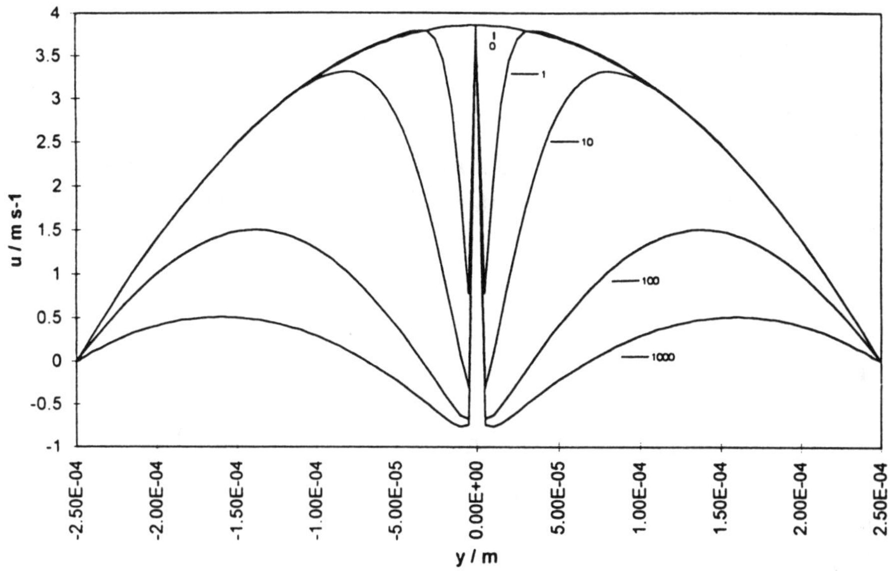

Fig. 3. Time sequence of velocity profiles following an ER step from Eq. (2.2.17). A plug forms between the maxima in real systems. Times in μs are indicated near the curve. From the top, t = 0.0, 1, 10, 100, 1000 μs.

2.3 Response to a step in pressure.

In this case we are interested in the response to a step in pressure at $t=0$

$$\Delta P(t) = \Delta P_o H(t) \quad , \tag{2.3.1}$$

where $H(t)$ is the Heavyside step function and ΔP_o is the size of the step. In this case we assume that the system is initially at rest so that

$$u(y,0)=0 \quad . \tag{2.3.2}$$

The equation to solve is then

$$\lambda \frac{\partial^2 U(y,s)}{\partial y^2} - sU(y,s) = -\frac{1}{s}\frac{\Delta P_o}{\rho l} \; . \tag{2.3.3}$$

The complementary function is again Eq.(2.2.6). A suitable particular integral U_{PI} can be found by seeking a form independent of y from

$$-sU_{PI}(s) = -\frac{1}{s}\frac{\Delta P_o}{\rho l} \tag{2.3.4}$$

giving

$$U_{PI}(s) = \frac{1}{s^2}\frac{\Delta P_o}{\rho l} \tag{2.3.5}$$

and so the general solution of (2.3.3) is

$$\begin{aligned} U(y,s) &= U_{CF} + U_{PI} \\ &= A\cosh\left(y\sqrt{\frac{s}{\lambda}}\right) + B\sinh\left(y\sqrt{\frac{s}{\lambda}}\right) + \frac{1}{s^2}\frac{\Delta P_o}{\rho l} \; . \end{aligned} \tag{2.3.6}$$

Applying the boundary conditions at the walls gives

$$A\cosh\left(\frac{h}{2}\sqrt{\frac{s}{\lambda}}\right) + B\sinh\left(\frac{h}{2}\sqrt{\frac{s}{\lambda}}\right) + \frac{1}{s^2}\frac{\Delta P_o}{\rho l} = 0 \; , \tag{2.3.7}$$

$$A\cosh\left(\frac{h}{2}\sqrt{\frac{s}{\lambda}}\right) - B\sinh\left(\frac{h}{2}\sqrt{\frac{s}{\lambda}}\right) + \frac{1}{s^2}\frac{\Delta P_o}{\rho l} = 0 \; . \tag{2.3.8}$$

From which $B = 0$ and

$$A = -\frac{1}{s^2}\frac{\Delta P_o}{\rho l}\left[\cosh\left(\frac{h}{2}\sqrt{\frac{s}{\lambda}}\right)\right]^{-1} \; . \tag{2.3.9}$$

The solution of Eq. (2.3.3) which satisfies the boundary conditions is then

$$U(y,s) = \frac{1}{s^2}\left(\frac{\Delta P_o}{l\rho}\right)\left[1-\frac{\cosh\left(y\sqrt{\frac{s}{\lambda}}\right)}{\cosh\left(\frac{h}{2}\sqrt{\frac{s}{\lambda}}\right)}\right] \quad . \qquad (2.3.10)$$

Laplace inversion [11] then leads to the time domain expression

$$u(y,t) = u_o\left[\left(1-\alpha\, y^2\right) + \frac{32}{\pi^3}\sum_{n=1}^{\infty}\frac{(-1)^n}{(2n-1)^3}e^{-t/\tau_n}\cos\frac{(2n-1)\pi y}{h}\right] , \qquad (2.3.11)$$

where the time constants are

$$\tau_n = \frac{h^2}{(2n-1)^2\,\pi^2\,\lambda} \quad . \qquad (2.3.12)$$

In an earlier paper [3] an approximate time constant τ_f for changes in flow profile was evaluated on the assumption of parabolic profiles. This took the form

$$\tau_f = \frac{3\rho h^2}{32\eta} . \qquad (2.3.13)$$

This gives a value of 334 μs for the parameters of Table 1 and agrees closely with τ_1 which dominates Eq. (2.3.11). The profiles generated by Eq. (2.3.11) are shown in Fig. 4.

In the same paper we also examined the pressure response to an ER change for an isolated valve at constant flowrate using a numerical

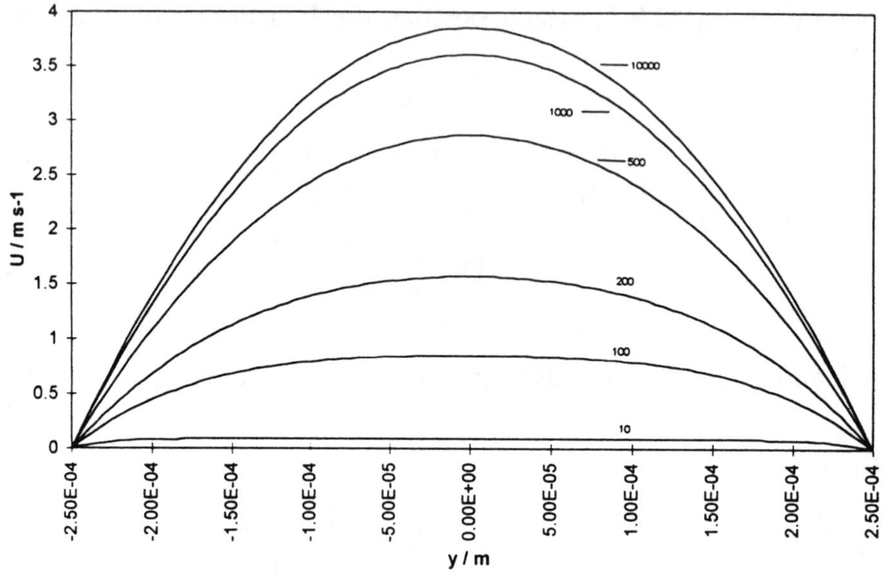

Fig. 4. Time sequence of velocity profiles following a pressure jump from Eq. (2.3.7). Times in μs are indicated near the curves. From the bottom, t = 10, 100, 200,500 μs and 1.0,10.0 ms.

model. The parameters used were comparable to those in Table 1, and the resulting response was fitted by two exponentials. The shorter time of 14 μs identified by this means was related to the model used for the ER response while the longer time of 50 μs was attributed to the fluid dynamic response and explained in terms of τ_f. It was suggested that this was reduced from the value of 334 μs by an effective increase in the viscosity η caused by the electro-stress. However, it can now be seen that the reasons for this faster response are more fundamental and that the yield stress plays no part in determining the response time.

3 Conclusions

We have examined the time dependent solutions for flow of a ER valve subject to a sudden change in the rheological properties of the fluid. The solutions give time constants that depend only on the gap size and kinematic viscosity of the fluid. Significantly, although the yield stress contributes to the magnitude of the change it does not influence the speed of response.

For comparison we have also examined the solutions for a sudden jump in pressure at zero field (i.e. the Newtonian case) and find that the fluid response to an electrorheological change is considerably faster than for a simple pressure jump. These findings parallel work in electrorheological clutches that we have recently reported[4,5]. Taken together they are potentially important for the design of high speed ER and MR machines and the optimisation of operating fluids.

References.

1. Electrorheological Fluids, Mechanisms, Properties, Technology and Applications. *Proceedings of the 4th International Conference Electrorheological Fluids* , Feldkirch, Austria. Ed. R. Tao and G.D. Roy. (World Scientific, Singapore, 1994).

2. A. Hosseini-Sianaki, W.A. Bullough, R. Firoozian, J. Makin and R. Tozer, *Proceedings of the International Conference Electrorheological Fluids,* Carbondale, Illinois. Ed. R. Tao. (World Scientific, Singapore, 1992). pp 219 - 235.

3. M. Whittle, R. Firoozian, D.J. Peel, and W.A. Bullough. Electrorheological relaxation times derived from pressure response experiments in the flow mode. *Jnl. Non-Newtonian Fluid Mech.* **57**, 1 - 25 (1995).

4. M. Whittle, R.J.Atkin and W.A. Bullough. Dynamics of a Radial ER clutch. To be submitted.

5. M. Whittle, R.J. Atkin and W.A. Bullough. Fluid dynamic limitations on the performance of an electrorheological clutch. *Jnl. of Non Newtonian Fluid Mechanics*, **57**, 61-81, (1995).

6. R.J. Atkin, , Xiao Shi, and W.A. Bullough. Solutions of the constitutive equations for the flow of an electrorheological fluid in radial configurations. *Jnl. Rheol.* **35**, 1441 - 1461, (1991)

7. M. Whittle, W.A. Bullough, and D. Peel, : The Dependence of Electrorheological Response Time on Conductivity and Polarization Time. *Phys. Rev. E*, **49**, 5249-5259 (1994).

8. D. Peel, D. Brookes and W.A. Bullough, Experiences in pumping electrorheological fluids. Proceedings 9th Conference on Fluid Machinery, Hungarian Academy of Sciences, Budapest, September 1991, p362-369.

9. For example, D.G. Shepherd. *Elements of Fluid Mechanics.* (Harcourt, Brace & World inc. 1965)

10. J.J. Eige, *ASME paper 63 MD* 1, 20/23 May 1963. See also L.Zheng, R.D. Ervin and F.E. Filisko, in *Proceedings of the 3rd International Conference on Electrorheological Fluids, Carbondale, Ill.* Ed. R. Tao, World Scientific, Singapore, 1992. For a comparable treatment in pipes see M. Reiner, *Deformation and Flow,* H.K. Lewis & Co. London , 1949.

11. M.R. Spiegel, *Mathematical Handbook of Formulas and Tables* (Schaum-McGraw Hill)

MATERIALS FOR ER FLUIDS

ROBERT BLOODWORTH AND ECKHARD WENDT
Central Research and Development
Bayer AG
D-51368 Leverkusen, Germany

ABSTRACT

Recent improvements in the physical understanding of ER fluids have led to the rational design of new ER materials with improved properties. This paper gives an overview of several recent developments in the formulation of ER fluids, concentrating on new particulate phases for ER dispersions. Examples of homogeneous ER fluids are also discussed. The trend leading to designed ER dispersions is demonstrated by a new class of electrorheological fluids based on non-aqueous polyurethane dispersions. The fluids exhibit an attractive combination of properties: low viscosity, high ER effect, and low conductivity. The dispersed phase consists of a specially developed polyurethane elastomer which solvates and stabilizes metal salts. The polymer network density influences the mobility of the dissolved ions, allowing a surprising degree of control over the ER effect. Properties such as the field strength dependence of the ER-effect, switching response, and conductivity of these fluids correlate directly with changes in the polymer structure. Electrorheological measurements in a couette viscometer (shear-mode) and in a model shock absorber (flow-mode) using a commercial polyurethane-based fluid show that the ER effect is also dependent upon the shearing geometry.

1. Historical Development of ER Fluids

The original ER fluids, dispersions of active powders in non conducting oils, were for the most part empirical inventions based in the chemical intuition of pioneers such as Winslow.[1] The mechanism of the ER effect in these dispersions was almost exclusively water-promoted ionic polarization. The next step in the evolution of ER materials was the recognition that the particle polarization necessary for the ER effect could be achieved by several different mechanisms. It is sensible to group the wide range of known ER dispersions based on the polarization mechanism of their electroactive phase.

2. Dispersed Phases in ER Fluids

The strong electrorheological fluids (ERF) known to date are dispersions of electrically polarizable particles in a non-conducting continuous phase. Overwhelming evidence suggests that the primary mechanism of the ER effect in these dispersions is the polarization of the particles in an electric field of sufficient strength. The following overview of ER materials will therefore concentrate on the chemistry of the particulate phase in these systems.

2.1. *Activated Ionic Phases*

The oldest and most widely studied ER fluids are based on water activated silica gels and zeolites,[2] including a number of fluids created in our laboratories.[3] Problems with the abrasiveness and sedimentation of these dispersions led to the development ERF's with polyelectrolyte gels as the dispersed phase.[4] All of these systems must be "activated" by the presence of a low molecular weight polar liquid, most frequently water. However, a number of reports suggest that anhydrous polar solvents (even aprotic ones like propylene carbonate) can also be used to activate these systems.[5,6,7,8] Poly(lithium methacrylate) is a well known dispersed phase in ER-fluids. As a dry powder, the lithium ions in the particles are strongly bound by the carboxylate groups. The addition of water or another substance which can solvate the lithium increases the ionic mobility and allows for a surface polarization of the particle in an electric field. The special role of water for the ER-effect lies primarily in its solvating ability, rather than in its inherent polarizability in an electric field. Water can therefore be replaced in most activated fluids by other polar liquids capable of solvating ions.

A number of commercial ER fluids are based upon a water-activated mechanism.

Poly(lithium methacrylate)	**Advanced Fluid Systems**
Versaflow™ ER-100	**Lord Corp.**
Derivatized Cellulose	**Lubrizol**
surface-sulfonated-	**Nippon Shokubai**
crosslinked polystyrene	**Asahi**

These fluids are still developmental products according to the manufacturers and may or may not represent current formulations.

2.2. *Intrinsically Polarizable Inorganics*

Filisko[9] has reported strong ER-effects with some scrupulously dried materials. Other than the purely dipolar dry systems, such as $BaTiO_3$ or TiO_3, which only work effectively with high frequency AC currents, the anhydrous aluminosilicates are ER active in DC fields. The ER activity of such highly dried materials was one of the first indications that intrinsically polarizable inorganic materials could be used as ER dispersed phases. A number of interesting systems based on anhydrous inorganic ion conductors such as zirconium doped aluminosilicates have also emerged in the patent literature.[10,11,12]

2.3. *Electronic Conductors*

More recently, a number of fluids based on semiconducting polymers have been reported. [13,14,15,16] Conductive polymers such as poly(aniline) have also received considerable attention.[17,18]

The activation energy for electron transport is much lower than that for ionic transport. Fluids based on electronic conductors would theoretically show ER activity down to low

2.4. Homogeneous ER Fluids

The problems associated with sedimentation and conductivity in dispersed ER fluids have spurred interest in homogenous ER fluids. A comparison of the relative strengths of the two types of formulations highlights the problems associated with the known homogeneous ER fluids (e.g. low strength, shear rate dependent ER effect)

Homogeneous Fluids	Dispersions
+ non-abrasive	? abrasion
+ no sedimentation	? sedimentation
? low conductivity	- moderate to high conductivity
? high zero-field viscosity	? low zero field viscosity
- low relative ER-effect	+ high ER-effect
? slow response times	+ fast response times
- not effective at high shear rates	+ effective at high shear rates

Table 2: Several examples of homogeneous ER fluids..

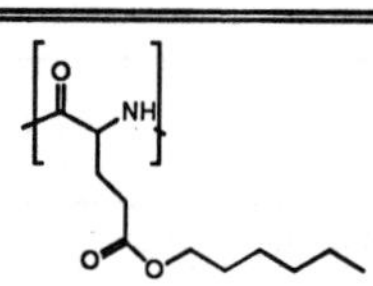	poly(n-hexylglutamate)[22] 3 kPa@0.5kV (g=0.1s^{-1})
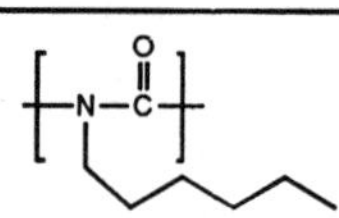	poly(n-hexylisocyanate) [23] 0.6 kPa@1kV (g=1s^{-1})
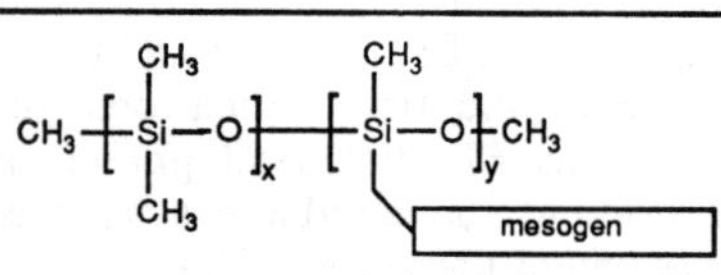	silicone liquid crystal side chain polymers. [24] 3kPa@2kV, t_r = 20-30 msec , h_{E0} = 10Poise (Asahi)
	aluminum soaps 400 Pa@8kV/mm g=200 s-1 , hE0 = 130 mPas (RWE-DEA)

Table 3: Examples of Homogeneous ER-Fluids

temperatures, as well as a drastically reduced temperature dependence of the conductivity. Most if not all of the organic conductors contain ionic species (some inadvertently). Unfortunately, the sensitivity of these systems to trace concentrations of the so-called "promoter liquids" often leads to high conductivity and poor reproducibility as well as instability of the ER effect.

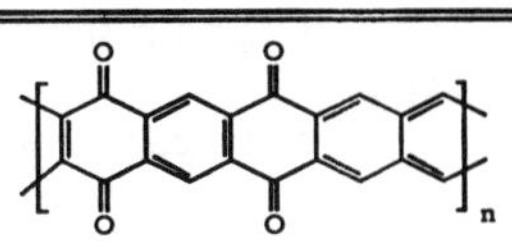	poly(acenquinone) radical (Block)
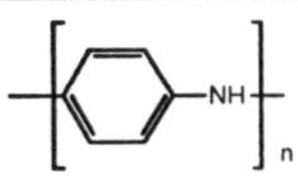	doped poly(aniline) (Block, Yamagata Univ., Lubrizol, Bridgestone, Toyota Jidosha, Kuramoto Noryuki) ionic dopants, conduction mechanism unclear
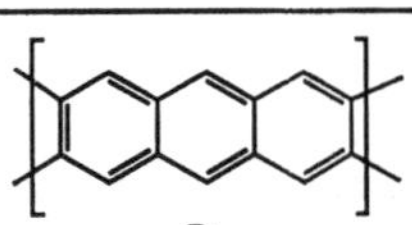 ?	carbonaceous particles (Bridgestone, Kawasaki Steel, Mitsubishi Kasei) poorly defined structure, characterized by C:H ratio (2.00-3.50)
	poly[2-(3-thienyl) ethanesulfonic acid (Toyota, Toyoda Chuo Kenkyusho) ionic doped organic semiconductor, conduction mechanism unclear

Table 1: Examples of semiconducting ER dispersed phases

New commercial ER fluids introduced by Bridgestone company[19] are based on specially synthesized carbonaceous particles. Although the magnitude of the field induced shear stress for the fluids is only moderate (1.5 kPa at 3 kV), the carbon-based fluid shows ER activity down to -40°C and a remarkably constant shear stress up to 140°C. This strongly suggests an electron polarization and may be the first example of an ER fluid based on a purely electronic conduction mechanism. The fluid conductivity, however, increases strongly with temperature, thus precluding power sensitive applications at sustained high temperatures. Other carbon particle based ER fluids have also appeared in the patent literature.[20]

The manufacture of such carbon-based particles has been described in several patents.[21] A carbonaceous powder of well-defined conductivity is produced by (a) subjecting a starting material comprising coal, coal tar, coal tar pitch, liquefied coal, coke, petroleum, petroleum tar, petroleum pitch or resin to heat treatment at a max. temp. of 300-800 deg.C to produce a carbonaceous material which is then pulverized and (b) subjected to an additional treatment at elevated temp. and/or reduced pressure. The controlled heat treatment is necessary in order to achieve the optimum particle conductivity.

Poly(isocyanate)s and the other liquid crystal ER-fluids function based on a field induced mesophase change. Although these solution show no sedimentation problems with sedimentation, the ER effect is relatively weak. Because a reorientation of the mesogens is necessary to achieve the ER effect in these systems, the response time of 10-30 msec makes these fluids an order of magnitude slower than standard ER dispersions. The liquid crystal ER fluids reported by Asahi are consist of a liquid crystalline silicone polymer with mesogenic groups on flexible side chains. This fluid shows a moderate ER effect and could be of interest for low shear robotics applications. Liquid crystal ER-fluids remain somewhat of a novelty in the ER field, but have definite advantages for some applications.

The exact mechanism of the ER effect in the RWE-DEA ER fluid is unclear, however, the shows a modest increase in its viscosity of the fluid in an electric field, as opposed to a Bingham-type shear stress.

3. Polyurethane-Based ER Fluids

In an effort to minimize the problems associated with activating liquids and abrasive particles, we have developed new commercial ER fluids using a polymeric solid ionic conductor as the dispersed phase.[25]

Polymers based on polyethers, as well as certain polyesters and polycarbonates can function as solid solvents for metal salts. The high solvating power of polyethylene oxide (PEO) segments for metal atoms stems from the interaction of the electron pairs on oxygen with the metal cation. The powerful cryptand 18-crown-6, an excellent complexing agent for lithium and sodium ions, is based on an advantageous circular arrangement of the oxygen atoms coordinating the metal cation. Similar structures, in which a central metal ion is enwrapped by a helical PEO-chain have been determined for crystalline complexes of ethylene glycol oligomers and alkali metal salts.[26]

Polyethylene glycol oligomers are especially interesting by virtue of their high solvating power and have been used as solid solvents in lithium-polymer batteries.[27] Unfortunately, these systems have poor mechanical properties. Multifunctional polyether polymers can be crosslinked to form elastomers which possess the required dimensional and thermal stability. The ER fluids based on these dispersed phases exhibit an attractive combination of properties: low viscosity, high ER effect, and low conductivity. The chemistry of the polymer can be widely varied to modify the polarity, Tg, and crosslink density, thus affecting the ionic mobility within the solid polymer. In addition to the practical advantage of allowing one to tailor the properties of these ER-fluids to particular applications, this well defined system provides a tool with which to examine the polarization process in the dispersed phase as a function of the ionic concentration and mobility.

A central question of ER phenomenon concerns the underlying mechanism of particle polarization. The purpose of this work was to investigate the effect of variations in fluid formulation such as the volume fraction, polymer structure and ionic strength of the dispersed phase on the ER properties of the fluid.

3.1. Experimental

The polyurethane-silicone oil dispersions were prepared using previously disclosed materials and methods.[28] Measurements of the ER effect in shear mode were made using a modified couette viscometer as described in reference 8. Measurements at constant shear rate were performed by incrementally increasing the field strength and monitoring the shear stress (τ) via torque measurements. The minimum field strength above which the shear stress increases (E_0) and the linear rate of increase in the shear stress with increasing field strength, e.g. the electrorheological effect (S) were determined from the intercept and slope respectively of the excess shear stress ($\tau - \tau_{0kV}$) versus field strength curve. Couette measurements were performed at 60°C. Measurements of the ER-effect in Poiseuille flow were made using a pneumatically driven piston assembly.[29] The fluid is forced through an electrified annulus valve by the piston of a modified shock absorber and the shear stress and shear rate calculated from the speed of the piston and pressure drop across the valve. The Poiseuille flow measurements were performed at 40°C.

3.2. Results

1. Volume Fraction of the Particulate Phase

The base viscosity of the fluid depends as expected upon the volume fraction of the particulate phase. Figure 1 indicates a limited shear rate dependence of the viscosity at high volume fractions at shear rates below 1000 s^{-1}, as well as surprisingly low zero field viscosities at higher volume fractions. A low base viscosity is critical to achieving a large switchable range of shear stresses in an ER device. The zero field viscosity scales approximately linearly with temperature, as would be expected for a silicone oil based fluid.

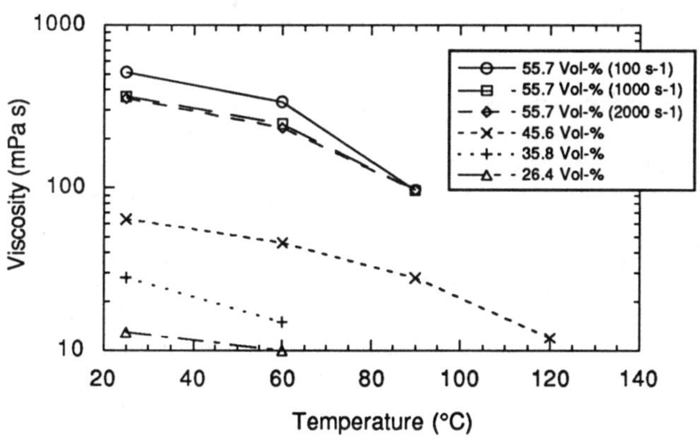

Figure 1: Effect of temperature, shear rate and volume fraction solids on the zero field viscosity.

2. Concentration of Ions in the Particulate Phase

The main factors contributing to the overall ER-effect are the concentration and mobility of the charge carriers within the particles. Modifying the number of charge carriers should directly influence the magnitude of the ER effect.

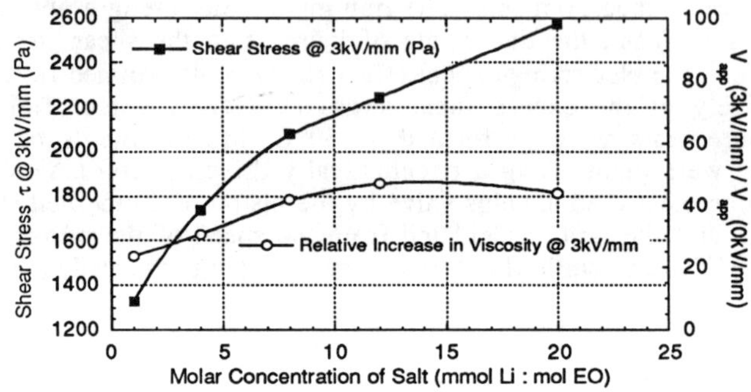

Figure 2: Effect of the Li^+ concentration on the shear stress (τ) and on the relative increase in the apparent viscosity (V_{3kV}/V_{0kV}) .

An AC voltage source was used to minimize any electrophoresis effects in the following experiments. Figure 2 shows the effect of increasing the concentration of an added salt ($LiNO_3$), which is dissolved in the dispersed phase, on the ER-effect as measured in a couette viscometer. The polyurethane dispersed phase is based on a trifunctional polyethylene oxide polyether (MW 675) crosslinked with 1 equivalent of toluene diisocyanate. The continuous phase is a polydimethylsiloxane oil with a viscosity of 5 mPa·s at 25°C.

After an initial strong increase in the ER effect, an approximately linear dependence of the shear stress on the concentration of Li^+ is observed. The switching factor, the ratio of apparent viscosity measured at 3kV/mm and at 0 kV/mm (η_{3kV}/η_{0kV}), is a measure of the fluids switchable viscosity range and increases only marginally above 10mmol Li^+/mol EO.

Examination of Figure 3 yields additional insight into the effect of the ion concentration on the polarization process. The threshold field strength for the onset of ER-activity, E_0, is high at low Li^+ concentrations but is much less sensitive to higher salt levels. This suggests that a minimum average concentration of ions is necessary for easy polarization of the dispersed phase, but that higher concentrations do not appreciably affect the barrier to polarization. The absolute magnitude of ER-effect, as well as the shear stress at 3kV/mm, appears to depend upon the total number of ions in the particles.

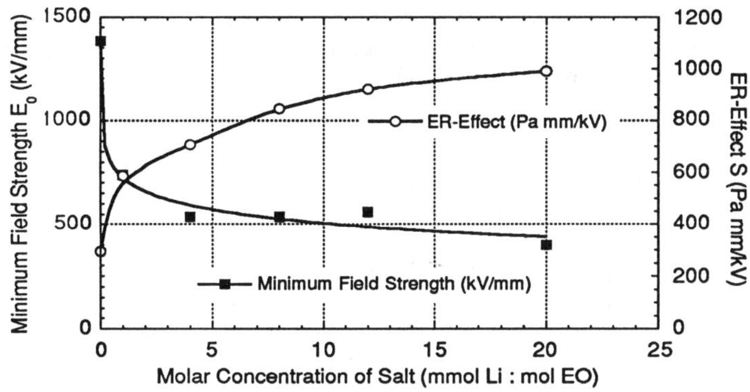

Figure 3: Minimum field strength for the ER-effect (E_0) and the absolute ER-effect (S) versus lithium ion concentration.

Figure 4 demonstrates the strict linear relationship between the concentration of dissolved ions and the conductivity of the ER fluid. This behavior is expected for solid ionic conductors at dilute concentrations. In the dilute regime, the ions are well solvated and do not agglomerate to form clusters. Although the shear stress and ER-effect continue to rise with the concentration of added salt, a strong increase in the bulk conductivity is associated with high ion concentrations. The concomitant increase in the power requirements of the fluid puts a practical limit on the amount of salt which may be incorporated into the dispersed phase.

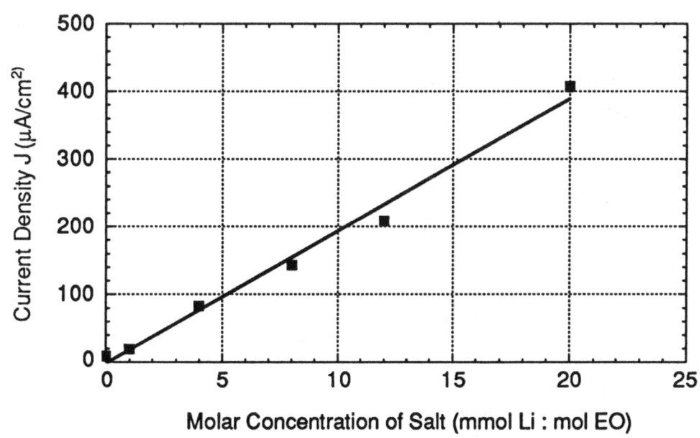

Figure 4: Current density (J) as a function of lithium ion concentration.

3. Crosslink Density of the Polyurethane

Figure 5 demonstrates the effect of varying the network structure of the dispersed polymer phase. Using mixtures of a trifunctional and a difunctional PEO polyol of the same equivalent weight (ca 225 g/equiv. OH) and crosslinking with toluene diisocyanate, the crosslink density of the polyurethane particles was varied while maintaining a nearly constant concentration of urethane groups within the polymer. The lithium ion concentration was held constant at 10 mmol Li: mole EO.

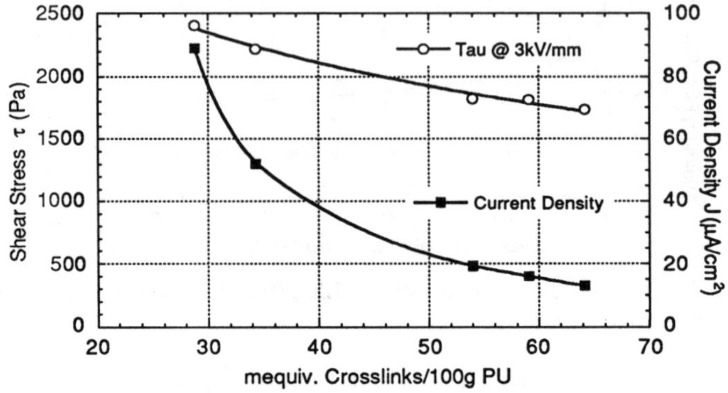

Figure 5: Shear stress (τ) and current density (J) versus crosslink density of the PU dispersed phase (ionic concentration of 10mmol Li: mol EO).

Increasing the crosslink density lowers the conductivity of the dispersed phase, indicating a decrease in the long range ion mobility. The magnitude of the shear stress (τ @ 3kV/mm), by comparison, decreases relatively slowly upon increasing the number of crosslinks, suggesting that the particle polarization is less affected.

An examination of Figure 6 reveals that the ER-effect falls approximately linearly with the crosslink density, as does τ. The current density falls off much more rapidly in a tighter network, scaling with the inverse of the crosslink density cubed. More surprisingly, the minimum field strength, E_0, remains relatively constant over the entire range of data.

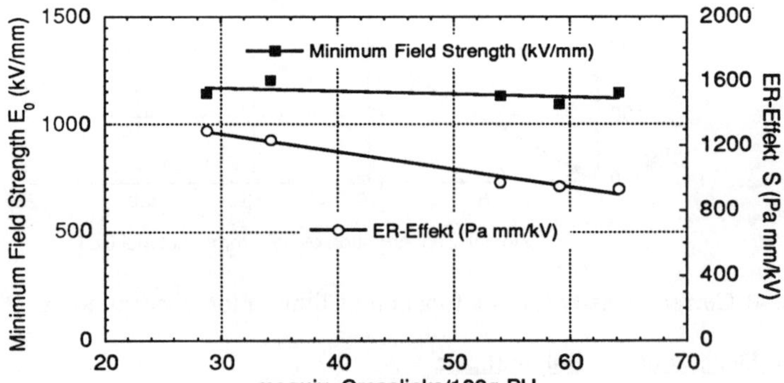

Figure 6: Minimum field strength (E_0) and ER-effect (S) versus crosslink density of the PU dispersed phase (ionic concentration of 10mmol Li: mol EO).

Additional experiments on related systems have shown that further increases in the crosslink density eventually lead to an increase in E_0, but that in a wide range of polymer networks, the polarity of the polymer matrix (as well as the concentration of ions) exerts a much greater influence on E_0 than does the crosslink density, or the mobility of the chain segments.

The data supports a model for particle polarization in which short range motions of the ions are responsible for an interfacial polarization[30] of the dispersed phase, and that ion transport *through the dispersed phase* controls the bulk conductivity of the ERF.

One consequence of these results for the design of improved ER-fluids lies in reducing the conductivity of the ERF without adversely affecting the ER response. Apparently, the bulk conductivity of the ERF, observable as the current density in Figure 5, can be reduced by restricting the movement of ions in the polymer matrix, albeit at a small cost to the maximum shear stress and the ER effect. This result suggests that it may be possible to make a strong ERF with a negligible power consumption.

3.3. Application of Polyurethane-Based Fluids

1. Poiseuille Flow versus Shear Flow

ER fluids behave differently depending upon the velocity profile of the fluid between the electrodes. In the shear mode, the fluid is sheared between electrodes moving relative to each other and the ERF experiences a linear shear gradient. The shear mode is prevalent in clutch applications of ER fluids. In Poiseuille flow (flow mode), the fluid is forced though an electrified annulus which functions as an electrorheological flow control valve. The Bingham character of the fluid in an electric field results in a complex velocity profile. The flow mode is common to a number of commercially attractive ER-applications, including hydraulics and automotive vibration damping.

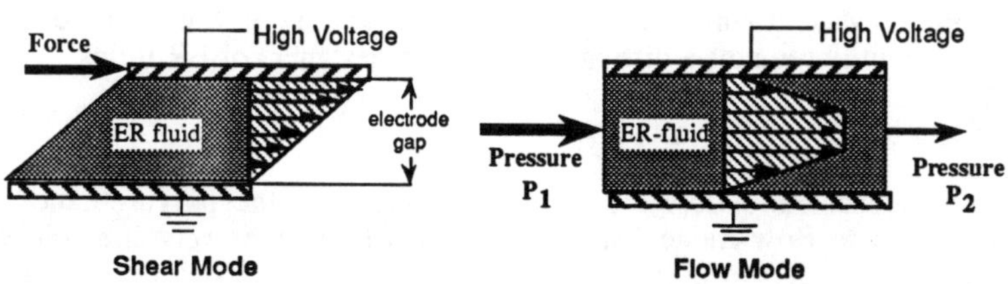

Figure 7: Comparison fluid velocity profiles in shear and flow mode applications.

Figure 8 shows the ER response of a commercial polyurethane-based ER fluid (RHEOBAY® VPAI 3565) in a flow mode testing apparatus based on an automotive shock absorber. Corresponding couette viscometer data is included for comparison. The couette viscometer used for the shear mode measurements has a maximum achievable shear stress of app. 2000 s^{-1}.The shear stress at 1000 s^{-1} extrapolated from flow mode measurements made at high shear rates lies significantly higher than would be expected from the couette measurement in shear mode. The reason behind this apparent discrepancy is the well known "chaining" phenomenon of the ER dispersed phase in an electric field. The pearl chains moving through the electrified valve do not experience a transverse shear and the fluid acts as a Bingham solid undergoing plug flow. This yields ER effects which are usually much higher at a given field strength than those estimated from standard couette measurements. Shear stresses of well over 10 kPa are possible with the correct selection of valve geometry and fluid.

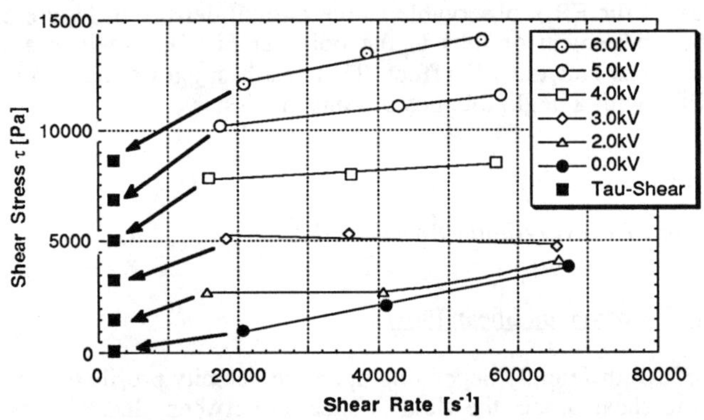

Figure 8: Shear stress versus shear rate data measured in flow mode using a modified ER shock absorber as compared to couette measurements (shear mode). Nominal shear rates calculated assuming Newtonian flow.

This complex behavior has implications for the design of ER fluids for flow mode applications. To optimize the plug flow within a valve, the initial polarization to form chains must be fast. However, once the plug has been formed, surface interactions likely dominate the interparticle forces. Maximizing the mechanical interaction of the particles while maintaining colloidal stability remains a difficult task for the designers of ER fluids.

2. ER Shock Absorber

The comparison of shear and flow mode data indicates that the polyurethane-based fluids are more effective in flow mode applications. An ER damper was constructed to demonstrate the application of the polyurethane fluids in an actual device. A standard automotive shock absorber was modified with an external ER valve (Figure 9). The restricted flow of the ER fluid through the external valve controls the damping force generated upon moving the piston.

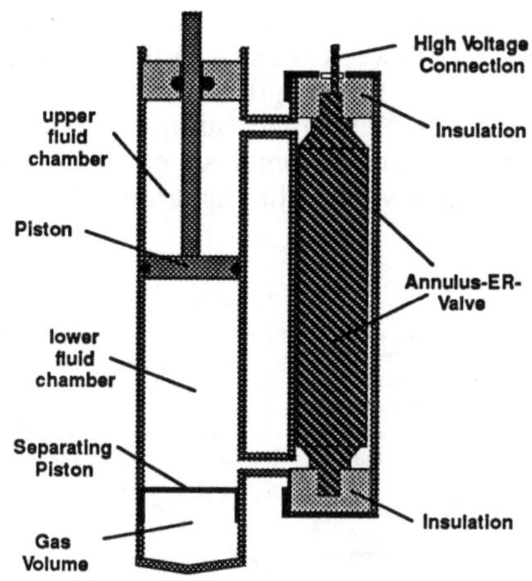

Figure 9: ER damper construction based on a modified shock absorber.

The damper was filled with a commercial polyurethane-based fluid (RHEOBAY® VPAI 3566, Bayer AG), and the damping force measured as a function of piston speed and electric field strength. The characteristic curves in Figure 10 demonstrate that the system responds symmetrically with a variable damping force between 2kN and 6kN. Forces of this magnitude are adequate for applications in passenger vehicles, thus demonstrating the utility of ER technology using these fluids.

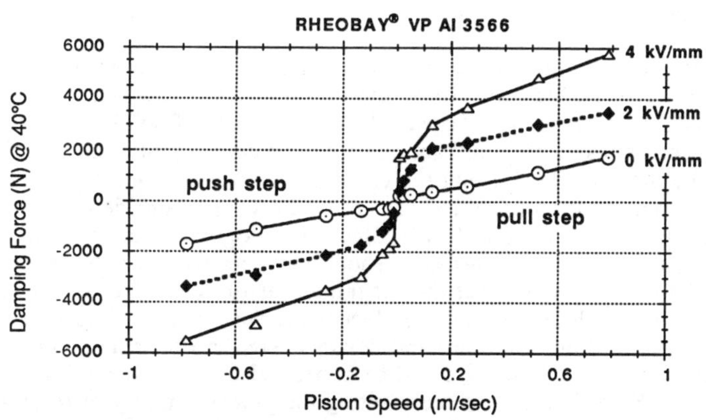

Figure 10: Characteristic damping curves of an ER shock absorber containing a commercial polyurethane-based ER fluid (RHEOBAY® VPAI 3566).

A major concern in the application of ER fluids is the long term stability of the ER effect under use. Figure 11 depicts the results of endurance tests on a polyurethane-based ER-fluid using a modified hydraulic apparatus. In this test, the polyurethane-based fluid is continuously recirculated through an electrified annulus valve and the electric field across the valve is varied continually between 0 and 9kV/mm. The curves indicate the change in pressure across the valve as the field strength is varied. The effective shear rate at the wall of the valve (as calculated for a Newtonian fluid) was app. 45,000 s^{-1}. Over a period of several thousand hours the ER effect remains nearly constant and the fluid properties unchanged. Neither the particles nor the apparatus show any signs of abrasion or degradation.

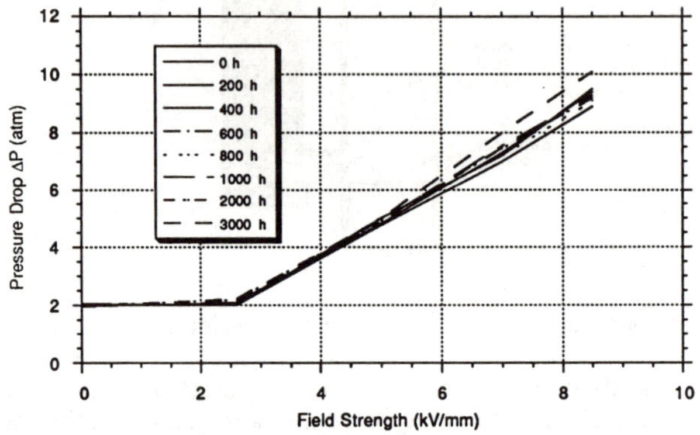

Figure 11: Endurance test of a polyurethane-based ERF in an hydraulic apparatus at a temperature of 50°C and a nominal shear rate of 45,000 s^{-1}.

4. Conclusions

New water free ER-fluids based on polyurethane dispersions exhibit a combination of properties: low viscosity, high ER effect, and low conductivity, which makes them attractive for commercial applications. The dispersed phase consists of solid solutions of metal salts dissolved in a crosslinked polyurethane matrix. The conductivity and polarizability of the particles can be modified by varying the crosslink density and polarity of the polyurethane as well as the concentration of added salt, thus allowing a considerable degree of control over the ER properties of the fluid. Electrorheological measurements in a couette viscometer (shear mode) and in a model shock absorber (flow mode) using a commercial polyurethane-based fluid show that the magnitude of the ER effect is dependent upon the velocity profile of the fluid between the electrodes. This result underscores the importance of carefully matching fluid properties and device design in ER applications. A continuously variable damping forces of 2-6 kN could be generated using a standard automotive damper modified with an ER valve, thus demonstrating the practical utility of these ER fluids. The fluids exhibit a stable ER effect which remain virtually unchanged even after prolonged use at elevated temperatures and high shear rates.

1. Winslow, W. M., J. *Appl. Phys.*, 20(0):1137(1949).

2. Klass, L. D., Martinek, Th. W., *J. Appl. Phys.*, 38(1):67(1966).

3. DE 3427499, DE 3517281, DE 3536934

4. UK 1570234

5. JP 1253110

6. EP 374525

7. JP 3074494

8. EP 509571

9. US 4,744,914

10. US 5,139,690

11. US 5,139,691

12. US 5,139,692

13. Block, Harry; Kelly, J. P.; Qin, A.; Watson, T., *Langmuir* 6(1): 6-14(1990).

14. Block, H., Kelly, J. P., Watson, T., *Spec. Publ. - R. Soc. Chem.* 87:51(1991).

15. Otsubo, Yasufumi; Sekine, Masahiro, *Nihon Reoroji Gakkaishi* 20(2): 51-5(1992).

16. Blackwood, K. M.; Block, H., *Trends Polym. Sci. 1(4)*: 98-105(1993).

17. EP 394005, WO 9307244, J 05239482, JP 06220419

18. Kuramoto, Noriyuki; Yamazaki, Mitsuo; Nagai, Katsutoshi; Koyama, Kiyohito; Tanaka, Katsufumi; Yatsuzuka, Kyoko; Higashiyama, Yoshio, *Thin Solid Films* 239(2): 169-71(1994).

19. Ishino, Yuichi; Maruyama, Takayuki; Ohsaki, Toshiyuki; Endo, Shigeki; Saito, Tasuku; Goshima, Norio, *Polymer Preprints*, 36(2): (1994).

20. EP 625566

21. EP 546552, JP 06234985

22. Tse and Shine, *Polym. Prep.*. 35(2) 383 (1994).

23. Yang and Shine *J. Rheol.* 36(6): 1079 (1992).

24. *Kagaku Kogoyo Nippo*, 3. Feb.1994.

25. Bloodworth, R., Electrorheological Fluids Based on Polyurethane Dispersions, in: *Electrorheological Fluids, Mechanisms, Properties, Technology and Applications*, R. Tao and G.D. Roy, ed., World Scientific, Singapore (1994)..

26. S. Okamura and Y. Chatini, *Polymer Preprints Japan,* 34:2089 (1985).

27. Wright, Perter V., *J. Makromol. Sci. Chem., A26*(2-3):519(1989)

28. DE 4026881, US 5,268,118

29. Wendt, E. and Büsing, K.W., *Proc. Actuator '94*, Bremen, Germany, in press.

30. Block, H., Kelly, J.P., Quin, A., Watson, T., *Langmuir,* 6(1):6(1990).

STEADY AND TRANSIENT RESPONSES OF ELECTRORHEOLOGICAL SUSPENSION PASSING THROUGH A RECTANGULAR CHANNEL

Masami NAKANO

Department of Mechanical Systems Engineering, Yamagata University
4-3-16, Jonan, Yonezawa, Yamagata, 992, JAPAN

Ryuji AIZAWA and Yoshinobu ASAKO

Tsukuba Research Laboratory, Nippon Shokubai Co., Ltd.
1-25-12, Kannondai, Tsukuba, Ibaraki, 305, JAPAN

ABSTRACT

The steady and transient responses of ER effect of ER fluids containing sulfonated polymer particles, passing through a rectangular channel between two fixed parallel electrodes, were investigated by ensuring pressure drops across the channel in which the flow established a Poiseuille flow geometry. Effects of the types and the strength of applied electric field and the mean flow velocity on the steady response were clarified, and compared with those in a Couette flow geometry using a rotary rheometer in order to investigate the dependence of ER effects on the flow geometry. As the mean flow velocity increased, the yield stress decreased exponentially under DC voltage, while, in the case of AC voltage, gradually increased and reached a peak level and decreased thereafter. Contrastively, the yield stress measured by the rotary rheometer decreased suddenly at low shear rate, and reached almost constant level. That is, the measured yield stress depends on the flow geometry and the type of applied electric field. The measured pressure response to a step input voltage can be approximated by a first order response, and the rising time constant of ER effect depend on the compositions of ER fluids and range from 5 to 40 msec.

1. Introduction

Electrorheological(ER) fluids show intriguing properties of yielding solids under an electric field and of rapid response to the electric field change, and thus, are anticipated some applications of industrial devices such as a damper, a valve, a clutch, and so on [1]-[5]. In most of these devices except the clutch, an ER fluid flows between two fixed parallel electrodes and its flow forms a Poiseuille flow geometry. So far, there are many researches on the ER effects in Couette flow geometries using a rotary rheometer having two concentric-cylindrical electrodes, but a few researches in Poiseuille flow geometries, and it seems that the details of ER effects in a Poiseuille flow geometry are not clarified. In the closed ER hydraulic systems, the steady and transient pressure responses of an ER fluid controlled by an ER valve are of significant importance in the estimation and the controllability of the induced force, respectively.

Therefore, we studied the steady and transient responses of RE effect in a Poiseuille flow geometry using a piston cylinder ER valve experimental system. Effects of the types (DC and AC) and the strength E of applied electric field and the mean flow velocity U on the steady response were clarified, and were compared with those in a Couette flow geometry using a rotary rheometer in order to investigate the dependence of ER effects on the flow geometries.

2. Experimental and Measurement

2.1 Piston Cylinder-ER Valve Experimental System

The exprimental apparatus was designed for testing ER effects in a Poiseuille flow geometry and consisted of a piston of 230 mm stroke, a cylinder of 60 mm or 20 mm in inner diameter and an ER valve shown in Fig. 1 [6]. The ER valve had a channel of 0.81 mm in height between two parallel plate electrodes of 20 mm in length and 10 mm in width.

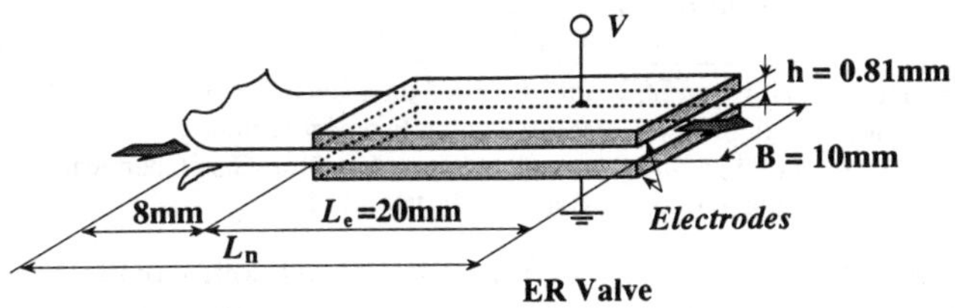

Fig. 1 Electrodes of ER valve

During testing, the piston was moved at a given constant speed by a speed controlled motor, so that the mean flow velocity between the electrodes should be constant. Under this condition, the step inputs of electric field were applied through a high voltage amplifier (Trek Model 664) of the frequency range from DC to 10 kHz. The resultant pressure responses were measured by a pressure transducer and were recorded on a digital recorder connected to a computer. From the saturated values and the time constants of the pressure responses obtained on the various conditions of mean flow velocity U and the electric field strength E, the characteristics of both steady and transient pressures were obtained as a function of U and E.

2.2 Rotary Rheometer

The ER effects in a Couette flow geometry was evaluated using a rotary rheometer (Rheometric Scientific Inc. Dynamic Analyzer RDA-II). The rheometer was modified to facilitate the supply of an electric field.

2.3 ER Fluid Samples

The characteristics of ERF samples used in this study were summarized in Table 1. Sulfonated polymer particles (NSP and SSD) were prepared according to the literature [7] and used as dispersed phase. NSP and SSD had the same average particle diameter of 5 microns. In NSP, each aromatic ring had 1.24 sulfonic acid substituents on average. The sulfonic acid substituents level of SSD was 0.88.

3. Steady Characteristics of ER Effects

3.1 DC Voltage

3.1.1 Piston Type In our preceding paper [6], we have theoretically derived, based on the Bingham Plastic Model of ER fluid, that the total pressure drop ΔP_T across the electrode

Table 1. The Characteristics of ERF Samples.

ERF Sample	Dispersed Particle	Disperse Medium	Water Content (wt.%)[1)]	Volume Fraction (vol.%)	Zero-Field Viscosity (mPas)[2)]
ER-1	NSP	Silicone Oil[3)] / DAIFLOIL#1[4)] = 1 / 3 (V/V)	2	20	30
ER-2	NSP	Silicone Oil	2	20	40
ER-3	SSD	Silicone Oil / DAIFLOIL #1 = 1 / 1 (V/V)	2	20	30
ER-4	NSP	Silicone Oil / DAIFLOIL #1 = 1 / 3 (V/V)	2	40	150
ER-5	NSP	Silicone Oil / DAIFLOIL #1 = 1 / 3 (V/V)	5	20	30

1) The water content of NSP and SSD was measured on a Karl-Fisher moisture meter.
2) The viscosity of ERF samples was measured on an E-type viscometer.
3) Shin-Etsu Silicone Oil KF96-20CS was manufactured by Shin-Etsu Chemical Co., Ltd.
4) DAIFLOIL #1 was manufactured by Daikin Industries, Ltd.

channel is divided into two components; one is a viscous component ΔP_0 proportional to the mean flow velocity U and the other is an electrorheological component ΔP_{ER} caused by ER effect, and we have reported that the approximate function of the experimental pressure drop ΔP_{ER} can be represented by the following Eq. 1[6],

$$\Delta P_{ER} = \Delta P_{ER,0} - (\Delta P_{ER,0} - \Delta P_{ER,S})(1 - e^{-\beta(E)U}) \qquad \cdots\cdots\cdots\cdots\cdots(1)$$

where $\Delta P_{ER,0}$ is the pressure drop at U=0, $\Delta P_{ER,S}$ is the pressure drop at sufficiently high mean flow velocity and β (E) [s/cm] is the coefficient indicating the dependence of ER effect on the mean flow velocity.

In this experiments, the ΔP_0 measured under non electric field is confirmed to increase in proportion to the mean flow velocity U without lack of the linearity. The variation of the ΔP_{ER} for ER-1 against the mean flow velocity are shown in Fig. 2 in terms of the electric field strength. Here, it should be stated that the data of the ΔP_{ER} at low U and high electric field strength E could not be measured with reliability due to violent and random fluctuations of the measured pressure induced by the occurrence of repeated blocking of ER fluid flow in the channel, and it was impossible to show these data points in the figure.

As the mean flow velocity increased, ΔP_{ER} tended to decrease almost exponentially from a maximum value $\Delta P_{ER,0}$ and to converge to a value $\Delta P_{ER,S}$ at any electric field strength. The increase of the electric field strength raised $\Delta P_{ER,0}$ and $\Delta P_{ER,S}$, weakening the decreasing rate to the flow velocity. The weakening of ER effect with increasing flow velocity suggests that the yield stress related to the ΔP_{ER} depends on the mean flow velocity and

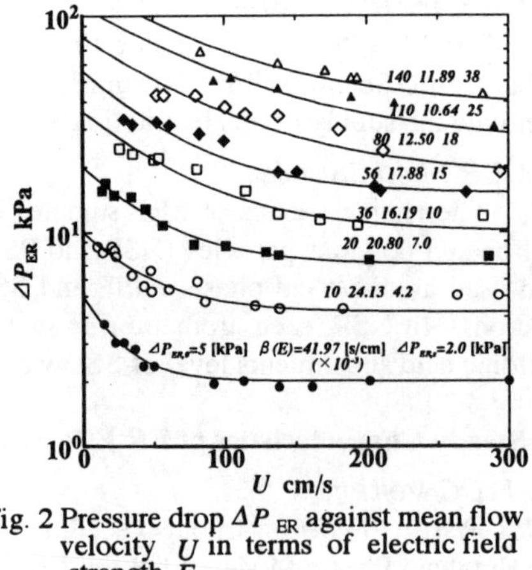

Fig. 2 Pressure drop ΔP_{ER} against mean flow velocity U in terms of electric field strength E

it may be due to the dependence of the cluster formation of dispersed particles on the mean flow velocity.

The application of a curve fitting method to the experimental data gave the values of $\Delta P_{ER,0}$, $\Delta P_{ER,S}$, $\beta(E)$, and fitted curves for each electric field strength as shown in Fig. 2. In the range of low U and high electric field strength E where there were no data points, the curves were extrapolated from the obtained data points shown in the figure. The fitted curves show a fairly good approximation of the experimental data, so it can be concluded that the function of Eq.1 is suitable for estimating the levels of ΔP_{ER}.

The yield stress τ_{ER} can be estimated from the values of ΔP_{ER}, according to the relationship of Eq. 2 as follows [6],

$$\tau_{ER} = (h/3L_e)\,\Delta P_{ER} \qquad \cdots\cdots(2)$$

where h is the channel height of electrodes and L_e is the channel length of electrodes.

The approximate function of τ_{ER} can be represented by the following Eq. 3,

$$\tau_{ER} = \tau_{ER,0} - (\tau_{ER,0} - \tau_{ER,S})(1 - e^{-\beta_p(E)U}) \qquad \cdots\cdots(3)$$

where $\tau_{ER,0}$ is the yield stress at U=0 and $\tau_{ER,S}$ is the yield stress at sufficiently high mean flow velocity.

Next, three parameters of $\tau_{ER,0}$, $\tau_{ER,S}$ and $\beta_p(E)$ for ER-1 should be arranged as a function of the electric field strength. These results are shown in Figures 3(a), 3(b), and 3(c), respectively. The parameters for ER-2, ER-3, ER-4 and ER-5 are indicated in the same figures. Both $\tau_{ER,0}$ and $\tau_{ER,S}$ increased in proportion to the square of electric field strength and the proportional coefficients $\alpha_{\tau p}$ [kPa·mm²/kV²] and $\gamma_{\tau p}$ [kPa·mm²/kV²] are as indicated in Fig. 3(a) and Fig. 3(b), respectively. Consequently, the approximate function of the τ_{ER} component is given as follows.

$$\tau_{ER}(E,U) = \alpha_{\tau p}E^2 - (\alpha_{\tau p}E^2 - \gamma_{\tau p}E^2)(1 - e^{-\beta_p(E)U}) \qquad \cdots\cdots(4)$$

The ER effect of ER fluids can be evaluated based on the coefficients of $\alpha_{\tau p}$, $\gamma_{\tau p}$ and $\beta_p(E)$. The $\alpha_{\tau p}$ decreased in the order of ER-4 > ER-5 > ER-1 > ER-2 > ER-3. The large values of $\alpha_{\tau p}$ of ER-4 and ER-5 can be explained on the high volume concentration

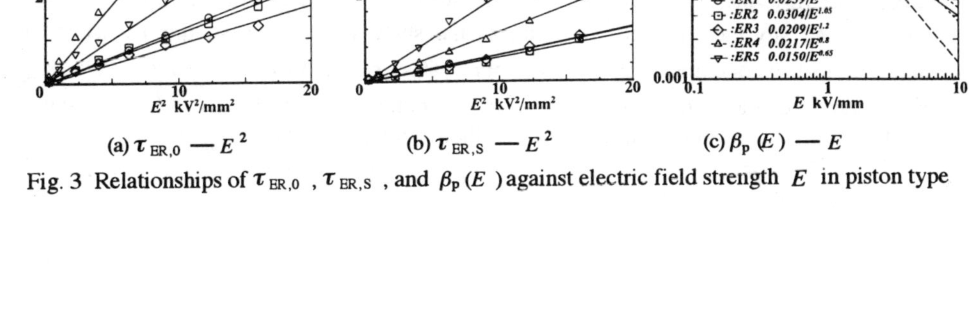

(a) $\tau_{ER,0}$ — E^2 (b) $\tau_{ER,S}$ — E^2 (c) $\beta_p(E)$ — E

Fig. 3 Relationships of $\tau_{ER,0}$, $\tau_{ER,S}$, and $\beta_p(E)$ against electric field strength E in piston type of dispersed particles and on the high water content in the dispersed particles, respectively. The $\beta_p(E)$ of ER-4 was larger than that of ER-5, so that the $\gamma_{\tau p}$ order of ER-4 and ER-5 was reversed.

Figure 4 shows the estimated yield stress τ_{ER} of ER-1 against the flow velocity in terms of the electric field strength. The fitted curves show a fairly good approximation of the experimental data, so it can be concluded that the approximate function of Eq. 4 is suitable for representing the levels of τ_{ER}.

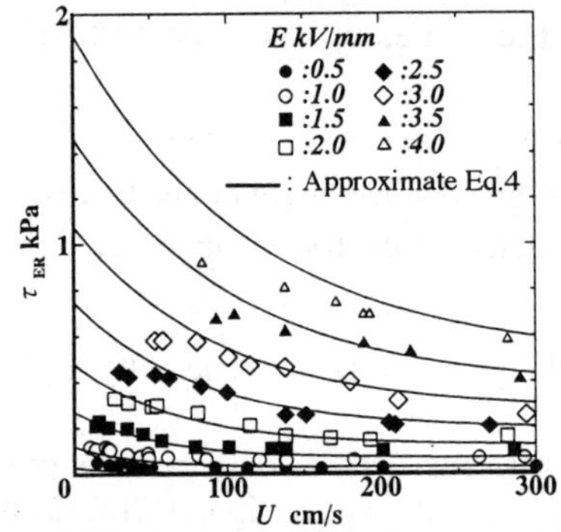

Fig. 4 Yield stress τ_{ER} against mean flow velocity U in terms of electric field strength E

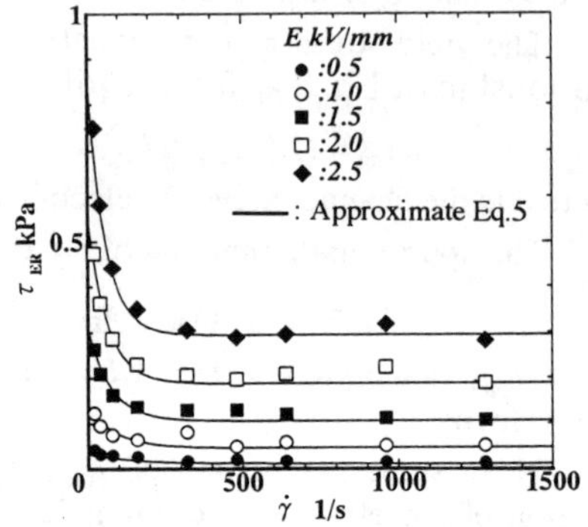

Fig. 5 Yield stress τ_{ER} against shear rate $\dot{\gamma}$ and electric field strength E

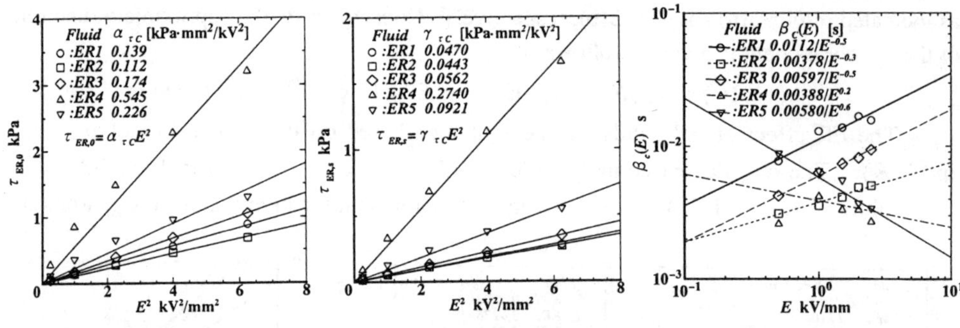

(a) $\tau_{ER,0}$ — E^2 (b) $\tau_{ER,S}$ — E^2 (c) $\beta_p(E)$ — E

Fig. 6 Relationships of $\tau_{ER,0}$, $\tau_{ER,S}$, and $\beta_C(E)$ against electric field strength E in rotor type

3.1.2 Rotor Type Figure 5 shows the change of the yield stress τ_{ER} for ER-1 with the shear rate in terms of the DC electric field strength, which are measured on the rotary rheometer. At any electric field strength, as the shear rate increased, the τ_{ER} tended to decrease almost exponentially from a maximum value $\tau_{ER,0}$ and to converge to a value $\tau_{ER,S}$ at shear rate $\dot{\gamma}$ of more than 200 s^{-1}. The approximate equation of τ_{ER} can be represented by the following Eq. 5, in the same manner as that of the piston type.

$$\tau_{ER}(E,\ \dot{\gamma}) = \alpha_{\tau C} E^2 - (\alpha_{\tau C} E^2 - \gamma_{\tau C} E^2)(1 - e^{-\beta_C(E)\dot{\gamma}}) \qquad \cdots\cdots(5)$$

The approximate curves of τ_{ER} are drawn in Fig.5. The $\tau_{ER,0}$ and $\tau_{ER,S}$ for ERF samples against the square of the electric field strength and the $\beta_c(E)$ [s] against the electric field strength are shown in Figures 6(a), 6(b), and 6(c), respectively. Both the coefficients $\alpha_{\tau C}$ [kPa·mm^2/kV^2] and $\gamma_{\tau C}$ [kPa·mm^2/kV^2] decreased in the order of ER-4 > ER-5 > ER-3 > ER-1 > ER-2. The orders were almost the same as that of the $\alpha_{\tau P}$ in the piston system.

3.2 AC Voltage

3.2.1 Piston Type Figure 7 shows the τ_{ER} of ER-1 against the mean flow velocity in terms of the AC (50Hz) electric field strength of r.m.s. value. As the mean flow velocity increased, the τ_{ER} gradually increased and reached a peak level and decreased thereafter.

Figure 8 shows the τ_{ER} of various ERF samples against the mean flow velocity at 2 kV/mm. ER-2 and ER-3 had similar dependence of ER effect on the flow velocity to ER-1. ER-4 induced the largest shear stress in the range of relatively low flow velocity but the shear stress decreased remarkably with the increase of the mean flow velocity.

Figure 9 shows the τ_{ER} of ER-1 against the frequency f of electric field at U = 22 cm/s in terms of the electric field strength. At the electric field of not more than 2 kV/mm, the τ_{ER} gradually decreased with the increase of the frequency. When the electric field of over 2 kV/mm was applied, however, the τ_{ER} gradually increased and reached a peak level and decreased thereafter as the frequency increased.

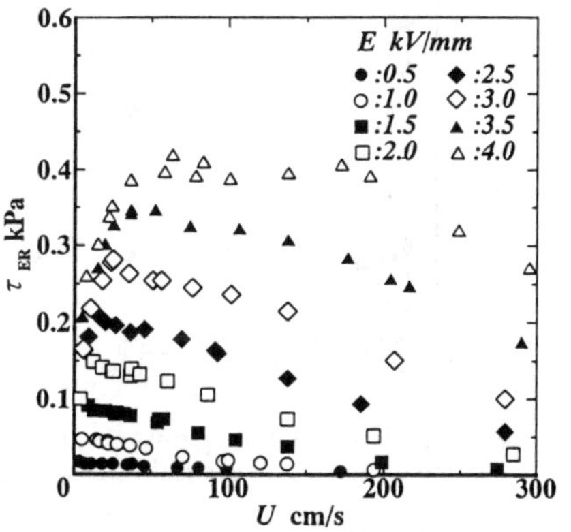

Fig. 7 Yield stress τ_{ER} of ER-1 against mean flow velocity U in terms of electric field strength E

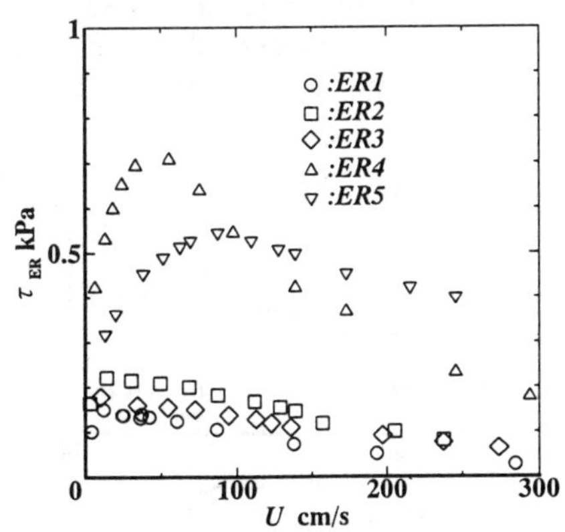

Fig. 8 Yield stress τ_{ER} of ERF samples against mean flow velocity U at 2 kV/mm

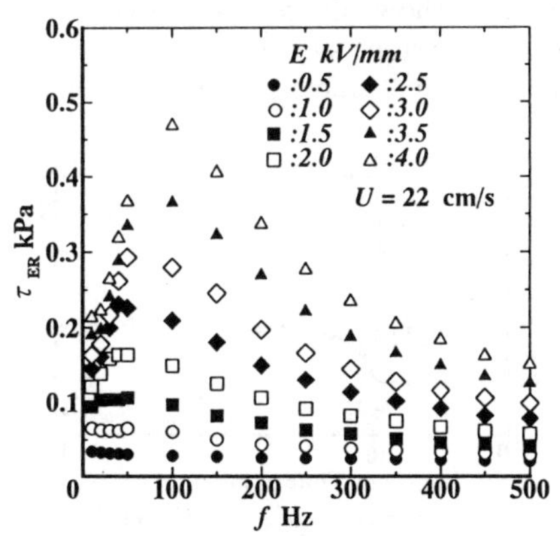

Fig. 9 Yield stress τ_{ER} of ER-1 against frequency of electric field at U =22 cm/s in terms of electric field strength E

3.2.2 Rotor Type Figure 10 shows the measured τ_{ER} of ER-1 against the shear rate $\dot{\gamma}$ in terms of the AC electric field strength. At any electric field strength, as the shear rate increased, the τ_{ER} tended to decrease almost exponentially from a maximum value of τ_{ER} at low shear rate and to converge to a value $\tau_{ER,S}$ in the range of the shear rate of more than 200 s^{-1}. The proportional coefficient $\gamma_{\tau C}$ between τ_{ER} at 960 s^{-1} and E^2 decreased in the order of ER-4 > ER-5 > ER-2 > ER-1 > ER-3, as seen in Fig. 11. The order was almost the same as that of $\gamma_{\tau C}$ in the rotor type under the DC electric field.

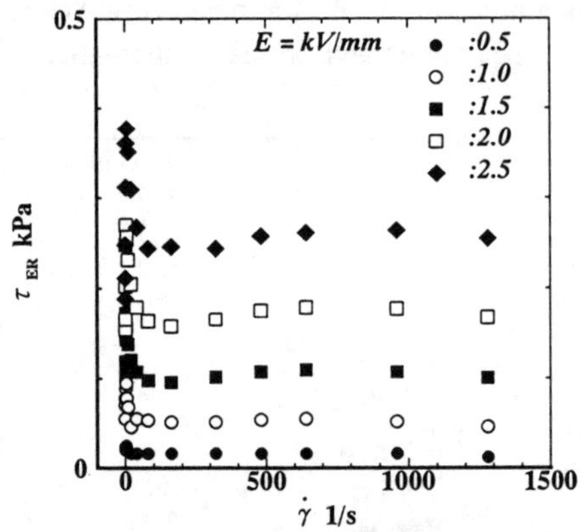

Fig. 10 Yield stress τ_{ER} against shear rate $\dot{\gamma}$ in terms of electric field strength E

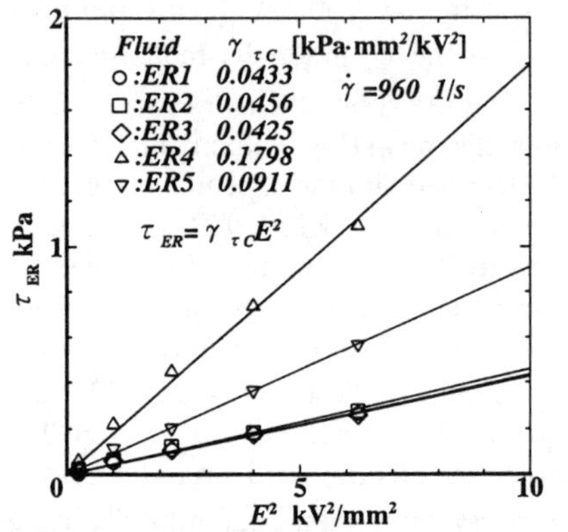

Fig. 11 Relationship between yield stress τ_{ER} at $\dot{\gamma}$=960 s^{-1} and the square of electric field strength E

4. Transient Pressure Response

The response time of the ERF samples in the ER valve illustrated in Fig. 1 was measured by the use of the piston system with a cylinder of 20 mm in inner diameter. Figure 12 shows the representative results of the transient pressure responses for ER-1 and ER-4 against the step input voltage. As the responses look as thought they have some second order character, the system phenomena related to the system parameters such as the

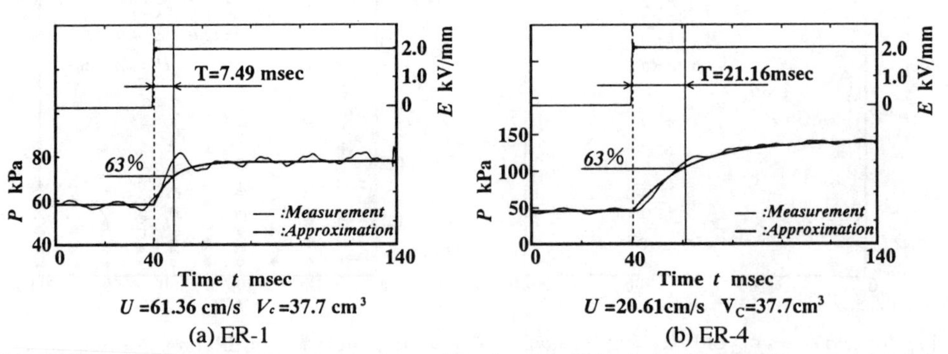

Fig. 12 Transient pressure responses to step voltage input to ER valve, which can be approximated to be first order, and the measured time constant at volume V_C of ER fluid in cylinder

compressibility of the ER fluid within a cylinder, a fluid inertia and a flow resistance within the ER valve may contribute to the measured transient pressure responses as described in our preceding paper[6]. In the almost all cases of this experiments, however, the second order character seems not to be remarkable, so that the transient pressure responses are approximated by first order responses to compare the response time of ERF samples easily. The responses can be approximated by a first order response after smoothing the experimental responses to remove noisy components.

Table 2 shows the time constants of the ERF samples. The time constant depended on the flow velocity, the electric field strength and the composition of the ERF samples. The measured time constants ranged from 5 to 40 msec. The tendency that the time constant increased with the increase of ER effect was observed.

Table 2. Measured Time Constant T [msec] of Transient Response of Tested ERF Samples.

ERF	U=20.61 cm/s		U=61.36 cm/s	
Sample	E=2.0 kV/mm	E=4.0 kV/mm	E=2.0 kV/mm	E=4.0 kV/mm
ER-1	6.66	7.17	7.49	9.54
ER-2	6.96	7.52	13.63	11.09
ER-3	6.92	12.93	5.90	9.63
ER-4	21.16	35.43	11.04	12.65
ER-5	19.97	40.43	10.41	13.73

5. Conclusion

The steady response of the ERF samples differed between the Poiseuille flow geometry and the Couette flow geometry. The flow geometry to evaluate an ERF should be selected according to that of the ER device to be developed. The ER effect in the Poiseuille flow geometry under DC electric fields was distinct from that under AC electric fields. On the other hand, The ER effect in the Couette flow geometry was hardly affected by the type of applied electric field.

6. References

1. D. L. Klass and T. W. Martinek, *Electroviscous fluids. I: Rheological properties, J. Appl. Phys.*, **38-1**(1967), 67.
2. H. Uejima, *Dielectric mechanism and rheological properties of electro-fluids, Jpn. J. Appl. Phys.*, 11-3 (1972), 319.
3. N. Sugimoto, *Winslow effect in ionic exchange resin dispersion, Bull. of JSME*, 20 (1977), 1476.
4. H. Block and J. P. Kelly, *Electro-rheology, J. Phys. D: Appl. Phys.*, 21 (1988), 1661.
5. A. R. Johnson, et. al., *Testing on a high speed electro-rheological clutch, Proc. 3rd Int. Conf. on ER fluids*, ed. R. Tao (World Scientific Publishing, 1992), 424.
6. M. Nakano and T. Yonekawa, *Pressure responses of ER fluid in a piston cylinder-ER valve system, Proc. 4th Int. Conf. on ER fluids*, ed. R. Tao & G. D. Roy (World Scientific Publishing, 1994), 477.
7. Y. Asako, S. Ono, R. Aizawa and T. Kawakami, *Properties of electrorheological fluids containing numerously sulfonated polymer particles, Polymer Preprints*, 35-2 (1994), 352.

Magnetic Field-Induced Phase Transitions in Ferrofluid Emulsion

Yun Zhu, M. L. Ivey, P. Sheaffer, J. Pousset, and Jing Liu
Department of Physics and Astronomy,
California State University, Long Beach, CA 90840

Abstract

Two field-induced phase transition regimes are found in a ferrofluid emulsion system, which are defined by three critical fields, $H_{c1} < H_{c2} < H_{c3}$. The first regime, which starts at H_{c1} and finishes at H_{c2}, is the transition from a gas to induced Nematic liquid crystal phase. In the second regime, or $H_{c2} < H < H_{c3}$, induced Nematic liquid crystal transits to columnar solid structure. A phase diagram is drawn and is explained well by a scaling calculation, which gives $H_{c1} \sim \Phi^{-1/2}$, $H_{c2} \sim \Phi^{-1/4}$, and $H_{c3} \sim \Phi^{\gamma} \exp\left(\pi G/\Phi^{\gamma/2}\right)$, where Φ is the volume fraction, $\gamma = 0.172$ and G=1.44.

The phase behavior in colloidal fluids is determined by the nature of the interaction between the colloidal particles and the particle volume fraction. The colloidal fluids characterized by isotropic interactions, such as hard-sphere and Coulomb interactions, have been understood very well[1]. However, much less understanding exists for colloidal fluids with anisotropic interactions such as dipole-dipole interactions. Magnetorheological (MR) fluids and electrorheological (ER) fluid are the two typical dipolar colloidal fluids which have attracted considerable attention recently [2,3].

The few studies dealing with the phase behavior of dipolar colloids have reported different results. Theoretically, Sano and Doi claim a "gas-liquid" like phase transition with a single critical field in their study of ferrofluid[4]. Recent numerical simulations by Tao suggest two critical fields for ER fluids, where the transition from the Induced Nematic liquid crystal phase to the columnar solid phase is a first-order phase transition[5]. Experimentally, Lemaire, et al, found one critical field in a polystyrene MR fluid[6], which was a linearly increasing function of the volume fraction Φ. A different measurement for a similar system was done by Hwang, et al, which gave two critical fields, $H_{c1} \sim \Phi^{-1/4}$, $H_{c2} \sim \Phi^{-1/2}$, but no theory could explain these results[7]. Furthermore, the most recent simulation shows that Peierls-Landau's instability promotes the formation of thick columns, and this needs to be confirmed directly by experiment. Thus, a complete picture of the field-induced phase transitions remains elusive in MR fluids as it does in ER fluids. Additional experiments and a good theoretical explanation are needed to clarify these fundamental problems in field-induced phase transitions of dipolar colloidal fluids.

In this letter, we report a systematic study of the magnetic field-induced phase behavior in a ferrofluid emulsion. Our primary results are as follows: (1) There are three critical fields, $H_{c1}<H_{c2}<H_{c3}$. Below H_{c1}, the system is a gas-like phase of randomly dispersed colloidal particles. Between H_{c1} and H_{c2}, the gas phase coexists with an

induced Nematic liquid crystal phase consisting of chains of colloidal droplets with random length and random positions. Between H_{C2} and H_{C3}, three phases coexist: a gas, an induced Nematic liquid crystal, and a columnar liquid phase. Above H_{C3}, all of the colloidal droplets have coalesced together into a solely columnar solid phase. (2) A phase diagram is constructed with our experimental data. A good fitting between our calculations and the phase diagram shows that $H_{C1} \sim \Phi^{-1/2}$, $H_{C2} \sim \Phi^{-1/4}$ and $H_{C3} \sim \Phi^{\gamma} \exp\left(\pi G / \Phi^{\gamma/2}\right)$ with $G = 1.44$ and $\gamma = 0.172$ for our sample. (3) The fluctuation coupling[8] is a dominant interaction in the second phase transition regime ($H_{C2}\, H < H_{C3}$) for driving the chains to coalesce. When $H > H_{C3}$, however, the short-range interaction of the chains, instead of the fluctuation coupling, plays a key role in finishing the induced Nematic liquid crystal $\rightarrow$ columnar solid phase transition.

In our study, light transmission, static light scattering and optical microscopy are used and have been described elsewhere[9]. The sample used is a ferrofluid emulsion consisting of colloidal sized ferrofluid droplets stabilized by surfactant and suspended in water. Ferrofluid droplets consist of small magnetic grains with permanent dipole moments. Brownian motion randomly orients these grains within the colloidal emulsion droplets so that the droplets have no net moment in the absence of applied field. Upon the

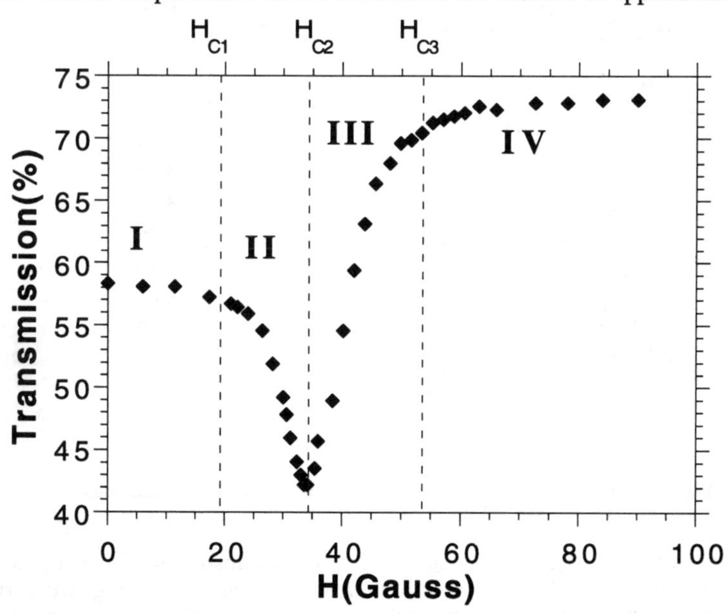

Fig.1. Transmission intensity vs. magnetic field at $\Phi = 0.9\%$, cell thickness $L = 100\mu m$. The critical fields H_{C1}, H_{C2} and H_{C3} are indicated by dashed lines.

application of an external magnetic field, the moments of the grains are preferentially aligned, inducing a large magnetic dipole moment in each droplet. In our experiment, the average diameter of the droplets is 0.3 μm with 35% polydispersity, as measured by dynamic light scattering. The sample is contained in a rectangular capillary cell with 100 μm in thickness. The measurements are taken as the magnetic field is increased with a rate of 0.013 Gauss per second, which is a slow enough for the system to reach an equilibrium state.

A typical light transmission measurement as a function of the magnetic field for Φ=0.9 is shown in Fig. 1. The light wavelength is 632.8nm. The field can be characterized by $\lambda = \beta^2 H^2 a^3 / 4k_B T$ where $\beta = (\mu_p - \mu_f)/(\mu_p + 2\mu_f)$, μ_f and μ_p are the permeability of the solvent and the droplet, respectively, a is the average radius of the droplets. In this case, the transmission intensity remains almost constant for small magnetic fields. At $H_{c1} \approx 18$ Gauss where $\lambda = 0.8$, the transmitted light intensity starts to decrease. The intensity continues to decrease with increasing field until an intensity minimum is reached at $H_{c2} \approx 34$ Gauss, where $\lambda = 2.6$. As the field continues to increase, the transmission intensity rises sharply and saturates after $H_{c3} \approx 53$ Gauss, where $\lambda = 7$. The transmission intensity as a function of the magnetic field is thus divided into four regions by the three critical magnetic fields. The results for various values of Φ are shown in Fig. 2.

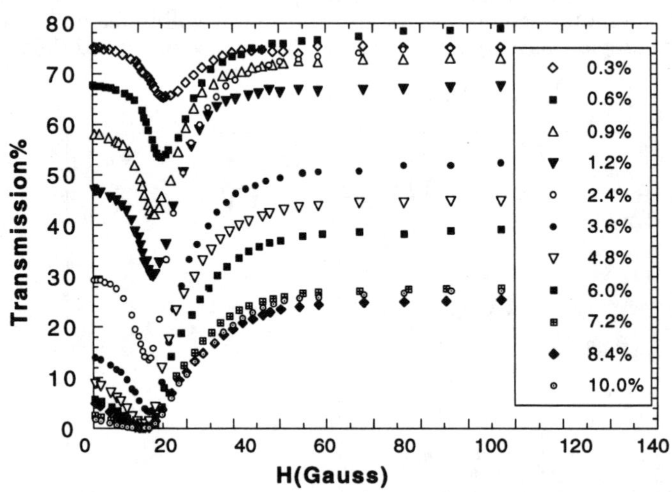

Fig.2. Transmission intensity vs. magnetic field at cell thickness $L=100\mu m$. for various values of Φ

In order to understand the dramatic changes in the transmission intensity at the three critical fields, we measure the small-angle light scattering intensity using two geometries. One geometry is done with the applied magnetic field parallel to the cell surface whereas the incident light is normal to the cell surface. As H increases and passes H_{c1}, the anisotropic scattering intensity pattern is observed to be an ellipsoid that evolves into a bright narrow band perpendicular to both the magnetic field and the incident light. This indicates chaining is parallel to the field direction. By rotating the magnetic coil by 90°, the magnetic field is parallel to the incident laser light. When the applied field reaches H_{c2}, an intense ring is observed that becomes sharper with an increasing field. Both the peak intensity of the ring, I_{max}, and the ring peak position in wave vector space, q_{max} (varies from 0.4 to 2 μm^{-1}) , are measured as functions of the magnetic field. I_{max} begins to increase dramatically around H_{c2}, and saturates at H_{c3}, providing clear evidence of increasing local order.

$D=2\pi/q_{max}$, the separation between columns, can be observed at H_{c2}, and declines very slowly thereafter.

The observation described above indicates that when $H>H_{c1}$, the chains are getting longer with the increasing field. The chains experience the Peierls-Landau's fluctuation. The displacement of a vibrating chain, which is an effective radius of a cylindrical scatterer, is $\propto (La)^{1/2} >> a$, where L is the average length of the chain. For a circular cylinder, Rayleigh-Gans scattering intensity is proportional to both square of length and quadratic of radius[10]. Hence, the decreasing of the transmitted intensity is not just caused by the increasing length of the chains, but also by the displacement of the chain fluctuation. When the field reaches H_{c2}, those completed chains span the cell and have about 300 droplets lined up, acting as long one-dimensional chains. The displacements of vibrating chains are limited due to the finite length of the chains which is restricted by the cell thickness. With further increasing field, the chains coalesce into columns, leading not only to a reduced number of scatters and decreased displacement amplitude, but also to a depletion zone around long chains. This opens up the gap between columns. It is therefore expected that the transmission intensity will increase. The dip in the transmission intensity at H_{c2} is a signal indicating that the coarsening of chains dominates the chaining of droplets. After H_{c3}, the columns are nearly equally spaced and tightened, and no single droplets or short chains are left. The transmission intensity reaches a saturation.

To confirm the intuitive physical picture of structural evolution mentioned above, microscopic images of the phase transitions are obtained by a CCD camera connected to a microscope and recorded on a VCR. Fig. 3 shows the typical structures of a ferrofluid emulsion at the different phases. For lower fields, the droplets are randomly distributed in space, and they move around like individual molecules in gas, as shown in Fig. 3.(a). The chaining commences at H_{c1} and is near completion by H_{c2}. Fig. 3(b) shows such an Nematic liquid crystal crystal-like phase, which we refer to as an induced Nematic liquid crystal phase. Notice that although most of the chains are completed, there are still some short chains, as well as single droplets. The polydispersity in the droplet size is an important contributing factor. In addition, the Peierls-Landau instability frequently breaks the chains which reform again with the itinerant short chains and single droplets around them. Such a coexistence of phases is inevitable even for a monodisperse sample. Because of the long-range attractive interaction between the chains produced directly by dipole fluctuation coupling[8], the induced Nematic liquid crystal phase will transit to columnar liquid and then to columnar solid. Fig. 3(c) shows a phase in the regime of $H_{c2}<H<H_{c3}$, which is a coexistence of induced Nematic liquid crystal and columnar liquid phases. The columnar solid phase of the ferrofluid emulsion is shown in Fig. 3(d). From the top view, a local order can be observed.

The above observations allow us to construct a quantitative phase diagram for the ferrofluid emulsions based on the transmission intensity measurements, which is shown in Fig. 4. In order to understand the mechanism for the phase diagram, we need to understand the dominant interactions in each phase, and the criteria for determining the critical fields.

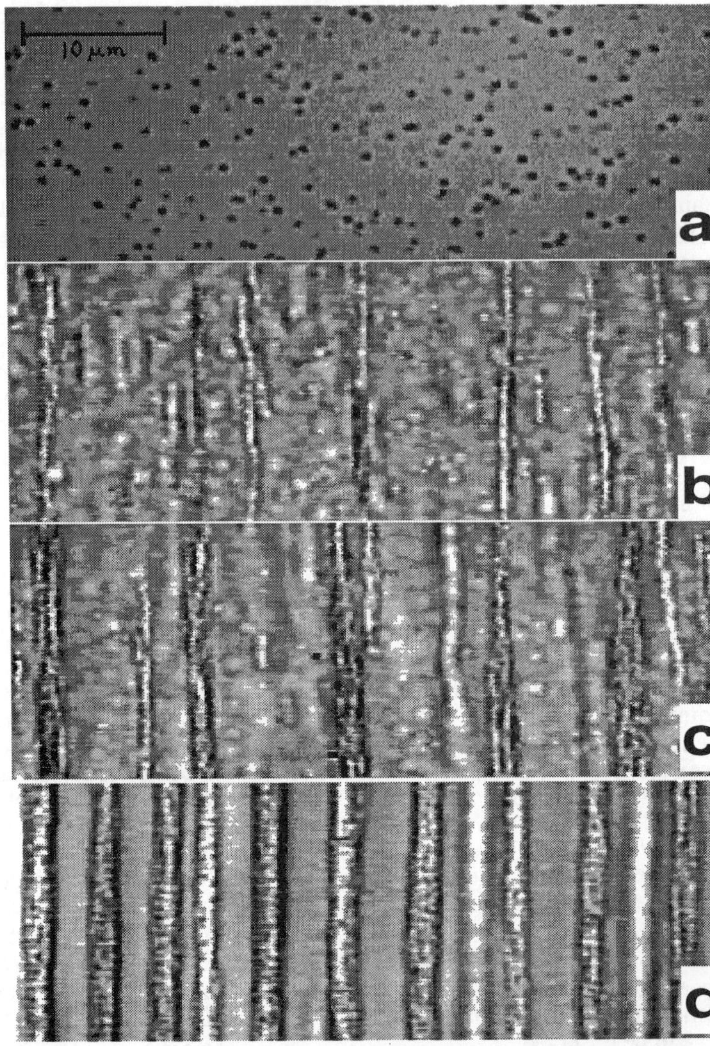

Fig.3. Microscope imaging showing the typical structures at the different phases with $\Phi=0.9\%$, $H_{c1}=18G$, $H_{c2}=33.8G$, and $H_{c3}=54G$. (a) The gas phase at $H=15\ G$. (b) the first coexistence of phases at $H=30\ G$. (c) the second coexistence of phases at $H=42\ G$ and (d) columnar solid phase at $H=60\ G$.

At H_{c1}, the transition from a gas to an induced Nematic liquid crystal phase is easy to understand. When the dipolar interaction strength exceeds the thermal energy, the droplets tend to form chains. Hence, H_{c1} is determined by $(2m^2)/d^3 \approx kT$, where $m=\beta\mu_f a^3 H$ is the dipole moment of each droplet, and d is the average distance between the droplets, $(a/d)^3 \sim \Phi$. Hence, H_{c1} is given by

$$H_{c1} \sim \frac{1}{\sqrt{2\mu_f}}\left(\frac{1}{s^3}\right)\frac{1}{\beta}\left(\frac{kT}{a^3}\right)^{1/2}\Phi^{-1/2}. \qquad (1)$$

Here $s=a^*/a$, a=0.15 μm and a*=0.2 μm, where a^* is the radius of the larger droplets which are the first ones to form chains because of their larger dipole moment. We use μ_f=2.7 and μ_p=1, Eq.(1) qualitatively fits the experimental data as shown in Fig.4.

Within the induced Nematic liquid crystal phase, nuclei will form which will initiate a new phase[11]. The nucleation is caused by the density fluctuation or impurities (in our sample, the bigger size droplets act as "impurities"). In microscopic observation, when the field just passes H_{c2}, we find that some chains are slightly thicker than others, and distributed randomly in space. They grow more rapidly than others with increasing field, and finally become solid columns. Hence, at H_{c2} we can observe these chains which will be nucleate centers. This nucleation mechanism of new phase formation is confirmed by a real-time light scattering study of the structural evolution of ferrofluid emulsions[12]. The nucleate columns capture the individual chains and recruit them for their growth. The basic coupling responsible for the growth process is the interaction between a stable nucleate

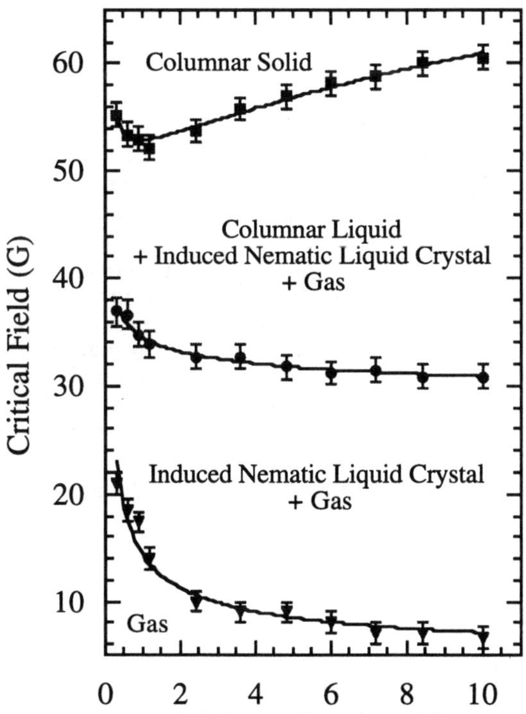

Fig.4. Volume fraction dependence of critical fields and phase diagram for the ferrofluid emulsion. The curves are the calculated results.

column and a single chain. The long-range dipole-dipole interactions are screened, thereby leading to a short-range interaction with an exponential decay $\exp(-\pi\rho/a)$, where ρ is the distance between a chain and its neighboring chain, typically $\rho/a \approx 10$. It is impossible to expect that such a short-range interaction could lead the chains to clump together. Whereas the long-range fluctuation coupling induced by Peierls-Landau instability becomes more important[8]. Assuming that chains in a nucleate column are correlated, the field at a transverse distance ρ from a nucleate column, induced by a fluctuating column, can be calculated approximately as [11]

$$H_f \approx (27\pi/128)^{1/2} N\left[(a)^2 k_B T/\rho^5\right]^{1/2}. \tag{2}$$

Where N is the number of chains in a nucleate column. Since N is proportional to the density of chains in Induced Nematic liquid crystal phase, so that $N \propto \Phi$ approximately[11]. The balance between the fluctuation coupling energy and the thermal energy, $(\beta a^3 H)\sqrt{<H_f^2>} \approx k_B T$, gives $H_{c_2} \sim (128/54\pi)^{1/2}\beta^{-1}(k_B T/a^3)^{1/2}\Phi\langle\rho\rangle^{-5/4}$ where $\langle\rho\rangle$ is a average separation between the nucleate columns and single chains. Since the size of nucleate columns at H_{c2} is small, $\langle\rho\rangle$ is equivalent to the average separation between chains and so $(a/\langle\rho\rangle)^2 \propto \Phi$. We then have

$$H_{c_2} \sim \left(\frac{128}{54\pi}\right)^{1/2} \frac{1}{\beta}\sqrt{\frac{k_B T}{a^3}}\Phi^{-1/4}. \tag{3}$$

As shown in Fig. 4., Eq.(3) qualitatively fits the experimental data very well.

Below H_{c3}, the columnar liquid coexists with the induced Nematic liquid crystal, above H_{c3}, it is the columnar solid phase. The main difference between columnar liquid and columnar solid is that the liquid-like column consists of chains and these chains are still fluctuating, it is like chain cluster; but in the solid column, the droplets have been re-arranged to be close packed such as BCT structure[14] and there is no longer a one-dimensional characteristic. During the transition from columnar liquid to columnar solid, the short-range direct interaction will increase as the chains approach each other, and it will be a dominant interaction for maintaining the body-centered tetragonal (BCT) lattice. The fluctuation of chains is favorable to a columnar liquid but not favorable to a solid BCT lattice. The vibration of chains tends to help the melting of a solid lattice. Hence, we define H_{c3} as a critical field where the liquid-like columns begin to tighten up and to transit to solid columns. H_{c3} can be obtained by balancing repulsive energy due to the fluctuation coupling and attractive energy due to the short-range interaction [11]. That is

$$(\beta a^3 H)\left(\frac{27\pi}{128}\right)^{1/2}\left(\frac{a^2 k_B T}{\rho^5}\right)^{1/2} = \left(\frac{m}{a}\right)^2 \pi^2 \left(\frac{2a}{\rho}\right)^{1/2} \exp(-\pi\rho/a)\cos(z\pi/a)$$

which gives

$$H_{c3} \approx \frac{1}{\beta\pi^2}\sqrt{\frac{k_B T}{a^3}}\exp\left(\pi\frac{\rho}{a}\right)\left(\frac{a}{\rho}\right)^{5/2} \frac{1}{|<\cos(z\pi/a)>|}. \tag{4}$$

Here the z-axis is in the direction of the magnetic field. With assuming a local hexagonal structure of chains in the induced Nematic liquid crystal phase, $(a/\rho)^2 \cong \Phi/(2\pi/3\sqrt{3}) \cong \Phi/(1.21)^2$, but for the columnar solid phase with the BCT structure[14], $a^2/\rho^2 = 1/3 = \Phi^0/3$. Hence, within the regime from Induced Nematic liquid crystal phase to solid column phase, it is reasonable to assume that

$$(a^2/\rho^2) = \Phi^\gamma/G^2, \tag{5}$$

with $1 \geq \gamma \geq 0$ and $1.21 < G < 1.73$. It is worthwhile to notice that γ reflects the sensitivity of volume fraction dependence of averaging distance between chains, and G is a geometrical factor which is related to the structure. Thus H_{c3} can be rewritten as

$$H_{c3} \approx \left(\frac{27\pi}{128}\right)^{1/2} \frac{1}{\beta\pi^2} \sqrt{\frac{kBT}{a^3} \frac{G^{-2}}{|< \cos(\pi z / a) >|}} \Phi^{\gamma} \exp\left(\frac{\pi G}{\Phi^{\gamma/2}}\right). \quad (6)$$

For a monodispersive sample, the ground state of columns gives the droplet position in a chain to be $z=(2j+1)a$, and $\cos(\pi z/a)=-1$. Our sample has 35% dispersity which implies that the chains match each other by 65%. On the average, we have $<\cos(\pi z/a)>=-0.65$. An excellent fit gives $\gamma = 0.172$, $G=1.44$, and $\cos(\pi z/a)= 0.65$, which fall exactly into the range that we expect. We don't need other adjustable parameters to obtain a good fit to the data, and the parameters G and γ have a definite physical meaning. The formula $(a^2 / \rho^2) = \Phi^{\gamma} / G^2$ is a scaling law which tells us how the separation between chains depends on the volume fraction Φ at H_{c3}. $\gamma = 0.172$, which is close to zero, indicates that the volume fraction dependence of separation becomes weak when the columns are becoming solid; $G=1.44$ implies that the structure transition from locally ordered structure of chains to BCT lattice of droplets is on the way. Obviously, H_{c3} is a critical point where the short-range direct interaction starts to take over the fluctuation coupling and to play a key role in the formation of the columnar solid phase. This result is consistent with Tao's results that BCT lattice structure in ER fluid is due to only the short-range direct interaction[14]; however, formation of thick columns would not be BCT structure without Peierls-Landau instability in single chains [5].

Another interesting result is that H_{c3} has the highest singularity and H_{c2} has the lowest as $\Phi \to 0$. This suggests that the columnar solid phase does not exist at very low volume fraction, and there is coexistence of only columnar liquid phase and some short chains.

We would like to thank K. Javier, Dr. J. Bibette, Prof. R. Sheng, G. Hu and Prof. L. Lerner for technical support and Dr. R. Tao, Professor W. Poon, and Dr. T. Hasey for helpful discussion. This work was partially supported by the NSF DMR #9321201, an award from Research Corporation, and CSULB University Research. Acknowledgment is made to the Donors of The Petroleum Research Fund, administered by the ACS, for partial support of this research.

REFERENCES

1. P.N. Pusey and W. van Megan, *Nature (London)* **320**, (1986) 340; N. A. Clark, A. J. Hurd and B. J. Ackerson, *Nature (London)* **281**, (1979) 57 .
2. For recent overview see, Electrorheological Fluids, edited by J. P. Carlson, A. F. Sprecher, and H. Conrad (Technomic Publishing, Lancaster, PA, 1990); M. Fermigier and A. P. Gast, J.of Colloid and Interface Science **54**, (1992) 522.
3. Jing Liu, et al, Phys. Rev. Lett. **74**, (1995) 2828.
4. K. Sano and M. J. Doi, Phys. Sco. Jpn. **52**,(1983)2810.
5. R. Tao, Phys. Rev. B **47**, (1993) 423; R.Tao and Qi Jiang, Phys. Rev. Lett. **73**, (1994) 205.
6. E. Lemaire, Y. Grasselli and G. Bossis, J. Phys. II France **2**, (1992) 359.
7. Y. H. Hwang and X. I. Wu, Phys.Rev.E **49**, (1994) 3102.
8. T. C. Halsey and W. Toor, J. Stat. Phys. **61**, (1990) 1257.

9. Eric M. Lawrence, Mark L. Ivey, George A. Flores and Jing Liu, International Journal of Modern Physics B **8**, (1994) 2765.
10. H. C. van de Hulst, *Light Scattering by Small Particles*, Dover Publications, Inc., 1981.
11. Yun Zhu, Mark Gross, and Jing Liu, to be published; L.D.Landau and E.M.Lifshitz,*Statistical Physics,* 2nd ed. (Addison-wesley), 1969.
12. Jing Liu, Tawei Mou, Yun Zhu, Jackie Pousset, Esmael Haddadian, and Sharlene R. Lim, to be published.
13. Yun Zhu and J. Liu, unpublished.
14. R. Tao and J. M. Sun, Phys. Rev. Lett. **67**, (1991) 398; Tian-jie Chen, R. N. Zitter,and R. Tao, Phys. Rev. Lett. **68**, (1992) 2555.

STRUCTURE-SENSITIVE LIQUID-FLOWING MEDIA BASED ON POLYVINYL ALCOHOL COMPOUNDS WITH CELLULOSE ESTERS

T. G. LAZAREVA and E. V. VASHUK

Institute of General & Inorganic Chemistry, Belarussian Academy of Sciences
9 Surganov St., Minsk 220072, Belarus

ABSTRACT

The class of structure-sensitive liquid-flowing media based on polyvinyl alcohol (PVA) compounds with water-soluble cellulose esters (carboxymethylcellulose, Na-CMC, methylcellulose, MC) have been investigated. PVA may be used as an excellent structure-sensitive matrix for films and gels. Introduction of active additives allows for the formation of liquids whose structures changes under the action of external fields. Introduction of additives, especially of Na-CMC and MC into PVA in the range of certain concentrations leads to the formation of the complexes of PVA with the active molecules linked by hydrogen bonds. Besides, due to the formation of macromolecular associates, a definitely ordered structure of the solution develops. Such changes in the structure of a solution result in abnormal properties, viz. rheological (characteristic flow curves), electrophysical and optical properties in the range of certain concentrations and temperatures. These fluids may be recommended for the formation of electric field-controlled anisotropic films and fibers as well as sensors and measuring elements sensitive to external fields.

1. Introduction

Structure-sensitive fluid media that change their structure under the action of electrical, magnetic and mechanical fields may find wide application for the production of anisotropic films and fibers and formation of gels. Such fluids may be used as different sensors and measuring devices. In this respect we have investigated a class of structure-sensitive fluid compounds based on polyvinyl alcohol (PVA) and water-soluble cellulose esters (Na-carboxyl methylcellulose (Na-CMC), methyl cellulose (MC)) and some others.

2. Experimental Results and Discussion

In investigations, we used PVA with a molecular mass of 40000 and degree of hydrolysis of 98%, Na-CMC with a polymerization degree of 86 and molecular mass of 600, MC with a substitution degree of 35. The compounds were prepared using 1.5 and 10% PVA solutions and 1% solutions of Na-CMC and MC upon mixing certain volumes of the corresponding solutions. Rheological properties were evaluated with the use of a REOTEST-2 rotational viscometer at shear rates (a rotational speed of the internal cylinder of the cell) ranged from 3 to 1312 s^{-1} , temperature from 20 to 90 °C and the solution concentration from 1 to 16 wt%.

As is known[1], for a polymer mixture without ionogenic groups three types of structures may be distinguished in the solution. At concentrations below C_1 the solution consists of a mixture of isolated macromolecules of the both polymers, at concentrations higher than C_2 the solution constitutes a two-phase emulsion, while in the range C_1-C_2 the system represents a single phase, however, it is a mixture of associates but not of isolated molecules. The effect of one polymer on the degree of association of the other polymer, *i.e.* on the short-range order of its macromolecules, causes the structural changes in each polymeric phase of the mixture to occur. This modifies the properties of the phases, in particular, the crystallization temperature, glass-formation temperature in mixtures[2,3], and affects the concentration dependence of the mixture density[4]. Somewhat different results are observed upon mixing of the polymers which can form complexes, *i.e.*, hydrogen-bonded structural elements of the mixed polymers. The complexes may form upon mixing the polymers with and without ionogenic groups, in particular, PVA with polyacrylic and polymethacrylic acids, and CMC in both basic and salt forms. Hydrogen-bonded complexes of different structures may form between the PVA molecules and polyacids. Some of the possible variants of their formation are described in literature[5].

Figure 1,a-c, shows the dependencies of dynamic viscosity on a ratio of polymers in a mixture for compounds based on 1.5 and 10% PVA solutions and the 1% Na-CMC solution.

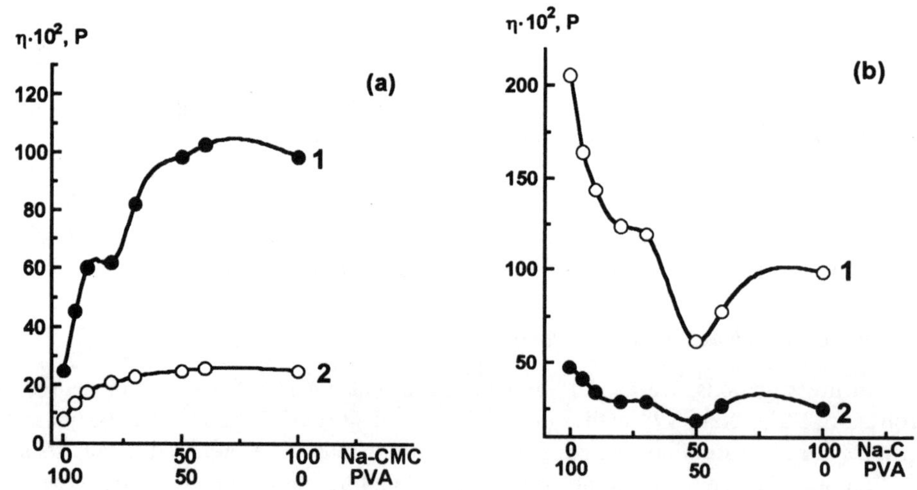

Fig. 1. Dynamic viscosity versus the polymer ratio in the mixture for 5% (a) and 10% (b) PVA solutions and 1% Na-CMC solution at shear rates 16.2 (curve 1) and 81 s^{-1} (curve 2) and temperature 30 (curve 2) and 70 °C (curve 2).

As is seen from Fig. 1, for all PVA concentrations the viscosity displays abnormal behaviour, *i.e.* departs from additivity, which is particularly pronounced at lower temperatures and shear rates. The abnormalities manifest themselves as inflections, maxima and minima in these curves. For the tested compounds, abnormal viscosity has been observed in the concentration ranges 10-20 and 50-60 wt% in the mixtures with Na-CMC. The occurrence of abnormal viscosity owes to the formation of hydrogen-bonded complexes of PVA and Na-CMC, which have different structures5. These complexes, as we have revealed, change electrophysical properties of the compounds. Figure 2,a,b,

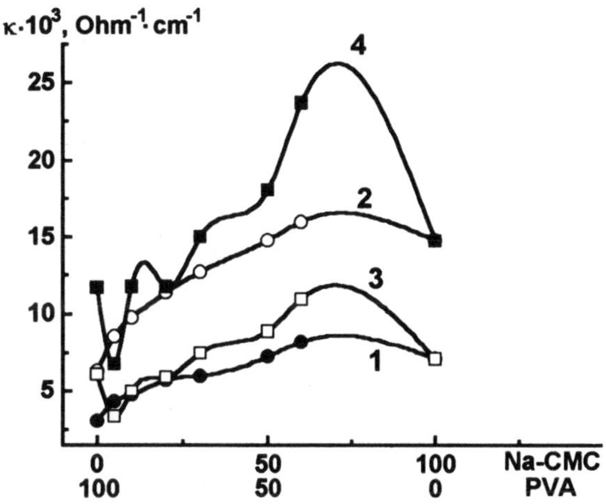

Fig. 2. Electric conductivity versus the polymer ratio in the mixture for 5% (curves 1 and 2) and 10% (curves 3 and 4) PVA solutions and 1% Na-CMC solution at the temperature 30 (curves 1 and 3) and 80 °C (curves 2 and 4).

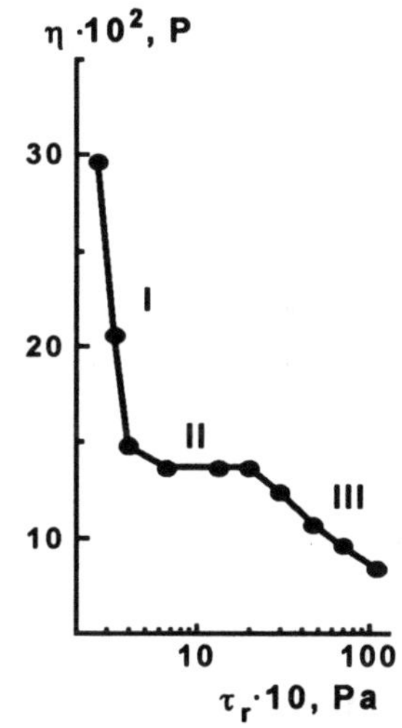

Fig. 3. Dynamic viscosity versus shear stress for composites obtained from 1% PVA and Na-CMC solutions at the temperature 30 °C. The polymer ratio [Na-CMC]:[PVA]=20:80.

shows a dependence of κ and tgδ for the tested compounds on the polymer ratio in the mixture. As in the case of the plot of dynamic viscosity versus composition, this dependence manifests abnormal electrophysical properties.

Figure 3 represents the dynamic viscosity as a function of a shear stress. As is seen, for the compounds based on 1 and 5% PVA solutions these dependencies display distinct inflections, the occurrence of which depends both on the PVA and Na-CMC concentrations in the compound and on temperature. This results may be related with the disintegration of macromolecular associates and the formation of anisotropic structures under the action of mechanical stress. Such abnormalities are typical for liquid-crystalline solutions of polymers and low-molecular liquid crystals[6,7]. Polymeric liquid crystals are characterized by a flow curve consisting of three segments, which reflects the abnormal behaviour of the media of this type[6]. Segment 1 pertains to the viscoplastic behaviour with yield strength (formation of slip planes, disintegration of a liquid-crystalline matrix into flow aggregates, an influence of the shear and thermal prehistory, a domain (polydomain) flow with a decrease of the domain sizes and an increase of the number of domains. Segment 2 represents the flow of oriented aggregates of an unchanged solution, the flow of a disintegrated domain structure in a continuous molecular phase. Section 3 corresponds to the molecular (monodomain) flow.

The above presented results are indicative, in our opinion, of the formation of anisotropic structures in the PVA solutions with Na-CMC under the action of mechanical fields. However, the formation of anisotropic liquid-crystalline structures in the PVA solution with Na-CMC, which is not subjected to mechanical forces, should not be ruled out. The PVA complexes with water-soluble polymers, which form hydrogen bonds, in particular, with Na-CMC, may be analogous, to some extent, to comb-shaped polymers[6]. However, in the case of complexes, the side off-shoots are linked with a main chain by hydrogen, not chemical, bonds.

In this aspect, the PVA-MC composites are, in our opinion, interesting. As is known[8], MC forms transparent viscous solutions possessing high binding, wetting and adhesive abilities and having good surface-active properties. MC macromolecules in the solution are covered with a hydrate envelope, which is hydrogen-bonded to the macromolecules. With a temperature rise, the hydrogen bonds break and macromolecule segments become movable. Then between the macromolecules, which are free of hydrate envelopes, the new bonds develop to form macromolecule associates.

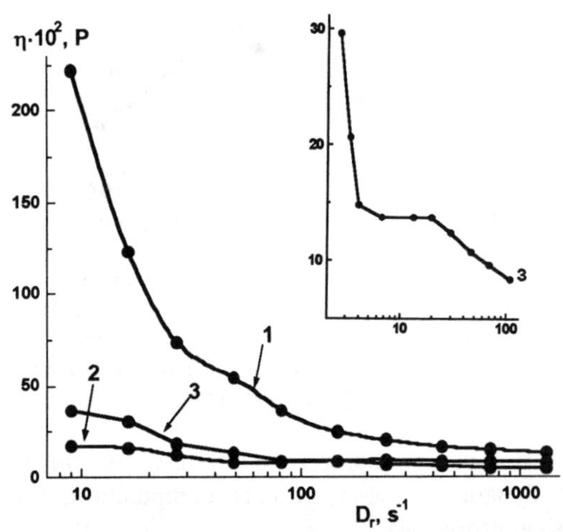

Fig. 4. Dynamic viscosity vs. shear rate at 50 °C for 1% MC solution (curve 1), 5% PVA solution (curve 2) and a compound based on it with the polymer ratio [PVA]: [MC]=61.5:38.5 (curve 3).

As it may be inferred from Fig. 4, for MC, as compared with the PVA solutions, the dependence of viscosity on a shear rate is more substantial. The composites based on 1 and 5% PVA solutions and on 1% MC solution are intermediates.

Figure 5 presents the viscosity of composites versus the MC concentration. For composites based on 5% PVA and 1% MC solutions, an extremal increase of viscosity is observed for the MC concentration of about 20 wt%. Abnormal viscosity is detected at the MC concentration of about 20 wt% at different shear rates (81 and 729 s^{-1}) and at the temperature of 30 and 60 °C (see Fig. 5). At 60 °C the extremum is more pronounced. For composites based on 1% PVA and 1% MC solutions, the viscosity increases with raising the MC content. The increase is particularly pronounced at a concentration of 20 wt%. For composites based on 10% PVA and 1% MC solutions, the viscosity, on the contrary, decreases drastically with increasing the MC concentration, the decline being most substantial below 10-20 wt%. The results shown in Fig. 5,b, may be connected with the formation of hydrogen-bonded associates of PVA and MC in the compound. It might be expected that anisotropic liquid-crystalline structures based on the associates may be formed under the action of a

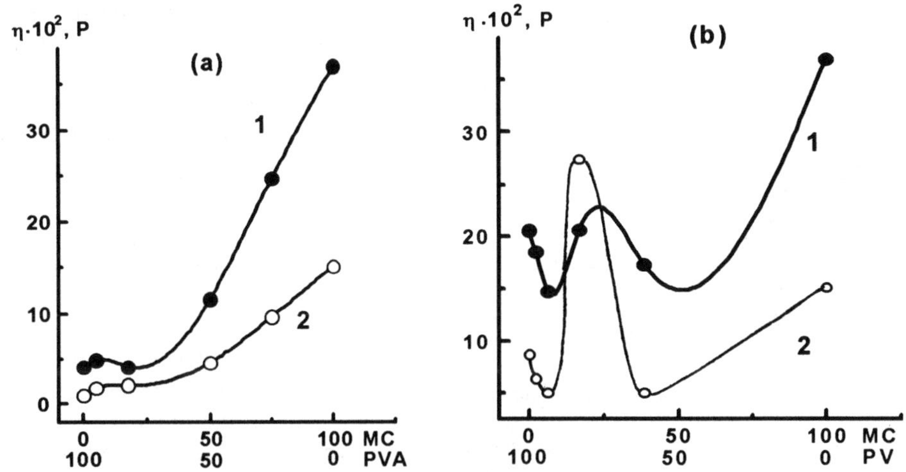

Fig. 5. Dynamic viscosity versus the polymer ratio in a mixture for compounds based on 1% MC solution with 1% (a) and 5% (b) PVA solution at 30 (curve 1) and 60 °C (curve 2), the shear rate being 81 (curve 1) and 729 s^{-1} (curve 2).

mechanical shear field. One of the features pointing to such structures is the presence of extrema in the viscosity versus concentration curve[6].

By Cogwell's definition[9], "liquid crystallinity is a rheological phenomenon, it manifests itself as a result of flow thus leading to a constant anisotropy in a liquid, which determines the properties of a product." According to the molecular theory developed by Doi[10], which provides a unified treatment of the properties of isotropic and anisotropic solutions of rigid-chain polymers, the concentration dependence of viscosity has a maximum in the concentration range that corresponds to the formation of a liquid-crystalline phase. A physical reason for such a phenomenon may be cooperative orientation processes in liquid-crystalline structures, as a consequence of which the total friction coefficient decreases. In this case, the functional dependence of viscosity on concentration depends on both the shear rate and shear stress, as well as on the range of temperature and concentration, in which the viscosity is measured. It is pertinent to note that the extremum on the concentration dependence of viscosity is still more pronounced when a temperature of the solution is higher than 50 °C.

As follows from Fig. 6, the temperature dependence of viscosity for such systems has an extremum (minimum) near 50 °C, above which the viscosity steeply increases. Apparently, the formation of associates is characteristic of the PVA-MC composites, as well as of MC solutions[9], and at a certain ratio of constituent components the formation of anisotropic liquid-crystalline structures is possible under the action of a mechanical field due to orientational transformations of the associates of such a type. These conclusions are supported by the results presented in Fig. 4, where the viscosity vs. shear rate curves display pronounced abnormalities.

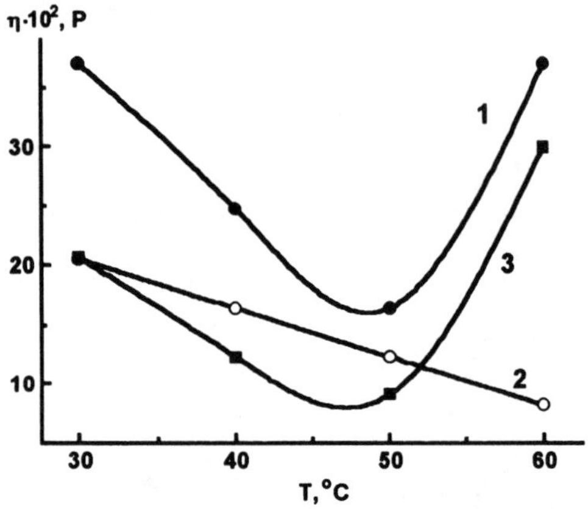

Fig. 6. Dynamic viscosity versus temperature at a shear rate 81 s^{-1} for the 1% MC solution (curve 1), 5% PVA solution (curve 2) and a compound with the ratio [PVA]:[MC]= 83.3:16.7 (curve 3).

The data on electrophysical properties of the tested compounds, which are shown in Fig. 7, reveal changes in the solution structure at the MC concentrations up to 30 wt%.

The dependence of tgδ on the MC concentration in the composite exhibits an inflection recorded at different temperatures. The dependence of κ on the composition of the polymer mixture has an extremum in the range 10 to 20 wt% MC, which is associated with the initial breakdown of the network of hydrogen bonds when MC is introduced in the

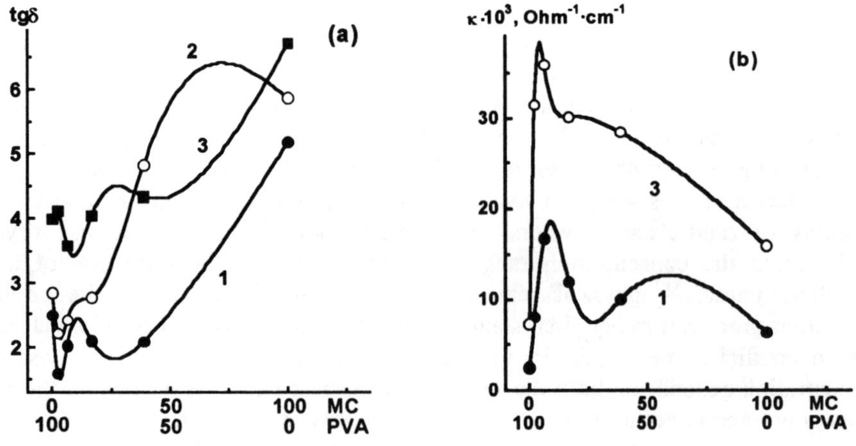

Fig. 7. Tangent of dielectric losses (a) and electric conductivity (b) versus the polymer ratio in mixture for compounds based on 1% MC and 5% PVA solutions at 30, 50 and 80 °C (curves 1-3).

solution, and with the subsequent formation of hydrogen-bonded complexes between the PVA and MC molecules. Moreover, we detected the occurrence of another relaxational transitions in such structures in the temperature range 50 to 70 °C (Fig. 8), which may be related with the formation of anisotropic associates. It should be noted that some information is known[11] on the possible formation of liquid-crystalline structures in polymeric composites, in particular, in those based on hydrogen-bonded ternary amines and some polyacids.

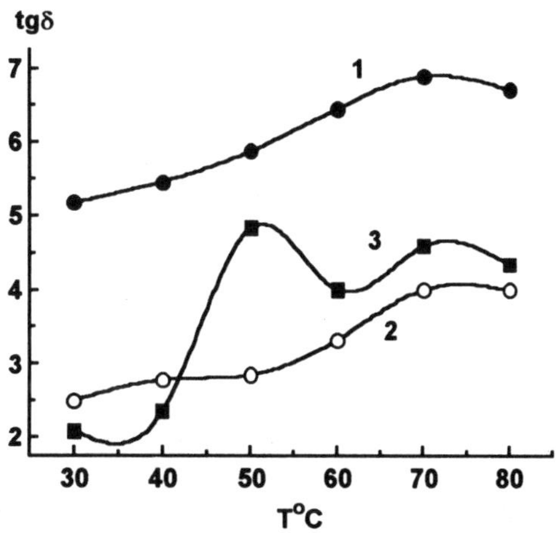

Fig. 8. Tangent of dielectric losses versus temperature for 1% MC (curve 1) and 5% PVA (curve 2) solutions and compounds on their basis with the ratio [PVA]:[MC]= 38.5:61.5 (curve 3).

In conclusion, the possibility is established of the formation, under the action of an orienting mechanical field, of anisotropic structures in the solutions based on compounds of PVA with some cellulose esters (MC and Na-CMC). This is connected with the formation of structure-sensitive associates in the form of complexes in the solution. The assumption is made that anisotropic liquid-crystalline structure may be formed in such solutions in a certain range of concentration without any additional orienting action.

3. References

1. V.N.Kuleznev, *Polymer Mixtures*, (Moscow, 1980).
2. G.V.Vinogradov, Yu.G.Yanovsky, V.N.Kuleznev and T.A.Ivabebko, *Kolloid. Zh.*, **28**, No. 5 (1966) 640.
3. S.Manahe, R.Marakoni and M.Takoynagi, *Jap. J. Polymer. Mater.* **1** (1971) 47.
4. V.N.Kuleznev and K.M.Igosheva, *Vysokomol. Soed.*, **4**, No 12 (1962) 1858.
5. S.K.Chatterjee and A.Malhotra. *Angew. Macromolec. Chem.*, **126** (1984) 153.
6. S.P.Papkov and V.G.Kuligikhin, *The Liquid-Crystalline State of Polymers*, (Moscow, 1977).
7. D.Marsh, J.Pochon and P.Erhardt, *J. Chem. Phys.*, **58** (1973) 5795.
8. N. Baiklse and N.Segal (eds.), *Cellulose and Its Derivatives*, Vol. **2** (1974).
9. F.N.Cogwell, in *Proc. Eur. Sci. Found. 6th Polym. Workshop Liq. Cryst. Polym. Syst.* (London, 1985) 165.
10. M.Doi, *J. Polym. Sci. Polym. Phys. Ed.*, **19** (1981) 229
11. R.N.Tal'roze and N.A.Plate, *Vysokomol. Soed.*, **36**, No 11 (1994) 1766.

THE CONDUCTION MODEL OF ELECTRORHEOLOGICAL EFFECT REVISITED

C. BOISSY, P. ATTEN and J.-N. FOULC
Laboratoire d'Electrostatique et de Matériaux Diélectriques
CNRS et Université Joseph Fourier - Grenoble 1
B.P. 166 – 38042 Grenoble Cedex 9 (France)

ABSTRACT

The conduction model for DC fields predicting the attraction force between slightly conducting spheres in contact is examined in the light of a refined picture of the conduction processes in the liquid phase. Taking into account the saturation of current density in the liquid and the injection of ions occurring at the solid/liquid interfaces, new expressions of the attraction force F between spheres in contact are derived. Predictions are compared with measurements performed using two half spheres of centimetre size immersed in a mineral oil. The case of injection occurring at the solid/liquid interfaces is also considered.

1. Introduction

At the last Conference on ER fluids (1993) a conduction model was presented which aimed at describing the interaction between particles of electrorheological (ER) fluids subjected to DC fields[1]. This model is based on the fact that the field and surface charge distributions for two particles in contact are determined by the conduction properties of both the solid (σ_S) and liquid (σ_L) phases, the passage of electrical current playing a fundamental role. In this model (spherical particles), two zones are distinguished : an external one where the surfaces of the spheres are approximately equipotential and the liquid resistance is higher than the solid resistance, and the contact zone of radius δ where, conversely, the solid has the greatest electrical resistance. By taking into account the enhancement of the conductivity of the liquid due to the field, it was possible to predict the attraction force to vary quasi linearly with the applied field E_0 for high enough values of E_0. The predictions concerning the variation laws as well as the order of magnitude of the force were succesfully tested on a scaled-up experiment[1,2] for values of the conductivity parameter $\Gamma = \sigma_S/\sigma_L$ up to about 10^3.

In the course of experiments examining the behaviour of the "model system" of two half-spheres for high Γ values (ranging from 10^3 to a few 10^5), it appeared that the assumption concerning the conduction current in the liquid was no longer valid. The same conclusion also holds in the case of ER fluids because of the very limited amount of dissociable species lying in the very small volumes between the interfaces of particles in contact. In both cases the conduction is non ohmic. The current density is proportional to the liquid thickness and also depends on the field. On another hand, the proposed conduction model[1,2] which appears to be sound from the electrical point of view fully disregards the physico-chemical processes responsible for the charge transfer at the solid/liquid interfaces. A recent work[3] suggests that electro-chemical processes occurring at the interfaces result in injections of ions into the liquid, which gives rise to a second component in the current density.

In this paper we examine closely how the phenomena of "saturation" of conduction current and of injection modify the previous predictions of the "conduction model". After a short recall on the electrical conduction of dielectric liquids, we derive new expressions of the attraction force F between spheres and compare these new predictions with measurements performed on two half spheres of centimetre size. Finally we consider the case of injections which leads to predicted laws very similar to those of the first conduction model.

2. Recall on bulk conduction in liquids

Two different processes create the ions which give rise to the current in insulating dielectric liquids[4] : i) dissociation of neutral species in the bulk and ii) creation of ions of the polarity of the electrode at the metal/liquid interface (injection). In this section we consider only the bulk conduction due to dissociation of electrolytic species. In the simplest case of a thermodynamic equilibrium between dissociation of a (unique) dissociable species (AB) and recombination of univalent ions A^+ and B^- :

$$(AB) \underset{k_D}{\overset{k_R}{\rightleftarrows}} A^+ + B^-$$

we have the following relation between the concentration c of neutral dissociable species and the charge density n_o of both positive and negative ions :

$$k_D\, c = k_R\, n_o^{\,2} \tag{1}$$

where k_D and $k_R = 2K/\varepsilon_L$ (ε_L : liquid permittivity, K : mobility of both positive and negative ions) are the dissociation and recombination kinetic rate constants respectively. Under a weak electric field E the equilibrium is only slightly perturbed and there is a ohmic behaviour (j : current density) :

$$j_{ohm} = \sigma_L E = 2Kn_oE \tag{2}$$

provided the field does not enhance the dissociation rate. In practice the equilibrium is modified near the electrodes in 2 layers of thickness[5] $\lambda = KE\tau$ (λ is the distance drifted by the ions during the relaxation time $\tau = \varepsilon_L/\sigma_L$). For strong fields such that the transit time $t_T = d^2/KV$ (d : electrode spacing; V : applied voltage) of ions from one electrode to the other is clearly lower than τ, then the system is fully out of thermodynamic equilibrium. After the first transit time, all ions initially present have been swept out and the current only arises from the ions created by dissociation :

$$j_{sat} = k_D cd \tag{3}$$

(this saturation current corresponds to the amount of electric charges created by dissociation in the liquid bulk per time unit). When the dissociation rate is constant, j_{sat} is independent of V and can be expressed in terms of the conductivity σ_L :

$$j_{sat} = \frac{\sigma_L^2 d}{2K\varepsilon_L} = \frac{\sigma_L d}{2K\tau} \tag{4}$$

The saturation field E_{sat} for which the ohmic current is equal to the saturation current is then :

$$E_{sat} = \frac{\sigma_L d}{2K\varepsilon_L} = \frac{d}{2K\tau} \tag{5}$$

We see on Table 1 the order of magnitude of the saturation field E_{sat} for various values of the liquid conductivity σ_L and of the distance d, in the case of mineral oil ($K \cong 10^{-9}$ m^2/Vs, $\varepsilon_L \cong 2\ 10^{-11}$ F/m). It clearly appears that for ER fluids with rather small particles (ϕ < 100 μm), the field between particles is usually much higher than the saturation field and the current is "saturated". The same conclusion holds for spheres of centimetre size when the liquid conductivity is small enough ($\sigma_L < (\sigma_L)_{max} \sim 10^{-12}$ S/m).

Table 1. The saturation field E_{sat} (in V/mm) for different values of the liquid conductivity σ_L and of the distance d.

σ_L (S/m)	d = 1 μm	d = 10 μm	d = 100 μm	d = 1 mm
10^{-13}	0.0025	0.025	0.25	2.5
10^{-12}	0.025	0.25	2.5	25
10^{-11}	0.25	2.5	25	250
10^{-10}	2.5	25	250	2500

In non-polar or weakly polar liquids the electrostatic interaction between ions plays the prime role[6,7] (electrostatic interaction energy on the order of 1 eV). There is an extremely low rate of dissociation of ions pairs into free ions and the electric field favours the dissociation. Onsager[8] gave a treatment of the problem and showed that the field enhanced dissociation is such that :

$$k_D(E) = k_D(0)\ f\left(\frac{E}{E_c}\right) \tag{6}$$

where the function f is expressed in terms of the modified Bessel function I_1 :

$$f(x) = \frac{2\ I_1\left(x^{1/2}\right)}{x^{1/2}} \tag{7}$$

and where the characteristic field E_c is $E_c = \pi\varepsilon_L U^2/e$ with $U = kT/e$, kT being the thermal energy and e the electron charge (for a non-polar liquid with $\varepsilon_L = 2.2\ \varepsilon_o$ we have $E_c \cong 0.25$ kV/mm at room temperature). Taking into account the field effect, the saturation current writes :

$$j_{sat} = \frac{\sigma_L d}{2K\tau} f\left(\frac{E}{E_c}\right) \tag{8}$$

3. Attraction force between spheres in the case of pure conduction

The basic element which makes it possible to estimate the attraction force in the conduction model of ER effect is the contact zone of radius δ. With plausible extra assumptions it is easy to obtain[1,2] :

$$F = \pi\varepsilon_L U^2 \left(\frac{R}{\delta}\right)^2 = 4\pi R^2 \varepsilon_L E_o^2 \left(\frac{R}{\delta}\right)^2 \tag{9}$$

where R is the radius of the spheres, U the potential difference between the spheres centres ($U = 2RE_o$) and E_o is the mean applied field. As the conductance C_S of the two half-spheres is on the order of that of two disks of radius δ, we write :

$$C_S = \alpha\, \sigma_S\, \delta \tag{10}$$

with $\alpha \sim 1$ (the value is 2 for two disks[1]). Eqs 9 and 10 lead to a relation expressing the correlation between the attraction force F and the current I_S flowing through the spheres :

$$F = \alpha^2 \pi\varepsilon_L (\sigma_S R)^2 \frac{U^4}{I_S^2} \tag{11}$$

Note that this relation is independent of the precise law of conduction in the liquid. Predictions for F are derived from the expressions of δ obtained through the condition $C_L = C_S$, C_L being the conductance of the liquid outside the contact zone[1]. Let us consider in this section the case of no injection at the solid/liquid interfaces. The previous derivation of the attraction force giving for high enough fields[1,2] :

$$F = 2\pi R^2 \varepsilon_L E_c E_o \left[\mathrm{Ln}\left(\frac{5\alpha\Gamma}{\pi}\sqrt{\frac{2E_o}{E_c}}\right)\right]^2 \tag{12}$$

is valid if the field E on the liquid is lower than the saturation field E_{sat} everywhere in the outer zone ($x > \delta$). This happens for rather high values of the liquid conductivity but is not really relevant to ER fluids. In the case where, conversely, $E > E_{sat}$ the conductance of the liquid layer between cylinders of radii x and x+dx is :

$$dC_L = \frac{di}{U} = 2\pi\, x\, dx \frac{j_{sat}}{U} = \pi \frac{\sigma_L^2\, y}{K\varepsilon_L U} f\left(\frac{E}{E_c}\right) x\, dx \tag{13}$$

Taking the asymptotic expansion[2]

$$f\left(\frac{E}{E_c}\right) \cong \sqrt{\frac{2}{\pi}} \left(\frac{E_c}{E}\right)^{3/4} \exp\left(\sqrt{\frac{E}{E_c}}\right) \tag{14}$$

for the Onsager function and integrating after a change of variables and a further approximation leads to :

$$C_L \cong \frac{\sqrt{2\pi}\,\sigma_L^2}{K\varepsilon_L E_c}\left(\frac{E_c}{2E_o}\right)^{9/4}\frac{\delta^{13/2}}{R^{9/2}}\exp\left[\sqrt{\frac{2E_o}{E_c}\frac{R}{\delta}}\right] \tag{15}$$

From the balance between C_L and C_S we obtain the following implicit expression for R/δ :

$$\frac{R}{\delta} \cong \sqrt{\frac{E_c}{2E_o}}\ \ln\left[\frac{\alpha\Gamma^2}{\sqrt{2\pi}}\frac{\varepsilon_L}{\sigma_S}\frac{KE_c}{R}\left(\frac{2E_o}{E_c}\right)^{9/4}\left(\frac{R}{\delta}\right)^{11/2}\right] \tag{16}$$

and relation 9 finally gives :

$$F \cong 2\pi R^2\varepsilon_L E_c E_o\left\{\ln\left[\frac{\alpha\Gamma^2}{\sqrt{2\pi}}\frac{\varepsilon_L}{\sigma_S}\frac{KE_c}{R}\left(\frac{2E_o}{E_c}\right)^{9/4}\left(\frac{R}{\delta}\right)^{11/2}\right]\right\}^2 \tag{17}$$

Again a quasi linear dependence of F on the field E_o is obtained. The dependence on Γ^2 of the logarithmic term indicates that the increase of the force should be more marked than predicted by relation 12. Note also that the ratio $F/(2\pi R^2)$ should increase when the radius R is decreased.

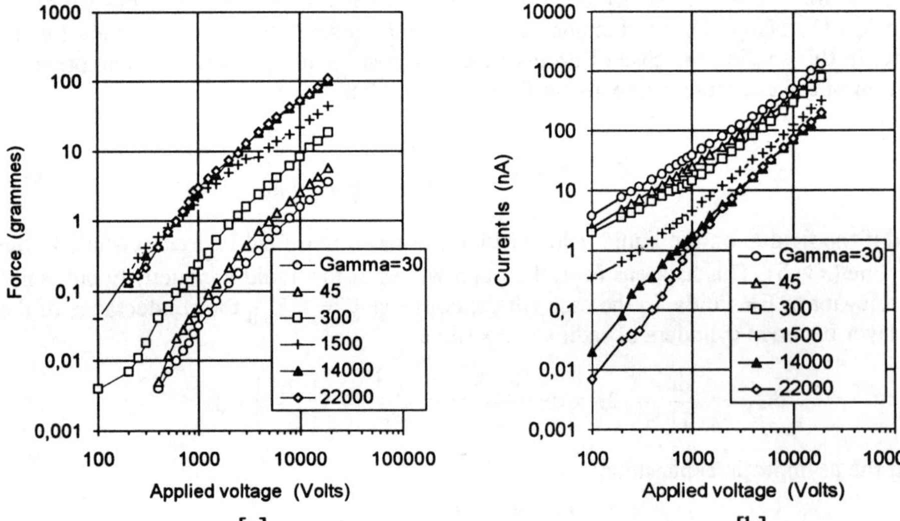

Fig. 1. Two half-spheres (radius R = 7 mm) immersed in mineral oil TF50. Variations versus applied voltage U for different values of Γ of : a) the attraction force F between spheres ; b) the current I_S through the spheres.

4. Experimental test on a scaled-up two-sphere model

In order to test the conduction model for high values of the conductivity parameter Γ, we worked again with the set-up consisting of two half-spheres of polyamid ($\sigma_S \cong 1.7\ 10^{-8}$ S/m, R = 7mm) immersed in a liquid[1]. The attraction force F and the current through the spheres I_S were measured for two liquids – mineral oil Elf TF50 and silicone oil RP 47V50 – the conductivity of which could be augmented by adding an ionic surfactant[1,2]. Typical results are shown on Fig. 1-a where the force is proportional to U^2 at low applied voltages and to U for high enough U values. Moreover F tends to decrease when the liquid conductivity σ_L is increased (and Γ is decreased). The current I_S exhibits first a linear variation with U for Γ values up to about 10^3 but a rather quadratic one for higher Γ values (Fig. 1-b). At high applied voltages there is a tendency towards a U^γ variation ($\gamma \cong 1.5$). For a given U value, I_S increases with σ_L, but not in proportion.

Fig. 2 shows the correlation between the attraction force and the ratio U^2/I_S for the two tested liquids. For Γ lower than a few thousands the different results fairly well define a unique curve with a slope 2 in the doubly logarithmic plot of Fig.2, consistently with Eq. 11. For high Γ values there is a clear departure for the points corresponding to low applied voltages (where $I_S \propto U^2$). This "universal" behaviour is a good test of the validity of the picture which distinguishes the contact and the outer zones. Therefore we can try to extract the coefficient α in formula 10 from the current I_S : $\alpha = C_S/(\sigma_S\delta) = I_S/(U\sigma_S\delta)$. Estimating δ from the measured force by applying Eq. 9, we see on Fig.3 that, to a first approximation, I_S is proportional to δ, which gives $\alpha \cong 0.6$ (for silicone oil we

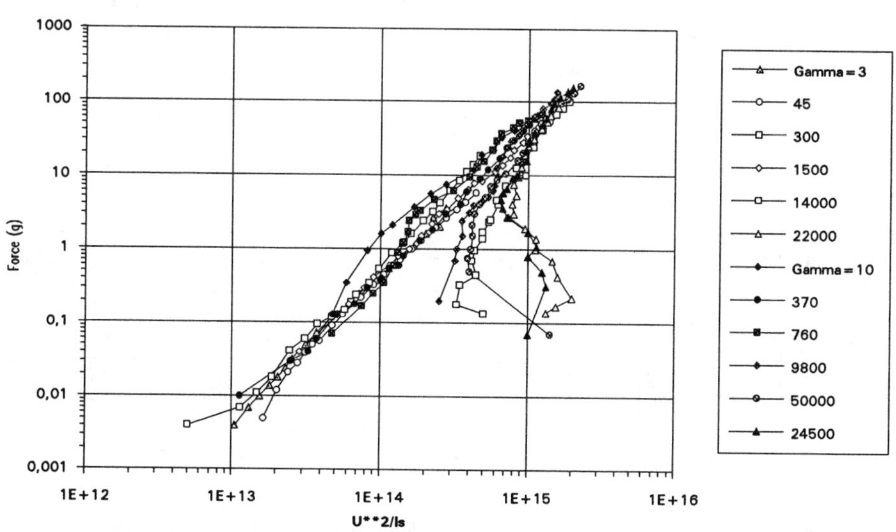

Fig. 2. Variations of the attraction force F between spheres as a function of the ratio U^2/I_S for different values of Γ in mineral oil (open symbols) and in silicone oil (dark symbols).

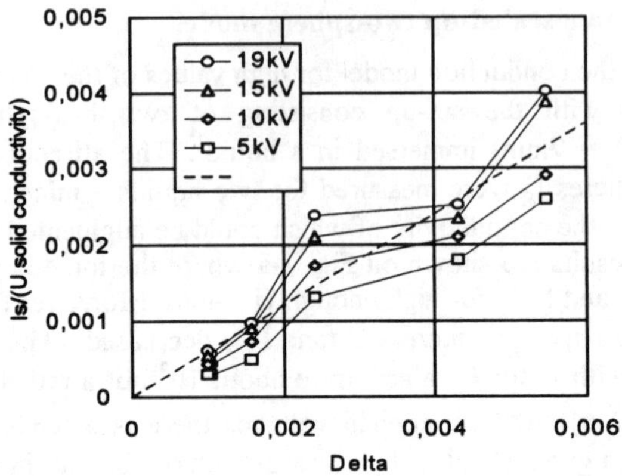

Fig. 3. Variations of the ratio $I_S/U\sigma_S$ versus the radius δ of contact zone determined from the force (through relation 9).

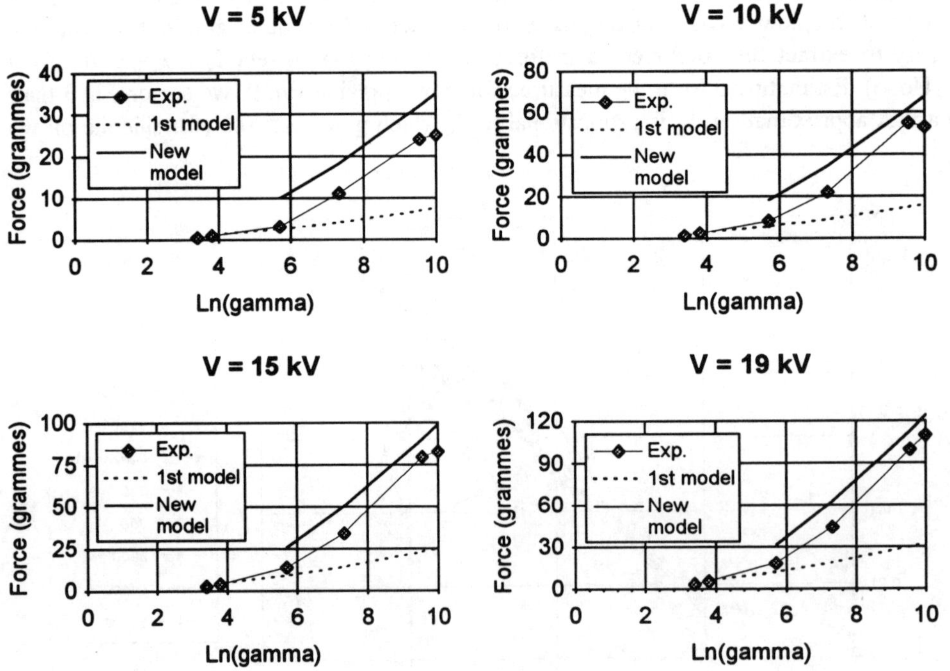

Fig. 4. Comparison of the experimental values of the force and the predictions as a function of the natural logarithm of Γ. Two polyamid spheres in mineral oil TF50.
First model (relation 12) : dotted curves; new model (relation 17) : full line curves.

obtained the value $\alpha \cong 0.65$).

Up to now no hypothesis has been made concerning the conduction mechanism. From the considerations of §3, it appears that the current in the liquid should be always "saturated" in the case of high Γ values and that only the predictions of relation 17 (and 16) might be correct. Fig. 4 shows the variations of the force with Γ along with the two predictions given by formulae 17 and 12. The agreement between the present theory and the measurements for $\Gamma \geq 10^3$ looks surprisingly good if we take into account the various assumptions and approximations (a similar agreement is obtained when working with silicone oil). This implies that the injection phenomena, which are fully ignored in the derivation of § 3, only play a minor role.

5. Attraction force between spheres in the case of dominant injection.

The results obtained on the scale-up two sphere model may be not fully relevant to the case of small particles. Indeed in ER fluids the particles are 3 orders of magnitude smaller than our spheres. At a given field E, the conduction current density given by (8) scales as R whereas the injected current density (see relation 18 below) is independent of R. Therefore for small values of R the injection process may provide the major contribution to the liquid conduction. It is therefore important to examine what is the expected behaviour in this case.

We assume the liquid conductivity to be very small so that the passage of current arises only from injections of ions at the interfaces. We consider a unipolar injection of ions of charge density $q_i(E)$ depending on the electric field E with fully negligible space charge effects. The current density then is :

$$j_i = Kq_i(E)\,E \tag{18}$$

(for an ambipolar injection there would be a second component). We proceed in the usual way to derive the expressions of δ and F. The elementary conductance of the liquid is :

$$dC_L \;=\; 2\pi\,K\,q_i\left(2E_o\,\frac{R^2}{x^2}\right)\,\frac{R}{x}\,dx \tag{19}$$

An explicit law for $q_i(E)$ is required to derive analytical expressions. By extrapolating here the relation valid for a metal/liquid interface[9], we write :

$$q_i(E) = q_{io}\,g\left(\frac{E}{E_c'}\right) \tag{20}$$

By taking a law of the form $g\left(\frac{E}{E_c'}\right) = \exp\left(\sqrt{\frac{E}{E_c'}}\right)$ we obtain after some calculations :

$$\frac{R}{\delta} \cong \sqrt{\frac{E_c'}{2E_o}}\,\ln\left(\frac{\alpha\Gamma_i}{2\pi}\sqrt{\frac{2E_o}{E_c'}}\right) \tag{21}$$

$$F \cong 2\pi R^2 \varepsilon_L E_c' E_o \left[\ln\left(\frac{\alpha \Gamma_i}{2\pi} \sqrt{\frac{2E_o}{E_c'}} \right) \right]^2 \tag{22}$$

Rel. 22 is very similar to rel. 12 with the parameter $\Gamma_i = \sigma_S/(\sigma_L)_{eq} = \sigma_S/(Kq_{io})$ characterising the injection level. If we take another law : $g\left(\frac{E}{E_c''}\right) = \exp\left(\sqrt{\frac{E}{E_c''}}\right)$, we obtain :

$$\left(\frac{R}{\delta}\right)^2 \cong \frac{E_c''}{2E_o} \ln\left(\frac{\alpha \Gamma_i}{2\pi} \sqrt{\frac{2E_o}{\pi E_c''}} \right) \tag{23}$$

$$F \cong 4\pi R^2 \varepsilon_L E_c'' E_o \ln\left(\frac{\alpha \Gamma_i}{2\pi} \sqrt{\frac{2E_o}{\pi E_c''}} \right) \tag{24}$$

The quasi linear dependence of F on E_o as found in relations 12, 17, 22 and 24 does not depend on the detail of the variations of the current density with the field E. It arises from the exponential character of the increase of j with E which, in practice, limits the value of the field in the liquid in the contact zone and near it. For a power law : $g\left(\frac{E}{E_c}\right) = \left(\frac{E}{E_c}\right)^m$, the variations of R/δ and F would be different (power laws) :

$$\frac{R}{\delta} \cong \sqrt{\frac{E_c}{2E_o}} \left[\frac{m\alpha\Gamma_i}{\pi} \sqrt{\frac{2E_o}{E_c}} \right]^{\frac{1}{2m+1}} \tag{25}$$

$$F \cong 2\pi R^2 \varepsilon_L E_c E_o \left[\left(\frac{m\alpha\Gamma_i}{\pi} \right)^2 \frac{2E_o}{E_c} \right]^{\frac{1}{2m+1}} \tag{26}$$

but there would also be a clear departure from the quadratic dependence of the force F on E_o which prevails at low applied fields.

6. Conclusion

In this paper we have extended the conduction model of ER effect under DC field to other situations of high field electrical conduction of the liquid, namely the "saturation" regime of bulk conduction and the injection. The tests on a scaled up two-sphere model confirm the predictions corresponding to saturated bulk conduction. Further work, in particular at even higher Γ values, has to be done in order to investigate the transition from the case where the current in the liquid is determined by the dissociation processes in the bulk to the case where the injection processes at the solid/liquid interfaces control the current level.

Acknowledgements

This work was performed with the joint financial support of the Centre National de la Recherche Scientifique and the company Elf-Aquitaine (under contract CNRS/SNEA No 50.8016). The authors wish to express their thanks to Prof. N. Felici for fruitful discussions.

References

1. P. Atten, J.-N. Foulc and N. Félici, *Int. J. Mod. Phys. B*, **8**, (1994) 2731.
2. J.-N. Foulc, P. Atten and N. Félici, *J. Electrostatics*, **33**, (1994) 103.
3. P. Atten, J.-N. Foulc and H. Benqassmi, in "*Progress in Electrorheology*", Plenum, 1995 (in press).
4. N. Felici, *Dir. Cur.*, **2**, (1972) 90.
5. J.J. Thomson and G.P. Thomson, "*Conduction of electricity through gases*", Cambridge Univ. Press, London, 1928.
6. N. Bjerrum, *Kgl. Danske Vid. Selskab, Math - fys medd.*, **7**, (1926) 9.
7. R.M. Fuoss, *J. Amer. Chem. Soc.*, **82**, (1960) 1013.
8. L. Onsager, *J. Chem. Phys.*, **2**, (1934) 599.
9. D.F. Blossey, *Phys. Rev. B*, **9**, (1974) 5183.

COMPARISON OF FREQUENCY CHARACTERISTICS IN A DAMPER USING MAGNETIC FLUID, ER FLUID DISPERSING SMECTITE, AND MIXED ER MAGNETIC FLUID

TOYOHISA FUJITA*, YASUHISA WADA*, GORO OBINATA*,
YOICHI AKAGAMI**, SHINZO NISHIMURA** AND YUJI OGASAWARA**

* Mining College, Akita University, 1-1 Tegata Gakuencho, Akita 010, Japan
**Akita Prefectural Industrial Technology Center, Akita 010, Japan

ABSTRACT

This paper presents a new type of damper using electro-rheological (ER) magnetic fluid for improving magnetic fluid damper and ER fluid damper. Silicone oil based magnetic fluid dispersing 10 nm size of magnetite, silicone oil based ER fluid dispersing 2 nm to 50 nm size of smectite and a mixture, that is, ER magnetic fluid have been used in a piston type damper. Dynamic visco-elasticities of these fluids have been measured by considering chain (cluster) formations. Next, frequency characteristics of transmissibility (magnification factor) and phase in damper have been compared with those fluids under magnetic field, electric field and both fields. The Voigt or Maxwell model can be applied in those dampers depending on amplitude. ER magnetic fluid has showed the decrease of amplitude ratios at resonance point and higher frequency at any amplitude under both magnetic and electric fields as low-pass filter.

1. Introduction

Recently various kinds of ER fluids whose apparent viscosities increase under electric field have been prepared[1]. Also, several solvent based magnetic fluids have been prepared[2]. If the fluid can be controlled by both magnetic and electric fields, new applications of this ER magnetic fluid can be expected, such as light scattering in a thin film[3]. However, when different solvent based magnetic fluid and ER fluid are mixed, the mixture often coagulates due to the changes of solvation force and steric interaction. In this experiment the mixture of same silicone oil based magnetic fluid and ER fluid have been used to prepare stable ER magnetic fluid.

As one application using magnetic fluid[4] or ER fluid[5], many kinds of damper performances have been reported. However, the relationship between chain formation and damper model has not been clarified well. Here a piston type damper whose inner cylinder is made from permanent magnetic circuit is constructed and the frequency characteristics on transmissibility of damper have been measured and compared by using magnetic fluid, ER fluid and ER magnetic fluid.

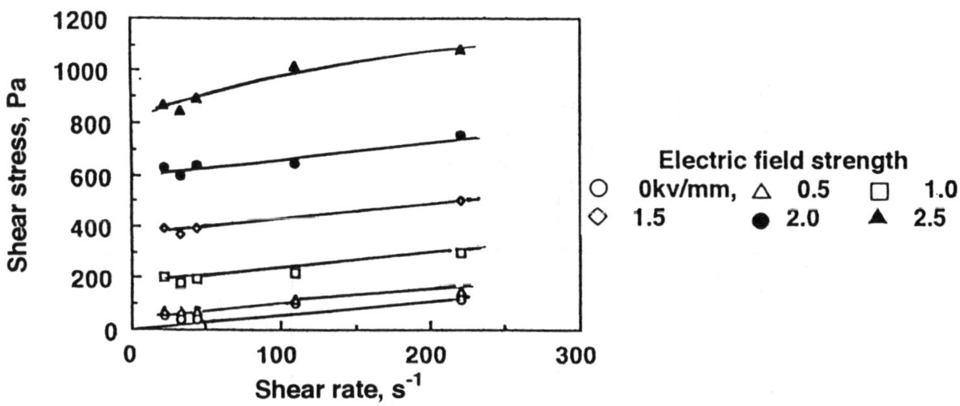

Fig.1 Effect of shear rate on shear stress of ER fluid containing 20wt% smectite under 50 Hz AC field.

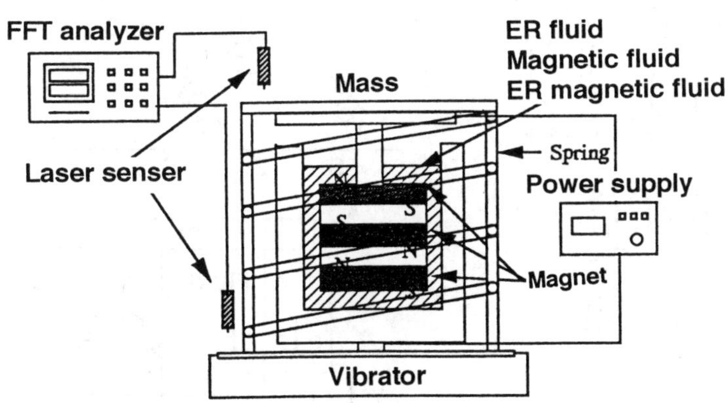

Fig.2 Experimental damper apparatus.
(Distance between cylinders 1mm, Spring constant 1.47 kN/m)

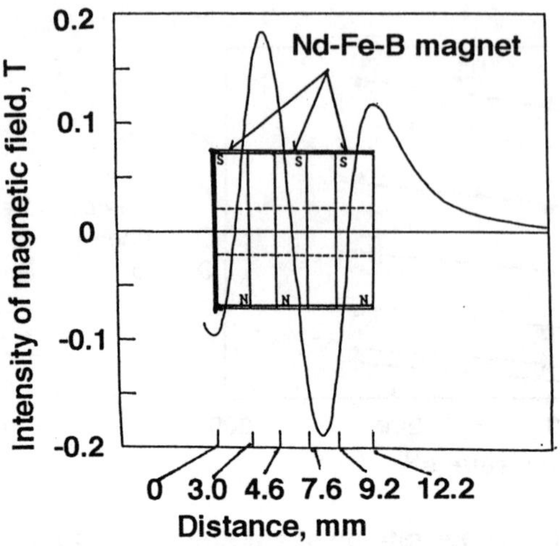

Fig.3 Magnetic field distribution on the surface of inner cylinder.

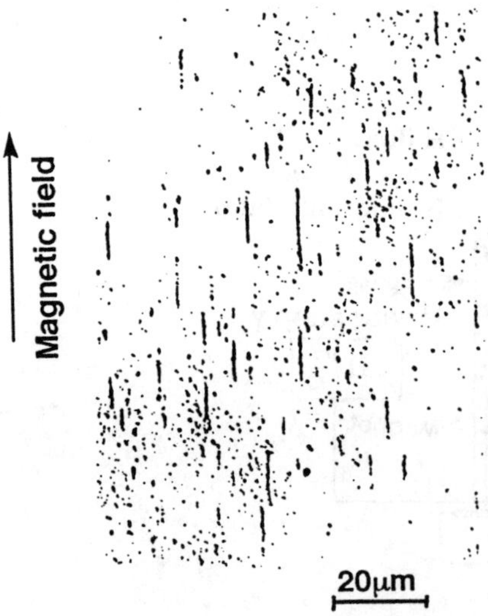

Fig.4 Magnetic clusters in magnetic fluid thin film under magnetic field by optical microscope observation.

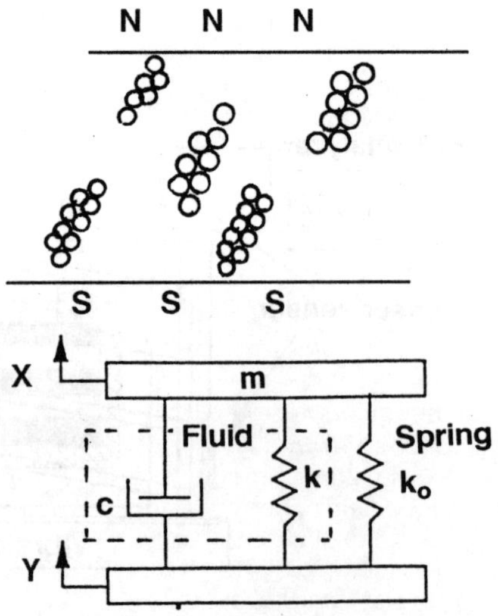

Fig.5 Concept of magnetic clusters in magnetic fluid and damper model using magnetic fluid.

2. Experiment

2.1 Fluids

A very stable water-free silicone oil based magnetic fluid contains 10 nm size of magnetite coated with surfactant in methyl phenyl silicone oil. The magnetization of fluid 0.027 T at 0.8 MA/m is used in a damper. The magnetic fluid viscosity increases about 4 times under magnetic field[6].

A water-free silicone oil based ER fluid dispersing 2 nm to 50 nm size of smectite (ion exchanged by organic ion) in methyl phenyl silicone oil does not show sedimentation at gravitational field[7]. The shear stress versus shear rate of the ER fluid under electric field measured by cylindrical rotation viscometer is shown in Fig.1. More than 1 kPa shear stress is caused at 2.5 kV/mm of AC field (50 Hz). This fluid is also used in a damper.

When above mentioned magnetic fluid and ER fluid are mixed, the magnetic and dielectric particles are dispersed very well in silicone oil. When 15 wt % of magnetite and 15 wt% of smectite are dispersed, the fluid viscosity is about 0.01 Pas under no field and shows 0.7 kPa yield stress at 2.5 kV/mm of AC field (50 Hz). The yield stress can be controlled easily by changing the particle concentration.

2.2 Experimental damper

An experimental damper apparatus is shown in Fig.2 The outer cylinder is made of brass and the spring connected with an inner cylinder are fixed on the vibrator. The vibrator is oscillated at the amplitude of 0.025 mm to 0.25 mm at a frequency of 1 Hz to 50Hz. The spring constant is 1.5 kN/m. A relative movement between the vibrator and the mass is measured by two laser sensors connected with FFT analyzer. The gap between inner cylinder and outer cylinder is 1 mm. AC electric field (50 Hz) is supplied between high voltage inner cylinder and ground outer cylinder. The inner cylinder is insulated with plastic plate from spring. The magnetic field distribution on the surface of inner cylinder is shown in Fig.3. Maximum intensity of magnetic field on the surface is about 0.2 T. The inner cylinder is 12 mm in diameter and 12.2 mm in length. Ne-Fe-B permanent magnet of 3 mm in length is set with facing same poles between brass spacer of 1.6 mm in length. The outside of inner cylinder is covered with brass of 0.1 mm thickness to supply uniform electric field. The hole of 4 mm diameter in the center of inner cylinder and mass is opened to decrease the fluid flow disturbance in the gap between cylinders. When magnetic fluid or ER magnetic fluid fills the cylinder gap, the inner cylinder is self- centered by magnetic force.

3. Experimental results and discussion

3.1 Damper using magnetic fluid

Magnetite chains of a few μm to 20 μm length in the direction of the magnetic field can be observed in a thin film (20 μm thickness) of magnetic fluid by the optical microscope as shown in Fig.4. If magnetic fluid is used in the experimental damper, the large relaxation time is needed to arrange the magnetite chain in the direction of magnetic field. Therefore, the Voigt model is considered in the damper model shown in Fig.5. At small strain, frequency dependence of dynamic modulus G' and dynamic loss G" of silicone oil based magnetic fluid is shown in Fig.6. G' and G" increased by applying a magnetic field. At the high frequency G' decreases, while G" is almost constant. When outer spring constant is k_o, fluid's viscous-damping constant is c and fluid's spring constant is k, the transfer function $G_1(s)$ is expressed as follows,

$$G_1(s) = X(s)/Y(s) = (cs + k + k_o)/(ms^2 + cs + k + k_o) \qquad (1)$$

where $s= j\omega$, ω is angular velocity.

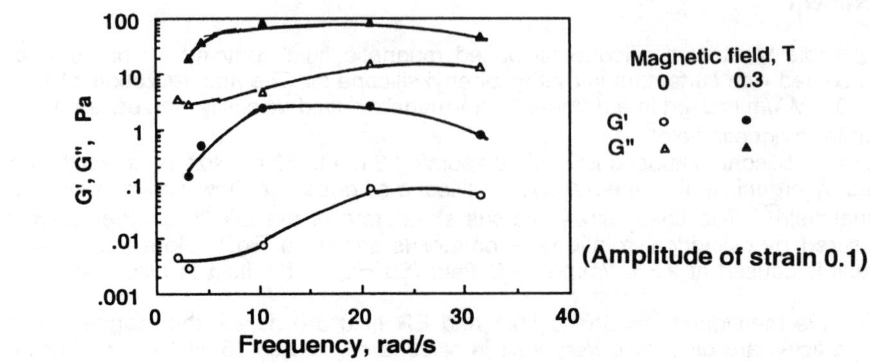

Fig.6 Frequency dependence of dynamic modulus G' and dynamic loss G" of silicone oil based magnetic fluid.

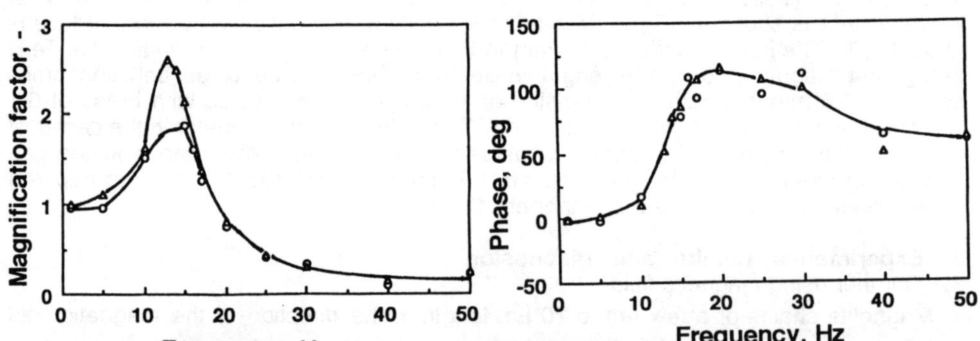

Fig.7 Effect of frequency on magnification factor and phase of damper using magnetic fluid. (Magnetization of fluid 0.03T)

Amplitude

—o— ± 0.025 mm

—Δ— ± 0.050

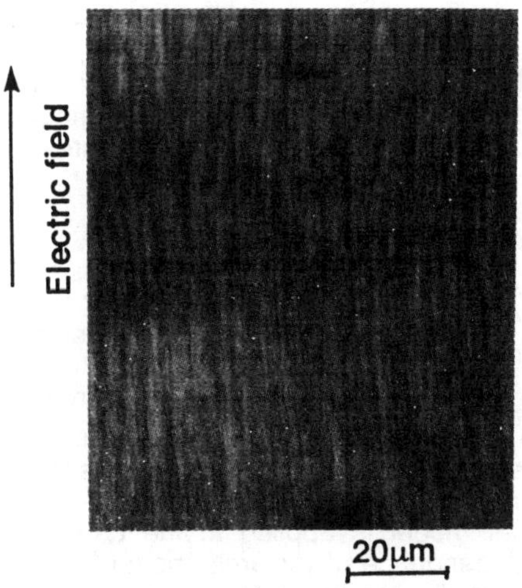

Fig.8 Dielectric particle clusters in ER fluid thin film under electric field by optical microscope observation.

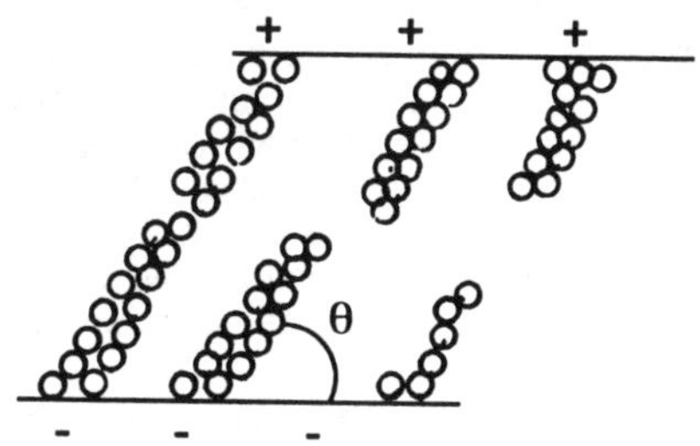

$$F = -\frac{3P^2}{4\pi\varepsilon r^4}(3\cos^2\theta - 1)$$

$\theta < 54.7$ attraction

P: Intensity of polarization

r : Distance between particle centers

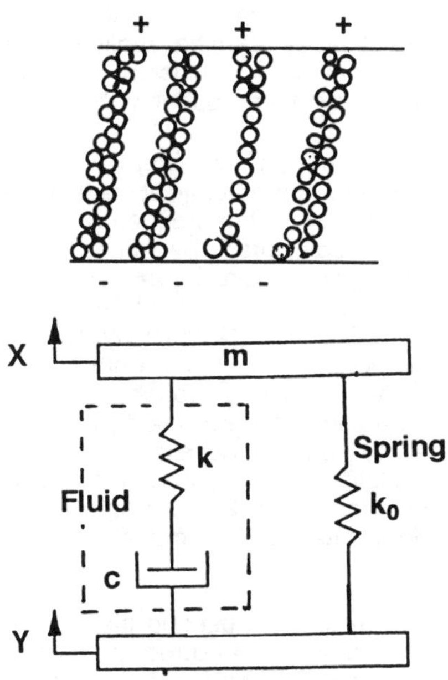

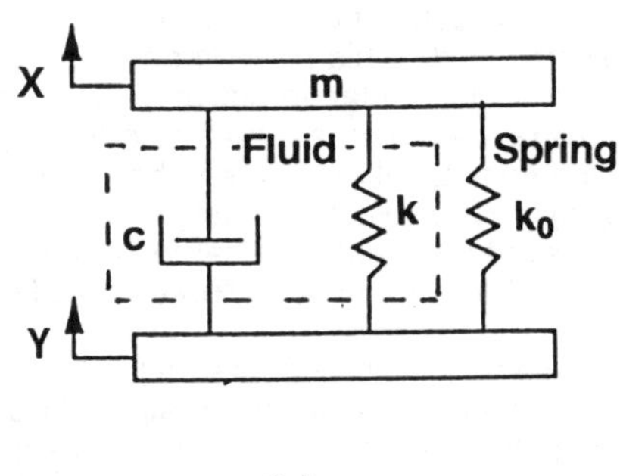

Fig.9 Concepts of dielectric particle clusters at small strain or displacement and damper model (a) and clusters at large strain or displacement and damper model (b).

The vibration transmissibility of the damper using very well dispersed magnetic fluid is shown in Fig.7. The resonance frequency does not change at different amplitudes, however, the magnification factor at resonance point increases as the amplitude is increased. At higher frequency magnification factor decreases to near zero. If magnetic particles are coagulated near the magnet surface, the resonance point changes to larger frequency as the decrease of amplitude.

3.2 Damper using ER fluid

As the smectite particles in silicone oil are colorless and transparent, almost same diameter (10 nm) particles of barium titanate are used to observe particle chains by applying an electric field. The barium titanate chain connecting between electrodes in the direction of electric field can be observed in a thin film of 20 μm thickness as shown in Fig.8. When ER fluid is used in the damper, the connected particle chains do not break under small displacements. Therefore, at the high frequency the elastic modulus effect becomes large and the Maxwell model is applied in the damper as shown in Fig.9 (a). At small strain, frequency dependence of dynamic modulus G' and dynamic loss G" of silicone oil based ER fluid is shown in Fig.10 When frequency increased, G" decreases while G' is almost constant. This relationship between G' and G" is opposite to the case of magnetic fluid. The transfer function $G_2(s)$ is expressed as follows,

$$G_2(s) = X(s)/Y(s) = [(k_o + k)cs + k_o k]/[mcs^3 + mks^2 + (k_o + k)cs + k_o k] \qquad (2)$$

On the other hand, when the displacement or disturbance of fluid is very large, particle chains will break at the moment like magnetic fluid chains. At the higher frequency, the viscosity effect increases, therefore, Voigt model can be applied to the damper as shown in Fig.9 (b). This model is qualitatively consistent with the observation that the elastic stress of ER fluid under an electric field suddenly decreases at a large strain[8].

Effect of frequency and amplitude on magnification factor of damper using ER fluid is shown in Fig.11. At the small displacement of ± 0.025 mm, the magnification factor at resonance point decreases to 1 and increases to 1 at the higher frequency as the applied electric field strength increases. The curves calculated by equation (1) expressed in the Maxwell model almost agreed with the observed values. In the calculation, spring constant k increased 4 and 8 times at 0.5 and 1.0 kV/mm, while damping constant c increased only 2 and 2.5 times at 0.5 and 1.0 kV/mm by applying electric field. This phenomenon is describable by Maxwell model. On the other hand, at a large displacement of ± 0.25 mm, the magnification factor approaches 1 at the resonance point as the electric field increases; however, it decreased at higher frequency at all field strengths. The curves calculated by equation (2) expressed in the Voigt model were almost same with the observed plots. In the calculation, spring constant k was almost same, while damping constant c increased 2 and 3 times at 0.5 and 1.0 kV/mm by applying electric field. This behavior is fitted to Voigt model.

3.3 Damper using ER magnetic fluid

Silicone oil based magnetic fluid and ER fluid were mixed and used in the damper. Magnification factor of damper using ER magnetic fluid was measured at different amplitude of ± 0.025 mm and 0.75 mm under electric field as shown in Fig.12. At both amplitudes, the magnification factor approaches 1 at resonance point which is similar to ER fluid performance and it decreases at even higher frequency which is similar to magnetic fluid performance.

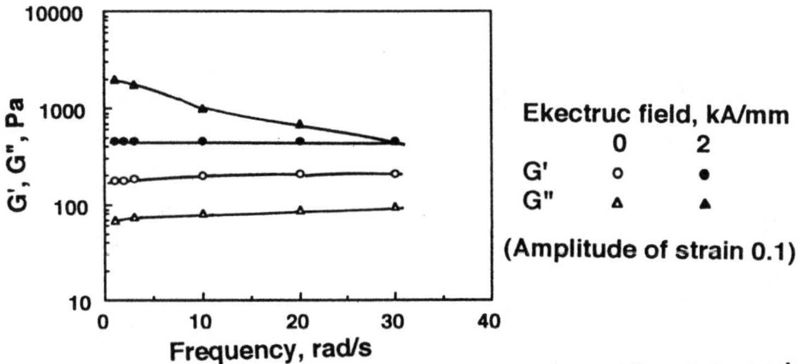

Fig.10 Frequency dependence of dynamic modulus G' and dynamic loss G" of silicone oil based ER fluid.

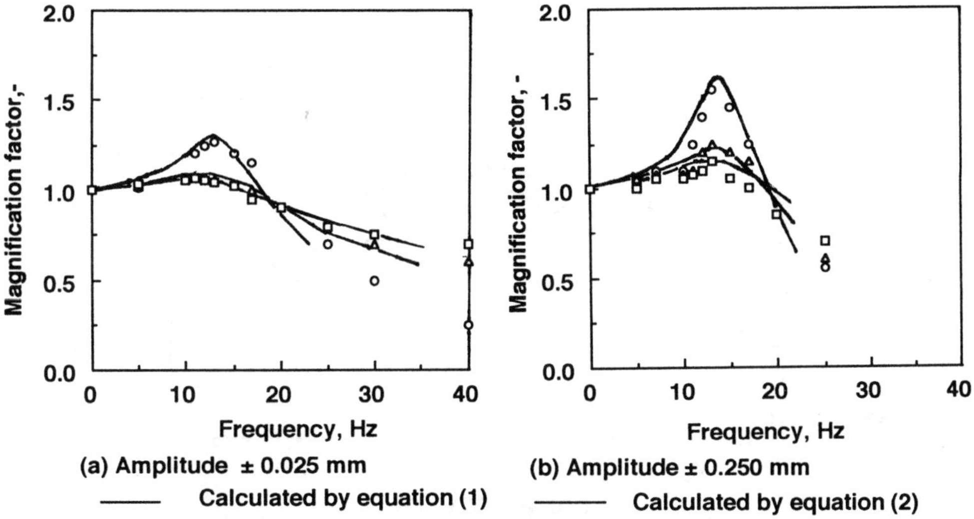

Fig.11 Effect of frequency and amplitude on magnification factor of damper using ER fluid. (Smectite concentration 15%)

Electric field, kV/mm ○ 0, △ 0.5, □ 1.0

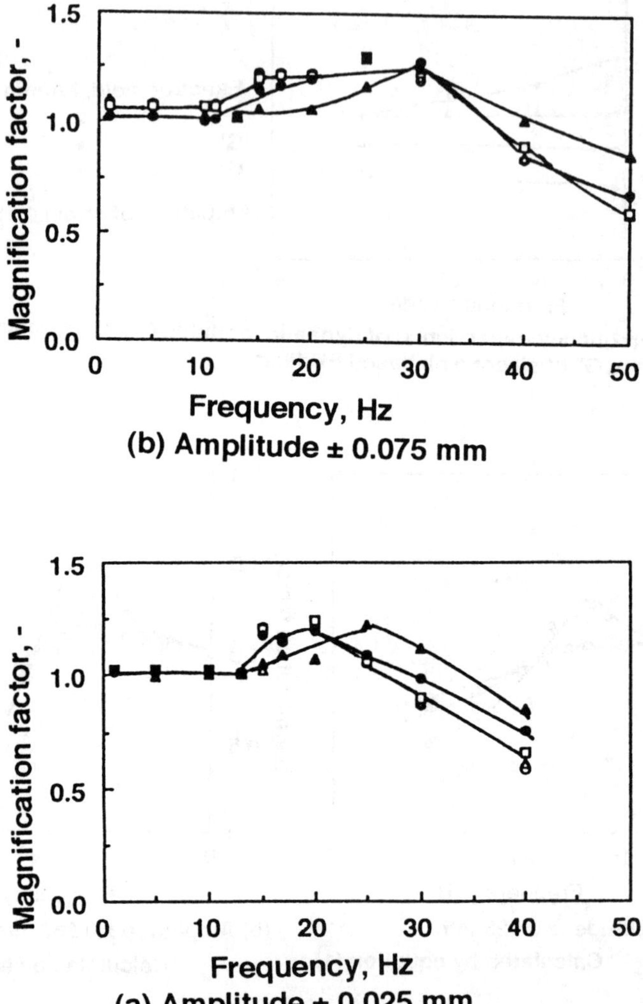

(b) Amplitude ± 0.075 mm

(a) Amplitude ± 0.025 mm

Fig.12 Effect of frequency and amplitude on magnification factor of damper using ER magnetic fluid.
(Smectite concentration 15 wt%, Magnetization of fluid 0.014T)
Electric field , kV/mm ○ 0, △ 0.5, □ 1.0, ● 2.0, ▲ 3.0

4. Conclusion

1. Magnetic fluid damper performance is described by the Voigt model because magnetic chains are not continuous at any amplitude, therefore, the magnification factor decreased at higher frequency. However, the damping factor is not enough at resonance point. When magnetic particles coagulate near the magnet surface, the resonance point changed to higher frequency and its point decreased at larger displacement.

2. As particle chains are connected between electrodes in ER fluid containing smectite under electric fields, the damper performance using ER fluid can be expressed as the Maxwell model. When electric field increased, magnification factor approaches 1 at any frequency. On the other hand, at large displacement particle chains are discontinuous; therefore, the damper performance can be expressed as the Voigt model like magnetic fluid. Even if electric field increased, magnification factor decreased at higher frequency at large displacement. However, at any strain or displacement the magnification factor approaches 1 at resonance point . The observed transmissibilities almost agreed with the calculated curves using these models.

3. When ER magnetic fluid produced by mixing silicone oil based magnetic fluid and ER fluid is used in damper, the magnification factor approaches 1 at resonance point and decreased at higher frequency at any displacement. This damper can be applied as a passive damper behaving like a low-pass filter.

Acknowledgments

The authors wish to thank Shin-Etsu Chemical Co., Ltd. for preparing magnetic circuit and Mr.H.Arima, Akita University, for helping the experiment. Financial support from the Education Ministry of Japan [Grant-in-Aid (05555276)] is gratefully acknowledged.

References

1. For example, H.Block and J.P.Kelly,J.Phys. D: Appl. Phys. **21**(1988)1661.
2. For example, B.Jeyadevan and T.Fujita, J. Soc. Materials Engineering for Resources of Japan, **3**,2(1990)100.
3. T.Fujita and K.Yamaguchi, J. Appl. Phys. **76**,6 (1994)3920.
4. For example, K.Nakatsuka, H.Yokoyama and J.Shimoiizaka, J.Magn. Magn. Mater. ,**65**(1987)359.
5. For example, D.L.Hartsck, R.F.Novak and G.J.Chaundy, J. Rheol. **35**(1991)1305.
6. R.E.Rosensweig, R.Kaiser and G.Miskolczy,J. Colloid and Interface Sci.,**29**,4(1969)680.
7. T.Fujita, H.Kogita, K.Yoshino, H.Saiki and G.Obinata, J. Mining and Materials Processing Inst. Japan, **111**(1995).
8. G.B.Thurston and E.B.Gaertner, J. Rheol. **35**(1991)1327.

ELECTRORHEOLOGICAL EFFECT OF REDUCING THE SUSPENSION VISCOSITY IN LAMINAR FLOWS BELOW THE VISCOSITY OF ANISOTROPIC DISPERSIVE MEDIUM

E.YU.TARAN, YU.V.PRIDATCHENKO
Department of Mechanics and Mathematics, Taras Shevchenko Kiev University, 6, Acad. Glushkov Ave., Kiev, 252127, Ukraine.

I.T.SELEZOV, D.E. TARAN
Wave Processes Department, Institute of Hydromechanics, Nat. Acad. Sci. of Ukraine, 8/4, Zhelyabov Str., Kiev, 252057, Ukraine.

ABSTRACT

Electrorheological features of a dilute suspension in Ericksen anisotropic fluid are analyzed on the basis of rheological constitutive equations obtained in the present paper. It is assumed that suspended particles possess a zero buoyancy and constant dipole moment, the anisotropic dispersive medium of suspension is electrically neutral. As a hydrodynamic model of suspended particles, the uniaxial dumbbell is used. The sizes of suspended particles are assumed to be much more than elements of microstructure which are responsible for anisotropy of dispersive medium.

The rheological equations of suspension under consideration are obtained on the basis of the structure–phenomenological approach. The rheological behaviour of suspension is investigated in a simple shear flow in the presence of external electric field. It is shown that the suspension behaves itself as anisotropic non–Newtonian fluid. The possibility of reducing the effective viscosity of suspension in laminar shear flows below the value of effective viscosity of anisotropic dispersive medium is predicted theoretically. It takes place at certain orientation of suspended particles of suspension under action of the external stationary electric field and the hydrodynamic forces.

The result obtained can be used for studying the suspension in liquid cristals.

On the basis of the structure–phenomenological approach [1,2] the rheological equations of dilute suspensions of rigid uniaxial dumbbells in non–Newtonian isotropic dispersive media [3–10] have been obtained. It enabled to investigate the influence of non–Newtonian properties of dispersive media on a rheological behavior of the suspensions.

In the present paper the structure–phenomenological approach is applied to construct the rheological equations of dilute suspensions of rigid uniaxial dumbbells in Ericksen anisotropic fluid [11,12].

1. Rheological model of dispersive medium.

The Ericksen anisotropic fluid [11,12] is the simplest phenomenological rheological model of the orientable fluid with nondeformable microstructure. To describe the behavior of the microstructure the unit vector n_i, being called a director, is used. It characterizes an orientation of the fluid particles in gradient flows. In this case the stress in the anisotropic fluid is the function of the strain rate tensor and director:

$$\sigma_{ij} + p\delta_{ij} = 2\mu\gamma_{ij} + \mu_1 n_{ij} + \mu_2\gamma_{lm}n_{lmij} + 2\mu_3(\gamma_{jl}n_{li} + \gamma_{il}n_{lj}). \tag{1.1}$$

In the Eq. 1.1 p is the isotropic pressure; δ_{ij} is the unit tensor; $n_{ij} = n_i n_j$; $n_{lmij} = n_l n_m n_i n_j$; μ, μ_1, μ_2, μ_3 are phenomenological constants; $\gamma_{ik} = (1/2)(v_{i,k} + v_{k,i})$; $v_{i,k}$ is the derivative of the velocity vector v_i in the direction of coordinate axis k.

The orientation of director n_i is determined by flow and, in general case, depends on the velocity gradient in fluid. In the case when the inertia of microstructure elements of anisotropic fluid is assumed to be negligible the constitutive equation for director n_i in the linear approximation by the velocity gradient is of the form

$$\frac{Dn_i}{Dt} = \lambda(\gamma_{il}n_l - \gamma_{lm}n_{lmi}), \tag{1.2}$$

where $Dn_i/Dt \equiv \dot{n}_i - \omega_{il}n_l$ is the Jaumann derivative with respect to time; a point over n_i denotes an individual derivative with respect to time; $\omega_{il} \equiv (1/2)(v_{i,l} - v_{l,i})$ is the tensor of the velocity vortex; λ is the phenomenological constant; $n_{lmi} = n_l n_m n_i$.

In accordance with Eq. 1.2, orientation of the director n_i depends esentially on the non–dimensional constant λ. At $|\lambda| < 1$ the director is periodically changed with respect to time [11], its orientation depends on the velocity gradients of the flow. The Eq. 1.1 and Eq. 1.2 in this particular case were used in [1] to construct the structure–phenomenological theory of stress state in a dilute suspension of rigid ellipsoids of revolution in the Newtonian dispersive medium.

In the present paper anisotropic fluids are considered for which $\lambda \geq 1$ and the stress at rest coincides with the isotropic hydrostatic pressure, that is fluids with $\mu_1 = 0$.

Director n_i at $\lambda \geq 1$ in steady state flows takes on a stationary orientation which is independent of the velocity gradients. Thus, the Eq. 1.2 in the steady simple shear flow

$$v_x = 0, \qquad v_y = Kx, \qquad v_z = 0 \qquad (K\text{— const}) \tag{1.3}$$

has a stationary solution

$$ctg^2\beta = \frac{\lambda - 1}{\lambda + 1}, \qquad \psi = \frac{\pi}{2}, \tag{1.4}$$

which is independent of the shear flow K. Here β, ψ are the angles, determining the director position in the laboratory set of coordinates $Oxyz$ ($n_x = cos\beta sin\psi$, $n_y = sin\beta sin\psi$, $n_z = cos\psi$); β is the angle between the axis Ox and projection of n_i onto Oxy plane; ψ is the angle between the axis Oz and vector n_i. In so doing, the orientation of the director depends upon the magnitude of the constant λ. The angle β in Eq. 1.4 varies within the limits $\pi/4 \leq \beta \leq \pi/2$ at $\lambda \geq 1$ and $-\pi/4 \leq \beta \leq 0$ at $\lambda \leq -1$.

When the geometry of the flow is changed, the orientation of director is changed too. Thus, in the steady uniaxial elongational flow ($v_x = -(q/2)x$, $v_y = -(q/2)y$, $v_z = qz$; q— const.), the director is oriented so to be collinear to the axis Oz at $\lambda \geq 1$ and parallel to Oxy plane at $\lambda \leq -1$. In this case the stationary orientation of the director does not depend on the elongational rate q.

In accordance with [13], the Eq. 1.1 may be presented in the form of the generalized Newtonian law

$$\sigma_{ij} + p\delta_{ij} = \eta_{ijkl}\gamma_{kl} \tag{1.5}$$

with the viscosity tensor of the form

$$\eta_{ijkl} = 2\mu I_{ijkl} + \mu_2 n_{ijkl} + 4\mu_3 n_{(i}\delta_{j)(k}n_{l)}, \tag{1.6}$$

where $4n_{(i}\delta_{j)(k}n_{l)} = n_i\delta_{jk}n_l + n_j\delta_{ik}n_l + n_i\delta_{jl}n_k + n_j\delta_{il}n_k$; on indexes in the brackets a simmetrization is held; $I_{ijkl} = (\delta_{ik}\delta_{jl} + \delta_{il}\delta_{jk})/2$ is the unit tensor of the forth rank.

In agreement with Eq. 1.6 the effective viscosity of anisotropic fluid defined by Eq. 1.1 and Eq. 1.2 at $\mu_1 = 0$ and $|\lambda| \geq 1$ in steady state flows is independent of gradients of the flow velocity. But unlike Newtonian fluids this one depends on the orientation of the director n_i. This allows to say about anisotropy of the fluid defined by Eq. 1.1 and Eq. 1.2 with respect to direction of the director n_i which occurs in fluid flows. Taking this into account the longitudinal $\eta_{\parallel}$ and transverse $\eta_{\perp}$ (with respect to the director n_i) viscosities of the fluid have been introduced in [13]. These viscous coefficients are of the basic significance. In terms of these coefficients the viscosities with respect to other directions are expressed.

The anisotropy of viscosity in the fluid defined by Eq. 1.1 and Eq. 1.2 leads to the anisotropy of other its characteristics. Thus, the basic viscosities $\eta_{\parallel}$ and $\eta_{\perp}$ determines the availability of dissimilar frictional coefficients $\zeta_{\parallel}$ and $\zeta_{\perp}$ of the spherical particle moving in the fluid defined by Eq. 1.1 and Eq. 1.2 along and across the director. According to [14] the drag force acting on the spherical particle moving with the velocity U_k in anisotropic fluid defined by Eq. 1.1 and Eq. 1.2 can be presented as $-\xi_{ik}U_k$ where

$$\xi_{ik} = \zeta_{\perp}\delta_{ik} + (\zeta_{\parallel} - \zeta_{\perp})n_{ik} \tag{1.7}$$

is the frictional drag tensor of the spherical particle in anisotropic fluid.

Basic viscosities $\eta_{\parallel}$ and $\eta_{\perp}$ may be determined in terms of rheological constants μ, μ_2, μ_3. For this purpose it is neccesary to present the rheological equation for the stress in the fluid defined by Eq. 1.1 and Eq. 1.2 according to [13] in the form

$$\sigma_{ij} + p\delta_{ij} = 2\eta_{\perp}\gamma_{ij} + 2(\eta_{\perp} - \eta_{\parallel})\gamma^*_{ij},$$

where $\gamma^*_{ij} = 2n_{ijlm}\gamma_{lm} - n_{il}\gamma_{lj} - n_{jl}\gamma_{li}$, and compare it with Eq. 1.1. Taking into account the relation $\mu_2 = -4\mu_3$ [13] enables to obtain $\eta_{\perp} = \mu$, $\eta_{\parallel} = \mu + \mu_3$.

Discription of the behavior of fluid medium using the director has been applied in continuum theories of liquid crystals [15]. The constitutive equations obtained in [16] and being used in description of dynamical properties of nematic liquid crystals are similar to the Eq. 1.1 and Eq. 1.2.

2. Structural theory.

As a hydrodynamic model of suspended particles possessing the axial and central symmetry the uniaxial dumbbell is used. This model consists of two point centers of the hydrodynamic interaction which are coupled by the undeformable junction (axis) of the length L. The point center of the dumbbell is assumed to interact with the dispersive medium like a spherical particle, that is under flowing the point center of the dumbbell by anisotropic medium of the velocity u_i, the point center is affected by a force determined by the relation $\xi_{ij}u_i$, where ξ_{ij} is the frictional drag tensor defined by Eq. 1.7.

On the one hand, the suspended particles, modelled by dumbbells, are assumed to have such sizes that the anisotropic medium interacts with them like with hydrodynamic bodies. On the other hand — the suspended particles have to be sufficiently small to make the velocity of dispersive medium v_i be a homogeneous function of coordinates ($v_i = \omega_{ik}r_k + \gamma_{ik}r_k$) in the vicinity of each particle (r_k is the radius–vector determining position of the point in vicinity of the particle in the laboratory set of coordinates $Oxyz$ the origin of which coincides with the middle of the dumbbell axis). It as assumed also that the suspended particles possess a zero buoyancy.

According to above mentioned assumptions hydrodynamic forces f_i, acting to the centers of the hydrodynamic interaction of the dumbbell from the side of the anisotropic homogeneous medium are determined by relations $f_i = \xi_{ij}(v_{j,k}R_k - \dot{R}_j - v_{0j})$ where v_{0j} is the migration velocity of the particle center relative to the dispersive medium; R_k is the radius–vector of the point center of hydrodynamic interaction of the dumbbell which takes the values $(L/2)\nu_k$ for one and $-(L/2)\nu_k$ for another point center; ν_k is a unit vector, characterizing the orientation of the dumbbell suspended particle in the chosen set of coordinates $Oxyz$. Then the main vector of hydrodynamical forces F_i acting to the dumbbell is of the form

$$F_i = -2\xi_{ij}v_{0j}. \tag{2.1}$$

The constitutive equation for a vector ν_i characterizing the orientation of the dumbbell particle is obtained similar to [2] at the vector multiplication of the equation of rotational motion of the suspended particle $\dot{\mathcal{L}}_i = M_i$ by the vector ν_i:

$$I(\ddot{\nu}_i + \dot{\nu}_k\dot{\nu}_k\nu_i) = \frac{1}{2}L^2(\xi_{ij}\gamma_{jk}\nu_k - \xi_{jk}\gamma_{kl}\nu_j\nu_l\nu_i - \xi_{ij}N_j + \xi_{jk}N_k\nu_j\nu_i). \tag{2.2}$$

Here $\mathcal{L}_i = I[\nu \times \dot{\nu}]$ is the angular moment of the particle; I is the moment of inertia of the dumbbell particle with respect to the axis passing through the center of the particle axis perpendicularly to it; $M_i = (1/2)L^2\epsilon_{ijk}\nu_j(\xi_{kl}v_{l,s}\nu_s - \xi_{kl}\dot{\nu}_l)$ is the principal moment of hydrodynamic forces with respect to the middle of the dumbbell axis; ϵ_{ijk} is the Levi–Chivita tensor; $N_i = \dot{\nu}_i - \omega_{ik}\nu_k$.

At $\zeta_{\|} = \zeta_{\perp} = \xi$ the Eq. 2.2 coincides with the constitutive equation for the vector ν_i obtained in [2] for suspensions with isotropic dispersive medium.

Inertial forces and moment of inertial forces of suspended particles in suspensions are very small and usually neglected in the rheology of suspensions. In such a case the equation of translational motion of suspended particles is $F_i = 0$, from where by Eq. 2.1 it is follows that $v_{0i} = 0$, e. g. a migration of suspended particles in dispersive medium is absent. In this case, the Eq. 2.2 takes the form

$$(\xi_{im} - \xi_{km}\nu_k\nu_i + \zeta_{\perp}\nu_m\nu_i)(\dot{\nu}_m - v_{m,k}\nu_k) + \zeta_{\perp}\gamma_{km}\nu_k\nu_m\nu_i \quad = \quad 0. \qquad (2.3)$$

The expression for the rate of dissipation of mechanical energy per unit volume of the suspension under consideration is obtained similar to [2] within the framework of the structure theory

$$\begin{aligned} W \quad &= \quad \tilde{W} + N_0(L^2/2)[\xi_{ij} < N_iN_j > -2\xi_{ij}\gamma_{jk} < N_i\nu_k > + \\ &+ \quad \xi_{ij}\gamma_{ik}\gamma_{jl} < \nu_l\nu_k >], \end{aligned} \qquad (2.4)$$

where $\tilde{W}$ is the rate of the mechanical energy dissipation per unit volume of the dispersive medium in the absence of the suspended particles; N_0 is the number of suspended particles per unit volume of suspension; the angular brackets denote the averaging with the use of the distribution function F of the angular positions of the vector ν_i, which is the solution of the equation

$$\frac{\partial F}{\partial t} + \frac{\partial}{\partial \nu_i}(F\dot{\nu}_i) \quad = \quad 0. \qquad (2.5)$$

3. Structure–phenomenological theory.

According to Eq. 2.4 the stress tensor Σ_{ij} in the suspension under consideration has to be determined by the relations of the form $\Sigma_{ij} = \sigma_{ij} + N_0 < t_{ij} >$, $t_{ij} = t_{ij}(\gamma_{km}, \xi_{ln}, \nu_s, N_p)$ (σ_{ij} is the stress tensor in dispersive medium in the absence of suspended particles, $N_0 < t_{ij} >$ is the stress cased by the presence of N_0 suspended particles per unit volume of the suspension). In this case the tensor t_{ij} has to be a polynomial function of matrices which should be linear with respect to its arguments $\xi_{km}\gamma_{ml}$ and $\xi_{km}N_m$. Besides, in view of the symmetry of the dumbbell model particles with respect to the middle of axis L, this function has to be invariant to the alternation of the direction of ν_i to opposite. Because of this the phenomenological equation for t_{ij} is obtained in a similar way to [12]

$$\begin{aligned} t_{ij} \quad &= \quad a_0\delta_{ij} + a_1\nu_i\nu_j + a_2\xi_{lk}\gamma_{lm}\nu_k\nu_m\nu_i\nu_j + a_3\xi_{in}\gamma_{nj} + a_4\xi_{jn}\gamma_{ni} + \\ &+ \quad a_5\xi_{ln}\gamma_{in}\nu_l\nu_j + a_6\xi_{ln}\gamma_{jn}\nu_l\nu_i + a_7\xi_{in}\gamma_{nk}\nu_k\nu_j + a_8\xi_{jn}\gamma_{nk}\nu_k\nu_i + \\ &+ \quad a_9\nu_i\xi_{jk}N_k + a_{10}\nu_j\xi_{ik}N_k. \end{aligned} \qquad (3.1)$$

The phenomenological constants a_i $(i = \overline{1,10})$ in Eq. 3.1 are being found comparing the rate of the mechanical energy dissipation per unit volume of the suspension defined by Eq. 2.4 determined in the structural theory whith the rate determined analogously to [2] by the formula $W = \tilde{W} + N_0 < t_{ij}\gamma_{ij} > -N_0 < N_i g_i >$ within the framework of phenomenological approach, where g_i is the right–hand side of Eq. 2.2. Taking into account the relation [2] $< t_{ij}\nu_j > - < t_{ji}\nu_j >=< g_i >$ such values can be obtained for rheological constants $a_1 = a_2 = a_3 = a_4 = a_5 = a_6 = a_8 = a_9 = 0$, $a_7 = L^2/2$, $a_{10} = -L^2/2$. Doing so permits to write finally the constitutive equation for the stress tensor Σ_{ij}

$$\Sigma_{ij} = \sigma_{ij} + (1/2)N_0 L^2(\xi_{im}\gamma_{mk} < \nu_k\nu_j > -\xi_{jk} < \nu_i N_k >). \qquad (3.2)$$

If the suspended particles modelled by dumbbells posseses a constant dipole moment $p_i = q\nu_i$, then on such particles in the external uniform electrical field of the intensity E_i will be exrted a couple of forces with the moment $M_i^e = q\epsilon_{ilm}\nu_l E_m$, tending to turn the particles in the direction of the vector E_i.

If the suspended particles are sufficiently small, then it is necessary to take into account their rotary Brownian motion induced by the Brownian couple of forces with the moment

$$M_i^B = -T\epsilon_{ilm}\nu_l \partial lnF/\partial\nu_m,$$

where T is the temperature in the energy dimension. The constitutive equation for the vector ν_i in the case when the moment of inertia force of the suspended particles is not taken into account is of the form

$$\frac{L^2}{2}(\xi_{im} - \xi_{km}\nu_k\nu_i + \zeta_\perp\nu_i\nu_m)(\dot{\nu}_m - v_{m,k}\nu_k) + \frac{L^2}{2}\zeta_\perp\gamma_{km}\nu_k\nu_m\nu_i - q(E_i - E_k\nu_k\nu_i) + T\left(\frac{\partial lnF}{\partial\nu_i} - \nu_i\nu_k\frac{\partial lnF}{\partial\nu_k}\right) = 0. \qquad (3.3)$$

Eq. 1.1, Eq. 1.2, Eq. 2.3 (or Eq. 3.3), Eq. 2.5 and Eq. 3.2 are the closed system of equations governing the stress state in the dilute suspension in anisotropic dispersive medium.

4. Steady simple shear flow.

The influence of the anisotropy of the dispersive medium on the behavior of the dilute suspension of dipole dumbbells in anisotropic fluid defined by Eq. 1.1 and Eq. 1.2 at $|\lambda| \geq 1$, $\mu_1 = 0$ in the simple shear flow defined by Eq. 1.3 in the presence of electric field $E_x = E$, $E_y = 0$, $E_z = 0$ (E — const) is investigated as an example of employment of the obtained rheological equations. The solution of Eq. 1.2 defined by Eq. 1.4 shows that electrically neutral director n_i of the anisotropic dispersive medium defined by Eq. 1.1 and Eq. 1.2 is oriented stationarily in the planes perpendicular to axis Oz, at the angle β to the coordinate plane Oxz, which is determined by the equation Eq. 1.4. It follows from Eq. 3.3 that the dipole suspended dumbbell particles take on the stationary orientation both in isotropic and anisotropic media in the planes perpendicular to axes Oz under the angle φ to the plane Oxz which is determined from the equation

$$cos^2\varphi - (\Delta - 1)cos\varphi sin\beta sin(\varphi - \beta) - \frac{1}{\sigma}sin\varphi \quad = \quad 0, \tag{4.1}$$

where $\sigma = K\gamma/(qE)$; $\gamma = \zeta_\perp L^2/2$; $\Delta = \zeta_\parallel/\zeta_\perp$.

It follows from the solution of Eq. 4.1 that with growing σ the angle of hovering of the dumbbell particle increases from $\varphi_0 = 0$ at $\sigma = 0$ to φ_∞ at $\sigma \to \infty$; the magnitude of φ_∞ is dependent of the anisotropy of the dispersive medium — at $\Delta \leq 1$ $\varphi_\infty = 90^o$, at $\Delta > 1$ $\varphi_\infty = arctg\{(1 + \Delta tg^2\beta)/[(\Delta - 1)tg\beta]\}$. At $\varphi < \beta$ the angle of hovering of the dumbbell particles in anisotropic dispersive medium with $\Delta < 1$ is less than in isotropic medium ($\Delta = 1$), while at $\varphi > \beta$ is larger; in anisotropic dispersive medium whith $\Delta > 1$ is vice versa.

In the case of stationary orientation of the suspended particles the distribution function of their angular positions, which in general case is found as a solution of the Eq. 2.5, turns into Dirac δ–function concentrated in the angle of hovering φ. Therefore, the effective viscosity of the suspension which is determined by using Eq. 3.2 takes the form

$$\mu_a \quad = \quad \mu_E + \frac{1}{2}N_0\gamma[cos^2\varphi + (\Delta - 1)cos\varphi sin\beta sin(\varphi - \beta)], \tag{4.2}$$

where $\mu_a \equiv (\Sigma_{xy} + \Sigma_{yx})/2K$; $\mu_E = \mu + \mu_2(\lambda^2 - 1)/4\lambda^2 + \mu_3$ is the effective viscosity of anisotropic dispersive medium defined by Eq. 1.1 and Eq. 1.2 at $|\lambda| \geq 1$, $\mu_1 = 0$ in the simple shear flow.

According to the Eq. 4.2 the anisotropic properties of the dispersive medium at $\Delta > 1$ lead to increasing of the increment $\nu = (\mu_a - \mu_E)/(N_0\gamma)$ of the effective viscosity μ_a of the suspension, and at $\Delta < 1$ to its decreasing in comparison with suspension with isotropic $\Delta = 1$ dispersive medium (Fig. 1, 2).

Fig. 1, 2 show the dependence $\nu = \nu(\sigma)$. Curves 1-5 of Fig.1 correspond to $\Delta = 0.2; 0.5; 1$ (isotropic dispersive medium); $1.5; 2$ at $\beta = 60^o$, curves 1, 2 of Fig.2 correspond to $\beta = 45^o, 75^o$ at $\Delta = 2$, curve 3 correspond to $\Delta = 1$ (isotropic dispersive medium), curves 4, 5 correspond to $\beta = 45^o, 75^o$ at $\Delta = 0.2$.

The presence of the ranges of σ has been discovered in which the increment ν of the effective viscosity μ_a increases with increasing σ for the suspension under consideration contrary the dilute suspensions of rigid particles in isotropic disperse medium ($\Delta = 1$). This effect takes place for $\Delta > 1$ at very small σ (curves 4, 5 on Fig.1 and 1, 2 on Fig.2) and for $\Delta < 1$ at great σ (curves 1, 2 on Fig.1 and 4, 5 on Fig.2). Increasing ν at small σ on Fig.1, 2 is slightly visible due to using the logarithmic scale for σ.

It is observed also (curves 1, 2 of Fig.1 and curves 4, 5 of Fig.2) the negative increment ν of the effective viscosity μ_a of suspension. It means that the effective viscosity of suspension in anisotropic fluid defined by Eq. 1.1 and Eq. 1.2 in laminar flows can be less than the effective viscosity of anisotropic fluid in the case of absence of suspended particles. It take place at the certain orientation of suspended particles under action of the external stationary electric field.

The effects mentioned above are not observed in the suspensions with Newtonian dispersive medium. These effects are first predicted theoretically in the present paper and it should be noted that they have not been observed in experiments earlier.

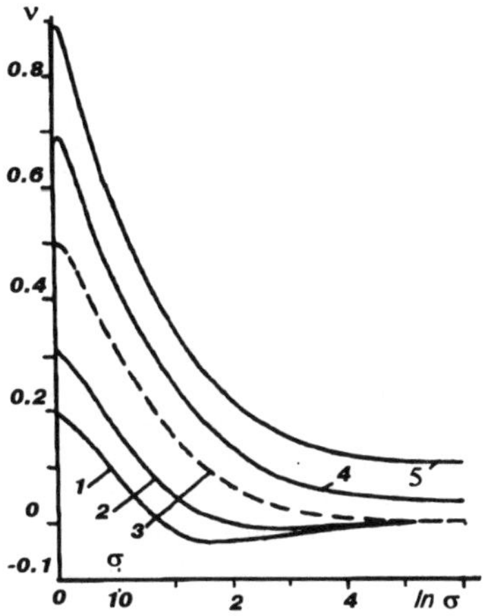

Fig. 1

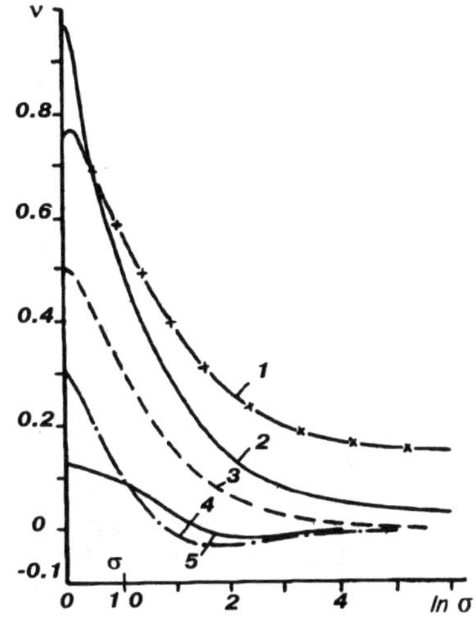

Fig. 2

The obtained results can be used to construct the theory of dynamic behavior of admixtures in the liquid crystals, as well as to investigate the influence of the admixtures on a rheological behavior of the liquid crystals.

5. References.

1. Yu.I.Shmakov and E.Yu.Taran, *Inzh.-fiz. Zhurn.* **18** (1970) 1019.
2. E.Yu.Taran, *Prikl. Mekhanika* **13** №4 (1977) 110.
3. E.Yu.Taran, *Inzh.-fiz. Zhurn.* **34** (1978) 622.
4. E.Yu.Taran, *Mekhanika Polimerov* №3 (1978) 519.
5. E.Yu.Taran, *Prikl. Mekhanika* **15** №9 (1979) 108.
6. E.Yu.Taran, *Doklady AN Ukr.SSR* Ser.A №7 (1984) 43.
7. E.Yu.Taran, *Doklady AN Ukr.SSR* Ser.A №3 (1986) 50.
8. E.Yu.Taran, *Vestnik Kiev. Univ. Matematika i Mekhanika* **19** (1987) 91.
9. E.Yu.Taran, *Gidromekhanika* **55** (1987) 55.
10. E.Yu.Taran, *Doklady AN Ukr.SSR* Ser.A №8 (1988) 43.
11. J.L.Ericksen, *Kolloid. Z.* **173** (1960) 117.
12. J.L.Ericksen, *Arch. Ration. Mech. Anal.* **4** (1960) 231.
13. V.S.Volkov and V.G.Kulichikhin, *J. Rheol.* **34** (1990) 281.
14. V.S.Volkov and G.V.Vinogradov, *Rheol. Acta.* **23** (1984) 231.
15. P.G. de Gennes, *The Physics of Liquid Crystals* (Clarendon Press, Oxford, 1974).
16. F.M.Leslie, *Arch. Rat. Mach. Anal.* **28** (1968) 265.

Surface Feedback for Virtual Environment Systems Using Electrorheological Fluids

P. M. Taylor, A. Hosseini-Sianaki, C. J. Varley

Department of Electronic Engineering
The University of Hull
HULL, HU6 7RX
U.K.

ABSTRACT

A tactile array system, when incorporated in an information system, provides a means of conveying tactile data to the user. Such arrays have many applications including simulators for training purposes and Braille systems for the blind. The use of ER fluids in these systems would provide an inexpensive alternative to the existing technologies because of the simple and flexible designs. An experimental study has been conducted which involves a simple 5x5 array of tactels, each tactel being independently energised in order to configure a variety of shapes. A series of tests are carried out on two different types of ER fluids under a range of electric fields and sensor speeds. The results show that for the given tactel spacing, the programmed shapes are easily detected for all conditions and are directly comparable to similar tests undertaken on a solid shape.

1. Introduction

Virtual Environment (VE) Systems, or Virtual Reality is a comparatively new technology using interactive simulations which aim to convince the user that he/she is in an artificial but realistic virtual world. Applications for this technology are plentiful, including simulators for training purposes, manufacturing processes and leisure activities.

The VE System technology presently available consists mainly of vision based elements and a major problem is the lack of complementary tactile and kinaesthetic feedback data. Research work has been carried out at Hull on a force feedback system[1], the 'Reactabot'(TM) and systems which give an illusion of shape through the creation of programmable surfaces which generate appropriate forces to a user's fingers as they hapticly search the surface. Two technologies, shape memory alloy[2] and electrorheological (ER) fluid[3], have been explored. It is the intention of this paper to demonstrate how the many properties of ER fluids can be used for such purposes.

At Hull, the application of ER fluids within VE systems was initially proposed by Monkman[4]. In a later publication, experimental work[3] carried out in order to verify the concept, was reported. This investigation included a simple method of detecting a hard wired square region of activated fluid. The results were extremely encouraging and have led on to further development described here in which the hard wired device has been replaced by a 5x5 array of tactels enabling a whole range of shapes to be programmed onto the surface. The device has been tested for different ER fluids under various field strengths and for a range of shape settings. The force measuring probe was set up in order to simulate human fingertips running across the display at different speeds.

2. Technical Equipment and Experimental Method

Figure 1 is a schematic diagram of the prototype tactile array including the force measuring probe. The principal part of the device consists of two parallel surfaces separated by a 2.5mm film of ER fluid. This nominal film thickness was chosen to give a depth of fluid which would promote the feeling of a real object rather than a 2-D area whilst still allowing the use of relatively low voltages. The upper surface - flexible membrane, is made of a thin conductive rubber and is grounded. The lower, solid surface is a complex electrode system (see Figure 2). It consists of a 5x5 array of tactels embedded in a perspex block which is solidly held within the frame of the device, each tactel being independently connectable to the high potential. Figure 2 shows the solid surface, it has dimensions 96mm x 96mm with the aforementioned 5x5 array centralised to it. Each tactel is a square of 11mm side and has a 2mm separation from adjacent tactels to minimise any coupling effects. It is also noted that in order to limit arcing between the tactels and also due to manufacturing restrictions, the corners of the tactels were rounded to a diameter of 2mm.

The high voltage was supplied by a unit capable of up to 10kV at 20mA current. The force measurements were carried out by a two axis strain gauge type force sensor accurate to 1 gram. An aluminium L-shaped brace was fitted to it allowing simultaneous force measurement in both directions. Attached to this brace is a cylindrical probe of cross-sectional area 7.85mm^2. This assembly was then mounted normal to the display device allowing the measurement of both horizontal(in the direction of motion) and vertical forces (see also Figure 1). The tactile display unit was mounted upon a high precision, computer controlled, X-Y table capable of a wide range of velocities. The force from both axes in addition to the current drawn, were logged by a data capturing system capable of up to 200kHz sampling with a resolution of 12 bits. Further details of this equipment is given in reference 3.

The shape configuration was achieved by a system of 25 mechanical switches, allowing each tactel to be individually energised, the non-energised tactels being instead connected to earth potential. Work is underway to replace the existing mechanical switches with a high voltage capability, electronic system enabling dynamic computer control.

The results from a total of three fluids are discussed, detailed descriptions of which are given here. For reference purposes, the fluids are named A, B and C. Fluid A consists of silicone oil with some 30% by volume of a Lithium salt of polyoxyethylenepolyester and 1% by volume of copolymer of polydimethylsiloxane and polythyleneglycol. The particles have a typical diameter of 3 microns. This fluid has a zero-voltage viscosity of 50 mPa s at a temperature of 20°C. Fluids B and C have also silicone oil for the continuous phase but contain inorganic materials ($Na_{12}Al_{12}Si_{12}O_{48}$) for the particulate phase. The average particle diameter in both cases is about 3 microns. Fluid B is made up of 24% by volume solid which contains 3.5% water whereas Fluid C has 22.5% by volume particles containing only 2% water. Another major difference is the steric stabiliser used in the chemical treatment to

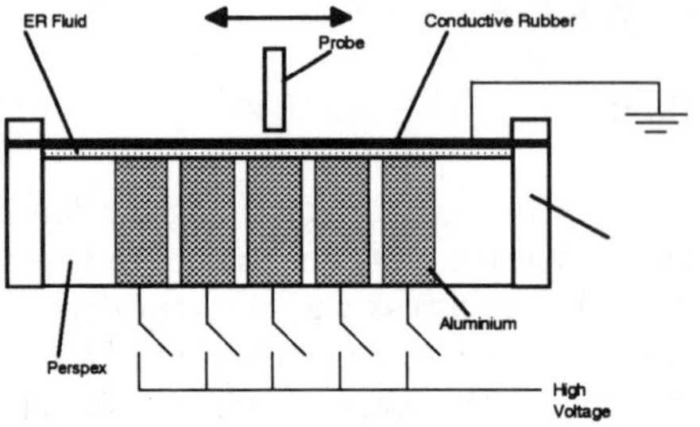

Figure 1. Cross section of the tactile array system in schematic form.

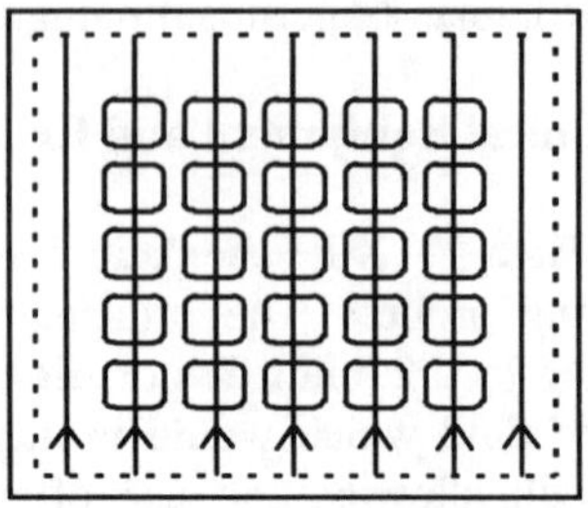

Figure 2. Top view of tactel arrangement showing lines along which tests were carried out. Arrows represent direction of the probe and the dotted line shows the tested area.

prevent settling out of the particles. 1% by volume of amino-substituted polysiloxane was grafted onto the particle surface for fluid B and 0.5% by volume of amino-substituted alkoxysilane was similarly employed for fluid C. The zero-voltage viscosities of the fluids are about 70 mPa s and 110 mPa s respectively at a temperature of 20°C.

For the force measurements taken, the probe arrangement was set up to closely simulate the action of a human finger. Some preliminary results and the experience gained from previous work[3] indicated that it was sufficient to examine the effect of the fluid as the probe moves along the surface i.e. 'drag' testing. Hence the point tests described in the earlier work[3] were not examined further. The surface of the display was explored systematically over a series of paths which are shown as lines in Figure 2.

Initially the probe was pressed onto the surface of the flexible membrane and the fine tuning of the vertical motion controller was used to set a gap of 1.4mm at the probe position between the conductive rubber and the solid bottom surface. This gap was chosen since it is large enough to prevent short circuiting whilst being small enough to produce significant force measurements. An initial vertical force is observed (typically 10 g to set this gap) which is equivalent to the reaction at the rubber surface. The procedure for a typical run, after the initial vertical setting and shape configuration was as follows: the high voltage was applied to the relevant tactels via the switching system and the probe set in motion using the Y-axis. The data capture system was instantaneously triggered, using a signal from the X-Y table, on commencement of probe motion. Each path was in turn tested for a selection of configured shapes. These tests were all conducted at room temperature (21°C), a voltage level of 2.4kV and a probe speed of 4mm s^{-1}. The fluid temperature was monitored regularly and no measurable change was observed.

Two other major sets of tests were conducted. The first examined the effect of varying voltage levels whereas the second set investigated the effect of varying probe speed along the same path. Again, these tests were carried out at room temperature.

3. Results and Discussion

The captured data from the tests were processed separately in terms of the horizontal and vertical forces. From these, a 3-D image of the apparent shape was depicted, showing a profile from above and from the side. Figures 3 and 4 show respectively the vertical and horizontal forces when the letter 'U' is configured. As the Tactile Array surface was divided into a finite number of lines (see Figure 2), and as the probe itself is of finite cross sectional area, for presentational purposes it was assumed that the reading from each line along which the probe is travelling, represents the area 6.5mm either side of the line. This was achieved by repeating the line data to cover the relevant area. Figures 5a and b show a more complicated configuration describing concentric squares. It is clear that a configured letter or shape can be detected equally well using either the horizontal or vertical force. However, in terms of magnitude, the vertical force is generally 3 to 4 times greater. The aberrations observed in the figures can be related to several factors including; high tension in the membrane at the clamped edges of the device; lower tension in the membrane at the centre of the device; the reluctance of the forces to die down after traversing a line of activated tactels due to either fluid retention or some other factors discussed later.

In the tests conducted, the current drawn from the apparatus was extremely small for all of the fluids tested. The consequent high level of noise upon the signal prevented any suitable pictorial representation within this paper. However, this point illustrates the low power, typically 100mW, required to operate such a device.

Since shapes can be detected with either vertical or horizontal forces, only the vertical forces will be shown from hereon.

Another objective of the present exercise was to verify earlier work[3] in which an array was not used to construct a square shape, but simply a square shaped block of aluminium was employed. The dimensions of the array has been chosen to enable an exact replica of the previous square to be constructed. Figures 6a and b compare the original solid square result and the reconstructed version using the tactel arrangement. The apparent reconstruction is clear, however a closer examination of a single line of data (see Figure 8) shows that for the case of the array, the spacing of each tactel is represented with a dip in the force and that these directly correspond to the spacing between each tactel. It should be noted that the forces displayed in Figure 6a are considerably greater since the gap between the electrodes at the probe position during the earlier work[3] was much less. To prevent the electrical shorting experienced in the earlier work, this gap was increased two fold leading to a lower electric field strength.

The result of comparing two different ER fluids, i.e. fluid A and fluid B is shown in Figure 7 when configuring a triangular shape. It is apparent that Fluid A produces a greater force although the edges of the shape are less sharp. This behaviour is equally true of both vertical and horizontal directions. Both fluids display the correct shape but some difficulties were encountered whilst experimenting with fluid A, caused by a reaction between the fluid and the rubber resulting in a gradual distortion of the rubber surface.

In order to investigate the effect of the voltage level, the system response was interrogated over one particular line of the array for a fixed probe speed. Figures 8a and b

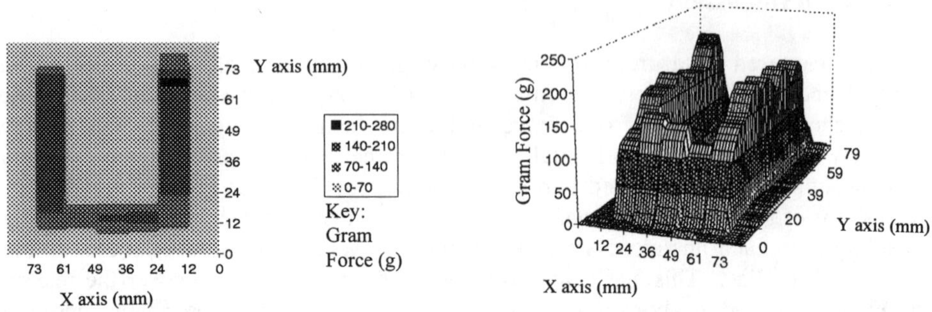

(a) Top view (b) Side view

Figure 3. Measured vertical forces when the letter U is configured

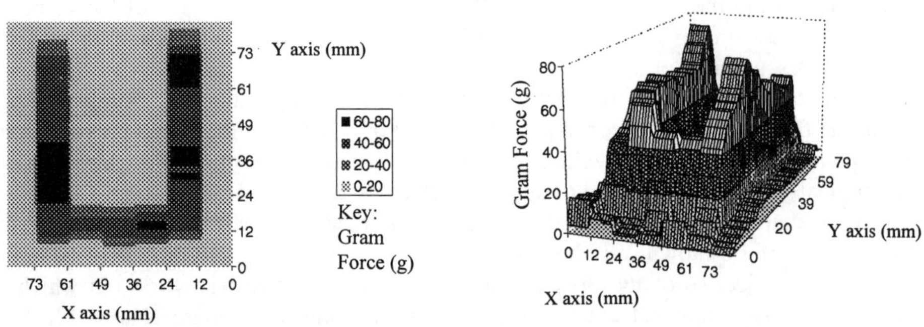

(a) Top view (b) Side view

Figure 4. Measured horizontal forces when the letter U is configured

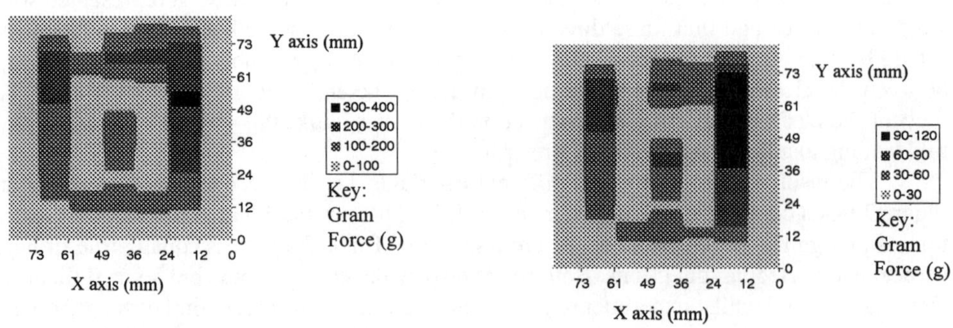

(a) Forces along vertical axis

(b) Forces along horizontal axis

Figure 5. Top view of the measured forces when concentric squares are configured

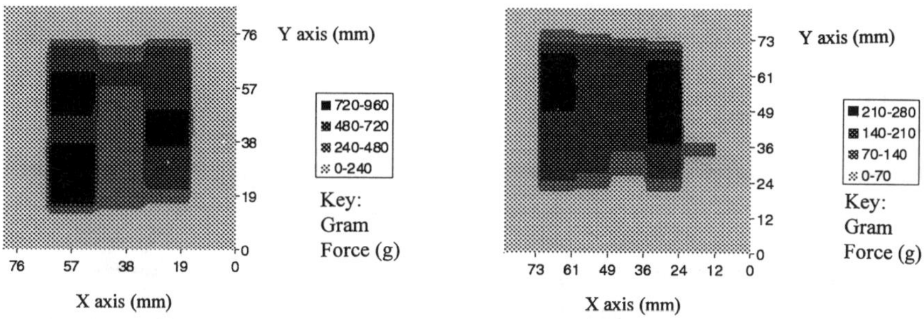

(a) Result from the solid block(fluid C)

(b) Result from the array(fluid B)

Figure 6. Comparison of a configured square from both solid block and array technique.

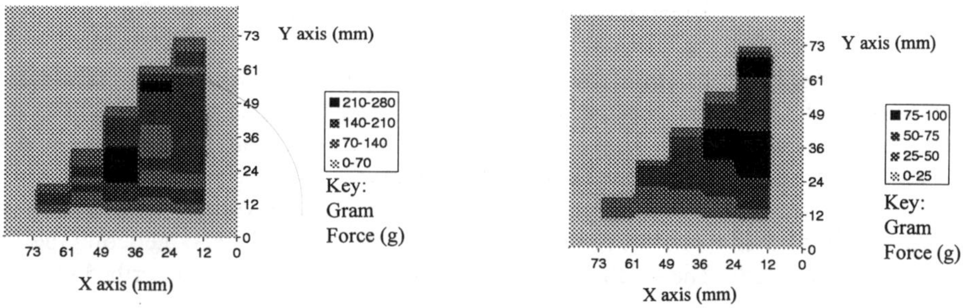

(a) Results from fluid A

(b) Results from fluid B

Figure 7. Results of fluids A and B for a triangle configuration. Top view of the measured vertical forces.

show the results for vertical and horizontal steady state forces respectively. Both figures clearly indicate that some response is apparent even with the voltage as low as 900V although the vertical force is more pronounced. There appears to be a general increase in the peak force level reaching a maximum at around the centre of the array. As mentioned earlier, the dips in the force as a consequence of the tactel spacing are evident and it is seen that as the voltage increases, so does the depth of these regions. Using the results from Figure 8 and removing the zero voltage response, an average peak force value from the three central tactels has been plotted against voltage and is shown in Figure 9. The observed behaviour is similar for both cases and resembles that for low speed Couette viscometer tests[5].

It is seen from the profile shown in Figure 8 that for both horizontal and vertical axes, as the probe moves over each subsequent tactel, the peak force increases to a limiting value. Two possible explanations for this behaviour are: an increasing strength of the fluid as the central body of the activated fluid is traversed; the fluid is being pushed into the

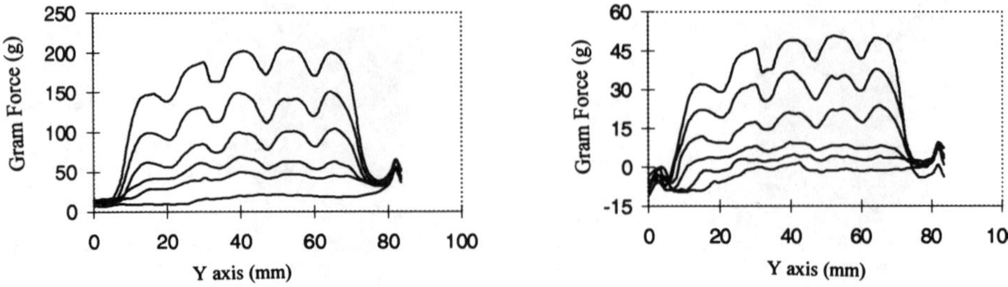

(a) Vertical axis reading (b) Horizontal axis reading

Figure 8. Line profile showing effect of increasing voltage on the measured forces. Note: from the lowest trace to the highest, V= 0, 600, 900, 1500, 2100, 2700.

active region due to the action of the probe, so increasing the mass of fluid in the local probe region. The latter effect will also increase the apparent effect of the ER phenomenon due to the greater concentration of fluid in the vicinity of the probe. This relationship is further verified for the vertical forces when analysing the various probe speeds (see Figure10). As the probe speed increases, the fluid apparently increases its resistance to the movement of the probe because the probe will push a larger volume of fluid into the neighbouring tactel region. However, it is seen that the horizontal forces react in a similar manner to those in the low speed Couette viscometer[5].

The effect of probe speed was examined as in the voltage case by choosing a particular line but in this case, fixing the voltage level to 2.4kV and conducting the tests for five different speeds. The results were processed in the same way as described above, i.e.

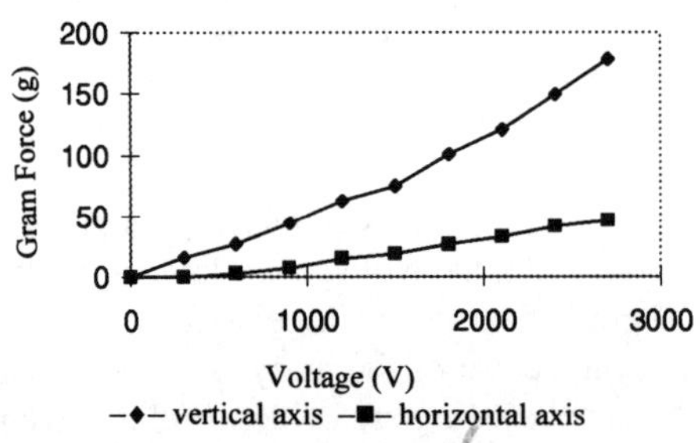

Figure 9. Variation of horizontal and vertical forces measured against the applied voltage.

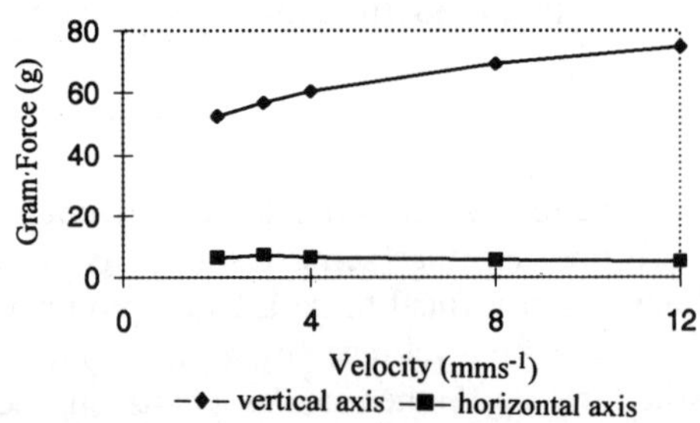

Figure 10. Horizontal and vertical forces measured for various probe speeds.

the average peak force level was calculated. The results can be seen in Figure 10. The profile of the forces are similar to those shown in Figure 8. However, within the range of tests performed, no significant relationship was observed between the dips due to the tactel spacing and the speed of the probe. The curves shown in Figure 10 show that the response force along the vertical axis increases with the probe speed whereas this force is decreasing with the probe speed along the horizontal axis, resembling the Couette viscometer measurements[5] and is also considerably smaller. For both cases, over the range of tests carried out, the response appears to be linear.

4. Conclusions

The work performed so far has shown that great potential exists for these fluids to be successfully employed within VE systems, not only for the purposes of complementing the visual systems already available, but also in the form of stand alone devices such as Braille displays. Such devices could also be used to enhance man/machine interface systems in general. The application of ER fluids in this field is particularly favourable in that they require very little power to operate and the active devices are of simple design comprising few moving mechanical components. These factors indicate that ER fluid based VE systems would be considerably less expensive than the other available options.

5. Acknowledgements

The work reported in this paper is funded by the European Union under Brite-Euram project no. BE-5072. The authors are grateful to Advanced Fluids Systems Ltd. and Metzeler Gimetall AG for supplying the ER fluids.

6. References

1. P.M. Taylor, K.G. Swift, D.R. Smith and R.S. Kalawsky, *The Reactabot: a tool for aiding the assessment of manual handling processes in assembly/disassembly,* Proc. 27TH ISATA Int. Dedicated Conf. on Mechatronics, Asachen, Germany, 31 Oct. - 4 Nov. 1994, p. 639-646.

2. P.M. Taylor et al, *A Dynamically Configurable Surface based on Shape Memory Alloy Wires for Virtual Reality Applications*, ibid., 617-625.

3. P.M. Taylor, A. Hosseini-Sianaki and C.J. Varley, *An Experimental Investigation into the Application of Electrorheological Fluids in Tactile Display Systems*, Proc. 5TH Int. Conf. on Adaptive Structures, 5-7 Dec. 1994, Sendai, Japan, 407-416.

4. G.J. Monkman, *Electrorheological Tactile Display,* Presence, MIT Press, 1(2), 1992.

5. H. Block and J.P. Kelly, *J. Phys. D: Appl. Phys.*, **21**(1988)1661.

THE ROTARY ELECTRORHEOLOGICAL EFFECT

THOMAS C. HALSEY
Exxon Research and Engineering, Route 22 East
Annandale, NJ 08801, U.S.A.
and
ROBERT A. ANDERSON and JAMES E. MARTIN
Advanced Materials Physics Division, Sandia National Laboratories
Albuquerque, NM 87185, U.S.A.

ABSTRACT

The viscous response of electrorheological fluids is usually manipulated through the use of DC or uniaxial AC electric fields. The result is that fibrillated structures parallel to the field form in a quiescent fluid; the distortion of such structures in a flow determines the enhanced viscous response, at least at low and moderate flow rates. We have conducted preliminary studies of electrorheological response in a different field configuration–a rotating electric field. With respect to the uniaxial AC case, there are two new developments in this type of field. The structures formed are disk-like, in the plane of the rotating field. Furthermore, the structures rotate either with or against the field, depending on the dielectric or conductivity contrast with the surrounding fluid.

1. Introduction

At the heart of the electrorheological (ER) effect is the ability to manipulate the interactions between colloidal particles through an applied electric field. While there has been considerable study of the way in which particle and fluid composition and response influence the ER response of the colloidal system[1], study of the influence of the electric field has been for the most part restricted to a study of the influence of the driving frequency of a uniaxial field.

In this contribution, it is our purpose to demonstrate that interesting effects arise by using more complicated electric field configurations. We focus upon circularly polarized electric fields. If the rotation frequency of the field is significantly lower than the characteristic response frequency of the polarization mechanism for the particles, then the interactions of the particles, while still uniaxial, are qualitatively reversed with respect to the normal ER case–particles attract if they are in the plane of the electric field, and repel along the direction perpendicular to this plane. This results in the formation of disk-like structures: we propose to call this the "Rotary" ER effect.

Any polarization mechanism will display some phase lag with the applied electric field. In the ordinary ER effect, the polarization is approximately collinear with the field even for appreciable phase lag, so while the strength of the ER effect may be altered by this phase lag, no torques develop on the particles. In the rotary ER effect, such torques play a significant role, and cause the disk-like structures to rotate at observable frequencies.[2]

2. Rotary Electrorheological Interactions

Consider a monodispersed suspension of spherical particles of radius r_d in an insulating fluid. For simplicity, we will assume that the particles have a simple dielectric response, so that their dipole moment obeys

$$\vec{p} = \beta r_d^3 \vec{E} \tag{1}$$

where $\vec{E}$ is the instantaneous electric field. We will further simplify matters by considering only dipolar interactions between the particles; we ignore non-electrostatic and multipolar interactions. As with ordinary ER fluids, we expect that this assumption will lead to a qualitatively correct and a quantitatively inadequate understanding of fluid behavior.

Let us suppose that although the frequency of the field ω is less than the characteristic frequency ω_p of the polarization response of the particles, that it is still greater than frequencies characteristic of particle movement in the fluid. To estimate these latter frequencies, we recall the Stokes relation for the steady velocity $\vec{v}$ of a particle of radius r_d subjected to a constant force $\vec{F}$ in a fluid of viscosity μ_0,[3]

$$\vec{F} = 6\pi\mu_0 \vec{v} r_d \tag{2}$$

Since we expect from dimensional analysis that the particles will be subjected to electrical forces $F_E \sim E^2 r_d^2$, we conclude that the field frequency ω should obey

$$\omega \gg \omega_a = \frac{\mu_0}{E^2} \tag{3}$$

so that particles move a distance much less than their diameter in one cycle of the applied field. Note that elsewhere we have termed ω_a the aggregation frequency, since its inverse gives the characteristic time scale for particle aggregation in the ordinary ER effect.[4]

We have identified a range of frequencies $\omega_a \ll \omega\omega_p$ over which we can approximately regard the particles as stationary over one cycle of the applied field, but with dipole moments following the instantaneous field. In this regime, to find the effective interaction between particles, we should average the instantaneous interaction over one cycle of the applied field. The instantaneous interaction energy u between two dipoles $\vec{p}^{(1)}$ and $\vec{p}^{(2)}$ separated by a distance $\vec{r}$ is

$$u(\vec{r}) = -p_i^{(1)} \left(\frac{3r_i r_j - r^2 \delta_{ij}}{r^5} \right) p_j^{(2)} \tag{4}$$

where the indices are Cartesian. We suppose that $\vec{p}^{(1,2)}(t) = \beta r_d^3 \vec{E}_0(t)$; thus the dipole moment is proportional to the applied field $\vec{E}_0(t)$–contributions of other dipoles to the local field are neglected. For a uniaxial field $\vec{E}_0(t) = \vec{E}_0 f(t)$, the resulting averaged interaction is

$$u(\vec{r}) = -\frac{\beta^2 r_d^6 E_0^2 \langle f^2(t) \rangle}{r^3} (3\cos^2\theta - 1) \tag{5}$$

where the brackets $\langle\rangle$ indicate a time average, and θ is the angle between $\vec{r}$ and the field direction. Equation 5 is simply the anisotropic dipolar interaction familiar from scores of ER fluid studies. We are concerned not with uniaxial fields, but with rotating fields

$$\vec{E}_0(t) = E_0(\hat{x}\cos\omega t + \hat{y}\sin\omega t) \tag{6}$$

where $\hat{x}$ and $\hat{y}$ are the normal three-dimensional basis vectors. Introducing this form for the field into Eq. (4) and averaging over time, we find that

$$u(\vec{r}) = -\frac{\beta^2 r_d^6}{2r^3}(1 - 3\cos^2\theta) \tag{7}$$

where θ is now the angle between the direction perpendicular to the plane of the rotating electric feld and $\vec{r}$. Note that the sign of Eq. (7) is the opposite of that of Eq. (5). Thus the particles repel in the direction perpendicular to the plane of the field, and attract in the plane of the field. Just as the ordinary ER anisotropy lends itself to the formation of chains of particles, so the rotary ER anisotropy will lend itself to the formation of disks or sheets of particles; these disks are seen in experiment (see Figure 1.)

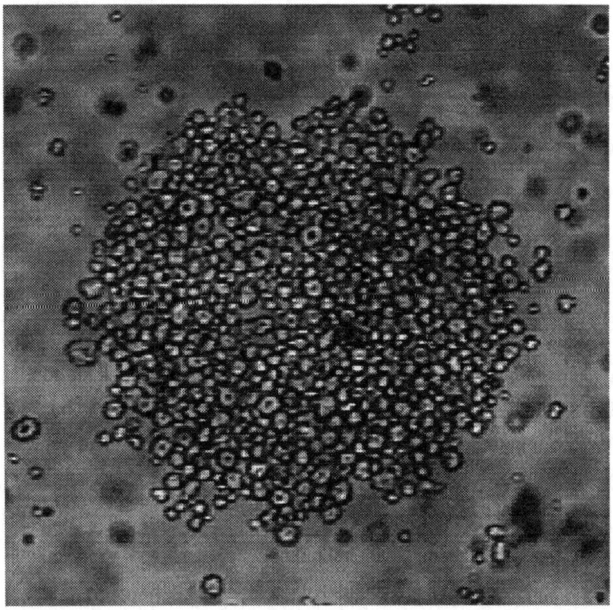

Figure 1. In a rotating electric field, particle disks form in the plane of the field. This photo shows a disk of 10 μm corn starch particles in such a field. (The experimental results displayed in Figures 2 and 3 were obtained with 90 μm corn pollen particles.)

3. Electrical Torques and Disk Rotation

One significant difference between the rotary ER effect and the ordinary ER effect is that significant torques develop on the particles in the former case. This effect is not to be confused with the torques that develop due to inter-particle interactions in, e.g., aqueous suspensions, as observed by Hu et al.[5] The torque T on a particle of dipole moment $\vec{p}$ in an applied field $\vec{E}_0$ is

$$\vec{T} = \vec{p} \times \vec{E}_0 \tag{8}$$

Suppose that a particle is subjected to a uniaxial harmonic field $\vec{E}_0 \cos(\omega t)$. If the polarization mechanism is linear, then the dipole moment of the particle will be $\vec{p}(t) = \beta(\omega) r_d^3 \vec{E}_0 \cos(\omega t - \delta(\omega))$, defining the polarizability β and the phase shift δ as functions of frequency. For the field given by Eq. (6), rotating at frequency ω, it is clear that the dipole moment will be

$$\vec{p}(t) = \beta(\omega) r_d^3 E_0 \left(\hat{x}\cos(\omega t - \delta(\omega)) + \hat{y}\sin(\omega t - \delta(\omega)\right) \tag{9}$$

and thus that the torque $\vec{T}$ will be

$$\vec{T} = \hat{z}\beta(\omega) r_d^3 E_0^2 \sin\delta(\omega) \tag{10}$$

For any polarization mechanism, we expect that the phase shift $\delta(\omega) \rightarrow 0$ for $\omega \rightarrow 0$. The behavior of $\beta(\omega \rightarrow 0)$ depends upon the nature of the polarization. If the effective dielectric contrast between the particle and the surrounding fluid is positive, then $\beta > 0$, and the torque for small phase shift δ will tend to rotate the particle in the same sense as the applied field. On the other hand, if the effective dielectric contrast is negative, then $\beta < 0$, and the torque will tend to rotate the particle in the opposite sense to that of the applied field.

As a concrete illustration of this latter case, consider an insulating particle suspended in a fluid of conductivity σ. For simplicity, we suppose that the particle and fluid can both be regarded as possessing underlying dielectric constants $\epsilon = 1$, so we need consider only the conductive response of the fluid. Determining the dipole moment of the particle in an arbitrary applied field is now an elementary problem–current will flow in the fluid around the particle until the charge buildup at the particle surface is sufficient to cancel the normal electric field at the exterior particle surface (Gauss' law insures that there is an electric field interior to the particle in this case). A simple computation then shows that in this case

$$\beta(\omega) = -\frac{1}{2}\frac{1}{1+(\omega/\sigma)^2} \tag{11a}$$

and

$$\delta(\omega) = \tan^{-1}(\omega/\sigma) \tag{11b}$$

where the $\omega \rightarrow 0$ result, $\beta = -1/2$, is that of a spherical hole in a medium of infinite dielectric constant.

A disk of particles, each subjected to a torque $\vec{T}$, will rotate at a frequency determined by the balance between this torque and the viscous drag. We model the disk as an oblate spheroid of semi-axes $(a, b, c) = (R, R, r_d)$, with $r_d \ll R$. The viscous torque L on such a spheroid rotating at frequency f in a fluid of viscosity μ_0 is given (for sufficiently low Reynolds number) by[6]

$$L = \frac{32}{3}\mu_0 R^3 f \tag{12}$$

The electrical torque on this disk will be given by the torque per particle from Eq. (10) times the number of particles N_p. For simplicity we write $N_p = \kappa R^2/r_d^2$ with $\kappa \sim 1$, and obtain

$$f \approx \frac{3\kappa\beta(\omega)\sin\delta(\omega)E_0^2}{32\mu_0}\left(\frac{r_d}{R}\right) \tag{13}$$

Note that the frequency is considerably reduced below its naive dimensional value of E_0^2/μ_0 by the factor r_d/R. This factor also insures that the viscous stresses at the edge of an isolated disk, no matter how large it is, are insufficient to cause yielding in the structure.

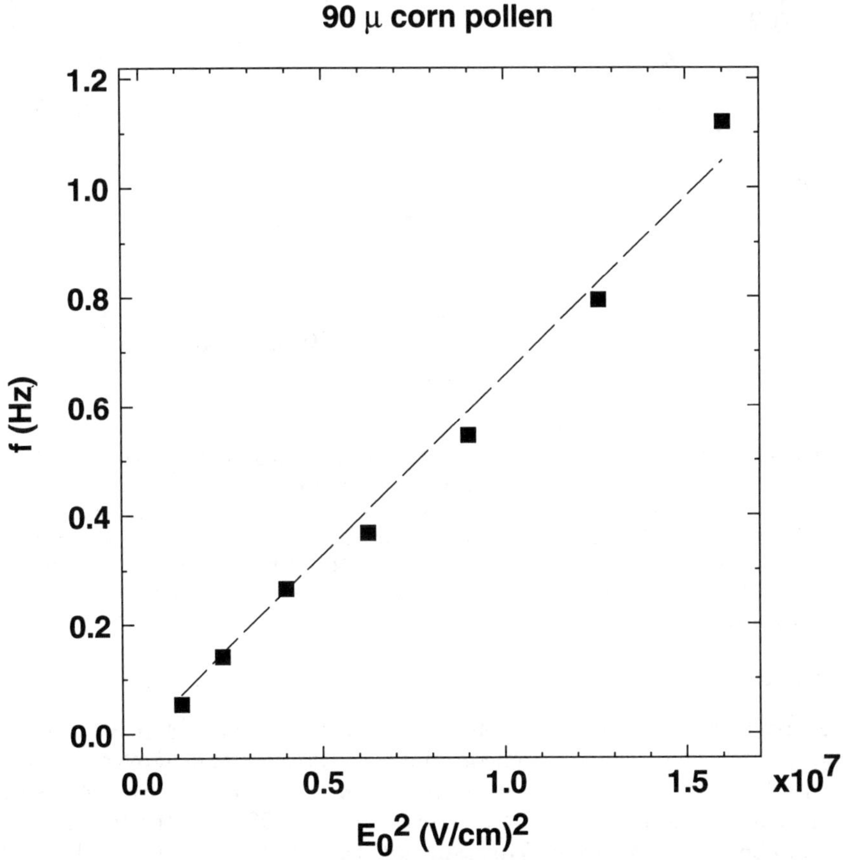

Figure 2. The disk rotation frequency f versus field magnitude E_0^2 for 90 μm corn pollen particles suspended in 4-methylcyclohexanol. The field rotates at a frequency of 1.0 kHz. Clearly $f \propto E_0^2$, as predicted by Eq. (13).

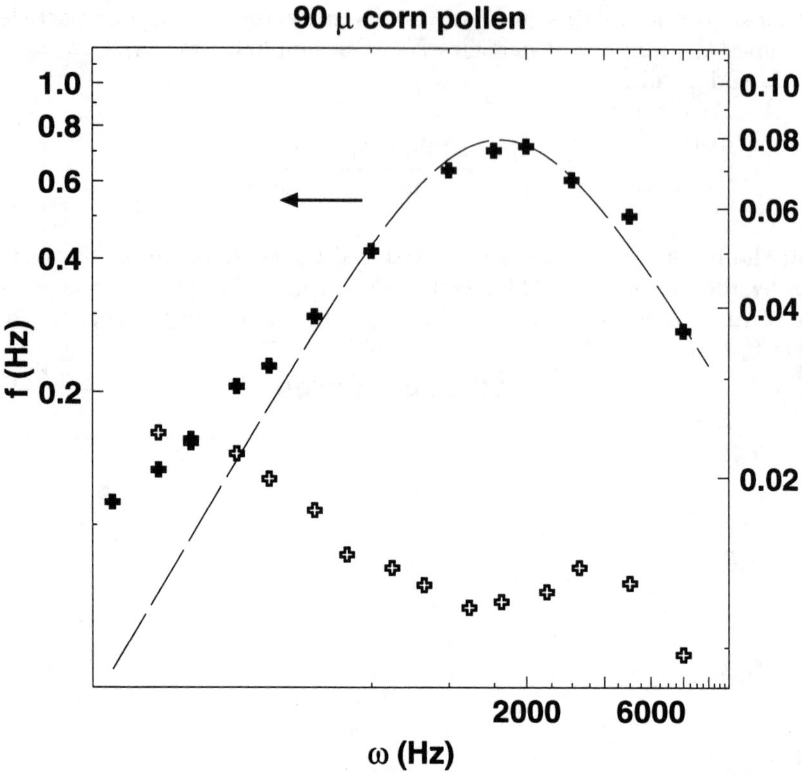

Figure 3. The disk rotation frequency f versus field rotation frequency ω for 90 μm corn pollen particles suspended in 4-methylcyclohexanol (solid crosses, left-hand scale) or mineral oil (open crosses, right-hand scale) at a field magnitude of 4 kV/cm. The conducting solvent (4-methylcyclohexanol) results are well modelled by the Debye form from Eq. (14), with $\sigma = 1.58 \times 10^3 \text{sec}^{-1}$ and $f_0 = 1.49Hz$. The disk rotation in the mineral oil was an order of magnitude slower, with a more complicated frequency dispersion.

4. Experimental Results

We have tested these ideas by constructing a cell bounded by phased electrodes, which generate a spatially homogeneous but rotating electric field over a large region in the interior of the sample. Details of the experimental procedure and results will be published separately.[7] Our ER fluid was made of 90 μm corn pollen particles; these could be easily visualized. We suspended these particles either in the relatively conducting solvent 4-methylcyclohexanol or in a much more insulating mineral oil.

In both solvents we were easily able to see disk formation at fields of from 0.3 to 4 kV/cm, and at field rotation frequencies between 0.5 and 10 kHz. For the conducting solvent, the disks counter-rotated, opposite to the sense of rotation of the field. This agrees with the prediction in section 3 above regarding the behavior of insulating particles in a conducting solvent. Figure 3 shows the dependence of the disk rotation frequency f on applied field; clearly the rotation frequency is $\propto E_0^2$, as predicted in Eq. (13).

The conducting model above predicts, using Eq. (11), that $f \propto \beta \sin \delta$, or equivalently that the dependence of f on field frequency ω is

$$f = f_0 \frac{(\omega/\sigma)}{1 + (\omega/\sigma)^2} \tag{14}$$

Figure 3 shows the disk frequency f versus the field frequency ω for each of the solvents. For the conducting solvent, a fit to the "Debye" form Eq. (14) is also shown; the agreement is good, although there is some additional dispersion at low frequencies. The insulating solvent behavior is not fit by a Debye form; in addition, the disks co-rotate with the field in these solvents, which suggests that the internal dielectric relaxation of the particles is determining the phase lag of the polarization in this case, leading to a positive effective dielectric contrast between the particles and the fluid.

Acknowledgements

T.C.H. is grateful to the NSF for the support of this research through the PYI program, grant DMR-9057156. The work of R.A.A. and J.E.M. was performed at Sandia National Laboratories and was supported by the U.S. Department of Energy under Contract No. DE-AC04-94AL85000.

References

1. See e.g. *Electrorheological Fluids: Mechanisms, Properties, Technology, and Applications,* eds. R. Tao and G.D. Roy, (World Scientific, Singapore, 1994).
2. There is a considerable literature on the torque experienced by a dielectric liquid in a circularly polarized field. See, e.g., E. Grossetti, Nuovo Cimento **10** (1958) 193 ; **13** (1959) 350; and **19** (1961) 1.
3. L. Landau and E.M. Lifshitz, *Fluid Mechanics*, (Pergamon, Oxford, 1959) pp. 63-68.
4. T.C. Halsey and W.R. Toor, *Phys. Rev. Lett.* **65** (1990) 2820; T.C. Halsey, Science **258** (1992) 761.

5. Y. Hu, J.L. Glass, A.E. Griffith, and S. Fraden, *J. Chem. Phys.* **100** (1994) 4674.

6. G.B. Jeffery, *Proc. Roy. Soc.* **A102** (1922) 761.

7. R.A. Anderson, T.C. Halsey, and J.E. Martin, unpublished.

TOWARDS UNDERSTANDING ER FLUIDS USING SALS/RHEOMETRY

Bryan J. Crosby*, Tom McLeish* and Harry Block±

*:Polymer Physics, University of Leeds, Leeds LS2 9JT, UK.
±: SIMS, University of Cranfield, Cranfield, Beds, MK43 OSU, UK.

Abstract

This paper details work in Cranfield and Leeds Universities of making a stock of transparent ER fluids, which could later be utilised in a new optical electro rheometer (OER) to be assembled at Leeds University.

Two basic routes were attempted. One was to use glass microspheres and the other was to use polymer spheres. In order to increase the strength of the ER effect, it was necessary to increase the volume loading while still maintaining sufficient transmission (about 75% over 2 mm). It was found to be possible to increase the ER effect quite substantially in some instances, and in others it was possible to get a near perfect refractive index match. It was not possible to combine both requirements in one fluid such that a high static yield stress was apparent in a transparent ER fluid. However one fluid was made which gave acceptable diffraction losses at high volume fractions, remained in suspension for extended periods and provide about 700 Pa yield stress at 4kV/mm and about 30% volume fraction viz: *untreated poly(ethylene vinyl acetate) microspheres in Cereclor/bromonaphthalene/polystyrene solution.*

The OER being assembled at Leeds University is intended to record small angle light scattering (SALS) profiles, electrical and mechanical properties of ER fluids simultaneously. The OER is based around a DSR 500 machine purchased from Rheometric Scientific with quartz tools coated with transparent indium tin oxide(ITO), which is capable of measuring both steady state (DC) and oscillatory (AC) material parameters.

Introduction

This work has been aimed at gaining a better understanding of the mechanism of how ER fluids work. It is assumed that the application of an electric field induces dipoles in one phase of the fluid. Consequently these electric dipoles align and form structures, which impede the flow of the liquid phase. After a longer interval, strings of dipoles are observed to grow which can traverse the gap between the electrodes[1,2]. The proportion of traversing strings would also be expected to depend on strain rate.

All of the different structures which can form after the application of an electric field should correspond to changes in the rheological properties of the fluid. As far as the authors are aware, no experimental apparatus has been used which measures the applied stress, strain rate, electric field, current and diffraction pattern of an ER fluid simultaneously. Although there are examples of measuring some of these parameters simultaneously[3,4,5,6] or all of them at different times on the same fluid[7]. An apparatus is being assembled at Leeds University to perform all of these measurements under conditions of both steady state and oscillatory, electric and stress fields.

It is intended to measure information about the structure factor by performing light scattering from a HeNe laser beam. Consequently it is necessary to reduce the scattering of 633 nm light in the medium under test so that at least 75% transmission is attained. Greater than 75% transmission should ensure that only single scattering events need to be considered during analysis. Most ER fluids are considerably more turbid than this. However non turbid fluids have been successfully made using silica or glass microspheres and adjusting the base fluids refractive index accordingly[6,8,9]. We have also tried to use glass microspheres, but have attempted to coat them with tin oxide to increase the strength of the ER effect. An increase in the ER effect appears to have been noted in one such fluid, but a greater increase in yield stress was noted by the addition of an anionic surfactant. We have also tested various polymeric particles and found that a commercial sample of poly(ethylene-vinyl acetate) (PEVAC) can be used to form a moderately weak, non-turbid ER fluid at high volume concentrations of solid.

Methods

Bromonaphthalene, aniline, $SnBr_4$, $SnCl_4$, surfactants and most specialist chemicals were AR grade and were purchased from Aldrich Chemical suppliers. Phenylated silicone oils: 550, 702, 704, 705 and 710 were supplied by Dow Corning. Cereclor LV50 is a partially chlorinated paraffin from ICI plc. Cross-linked polystyrene beads were supplied by Polymer Laboratories(10μm diameter, putatively monodisperse). Polymethylene(3-45 μm) and PEVAC(8-30 μm, mean>20 μm) particles, were purchased from Polysciences. Soda-lime glass microspheres(CP5000, 90% between 0.5 and 19.3μm,) were supplied by Croxton and Garry, while the boro-silicate glass microspheres were supplied by Potters-Ballotini(0.5 to 40μm). TP Al 3565 ER fluid was supplied by Bayer AG.

Liquid phase of transparent ER fluid.

Table 1 Fluids refractive indices

liquid	n_d	ρ(g/cm^3)
Cereclor (chlorinated hydrocarbon oil)	1.496	1.2
Dow Corning 702 Diff. pump fluid	1.517	1.09
Dow Corning 704 Diff. pump fluid	1.560	1.07
Dow Corning 705 Diff. pump fluid	1.579	1.09
Dow Corning 710 fluid	1.536	1.11
Dow Corning 550 fluid	1.498	1.07
Bromonaphthalene	1.656	1.489
MW=48K polystyrene: BN = 1:10w/w	1.651	~1.43
MW=218Kpolystyrene:BN=1:13.8w/w	1.654	~1.45

Seven fluids were found to be suitable as suspending media for use in transparent ER fluids. Each of these fluids is miscible in the other six. This is an important property for refractive index(RI) matching as any liquid of an intermediate RI can be made by mixing 2 of the seven fluids. In addition polystyrene is soluble in the fluids and so can be used to increase the viscosity, thus decreasing sedimentation rate of suspensions. The range of

RI's by mixing the viscous fluids is between 1.496 and 1.654. The breakdown potential of the Cereclor and silicone oils is very high and the loss tangent of the bromonaphthalene: polystyrene mixtures is very low indicating that these fluids of themselves should also be able to withstand the high electric fields expected within an ER fluid test rig.

The Bromonaphthalene-polystyrene solution was used to test refractive indices of the solids as it enabled us to access the largest range. It was made by adding weighed quantities of the constituents to a brown, sealable bottle and rotating on a roller mill until the polystyrene was dissolved, forming a clear solution. Heating above room temperature was avoided as this tended to form a cloudy mixture.

Refractive Index matching

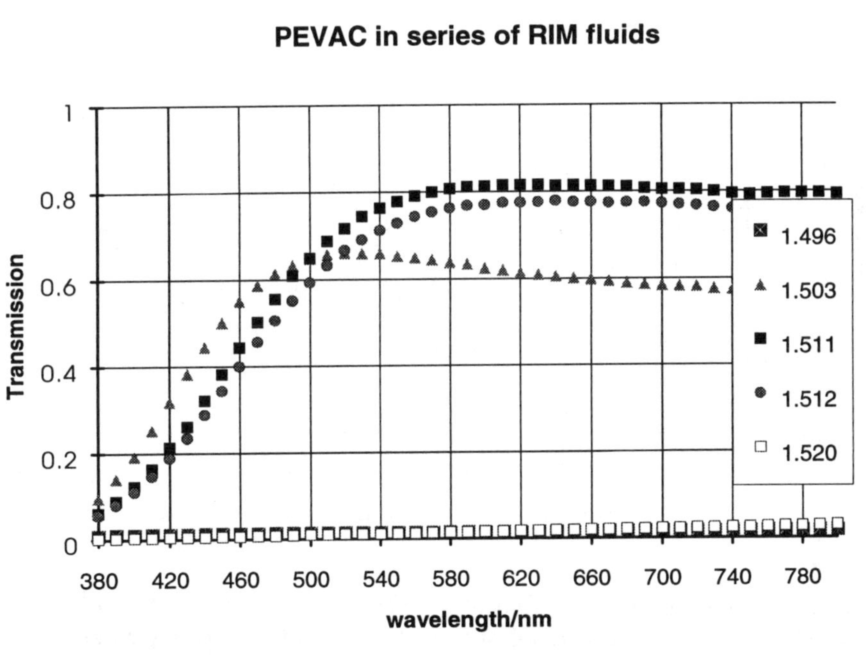

Figure 1. Transmission of suspension of PEVAC (5 volume%) in refractive index matching fluids, path length = 10 mm.

This was done by preparing two solutions with RIs such that one was higher and one lower that the solid to be matched. The solid was then weighed at constant volume loadings into each of the two solutions. The RI of a solution made from two liquids mixed homogenously was assumed to follow a logarithmic law of mixtures[10] :

$$\log(n_{12}) = \varphi_1 . \log(n_1) + (1-\varphi_1) . \log(n_2) \qquad (1)$$

where n_{12}, n_1 & n_2 are the refractive indices of the mixture, liquid 1 and liquid 2, while φ_1 is the volume fraction of solution 1 in the mixture. The transmission was measured in a UV/Vis spectrophotometer, of the mixtures of solid particles in fluid, and weighed mixtures of the two until the best match was determined. Measuring the RIs of some mixtures confirmed Eq.1 within experimental error. The RI of a good match(near maximum transmission) can be determined directly on the Abbe refractometer, or it can be calculated from the volume fractions of each component of the matching liquid using Eq. 1. When done for some mixtures it confirmed the applicability of Eq. 1. A series of graphs was collected as shown in figure 1. The figure shows maximum transmission when the best RI match between solid and liquid phases occur. If the solid phase has a homogeneous RI, then the maximum transmission is nearly 100%. In fact the transmission can appear to be better than that for air if the reference path includes empty cuvettes in the spectrophotometer! This is simply due to there being lower reflectivity at a glass-liquid interface than at a glass-air interface. Figure 2 shows the transmission properties of 6 tested suspensions for light of 633 nm wavelength.

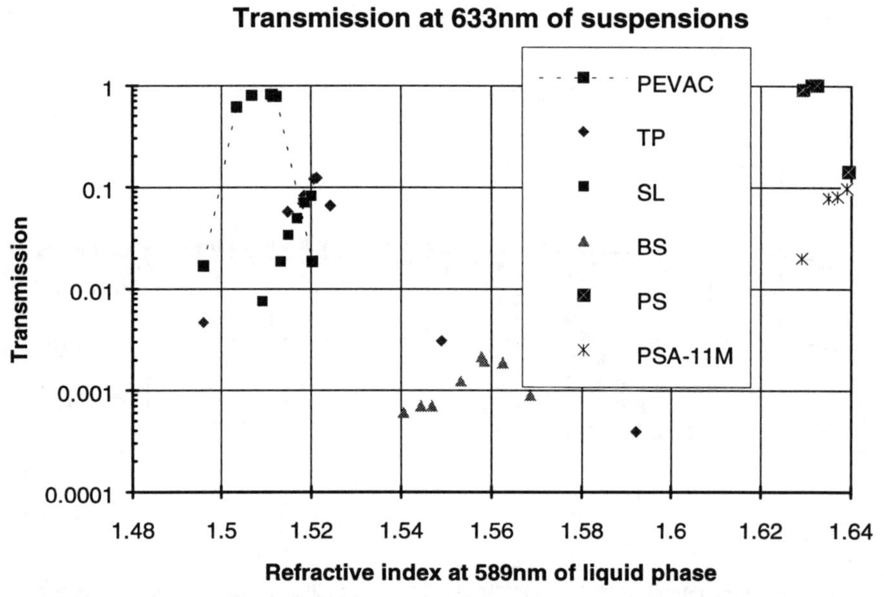

Figure 2. Transmission characteristics of 6 solids suspended in mixed solutions of Cereclor, bromonaphthalene and polystyrene. PEVAC refers to poly(ethylene-vinyl acetate), TP refers to the solid phase of TP AL 3565, SL to soda-lime glass microspheres(CP5000), BS to borosilicate glass microspheres, PS to uncoated cross-linked polystyrene and PSA 11M to a PS sample after being lightly coated with poly-aniline

Tin tetrabromide treated (TTBT) glass microspheres

In order to make the surface of glass spheres electronically conducting, attempts were made to coat glass microspheres with tin oxide. The initial attempts by hydrolysis of a surface layer of $SnCl_4$ did not appear to enhance the ER effect or to produce a coat of SnO_X. It appears that the stage of hydrolysis by water of bound $SnCl_4$ also etched any SnOx off the glass due to the production of HCl. In order to overcome this problem a method was used whereby, hydrolysis was done by steam during coating with $SnBr_4$.

$SnBr_4$ (ρ=3.34)was used as a source of tin because it is a high melting point liquid which is denser than glass(ρ=2.5). Similarly $CHBr_3$(ρ=2.7) also has a density higher than glass. Consequently $SnBr_4/CHBr_3$ mixture was used to float the glass spheres. The $SnBr_4$ was then expected to react at the glass surface and with moisture present near the solution surface because the reaction was performed standing in a heated water bath as indicated in figure 3.

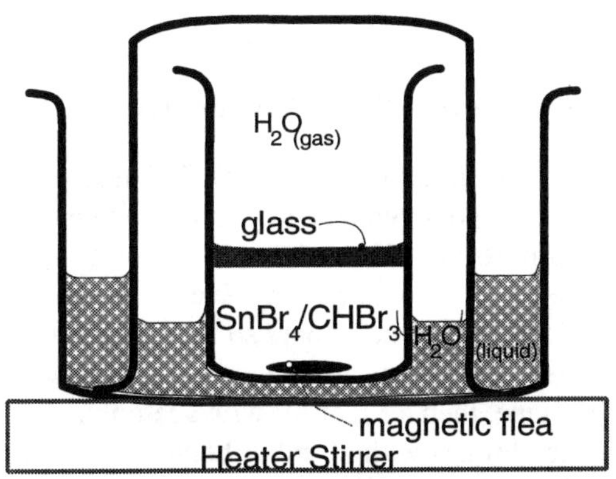

Figure 3. Schematic of reaction vessel for tin tetrabromide treatment (TTBT) of microspheres.

Before use, the soda lime microspheres (CP5000) were sedimented in water for about 15 hours to remove spheres with a radius less that 2μm. Dried spheres were mixed with 9 wt% $SnBr_4/CHBr_3$ (7.1g:70.6g) solution. Glass spheres (13.6g) weighed about twice the weight of $SnBr_4$ used (7.1g). A magnetic stirring flea was placed in the reaction beaker while it was placed in a larger beaker with some water(taking care not to mix the contents of the two beakers). An intermediate size beaker was placed over the reaction beaker. The assembled apparatus was mixed with ultrasonics or magnetic stirrer until reaction appeared to be complete(about 5 hours). A green coloured, viscous mixture ("slime") floated on top of the $CHBr_3$ containing the glass. Initially the glass microspheres tended to aggregate into a viscous 'blob' on top of the $SnBr_4$ solution. After the apparatus was heated until the water was boiling or nearly so, the glass blob spread over the surface of the suspending liquid. The reaction appeared to be controlled, however it was performed in a fume hood.

A glass bubbler was made heat the green slime to a high temperature, so that the coating on the glass could be converted to SnO_x. This bubbler was heated to about 300°C in a fluidised bed bath with flowing nitrogen gas. The heating regime took 2 hours to reach near 300° and was held near that temperature for about 2 more hours.

A brown crusty material resulted from the heat treatment which was fairly easily broken up and removed from the bubbler with a strong glass rod. This material was washed with ethanol and acetone and dried. 5.1 g of this material was mixed with 8 grams of dry Cereclor for measurement of static yield stress and dielectric loss measurements. Later the treated beads were also treated with dichloro-dimethyl-silane by mixing the treated powder with the silane diluted in CH_2Cl_2. The resulting ERF suspended easier in Cereclor than untreated beads and was not observed to flocculate in the absence of electric field.

Surfactant Coating of Glass Microspheres

0.5 g of trimethylcetyl ammonium bromide was dissolved in 9.5 g of water and 20 g of ethanol. To this was added 24 g of CP5000. The mixture was shaken thoroughly then poured carefully onto 80 ml of propan-2-ol. There appeared to be convective mixing at the interface of the two liquids, however the mixture was allowed to sediment for a while and the sediment collected. The unwanted liquid was mostly siphoned away. The spheres were collected by filtration and then washed 2x in propan-2-ol, air dried and suspended in Cereclor, ready for testing dielectric properties and static yield stress.

Poly-aniline Coating of Cross-linked Polystyrene and Other Polymer Spheres

Poly-aniline is a conducting polymer which has been used in ER fluids previously. In this procedure it was used to coat polystyrene beads in order to increase their polarizability and hence potentially form strong ER fluids.

PSA 11N : 53 g of aniline was mixed with 5.2 g of polystyrene microspheres (PSM) until a dispersion formed. It was stirred for about 2 hours so that aniline would soak diffuse into surface of the microspheres. The excess aniline was filtered away by washing in 43g of 6.3%w/w acetone in heptane, then 42 g of 5%w/w acetone in heptane, then 34g of 1% acetone in heptane, 2 times with 20 ml of heptane, resuspended in 1%w/w acetone/heptane, filtered and rinsed with 20 ml of heptane. It was dried in vacuum oven at 40°C for 1 hour with vacuum pumping. It was necessary to remove heptane so that PS can be dissolved in HCl for next reaction. Powder was added to 50 ml of 2M HCl, stirred until mixed and the reaction beaker placed in ice bath for 10 minutes, before acid ammonium per sulphate is added drop wise with continual stirring to polymerise the aniline. The microspheres were then filtered, rinsed with 2M HCl and then propan-2-ol and dried in vacuum oven at 40°C overnight. When PSA 11N was mixed with Cereclor, an emerald green fluid resulted. PSA 11M was made in a similar fashion to this except that washing of aniline was more thorough and colour of powder was a much lighter green.

PSA 11P : 1.5g of PSA 11N was mixed with pH 7.05 buffer, made from buffer tablets. In order to facilitate mixing it was necessary to add about 20 ml of propan-2-ol. The colour changed to light blue. The solids were filtered then rinsed 2x in water and washed 1x in water, then the same rinse wash cycle was repeated for propan-2-ol then acetone. The powder was then dried overnight in a vacuum oven. 1.02 g was mixed with 5.54 g of Cereclor and ultrasonicated to form a blue suspension.

Coating of PEVAC and Poly-methylene spheres was attempted in a similar fashion to PSA 11N above.

Static Yield Stress and Dielectric Loss Measurements

Yield stress was measured on a rig designed at Cranfield University using plates of 4 cm^2 area each and separated by 0.5 mm. The measurement process and rig is described by Block *et al.*[11]. The dielectric loss at frequencies between 12 Hz and 100 kHz was measured using an impedance bridge connected to a parallel plate capacitor with a guard ring and an adjustable spacing.

Results and Discussion

The dielectric loss measurements often indicated when suspensions were too moist and lossy, but once the loss tangent had been reduced by drying the fluid over 4Å molecular sieve, these measurements did not provide conclusive results on ER effectiveness. The static yield stress measurements on glass based ERFs showed that the ER effect could be improved with TTBT and silanizing. The silanizing had the effect of increasing the dispersibility of the spheres and may also have had the effect of reducing the water absorption. However the ER effect was not conclusively shown to be due to SnO_x formation on the surface. Figure 2 shows that the glass spheres used here do not have a homogeneous RI because the maximum transmission was less than 0.1. Attempts to find a source with a homogeneous RI have so far been unfruitful. However two methods have been found to enhance the static yield stress under electric fields, so that if a source of glass spheres with homogeneous RI is obtained, then the methods described here would be expected to enhance their ER effectiveness.

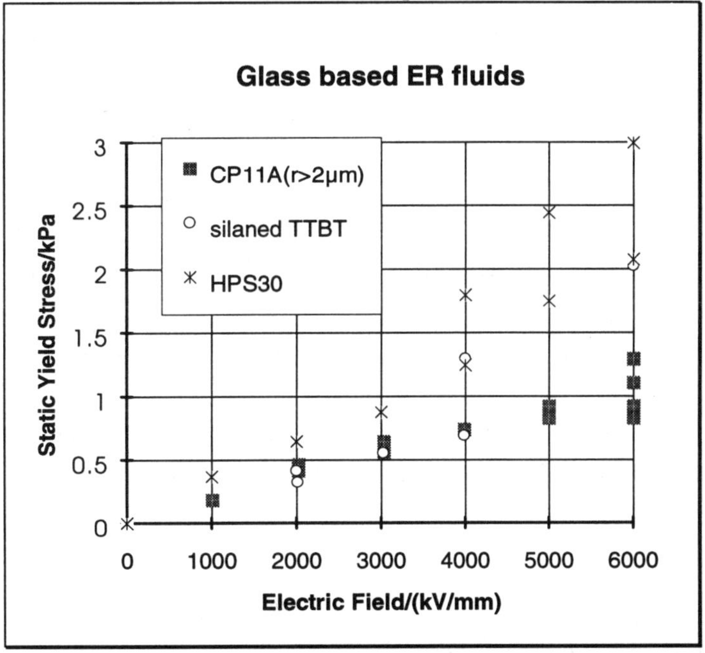

Figure 4: Static yield stress on 30 volume%. "CP11A" is CP5000 with radius greater than 2μm, in Cereclor. "HPS30" has been treated with an anionic surfactant, and "silanized TTBT" has been treated with $SnBr_4$ and a dichloro dimethyl silane as indicated in the methods section.

The static yield stress for polymer based systems are presented in figure 5. This shows that poly-aniline coating of polymer spheres can increase their static yield stress significantly. PSA 11P gave a static yield stress of 2.5 kPa at 6 kV/mm field. However for samples that transmitted more than 10% of 633 nm light, such as PSA 11M, then no ER effect was measured. Thus poly-aniline coating of transparent spheres did not produce ER fluids which were transparent, as its specific absorptivity was too high.

It was found however that PEVAC based ERFs were sufficiently active (possibly surfactant mediated) without coating and figures 1 and 2 showed that its transmission properties were also adequate under high volume loadings when suspended in a liquid with RI ~ 1.51. Figure 2 shows that the solid phase in TP Al3565 ER fluid has a RI of 1.52. Its maximum transmission was just over 10%. However, part of the scattering was due to agglomeration of particles in the RI matching fluids. The application data sheets for TP Al3565 state that it is a strong ER fluid.

Consequently if the ER properties can be maintained while resuspending it homogeneously in an index matching medium, then it will form the basis of another model transparent ER fluid. The other suspensions were not observed to flocculate in absence of electric field, although this was not rigorously tested.

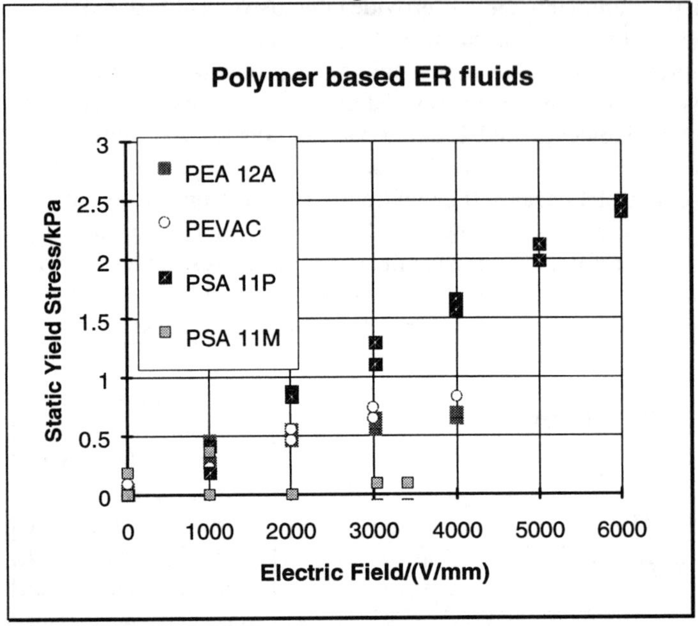

Figure 5 : Static yield stress for 4 polymeric based ER fluids at ~30%v/v in Cereclor. PEA 12A refers to poly-methylene spheres coated lightly with poly-aniline, PEVAC to uncoated PEVAC, PSA 11M to poly-styrene spheres coated lightly with poly-aniline (the 633 nm transmission for PSA 11M is shown in figure 2) and PSA 11P to the PS spheres prepared as described in Methods section

The next stage of this project is to study the transparent ER fluids in an optical electro rheometer, where we will be able to accurately measure stress, strain, electrical and structural properties and can begin as soon as construction has been completed.

Acknowledgments : This work has been funded by an EPSRC grant, and aided by Andrew Stollard, Davinda Riat and discussions with Geoff Ashwell.

References

[1]W.M. Winslow, J.Appl.Phys., **20**: 1137 (1990).

[2]D.J. Klingenberg and C.F. Zukoski, *Langmuir*, **6**: 15 (1990).

[3]K.L. Smith and G.G. Fuller, *Journal of Colloid and Interface Science*, **155**:183-190, 1990

[4]W.T. Culberson and M.R. Tant, *J. App.Polym.Sci.*, **47**: 395-405

[5]R.H. Ottewill, A.R. Rennie and A.Scholefield, *Progr.Colloid.Polym.Sci.*,**81**: 1-5, 1990

[6]H. Block and P. Rattray, "Recent developments in ER fluids" in *Progress in Electrorheology*, Ed. K Havelka and Philisko, Plenum Press in Press.

[7]J.E. Martin, "Light scattering studies of an electrorheological fluid in steady and oscillatory shear", 66th annual meeting of Society of Rheology, Philadelphia October, (1994).

[8]T.C. Halsey, in "Electrorheological Fluids-Mechanisms, Porperties, Structure, Technology and Applications" R.Tao. ed., World Sci. Pub. Co., (1992).

[9]T-j. Chen, R.N. Zitter and R. Tao, *Phys. Rev. Lett.* **68**: 2555 (1992).

[10]K. Lichtenecker, *Physik. Z.* **10**: 105 (1909).

[11]H. Block, J.P. Kelly, A. Qin and T. Watson, *Langmuir*, **6**:6-14 (1990).

NUMERICAL STUDY OF THE RESPONSE TIME OF ER SUSPENSIONS

JUN-ICHI TAKIMOTO, KEIJI MINAGAWA and KIYOHITO KOYAMA
Department of Materials Science and Engineering
Yamagata University, Yonezawa 992, Japan

ABSTRACT

Transient response of model ER suspensions to step-wise electric field under steady shear flow is studied by numerical simulations. The electrically induced stress increases almost linearly with time up to the steady-state value, and this increase directly reflects the growth of the chain-like clusters of the dispersed particles. As the shear rate increases, the response becomes faster since the growth rate of the clusters increases and the steady-state length of the clusters decreases. The response also becomes faster by increasing the particle density or electric field, but the dependence on the field strength is rather weak.

1. Introduction

Electrorheological (ER) suspensions[1] generally consist of polarizable particles and insulating oil, and their rheological properties can be reversibly changed by an external electric field. It is well known that most ER suspensions behave as Bingham fluids under the electric field, and the yield stress (or electrically induced stress) is proportional to the square of the field strength. Under the electric field, dispersed particles form chain-like clusters in the field direction, and this structure is widely believed to be the main origin of the induced stress. In our previous works[2,3], we have shown, by theory and numerical simulations, that the length of the clusters becomes shorter as the shear rate increases, while the induced stress is independent of the shear rate.

Since the rheological properties of ER suspensions can be easily controlled by electric field, many mechanical applications of ER suspensions have been proposed. From application point of view, one of the advantage of the ER suspensions is their fast response to the applied electric field; typically response time of the order of 10 ms has been reported[4-6]. Under shear flow, a few reports[5,6] indicate that the response becomes faster as the shear rate increases. However, detailed relations between the response time of the shear stress and other physical parameters such as the particle density are not studied yet.

In this paper, we study, by numerical simulation, the transient response of model ER suspensions to the step-wise electric field under steady shear flow. Our aims are (i) to clarify the dependences of the response time of the shear stress on the shear rate, field strength and particle density, and (ii) to understand the physical origin of these dependences. On of the advantage of computer simulation over experiment is that the former can give detailed information about the motion of the particles. Using this information, we will calculate the temporal growth of the clusters, and study the relation of the growth of the clusters with the response time of the shear stress.

2. Method of Simulation

The method of simulation is the same as that used in the previous studies[2,3], and similar to the method used by other researchers[7-9].

We approximate the electrostatic interaction between two particles by the following dipole-dipole interaction:

$$\mathbf{F}(\mathbf{r}) = -\frac{3}{16}G\left[(5\cos^2\theta - 1)\mathbf{e}_r - 2\cos\theta\mathbf{e}_e\right]\left(\frac{2a}{r}\right)^4. \tag{1}$$

Here $\mathbf{r}$ is a radial vector between the two particles, $\mathbf{e}_r$ and $\mathbf{e}_e$ is unit vectors in the direction of $\mathbf{r}$ and the electric field $\mathbf{E}$, respectively, θ is the angle between $\mathbf{e}_r$ and $\mathbf{e}_e$, and a is the radius of the particle. G is a parameter characterizing the strength of the interaction; in the simplest approximation, it is given by

$$G = 4\pi\varepsilon_s\beta^2E^2a^2, \tag{2}$$

with $\beta = (\varepsilon_p - \varepsilon_s)/(\varepsilon_p + 2\varepsilon_s)$. Here ε_p and ε_s is the dielectric constant of the particle and the dispersing fluid, respectively.

The hydrostatic force $\mathbf{f}$ acting on a particle is approximated by simple Stokes resistance

$$\mathbf{f} = -6\pi\eta_s a(\mathbf{v} - \mathbf{v}_s), \tag{3}$$

where η_s is the viscosity of the dispersing medium, $\mathbf{v}$ is the velocity of the particle, and $\mathbf{v}_s$ is the unperturbed velocity of the fluid. The velocity $\mathbf{v}$ at a given instant t is obtained by balancing this resistance force $\mathbf{f}$ with the sum of the electrostatic forces acting on the particle. The position of the particle at the "next" instant $t+dt$ is obtained by moving the particle with the velocity $\mathbf{v}$ for the interval dt. In the present simulation, we take $dt = 0.01t_0$, where t_0 is defined by

$$t_0 \equiv \frac{6\pi\eta_s a^2}{G} = \frac{3\eta_s}{2\varepsilon_2\beta^2E^2}. \tag{4}$$

In the following, the time t and the shear rate γ are measured in unit of t_0 and $1/t_0$, respectively. In the typical case of $\varepsilon_s \sim \varepsilon_0$ (dielectric constant of the vacuum), $\beta \sim 1$, $\eta_s \sim 0.01$ Pa·s and $E \sim 1$ kV/mm, we get $t_0 \sim 1$ ms.

Simulations are carried out in 2-dimension; the center of each particle is restricted to be on a single plane. The flow velocity is in the x-direction, and the velocity gradient and the electric field are in the y-direction. Number of the particles is 168. The ratio of the length of the system in x and y-direction is 6:7, and the sheared periodic boundary condition is used. The initial configuration of the particles is generated by Monte Carlo simulation of the hard disk system. At t=0 the electric field is turned on, and the following increase of the shear stress is calculated as a function of time t.

3. Summary of the Results for Steady Electric Field

Before discussing the response of the fluid to step-wise electric field, let us summarize the results of our previous studies[2,3] for steady electric field.

Under electric field, the dispersed particles form chain-like clusters along the field direction. If shear flow is applied, these clusters are broken into short segments. The maximum length n_c of the clusters which can exist under the electric field E and the shear rate $\dot{\gamma}$ is given by

$$n_c = \sqrt{\frac{2F_c}{3\pi\eta_s a^2\dot{\gamma}}}, \tag{5}$$

where F_c is the maximum of the x-component of the attractive force between two particles. With the interaction given by Eq.1, F_c is given by

$$F_c = \frac{G}{\sqrt{15}}. \tag{6}$$

As the shear rate $\dot{\gamma}$ increases, the clusters become shorter in proportion to $1/\sqrt{\dot{\gamma}}$. The electrically induces stress σ_e, on the other hand, is almost independent of the shear rate, and given by

$$\sigma_e \approx 0.2\phi_2 \frac{F_c}{a}. \tag{7}$$

Here ϕ_2 is the "area fraction" of the dispersed particles. Note that stress is defined as a force per unit length in 2-dimensional system. If we assume that the total shear stress is given by the sum of σ_e and the stress due to the viscosity of the fluid, the independence of σ_e on $\dot{\gamma}$ means that the fluid behaves as a Bingham fluid, as experimentally observed.

4. Response to Step-wise Electric Field

Now let us study the response of the fluid to sudden application of the electric field. Fig.1 shows the time dependence of the electrically induced stress σ_e for various

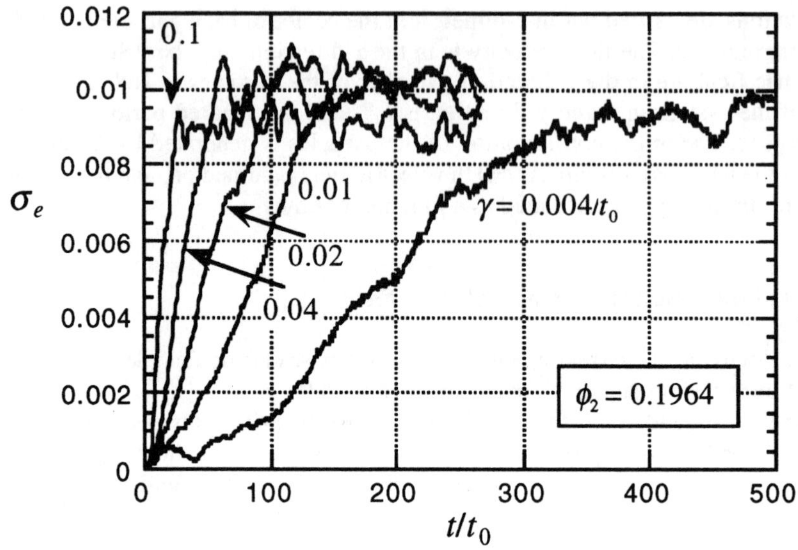

Fig.1 Transient response of the electrically induced stress σ_e at various sherar rate γ. σ_e is measured in units of G/a. Particle density is $\phi_2 = 0.1964$.

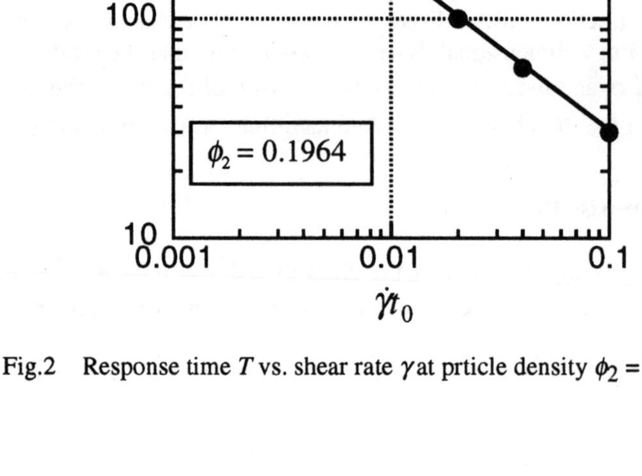

Fig.2 Response time T vs. shear rate γ at prticle density $\phi_2 = 0.1964$.

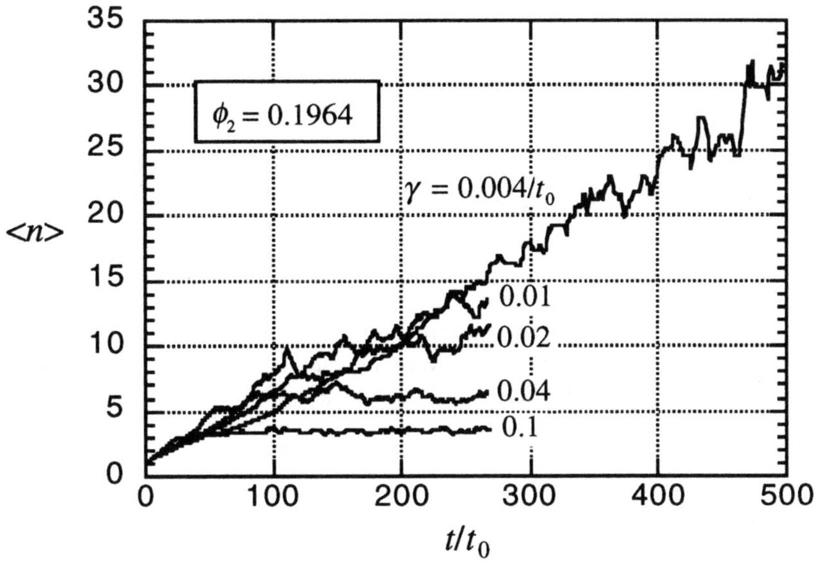

Fig.3 Time dependence of the average size of clusters at various sherar rate γ. Particle density is $\phi_2 = 0.1964$.

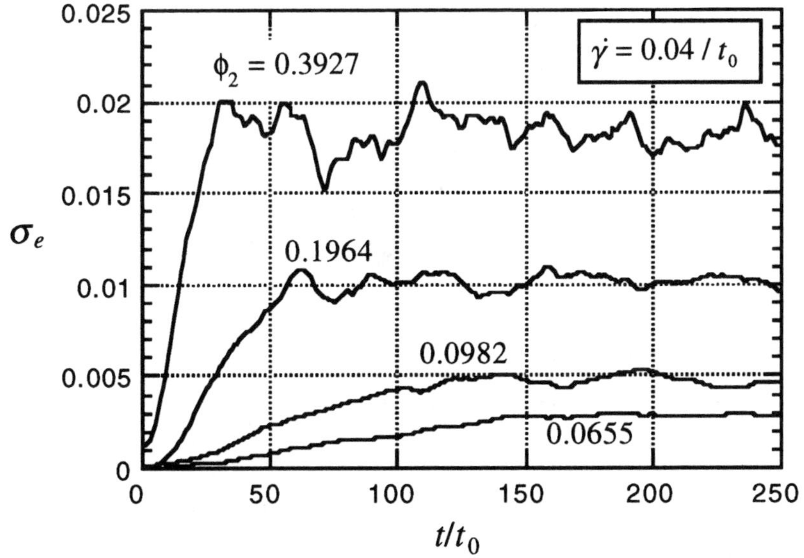

Fig.4 Transient response of the electrically induced stress σ_e at various particle density ϕ_2. σ_e is measured in units of G/a. Shear rate is $\gamma = 0.04/t_0$.

shear rates. Each curve in this figure is obtained by the average of four independent simulations. Three points should be noted here: (i) the steady-state values of the stress do not depend the shear rate as predicted by Eq.7, (ii) the stress increases up to the steady-state value almost linearly with time, and (iii) the response becomes faster with increasing shear rate.

To discuss the last point more quantitatively, we define the response time T as follows: the increasing part of the stress is fitted by a straight line $\sigma_e \propto t$, and the steady-state value of the stress is fitted by a line σ_e=const. The response time T is defined as the time at which these two lines cross. Fig.2 shows the dependence of the response time T on the shear rate. The straight line in the figure corresponds to $T \propto \dot{\gamma}^{-3/4}$.

To see what is determining this response, the average size <n> of the clusters is calculated as a function of time. Here, the size of a cluster is defined as the total number of particles in the cluster. The results are shown in Fig.3. Note that the steady-state size of the clusters decreases as the shear rate increases, as predicted by Eq.5. By comparing this figure with Fig.1, we can see that the response time of the stress corresponds to the time necessary for the cluster to form.

Next we study the dependence of the response time on the density of the particles. Fig.4 shows the transient response of the induced stress for various area fraction ϕ_2. Note that the steady-state stress linearly increases with ϕ_2 as predicted by Eq.7. Also note that the response becomes faster as ϕ_2 increases. This can be easily understood by considering the increase of the collision frequency with ϕ_2. Fig.5 shows the dependence of the response time T on ϕ_2. We can see that T is inversely proportional to ϕ_2.

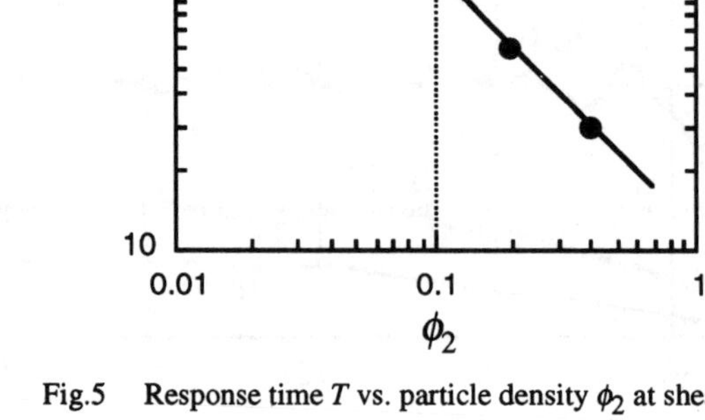

Fig.5 Response time T vs. particle density ϕ_2 at shear rate $\dot{\gamma}$= 0.04 / t_0

The results of Fig.2 and Fig.5 can be summarized as

$$\frac{T}{t_0} = \frac{c}{\phi_2^{\alpha} \cdot (\gamma t_0)^{\delta}} \tag{8}$$

with

$$\alpha = 1, \qquad \delta = 3/4 \qquad \text{and} \qquad c \approx 1.0. \tag{9}$$

The present simulations are carried out in 2-dimension. In 3-dimensional system, the values of the exponents α and δ (and the constant c) may somewhat differ. But the general tendency that the response becomes faster as γ or ϕ_2 increases is expected to hold also in 3-dimension. By combining Eqs.4, 8 and 9, we get $T \propto E^{-1/4}$. Thus the response becomes faster as the electric field E increases, but the dependence on E is rather weak. Also note that the particle size has no effect on the response time if ϕ_2 is kept constant.

The reason that the response becomes faster with increasing γ is twofold: one is that the collision frequency among particles (or among clusters) increases with γ, and thus the growth rate of the cluster size increases. Another is that the length of the clusters at the steady state (Eq.5) becomes shorter as γ increases, and it takes shorter time for clusters to reach steady-state length even if the growth rate is constant. As the particle density ϕ_2 increases, the collision frequency increases, and the growth rate of the cluster increases. Thus the response becomes faster as ϕ_2 increases. As the strength of the electric field E increases, the growth rate becomes faster, but the steady state length of the cluster becomes longer. Thus the response time depends on E only weakly.

Experimentally, Ginder and Ceccio[6] have found that the response time is approximately proportional to $1/\gamma$ at high shear rate and depends on the field strength only weakly. The results of our simulation qualitatively agree with their results. If we take $t_0 \sim 1$ ms (see Eq.4 and the paragraph following it), Eq.8 gives $T \sim 20$ ms at $\gamma =$ 100/s and $\phi_2 = 0.3$. Experimental values[6] for zeolite and aluminosilicate suspensions under similar condition are about 2 ms and 20 ms, respectively.

Before closing this section, let us briefly discuss the limitation of the present simulation. Although order of magnitude of T given by the simulation is consistent with the experiments, quantitative prediction of T by the present simulation is rather difficult, considering the following points; (i) The dipole-dipole interaction (Eq.1) are based on the assumption that the polarization on each particle is fixed to that of an isolated particle. When two particles are close to each other, however, there should be a mutual enhancement of the polarization. Thus Eq.1 underestimates the interaction strength, and overestimates the response time. (ii) Eq.2 ignores the hydrodynamic interaction among particles. Since this interaction is repulsive when two particles are approaching each other, use of Eq.2 may underestimate T. (iii) Simulations are carried in 2-dimension.

In the limit $\gamma \rightarrow 0$, Eq.8 predicts infinitely long response time. This is because the steady-state length of the clusters becomes infinitely long in this limit. In actual experiments, however, system size in the y-direction (the direction of $\mathbf{E}$) is limited by electrodes, and if the length of the clusters reaches this limit, the response time would show different shear rate dependence. Experiments by Ginder and Ceccio[6] show that the response time approaches constant value in the small shear rate limit. Simulations which take account of the effect of the electrodes, as well as simulations in 3-dimensional system, are left for future studies.

In the high shear rate limit ($\dot{\gamma} \rightarrow \infty$), Eq.8 predicts infinitely rapid response. Actual response time, however, can not become arbitrary short but is limited by the dielectric response time of the particles.

5. Conclusion

Transient response of the electrically induced stress of ER suspensions directly reflects the growth of the clusters. As the shear rate $\dot{\gamma}$ increases, collision frequency among particles increases, and thus the growth rate of the clusters increases. Moreover, the steady-state length of the clusters becomes shorter with increasing $\dot{\gamma}$. Thus the response time (=the time necessary for the clusters to develop) becomes shorter as $\dot{\gamma}$ increases. The response also becomes faster with increasing particle density due to the increase of the collision frequency.

Fast response to the electric field is one of the advantages of ER fluids. In order to utilize this merit in practical applications of ER fluids, it is preferable to design a system so that the shear rate in the device is sufficiently high. As pointed out by Ref.6, however, the shear stress under no electric field increases with $\dot{\gamma}$, and the dynamic range of the shear stress decreases. Thus the shear rate in the device should be carefully chosen in order to optimize the performance of the system.

References

1. H. Block and J. P. Kelly, *J. Phys. D: Appl. Phys.* **21** (1988) 1661.
2. J. Takimoto, in*Proc. 3rd Int. Conf. ER Fluids*, ed. R. Tao (World Scientific, 1992) p.53.
3. J. Takimoto, *Nihon Reoroji Gakkaishi* **20** (1992) 95 (in Japanese).
4. K. Tanaka, T. Yoshida and K. Koyama, in *Proc. 3rd Int. Conf. ER Fluids*, ed. R. Tao (World Scientific, 1992) p.289.
5. A. Hosseini-Sianaki, W. A. Bullough, R. Firoozian, J. Makin and R. C. Tozer, in *Proc. 3rd Int. Conf. ER Fluids*, ed. R. Tao (World Scientific, 1992) p.219.
6. J. M. Ginder and S. L. Ceccio, *J. Rheol.* **39** (1995) 211.
7. D. J. Klingenberg, F. van Swol and C. F. Zukoski, *J. Chem. Phys.* **94** (1991) 6160.
8. H. See and M. Doi, *J. Phys. Soc. Jpn.* **60** (1991) 2778.
9. H. See and M. Doi, *J. Rheol.* **36** (1992) 1143.

ORIENTATIONAL DYNAMICS OF VISCOELASTIC MAGNETIC SUSPENSIONS

YURI L. RAIKHER, VICTOR V. RUSAKOV

Institute of Continuous Media Mechanics
Urals Branch of the Russian Academy of Sciences, Perm, 614061, Russia
E-mail: ylr@icmm.perm.su

and

IGOR G. SEMAKIN

Mathematical Department, Perm State University
Perm, 614005, Russia

ABSTRACT

We present a theoretical model to describe a dynamic orientational response of a suspension of single-domain magnetic particles in a viscoelastic liquid carrier of the Maxwell's fluid type. Introducing thermal rotary diffusion, a pertinent Fokker-Planck equation is obtained and solved approximately. Absorption spectra of initial dynamic magnetic susceptibility display a complex comb-like structure the origin and properties of which are discussed.

1. Introduction

In this paper we derive and analyze the dynamic susceptibility of a statistical assembly of fine ferroparticles in a Maxwellian fluid. The latter is a simple conventional model of a polymeric solution. Of course, due to its simplicity, it gives up a lot of details relevant for the behavior of a particular polymer. The gain, though, is the opportunity to get acquainted with the effect of retardation—probably the most general non-Newtonian feature of polymeric solutions—on the frequency-dependent magnetic response. Our present study continues the work started in [1,2]. Its long-range objective is to develop a theoretical basis for microrheological measurements. The idea of the latter is to use the dynamic response of ferromagnetic nanoparticles embedded in a bulk of a complex fluid as a method of non-destructive probing of its fine structure.

2. Magnetic particle in a viscoelastic fluid

Let us consider a two-dimensional orientational motion of a single-domain magnetically rigid particle in a viscoelastic fluid. We choose the 2D representation for its utter simplicity. The 3D one, though far more cumbersome in mathematical treatment, would provide the results differing from ours but by numerical coefficients of the order of unity. This fact, though maybe lacking a fundamental proof, is widely used as a practical way to shorten calculations in the theory of rotary Brownian motion since its foundation by Debye—see[3], for example.

In the adopted framework a magnetic particle is a solid disk of radius a confined in the arbitrary plane passing through the direction of the applied magnetic field $\boldsymbol{H}$, which we take for the polar axis of the coordinate system. The magnetic moment $\boldsymbol{\mu}$ of the particle is constant in its absolute value, its direction inside the plane of the disk is fixed. Orientation of the vector $\boldsymbol{\mu}$ with respect to the polar axis is defined by the angle ϑ. The equation of rotational motion then writes

$$I\ddot{\vartheta} + Q(t) + \mu H \sin\vartheta = y(t)\,. \qquad (1)$$

Here the dots denote differentiation with respect to time, I is the moment of inertia of a particle when rotated about its center, $Q(t)$ is the resistance torque, and the noise torque $y(t)$ accounts for thermal motion in the fluid.

For the Maxwellian model the resistance torque exerted by the fluid carrier on a rotating particle is given by equation

$$\dot{Q} = -\tfrac{1}{\tau_M}(Q - Q_0) \qquad \text{with} \quad Q_0 = \zeta\dot{\vartheta}\,, \qquad (2)$$

where ζ is the friction coefficient proportional to the fluid viscosity η and τ_M is the characteristic time of elastic stress relaxation. The Maxwellian time τ_M, as well as the viscosity, is the characteristics of the fluid proper and as such is insensitive to the presence of particles.

To clarify the meaning of the relaxational equation (2) let us substitute there the angular displacement in the form $\vartheta = \vartheta^{(0)}\exp(-i\omega t)$. For the torque amplitudes one finds

$$\mathrm{Re}Q^{(0)} = \frac{\zeta\omega^2\tau_M}{1+\omega^2\tau_M^2}\,\vartheta^{(0)}\,, \qquad \mathrm{Im}Q^{(0)} = \frac{\zeta\omega}{1+\omega^2\tau_M^2}\,\vartheta^{(0)}\,.$$

At low frequencies ($\omega\tau_M \ll 1$) only the imaginary part of the resistance is relevant, which leads to $Q \approx -i\mathrm{Im}Q \propto \dot{\vartheta}$, meaning that the torque is predominantly viscous. The case of the Newtonian fluid is recovered at $\omega \to 0$, where $Q = Q_0$—see Eq. (2). At high frequencies one has $Q \approx \mathrm{Re}Q \propto \vartheta$, as if the fluid were transformed in a Hookean elastic body with the effective strain modulus η/τ_M. In this case the response has a reversible, non-dissipative, character. In the intermediate range the reaction of the system is a superposition of viscous (fluid-like) and elastic (solid-like) contributions.

As it has been shown[1], the random torque in Eq. (1) is not a white noise. It is correlated with a finite time τ_M, and evolves according to equation

$$\widehat{R}\,y \equiv \left(1 + \tau_M\frac{d}{dt}\right) y = f(t), \qquad \big\langle f(t)f(0)\big\rangle = 2\zeta k_B T\delta(t)\,, \qquad (3)$$

Adding this relationship to Eqs. (1) and (2), one obtains a closed set of stochastic equation with a standard white noise $f(t)$. Note that the linear retardation operator

$\widehat{R}$ entering the lhs of Eq. (3) coincides with that of Eq. (2) for Q. That means that the thermal and resistance torques have the same type of frequency dispersion. Eliminating Q and y, Eqs. (1)–(3) unite in the Langevin equation of the problem, that reads

$$\dddot{\vartheta} + \frac{1}{\tau_M}\ddot{\vartheta} + \frac{1}{I}\left(\frac{\zeta}{\tau_M} + \mu H \cos\vartheta\right)\dot{\vartheta} = -\frac{\mu}{I\tau_M}\sin\vartheta\,\widehat{R}\,H + \frac{1}{I\tau_M}f(t)\,. \tag{4}$$

With Eq. (4) it is easy to proceed to any statistical description of the particle assembly, e.g., obtain the Fokker–Planck equation (FPE).

3. Dynamic susceptibility from the Fokker-Planck equation

The corresponding FPE has been derived in[2]:

$$\begin{aligned}&\left(\frac{\partial}{\partial t} + \dot{\vartheta}\frac{\partial}{\partial\vartheta} + \ddot{\vartheta}\frac{\partial}{\partial\dot{\vartheta}}\right)W \\ &\quad = \frac{1}{I\tau_M}\frac{\partial}{\partial\ddot{\vartheta}}\left\{\left[I\ddot{\vartheta} + \zeta\dot{\vartheta} + \mu\sin\vartheta\left(H + \tau_M\frac{dH}{dt}\right)\right]W\right\} + \frac{\zeta k_B T}{I^2\tau_M^2}\frac{\partial^2}{\partial\ddot{\vartheta}^2}W\,.\end{aligned} \tag{5}$$

Here the function $W(\vartheta, \dot{\vartheta}, \ddot{\vartheta}, t)$ is the probability density for a particle to be at the point $(\vartheta, \dot{\vartheta}, \ddot{\vartheta}, t)$ of the phase space at the moment t, and is normalized to unity.

We remark that all the combinations of material parameters of the system entering Eq. (7) may be expressed in terms of three characteristic times. Namely, the already introduced Maxwellian time τ_M, the Debye time $\tau_D = \zeta/k_B T$ and the inertial time $\tau_I = I/\zeta$. The detailed estimates[4] show that in colloids and suspensions the inertial time is always much less than the orientational diffusion (Debye) one, i.e., $\tau_I \ll \tau_D$. In a number of cases this condition allows to achieve a considerable simplification of the kinetic equation, and to treat it in the so-called *non-inertial* approximation. Then the only relevant variable is the angle (or the set of angular coordinates, see[4]). However this is valid only for low-frequency phenomena. As soon as one intends to study the processes affected by the dynamic elasticity of the fluid carrier, the non-inertial approach becomes insufficient and the problem must be described in full, with the aid of Eq. (7) or its 3D version, if necessary.

In the absence of the external field the equilibrium solution of Eq. (7) is

$$W_0(\dot{\vartheta}, \ddot{\vartheta}) = Z^{-1}\exp\left[-I\left(\dot{\vartheta}^2 + \tau_I\tau_M\ddot{\vartheta}^2\right)/2k_B T\right]\,. \tag{6}$$

Here Z is a partition function defined as usual. The way to obtain the distribution (8) is as follows. Transforming the Langevin equation (4) to the equivalent chain of moment equations, one finds that at $H = 0$ the stationary solution of the latter is rendered by $\langle\dot{\vartheta}^2\rangle = k_B T/I$ and $\langle\ddot{\vartheta}^2\rangle = k_B T/I\tau_I\tau_M$. Since the only distribution function entirely defined by its second moments is a Gaussian, this yields Eq. (8).

To shorten notations, let us introduce dimensionless variables defining the angular velocity and acceleration as

$$v = \dot{\vartheta}/\dot{\vartheta}_T, \qquad w = \ddot{\vartheta}/\ddot{\vartheta}_T. \tag{7}$$

Here $\dot{\vartheta}_T = \sqrt{2k_BT/I}$ and $\ddot{\vartheta}_T = \dot{\vartheta}_T/\sqrt{\tau_I\tau_M}$ are the characteristic thermal angular velocity and angular acceleration of the particles, respectively.

To evaluate the linear response of an assembly of independent particles with a number density n to the probing external field, i.e., the dynamic magnetic susceptibility

$$\chi = n\mu\,\partial\langle\cos\vartheta\rangle/\partial H, \tag{8}$$

it is sufficient to linearize Eq. (7) with respect to H. Using dimensionless variables (9), one gets

$$\left[\tau_M\frac{\partial}{\partial t} + \alpha v\frac{\partial}{\partial\vartheta} + \beta\left(w\frac{\partial}{\partial v} - v\frac{\partial}{\partial w}\right)\right] W(\vartheta, v, w, t)$$
$$= -\gamma w\left[(\xi + \tau_M\dot{\xi})\sin\vartheta + \alpha\xi v\cos\vartheta\right] W_0 + \frac{\partial}{\partial w}\left(w + \frac{1}{2}\frac{\partial}{\partial w}\right) W, \tag{9}$$

$$\alpha = \dot{\vartheta}_T\tau_M = \sqrt{2\tau_M^2 k_BT/I}, \qquad \beta = \sqrt{\tau_M/\tau_I}, \qquad \gamma = \sqrt{2\tau_M/\tau_D}.$$

Here $\xi = \mu H/k_BT$ is a dimensionless field strength, and under the adopted definitions of the relaxation times the equality $\alpha = \beta\gamma$ holds. The first term in rhs of Eq. (11) has been transformed taking into account the explicit form of the equilibrium distribution (8).

Since the variation range of v and w—see Eq. (9)—is unbounded, and function W_0 is a Gaussian form in them, it is natural to expand the non-equilibrium solution of the FPE (11) in the Hermit polynomial series as

$$W(\vartheta, v, w, t) = W_0(v, w)\left[1 + \sum_q\sum_{m,n} A_{mn}(q,t)e^{iq\vartheta}H_m(v)H_n(w)\right]. \tag{10}$$

where $q = \pm1, 2, \ldots$ and $m, n = 0, 1, 2, \ldots$. Then the sought for macroscopic quantity, i.e., dynamic susceptibility (10) may be obtained with the aid of relation

$$\langle\cos\vartheta\rangle = \tfrac{1}{2}\langle e^{i\vartheta} + e^{-i\vartheta}\rangle = \tfrac{1}{2}\left[A_{00}(1,t) + A_{00}(-1,t)\right], \tag{11}$$

where the angular brackets denote averaging over the distribution function W. Substituting formula (12) into Eq. (11), after standard transformations one arrives at a set of algebraic equations for the amplitudes

$$(n - i\omega\tau_M)A_{mn}(q,\omega) + \tfrac{1}{2}i\alpha q\left[A_{m-1,n}(q,\omega) + 2(m+1)A_{m+1,n}(q,\omega)\right]$$
$$+\beta\left[(m+1)A_{m+1,n-1}(q,\omega) - (n+1)A_{m-1,n+1}(q,\omega)\right] \tag{12}$$
$$= \tfrac{1}{4}\gamma\xi\delta_{n,1}\left\{\delta_{q,1}\left[(i+\omega\tau_M)\delta_{m,0} - \tfrac{1}{2}\alpha\delta_{m,1}\right] - \delta_{q,-1}\left[(i+\omega\tau_M)\delta_{m,0} + \tfrac{1}{2}\alpha\delta_{m,1}\right]\right\}.$$

Here we assume that a response to a probing oscillating magnetic field $H(t) = H\exp(-i\omega t)$ may be presented as $A(q,t) = A(q,\omega)\exp(-i\omega t)$.

The infinite three-index set (14) is rather difficult to solve. As a first step, let us consider the simplest approximation setting all the amplitudes A_{mn} with the sum of indices $m+n>1$ equal to zero. This reduces the number of equations to three:

$$\begin{aligned} \omega\tau_M A_{00}(1,\omega) - \alpha A_{10}(1,\omega) &= 0\,, \\ \omega\tau_M A_{10}(1,\omega) - \tfrac{1}{2}\alpha A_{00}(1,\omega) - i\beta A_{01}(1,\omega) &= 0\,, \\ (1-i\omega\tau_M)A_{01}(1,\omega) + \beta A_{10}(1,\omega) &= \tfrac{1}{4}\gamma\xi\,(i+\omega\tau_M)\,. \end{aligned} \tag{13}$$

The solution yields one of the amplitudes needed to for evaluation of $\langle \cos\vartheta \rangle$, namely

$$A_{00}(1,\omega) = \tfrac{1}{2}\xi\,\frac{1-i\omega\tau_M}{(1-i\omega\tau_M)\,[1-2(\omega\tau_M/\alpha)^2] - 2i\omega\tau_M(\beta/\alpha)^2}\,. \tag{14}$$

The complementary term, instead of deriving and solving a set similar to (15), may be found immediately using a general symmetry relation

$$A_{mn}(q,t\,|\,\alpha,\beta,\gamma) = A_{mn}(-q,t\,|\,-\alpha,\beta,-\gamma)\,,$$

following from Eqs. (14). Thence $A_{00}(1,\omega) = A_{00}(-1,\omega)$ leading to $\langle \cos\vartheta \rangle = A_{00}(1,\omega)$, and from formula (10) for the imaginary part χ'' of the susceptibility, i.e. absorption, one gets

$$\frac{\chi''(\omega)}{\chi_0} = \frac{2\omega\tau_M(\beta/\alpha)^2}{[1-(2\omega\tau_M/\alpha)^2]^2 + \omega^2\tau_M^2\,[1+(2/\alpha)^2(\beta-\omega^2\tau_M^2)]^2}\,. \tag{15}$$

where $\chi_0 = n\mu^2/2k_BT$ is the static susceptibility for a 2D system. The last expression recovers the result[1] obtained directly from the Langevin equation (4) using the method proposed by Coffey[5]. It has a plausible form and incorporates all the reference times of the model. Formula (18) may be rigorously expanded in a sum of: *(i)* relaxational quasi-Debye oscillation with a characteristic frequency $\sim 1/\tau_D$ and *(ii)* a resonant Lorenzian line centered at $\omega_0 \sim [(\tau_M+\tau_D)/\tau_I\tau_M\tau_D]^{1/2}$. However, the high-temperature limit of the resonance at $\tau_M > \tau_D$ (pronounced viscoelasticity) looks somewhat strange: one gets $\omega_0 \sim \sqrt{k_BT/I}$ as if viscoelasticity had no effect on oscillations.

4. Higher approximations

Though difficult to prove directly, one may suppose that the truncation of the set (14) yielding expression (18) for the dynamic susceptibility, is somewhat equivalent to an expansion of the exact result with respect to parameter τ_M/τ_D. Therefore, justification of the range of validity of formula (18) is necessary.

As a test, we had shifted, the cutoff value of the sum of indices in Eq. (14) to $m+n>2$. The resulting set comprised six equations and yielded for $\chi(\omega)$ a formula too cumbersome to present here. However, when analyzing it, one encounters a qualitative change of the form. In particular, in the absorption spectrum $\mathrm{Im}\chi(\omega)$ besides the resonance maximum at ω_0 predicted by equation (18), another maximum at $\omega \approx 2\omega_0$ turns up. Moreover, the ratio of their heights is $\mathcal{A}(\omega_0)/\mathcal{A}(2\omega_0) \sim \tau_M/\tau_D$. This fact is a serious argument in favor of the assumption proposed at the beginning of this section, and it definitely reduces the validity range of Eq. (18) to $\tau_M \ll \tau_D$.

To achieve satisfactory accuracy, we have carried out numerical calculations increasing the number of amplitudes A_{mn} involved in the sets constructed alike Eq. (15). It is easy to show that the size of the functional basis, if $m+n=N$ is the cutoff value, is $(N+1)(N+2)/2$. Ascending in N, we have obtained the results for $N \leq 13$ that means the basis of up to 251 Hermit polynomials.

5. Results and discussion

To present and discuss the novel results, we pass to a time scale based on τ_I thus introducing dimensionless parameters

$$M=\tau_M/\tau_I\,, \qquad D=\tau_D/\tau_I\,, \qquad \Omega=\omega\tau_I\,. \tag{16}$$

As it is shown in[4], in a suspension the condition $\tau_I \ll \tau_D$ always holds. Because of that, the values of M and D for real systems are high, whereas the relevant frequency ranges are $\Omega \ll 1$—see below for numerical estimations.

An example of a susceptibility pattern $\mathrm{Im}\chi(\Omega)$ obtained for $N=10$, i.e., with 66 functions, is shown in Fig. 1. Note that due to the fact that the width of peaks is much smaller than inter-peak distances, only the essential parts of the curve are plotted. All over the ranges in between the framed peaks, the susceptibility is very small in the adopted vertical scale. The same applies to the fourth, etc. peaks expected to the right of the third one.

In general, the numerical study of $\chi(\Omega)$ may be described as follows. At $M \ll D$ (weak viscoelasticity) the susceptibility is given by formula (18). At pronounced viscoelasticity ($M \sim D$) in the high-frequency range a sequence of equidistant peaks positioned at $\Omega_k = k\Omega_M$ with $k=1,2,\ldots$ emerges. This set is based on the reference value $\Omega_M = 1/\sqrt{M}$, and the peak widths obey the relation

$$\Delta\Omega_k = 1/D + k/2M\,. \tag{17}$$

The heights $\mathcal{A}_k$ of the peaks behave differently depending on the ratio M/D. For $M/D<1$ the sequence $\{\mathcal{A}_k\}$ decreases monotonically. For $M/D>1$ it first grows, passes a maximum at $\Omega \sim \Omega_T = \sqrt{2/D}$ and finally goes down to zero. The curves of Fig. 1 illustrate the behavior of the latter type.

As for the low-frequency ($\Omega \sim 1/D$) relaxational maximum of $\mathrm{Im}\chi(\Omega)$ which the system "inherits" from the classical case, i.e., Newtonian fluid, with M growing it becomes less pronounced, and eventually disappears.

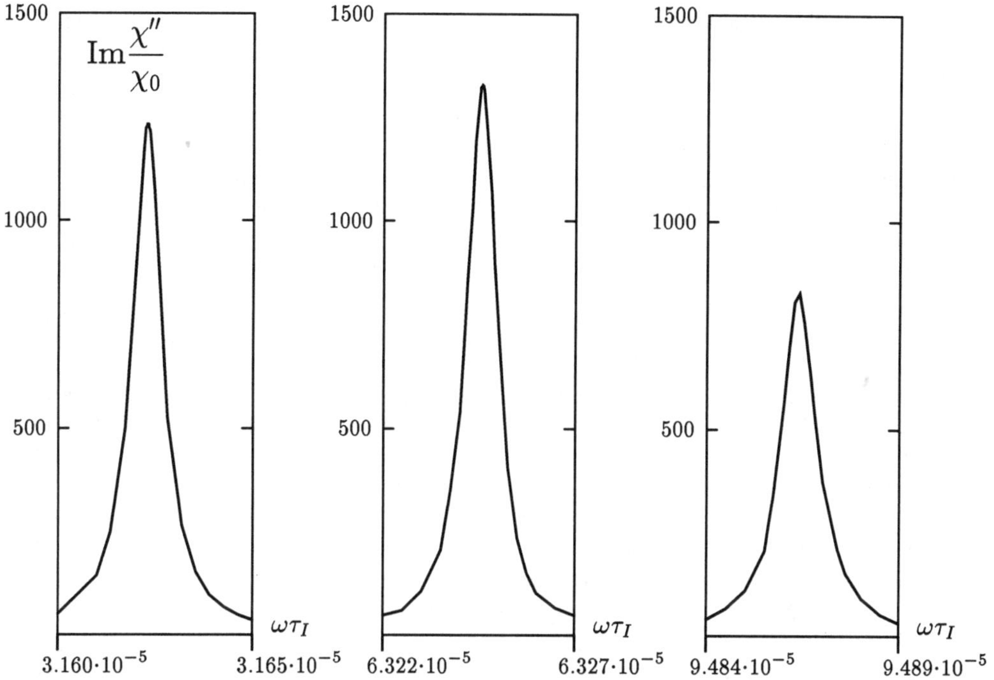

Figure 1: Absorption line for a viscoelastic magnetic suspension. The set of parameters corresponds to that of Eq. (18). Due to the tiny width of lines in comparison to the inter-peak distances, the latter are contracted—see the marks on the horizontal axis.

With the available numerical tools we were not able to proceed to the limit of high viscoelasticity which should take place at $\sqrt{M} \ll D$. The corresponding work is in progress and would be published elsewhere.

The obtained resonant structure of the high-frequency absorption spectra in a system of dipolar rotators is due to the interplay of inertial and elastic forces acting on a particle. The estimate for the basic resonant frequency Ω_M may be easily recovered right from Eq. (1) if to set the inertial term written as $I\Omega_M^2/\tau_I^2$, equal to the high-frequency limit $\zeta/\tau_M = I/\tau_I\tau_M$ of the resistance torque. However, the thermal motion is essential for "propagation" of resonances, i.e., $\Omega_k = k\Omega_M$. The matrix representation of the FPE (14) shows how the chain-linking of different (m,n)-modes occurs.

We remark that a similar situation has been previously dealt with in the molecular spectroscopy theory when analyzing linear non-Markovian stochastic processes[3]. In particular, a multi-peak spectral pattern has been predicted when a molecule was schematized as a set of two co-centered inertial rotators coupled through a torque-retarding elastic link (the itinerant oscillator model)[6]. However, in molecular liquids the range of material parameters where the multi-peak structure might occur, turned out to be on the edge of the physical realm, and due to that pronounced rather poorly [7].

Let us show that with composite fluid media like viscoelastic magnetic suspensions, the multi-peak patterns are achievable, at least in principle. For estimates we take a spherical solid particle of a radius $a \sim 0.1\,\mu$m in a Maxwellian fluid with the viscosity $\eta \sim 1$ P and the stress relaxation time $\tau_M \sim 0.1$ s. Then we get

$$\tau_I \sim a^2/\eta \sim 10^{-10}\,\text{s}\,, \qquad \tau_D \sim a^3\eta/k_BT \sim 3\cdot 10^{-2}\,\text{s}\,,$$

which in dimensionless units corresponds to

$$M = 10^9\,, \quad D = 3\cdot 10^8\,, \quad \Omega_T = 10^{-4}\,, \quad \Omega_M = 3\cdot 10^{-5}\,, \quad \Delta\Omega_k \sim 10^{-9}\,. \tag{18}$$

The last set renders the actual values used for numerical plotting of the curves in Fig. 1. Apparently, the positions and widths of the peaks shown, fairly well agree with the given estimations of Ω_M and $\Delta\Omega_k$. Transforming the characteristic frequencies back to a dimensional form, one obtains $\omega_M \sim 3\cdot 10^5$ and $\Delta\omega_k \sim 10-10^2\,\text{rad}^{-1}$ that is rather usual for magnetic susceptibility measurements. However, prior to an attempt of experimental verification of the given predictions, one essential point must be clarified. It is: How does the ever present in any suspension distributions of stress relaxation times and particle sizes affect the fine structure of the spectral patterns.

6. Acknowledgments

We are grateful to I. I. Wertgeim and P. A. Sosnin for their help in the preliminary numerical studies, and to W. T. Coffey for bringing Refs. 6 and 7 to our attention.

This work was partially supported by grants: RMD 300 from the International Science Foundation and Russian Government, and 95–02–03953 from the Russian Basic Research Foundation.

7. References

[1] Yu. Raikher and V. Rusakov, *J. Magn. Magn. Mater.* **122** (1993) 172.
[2] Yu. Raikher and V. Rusakov, *Abstr. VI Int. Conf. on Magnetic Fluids* (Paris, 1992) 340; *Physica A* (1995), submitted.
[3] W. T. Coffey, M. Evans and P. Grigolini, *Molecular Diffusion and Spectra* J. Wiley & Sons, New York, 1984. Ch.4.
[4] Yu. L. Raikher and M. I. Shliomis, *Adv. Chem. Phys.* **87** (1994) 595.
[5] W.T. Coffey, *J. Chem. Phys.* **93** (1990) 724; **95** (1991) 2026.
[6] W. T. Coffey, P. M. Corcorane and M. Evans, *Mol. Phys.* **61** (1987) 15.
[7] W. T. Coffey, P. M. Corcorane and J. K. Vij, *Proc. Roy. Soc. A* **412** (1987) 339.

HIGH SPEED JOURNAL BEARINGS LUBRICATED WITH ELECTRO-RHEOLOGICAL FLUIDS. AN EXPERIMENTAL INVESTIGATION

P. G. NIKOLAKOPOULOS
Mechanical Engineering Dept., University of Patras, Machine Design Lab
Patras, 265 00, Greece

and

C. A. PAPADOPOULOS
Mechanical Engineering Dept., University of Patras, Machine Design Lab
Patras, 265 00, Greece

ABSTRACT

It is well known that the imposition of an electric field on an Electro-Rheological (ER) fluid alters the viscosity and as a consequence the flow properties of the fluid. If such a fluid is used to lubricate a journal bearing system, it is expected that the imposition of an electric field between the rotor and the stator will cause the alteration of the dynamic properties of the journal bearing. For the present, it has been proved that this is valid only for low speeds and high radial clearances of Couette type viscometers. In this paper an experiment in a high speeds (16000 to 65000 s^{-1}) journal bearing with small radial clearance is presented. The experiment performed has showed the phenomenon and has proved that the ER FLUID at high shear rates under constant temperature, follows the Bingham model in realistic bearings. Properties such as wall shear stress, dynamic yield stress relative viscosity are experimentally determined as functions of the electric field, for different particle concentrations and the shear strain rate under constant or free to vary temperature (due to operating conditions, angular velocity, friction). Concluding the ER fluids can be used to create "smart" journal bearings. Vibration controllers can be constructed to control the stability of the ER fluid lubricated bearings.

1. Introduction

Smart materials and structures have sensing, actuating, or controlling capabilities, either by themselves or throughout embedded sensors, while they are also capable of adapting themselves to changes of the external conditions to maintain a specified goal. Such materials are the optical fibers, the piezoelectric crystals, the memory alloys, the magneto-rheological (MR) and the electro-rheological (ER) fluids. The ER fluids consist of dielectric particles (mainly hydrophilic) dispersed in electrically non-conducting liquids. Externally applied voltage rapidly transforms these fluids from Newtonian to pseudo-plastic materials with higher apparent viscosity.
The phenomenon of electro-rheology, was first reported by Winslow[1] and occasionally bears his name, i.e. Winslow effect. It appears as an increased resistance in flow, and in some cases, the fluid converts itself to a solid when an increase of the electric field take place[2].
Klass and Martinek[3, 4] have reported experiments with silicone-based ER fluid for shear

strain rates ranging from 0 to 2300 s^{-1}. Their data show that a threshold value of shear stress, the "static yield stress", is required to initiate the flow at zero shear strain rate. This excessive shear stress was defined by Kraynic[5] as dynamic yield stress since it refers to dynamic shearing where the hydrodynamic forces dominate. Constant dynamic yield stress under high shear strain rates implies pseudo-plastic behavior described by the Bingham model. Bullough[6] reported tests for a 28% volume fraction of starch in transformer oil for shear strain rates in the range from 0 to 350 s^{-1}. He also observed anomalous behaviour at low shear strain rates and dynamic yield stress reduced with shearing. Moisture effects have been investigated by Filisko and Radzilowski[7], who mention that Zeolite suspensions exhibit ER effects even with small amounts of water. A complete study was also reported by Sprecher et al[8] and the electromechanical properties of zeolites were investigated for small shear rates approximately to 10 s^{-1}.

Kollias and Dimarogonas[9], in their experimental work, have examined the bulk properties of Zeolite and corn starch based ER fluids at shear rates from 0 to 20000 s^{-1}.

The influence of temperature is very complex, since it affects the fluid dielectric properties and the surfactant action. For example, a fourfold increase of the shear force was observed when temperature increased from 25°C to 65°C[3, 4].

Nelson and Suydam[10] presented a series of experimental results for the influence of temperature on the electro-rheological effect and the other properties of mechanisms that utilised ER fluids.

Oppermann et al[11] presented experimental results using Bayer electro-viscous fluid (ER fluids). The shear stress and the dynamic viscosity were measured for various temperatures between 25°C and 110°C.

The principle of ER fluid based devices is the wide variation of apparent viscosity (from fluid to solid) and the fast response time (order of several milliseconds)[12]. One very promising application is the use for vibration control of high speed lubricated rotor bearing systems, based on the electrical alteration of stiffness and damping.

Many applications have been reported in the literature. Nikolajsen and Hoque[13] investigated the performance of ER squeeze film damper and recognised its efficiency. Morishita and Mitsui[14] have presented their results on the successful application of the ER fluid to a squeeze film damper which controls vibration amplitude of a flexible rotor in a wide range of rotating speeds. Tichy[15] applied the Wada et al[16], method for Bingham materials, for an one dimensional squeeze film damper bearing. Leek et al[17] presented an experimental investigation for the transport properties of unexcited and excited electro-rheological fluid in a Rayleigh step bearing.

Wada, Hayashi and Haga presented the three dimensional general theory for a Bingham solid in hydrodynamic lubrication in a finite journal bearing[16]. They have also presented some analytical and experimental results using Bingham solid as a lubricant to step bearings[18]. Tichy[19] presented the two dimensional modified Reynolds equation for Bingham materials for use in journal bearings.

Kollias and Dimarogonas[20] tested low speeds (28 rpm) partial journal bearings. They have also measured the produced hydrodynamic pressure and compared their results with analytical models using the theory of Wada et al. for Bingham materials and observed good agreement. Also in their papers[21] the possibility to stabilise rotor bearing systems using electro-rheological fluids is shown.

In the experimental work of this paper two procedures are presented. a) In the first one the temperature of the lubricant was not kept constant. It was increesing during the experiment because of the operating speed resulting in different friction coefficients

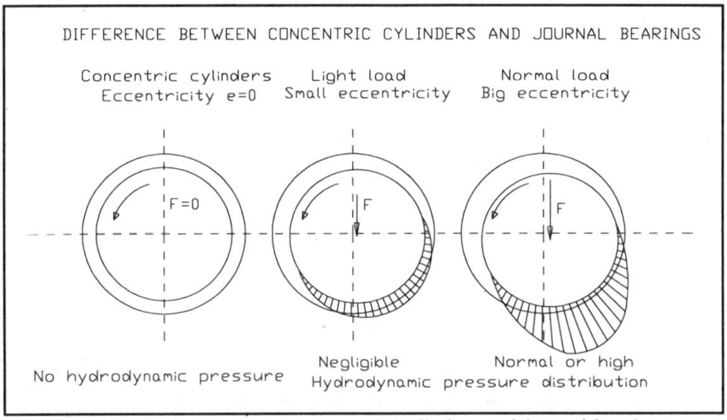

Fig. 1. Difference between concentric cylinders and journal bearings.

and the limited volume of ER fluid used (one litre). But, for each operating condition the measurements were taken at the steady state. b) In the second procedure the temperature was kept constant, within the limit of some degrees Celsius, using air and/or water heat exchangers. The bulk hydrodynamic properties of zeolite based ER fluids at high shear rates were investigated for both procedures. The contribution of this paper is the test of an ER fluid for usage in high speed journal bearings having small radial clearance. The shear rates were between approximately 25000 and 65000 s^{-1} and 16000 to 35000 s^{-1} with the ratio of the journal radius over the radial clearance r/c = 264.5 for c=100 μm and r/c = 176.5 for c=150 μm. These ratios show that the experimental machine is not only a common concentric viscometer but a real journal bearing. The other goal achived in the present paper is the use of common materials, for the construction of the ER fluid which is suitable for journal bearing operation. Zeolite particles are dispersed in SAE 30 oil which have very good lubrication properties for such small clearances.

2. ER-Fluid Lubricated Journal Bearing and Rheology

In the rotor bearing systems (journal bearings, squeeze film bearings), where the ER fluids could be applied, the shear strain rates are in general much higher than the maxima of most of the published experimental results of ER fluid applications. Moreover for high shear rates the validity of the Bingham model was questioned by many authors without a specific experimental verification.

When no load is applied on the bearing no eccentricity appears and thus the journal and bearing centers coincide. In this case the film thickness is constant and equal to the radial clearance of the bearing and there is no hydrodynamic pressure (Fig. 1). The operation of the bearing in this case is similar to a Couette type viscometer. If a light or normal load is applied on a journal bearing system then a small or normal eccentricity appears and the film thickness and the hydrodynamic pressure distribution is shown in Fig. 1. The first case of light loaded bearings, which is the case presented in this paper, could be approximated by the analysis valid for concentric cylinders, or by the Petroff's law. The light loads were due to the bearing self weight (2 kp).

In the experimental work described below the main goal is the investigation of the ER

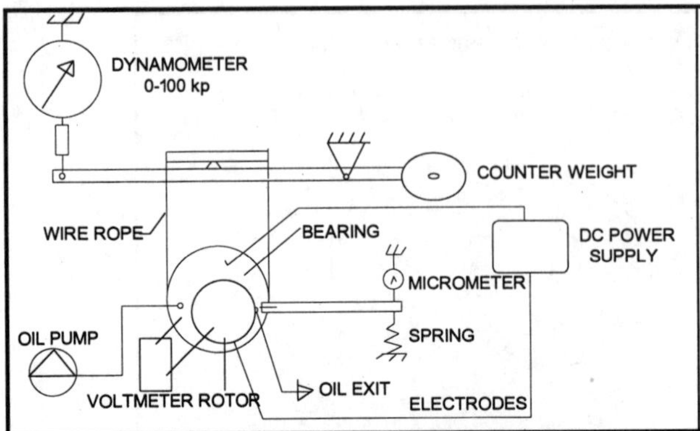

Fig. 2. General view of the experimental device

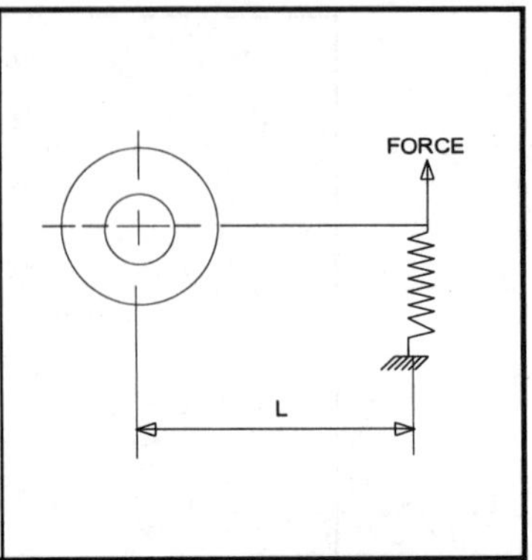

Fig. 3. Torque measurement system

fluid in a real journal bearing machine with true ratio r/c and shear strain rate approximately between 16000 to 65000 s^{-1}. In this experiment it was observed that the tested ER fluids behaved like Bingham fluids, since there was linear relation between torque and speed, under constant temperature during the course of the measurements. The Bingham model in one dimensional flow could be written as follow:

$$\tau = \tau_o + \mu_o \dot{\gamma} \quad (1)$$

where τ is the shear stress, μ_0 is the plastic viscosity, $\dot{\gamma}$ is the shear rate, $\dot{\gamma} \cong \omega r / c$, ω is the angular velocity of the rotor and τ_0 is the dynamic yield stress. The shear rate can approximately be given by the above relation because of the light bearing load assumption where the film thickness is almost constant. At high speeds the material completely yielded and sheared. The shear stress is related with torque by,

$$\tau = \frac{T}{2\pi L r_1^2}, \quad \text{where } r_1 = \frac{R+r}{2} \quad (2)$$

where L and R are the bearing length and radius respectively and r is the rotor radius. The dynamic yield stress at zero speeds can be calculated as[22],

$$\tau_o = \tau_1 \frac{1 - 1/s^2}{2\ln(s)} \quad (3)$$

where $s = R / r$, R the bearings radius, $\tau_1 = T_0 / (2\pi L r^2)$ and T_o is the dynamic yield torque which can be computed from the extrapolation of the linear correlation of the experimental data (torque versus speed).
Finally the relative viscosity should be considered, which is defined as the viscosity with electric field to zero field viscosity ($\mu_R = \mu_a / \mu_o$).
The viscosity can be calculated from the following equation[22],

$$\mu_a = \frac{\tau_r - \tau_R - 2\tau_o \ln(s)}{2\omega} \quad (4)$$

where $\tau_r / \tau_R = (R / r)^2 = s^2$ or using the Petroff's equation[23]

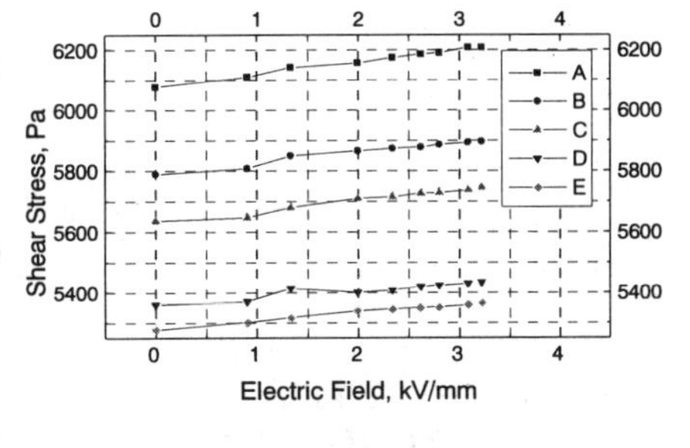

Fig. 4: Shear stress versus electric field. ERF-1.

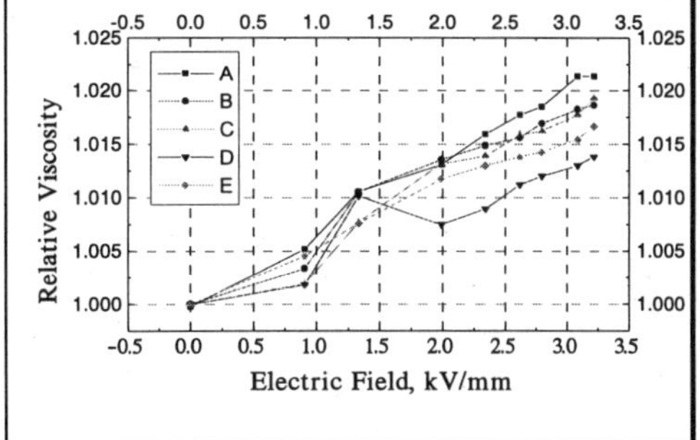

Fig. 5: Relative viscosity versus electric field. ERF-1.

$$\mu_a(\dot{\gamma}) = Tc / (\pi^2 d^3 LN) \tag{5}$$

where N are the revolutions per unit time, L is the bearing length and c is the radial clearance.

3. The ER fluids used.

Oil SAE-30, a common oil for lubrication, (Rotella sx 30 by Shell with kinematic viscosity 91.3 cSt at 40°C and density 0.894 Kg/lt at 15°C) and zeolite particles (Z3125 by Sigma Chemical Co. with Chemical formula $Na_{86}[(AlO_2)_{86}(SiO_2)_{106}]xH_2O$, nominal pore diameter 9-10Å and particle diameter 10 μm) are used to prepare the ER fluid. For the present work four mixtures were prepared, ERF-n, n=1, 2, 3 and 4 to be used in the corresponding experiments. In Table 1, the percentage by weight of zeolite is presented, together with the corresponding hours of drying process and degrees Celcius. The three first mixtures were used in a journal bearing with 150μm radial clearance and the fourth one in the same journal bearing but with 100μm radial clearance.

Table 1 : ER fluids prepared to be used in the experimental procedures

		Drying Process			
Mixtures of ERF's	Zeolite, %	Hours	^{0}C	Loss of weight	Radial Clearance
ERF-1	25%	5h	240°C	9.5%	150μm
ERF-2	30%	5h 45min	220°C	2.3%	150μm
ERF-3	25%	5h	220°C	7.6%	150μm
ERF-4	25%	5h	220°C	5.2%	100μm

4. Experimental setup

A journal bearing system (see Fig. 2), with rotor and bearing diameters $d = 52.9mm$ and $D = 53.2mm$ respectively and radial clearance $c = 150\,\mu m$ was used for the three first experiments. During the fourth experiment another bearing with diameter $D = 53.1mm$ and resultant radial clearance $c = 100\,\mu m$ was also used. The rotor can be rotated from 0 to 2590 rpm driven by a 5 HP AC electric motor with mechanically variable belt transmission. The ER fluid is pumped to the journal bearing by a gear oil pump. The

ERF-2

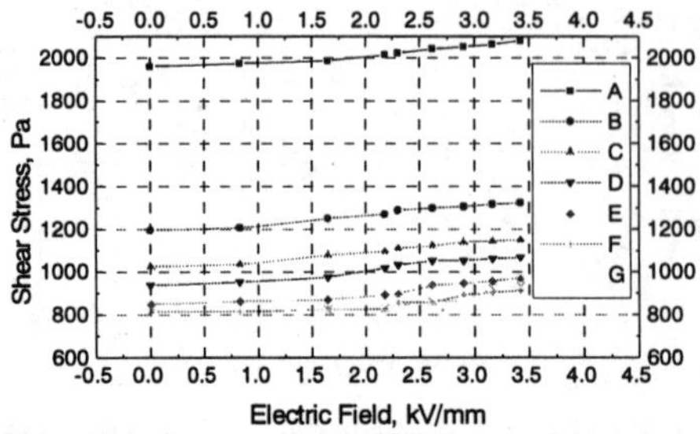

Fig. 6: Shear stress versus electric field. ERF-2

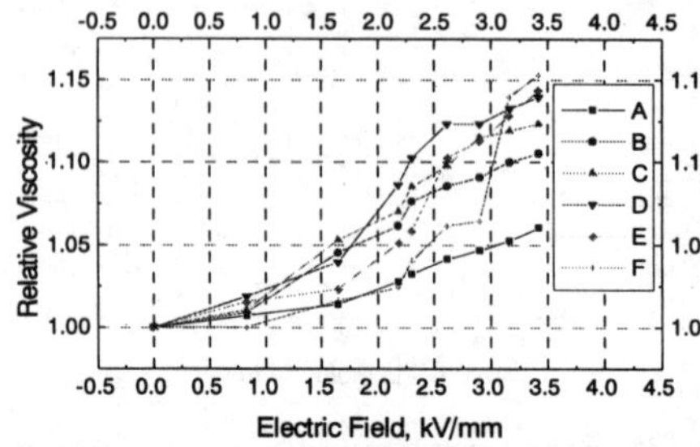

Fig. 7: Relative viscosity vs electric field. ERF-2.

possibility to load the bearing with static forces using the mechanism shown in Fig. 2 also exists. An electric field of some kV/mm, produced by a DC power supply (0-3500V DC), is then imposed on the ER fluid. All the elements of the rotor are perfectly electrically insulated from those of the bearing. The current observed during the experiment was of the order of 1 - 4 mA. Due to the small clearance some hundred of volts are capable of creating high electric fields. For example 200V into the bearing with c=0.1 mm produced 2 kV/mm field density. Every change in the operating conditions (speed, load, temperature, or electric field) causes finally a respective change at the deflection of a spring, measured by a TESA sensitive micrometer with accuracy 1μm.

5. Experimental procedure

Two different experimental procedures (1. Without and 2. With cooling system) are take place in this experimental investigation. In the first one, two similar experiments using ERF-1 and ERF-2 were performed. The shear rates varied from 16000 to 35000 s^{-1}, the applied loads for each shear rate from 0 to 25 kp and the imposed electric field for each combination of shear rate and load from 0 to 3 kV/mm. For each combination of speed, load and electric field the deflection caused at the spring (Fig. 2 and 3) by the different torques was measured. No cooling system was present in this experiments and the temperature varied from 35°C to 55°C during the measurements.

In the second experimental procedure two experiments are also taking place. In the first one the ERF-3 was used and the shear rates changed from 16000 to 35000 s^{-1}. The temperature was kept constant (within some limits) with a water heat exchanger. The system of the temperature stabilization in this experiment consisted of a temperature sensor, an electric valve and a temperature controller in order to control the water flow into the exchanger. The load applied to the bearing was varied from 0 to 25 kp and the temperature was controlled during the measurements between 32°C to 37°C.

In the last experiment the ERF-4 mixture was tested between 25000 and 65000 s^{-1}. The temperature was kept constant using a fan cooling system. In this experiment no load was applied to the bearing except its own weight (2 kp).

Some general comments on the procedure are the follows:

- For both experiments one litre of ER fluid was used.
- For each change of the rotational speed the rotor operated enough time in order to

ERF-3

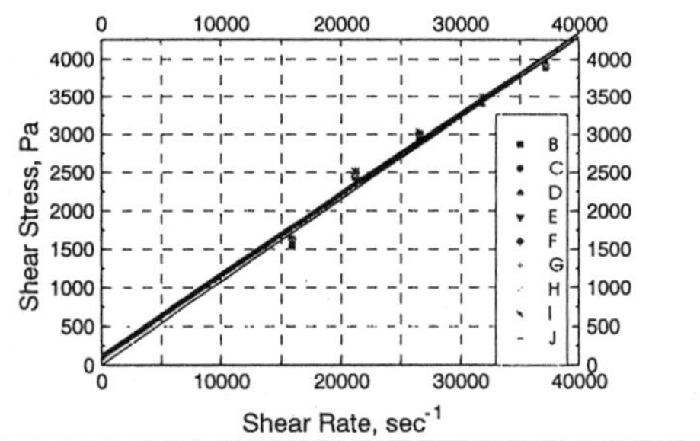

Fig. 8. Shear stress versus shear rate, T=34±2°C

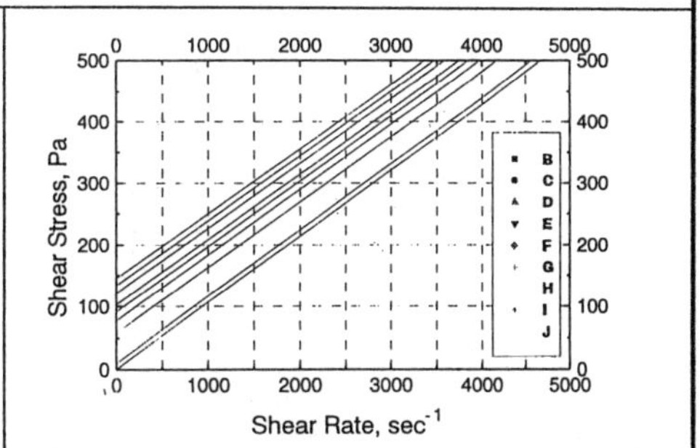

Fig. 9 . Zoom of previous Fig. 8

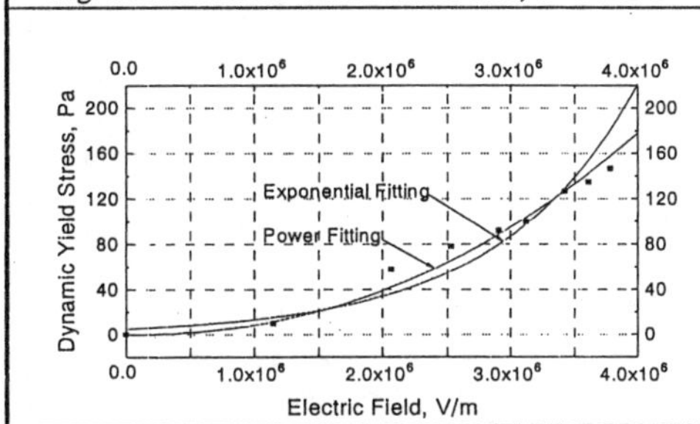

Fig. 10. Dynamic yield stress versus electric field

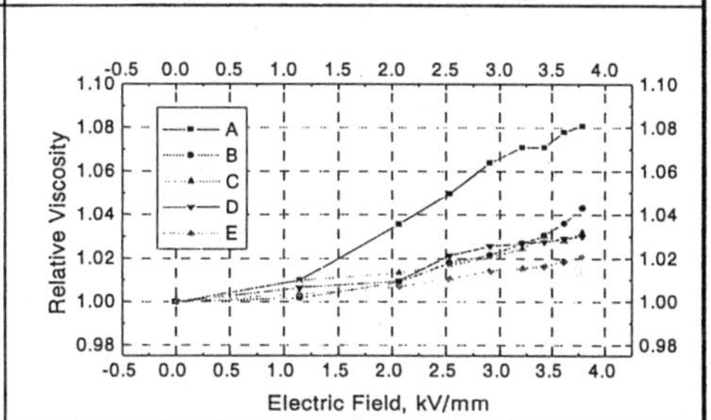

Fig. 11. Relative viscosity stress versus electric field

stir the ER fluid and to stabilize the temperature of the fluid.

- The effect was stronger at lower shear rates than the higher shear rates.
- The measurements were repeatable in all ranges of shear rates.
- For each rotational speed the torque was measured at least five times for every set of electric field and the collected data was in agreement to within a few percent.

6. Results - discussion

6.1 First experimental procedure- ERF-1 (without cooling system)

In Fig. 4 the total shear stress for every rotational speed is shown (see table 2). During this procedure, no cooling system was used. So the temperature rose during the operation but was constant for each speed value. It is obvious that the shear stress tended to increase with the electric field. In Fig. 5 the relative viscosity versus the electrical field are illustrated. The increased tendency of the viscosity versus the electric field is obvious.

Table 2 : Shear rates and temperatures corresponding to Fig. 4 and 5

	A	B	C	D	E
$\dot{\gamma}$, sec^{-1}	15926	19111	22297	25482	28667
t_1, °C	35.3	37.7	41.4	45.2	48

ERF-4

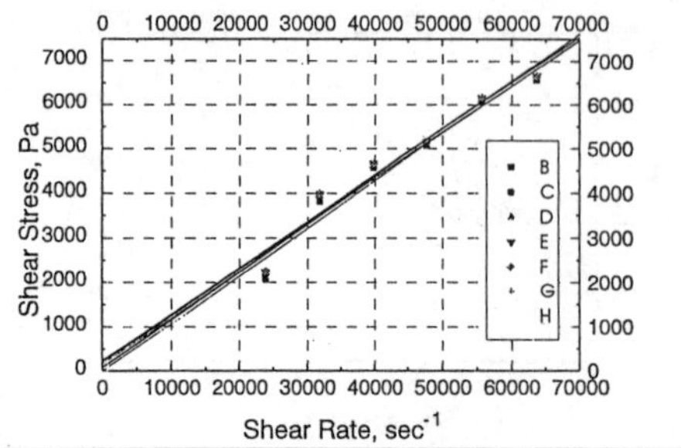

Fig. 12. Shear stress versus shear rate, T=45±3°C

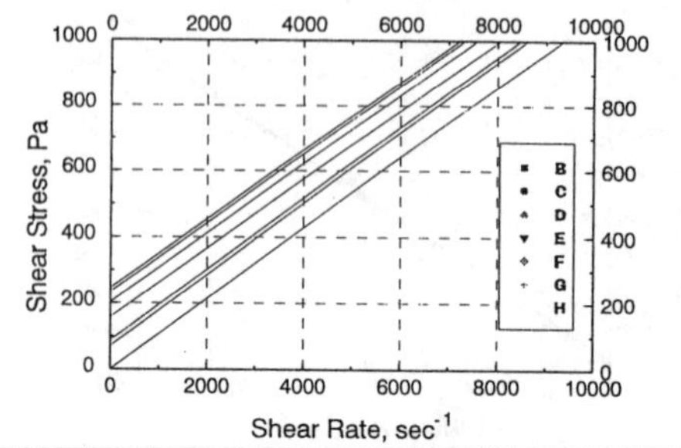

Fig. 13. Zoom of previous Fig. 12

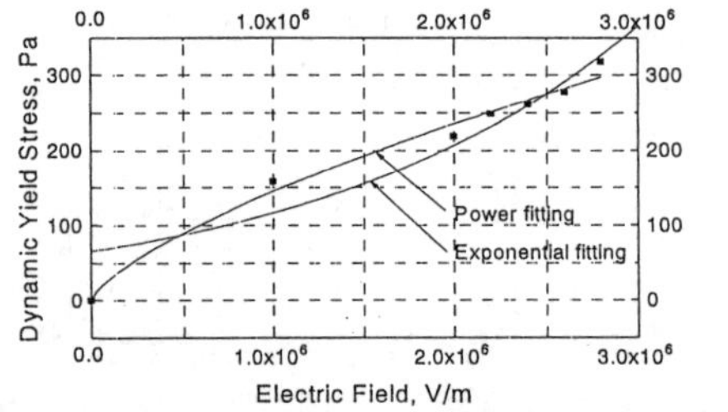

Fig. 14 . Dynamic yield stress versus electric field

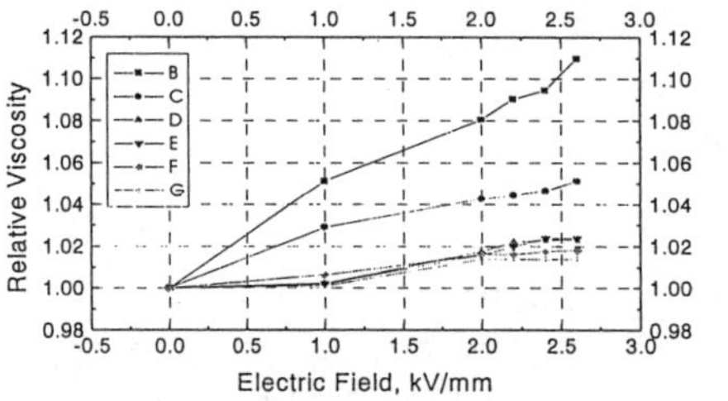

Fig. 15. Relative viscosity versus electric field

6.2 First experimental procedure - ERF-2 (without cooling system)

In Fig. 6 and 7 the curves correspond to the following shear rates and temperatures.

Table 3 : Shear rates and temperatures corresponding to Fig. 6 and 7

	A	B	C	D	E	F	G
$\dot{\gamma}$, sec^{-1}	15926	19111	22297	25482	28667	31853	35020
t_1, °C	34.3	38.4	42	44.8	47.6	50.8	55.8

6.3. Second experimental procedure. ERF-3 (Cooling with heat exchanger)

In Fig. 8 the total shear stress versus shear rate is illustrated, which take place for 25% b/w. zeolite and 150 μm radial clearance. Zooming in Fig. 8 at low shear rates the curves for shear stress and shear rate were obtained in Fig. 9.

The temperature was stabilized using a water heat exchanger. Despite of the existence of cooling system the stabilization of the temperature at a discrete value was not obtainable. So, during the course of the measurements, the temperature ranged of a range of 34 ± 2 °C. It seemed by the same Fig. that in this range of shear rates and this range of temperature difference, the material behaves as a typical Bingham fluid. In Fig. 11 the relative increment of the viscosity is illustrated for this mixture.

Fig. 10 shows the power approximation of dynamic yield stress. This approximation is $\tau_o = AE^p$ with A=6.69E-13 Pa/(m/V)p, p=2.185. In the same Fig. the exponential approximation of dynamic yield stress is also illustrated. This approximation is $\tau_o = Ae^{qE}$ with A=5.317104 Pa/exp(V/m), q=9.32E-7, where E is the electric field density.

6.4 Second experimental procedure, ERF-4 (With air cooling system)

In Fig. 12, shear stress versus shear rate is illustrated for ER fluid consisting of 25% zeolite b/w. tested between 25000 and 65000 s^{-1}. In this experiment the temperature was kept at $45 \pm 3\ ^oC$ by an air fan. In Fig. 13 the zoom of Fig. 12 at low shear rates is presented

Table 4 : Correspondence between curves and electric field (kV/mm) in fig. 8 and 12

	B	C	D	E	F	G	H	I	J
Fig. 8	0	1.1	2.0	2.5	2.9	3.2	3.4	3.6	3.8
Fig. 12	0	1.0	2.0	2.2	2.4	2.6	2.8	-	-

From the Fig. 8 and 12 it is proven that the ER fluid in high speed real journal bearings behaves as Bingham material.
In Fig. 14 is illustrated the exponential approximation of dynamic yield stress. This approximation is $\tau_o = Ae^{qE}$ with A=65.9717 Pa/exp(V/m), q=5.71E-7. In the same Fig. the power approximation of dynamic yield stress is also illustrated. This approximation is $\tau_o = AE^p$ with A=0.098Pa/(m/V)p, p=0.695. It is noticeable that this mixture does not follow the quadratic low approximation by the experimental data, in power approximation.

7. Conclusions

Under constant temperature this zeolite ER fluid proved to be Bingham like fluid under such small clearances and high shear rates (25000 to 65000 s^{-1} Fig. 8 and 12) . Their response during the experimental procedure was very fast and well repeatable in all ranges of the shear rates.
The effect was stronger at lower shear rates. From Fig. 5, 7, 11 and 15 it is concluded that the viscosity increases with the electric field at every shear rate. The viscosity did not increased dramatically but the increase affects the dynamic properties of the bearing (eccentricity, film thickness, pressure distribution and stiffness and damping coefficients).
It is noticeable that the chains of particles of the ER fluid with maximum ten particles for the bearing with 100 μm radial clearance and 15 particles for the bearing with 150 μm radial clearance induced an important effect. It is possible other fluids with higher yield stress to induce more effect.
From Fig. 4 and 6 it is clear that the shear stress increases with the electric field and decreases with the increment of temperature for every shear rate. This means that the effectiveness of ER fluid drops when the temperatures rise for different operational conditions. Finally Oka's [22] formula and Petroff's law give the same results for apparent viscosity.

8. References

1. Winslow, W. M., "Methods and means of translating Electrical Impulses in to Mechanical forces", U. S. Patent **2**, 147-150 (1947).
2. H. Block and J. P. Kelly, Review Article Electro-rheology. Journal of Physics D: Ap-

plied Physics, **Vol. 21** (1988) 1661-1677.

3. Donald L. Klass and Thomas W. Martinek, "I. Rheological properties", Journal of Applied Physics, **Vol. 38**, No. 1, (January 1967).
4. Donald L. Klass and Thomas W. Martinek, "II. Electrical properties", Journal of Applied Physics, **Vol. 38**, No. 1, (January 1967).
5. KrayniK, A. K., "Comments on ER fluid rheology", Porch. 2nd Int. Conference on ER fluids, Ed. Carson, J. D., Sprecher, A. F. Conrad, H. (Technomic Publ. Co., 1989) p. 445.
6. Bullough W. A. "Miscellaneous Electro-rheological phenomena, Part II", Proc. 2nd Int. Conference on ER fluids, Ed. Carlson, J. D., Sprecher, A. F. Conrad, H. (Technomic Publ. Co. 1989), p. 445.
7. Filisko F. E., and Radzilowski L. H, "An Intrinsic mechanism for the activity of alumino-silicate based electro-rheological materials", Journal of Rheology **34**(4), (1990), pp. 539-552
8. A. F. Sprecher, J. D. Carlson and H. Conrad, "Electro-rheology at Small Strains and Strain Rates of Suspensions of Silica Particles in Silicone Oil", Materials Science and Engineering, **95**(1987) 187-197.
9. A. Kollias and A. D. Dimarogonas, "Rheology of Zeolite and corn Starch based ER Fluids at High shear strain rates", Proc. of the conf. on Recent Advances In Adaptive and Sensory Materials and their Applications, ed. by C.A. Rogers and R. C. Rogers. (1992), pp. 548-558.
10. D. A. Nelson, E. C., Suydam, "The Thermal Aspects of the Electro-rheological Effect and its Impact on Application Design", FED-**Vol. 164**, Electro-Rheological Flows ASME 1993, pp. 71-84.
11. G. Oppermann, G. Penners, M. Schulze, G. Marquardt, R. Flindt and T. H. Naumann, "Applications of Electro-viscous Fluids as Movement Sensor Control Devices in Active Vibration Dampers", Proc. 2nd Int. Conference on ER fluids, Ed. Carlson, J. D., Sprecher, A. F. Conrad, H. (Technomic Publ. Co. 1989), P. 287.
12. John J. Hill and Thomas H. Van Steehkiste, "Response Times of Electro-rheological fluids", Journal of Applied Physics, **70**(3), 1 (August 1991).
13. J. L. Nikolajsen and M. S. Hoque, "An Electro-viscous Damper for Rotor Applications", Journal of Vibrations Acoustics, **Vol. 112**, (October 1990), pp. 440-443.
14. S. Morishita, J. Mitsui, "Controllable Squeeze Film Damper (An Application of Electro-rheological Fluid)" Journal of Vibrations and Acoustics, **Vol. 114**, (July 1992), pp. 354-357.
15. John A. Tichy, "Behavior of Squeeze film Damper with an Electro-rheological Fluid". Tribology Transactions, **Vol. 36** (1993) 1, 127-133.
16. H. Wada, H. Hayashi and K. Haga, "Behavior of Bingham Solid in Hydrodynamic Lubrication" Part 1, General Theory. Bulletin of the JSME, **Vol. 16**, No 92, (Feb. 1973), pp. 422-431.
17. T. H. Leek, S. Lingard, R. J. Atkin and W. A. Bullough, " An experimental investigation of the flow of an Electro-rheological fluid in Rayleigh step bearing", J. Phys. D: Applied Physics, **26**(1993) 1592-1800.
18. H. Wada, H. Hayashi and K. Haga, "Behavior of Bingham Solid in Hydrodynamic Lubrication" Part 2, Application to Step Bearing. Bulletin of the JSME, **Vol. 16**, No 92, (Feb. 1973), pp. 432-440.
19. J. A. Tichy, "Hydrodynamic Lubrication Theory for the Bingham Plastic Model" Journal of Rheology **35**(4), May 1991. pp. 477-496.
20. A. Kollias and A. D. Dimarogonas" Electro-rheological fluid flow In Partial Journal

Bearings", ASME Winter annual meeting, Electro-rheological flows, (1994), Chicago.
21. A. D. Dimarogonas and A. Kollias, "Electro-Rhelogical Fluid-Controlled "Smart" Journal Bearings", Tribology Transactions (1992).
22. Oka. S, "The principles of Rheometry", Rheology-Theory and applications, (Eirich F. R., Academic Press), **Vol. 3**, P.3.3.
23. P.R. Trumbler, "Design of Film Bearings", (The Macmillan company, New York 1966).

STATIC AND DYNAMIC LIGHT SCATTERING OF DILUTE MAGNETORHEOLGICAL EMULSIONS

M. HAGENBÜCHLE, P. SHEAFFER, Y. ZHU, AND J. LIU
Department of Physics and Astronomy,
California State University, Long Beach, CA 90840.

ABSTRACT

The structure formation and dynamics of a dilute, monodisperse ferrofluid emulsion under the influence of an external magnetic field have been investigated using static and dynamic light scattering techniques. In the absence of an external field, the spherical emulsion particles behave as a dilute system of hard spheres where the static light scattering is dominated by the droplet form factor and the dynamic light scattering provides the diffusion coefficient due to their Brownian motion. An applied magnetic field induces a magnetic dipole moment in each droplet. Static light scattering is used to study the time dependence of the structure formation caused by the dipole-dipole interaction as a function of the field strength. The change of the dynamic properties due to chain formation is studied by dynamic light scattering.

1. Introduction

Static and dynamic light scattering are well established techniques in the field of complex fluids. They were, for example, widely used to study colloidal suspensions of both hard spheres and charged spheres[1-3]. In these systems, the Coulomb interaction between the suspended particles or entropy causes them to order into gaslike, liquidlike or crystalline structures depending on the ionic strength and the particle concentration. Since the interparticle distances are usually comparable to the wavelength of light, static light scattering (SLS) is an ideal tool to gain structural information. On the other hand, dynamic light scattering (DLS) has shown to be a powerful instrument to examine the dynamics of these colloidal spheres. This is due to the fact that DLS is sensitive to the interaction between colloidal particles and also that the relevant time scales of the particle dynamics is within the limits of digital correlators used in DLS. Therefore, DLS is employed in many studies of colloidal suspensions and great understanding has been achieved.

Electrorheological (ER) and magnetorheological (MR) fluids are complex fluids which have attracted much interest in recent years, mainly because they promise to have a broad spectrum of possible applications[4]. They consist of polarizable or magnetizable particles suspended in a suitable liquid. Their technological power is based on the extraordinary rheological behavior they exhibit. Applying an external electric or, in the case of MR fluids, magnetic field changes them from a low viscosity Newtonian fluid into a high viscosity Non-Newtonian fluid, which is often empirically described as Bingham plastic. The reason for these unusual rheological properties is found in changes of the microstructure which are caused by the external field. The field induces a dipole moment in each particle. When the strength of the dipole-dipole interaction between the particles exceeds the thermal energy, they adhere and form fibrous structures along the field lines. Compared to the above mentioned spheres with Coulomb or hard sphere interaction, ER and MR fluids are certainly more complex. Mainly because of the anisotropy of the dipole-dipole interaction, but also because they often contain additional components, for example surfactants. Nevertheless, the encouraging example of Coulomb spheres and hard spheres indicates that the key for better understanding and technological improvements is a detailed knowledge of their structural and dynamic properties.

In this preliminary study we present static and dynamic light scattering measurements on a dilute MR fluid. Due to the high dilution and the weak applied fields we observe only chain formation and no coarsening of the chains into columns. The chain formation strongly affects both static and dynamic light scattering. In some respect, this work provides the basis for the understanding of the equilibrium structure and phase transitions of MR fluids at high volume fractions[5].

2. Experimental

We employed a ferrofluid emulsion[6] as a model magnetorheological fluid. It consists of small ferrofluid droplets dispersed in water. The droplets are stabilized against irreversible aggregation and coalescence by the surfactant *sodium dodecyl sulfate (SDS)*. The ferrofluid itself is based on kerosene and iron oxide grains which have an average diameter of 90 Å. Each of these grains possess a permanent dipole moment. Yet, there is no net dipole moment of the whole droplet since thermal motion keeps the orientation of the grains random. Only when an external magnetic field is applied, the grains orient and thus induce a dipole moment in each droplet which depends on the strength of the applied field. The distribution of droplet sizes in an emulsion is generally very broad. We obtained a monodisperse emulsion by applying the fractionated crystallization scheme which is described elsewhere[7]. Size and monodispersity of the droplets were measured by dynamic light scattering. The particle radius a was found to be 233 ± 17 nm. The sample was diluted to a volume fraction of $\phi = 7.65 \times 10^{-5}$ to avoid multiple light scattering.

We used an air-cooled Argon ion laser at a wavelenghth of 514.5 nm for our light scattering experiments. A lens was used to focus the laser beam into a cylindrical glass tube of 20 mm outer diameter which contained about 10 ml of the sample. This sample cell was surrounded by an index-match bath to reduce reflections and to keep the temperature of the sample constant. A Helmholtz coil with the sample in its center provided a homogeneous magnetic field perpendicular to the scattering plane. After passing the detector optics, scattered light was detected by a photomultiplier tube at a fixed scattering angle of 90°. The photon counts were then sent to a digital correlator (BI-9000AT, Brookhaven Instruments Corporation) for processing, from which the averaged scattered light intensity I and the homodyne intensity autocorrelation function $<I(\tau)I(0)>$ were obtained. Using the Siegert relation[8] we converted the intensity autocorrelation function into the normalized field autocorrelation function $g^{(1)}(q,\tau) = \langle E(\tau)E(0)\rangle / \langle |E(0)|^2 \rangle$ from which the diffusion coefficients were derived as described below.

3. Results

3.1. Static light scattering

Figure 1 shows schematically the scattering pattern of our sample on a cylindrical screen with and without applied magnetic field. Without magnetic field the emulsion can be described as a dilute suspension of non-interacting spheres. The angular dependence of the scattered light of such a system is determined by the formfactor[9] $P(a,q)$ which depends only on the radius a of the spheres and the scattering vector $q = 4\pi n \cdot \sin(\Theta/2)/\lambda$, where n is the refractive index of the solution, λ the wavelength of light, and Θ the scattering angle. $P(a,q)$ is a smooth function which decreases only slowly with increasing q. Therefore, the scattering pattern in Fig. 1a shows no distinct structure, and the intensity on the screen decreases slowly with increasing distance to the spot where the incident beam hits the screen. Fig. 1b demonstrates qualitatively the changes when a magnetic field of about 40 G is applied. A bright band is found in a narrow zone around the equatorial plane. Nearly no scattering is observed outside this band. Martin et al.[10] have reported similar patterns for more concentrated samples of an ER fluid subjected to an electric field.

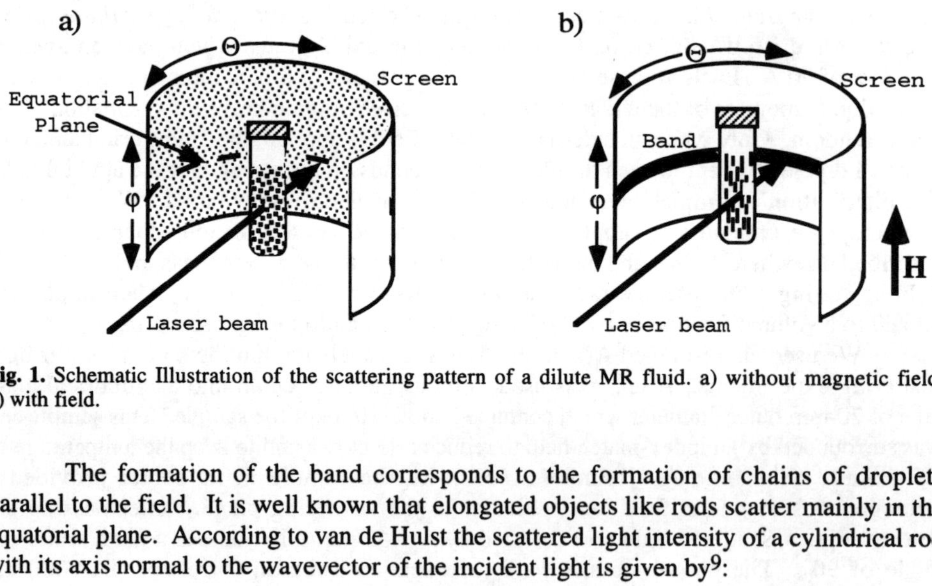

Fig. 1. Schematic Illustration of the scattering pattern of a dilute MR fluid. a) without magnetic field. b) with field.

The formation of the band corresponds to the formation of chains of droplets parallel to the field. It is well known that elongated objects like rods scatter mainly in the equatorial plane. According to van de Hulst the scattered light intensity of a cylindrical rod with its axis normal to the wavevector of the incident light is given by[9]:

$$I \propto a^4 l^2 I_0 \cdot F(a,k,\Theta)^2 \cdot E(l,k,\varphi)^2. \qquad (1)$$

Here, a denotes the radius of the rod, l its length, k and I_0 the wavevector and the intensity of the incident light, respectively. Θ is the angle in the plane perpendicular to the axis of the rod, i.e. the equatorial plane, and φ is the angle between this plane and the axis of the rod. Both $F(x)$ and $E(x)$ are Bessel functions which are listed in reference 9. Note that $F(a,k,\Theta)$ which describes the angular dependence of the scattered light in the equatorial plane (along the direction of the band) is a function of the radius of the rods. This was exploited by Martin et al.[10] to measure the segregation of chains into columns which is known to occur in more concentrated samples. On the other hand $E(l,k,\varphi)$, describing the scattering perpendicular to the equatorial plane (across the band), depends only on the length.

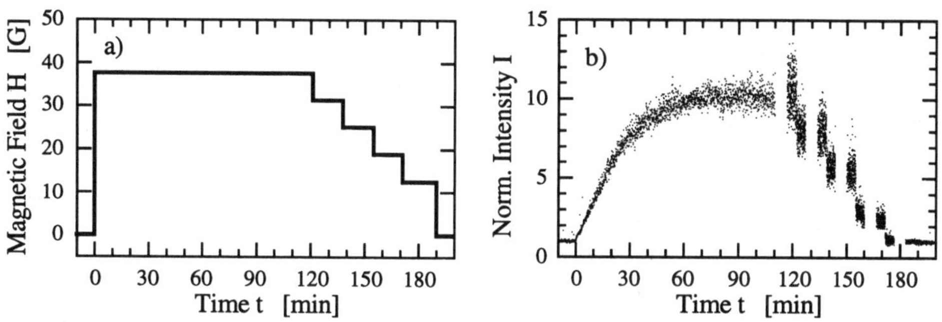

Fig. 2. Time dependence of static light scattering at Θ = 90°, φ = 0°. a) Applied field $H(t)$ b) Corresponding scattered light intensity$I(t)$. $I(t)$ is normalized by the zero-field intensity.

Fig. 2 illustrates the formation and disintegration of the droplet chains. In our experiment the detector is placed in the equatorial plane. In terms of the angular coordinates used in Eq. 1, the detector is located at $\Theta = 90°$ and $\varphi = 0°$. As shown in Fig. 2a, a magnetic field of 37.5 G is applied at time 0 and kept constant for 120 minutes. Then the field is lowered step by step until it reaches 0 G after another 70 minutes. The corresponding intensity signal is presented in Fig. 2b. The scattered light intensity I is normalized to the zero-field intensity. After the magnetic field is turned on, the chain formation starts instantly as demonstrated by the increasing intensity. Initially, there is a steep increase indicating a fast chain growth, but after approximately 30 minutes the increase in intensity slows down and finally an asymptotic value, about ten times higher as the off-field intensity, is reached. Obviously, chain formation is a very slow process in this sample since it takes more than an hour before the scattered intensity saturates. On the other hand the time scale for chain formation in concentrated samples of the same emulsion has been reported to be milliseconds[11]. Hence, the high dilution must be responsible for the slow chain formation.

This conclusion is supported by a Computer simulation by Fermigier and Gast[12]. The authors traced the paths of two initially well separated Brownian particles which are subjected to dipole-dipole interaction. The simulation demonstrates that most of the time the motion of the particles is diffusive. Only within a threshold distance r_c the motion ceases to be random. The particles then approach each other on deterministic trajectories and finally adhere. The threshold distance r_c is reported to be proportional to $\lambda^{1/3}$, where λ is the coupling constant defined by the ratio of the magnetic interaction dipole interaction energy to the thermal energy. In consequence, we conclude that chain formation in dilute emulsions is a diffusion-limited process.

In our experiment, caused by the very low volume fraction, the average distance between two particles is very large and, in addition, the threshold distance r_c is small due to the low applied field strength. Both reasons account for the very slow chain formation displayed in Fig. 2. After the initial stage, where mainly single particles adhere, there is an additional slowing down of the chain growth because the diffusion coefficient of a chain decreases with increasing chain length.

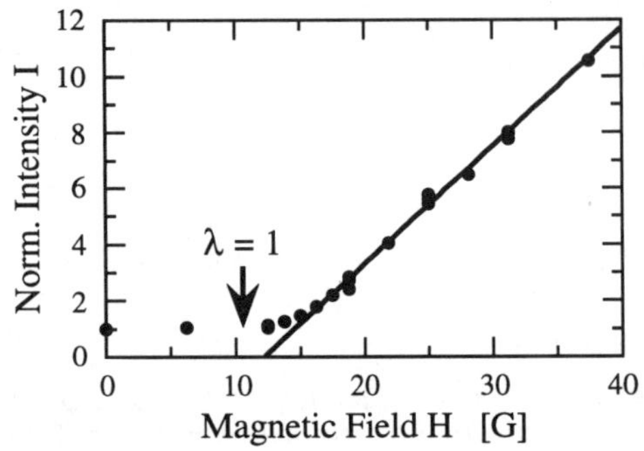

Fig. 3. Dependence of the saturation intensity I_{sat} on the magnetic field strength *H*.

The saturation of the *I(t)* in Fig. 2b indicates that at equilibrium chain length is finite for a given field strength. This is clearly confirmed by the descending part of Fig. 2b, where the magnetic field is lowered step by step. Responding almost instantaneously to the change of the magnetic field, I_{sat} decreases also step by step. This means that at each step longer chains (higher intensity) break apart into smaller fragments (lower intensity). It is evident from Fig. 2b that the intensity, i.e. the chain length, depends on the strength of the applied field.

To illustrate more clearly this, the saturation intensity I_{sat} is plotted as a function of the applied magnetic field in Fig. 3. No chain formation occurs up to a field strength of 12.5 G. In this region the dipole interaction energy is smaller than the thermal energy i.e. $\lambda < 1$. Above 15 G we find a linear increase of the intensity with increasing field. To find out, how the chain length l itself depends on the magnetic field H, we take Eq. 1 as a good estimate for the scattering of the droplet chains. As mentioned before in our experiment φ in Eq. 1 equals zero, therefore, $E(x) = 1$. $F(x)$ is a constant factor because θ is fixed and the radius of a chain of droplets is the same as that of a single droplet. Thus Eq. 1 yields $I_{sat} \sim l^2$. But this is only valid if the number of chains N_c remains constant while the length is increased. If the number of single particles is a conserved quantity then the number of chains decrease with increasing chain length l according to $N_c \sim 1/l$. Since I_{sat} is proportional to N_C, this leads to $I_{sat} \sim l$. Based on this result and on Fig. 3 we expect that for dilute emulsions at low field strengths the chain length l is proportional to the field strength H.

3.2. Dynamic light scattering

The normalized field autocorrelation function $g^{(1)}(q,\tau)$ is related to the dynamic structure factor $S(q,\tau)$ by[13]

$$g^{(1)}(q,\tau) = \frac{S(q,\tau)}{S(q)}, \tag{2}$$

where $S(q) = S(q,0)$. The dynamic structure factor is a measure of the correlations between the position $r_i(\tau)$ of particle i at time τ and the position $r_j(0)$ of particle j at time 0:

$$S(q,\tau) = \frac{1}{N}\sum_{i,j}^{N}\left\langle e^{iq(r_i(\tau)-r_j(0))}\right\rangle. \tag{3}$$

$S(q,\tau)$ is most important since it includes all the information on the dynamics.

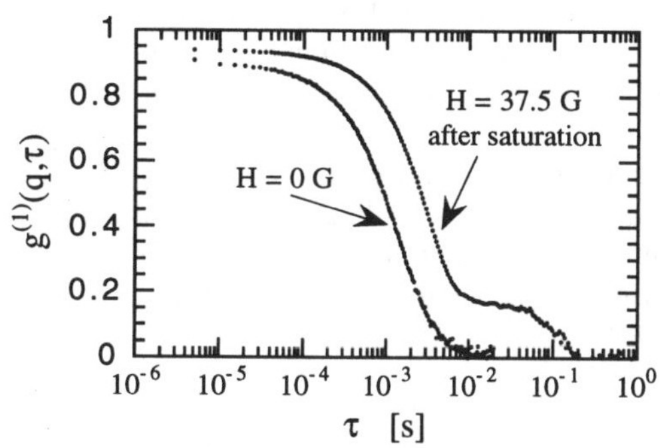

Fig. 4. Normalized field autocorrelation function $g^{(1)}(q,\tau)$ without magnetic field (H = 0) and after saturation (H = 31 G).

Fig. 4 shows the comparison of the normalized field autocorrelation function $g^{(1)}(q,\tau)$ at H = 0 G and H = 31 G. The correlation function with applied magnetic field was measured after the chain formation had stopped which means after the scattered intensity $I(t)$ had saturated. We used static light scattering to monitor the chain formation. There are apparently enormous differences between the two correlation functions shown in Fig. 4. The correlation function at zero field can essentially be described by a single exponential. Due to the high dilution the emulsion behaves like a system of non-interacting spheres. The small nonexponentiallity is caused by the small size polydispersity of the droplets. In fact, particle sizing by analyzing the deviation from the single exponential behavior is a standard application for dynamic light scattering[14]. In contrast, $g^{(1)}(q,\tau)$ with applied magnetic field decays generally slower and shows two distinct decay times. The main decay time in earlier time is of the same order of magnitude as the zero-field decay time. Compared to this short-time decay the long-time decay is about two orders of magnitude slower.

In the following we focus only on the short time behavior of $g^{(1)}(q,\tau)$. The short-time dynamics is usually described in terms of the first cumulant $\Gamma_1(q)$[13]:

$$\Gamma_1(q) = -\frac{\partial}{\partial\tau}\ln g^{(1)}(q,\tau)\Big|_{\tau=0}. \tag{4}$$

The first cumulant defines an apparent diffusion coefficient through:

$$\Gamma_1(q) = q^2 D_{app}(q). \tag{5}$$

In Eq. 3 the $i = j$ terms account for single particle diffusion which is often called self diffusion, whereas the $i \neq j$ terms describe collective diffusion. Therefore, in general $D_{app}(q)$ contains both self and collective diffusion. However, there are two arguments that in our experiment $D_{app}(q)$ is solely determined by the self diffusion of the particles. First, since the threshold distance, which is a measure of the magnetic interaction range, is very small and the dilution is very high, we believe that, after saturation, we have essentially a system of non-interacting chains. Without interaction there are no correlations between different particles i and j. Thus, $D_{app}(q)$ reduces to $D_{app}^{self}(q)$. Second, it has been pointed out by several authors[13,15] that, for $q >> q_{max}$, $D_{app}(q)$ reduces to $D_{app}^{self}(q)$ even in interacting colloidal suspensions. Here, q_{max} is given by the nearest neighbor distance $d_{max} = 2\pi/q_{max}$. The distance corresponding to the scattering vector q employed in our measurements is approximately the size of the particles. Due to the high dilution, d_{max} is definitely much larger than the particle size and therefore the above mentioned condition $q >> q_{max}$ is well fulfilled.

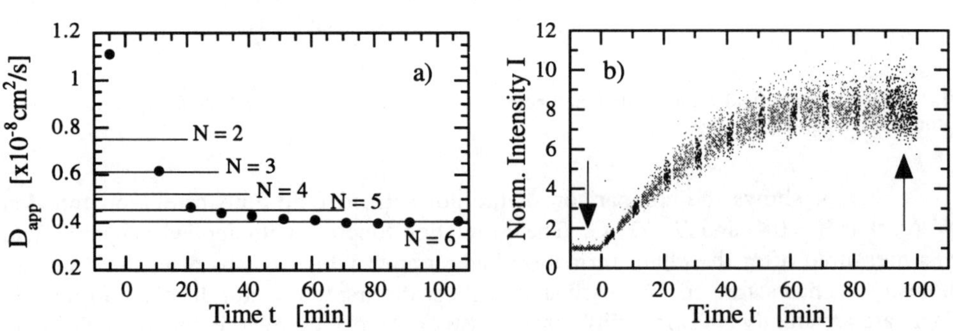

Fig. 5. a) Short-time self diffusion coefficient during chain formation at a magnetic field strength of 31 G. The constant levels are calculated diffusion coefficients for a chain consisting of N spheres. b) Corresponding static light scattering signal. The dark parts indicate the time when a correlation function was measured. The arrows correspond to the correlation functions shown in Fig. 4.

Figure 5a shows $D_{app}^{self}(q)$ during the chain formation. For comparison, the corresponding static light scattering signal is plotted in Fig. 5b. The magnetic field was switched on at time 0 and kept constant at 31 G. The time to measure a correlation function was two minutes as indicated by the dark parts of $I(t)$ in Fig. 5b. The first and the last point in Fig. 5a correspond to the two correlation functions presented in Fig. 4. The diffusion coefficient of a chain of particles decreases with increasing number of particles in the chain. Thus, Fig. 5a directly reflects the chain formation. Due to our scattering geometry we are mainly sensitive to the motion of the chains perpendicular to their axes. Using the Stokes Einstein relation and the appropriate friction coefficient $F_{chain}^{\perp}$[16], we can calculate the diffusion coefficient $D_{chain}^{\perp}$:

$$D_{chain}^{\perp} = \frac{k_B T}{F_{chain}^{\perp}}$$

$$F_{chain}^{\perp} = \frac{4N}{3\ln 2N + \gamma^{\perp}} \cdot F_{sphere}. \tag{6}$$

Here, F_{sphere} is the friction coefficient of a single sphere, N is the number of spheres in the chain, and for $\gamma^{\perp}$ a value of -.418 was reported[17]. The calculated values for $N = 2$ to 6 are presented in Fig. 5a as constant levels. We find that at saturation the chain consists of approximately six particles. However, some remarks have to be made on this result. First, we do not expect one single chain length but a distribution of chain lengths to be created by the applied field. Therefore the diffusion coefficient we measure is actually an averaged diffusion coefficient corresponding to an average chain length. Second, dynamic light scattering is sensitive to all kinds of motion. Thus, if there are bending modes or chain fluctuations, they will certainly be contained in $D_{app}^{self}(q)$. Further q-dependent measurements will clarify these points and provide more insight in the dynamics of chains.

Figure 5b shows that the scattered light intensity has increased by roughly a factor of eight at saturation. If we assume that $I \sim l$, as explained earlier, then according to static light scattering the equilibrium chain consists of approximately eight particles. This is in remarkable agreement to the dynamic light scattering result regarding all the assumptions and uncertainties involved in the determination of the chain length by both dynamic and static light scattering.

4. Conclusion

In summary, we have presented initial static and dynamic light scattering measurements on a dilute magnetorheological emulsion. Field induced chain formation was observed for $\lambda > 1$ and was found to be a diffusion-limited process. Our measurements show clearly that in these dilute samples an equilibrium chain length exists, which depends on the applied magnetic field. With dynamic light scattering the short time self diffusion coefficient $D_{app}^{self}(q)$ was measured. $D_{app}^{self}(q)$ decreases with increasing chain length. Overall, the results of our measurements demonstrate the usefulness of static and dynamic light scattering as a non-invasive tool to study the structure formation in MR and ER fluids.

5. Acknowledgments

This work was partially supported by an award from the Research Corporation, and by CSULB University Research. We also gratefully acknowledge the support by NASA, NAG 3-1634.

6. References

1 R. Klein, in *Structure and Dynamics of Strongly Interacting Colloids and Supermolecular Aggregates in Solution* , eds. S. H. Chen, J. S. Huang, P. Tartaglia (Kluwer Academic, Dordrecht, 1992).
2 P. N. Pusey and W. van Megen, *Nature* **320** (1986) 340.
3 J. Liu, D.A. Weitz, B. J. Ackerson, *Phys. Rev. E* **48** (1993) 1106.
4 *Proceedings of the Conference on Electrorheological Fluids*, Feldkirch, AUSTRIA, 20 - 23 July, 1993, R. Tao and G. D. Roy, Eds., (World Scientific, Singapore, 1994).
5 *Proceedings of the 5th Int. Conf. on Electro-Rheological Fluids, Magneto-Rheological Suspensions and Associated Technology*, University of Sheffield, UK, 10 - 14 July 1995.
6 J. Bibette, *J. Magnetism and Magn. Materials* **122** (1993) 37.
7 J. Bibette, *J. Coll. Int. Sci.* **147** (1991) 474.
8 B. J. Berne and R. Pecora, *Dynamic Light Scattering* (Wiley, New York, 1976).
9 H. C. van de Hulst, *Light Scattering by small particles* (Dover, New York, 1957).
10 J. E. Martin, J. Odinek, and T. C. Halsey, *Phys. Rev. Lett.* **69** (1992) 1524.
11 T. Mou, G. A. Flores, J. Liu, J. Bibette, and J. Richard, *Int. J. Mod. Phys. B* **8** (1994) 2779.
12 M. Fermigier and A. P. Gast, *J. Coll. Int. Sci.* **154** (1992) 522.
13 P. N. Pusey, *J. Phys. A: Math. Gen.* **11** (1978) 119.
14 B. E. Dahneke *Measurements of suspended particles by quasi-elastic light scattering* (Wiley, New York, 1983).
15 K. J. Gaylor, I. K. Snook, W. J. van Megen, and R. O. Watts, *J. Phys. A: Math. Gen.* **13** (1980) 2513.
16 K. Zahn, R. Lenke, and G. Maret, *J. Phys. II France* **4** (1994) 555.
17 H. Yamakawa, *J. Chem. Phys.* **53** (1970) 436.

RHEOLOGICAL SYNERGISTIC EFFECTS OF ELECTRIC AND MAGNETIC FIELDS IN IRON PARTICLE SUSPENSION

KIYOHITO KOYAMA

Department of Materials Science and Engineering, Yamagata University, Yonezawa 992, Japan

ABSTRACT

The effects of electric field and/or magnetic field on the rheological property of iron suspension were studied by use of parallel-plate rheometers developed for electro-magneto-rheological (EMR) measurements. The EMR effect, which is defined as an increase in the shear stress under combined fields, was dependent on the strength and direction of the fields. In particular, significant synergism was found in the EMR effect when the electric and magnetic fields were applied in parallel. Observation of the suspensions solidified in parallel field and crossed field conditions revealed different structures of aggregated particles for these combined fields. The parallel field induced many chains along the field, while the crossed field induced a network structure. It is concluded that the synergistic EMR effect in parallel field is due to the increase in the number of particle chains bridging the plates.

1. Introduction

The ER and MR effects of suspensions, which are typically observed as an increase of shear stress in response to an electric or a magnetic field, are based on reversible structuring of particles dispersed in the fluids [1-4]. There have been some attempts to combine these effects in order to improve the property of the ER and MR fluids [5-7]. Kordonsky et al. found that combined electric and magnetic fields induce larger stress increase when appropriate materials responsive to the both fields are used [6]. They indicated that the shear stress induced by the combined field is larger than that expected from linear additivity of ER and MR effects. Recently we also reported that a suspension of titania-coated iron particles exhibits a significant synergistic EMR effect [8]. Though the mechanism of the synergistic effect can be speculated on the basis of the cluster model for ER fluids, there have been little information on the cluster structures in the combined field. It is necessary to clarify the structure induced by the electric and/or magnetic fields.

Here we carried out shear stress measurements of an iron suspension under electric and magnetic fields superimposed in parallel or in perpendicular. We also observed the structure of the suspensions solidified in the presence of the fields with an optical microscope. The synergistic EMR effect is discussed in terms of the induced stress and the structure of particle clusters.

2. Experimental Method

2.1. Rheometer

Fig. 1 schematically illustrates the parallel-plate rheometer modified for EMR measurements. A d.c. or a.c. electric field is applied to the fluid sandwiched by the parallel plates. The gap between the two plates is 0.5mm. The bottom plate slides at a constant rate and the displacement of the upper plate is detected by a strain gauge. A magnetic field is generated in parallel or perpendicular direction to the electric field with one or two electromagnetic coils as shown in parts (a) and (b) of Fig.1. Here the shear deformation is represented as γ_{xz} and the field directions as E_z/H_z (a) and E_z/H_x (b).

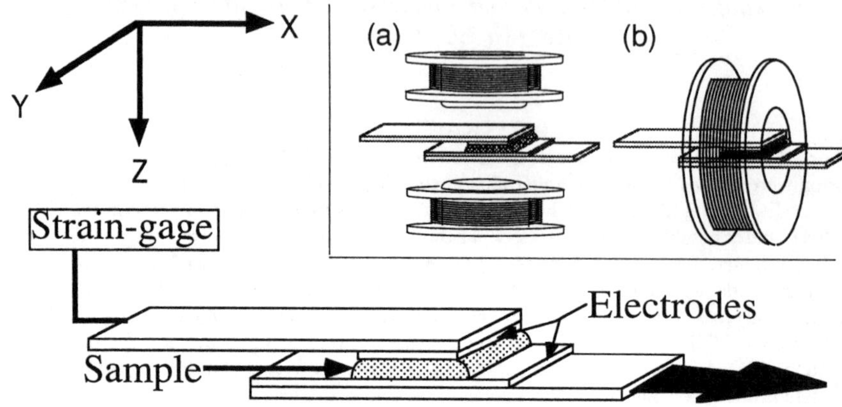

Fig. 1 Parallel-plate rheometer; (a) parallel field (E_z/H_z), (b) crossed-field (E_z/H_x).

2.2. Materials and Measurement Condition

The EMR suspension, supplied by Nippon Oil Co. Ltd., is composed of iron particles of diameter 5~10μm and a liquid silicone rubber (RTV rubber; Shinetsu Silicone KE-103). The surface of the particles is coated with titania. The concentration of the particle is 40 wt%. The shear stress response was examined under an electric field (0~2kV/mm) and/or a magnetic field (0~2 kOe) for constant shear rate 2.0 sec^{-1}. All measurements were carried out at room temperature (ca. 25℃).

2.3. Observation of Structure

The EMR suspension was mixed with a small amount of hardening agent (Shinetsu Silicone CAT-103). The particle concentration was reduced (10wt%) to obtain clear images of the structure. The mixture was solidified by heating in the presence of electric and/or magnetic fields applied with the EMR rheometer. The solidified samples were sliced into 50~100 μm-thick fragments. The cross section of the sample was observed with an optical microscope.

3. Results and Discussion

3.1. Shear Stress

Both electric and magnetic fields were found to increase the shear stress of the iron suspension. The ER, MR, and EMR effects are evaluated as the increment of the stress, i.e., difference between stress values in the presence and absence of the fields. The induced stress values are plotted against the strength of electric and magnetic fields in Figs. 2 and 3, respectively. The EMR effect is compared with MR effect in Fig. 3.

Here the stress value measured under no external field is defined as 0 Pa. The stress increment is proportional to the square of electric field strength and linearly proportional to the magnetic field strength.

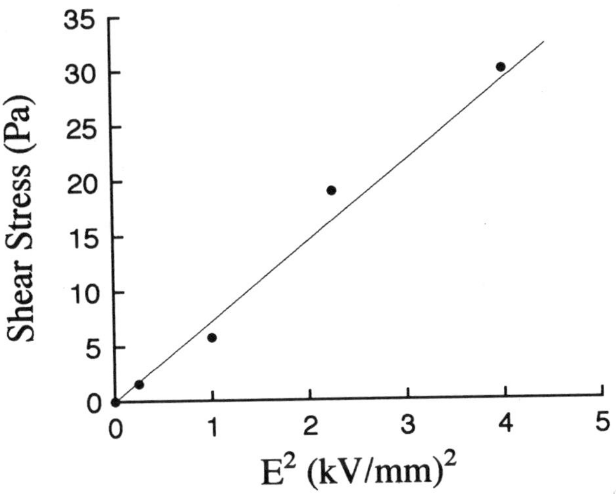

Fig. 2 ER effect of the iron suspension.

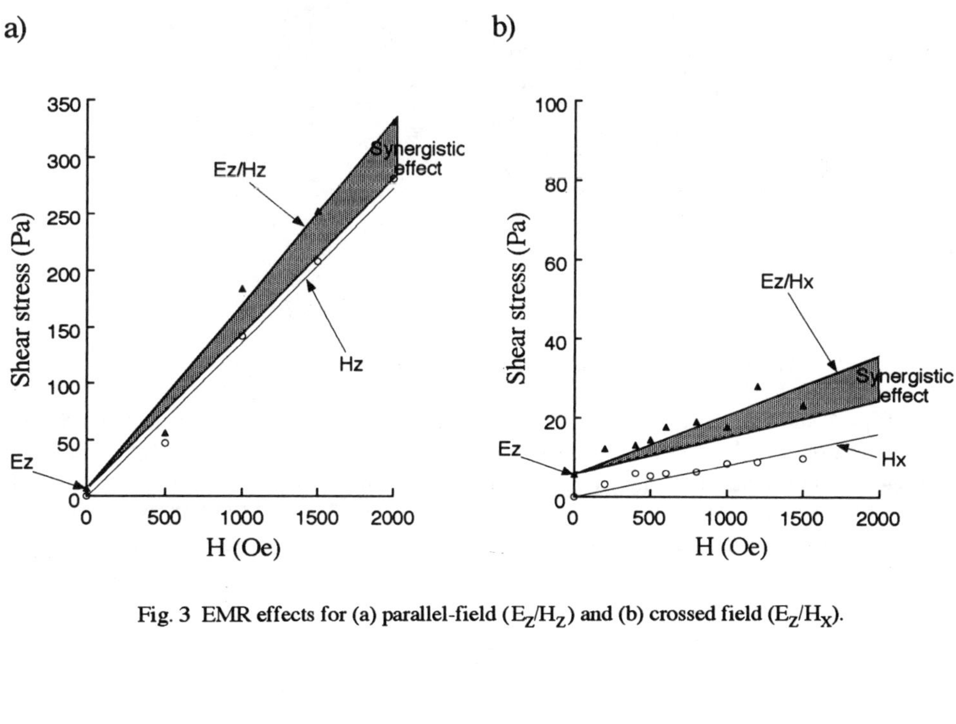

Fig. 3 EMR effects for (a) parallel-field (E_Z/H_Z) and (b) crossed field (E_Z/H_X).

Under the field condition examined, the effect of magnetic field is predominant for the increase of shear stress. The electrically induced stress increment is less than 10Pa for E_z = 1 kV/mm, while the magnetic field induces much larger stress when the direction is perpendicular to the shearing plane (H_z = 0.5~2 kOe). On the other hand, the H_x field with the same strength induces small increase of the stress. The difference between stress increments by H_z and H_x fields indicates that some clusters with different structures are formed in the parallel and crossed fields.

The total stress increment induced by the superimposed field is larger than that expected from the linear additivity of ER and MR effects. We call this nonlinear stress increase a rheological synergistic effect of electric and magnetic fields [8]. The synergistic effect is observed under both parallel-field and crossed-field conditions, and the synergism is very significant in the parallel-field. This effect is probably due to a change in the cluster structure under the combined field. The information on the cluster structure is needed for further understanding of the stress change. The following experiments concern observation of the cluster by using solidified samples.

3.2. Cluster Structure

Fig. 4 Structure of induced clusters formed under a combined field.

Fig. 4 shows a micrograph of cross section for the EMR suspension solidified under a combined E_z/H_z field (E_z= 1 kV/mm, H_z= 1 kOe) applied with the parallel-field rheometer. The thickness of the sample is about 0.5 mm. Many fibrous clusters parallel to the field are observed, which is similar to those often observed for ER fluids. It was found that the combined parallel field results in an increase of the number and length of the clusters.

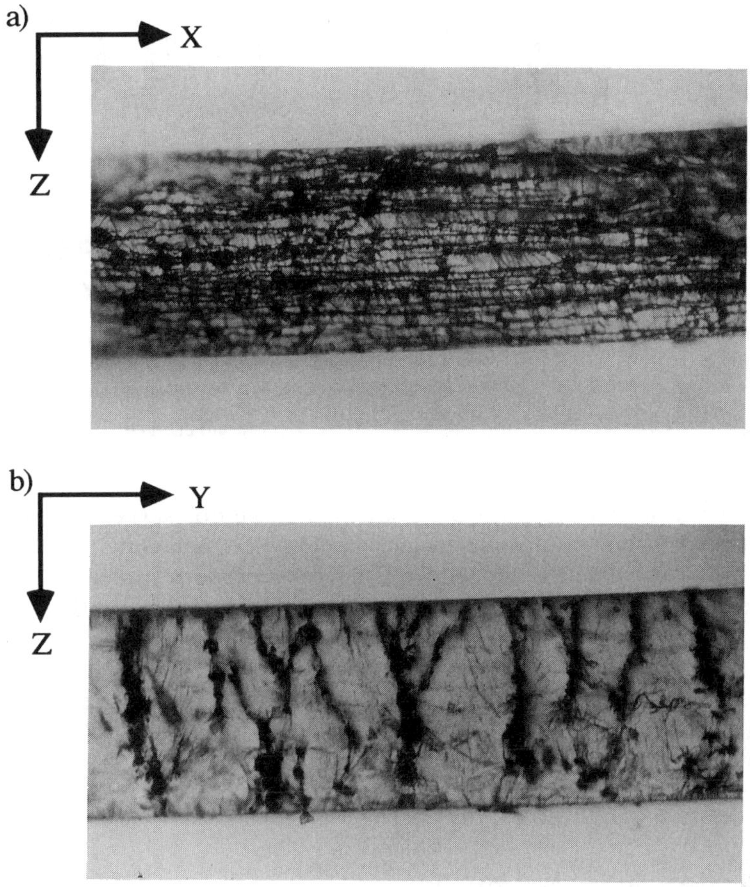

Fig. 5 Structure of induced clusters formed under a combined field; cross sections of (a) X-Z plane, and (b) Y-Z plane.

Fig. 5 (a, b) shows micrographs indicating another cluster structure formed under a crossed field condition (E_Z/H_X; E_Z = 1kV/mm, H_X = 1 kOe). From the observation of X-Z plane (a), it is found that long clusters orient along the X-axis, i.e., along the applied magnetic field. It is also found from the photograph of Y-Z plane (b) that there is a network structure formed by the fibrous clusters. However, this network structure is not effective for increasing the shear stress.

The cluster structures are schematically represented in Fig. 6. Considering only the effects of the clusters on the shear stress, it is concluded that the synergistic EMR effect is generated by the increase of the number of particle chains which bridge the plates and that the network structure has little effect on the stress increase.

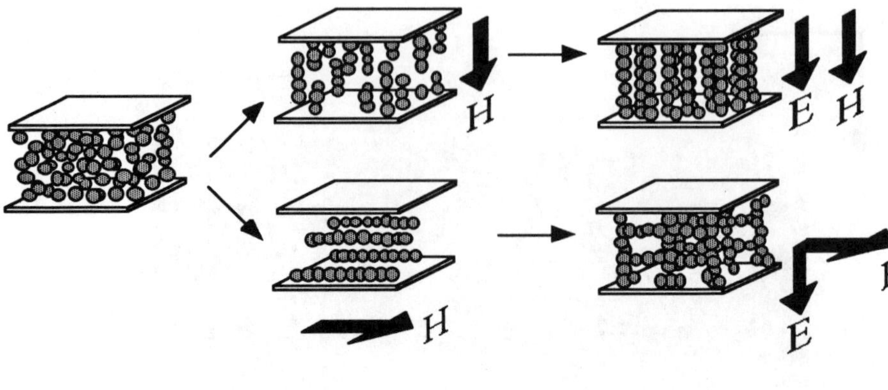

Fig. 6 Model of the cluster structures formed under electric and magnetic fields.

4. Conclusion

The induced shear stress and cluster structure of the iron suspension depend on the strength and direction of electric and magnetic fields. The parallel field induces many chains along the field, while the crossed field induces a network structure. The shear stress increases with an increase in the number of clusters bridging the parallel plates.

5. Acknowledgements

The author is grateful to Nippon Oil Co. Ltd. for providing the samples, and to his colleagues in Yamagata University for valuable help and discussion. This work was partly supported by the Grant-in-Aid from the Ministry of Education, Science and Culture of Japan.

6. References

1 W. M. Winslow, *J. Appl. Phys.* **20** (1949) 1137.
2. H. Block and J. P. Kelly, *J. Phys. D* **21** (1988) 1661.
3. Z. P. Schulman, V.I. Kordonsky, E.A. Zaltsgendler, I. V. Prokhorow, B. M. Khusid, and S. A. Demchuk, *Int. J. Multiphase Flow* **12** (1986) 1935.
4. E. Lemaire and G. Bossis, *J. Phys. D* **24** (1991) 1473.
5. T. Fujita, J. Mochizuki, and I. J. Lin, *J. Magn. Magn. Mater.* **122** (1993) 29.
6. W. I. Kordonsky, S. R. Gorodkin, and E. V. Medvedeva, in *Proceedings of the 4th International Conference on ER Fluids,* eds. R. Tao and G. D. Roy (World Scientific, Singapore, 1994) p. 22.
7. K. Minagawa, T. Watanabe, M. Munakata, and K. Koyama, *J. Non-Newt. Fluid Mech.* **52** (1994) 59.
8. K. Minagawa, T. Watanabe, K. Koyama, and M. Sasaki, *Langmuir* **10** (11) (1994) 3926.

THE FREQUENCY DEPENDENCE AND NON-LINEAR DIELECTRIC PROPERTY OF ER FLUID

KUNQUAN LU, WEIJIA WEN, CHENXI LI
Institute of Physics, Chinese Academy of Sciences
Beijing 100080, P.R.China

ABSTRACT

The frequency dependence of the shear stress in ac field and the non-linear dielectric property of ER fluid have been studied. We find that the shear stresses of some water-free ER fluids increase monotonously with the frequency and tend to reach saturated values at high frequency. The measurements on $KNbO_3$/silicone ER fluid show that the shear stresses under 10^3 Hz frequency a.c. field are several times or even an order larger than that under d.c. field for the same field strength. The studies of non-linear dielectric properties of ER fluids show that the permittivity of ER fluid increases linearly with increasing field strength when the electric field exceeds a threshold value E_1 and tends to a saturated constant beyond a high field strength E_2. Correspondingly the current density follows linear behavior no longer in the region between E_1 and E_2. A model based on the rearrangement of the particles under the electric field, which causes the variation of the dielectric property of the ER fluid, is proposed and the analysis is consistent with the measured results.

1. Introduction

The effect of the frequency of the applied ac electric field on ER fluids has not been sufficiently studied although some inconsistent observations have appeared in the literature[1-7].

Davis [8-10] analyzed the conductivity effect in ER fluids and concluded that the best performance should be obtained at high frequency for a metal-particle/insulating oil system which was demonstrated experimentally by Inoue [11].

In this paper a study of the frequency dependence of the shear stress in an a.c. electric field is reported. We find that when an electric field of a constant amplitude is applied to some water-free ER fluids the apparent viscosities (or shear stresses) increase monotonically with ac frequency and tend to saturated values at high frequency. The shear stresses of ER fluids can be several times or even one order of magnitude larger at higher frequency than that under dc field.

The non-linear dielectric properties has been observed in ER fluids for a long time [7], most commonly noticed one of which is that the current density increases more rapidly with the field strength than linear behavior. Up to date the reliable explanation about this phenomenon is still absent. We find that the non-linear behavior is mainly caused by the arrangement of the particles in the fluid. The permittivity of ER fluid increases linearly with increasing the field strength when electric field exceeds a threshold value E_1 and tends to a saturated constant beyond a high field strength E_2. Correspondingly the current density follows linear behavior no longer in the region between E_1 and E_2. The measured results are well consistent with the analysis.

2. Experimental

For measuring the frequency dependence of ER fluids in ac electric field, a high voltage a.c. power supply was designed and set up. The sinusoidal voltage output and its frequency are continuously adjustable. The effective ranges of the voltage and the frequency are 0-5KV and 10Hz-40KHz respectively. However, in most of the measurements of ER fluids reported herein the current output limited the apparatus to be applied to about 3KHz. Beyond that frequency the power supply is unable to withstand the large current caused from the capacitance effect.

The frequency dependencies of shear stresses in several kinds of ER fluids have been studied. The ER fluids consist of silicone oil with suspended $KNbO_3$, $BaTiO_3$, $SrTiO_3$, $LiNbO_3$ or Si particles. All particles were prepared by grinding the single crystals and then drying in an oven above 200°C to evaporate any water. As an example, we present the results on the $KNbO_3$/silicone oil ER fluid.

Particles of ferroelectric $KNbO_3$ crystal (average size 35 μm) were suspended in a silicone oil. The volume fractions are 0.025, 0.05, 0.075 and 0.1 respectively. The rheological measurements were performed on a rotating cylinder rheometer.

In order to keep a constant magnitude of the electric field between the electrodes it is necessary to adjust the voltage of the power supply carefully when the frequency is changing. In other words, special care must be taken to avoid false appearances coming from the resonance effect of ac electric circuit.

Field dependences of the permittivity of two kinds of ER fluids, $KNbO_3$/silicone and $SrTiO_3$/silicone, were measured. The container of ER fluid was installed in a silicone oil bath of which the temperature can be controlled. All the measured current densities and field strength were mean effective values and the experiments were performed under ac field with 50 Hz at room temperature.

3. Results and Discussions

A. The frequency dependence of ER fluid in ac field

When we keep a fixed mean effective voltage across ER fluid at different frequencies by adjusting the output of the power supply the measured mean effective current passing through ER fluid is linearly with respect to the frequency. Fig. 1 shows the frequency dependence of the shear stresses for different electric fields; while Fig. 2 shows how the shear stresses change with the field strength at different frequencies.

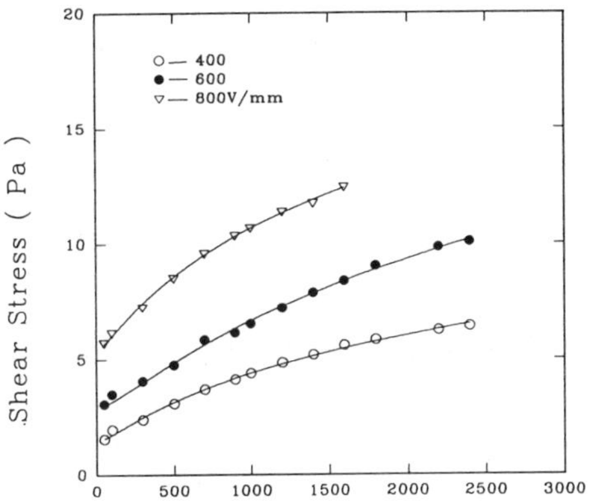

Fig.1 Variations of shear stress of $KNbO_3$/silicone oil ER fluid (ϕ=0.1) with the frequency as E=400, 600, 800V/ mm respectively (shear rate: $15.50s^{-1}$)

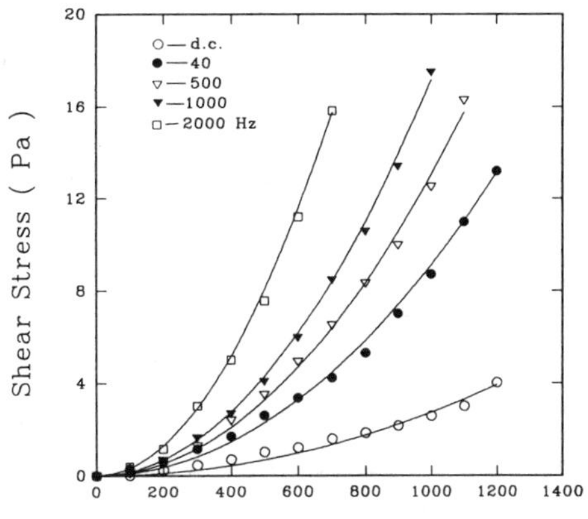

Fig.2 Variations of shear stress of $KNbO_3$/silicone oil ER fluid (ϕ=0.1)with the electric field for f=0,40,500,1000,2000 Hz respectively (shear rate:$15.50s^{-1}$)

The shear stress of ER fluid increases monotonously with the frequency at a fixed field as shown in Fig. 1. It can be seen that the shear stresses at a higher frequency are several times or even an order of magnitude larger than that in a dc field for a same field strength. In addition to the $KNbO_3$/silicone oil ER fluid, some other water-free ER fluids such as $BaTiO_3$, $SrTiO_3$, $LiNbO_3$ or Si particles suspended in the silicone oil exhibited the same phenomena, that is, the shear stresses increase monotonically with the frequency up to 3KHz.

According to the conventional expression [1, 10], the force between two spherical particles in ER fluid and further the shear stress is proportional to β^2. β is given by

$$\beta = \frac{\tilde{\varepsilon}_p(\omega) - \tilde{\varepsilon}_f(\omega)}{\tilde{\varepsilon}_p + 2\tilde{\varepsilon}_f(\omega)} \quad (1)$$

The complex dielectric constant $\tilde{\varepsilon} = \varepsilon + \dfrac{\sigma}{i\varepsilon_0\omega}$, σ is the conductivity, angular frequency $\omega = 2\pi f$. The subscripts p and f indicate the particle and the fluid, respectively. Then

$$\beta^2 = \frac{(\sigma_p - \sigma_f)^2 + \omega^2\varepsilon_0^2(\varepsilon_p - \varepsilon_f)^2}{(\sigma_p + 2\sigma_f)^2 + \omega^2\varepsilon_0^2(\varepsilon_p + 2\varepsilon_f)^2} \quad (2)$$

It is obvious, that when the condition $\frac{|\varepsilon_p - \varepsilon_f|}{\varepsilon_p + 2\varepsilon_f} > \frac{|\sigma_p - \sigma_f|}{\sigma_p + 2\sigma_f}$ is satisfied, β^2 should monotonously increase with increasing frequency and reach $\left(\frac{\varepsilon_f - \varepsilon_p}{\varepsilon_p + 2\varepsilon_f}\right)^2$ at high frequency. Therefore the shear stress of ER fluid should be enhanced by increasing the frequency of ac field and tend to a saturated value as the frequency being high enough.

Although the tendency of the measured shear stress increasing with the frequency is consistent with the analysis described above, the quantitative agreement of the behavior between the calculation and the measurement is unable to be achieved. The reason of this disagreement is still poorly understood.

All of our results were obtained with the anhydrous ER fluids. In the case of water-activated ER fluids the frequency responses greatly depend on the water content and usually a maximum of the shear stress appears in the frequency range below 10^3 Hz [12].

B. The non-linear dielectric properties

The current density and permittivity dependence of $KNbO_3$/ silicone oil ER fluid on electric field are shown in Fig.3(a) and Fig.3(b). In Fig.3(b), the insertion plots the behavior under low electric field. It is obvious that in a certain range of the electric field the permittivity increases linearly with field strength and correspondingly the current density behaves non-linearity in this region. In Fig.3 this region is from about E_1=50V/mm to E_2=2800V/mm for the case of $\phi = 0.2$, while E_1 and E_2 are lower in the case of $\phi = 0.05$. Below E_1 the permittivities are constants and beyond E_2 that tend to saturated values. The same results were obtained with a $SrTiO_3$/silicone oil fluid which are not shown here. The physical origin of the variation of the permittivity is due to the arrangement of the particles in the oil. When the field strength is lower than E_1, the particles in ER fluid are randomly dispersed without rearranging. The permittivity of the system is a constant $\varepsilon = \varepsilon_1$, which can be presented with Looyenga formula [13].

$$\varepsilon_1 = \left[\varepsilon_f^{1/3} + \phi(\varepsilon_p^{1/3} - \varepsilon_f^{1/3})\right]^3 \tag{3}$$

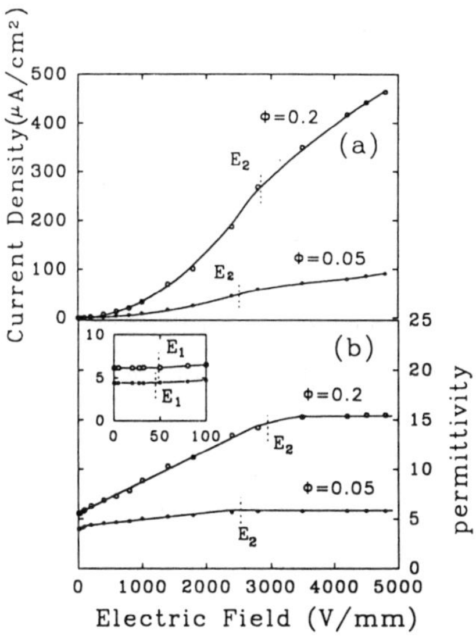

Fig.3 Relation of the current density (a) and permittivity (b)vs. field strength in $KNbO_3$/silicone oil ER fluid (ϕ=0.05 and 0.2)

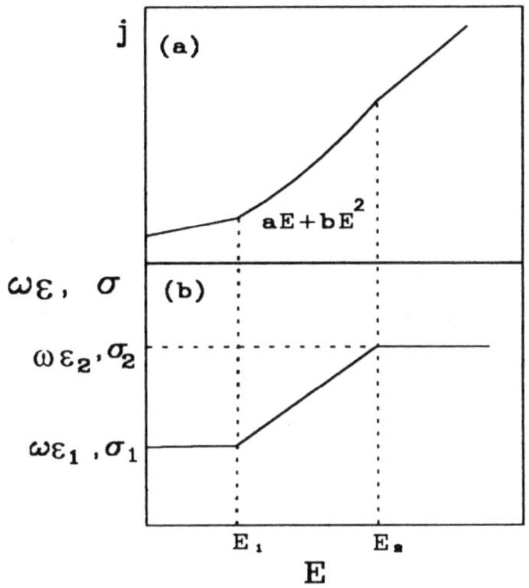

Fig.4 Relation of j (a) and ωε or σ (b) vs. the field strength according to Eq.(5)- (8)

Any aggregation of the particles along the direction of the electric field must cause the permittivity of the system increasing. As the electric field $E>E_1$, the particles are gradually aligned in the direction of the field forming chains and columns. Finally all particles are arranged in chains and columns at E_2 beyond which the pattern would be fixed and no more change could happen. In this case the permittivity of the system can be derived by regarding the total capacitance as the sum of all capacitancies of solid chains and liquid in parallel. The permittivity

$$\varepsilon_2 = \phi\varepsilon_p + (1-\phi)\varepsilon_f \ . \tag{4}$$

It is obvious that for a certain volume fraction ϕ, ε_2 is always larger than ε_1.

The relationship of the permittivity and electric field can be expressed as:

$$\varepsilon = \begin{cases} \varepsilon_1, & \mathrm{E} \le \mathrm{E}_1 \\ \varepsilon_1 + \kappa(\mathrm{E}-\mathrm{E}_1), & \mathrm{E}_1 < \mathrm{E} < \mathrm{E}_2 \\ \varepsilon_2 = \varepsilon_1 + \kappa(\mathrm{E}_2 - \mathrm{E}_1), & \mathrm{E} \ge \mathrm{E}_2 \,. \end{cases} \tag{5}$$

where k is a constant.

Then the current density in high frequency ac field limit can be written as

$$j = \omega\varepsilon E$$
$$= \begin{cases} \omega\varepsilon_1 E, & E \le E_1 \\ [\omega\varepsilon_1 + \kappa\omega(E - E_1)]E, & E_1 < E < E_2 \\ \omega\varepsilon_2 E = [\omega\varepsilon_1 + \kappa\omega(E_2 - E_1)]E, & E \ge E_2. \end{cases} \quad (6)$$

Because of the non-linear polarization of ER fluid in the region between E_1 and E_2, the field dependence of the current density behaves a form $j=aE+bE^2$ where $a=\omega\varepsilon_1 - \kappa\omega E_1$ and $b=\kappa\omega$.

It is also reasonable to substitute the conductivity for the permitivity in Eq.(3)-(5). Then the conductivity and the current density should exhibit the same non-linear behavior in a dc field. They can be expressed as:

$$\sigma = \begin{cases} \sigma_1, & E \le E_1 \\ \sigma_1 + \kappa'(E - E_1), & E_1 < E < E_2 \\ \sigma_2 = \sigma_1 + \kappa(E_2 - E_1), & E \ge E_2 \end{cases} \quad (7)$$

and

$$j = \sigma E$$
$$= \begin{cases} \sigma_1 E, & E \le E_1 \\ [\sigma_1 + \kappa'(E - E_1)]E, & E_1 < E < E_2 \\ \sigma_2 E = [\sigma_1 + \kappa'(E_2 - E_1)]E, & E \ge E_2. \end{cases} \quad (8)$$

κ' is a constant being different from κ. In the regions below E_1 and above E_2 the current densities still follow Ohmic rule but with different slopes which are $\omega\varepsilon_1$ or σ_1 and $\omega\varepsilon_2$ or σ_2 respectively. Obviously $\omega\varepsilon_2$ or σ_2 is usually much larger than $\omega\varepsilon_1$ or σ_1, i.e. the current density increases with E much steeper beyond E_2 than below E_1. Fig. 4(a) and (b) plot the results described in Eq. (5) and (6) or Eq.(7) and (8). With this analysis the measured behaviors of the permittivity and current density shown in Fig.3 can be well understood.

In our experiment, because the anhydrous ER fluids were mearsured, the relaxation is not important in the case of that the field frequency is not very high. The Looyenga equation and the model we proposed are valid. However, if ER fluids contain water or mobile ions, the relaxation problem would be serious and should be taken into account. Nevertheless, the structure induced non-linear dielectric property is a general behavior in ER fluids.

4. Conclusions

In conclusion, it is rather popular behavior that the shear stresses of anhydrous ER fluids monotonously increase with the frequency and tend to reach saturated values at high frequency under a fixed ac field strength. This phenomenon depends on the mismatch of the conductivity and the permittivity of particle and fluid. As an example the detailed results of the $KNbO_3$/silicone fluid have been reported, in which remarkable enhancement of the shear stresses has been observed in the frequency range of 10Hz-3KHz comparing with that in dc field. The frequency response of ER fluid indicates that under ac electric field the polarization current of ER fluids, which is proportional to the frequency, must strengthen the interaction between the particles. A further study about the mechanism of the frequency effect on ER fluids should be carried out.

The electric field dependence of the permittivity and the current density of ER fluid was measured and analyzed. The permittivity of ER fluid increases linearly with increasing field strength in a region between a threshold field E_1 and a critical field E_2. The alignment of the particles causes the permittivity or the conductivity of the system increasing. Below E_1, where the particles are randomly dispersed, and beyond E_2, where the particles form chains and columns, the permittivities do not vary with E and have a lowest and a highest values respectively. Correspondingly the current density of ER fluid follows Ohmic rule no longer in between E_1 and E_2, while it still shows Ohmic behavior in the regions below E_1 and beyond E_2.

Acknowledgments

This research is supported by the Chinese National Science Foundation and a grant for Key Research Project in Climbing Program from the State Science and Technology Commission of China.

References

[1]. H.Conrad, in *Particulate Two-phase Flow* ed. M.C.Roco (Butter-worth, Boston, MA, 1992) p.355.

[2]. J.E.Stangroom, *J.Stat.Phys.*, **64**(1991), 1059.

[3]. H.Conrad and A.F.Sprecher, *J.Stat.Phys.*, **64**(1991), 1073.

[4]. D.L.Klass and T.W.Martinek, *J.Appl.Phys.*, **38**(1967), 67.

[5]. T.b.Jones, in *Electrorheological Fluids, Proc. of the Second International Conference on ER Fluid,* ed. J.D.Calson, A.F.Sprecher and H.Conrad, (Technomics Publishing Co.,1990), p.14.

[6]. T.Garino, D.Adolf and B.Hance, in *Electrorheological Fluids, Proc. of the International Conference on Electrorheological Fluids: Mechanisms, Properties, Structure, Technology and Applications*, ed. R.Tao, (World Scientific Publishing Co.,1992), p.167.

[7]. M.Whittle, W.A.Bullough, D.J.Peel and R.Firoozian, *Phys. Rev.* **E, 49**(1994), 5249.

[8]. L.C.Davis, *Appl. Phys. Lett.*, **60** (1992),319.

[9]. L.C.Davis, *J. Appl. Phys.*, **72**(1992),1334.

[10]. L.C.Davis, *J. Appl. Phys.*, **73**(1993),680.

[11]. A.Inoue, in *Electrorheological Fluids, Proc. of the Second International Conference on ER Fluid*, ed. J.D.Sprecher and H. Conrad, (Technomics Publishing Co., 1990)p.176.

[12]. T.Y.Chen, P.F.Luckham, *Colloids and Surfaces* **A, 78**(1994),167.

[13]. H.Looyenga, *Physica*, **31**(1965), 401.

A COMPARISON OF THE PERFORMANCE OF

ER FLUIDS IN SQUEEZE

J L Sproston and A K El Wahed
Department of Mechanical Engineering
E W Williams
Department of Applied Mathematics and Theoretical Physics
University of Liverpool
UK

and

R Stanway
Department of Mechanical and Process Engineering
University of Sheffield
UK

ABSTRACT

This paper is concerned with an experimental and theoretical study of the characteristics of two ER fluids and their application to vibration control when used in squeeze flow.

It is seen that when the fluids are sandwiched between two electrodes, the lower one oscillating and the upper one fixed, displacement control of the lower electrode can be achieved by control of the applied field. Of particular interest is the dependence of the force transmitted across the fluids on oscillation frequency and applied field. A time-dependent theoretical model is seen to satisfactorily predict the device performance when allowance is made for a non-linear dependence of the post-yield shear stress on shear rate.

INTRODUCTION

In recent papers by Stanway et al (1993), Williams et al (1993), Sproston et al (1994) and Monkman (1995) it was suggested that ER fluids could be used more effectively in squeeze rather than in shear mode. In applications using ER fluids in shear where the available yield stresses were insufficient to accommodate the required engineering performance criteria, it appeared that this alternative squeeze-mode might prove beneficial. The most obvious application of fluids in squeeze might be concerned with vibration control, particularly in short range devices where the maximum stroke is a few millimetres.

The present investigation addresses the determination of the performance of two ER fluids in squeeze under various mechanical and electrical conditions in an experimental rig and then compares the resulting data with predictions made by a quasi-steady, time dependent theoretical model.

EXPERIMENTAL ARRANGEMENT

The experimental rig (Fig 1) consists of a Ling Dynamic Systems electromagnetic shaker (Model No 406) which is capable of providing vertical oscillatory motion with a maximum amplitude of ± 7 mm over a frequency in the range DC to 9 kHz. The shaker head is attached rigidly to a Kistler (Model No 9311A) piezoelectric force link with an earthed brass electrode having a recessed cylindrical cavity of diameter 95 mm which provides the reservoir for the ER fluid. The high voltage upper electrode is a circular brass disc of diameter 56 mm, its circumferential edge and rear face surrounded and supported by a PTFE collar. This is rigidly attached to a second identical force link and positioning assembly to the supporting frame.

The instantaneous displacement y of the lower electrode is determined using a RDP (type GTX2500) LVDT and the acceleration $\ddot{y}$ using an accelerometer Endevco (type 7254-100) attached to the upper surface of the lower electrode.

Electrical excitation of the ER fluid is achieved by means of a Brandenburg high voltage DC generator (Model 807R). Data acquisition and processing of the two force magnitudes together with the displacement and acceleration of the lower electrode is achieved using a Measurement Group (type Esam) electronic signal acquisition module linked to an IBM compatible personal computer. Feedback control of displacement amplitude of the lower electrode can be achieved using a Ling Dynamic Systems (Model DSC4) Digital Sine controller employing as input the signal from the accelerometer. This control is available when the amplitude of the input displacement produced by the shaker was reduced due to the increasing viscous resistance of the ER fluid with increasing potential difference across the electrodes.

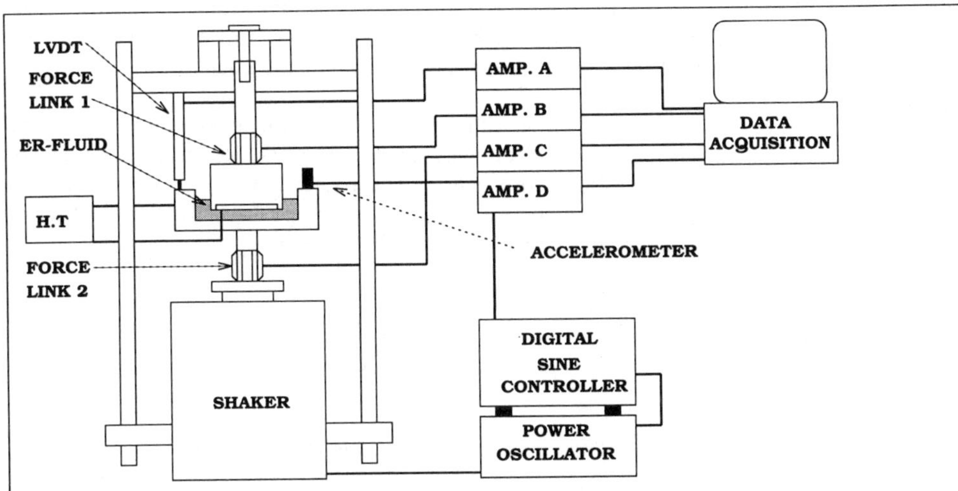

FIGURE 1 EXPERIMENTAL ARRANGEMENT

TESTS CARRIED OUT

For various displacement amplitudes of the lower electrode in the range (0.4—1.0 mm) and for excitation frequencies in the range (1—50 Hz) the input force F_1 to the ER fluid and the transmitted output force F_2, together with the displacement y and acceleration $\ddot{y}$ of the lower electrode were recorded over an interval of about one second at a sampling frequency of 5 kHz. These results were taken for DC excitation of the fluid in the range (0—2.5 kV) and for the case when the mean separation of the electrodes was 2 mm.

ER FLUIDS

The fluids employed here were as follows:

ER FLUID 1: a suspension of calcium alumina-silicate in silicone oil. Oil viscosity is 50 cSt and solid phase in size-range (0.5 to 5 μm)

and

ER FLUID 2: a suspension of polyurethane in silicone oil (Bayer plc RHEOBAY TP AC 3565). Oil viscosity is 5 cSt and solid phase in size-range (5 to 6 μm).

THEORETICAL ANALYSIS

Denoting the displacement of the lower electrode as y(t) the theoretical modelling is directed at the determination of the input force F_1 (t) and the output force F_2 (t) from a set of experimental data for y(t), by using the analysis outlined in Williams et al (1993). These forces can then be compared with the values taken from the experimental rig.

The analysis assumes a quasi-steady flow with a bi-viscous characteristic for the ER fluid in which the ratio γ represents the ratio of the viscosities in the 'yielded' to the 'unyielded' regions. The post-yield region also incorporates a shear-rate dependence obtained from additional experimental testing of the fluids. This was accomplished, for the electrically unstressed fluids, using a Carrimed Controlled Stress Rheometer.

The non-dimensional pressure gradient G given by

$$G = \frac{-(h-y(t))}{2\tau_1}\left[\frac{dp}{dr}\right] \tag{1}$$

and is determined as a function of the non-dimensional radius S given by

$$S = \frac{-\dot{y}(t)\ \eta r}{(h-y(t))^2\ \tau_1} \tag{2}$$

where h is the mean separation of the electrodes, τ_1 is the fluid yield stress, $\dot{y}(t)$ is the velocity of the lower electrode and r, θ, z are appropriate cylindrical polar coordinates.

Integration over the upper electrode then gives the output force F_2 (t) in the form

$$F_2 = \frac{2\pi\tau_1 a^3\Phi(X)}{h-y(t)} \tag{3}$$

where a is the radius of the upper electrode and X is the value of S at r = a.

The non-dimensional output force Φ may be expressed as

$$\Phi = \frac{1}{108}\left[\frac{\gamma}{X}\right]^3 + \frac{1}{X^3}\int_{S=\frac{\gamma}{3}}^{X} S^2 G\, dS \tag{4}$$

and the integration, once G is determined (numerically) as a function of S, is evaluated using simple quadrature.

Once F_2 is obtained the input force F_1 is then obtained using Newton's second law applied to the lower electrode in the form

$$F_1 - F_2 = M\ddot{y} \tag{5}$$

where M is the mass of the lower assembly.

RESULTS

Comparisons between the experimental results for the input and output forces and the predicted values from the theoretical model are shown in Figures 2 and 3 for ER fluid 1 for a range of applied fields. It is seen that agreement is fair over the middle range of oscillation frequencies but not so convincing at lower and higher frequencies. At the lower frequencies any difference is likely to be due to the inaccuracy in measuring the acceleration of the lower assembly whereas at the higher frequencies the assumption of a quasi-steady state may be responsible.

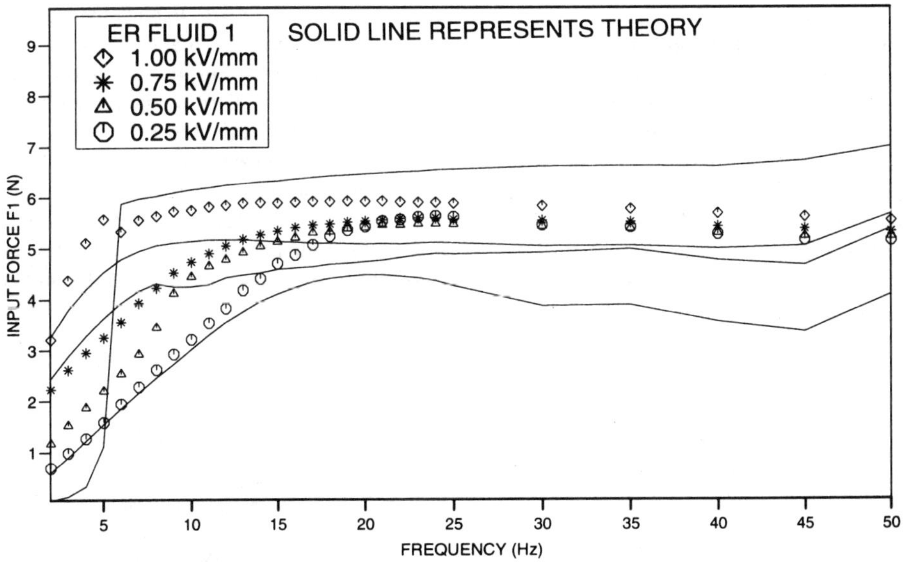

FIGURE 2. VARIATION OF INPUT FORCE WITH FREQUENCY

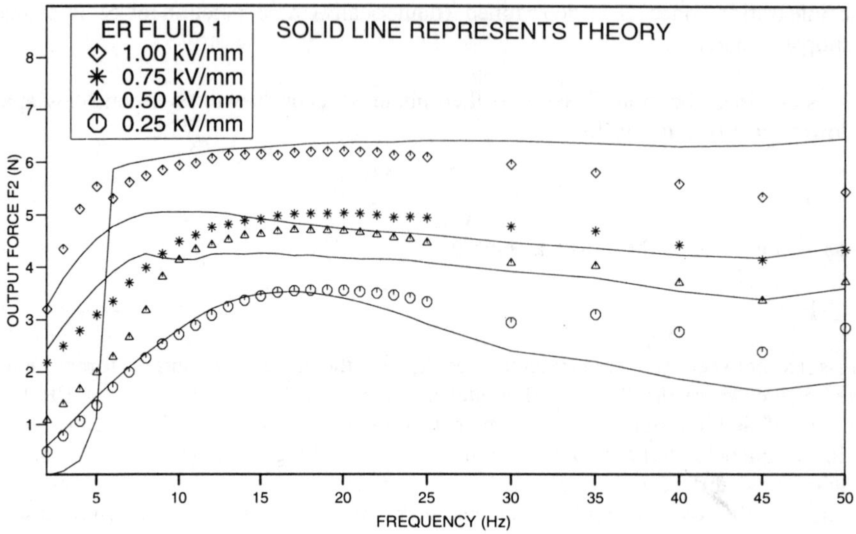

FIGURE 3. VARIATION OF OUTPUT FORCE WITH FREQUENCY

For ER fluid 2 the experimental results for peak-to-peak values of displacement of the lower assembly, input force and output force are shown in Figures 4a, b and c, respectively, for a range of frequencies and applied voltages. It is evident from Fig 4a that when a field of 1 kV/mm is applied the vibration is completely arrested for all frequencies within the range considered. At the lower values of 0.5 and 0.75 kV/mm only the higher frequency oscillations (above about 25 Hz) are arrested. In Fig 4b it is seen that the variation of input force with frequency follows the expected trend in which the force increases with applied field. However for the cases of 0 and 0.25 kV/mm, not only do the characteristics display a marked difference in behaviour to those for the higher fields at the lower frequency end, but for the electrically unstressed case the force generally exceeds the value of the stressed cases in the frequency range 15 to 30 Hz.

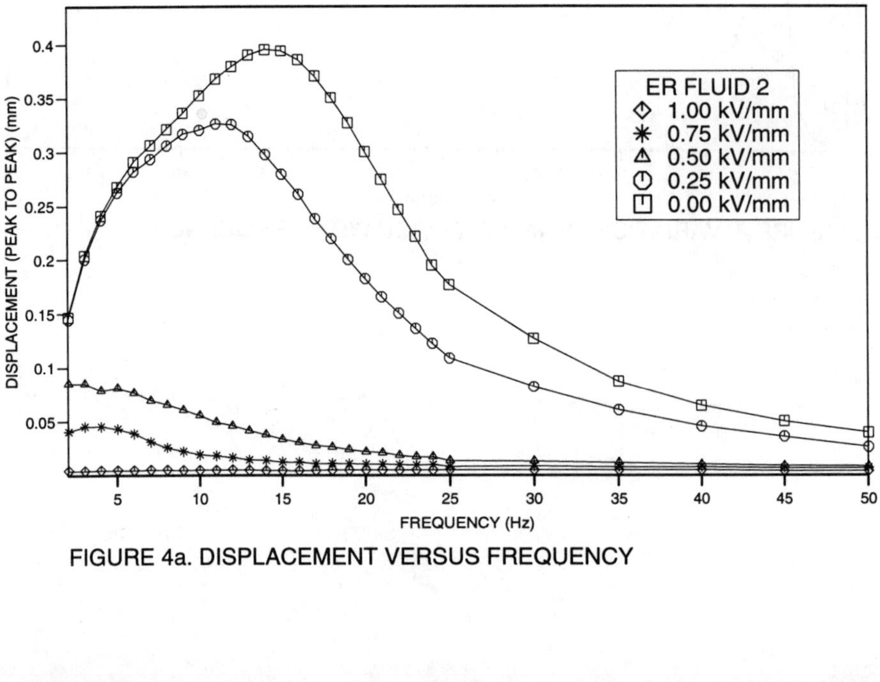

FIGURE 4a. DISPLACEMENT VERSUS FREQUENCY

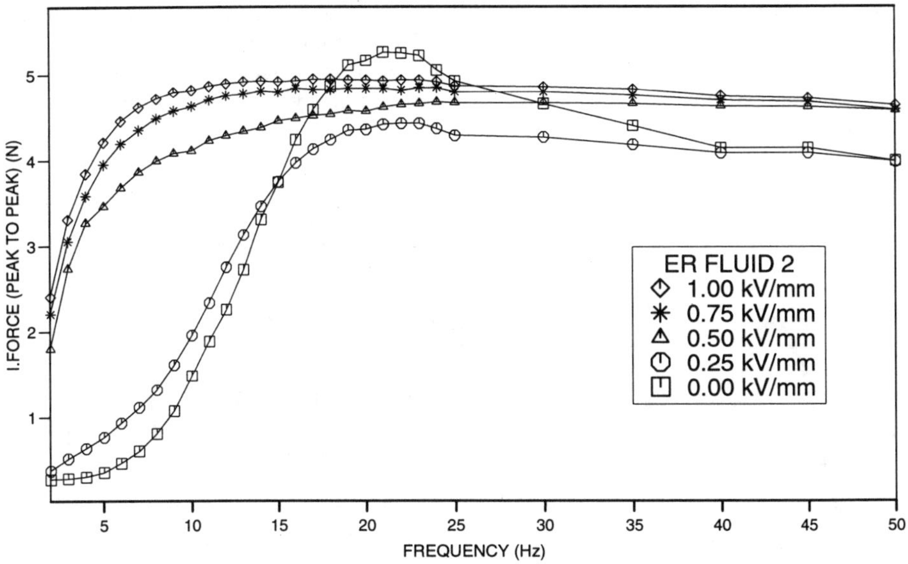

FIGURE 4b. INPUT FORCE VERSUS FREQUENCY

The dependence on frequency of the output force is shown in Fig 4c, where, for the higher fields of 0.5, 0.75 and 1 kV/mm the force asymptotes to a value independent of frequency for frequencies greater than about 20 Hz.

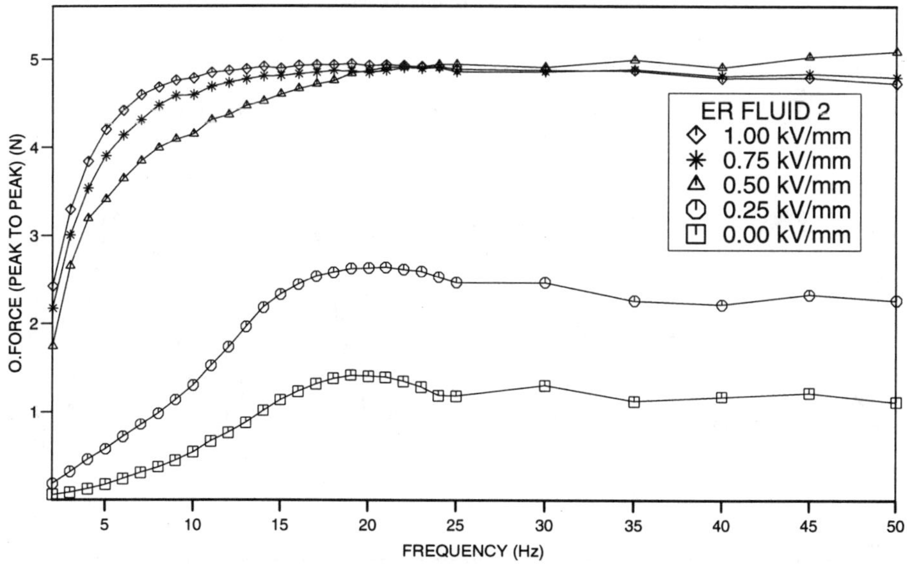

FIGURE 4c. OUTPUT FORCE VERSUS FREQUENCY

A comparison between the experimental and theoretical values of the temporal variation of input and output forces is shown in Figs 5 and 6 for ER fluid 1 at a frequency of 5 Hz and an applied field of 0.25 kV/mm. Satisfactory agreement is evident although the theoretical model overestimates the peak values of the output force by between 10 and 20%, even though the input force is closely predicted.

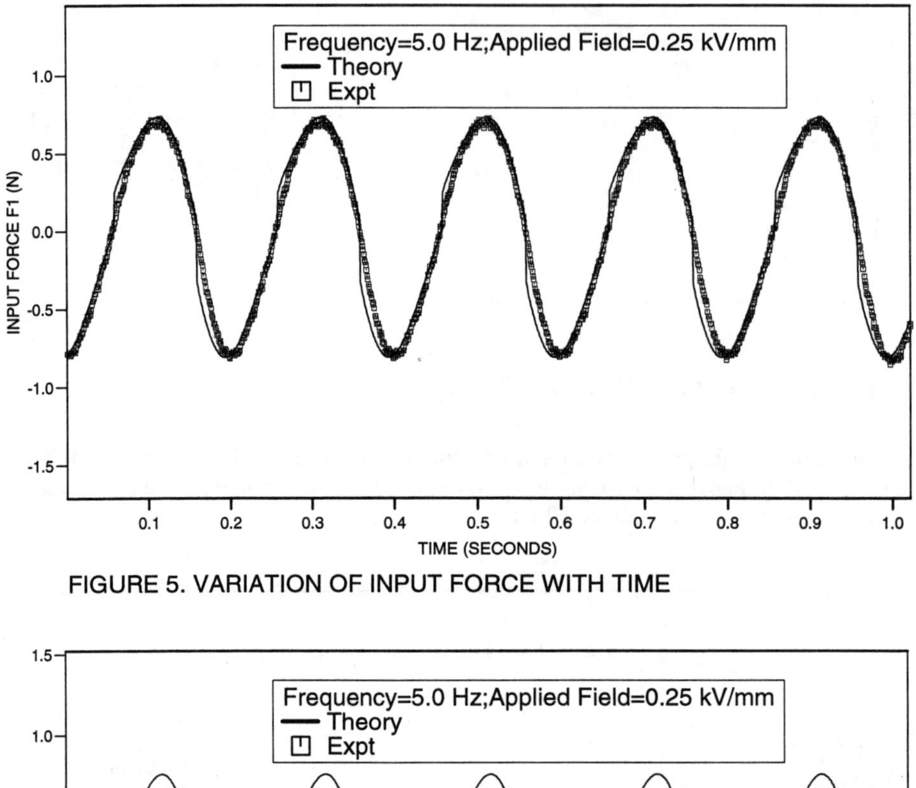

FIGURE 5. VARIATION OF INPUT FORCE WITH TIME

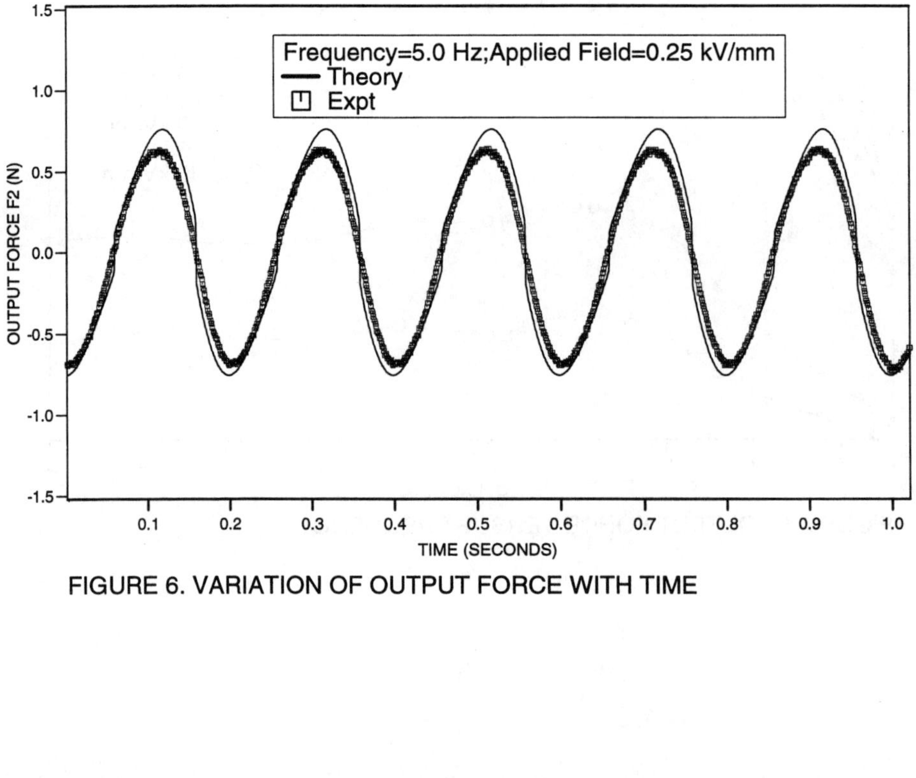

FIGURE 6. VARIATION OF OUTPUT FORCE WITH TIME

Of particular interest in the application of ER fluids to vibration control is the behaviour of the experimental rig expressed in terms of the force transmissibility defined as the ratio of the output force to the input force. In Fig 7 force transmissibility for ER fluid 1 is shown plotted as a function of the ratio of vibration frequency to resonant frequency for a range of applied fields. For the highest value of 1 kV/mm it is evident that the transmissibility is close to unity, confirming that the fluid is exhibiting 'solid body' characteristics. For the other values of voltage there is a clear distinction between the corresponding values of transmissibility and these generally reduce in magnitude for frequencies beyond resonance.

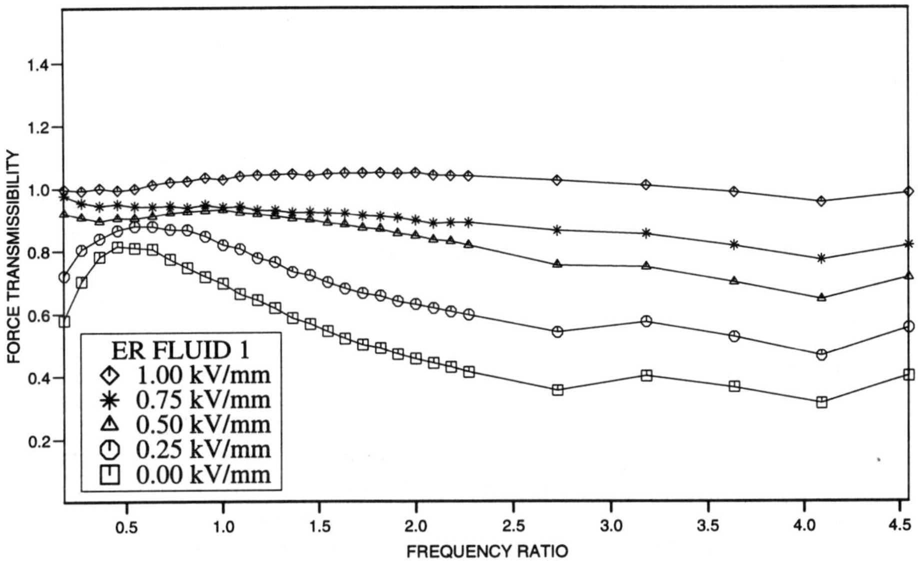

FIGURE 7. FORCE TRANSMISSIBILITY VERSUS FREQUENCY RATIO

The corresponding results for ER fluid 2 seen in Fig 8 show very different characteristics for the higher fields and it is evident that the 'solid body' state is exhibited even for a field of 0.5 kV/mm.

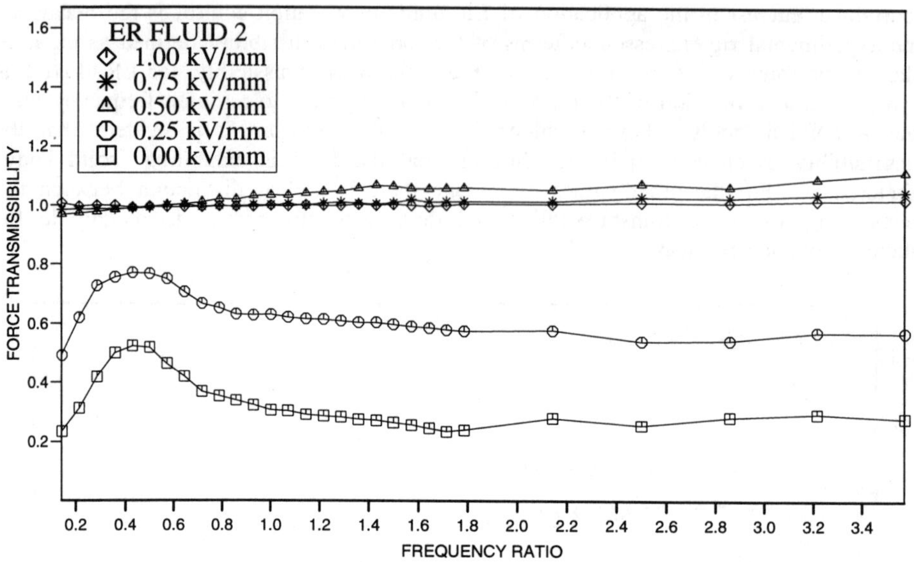

FIGURE 8. FORCE TRANSMISSIBILITY VERSUS FREQUENCY RATIO

CONCLUDING REMARKS

In this paper the authors have presented work which is directed towards an assessment of the applicability of two ER fluids in squeeze flow to vibration control. A time-dependent quasi-steady theoretical model, in conjunction with a fluid viscosity characteristic which incorporates a shear-rate dependence, was seen to satisfactorily predict the temporal variation of both the input and output forces and their dependence on oscillation frequency. For both fluids the vibration could be arrested through the application of a sufficiently high value of applied field across the fluids, and that the second fluid (Bayer RHEOBAY TP AC 3565) exhibited rather more useful characteristics in squeeze at lower fields than the first fluid.

Further work is to be carried out on the fluids when subjected to AC voltages and this is to be the subject of a future paper.

REFERENCES

Sproston, J.L., Rigby, S.G., Williams, E.W. and Stanway, R., 1994, "A Numerical Simulation of ER Fluids in Oscillatory Compressive Squeeze Flow", Journal of Applied Physics, D, Vol 27, pp 338-343.

Stanway, R., Sproston, J.L., Prendergast, M.J., Case, J., and Wilne, C.E,, 1992, "ER Fluids in Squeeze Flow Mode: An Application to Vibration Isolation", Journal of Electrostatics, Vol 28, pp 89-94.

Williams, E.W., Rigby, S.G., Sproston, J.L. and Stanway, R., 1993, "ER Fluids Applied to an Automotive Engine Mount", Journal of Non-Newtonian Fluid Mechanics, Vol 47, pp 221-238.

Monkman, G.J., 1995, "The Electrorheological Effect Under Compressive Stress", Journal of Applied Physics, D, Vol 28, pp 588-593.

ACKNOWLEDGEMENT

The authors would like to acknowledge the provision of ER fluid 2 by Dr Eckhardt Wendt of Bayer AG.

PHASE SEPARATION AND TURBIDITY OF ELECTRORHEOLOGICAL FLUIDS

S. CUTILLAS, A. MEUNIER, E. LEMAIRE, G. BOSSIS
Laboratoire de Physique de la matière Condensée
Université de Nice Sophia Antipolis, Parc Valrose, 06108 Nice Cedex 2-France

and

J.PERSELLO
† Laboratoire d'Electrochimie des solides
Université de Besançon-25030 Besançon-France

ABSTRACT

The turbidity of a suspension of silica particles in a silicone oil is studied in the presence of an electric field. We present experimental results which show a very strong attenuation of a laser beam when the electric field is switched on. We give a theoretical interpretation of these results with the help of a Rayleigh Gans Debye theory applied to cylinders. This model allows us to obtain the average length of the aggregates as a function of the coupling parameter λ which is the ratio of the electrostatic energy to kT. These results agree with a theoretical prediction and also with the results of Monte-Carlo simulations. The kinetics of the beginning of field induced aggregation can be obtained from the change of the turbidity with time after the application of a voltage step.

1. Introduction

Suspensions of solid dielectric particles can aggregate in the presence of an electric field. This aggregation is due to the electrostatic interactions between the polarized particles. When these electrostatic forces are high enough, the transient aggregates which are formed can resist to a mechanical stress of several thousand Pascal and then strongly modify the rheology of these suspensions. Most of the studies on electro-structured suspensions deal with the effect of this electrostatic stress on the flowing properties of the suspension. In this paper we are not going to consider the mechanical properties but the change of optical properties due to the application of an electric field. More precisely we are going to focus on the change of turbidity which is induced by the application of an electric field. This optical study allows to obtain information both on the state of aggregation and on its dynamics[1]. These information are then useful to understand the rheology in transient regimes (for instance the increase of apparent viscosity when the field is switched on). Apart from possible applications in optical devices, this optical study of field induced aggregation can also help to characterise the phase separation which is obtained in the presence of an electric field. The understanding of phase separation in the presence of anisotropic forces [2, 3] is the straight continuation of the study of phase transition in colloidal suspensions with radial forces. Here the controlling parameter for the phase transition is not the temperature nor the ionic strength (which changes the range of repulsive forces) but the ratio of electrostatic energy to kT :

$$\lambda = \frac{m_p^2}{4\pi\, \varepsilon_o\, \varepsilon_f d^3 kT} \qquad (1)$$

In Eq. (1) m_p is the dipolar moment of the particle, d its diameter and ε_f the relative permittivity of the suspending fluid. The electrostatic energy appearing in Eq. (1) corresponds to the situation of two particles at a distance d and with the line joining their centres perpendicular to the direction of the field.

In the second section of this paper we present the experimental results concerning the change of the turbidity due to the application of an electric field together with an interpretation of these results based on Rayleigh - Gans - Debye theory of scattering. In the third section we use a model for the kinetics of aggregation in order to model the time dependence of the extinction of a laser beam after the application of a voltage step. In the fourth section we compare our experimental determination of the average size of the aggregates to some theoretical predictions and also to results obtained by Monte-Carlo simulations.

2. Turbidity dependence on the electric field

The suspension used was made of silica particles synthetised by the Stober method and dispersed in a silicone oil of viscosity η = 20 mPa.s . The particles are monodisperse with a radius a = 0.47±0.05 µm. The suspension is placed between two glass disks covered with transparent electrodes. A laser beam is directed perpendicularly to the plane of the disks (and so parallel to the electric field) and the light transmitted through the sample is recorded on a photodetector. In fig. 1 we present the result of an experiment where we apply a voltage step and we record the change of transmission of the suspension versus the time. The different curves correspond to different voltage steps. What we first notice is a very strong decrease of the transmission after the application of the field.

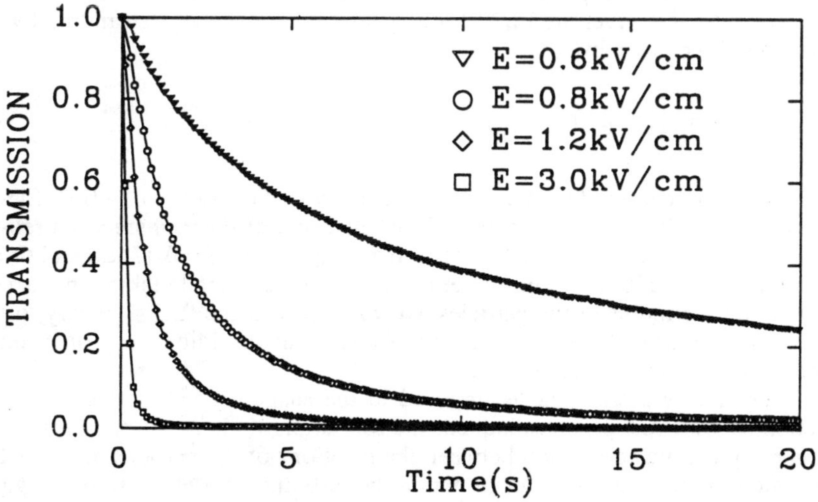

Fig 1. Transmission versus time after the application of different voltage steps

For a cell of thickness 100 μm the transmission decreases by more than two orders of magnitude and , as expected, the higher the applied voltage the faster the decrease of the transmission. In this experiment we have applied an alternative field at a frequency of 1 kHz. In order to explain this increase in turbidity with the field we are going to use a simple model of turbidity based on the Rayleigh - Gans - Debye approximation. In this model we assume single scattering and also that the local electromagnetic field is identical to the incident field.

The usual criteria of applicability of R.G.D. theory are given by $m-1 \ll 1$ and $2\pi\,(L/\lambda)\,(m-1) \ll 1$, m is the ratio of the refractive index of the particle to the one of the solvent. The refractive index of the particles has been obtained by measuring the refractive index of the suspension as function of the volume fraction. Then extrapolating at zero density and using Maxwell-Garnett theory [4] we obtain m = 1.444/1.4023 at T = 20°C. As the diameter of a particle is of same order as the wavelength ($\lambda = 0.63\mu$), a phase shift of one is obtained for a length corresponding to five particles. Actually the most important criterion ($m - 1 \ll 1$) is well verified and taking into account that the average electromagnetic field in the suspension plays the role of the incident field, the second condition which states that the phase of the field inside the particle must be equivalent to that of the incident field, is certainly less drastic for an assembly of aggregates than for a single aggregate. In the frame of this theory the turbidity which is the ratio of the scattered intensity to the incident intensity per unit length, is given for a polarized beam by :

$$\tau = \frac{1}{3}\,\Phi(m-1)^2\,k^4a^3 \int_0^{\pi} (1+\cos^2\theta)\,F(\theta)^2 \sin\theta\,d\theta \qquad (2)$$

with

$$F(\theta) = \frac{3}{(qa)^3}(\sin qa - qa\cos qa)\;;\; q = \frac{4\pi n}{\lambda}\sin\frac{\theta}{2} \qquad (3)$$

In Eq (2) $k = 2\pi n/\lambda$ with n the refractive index of the suspension, Φ is the volume fraction of the suspension and a the radius of the particles. $F(\theta)$ is the so-called form factor, which is given by Eq. 3 for a sphere. The intensity transmitted by a cell of thickness h will be given by :

$$I_t(h) = I_o\,e^{-\tau h} \qquad (4)$$

We have measured the transmitted intensity in the absence of a field for different thickness of the cell. Taking the slope of the logarithm of the transmission relatively to the thickness we get $\tau_{exp} = 21500\ m^{-1}$ and from Eq. 2 with a = 0.47 μm and $\Phi = 14.5\%$ we obtain : $\tau_{th} = 16500\ m^{-1}$. The agreement is correct, especially taking into account the uncertainty on the radius of the particles, so we shall trust the R.G.D. theory to calculate the turbidity. We have also verified that the value of the turbidity was proportional to the volume fraction at lower densities.

When an electric field is imposed to the suspension the effect of the dipolar interaction between the particles induces the formation of chains of particles and we can no longer neglect the correlation between the positions of the particles in the calculation of the scattered intensity. In order to obtain the turbidity of a given state of aggregation we can apply the R.G.D. model to cylinders of diameter d = 2a and length L = N.2a with

For a cell of thickness 100 μm the transmission decreases by more than two orders of magnitude and , as expected, the higher the applied voltage the faster the decrease of the transmission. In this experiment we have applied an alternative field at a frequency of 1 kHz. In order to explain this increase in turbidity with the field we are going to use a simple model of turbidity based on the Rayleigh - Gans - Debye approximation. In this model we assume single scattering and also that the local electromagnetic field is identical to the incident field.
The usual criteria of applicability of R.G.D. theory are given by $m-1<<1$ and $2\pi\ (L/\lambda)(m-1) << 1$, m is the ratio of the refractive index of the particle to the one of the solvent. The refractive index of the particles has been obtained by measuring the refractive index of the suspension as function of the volume fraction. Then extrapolating at zero density and using Maxwell-Garnett theory [4] we obtain m = 1.444/1.4023 at T = 20°C. As the diameter of a particle is of same order as the wavelength ($\lambda = 0.63\mu$), a phase shift of one is obtained for a length corresponding to five particles. Actually the most important criterion ($m - 1 << 1$) is well verified and taking into account that the average electromagnetic field in the suspension plays the role of the incident field, the second condition which states that the phase of the field inside the particle must be equivalent to that of the incident field, is certainly less drastic for an assembly of aggregates than for a single aggregate. In the frame of this theory the turbidity which is the ratio of the scattered intensity to the incident intensity per unit length, is given for a polarized beam by :

$$\tau = \frac{1}{3}\Phi(m-1)^2 k^4 a^3 \int_0^{\pi} (1+\cos^2\theta)\, F(\theta)^2 \sin\theta\, d\theta \qquad (2)$$

with

$$F(\theta) = \frac{3}{(qa)^3}(\sin qa - qa\cos qa)\ ;\ q = \frac{4\pi n}{\lambda}\sin\frac{\theta}{2} \qquad (3)$$

In Eq (2) $k = 2\pi n/\lambda$ with n the refractive index of the suspension, Φ is the volume fraction of the suspension and a the radius of the particles. $F(\theta)$ is the so-called form factor, which is given by Eq. 3 for a sphere. The intensity transmitted by a cell of thickness h will be given by :

$$I_t(h) = I_o\, e^{-\tau h} \qquad (4)$$

We have measured the transmitted intensity in the absence of a field for different thickness of the cell. Taking the slope of the logarithm of the transmission relatively to the thickness we get $\tau_{exp} = 21500\ m^{-1}$and from Eq. 2 with a = 0.47 μm and Φ = 14.5% we obtain : $\tau_{th} = 16500\ m^{-1}$. The agreement is correct, especially taking into account the uncertainty on the radius of the particles, so we shall trust the R.G.D. theory to calculate the turbidity. We have also verified that the value of the turbidity was proportional to the volume fraction at lower densities.

When an electric field is imposed to the suspension the effect of the dipolar interaction between the particles induces the formation of chains of particles and we can no longer neglect the correlation between the positions of the particles in the calculation of the scattered intensity. In order to obtain the turbidity of a given state of aggregation we can apply the R.G.D. model to cylinders of diameter d = 2a and length L = N.2a with

N the number of spheres in the chain. Assuming that all the cylinders have the same length and are aligned on the direction of the field we obtain for a polarized beam[5] :

$$\tau = \Phi\,(m^2-1)^2\,k^4\,a^3\,N\left[I_1 - \frac{I_2}{2}\right] \qquad (5)$$

with

$$I_1 = \int_0^{\pi} \left(j_0(\beta N)\,\frac{J_1(\gamma a)}{\gamma a}\right)^2 \sin\theta\,d\theta$$

$$I_2 = \int_0^{\pi} \left(j_0(\beta N)\,\frac{J_1(\gamma a)}{\gamma a}\right)^2 \sin^3\theta\,d\theta$$

$$\beta = ka(1 - \cos\theta) \;\; ; \;\; \gamma a = ka \sin\theta \cos(\theta/2)$$

In Eq. 5 $j_0(x) = \sin x / x$, J_1 is a first order Bessel function and we have taken into account the difference between a chain of spheres and a cylinder by multiplying the result for a cylinder by 2/3 which is the ratio of the volume of a sphere to the one of the cylinder. We see that the turbidity is now proportional to N the number of spheres which is contained in the cylinder of length L. This factor N expresses the fact that for N particles which would scatter in phase, the scattered field would grow like N and so the intensity like N^2 or ΦN. Using Eq. 5 we obtain (1, 1.81, 2.41, 2.86, 3.20) for the ratios of turbidity of chains of spheres with the number of spheres increasing from 1 to 5. An other model where we assume that the field scattered by each sphere comes from its centre (so the phase shifts are only considered between the centres of the spheres) gives very similar results.

As the transmitted intensity decreases exponentially with τ, it is readily seen that the transmitted intensity can be a very sensitive probe of the beginning of aggregation. On the other hand we must keep in mind that in our experimental conditions it is not clear to which extent the R.G.D. model can apply to chains of spheres longer than a few units.

3. - Kinetics of Aggregation

In order to obtain a quantitative representation of the time evolution of the transmitted intensity we have to know what is the time evolution of the average size of the aggregates. A simple analytical model for field induced aggregation has been proposed by See and Doi [6]. This model applies for irreversible aggregation and assumes that at a given stage of aggregation all the chains have the same length L. The following step, the length will be 2L and the time needed to perform this step is obtained from the velocity of two chains aligned on the direction of the field and separated by a distance corresponding to an homogeneous distribution of chains of length L. The force acting between two chains is obtained from two linear dipolar distributions and the hydrodynamic mobility is inversely proportional to the length. With these assumptions they obtain :

$$\langle N(t)\rangle = \left(\frac{1}{2\pi\varepsilon_0\varepsilon_f\eta_F}\right)^{1/2} \frac{3m_p}{16\,\pi\, a^3}\, t^{1/2}\,\Phi^{1/2} \qquad (6)$$

In Eq. 6 η_F and ε_F are respectively the viscosity and the permittivity of the suspending fluid (we have measured $\varepsilon_F = 2.69$ and $\eta_F = 20$ mPa.s for T = 20°C).

Combining Eqs (5) and (6) we are able to predict the time evolution of the transmitted intensity. The comparison between this model and the experimental results for two voltage steps E = 120V/mm and E=300V/mm is made in Fig. 2. We see that for the higher field we have a fair agreement between the theoretical and the experimental curve. For the lower field the theory of See and Doi gives an aggregation rate which is noticeably too fast. One reson for this discrepancy could be that this theory only deals with irreversible aggregation whereas at the lower values of the field we must also take into account the rupture of some aggregates in the kinetics which will slow down the rate of aggregation.

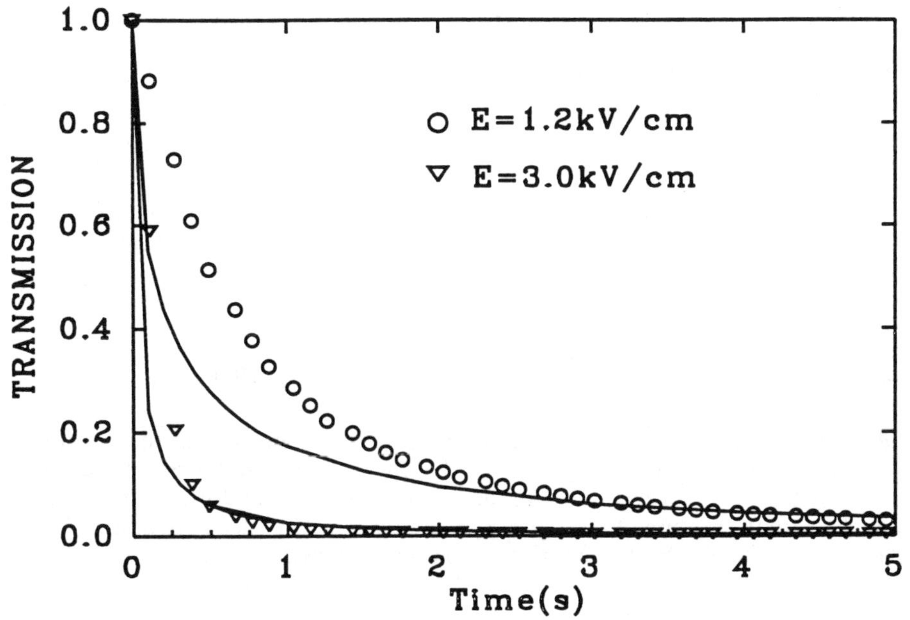

Fig.2 Prediction of the change of transmission with time after a voltage step. The open dots and the triangles are the predictions of the model. The continuous lines are the theoretical predictions

Nevertheless this approach can be useful for instance to obtain a good estimation of the force between the particles which drives the first stages of aggregation. Fig. 2 concerns irreversible aggregation, where the field corresponds to a value of λ (Eq 1) much larger than the critical value corresponding to the phase separation. We are now trying to use these experiments in order to obtain the average size of the aggregates as a function of λ.

4. Equilibrium length of the aggregates

The coupling parameters expresses the balance between electrostatic and thermodynamic forces. Then we expect that above a critical value λ_c we shall observe a first order phase transition equivalent to a gas-liquid transition. The essential difference comes from the anisotropy of the attractive force of dipolar origin compared to the usual radial Van Der Waals force. Instead of having the growth of spherical droplets at the phase transition we observe the growth of elongated aggregates or, at low density, of linear chains of spheres. Some attempts to predict the average length of these aggregates have been made in the context of ferrofluids[7-9]. For suspensions at low density, where pair interactions are dominant, the total pressure can be approximated either by a virial expansion of the density or by the sum of the partial pressures corresponding to different species of aggregates. Equating these two expressions of the pressure we have[8] :

$$\langle N \rangle = 1/(1-\rho B_2) \qquad (7)$$

with ρ, the number density, and B_2, the virial coefficient:

$$B_2 = -d^3 \int_1^{\infty} \int_0^{+1} \left(e^{+\frac{\lambda}{r^3}(3u^2-1)} - 1 \right) 2\pi r^2 du\, dr \qquad (8)$$

If λ is much larger than one we obtain $B_2 \approx -\pi d^3 e^{2\lambda} / 18\lambda^2$.
A similar expression was derived by De Gennes and Pincus[9] for an application to ferrofluids.
An other approach has been followed by Tsebers[8] who calculates the equilibrium distribution of sizes from the equality of the chemical potentials associated with a "chemical reaction" between any two aggregates which can form a third one (or reciprocally with the rupture of one aggregate into two smaller ones). Assuming the same association constant $K=2B_2$ for every "reaction", he obtains:

$$\langle N \rangle = 1/(1-2\rho_1 B_2) \qquad (9)$$

with

$$2\rho_1 B_2 = 1 + \frac{1}{4\rho B_2} - \frac{\sqrt{8\rho B_2 + 1}}{4\rho B_2}$$

In Eq.(9) ρ_1 is the average density of isolated particles . From the measurement of the turbidity for different values of λ we can obtain with the help of Eq. 5 the evolution of the average size of the chains with λ . In this aim we need to relate λ to the value of the field. We have obtained the dielectric constant of the suspension by measuring the change of the capacity with the spacing between the electrodes. We obtain $\varepsilon = 5.1$ at a frequency of 1 KHz and with $\Phi = 0.145$. The dipolar moment, m_p, of the particle which appears in Eq. (1) is obtained from :

$$\rho\, m_p = \varepsilon_0 (\varepsilon - \varepsilon_F) E \qquad (10)$$

We find with these values $\lambda = 4.07\ 10^{-9}\ E^2$ where E is the root mean squared field. We have plotted in fig. (3) the experimental results (filled squares) for the average size of the chains versus λ together with the predictions of Eqs (7) (hexagons) and (9) (triangles). The last curve (empty square) is the result of a Monte Carlo simulation of hard spheres interacting through a dipolar potential. The basic cell was tetragonal with either 125 or 320 particles and an Ewald sum[10] was used to sum the dipolar interactions on an infinite periodic medium based on the vectors defining the unit cell. Two particles were considered to belong to the same aggregate if their distance was less then 0.5a. This criterion is somewhat arbitrary but was chosen on the observed invariance of the results for the choice of separations ranging from 0.3 to 0.7. We can see on this figure that the result obtained by Monte-Carlo simulation agrees quite well with the prediction of Tsebers[8] but rules out the theory given in Eq.7 .The experimental points (filled squares) are still closer to the theoretical values and confirm very well this conclusion. It must be noted that special care is taken during the experiments in order to do the measurements in a time small compared to the sedimentation time(typically 1 hour) but still large enough to reach the equilibrium state of aggregation.

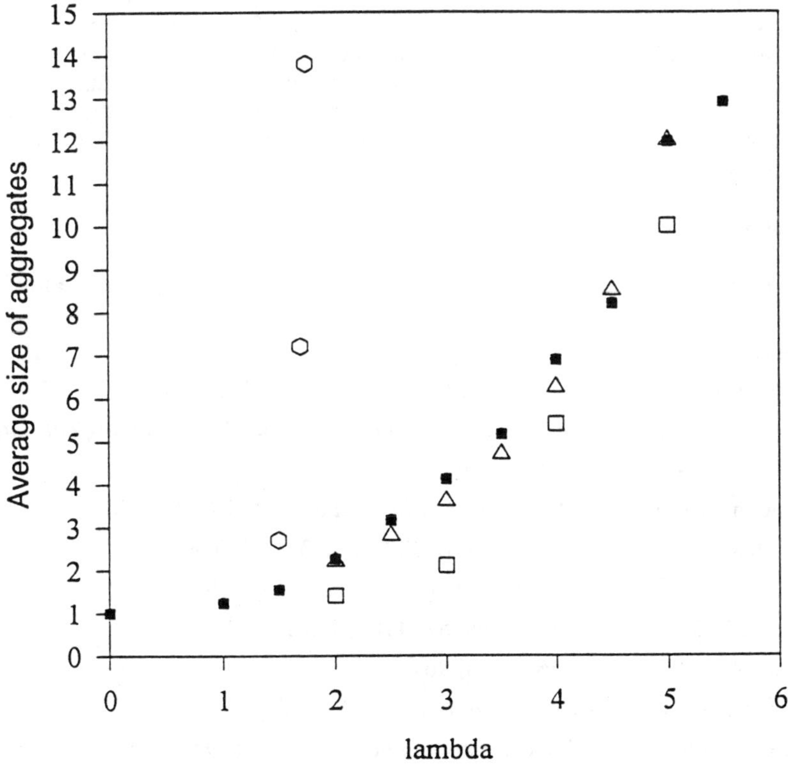

Fig.3 Average size of the aggregates versus λ (Cf Eq.1). ■ ■ ■ Experimental results
○ ○ As given by Eq. 7 △ △ △ As given by Eq.9 □ □ □ M.C. simulations

5. Discussion

We have shown that using micronic particles with a refractive index close to that of the suspending fluid we can strongly attenuate the transmission of a laser beam by imposing a voltage step of a few Volts on a thin cell containing the electrorheological fluid. This effect has been interpreted on the basis of Rayleigh-Gans-Debye theory applied to spheres and cylinders. This theory seems to work quite well and we are able to predict the time evolution of the transmitted intensity as well as the equilibrium length of the aggregates without using any free parameter. Both experiments and simulations show that the model of aggregation based on chemical equilibrium between the different species of aggregates is appropriated at least for volume fractions up to 14%. If we follow this model we have aggregates with a length equals to the thickness of the cell for λ=6. Experimentally this chain structure can be kept during several days with a field corresponding to $6<\lambda<10$ without any change of the turbidity. Nevertheless it does not seem to be an equilibrium structure since small mechanical vibrations are able to induce the phase separation which manifests by a strong increase of the transmitted light. At higher values of λ we can observe a phase separation even without the help of mechanical vibrations but we still have to increase the field slowly: a quick increase of the field produces a glassy like structure which does not evolve towards phase separation. This is still more evident at higher volume fractions. A complete study of phase separation in these conditions is not easy since the sedimentation time is often lower than the typical times needed to reach a true thermodynamic equilibrium. This study is easier in the presence of a small amplitude vibration which allows to reach quickly the structures of minimum energy. As long as the Péclet number remains smaller than unity we can assume that the critical value of λ will not be modified by the presence of the oscillating flow. We believe that such a study should give us the experimental phase diagram of this field induced transition.

Acknowledgments

Computer time for the M.C.simulations was kindly provided by the computer centre IMT in Marseille

6. References

[1] K.L.Smith, G.G. Fuller, J. Colloid Interface Sci.**155** (1993), 183

[2] R. Tao, Qi Jiang in *Proceedings of the Conference on Electrorheological fluids*, ed. R. Tao, (World Scientific, Singapore, 1994)

[3] S. Fraden, A.J. Hurd, R.B.Meyer *Phys.Rev.Letters* **63** (1989) 2373

[4] J.C. Maxwell-Garnett *Philos. Trans. R. Soc.* **203**, (1904)385

[5] A.R.Jones, H. Savaloni, *Part. Part. Syst. Charact.* **6** (1989) 110

[6] H. See, M. Doi, *J. Phys. Soc. Jap.*, **52** (1983) 2810

[7] P.C. Jordan, *Mol. Phys.*, **25** (1973) 961

[8] A.O. Tsebers, *Magnitmoya. Gidrodinamika*, **2** (1974) 36

[9] P.G. De Gennes ,P.A. Pincus, *Phys.Kondens.Mat.*, **11** (1970) 189

[10] D.J. Adams I.R.McDonald *Molecular Physics* **32** (1976) 931

A GENERALISED PRESENTATION OF VALVE AND CLUTCH DATA FOR AN ER FLUID, AND PRACTICAL PERFORMANCE PREDICTION METHODOLOGY

D J PEEL, R STANWAY AND W A BULLOUGH

*Department of Mechanical and Process Engineering,
The University, Sheffield, S1 3JD, UK*

Abstract

In previous papers a technique for comparing the performance of a particular electro-rheological (ER) fluid in different situations was derived from steady, laminar, shear and flow mode experimental data and static cell tests. By plotting non-dimensional Hedström Number (He) and Friction Coefficient (Cf) against Reynolds Number (Re) via use of the Bingham plastic constitutive equation, one continuous and overlapping relationship was shown for constant field excitation over a range of operating conditions and geometries. In the present work the same treatment is extended to a different fluid, of higher solid volume fraction and different constituents, which was tested on different instruments at a different uniform temperature, so as to attest the generality of the technique. In doing this the methodology of converting low shear rate Couette viscometer data to practical flow valve performance predictions is illustrated.

1. Introduction

Electro-rheological test data has generally been derived from steady flow tests on concentric clutch (shear mode) or effectively parallel plate valve (flow mode) configurations, having a large radius to electrode separation gap ratio. Initially, the results were plotted respectively as T v. Ω v. E or Δp v. $\dot{q}$ v. E, for a range of lower than practical conditions of device operation. Later on, in order to make it applicable to a wide range of device sizes, the data was reduced to τ v. $\dot{\gamma}$ v. E [1], for both test modes and more intense operational areas, e.g. see Figs. 1 and 2. For simplification towards providing a convenient and approximate design/sizing up method, uniform conditions were assumed. From the Couette formulae for a concentric clutch, $\tau_w = T/2\pi r_1^2 b$, $\dot{\gamma}_n = \Omega r_2/h$ and $E = V/h$ - (1). For the valve, $E = V/h$, $\dot{\gamma}_n = (du/dy)_w = 6\dot{q}/bh^2$ - (2) from the Poiseuille flow formulae at zero volts, and $2\tau_w b\ell = \Delta p\ bh$ - (3) The shear stress data from the two arrangements did not truly correspond with one another, nor with static shear test results.

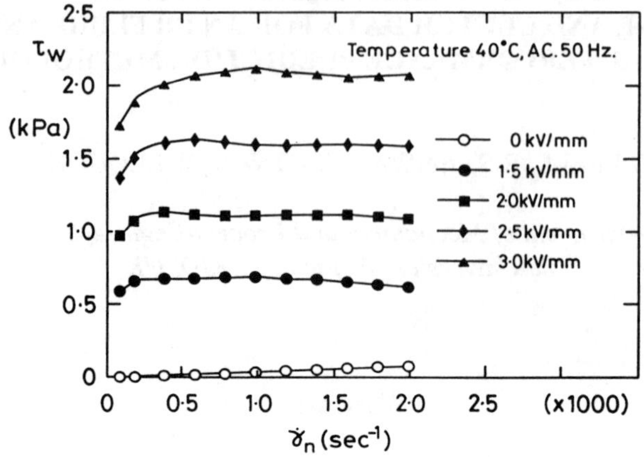

Figure 1. Couette Rheometer Experimental Data - after use of equation (1) [by courtesy of Bayer AG]

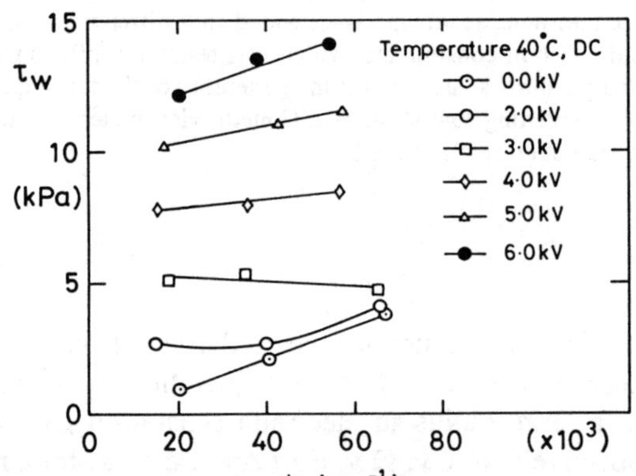

Figure 2. Poiseuille Flow Experimental Data - after use of equations (2) and (3). [by courtesy of Bayer AG]

Subsequently, ER fluids were used in the squeeze mode, in two-dimensional flow in radial clutches, and in hydrodynamic bearings etc. If an "ideal" Bingham plastic constitutive model were chosen, a best choice of τ_e and μ had to be made from the test device characteristics. Clearly the time had come to try to establish common parameters of ER fluid operation on a continuum property basis, otherwise for any one fluid a plethora of test data would be forthcoming.

As a first step the flow mode test derived fluid property data for one valve type, of different sizes, was shown to be consistent over a range of flow rates and operational voltages applied to the fluid [2]. The approach was consolidated when static shear test data was shown to form part of a continuous function derived therefrom [3]. More recently, using a similar fluid, the technique has been extended [4] to directly relate concentric clutch and plane valve data. The present work now applies the same technique to data from a completely different fluid, tested independently on differing sizes of shear and flow mode apparatus, and over a much wider overall range of operation.

In these tests the fluid was Bayer Rheobay AI3565 x 50% volume fraction. The particles are of mixed size of nominal average 6 μm (to keep the bulk mixture zero volts viscosity low) and are covered by the polymers within patent [5]. There is no added water and the base oil is Baysilone M5. The mixture viscosity was ~ 60 mPas at 40°C (at which all calculations are carried out) and the density ~1000 kg/m^3. At 8 kV/mm field strength the current density is nominally 100 μA/cm^2.

Kinematic shear mode testing [1] was carried out in a cylindrical clutch device of outer (stationary) cup 22.4 mm dia, channel width 0.55 mm, and effective axial length of electrode 32.3 mm. The steady test data achieved at 50 Hz binary square wave field excitation is shown in Fig. 1, evaluated with equation (1). Flow mode tests [1] were carried out in a single annular channel valve of electrode separation h = 0.75 mm, the outside diameter 13 mm (mean circumference = 38.5 mm = b) and length ℓ = 100 mm. The sample is driven through the valve by a piston which advances at a known speed, the pressure in front of the valve being measured relative to the reservoir (atmospheric) pressure at valve outlet. The field was DC and the steady test data shown in Fig. 2 is derived using equations (2) and (3).

2. The Technique - A Summary

Taking a fully developed flow situation in an arbitrary device to be characterisable by shear stress τ, electrode separation h, an average velocity $\bar{u}$, and continuum fluid properties τ_b, μ and ρ, and using the method of dimensions,

$$Cf = f(Re, He) \qquad (4)$$

$$\text{when} \quad Cf = \tau/\rho\bar{u}^2, \quad Re = \rho\bar{u}h/\mu \quad \text{and} \quad He = \tau_b\, \rho h^2/\mu^2$$

To derive the dimensionless parameters via the experimentally determined τ_w and $\dot{\gamma}$, in regions where a plug flow ($\tau < \tau_b$) accompanies the shearing flow, requires the use of a further equation. Previous work [2,3,4] suggests that this can be based on a Bingham plastic constitutive form, with a yield stress τ_b which varies with field excitation E and with the flow, and μ a (constant) zero volts viscosity.

The next step is to demonstrate that values of Cf plotted against Re for arbitrarily selected values of constant He approximate the constant field excitation experimental lines of Cf vs. Re, for both valve and clutch.

Beyond this, it remains to show that a manageable empirical relationship exists between clutch and valve situations, and to establish the basic parameters (uniform field, velocities etc) in the context of the general applicability of the result for different devices, fluids, geometry, excitation and operating conditions. This is done by taking clutch derived data and presenting it for a prediction of valve flow for the same fluid, for comparison then with experimental valve flow results.

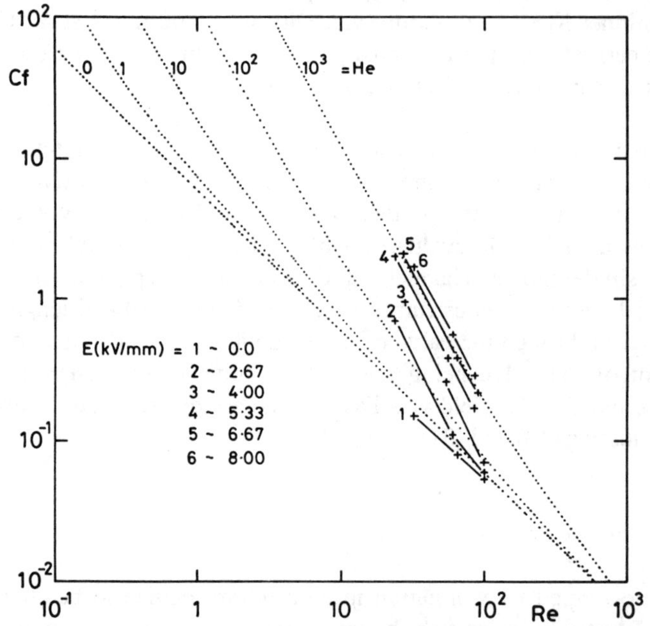

Figure 3. Cf v Re in the valve: lines of arbitrary constant He (...) and lines of constant field strength experimental data(—)

3. Valve Data

In Fig. 3, taking the experimental data from Fig. 2, the experimental Cf for a constant field excitation (solid line) is calculated from $\tau_w/\rho\bar{u}^2$ using equation 3, with $\bar{u} = \dot{q}/bh$, and Re from $\rho\bar{u}h/\mu$, where μ is derived from the zero volts line of Fig. 2. Lines of constant He (the dotted lines) are obtained using the Bingham plastic equation, viz.

$$(\tau_w/\tau_b)^3 - [3/2 + 6\mu\bar{u}/h\tau_b]\,(\tau_w/\tau_b)^2 + ½ = 0 \tag{5}$$

which is $$Cf^3 - [3/2 + 6\,Re/He]\,(He/Re^2)\,Cf^2 + (He/Re^2)/2 = 0 \tag{6}$$

in terms of the dimensionless parameters. The lines are constructed by selecting He, then solving equation (6) for Cf over a range of values of Re. It can be seen that the constant field strength experimental data lies in similar fashion to the constant He characteristic.

In Fig. 4, the experimental constant field excitation lines of Fig. 3 are plotted in the form of He v Re for each field strength value. The value of He for each experimental point is calculated by using equation (5) to give an effective uniform yield stress τ_b corresponding to the experimental τ_w, $\dot{q}$ values, then forming He with that τ_b. The dotted lines are best fits to the data.

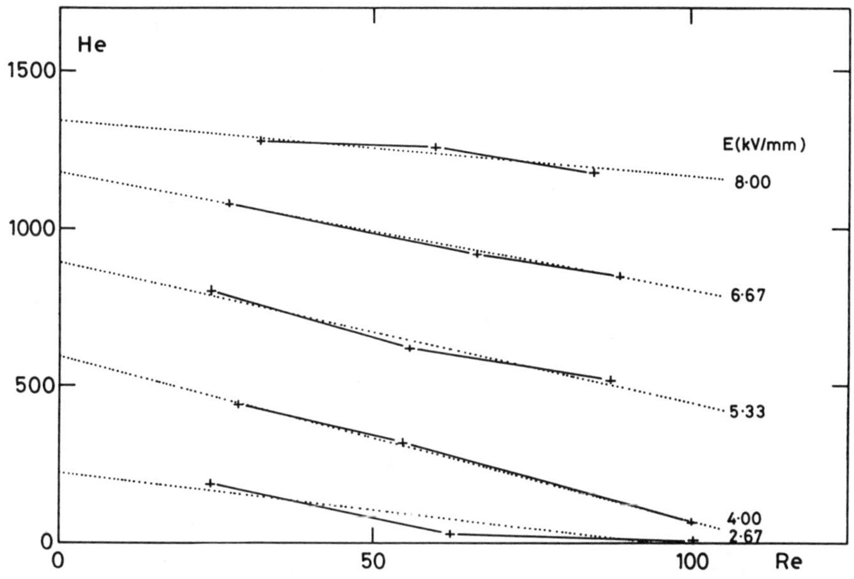

Figure 4. He v Re for valve - derived from experimental constant field strength data points of Fig. 2 and equation (5).

4. Clutch Data

Using a technique similar to that for the construction of Figs. 3 and 4, Figs. 5 and 6 are formed. However, in this case, superimposing an angular velocity $-\Omega$ to treat the motion as one with rotating outer cylinder and stationary inner cylinder, the velocity $\bar{u}$ is derived, for each experimental test point, from an integration (and average) of assumed velocity profile based on the Bingham plastic equations, viz,

$$u/r = T\,(1/r_1^2 - 1/r^2)/4\pi\mu b - \tau_b \ln(r^2/r_1^2)/2\mu \quad (7)$$

$$\text{with} \quad u/r = \Omega \ \text{ for } r > \sqrt{T/2\pi b\tau_b} \quad (8)$$

Equations (7) and (8) lead to a criterion for the presence of plug flow. The dashed line in Fig. 5 separates plug flow (to right and top) and uniform velocity profiles. The equations found by integration of (7) and (8) in terms of the non-dimensional parameters, for uniform and plug flow velocity profiles respectively, are, where $r_2/r_1 = R$,

$$Cf.\,Re^2/He = [4(R-1)^2\,Re/He + R^2 \ln R^2 - R^2 + 1] \,/\, [R^2 - 1 - \ln R^2] \quad (9)$$

$$0 = [1 + R^2](Cf.\,Re^2/He - 1) - [Cf.\,Re^2/He + R^2] \ln(Cf.\,Re^2/He) - 4(R-1)^2\,Re/He \quad (10)$$

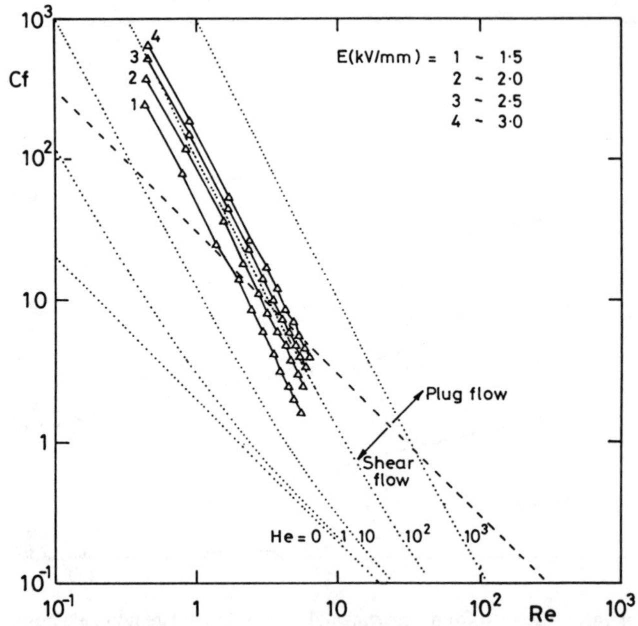

Figure 5. Cf v Re in the clutch: lines of arbitrary constant He (....) and lines of constant field strength experimental data(—).

In Fig. 5, the experimental values of Cf for a constant field excitation (solid lines) are calculated from τ_w and with $\bar{u}$ determined via equations (7) and (8) by integration, μ being taken as for the valve flow. Lines of constant He (the dotted lines) are determined with equations (9) and (10) by selecting He and calculating Cf over a range of values of Re. As for the valve, there is similarity between the experimental data at constant field excitation and the lines of constant He.

Figure 6 shows experimental He v Re for constant field excitation, with the best fit of data in the higher Re region, in similar fashion to Fig. 4 for the valve.

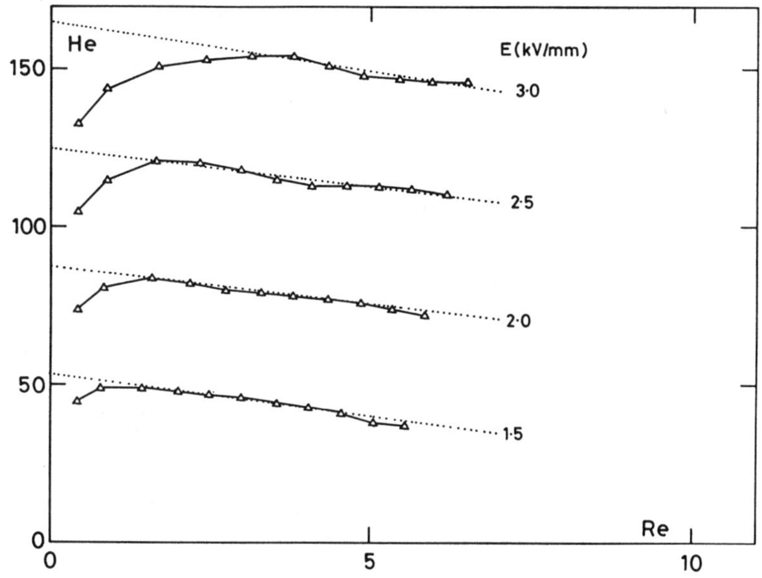

Figure 6. He v Re for clutch - derived from experimental constant field strength data points of Fig. 1 and equation (7).

5. Fluid Data

The slopes of the near parallel constant field lines of Figs. 4 and 6 average -3.503 for the valve and -2.678 for the clutch, respectively. Given this linearity and close similarity, the values of He (Re = 0) for the extrapolated lines of best fit are plotted on Fig. 7. The full line is a second order polynomial fit derived from the clutch data Δ (from Fig. 6) whilst the dotted line is a linear fit to the valve data + (from Fig.4).

Fig. 7 is concerned primarily with the valve. Since E (through τ_b) and h are the two variables which most influence He (Re = 0), the clutch data of Fig. 6 for He (Re = 0) is transformed to a prediction of He (Re = 0) in the valve, viz.

$$\text{He (Re = 0, valve)} = \text{He (Re = 0, clutch)} \times [\text{h(valve)} / \text{h(clutch)}]^2 \qquad (11)$$

for each of the test field excitation values of the clutch data of Fig. 6. The valve experimental data from Fig. 4 is then entered in Fig. 7 for comparison.

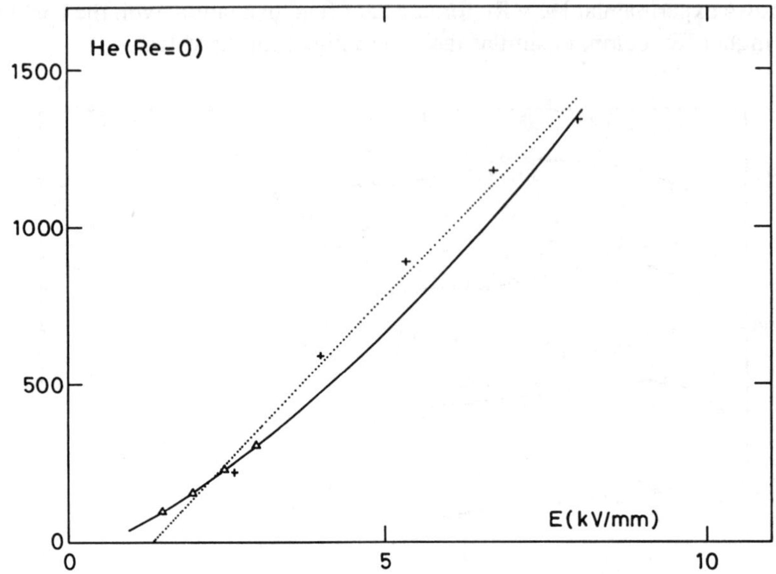

Figure 7. He for valve at Re = 0, v. E, prediction from Fig. 6 (Δ—) and comparison with Fig. 4 (+....).

Thus the He(Re = 0) and He vs. Re at constant field excitation E data is interchangeable between clutch and valve. The predictions based purely on the clutch data in Figs. 6 and 7 are

$$\text{He (Re = 0, valve)} = -59.89 + 89.13\,E + 10.99\,E^2 \quad (E\ \text{kV/mm} > 1)$$

and
$$\text{He (valve)} = \text{He (Re = 0, valve)} - 2.678\,\text{Re} \qquad (12)$$

6. Conversion of Data

The confirmation of the technique is shown in Figs. 8 and 9. In Fig. 8 the dotted lines are predictions for the valve performance from equations (12) with the full line showing the original valve data from Fig. 4.

In Fig. 9, for each of the valve constant field excitation values of Fig. 8, over a range of values of Re, He in the valve is predicted with equations (12), and Cf found by solving equation (6). This data is plotted in the dotted lines of Fig. 9 as wall shear stress τ_w vs. nominal shearing rate $\dot{\gamma}_n$ as in Fig. 2, by means of equations (4), (3) and (2). The actual experimental valve data from Fig. 2 is shown in the full lines of Fig. 9 for comparison.

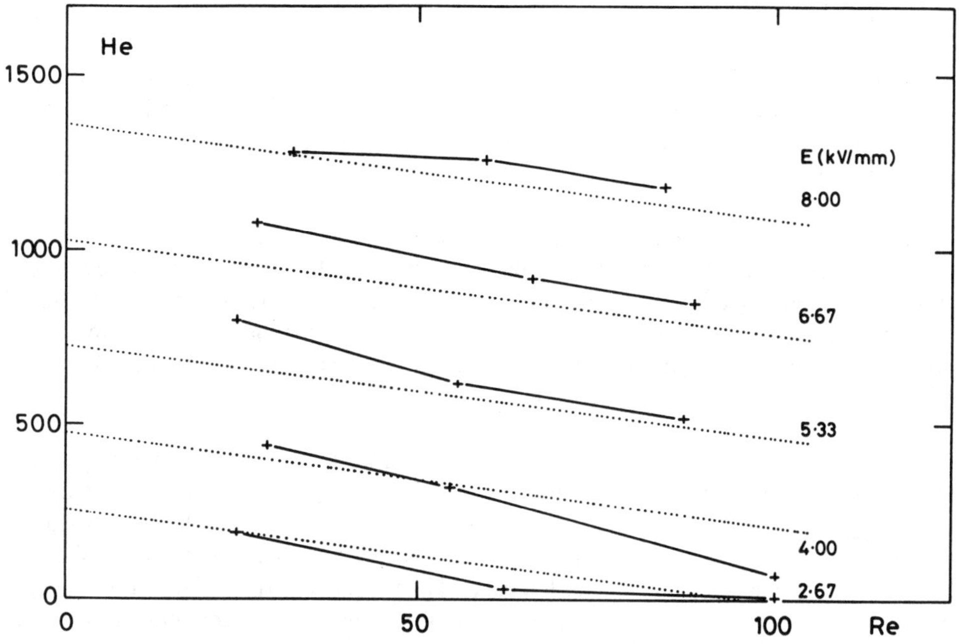

Figure 8. He v Re for valve, as predicted from clutch (....) and by experiment (—) for given values of E.

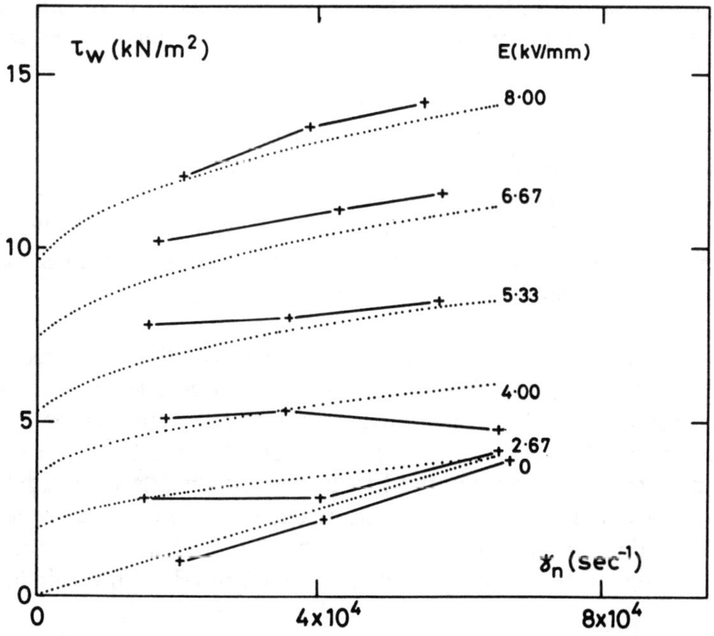

Figure 9. Comparison of original valve experimental data of Fig. 2 with predictions from the clutch experimental data of Fig. 1.

7. Discussion

The flow of a fluid having a yield characteristic in a plane valve gives rise to a central 'plug' region in the flow where no shear takes place, and shearing rate will vary between zero at the edge of the plug and a maximum at the wall of the flow channel. The plug is always present (e.g. reference [6]).

In the concentric cylindrical clutch, a plug may be present at the outer cylinder wall, if the fluid yield effect is sufficiently great or if the angular velocity of the cylinder is sufficiently small. Otherwise in the clutch no plug region will be present, and all fluid in the flow channel will be in shear. Then, with increasing angular velocity and diminishing significance of the yield effect, the shearing rate will tend to a uniform value across the flow channel (e.g. reference [7]).

The Bingham plastic 'ideal' constitutive form is a uniform yield stress superimposed on the underlying Newtonian shear characteristic. The analysis of the clutch test data, in Figs. 5 and 6, shows that this ideal is met only to an approximation - otherwise, assuming a uniform excitation field across the flow channel, the yield parameter of He shown in Fig. 6 would be constant. This is not the case. The yield stress τ_b, calculated by means of the equations (5) and (7) for each data point of the valve and clutch tests, and incorporated into He (experimental) represents an effective uniform or mean value across the channel for the prevailing flow and field conditions. This mean yield stress relates the clutch and valve modes, in plug or shear flows, which is shown by the connection of valve with clutch test data in Fig. 8, in terms of mean velocity, mean field strength [7] and channel width. The yield stress is thus shown to be a fluid rather than a device characteristic.

The results of the analysis seem to indicate that the ER effect is not significantly influenced by the choice of binary-square wave or dc excitation. Also for the clutch, the mean velocity as determined for the plug flow differs appreciably from the value derived as $\bar{u} = \frac{1}{2} r_2 \Omega$ - this could be shown by comparing Reynolds Numbers from Fig. 7 with those shown from Fig. 9. The effect of this on the slopes of the relationships of Fig. 9 is however much smaller than might be expected.

Although there is no overlap of much significance between the clutch and valve data in Figs. 6 and 4, the two sets of data have nevertheless been related. A more extensive range in the test variables for one or other of the modes would be expected to give a clearer connection between the two. The polynomial curve fit to the clutch data of Fig. 7 is made with weighting (use of) x 1 to the two outer points and x 2 to the two inner points. The valve data is shown (for comparison) with a linear least-squares fit. In a similar vein, the linear relationships of Figs. 4 and 8 (dotted line) have some limitation, viz. the field strength E = 2.67 kV/mm line meets the axis (He = 0) within the span of the test data. Over the test range otherwise, the choice of linear variation gives a satisfactory result. Different operating temperatures, or different fluid compositions, will show some difference of characteristic (e.g. reference [4]).

To improve upon the technique will require the use of local rather than averaged values of the variables. This in turn would involve, for annular rheometers, a detailed knowledge of the variation of electric field with radius and a very accurate model for the dependence of yield stress on electric field and, say, shear rate for any given electrode separation gap. Impinging upon this area would be the questioning of the fluid homogeneity assumption which is implicit in the above work. Perhaps the next step should rather be in the use of the present approach to correlate data in the more complex motions of squeeze film and Raleigh step bearing apparatus.

8. Conclusion

The dimensional analysis used here has been applied to independent steady flow valve and clutch tests carried out with independently developed and manufactured ER fluids of different types, in each case with a satisfactory result. The value of the approach lies in the fact that a relatively small amount of initially device-specific test data can be interpreted as generalised fluid data, i.e., for the fluid concerned. The procedure offers a design tool or method whereby ER fluid test data from one laboratory scale device can be incorporated into the modelling of a range of different industrial scale types of device.

Acknowledgements

The authors wish to thank Bayer AG for permission to use and publish experimental data, and the Advanced Railway Research Centre (UK) for their financial support of the analysis.

References

1. Bayer AG. *Provisional Product Data - TP.AI.3565*; **AI 12601e 6/94.**
2. D.J. Peel and W.A. Bullough, *Int. Jnl. Mod. Phys. B*, **Vol. 8, Nos. 20, 21**, pp. 2967 - 2986, Sept. 1994.
3. D.J. Peel and W.A. Bullough, *Proc. Inst. Mech. Eng., Pt.C, Jnl. Mech. Eng. Sc.*, **Vol. 208**, p. 253-266, 1994.
4. D. J. Peel and W. A. Bullough, *A Technique For The Normalisation Of Electro-Rheological Fluid Data In Cylindrical Clutch And Plane Valve Modes Of Steady Operation*, to be submitted for publication.

5. R. Bloodworth, G. Penners, G. Opperman and R. Flindt, *'Electroviscous Liquids based on Polymer Dispersions with an Electrolyte-Containing Disperse Phase'.* **USA Patent 5, 268, 118** (07.12.93). See also **EU Patent 0432 601 A1** (01.12.90)
6. W. L. Wilkinson, *Non Newtonian Fluids*, 1960 (Pergamon Press, Oxford).
7. R. J. Atkin, Shi Xiao and W.A. Bullough, *J. Rheol,* **35(7)**, pp. 1441 - 1461, Oct. 1991.

Nomenclature

b	electrode width, of flow aperture in valve, axial span in clutch
E	electric field strength = V/h
$f(\,)$	a function
h	electrode separation gap
ℓ	valve electrode length in flow direction
Δp	valve pressure drop
$\dot{q}$	volume flowrate
r	radius in clutch
r_1	radius of clutch inner wall
r_2	radius of clutch outer wall
R	ratio r_2/r_1
T	torque applied to clutch
$\overline{u}$	mean flow velocity between electrodes
u	general value of flow velocity
$(du/dy)_w$	wall shearing rate
V	potential between adjacent electrodes
Cf	Friction Coefficient
He	Hedström Number
Re	Reynolds Number
$\dot{\gamma}_n$	shearing rate in fluid
μ	Newtonian dynamic viscosity coefficient
ρ	fluid density
τ	general value of shear stress in fluid
τ_b	yield stress value
τ_e	electro-stress component of wall shear stress
τ_w	shear stress, wall value
Ω	angular velocity in clutch

A Non-Linear Model for Dynamic Response of Electro-Rheological Fluid in Flow Mode

by

R. Firoozian, B.Sc., M.Sc., Ph.D.
Department of Mechanical Engineering
Sharif University of Technology,
Tehran-Iran
Fax: (010) 98 21-918-2366

ABSTRACT

In this paper a model is presented for the electrical and mechanical properties of ER fluid in flow mode and it is concluded that a similar model must also govern the properties of ER fluid in shear mode of operation. A settling time constant is defined for the fluid which determines the speed of response. The model has been derived from several years of experience gained working with ER fluid and has not been verified experimentally. The proposed model is based on the theory that the electrical properties of ER fluid such as resistance and capacitance are functions of the applied voltage and time. This function is characterised by a time constant (τ_R) for a simple model and may be characterised by a natural frequency and a damping ratio for more complex model. The numerical solution of these functions when compared with the experimental results obtained from large step input confirm the viability of the proposed model. Further work, however, is required to identify the coefficients numerically from data obtained on a special test rig.

Nomenclature

C, C_o, C_∞	Capacitance of ER fluid as defined
I	Current
R, R_o, R_∞	resistance of ER fluid as defined
s	Laplace Transform Operator
t	time
V	voltage applied across the electrode
z	constant of equation
$\frac{d}{dt}$	Derivative with respect to time
τ_R	time constant
ω_n, ξ	natural frequency and damping ratio respectively
σ, σ_y	Instantaneous and maximum yield stress for a given voltage respectively.

Introduction

Electro-Rheological (ER) fluid [1, 2] in which an electrical voltage across a gap of approximately ½ mm produces variation in fluid yield stress has many potential applications in industry. At present because of fluid sensitivity to temperature and low overall yield stress have limited its application to only laboratory test equipment. The emergence of fluid with high yield stress is changing this and new applications are emerging in industry. The fluid property changes rapidly in order of less than one millisecond, and hence may be used in the area of high speed actuation. Since the electrode can be made of very thin conducting material, the moving parts can be made to be very lightweight and therefore potentially high acceleration and deceleration is realizable. Electro-rheological actuator [3, 4] can be made to operate an order of magnitude faster than electrical servo motors.

For design purposes the dynamic behaviour of the fluid must be well established so that various aspects of response characterististics can be studied and a suitable power supply can be made to meet the response requirements. The static behaviour of the fluid such as voltage-yield stress and voltage-current characteristics can be studied from simple experimentation but the dynamic behaviour requires special test equipment and correct experimental test procedures must be followed so that the collected data may be processed at a later stage.

As mentioned above the response is very fast and several investigators [5, 6] have reported the difficulties of measuring such a fast response. Often the true response is superimposed on the dynamic response of the test equipment and instruments used in measuring the response [6, 7]. Therefore, the design of the test rig requires special attention and equipment with known dynamic behaviour must be used. Data analysis such as identification method may be used to obtain the true dynamic responses from the measured data [7, 8]. The electrical properties of the ER fluid have been studied and various electrical models are proposed. In this paper the proposed models are studied and a different non-linear model which describes both small amplitude and large amplitude excitation test signals is proposed.

It is known that ER fluid responds quickly to an applied voltage but little is known about the mechanisms by which this transition occurs. In this paper a model is proposed for the electrical behaviour of ER fluid and by analogy a model for the mechanical properties of the fluid is proposed. Further work would be required to identify the coefficients proposed in the model from experimentally measured data.

The results presented in this paper are obtained from data stored previously on a concentric valve of 100 mm in length and 60 mm in diameter with a gap of 0.5 mm. Biased sinusoidal excitation tests of 600 and 1200 volts base voltage with harmonic amplitude of ±50 volts added on the base voltage were available from 2-3 Hz to 200 Hz. In addition large step input of 600 and 1200 volts were available so that the identified model could be tested for accuracy. All tests were repeated for flow rates of 3, 9 and 15 lit/min.

Another important consideration in design of ER devices is the power supply that must have high frequency bandwidth. The power supply used in the experiment was designed purposely to have a fast response in step change of voltage. To achieve this a 2 μF capacitor was attached to the power supply in parallel. The combined system had a rise time of less than 0.1 millisecond for 1200 volts of step change. The power supply itself was operated in open loop and for biased sine wave excitation had a large frequency bandwidth.

Electrical Model

(a) *Steady state*

At steady state the current drawn across the electrodes is a non-linear function of voltage, flow rate and valve dimensions. For a given valve dimension, the current-voltage curve (Fig. 1) can approximately be modelled by a non-linear function as [9],

$$I = Z(V - V_T)^2 = z\left(1 - \frac{V_T}{V}\right)^2 V^2 \qquad V > V_T \qquad (1)$$

$$I = 0 \qquad V < V_T$$

Comparing Eq. 1 with Ohm's law of V = RI, we can define a non-linear resistance as,

$$R = \frac{1}{Z\left(1 - \frac{V_T}{V}\right)^2 V} \qquad (2)$$

The above equation shows that the equivalent resistance is inversely proportional to the applied voltage and it reduces from a large value (in theory ∞) when $V < V_T$ to a finite value at the applied voltage of V. This indicates that resistance changes rapidly when a large step voltage is applied and the assumption of constant resistance is not valid. This also indicates that both resistance and capacitance must also be functions of time.

(b) *Transient Response*

To determine the dynamic behaviour of ER fluids two standard tests of sinusoidal excitation and step change of voltage may be used. As discussed earlier, ER fluid properties are highly non-linear functions of voltage and hence only small amplitude excitation levels should be used to identify properties of ER fluid when a linear model is used and large step input change should only be used to identify non-linear models.

A series of tests measured previously in flow mode consist of small amplitude harmonic excitation over a frequency range of 0-200 Hz and large step voltage change of 600 and 1200 volts. A simple RC network (Fig. 2) was used to model the dynamic behaviour of ER fluids. The governing differential equation of Fig. 2 may be derived as,

$$I = \frac{V}{R} + C\frac{dV}{dt} = \frac{1}{R}\left[V + RC\frac{dV}{dt}\right] \tag{3}$$

Using Laplace Transform Eq. 3 can be written in standard form of lead characteristics as,

$$\frac{I}{V} = \frac{1}{R}(RCs+1) \tag{4}$$

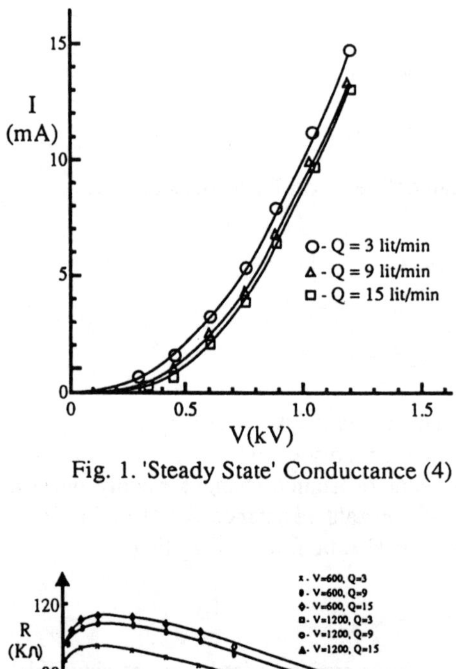

Fig. 1. 'Steady State' Conductance (4)

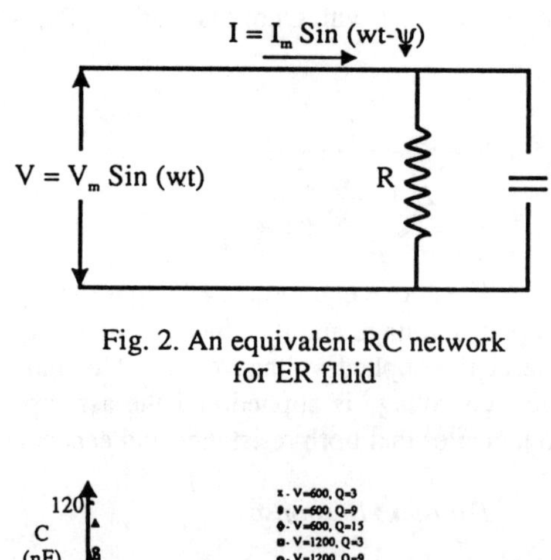

Fig. 2. An equivalent RC network for ER fluid

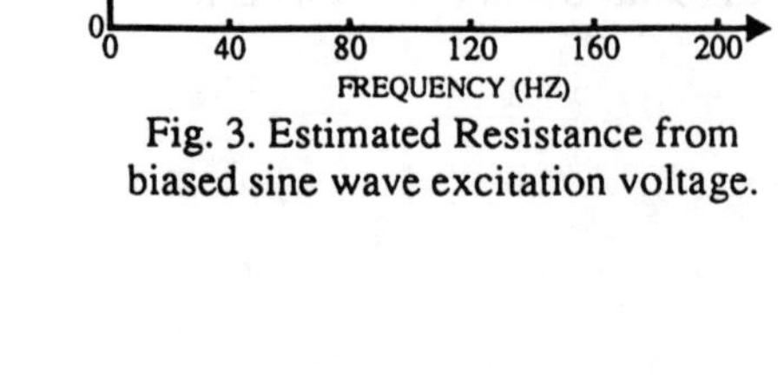

Fig. 3. Estimated Resistance from biased sine wave excitation voltage.

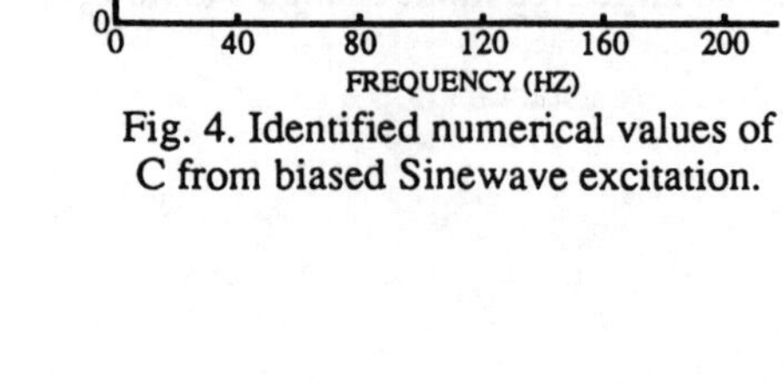

Fig. 4. Identified numerical values of C from biased Sinewave excitation.

Both the experimental data measured on the test rig and Eq. 4 shows that the current leads the voltage and this confirms the accuracy of RC network model for small amplitude excitation test signals. The purpose was thus to identify numerical values of R and C from small amplitude sinusoidal test signals. Identification algorithm was then developed for this purpose [8] and Fig. 3 and Fig. 4 show the identified numerical values of R and C over the frequency range of 0-200 Hz and base voltage of 600 and 1200 volts. As suspected both R and C are functions of the base voltage and excitation frequency. Unfortunately high frequency test signals were not available to continue the identification exercise. However, the trend is that at high frequency both R and C remain constant.

Although Eq. 4 gives a satisfactory model for biased harmonic excitation, but for large input it shows that initially the current rises to a large value (∞ in theory) and the current should reach a constant value immediately after the voltage has reached its steady state value. Experimentally measured data, however, shows that the current rises to a large value when the voltage is applied but there is an exponential decay of current (Fig. 5). Due to the limitations of data acquisition system maximum value of the current has not been measured.

As explained earlier both resistance and capacitance are functions of voltage and time. In this paper a non-linear model is proposed for large step input. The steady state numerical values of resistance and capacitance at various voltages are obtained from Figs. 1, 3 and 4. It is assumed that both resistance and capacitance change exponentially characterised by a time constant τ_R. When a large step input of voltage is applied the equivalent resistance changes from a large value to a finite value (Fig. 1) and that the capacitance changes from zero to a finite value (Fig. 4).

With the assumption that both resistance and capacitance are functions of time, the governing differential equation of Fig. 2 may be derived as,

$$I = \frac{V}{R} + C\frac{dV}{dt} + V\frac{dc}{dt} \tag{5}$$

In simplest form, we can assume that resistance and capacitance change with time exponentially, i.e.

$$R = R_\infty + (R_o - R_\infty)\, e^{-t/\tau_R} \tag{6}$$

and

$$C = C_\infty + (C_o - C_\infty)e^{-t/\tau_R} \tag{7}$$

Where R_∞ and C_∞ are the steady state resistance and capacitance at voltage V and R_o and C_o are the resistance and capacitance at V = 0 respectively.

Substituting Eqs. 6 and 7 in Eq. 5 gives,

$$I = \frac{V}{R_\infty + (R_o - R_\infty)\, e^{-t/\tau_R}} + \left[C_\infty + (C_o - C_\infty)e^{-t/\tau_R}\right]\frac{dV}{dt} - (C_o - C_\infty)\frac{V}{\tau_R}e^{-t/\tau_R} \quad (8)$$

For large step input and at t = 0, $\frac{dv}{dt} = \infty$ and $C_o = 0$ and at $t = 0^+$, $\frac{dv}{dt} = 0$. Eq. 8 therefore reduces to,

$$I = \frac{V}{R_\infty + (R_o - R_\infty)\, e^{-t/\tau_R}} - (C_o - C_\infty)\, Ve^{-t/\tau_R} / \tau_R \quad (9)$$

Eq. 9 was then solved for the following numerical values of the parameters,

$R_o = 100 \times 10^4 = 1 \times 10^6\ \Omega$
$R_\infty = 100 \times 10^3 = 1 \times 10^5\ \Omega$
$\tau_R = 1$ ms $= 1 \times 10^{-3}$ sec
$C_\infty = 80 \times 10^{-9}$ F
$C_o = 0$

A typical response of the current to a step input voltage of 1200 volts is shown in Fig. 6. The current rises to a large value and then it decays exponentially to a minimum giving an undershoot and then it reaches to a steady state level. This is consistent with the measured current (Fig. 5) on the test rig.

The assumption that both resistance and capacitance change in value exponentially gives a satisfactory step input response. Further work however, would be required to identify the parameters from the measured current response. A second order model characterised by a natural frequency (ω_n) and a damping ratio (ξ) might give a better model giving a better agreement between the response predicted by the resulting non-linear model and experimentally measured current.

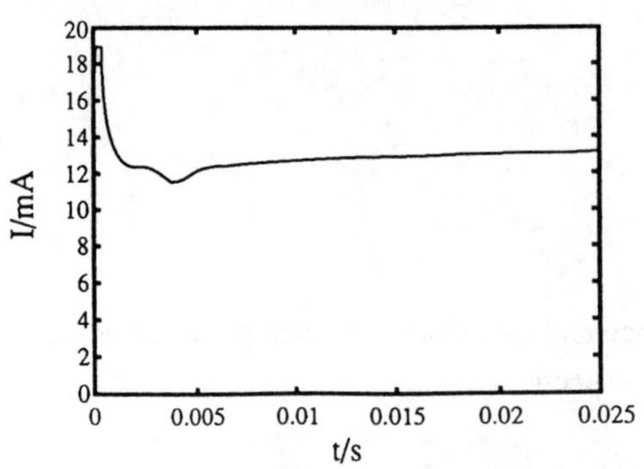

Fig. 5. Typical current response to a voltage step input. Q = 3 Lit min-^{1}, V = 1.2 kV : step input

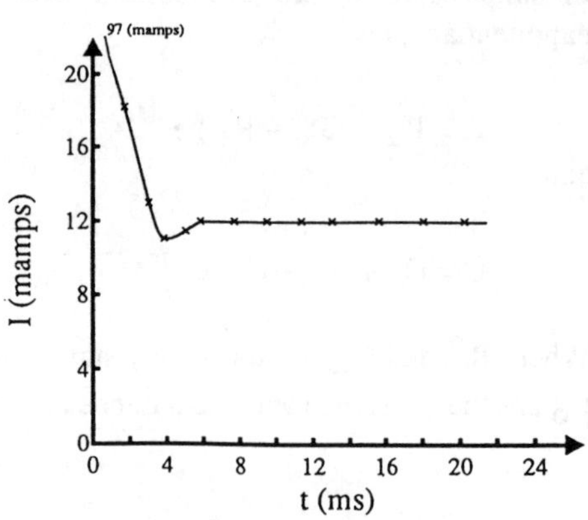

Fig. 6. Current response from the non-linear model to a step input of 1200 volts.

A Model for the Yield Stress

One of the most important parameters of ER fluid is the yield stress. When the voltage is applied across the electrodes the yield stress changes from zero to a finite value. The magnitude of the yield stress is a function of the fluid properties, temperature and applied voltage. Various theories are proposed for this event but little is known about the mechanism by which this event takes place. It is known that the yield stress is a function of time and the response is very fast in order of less than one millisecond. In simplest form we can assume that the yield stress reaches its final value exponentially, i.e.

$$\sigma = \sigma_y \,(1 - e^{-t/\tau_R}) \tag{10}$$

where σ is the instantaneous yield stress and σ_y is its final value.

The time constant τ_R determines the speed of response and in theory should have the same numerical value as the time constant for the resistance and capacitance. All the experimental work shows that τ_R is less than three milliseconds.

In a more complex form we can assume that the yield stress reaches its final value in second order form, i.e.

$$\sigma = \sigma_y \left\{ 1 - \frac{e^{-\xi\omega_n}}{\sqrt{1-\xi^2}} \sin\left(\omega_n\sqrt{1-\xi^2} + \psi\right) \right\} \tag{11}$$

where $\psi = \cos^{-1}\xi$

In equation (11) the damping ratio $\xi<1$ and ω_n is the natural frequency. Modelling the dynamic response of the yield stress in this manner indicates that the properties of ER fluid is characterised by a natural frequency and a damping ratio in which the damping ratio is less than one. There are some experimentally measured data [7] which shows that equation (11) gives a better representation of the dynamic behaviour.

The response is so fast that it can not be measured experimentally and it can only be confirmed from secondary effect such as pressure in flow mode and torque in shear mode. However the true response of ER fluid is often superimposed on the dynamic response of the system set up to measure the event [7]. Special data processing techniques must be developed so that the true response is separated from the system's effect.

Conclusions

A simple RC network is used to model the electrical behaviour of ER fluid. The numerical values of R and C have been identified from biased sinewave excitation of ER fluid and the results confirm that both R and C are function of voltage, frequency and time. By adding a time function in the model the RC network presented in this paper can also describe large step input voltage response. It can therefore be concluded that a simple RC network can approximately model the dynamic behaviour of the electric response of ER fluid and may be used in design of the power supply. The dynamic behaviour of the yield stress that cannot be measured directly can be modelled by a simple first order term characterised by a time constant or by a second order term characterised by a natural frequency and a damping ratio. Further work would be required to confirm the model presented in this paper and to identify the numerical values of the parameters from experimentally measured data. For this purpose a special test rig has to be design so that the system effect is minimum allowing the true response to be measured directly.

Acknowledgement

The data used in the identification programme was donated by Mr. W. A. Bullough and Dr. D. J. Peel of the Department of Mechanical and Process Engineering, University of Sheffield, on a test rig built for analysing ER fluid in flow mode.

References

1. Bullough, W. A. and Peel, D. J., "Progress Towards a Hydraulic Semi-Conductor for Vehicle Application", Society of Automotive Engineers, paper No. 881786, presented at the 1988 SAE passenger Car Meeting, Dearbon, Michigan, November 1988.

2. Stangroom, J. E., "Improvements (further improvements) in or Relating to Electric Field Responsive Fluids", UK Pat. 1,570, 234, published 25 June 1980.

3. Firoozian, R., et al., "Dynamic Torque Responses of an Electro-Rheological Clutch", ASME, DE Vol 35, Rotating Machinery and Vehicle Dynamics, Miami, 1991, pp. 295-301.

4. Bullough, W. A., et al., "The Electro-Rheological Catch/Latch/Clutch, Implementation and Prospects", paper C414/070, Proc. I.Mech.E., Eurotech Direct, 91 Conference, NEC, Birmingham, 1991, pp. 129-136.

5. Bullough, W. A., et al., "Quick-Acting Connections", patent application.

6. Hosseini-Sianaki, A., et al., "Operational Considerations in the Use of an Electro-Rheological Catch Device", Proc. China Fluid Power Society, 1st Fluid Power Trans. and Control Symposium, Beijing, 1991, pp. 591-595.

7. Whittle, M., Firoozian, R., Peel, D. J. and Bullough, W. A., "Decomposition of the Pressure Response in an ER Valve Control System", Journal of Intelligent Material, Systems and Structures, Oct. 1993.

8. Firoozian, R., Bullough, W. A. and Peel, D. J., "Time Domain Modelling of the Response of an Electro-Rheological Fluid in the Flow Mode", ASME Publication DE-Vol. 18-4, 1989, pp. 45-50, ASME.

9. Whittle, M., Firoozian, R., Peel, D. J. and Bullough, W. A., "A Model for the Electrical Characteristics of an ER Valve", Proc. ER Conf., Carbondale, Oct. 1991, also in Journal of Modern Physics B.

Transfer Function Identification of an Electro-Rheological Actuator

Brookfield, D.J.
and
Dlodlo, Z.B.

Department of Mechanical Engineering,
University of Liverpool,
Liverpool L69 3BX, UK.

Abstract

A fluid clutch utilising an Electro-Rheological (ER) suspension provides a controlled torque coupling between input and output through the control of the applied electric field. If the input is driven at constant speed the device can be considered as an ER torque actuator and thus be used to drive robot links or other mechanisms requiring precise positioning. Such an ER torque actuator can replace a DC servo-motor in robotic applications with the benefits of low time constant and smooth output torque unaffected by cogging (i.e. variation in torque of a DC motor as the magnetic reluctance of the armature-stator path changes with rotation).

Although the ER actuator has many benefits, it suffers from a non-linear and time varying relationship between input voltage and output torque. These undesirable characteristics can be mitigated by providing a local closed loop controller around the system. The design of such a controller requires a knowledge of the relationship between the applied voltage and output torque; i.e. the transfer function of the actuator. This transfer function has been determined by observing the response of an ER torque actuator in the frequency domain. It is shown that a linear transfer function model reasonable represents the actuator behaviour, that the actuator is a stable second order system and that the time constant of the clutch studied is sufficiently short to hold considerable promise for robotic applications. Furthermore, the maximum torque capability is shown to be sufficient for many medium scale industrial robots.

Notation

a	Break frequency (rad s^{-1}).
d	Clutch plate spacing (m)
E(s)	Field applied to ER fluid expressed in frequency domain (V m^{-1}).
f_1	DC Motor shaft speed (rad s^{-1}).
f_2	Actuator output shaft speed (rad s^{-1}).
G(s)	General transfer function.
$G_A(s)$	Transfer function of ER actuator, i.e. of clutch and power supply.
$G_{C1}(s)$	Arm position controller transfer function.
$G_{C2}(s)$	Torque controller transfer function.

$G_E(s)$	Transfer function of ER clutch.
$G_R(s)$	Transfer function of robot arm (i.e. a model of the arm dynamics).
$G_V(s)$	Transfer function of computer controlled power supply.
$H(s)$	Transfer function of feedback transducer.
J	Rotational inertia of robot arm (kg m^2).
k	General transfer function gain.
k_1	Viscous friction in robot joint (Nm s).
T	Output torque of actuator (Nm).
T_0	Mean output torque of actuator (Nm).
T_1	Amplitude of fundamental frequency output torque from actuator (Nm).
T_3	Amplitude of third harmonic output torque from clutch (Nm).
U	Power dissipated in actuator (Watts).
V	Voltage applied to clutch (V).
$V(s)$	Voltage applied to clutch, expressed in frequency domain (V).
V_0	Mean voltage applied to clutch (V).
V_1	Amplitude of fundamental frequency voltage applied to clutch (V).
V_3	Amplitude of third harmonic voltage applied to clutch (V).
V_A	Amplitude of voltage applied to high voltage power supply (V).
V_H	Input voltage to high voltage power supply (V).
$X(s)$	Position of arm in robot arm positional controller.
$X_D(s)$	Desired position of arm in robot arm positional controller.
ϕ_{V1}	Phase shift on fundamental in high voltage power supply (deg.).
ϕ_{V3}	Phase shift on third harmonic in high voltage power supply (deg.).
ϕ_{T1}	Phase shift on fundamental in clutch (deg.).
ϕ_{T3}	Phase shift on third harmonic in clutch (deg.).
ω_1	Test frequency (rad s^{-1}).

Introduction

The Electro-Rheological (ER) effect in a dielectric fluid containing suspended silica gel particles was first described by Winslow[1]. The main feature of the effect is a change in the mechanical properties of the suspension induced by the application of a high electric field, typically of the order of 1-5 kV mm^{-1}. Two groups of ER suspensions are now common; those in which water adsorbed onto the particles is essential (such as Winslow's fluid) and water free suspensions such as those described by Block and Kelly[2] and by Filisko and Armstrong[3].

Although initially seen more as a scientific curiosity, the ER effect was later applied to the design of variable damping systems where the shear properties of a suspension in a damper were controlled by an applied electric field. Such dampers have applications in automotive engine mounts[4,5] and in many other areas of vibration control.

Control of the suspension shear properties has also allowed the design of ER valves. In such valves the pressure drop across the valve is controlled by the applied potential and four ER valves, in a bridge configuration, can be used to control a hydraulic ram type actuator to provide translational movement[6]. A bridge configuration is necessary because each valve cannot form a perfect seal against fluid flow and thus a pressure balance on each side of the ram piston is maintained to control the position.

A further example of the use of ER suspensions in shear is the ER clutch in which the transmitted torque is controlled by the applied field[7]. If such a clutch is driven by a constant speed motor it can be considered as a rotational actuator and thus used for positional control of rotating machinery. An obvious example of a torque driven system is a rotational robot arm joint and such an application was considered by Monkman[8]. When compared to conventional robot designs, where the actuator is a DC servo-motor, the ER rotational actuator has a number of potential advantages. The ER actuator gives very smooth and continuous control of torque when compared to DC motors. In such motors the output torque varies as a function of the shaft angle due to the changing reluctance of the motor magnetic circuit as the armature rotates and due to the effect of the commutator[9]. This torque ripple, or "cogging" complicates the control of the robot particularly when a model reference strategy is used. In addition to providing a smooth torque, the ER rotational actuator can also provide rapid control of the torque with a typical response time of a few milliseconds.

Whilst the transfer function of a DC servo-motor can be reasonably well approximated by a constant linear relationship between input current and output torque, it is unfortunate that the relationship between applied field and torque for an ER rotational actuator is more difficult to model. Tao, et al,[10] observed that a minimum threshold field is required to cause any ER effect. Furthermore, ER suspensions are non-Ohmic and the current density is an approximately parabolic function of the applied field. The shear stress at a particular shear rate approximately follows the current density and this non-linearity complicates the implementation of a simple torque control strategy.

Of greater importance in the design of a controller is the change in output torque caused by changes in temperature of the suspension. The ER rotational actuator operates by providing controlled slip between the constant speed driving motor and the output shaft. This slip necessarily causes a power U, given by $U = T(f_1 - f_2)$, to be dissipated in the actuator and this power input raises the temperature of the suspension. The increase in temperature manifests itself through two effects; a decrease in the zero field viscosity of the suspension and an increase in the conductivity of the suspension. This conductivity increase consequently increases the current density and thus the torque that can be obtained with a particular applied field. The response of the actuator is thus a time varying function of the field.

In any system, non-linearity or a time varying characteristic can be mitigated by providing a local closed control loop around the system. For the ER rotational actuator, such a controller can be implemented by varying the applied voltage and thus the field in response to the error between the desired output torque T_D and the actual output torque T measured by a torque load cell located on the actuator output shaft.

The proper design of a torque controller requires some knowledge of the relationship between the applied field E and the output torque T. It is convenient and will be justified from experimental results to describe the actuator using a linear model where the non-linearity effectively appears as a disturbance term in the controller. The input-output relationship can be specified as the clutch transfer function $G_E(s)$ defined by:

$$G_E(s) = \frac{T(s)}{V(s)} \tag{1}$$

$$E(s) = \frac{V(s)}{d} \tag{2}$$

In practice, control of the actuator requires the control of the applied voltage V using a computer controlled power supply. The effective transfer function $G_A(s)$ of the actuator is then product of the clutch transfer function $G_E(s)$ and the transfer function $G_V(s)$ of the computer controlled power supply.

$$G_A(s) = G_E(s)\, G_V(s) \tag{3}$$

The design of a torque controller and the subsequent design of a position controller for a rotational robot joint driven by the actuator thus requires a knowledge of $G_A(s)$. The aim of the present paper is to report on the identification of this transfer function through observation of the actuator behaviour in the frequency domain.

The ER Torque Actuator

Fig. 1 shows an ER clutch adapted as a torque actuator. A variable speed DC motor was used to drive the clutch centre plate via an insulated (nylon) coupling. The high voltage was applied to the left hand bearing pedestal and thus to the centre plate. This plate was enclosed by a housing coupled to the output shaft such that a 0.002m gap existed between either side of the plate and the housing. This gap contained the ER suspension. In the present tests this fluid was a suspension of starch particles of a maximum diameter 40 x 10^{-6} m in 100 cS silicon oil. The volume fraction of particles in the suspension was 1:3.

A torsion load cell comprising a thin tubular section carrying four resistance strain gauges measured the output torque. The gauges were aligned at 45° to the load cell axis, that is, in the principal stress directions for torsion. The resistance changes were read using a

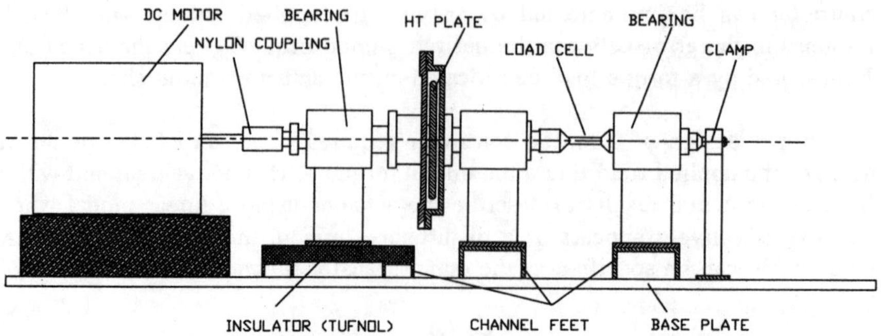

Fig. 1 Basic ER Actuator Configuration

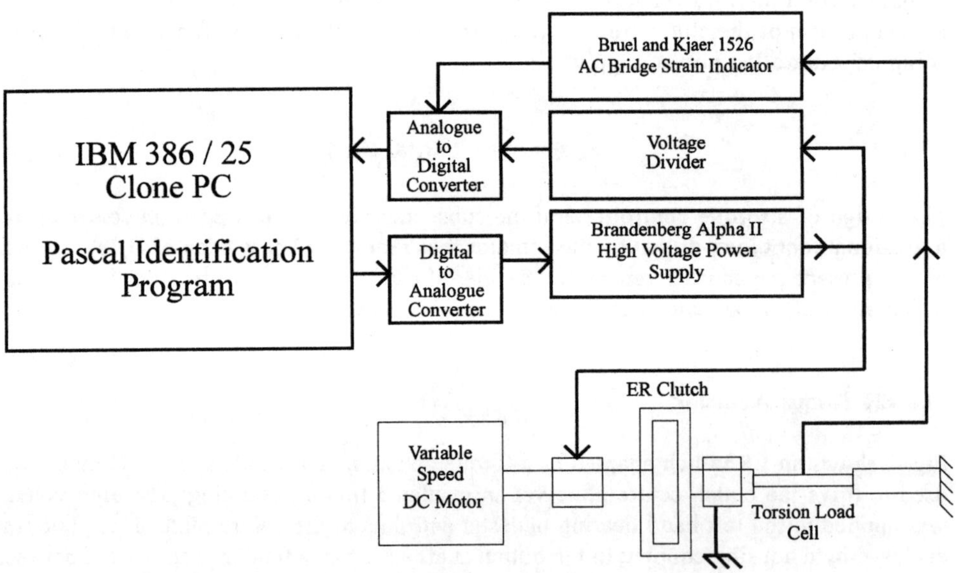

Fig. 2 ER Actuator Torque Control or Identification

Bruel and Kjaer 1526 strain indicator which is an AC bridge instrument having a cut-off frequency of 1885 rad s^{-1}. For the identification of the actuator transfer function the load cell output shaft was clamped (f_2 = 0).

Transfer Function Identification

(i) Experimental Arrangement

The technique used to determine the actuator transfer function was the plotting of the amplitude and phase responses of the actuator to an offset sine wave input as a function of frequency; that is, Bode plotting. The basic technique was derived by Bode[11] when investigating feedback amplifiers and is described in any standard control engineering textbook[12]. The Bode plots comprise two graphs: the amplitude response expressed in dB against log(frequency) and the phase shift in degrees against log(frequency).

The responses were determined using the experimental arrangement shown in Fig. 2. A 386/25 clone PC coupled to high speed 12 bit Analogue to Digital (A-D) and Digital to Analogue (D-A) converters was used to generate an offset sine wave input to the power supply and to record the power supply output voltage and the actuator output torque. The offset sine wave output from the D-A was of the form:

$$V_H = V_A(1 + \sin(\omega t)) \qquad (4)$$

The voltage thus varied from zero to 2 V_A . The D-A output V_H was applied to a Brandenberg Alpha II high voltage power supply. In the present tests a maximum field E = 3 x 10^6 V m^{-1} (3 kV mm^{-1}) was selected to ensure a reasonable ER effect with no risk of breakdown. With a plate gap of d =0.002m this field required a maximum voltage of 6 kV. For the power supply used this corresponded to a voltage V_A = 0.63 V.

The mechanical input to the actuator was provided by a speed controlled DC motor set to operate at a shaft speed of f_1 = 31.4 rad s^{-1} (300 rpm). This motor was large (400 W power capability) and of high inertia and thus provided a steady input angular velocity to the actuator. An identification program written in Pascal generated V_H according to Eq. (4) and collected the voltage and torque data producing a disk file of 2000 observations per test. Each test comprised 10 cycles of the offset sine wave input although the first two cycles (400 observations) were ignored in processing the data to allow initial transients to decay.

(ii) Determination of the Amplitude and Phase responses

In order to produce the Bode plots it was necessary to know the amplitude and phase of the high voltage power supply output voltage and the actuator output torque. Whilst these could have been determined graphically, it is generally difficult to obtain by such a method good estimates of the values in the presence of any experimental noise. Furthermore, as the actuator was known to be somewhat non-linear, it was necessary to determine the amplitudes and phases in the presence of some waveform distortion and to estimate the extent of the non-linearity.

By determining both the fundamental and third harmonic amplitudes and phases it was possible to produce Bode plots from the fundamental amplitudes and phases and to investigate the non-linearity in terms of the percentage third harmonic distortion.

A least squares technique was developed to determine the following parameters from each data set recorded on disk: V_0 , V_1 , V_3 , ϕ_{V1} , ϕ_{V3}, T_0 , T_1 , T_3 , ϕ_{T1} and ϕ_{T3} . The first five of these parameters appear in the voltage data according to Eq. 5 and the latter five in the torque data according to Eq. 6.

$$V = V_0 + V_1 \sin(\omega t + \phi_{V1}) + V_3 \sin(3\omega t + \phi_{V3}) \tag{5}$$

$$T = T_0 + T_1 \sin(\omega t + \phi_{T1}) + T_3 \sin(3\omega t + \phi_{T3}) \tag{6}$$

Eqs. 5 and 6 can be re-written using the standard angle addition formula as Eqs. 7 and 8.

$$\begin{aligned} V = {} & V_0 + V_1 \left[\cos(\phi_{V1}) \sin(\omega t) + \sin(\phi_{V1}) \cos(\omega t) \right] \\ & + V_3 \left[\cos(\phi_{V3}) \sin(3\omega t) + \sin(\phi_{V3}) \cos(3\omega t) \right] \end{aligned} \tag{7}$$

$$\begin{aligned} T = {} & T_0 + T_1 \left[\cos(\phi_{T1}) \sin(\omega t) + \sin(\phi_{T1}) \cos(\omega t) \right] \\ & + T_3 \left[\cos(\phi_{T3}) \sin(3\omega t) + \sin(\phi_{T3}) \cos(3\omega t) \right] \end{aligned} \tag{8}$$

Solutions for the ten parameters were obtained by minimising the sum of the squares of the errors between the voltages and torques V and T predicted by Eqs. 7 and 8 and corresponding values from the experimental data set. Application of a conventional least squares approach thus generated two matrices of summations from which solution matrices were obtained using Cramer's rule. These solution matrices represented each of the terms in Eqs. 7 and 8 and the parameters values were found from these using Eqs. 9-16.

$$V_1 = \left[(V_1 \cos(\phi_{V1}))^2 + (V_1 \sin(\phi_{V1}))^2 \right]^{1/2} \tag{9}$$

$$\phi_{V1} = \tan^{-1}\left[\frac{V_1 \sin(\phi_{V1})}{V_1 \cos(\phi_{V1})}\right] \tag{10}$$

$$V_3 = \left[(V_3 \cos(\phi_{V3}))^2 + (V_3 \sin(\phi_{V3}))^2 \right]^{1/2} \tag{11}$$

$$\phi_{V3} = \tan^{-1}\left[\frac{V_3 \sin(\phi_{V3})}{V_3 \cos(\phi_{V3})}\right] \tag{12}$$

$$T_1 = \left[(T_1 \cos(\phi_{T1}))^2 + (T_1 \sin(\phi_{T1}))^2 \right]^{1/2} \tag{13}$$

$$\phi_{T1} = \tan^{-1}\left[\frac{T_1 \sin(\phi_{T1})}{T_1 \cos(\phi_{T1})}\right] \tag{14}$$

$$T_3 = \left[(T_3 \cos(\phi_{T3}))^2 + (T_3 \sin(\phi_{T3}))^2 \right]^{1/2} \tag{15}$$

$$\phi_{T3} = \tan^{-1}\left[\frac{T_3 \sin(\phi_{T3})}{T_3 \cos(\phi_{T3})}\right] \tag{16}$$

Further details of the identification procedure are given by Brookfield[13].

Bode amplitude plots were produced by converting the fundamental amplitudes to deciBel using Eqs. 17 and 18 and the phases were plotted from ϕ_{V1} and ϕ_{T1} .

$$V_1\ (db) = 20 \log_{10}\left[\frac{V_1}{V_A}\right] \tag{17}$$

$$T_1\ (db) = 20 \log_{10}\left[\frac{T_1}{V_1}\right] \tag{18}$$

iii) Transfer function determination from Bode plots

The standard Bode approximations are based on the assumption s = jω and study of the behaviour of the transfer function at low and high driving frequencies after this assumption has been made. The transfer functions for the high voltage power supply and the clutch were both found to be first order poles having the general form:

$$G(s) = \frac{k}{(s + a)} \tag{19}$$

Eq. 19 is replaced by Eq. 20 after making the assumption s = jω.

$$G(s) = \frac{k}{(j\omega + a)} \tag{20}$$

The behaviour is then determined at low frequency (that is, $\omega < a$) and high frequency ($\omega > a$). At low frequency the amplitude ratio is k/a and the phase shift zero. At high frequency the amplitude ratio is k/ω and the phase shift -90°. This behaviour is shown on the Bode plots as asymptotic approximations, i.e straight line segments, with a transition in the amplitude plot at the break frequency (ω=a) from a constant amplitude $20 \log_{10}$ (k/a) to a decreasing amplitude $20 \log_{10}$ (k/ω). The phase shift is assumed to be 0° for frequencies $\omega < (a/10)$, -45° at frequency ω=a and -90° for frequencies $\omega > 10a$. By sketching these approximations onto experimentally obtained Bode plots estimates of 'k' and 'a' can be obtained for the power supply and clutch transfer functions.

Results

High Voltage Power Supply

22 Tests were undertaken over a 1000:1 frequency range ω = 0.37 - 370 rad s^{-1} . In each test the amplitudes and phases were determined using the least squares method already described. Fig. 3 shows the Bode amplitude plot and Fig. 4 the Bode phase plot for the power supply. The amplitude data has been normalised using the specified power supply calibration. Also shown on these plots are Bode asymptotic approximations chosen to match the experimental data. These approximations were then used to determine the power supply transfer function $G_V(s)$ given in Eq. 21.

$$G_V(s) = \frac{9.32}{(s + 8.6)} \tag{21}$$

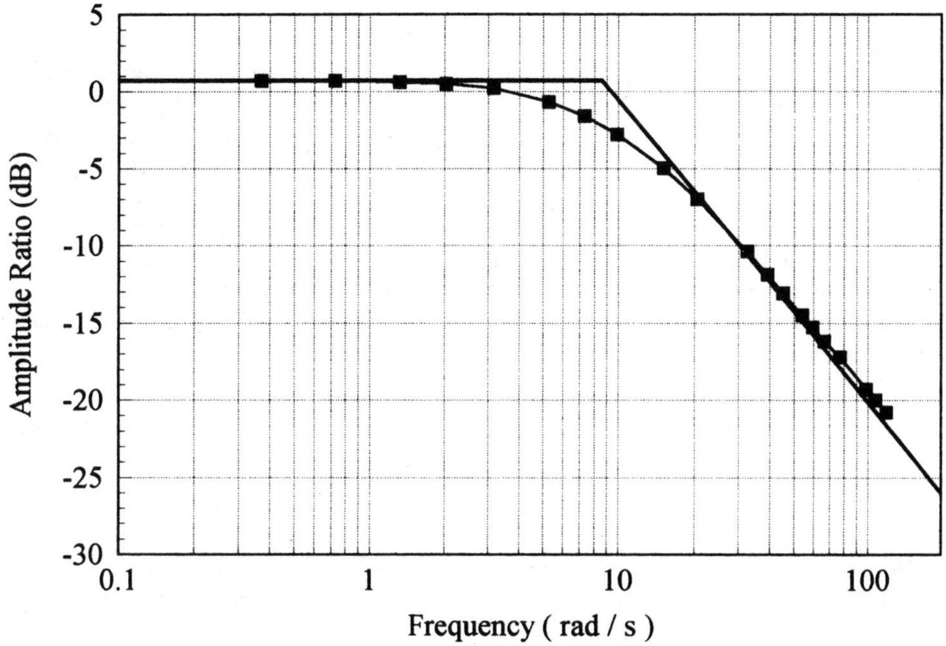

Fig. 3 Bode Amplitude Plot - Brandenberg PSU

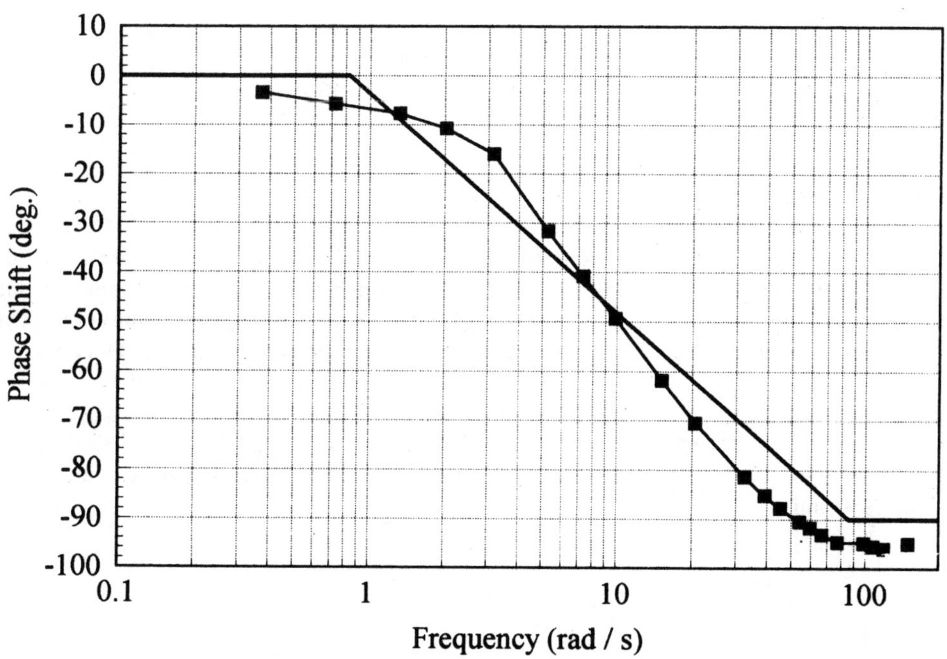

Fig. 4 Bode Phase Plot - Brandenberg PSU

ER Clutch

In each of the 22 tests described above, the amplitude ratio for the clutch was determined from Eq. 18 and the phase shift as the difference between ϕ_{T1} and ϕ_{V1}, (that is, the additional phase shift introduced by the clutch over and above that due to the power supply). Figs. 5 and 6 show the amplitude and phase plots respectively.

The amplitude ratio in Fig. 5 shows a clear decreasing trend with frequency despite the small range of amplitude ratios observed. Fig. 6 shows a phase lag at all frequencies and the behaviour of this phase lag, combined with the low frequency amplitude ratio, allowed the determination of the clutch transfer function $G_E(s)$ given in Eq. 22 and shown as an asymptotic approximation in Fig. 6.

$$G_E(s) = \frac{0.0423}{(s + 150)} \tag{22}$$

Eq. 22 is of first order form. The absence of any form of resonance peak in the amplitude response shown in Fig. 5 justifies the assumption that damping, rather than inertial, terms predominate. Furthermore the phase shift in Fig. 6 does not correspond to that expected from a system where inertia is significant.

Actuator Transfer Function

The combination of Eqs. 21 and 22 according to Eq. 4 gives the actuator transfer function $G_A(s)$ in Eq. 23.

$$G_A(s) = \frac{0.394}{(s^2 + 158.6\,s + 1290)} \tag{23}$$

Actuator Linearity

In electronic instrumentation linearity is conventionally measured in terms of the percentage third harmonic distortion; that is, the ratio between the third harmonic and fundamental amplitudes at the output when a pure sine wave is applied to the input. The least squares technique reports the third harmonic amplitudes as V_3 in Eq. 11 and T_3 in Eq. 15.

Averaged over the 22 tests conducted, the mean percentage third harmonic distortions in the power supply and the clutch were 3.3 % and 3.7% respectively. The maximum values were 5.8% and 6.4%.

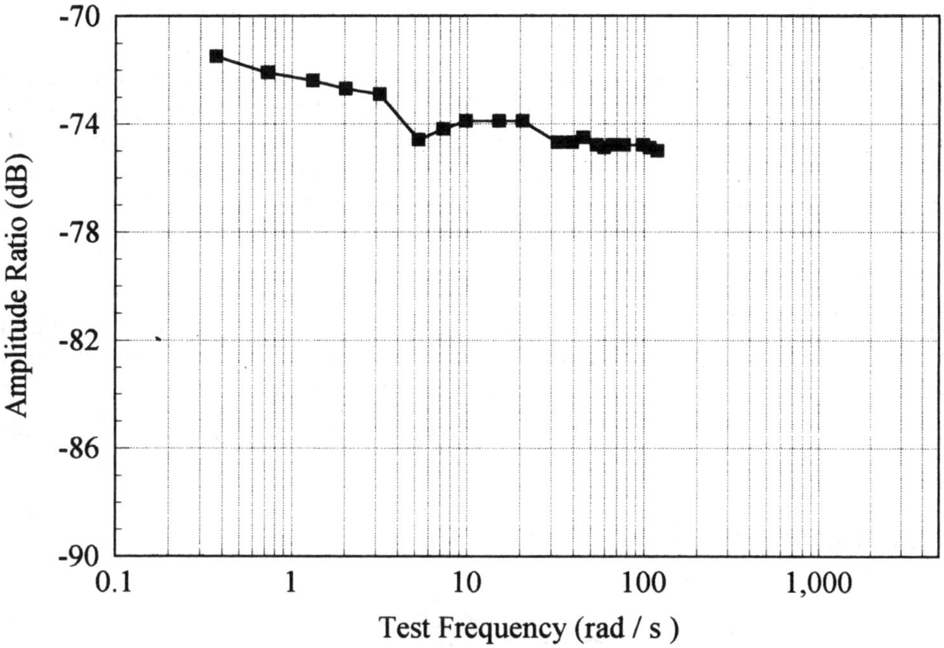

Fig. 5 Bode Amplitude Plot - ER Rotational Actuator

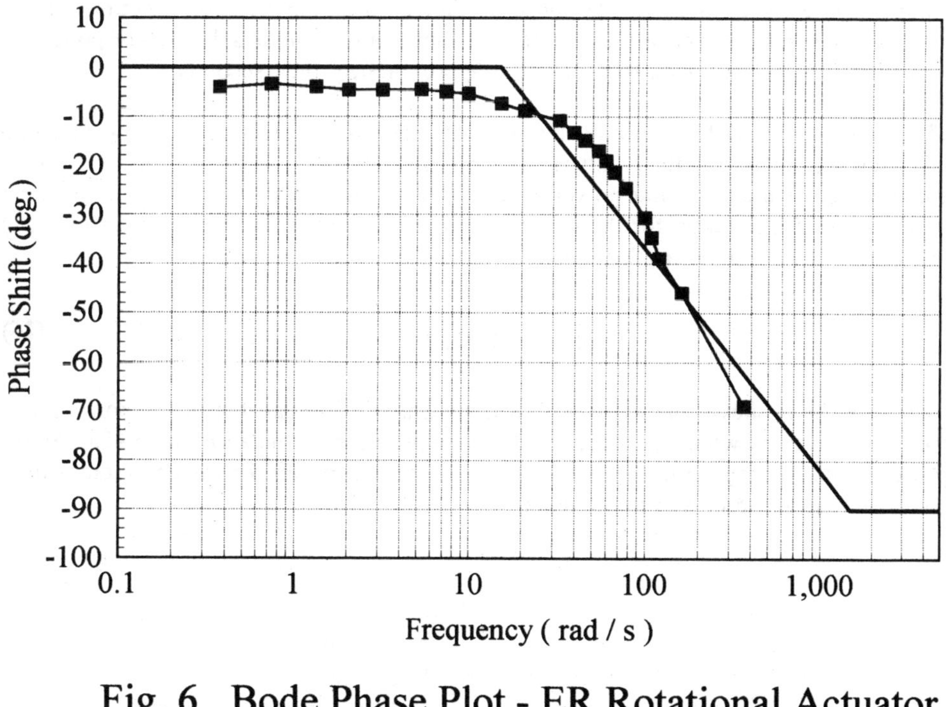

Fig. 6 Bode Phase Plot - ER Rotational Actuator

Discussion

The Bode plotting identification method allowed the identification of the actuator transfer function $G_A(s)$. This was the product of the power supply transfer function $G_V(s)$ and the ER clutch transfer function $G_E(s)$. It was noticeable in $G_V(s)$ that the low frequency gain was 9.32 / 8.6. This corresponded to a power supply calibration error of 8% which resulted in the voltage applied to the clutch having a maximum value of 6500 V and not 6000 V as intended. This did not represent any difficulty in the experiments and breakdown was not observed.

The combined actuator transfer function found in Eq. 23 represents a stable system; the roots of the equation are negative real. The time response of the actuator to a step input can be estimated using the product of $G_A(s)$ and a suitable step input.

$$T(s) = V(s)\, G_A(s) \tag{24}$$

Considering a 6 kV step the inverse transform (that is, the time domain solution) is obtained from Eq. 24 as Eq. 25.

$$T(t) = 1.832 \left[1 - \left(\frac{150\, e^{-8.6\, t}}{141.4} \right) + \left(\frac{8.6 e^{-150\, t}}{141.4} \right) \right] \tag{25}$$

Fig. 7 shows the torque from Eq. 25 plotted as a function of time (marked points). The curve shows results from an experimental step test on the actuator. The zero field torque from the experimental test has been added to the torque from Eq. 25 to allow direct comparison between the step response predicted from the transfer function and the experimental data. It is clear that the transfer function well models the step response with a maximum torque error of 16% of fsd at approximately 0.04 seconds.

The results shown in Fig. 7 imply that the actuator has an effective time constant (time taken to reach 0.63 of final value) of approximately 0.13 seconds. This is simply a consequence of the particular power supply used in the identification tests. Considering only the transfer function for the clutch, the inverse transform of the product of Eq. 22 and a 6 kV step is given by Eq. 26.

$$T(t) = 1.69 \left[1 - e^{-150t} \right] \tag{26}$$

This response is shown plotted in Fig. 8 from which it is clear that the response time is 0.0067 seconds as would be expected for a single pole of the form 1/(s+150). With a power supply having a more rapid response the time constant for the complete actuator could be reduced to close to this value.

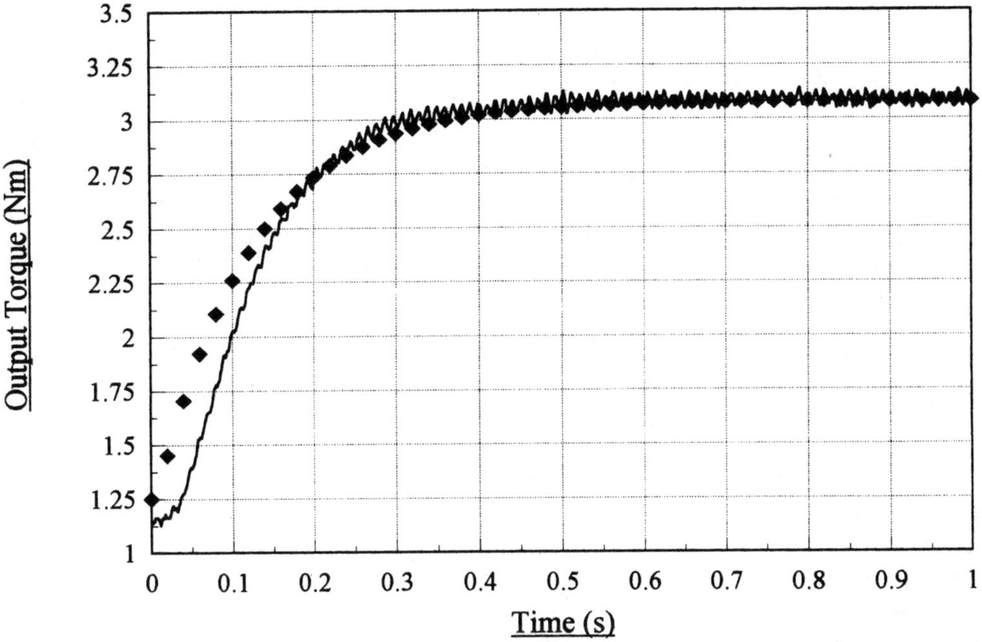

Fig. 7 Step Response of Power Supply and Actuator
1.26 Volts Step Input at Time t = 0

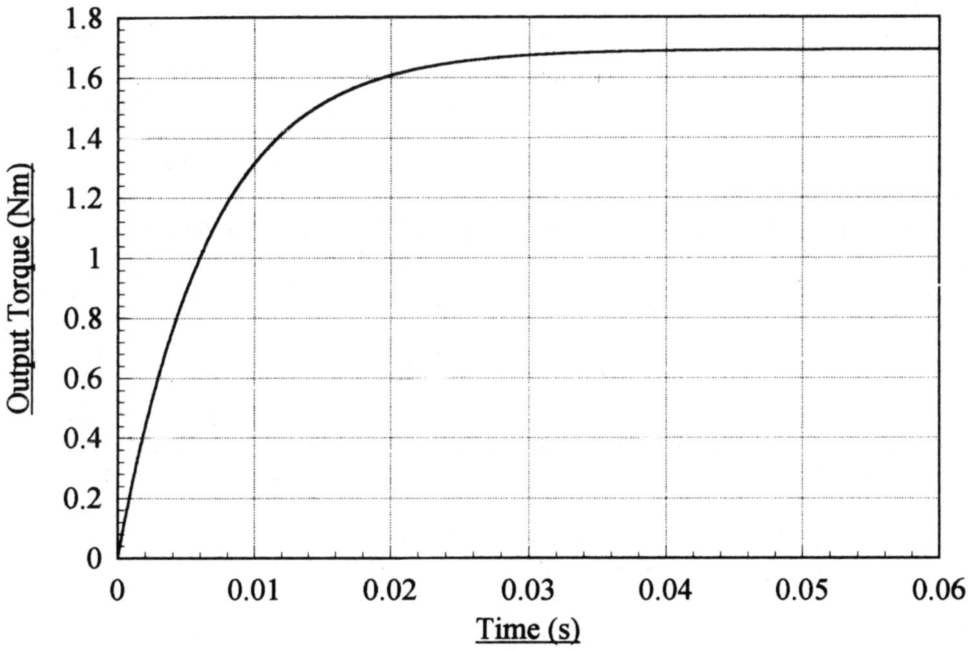

Fig. 8 Simulated Step Response of Actuator

The response time obtained includes the strain indicator and torsion load cell time constants. The time constant of the strain indicator was approximately 0.0005 seconds and thus represents less than 8% of the observed clutch time constant. The load cell time constant is primarily the settling time of the second order system formed by the clutch disk inertia, the damping of the disk in the ER fluid and the load cell stiffness. It is impossible to separate this settling time from the clutch behaviour in that the damping is clearly a function of the electrically stressed ER fluid behaviour. It is thus argued that, provided the torsion load cell is reasonably torsionally stiff, the observed time constant provides a reasonable prediction of the torque response time when driving an inertial mechanical system such as a robot link.

Bullough, et al.[14], investigated the step response of a concentric cylinder type clutch. Having compensated for the time constant of the electro-magnetic torque transducer the time constant for the clutch was found to be of the order of 0.001 seconds. This provides further justification for the use of an ER clutch as an actuator in robotics.

The ER actuator is accepted to be a non-linear system and the use of a linear model, as implied by the determination of a transfer function, can only be justified if the non-linearity is sufficiently small that it does not significantly affect the output torque. The agreement between experiment and asymptotic approximations shown in Figs. 3, 4 and 6 is good and the deviation of the points from the lines is primarily due to the nature of the asymptotic approximations. This good agreement provides justification for the linear model. The step response data plotted in Fig. 7 also show good agreement between predictions from the linear model and experimental data. Finally the low observed third harmonic distortion (mean values of 3.3% and 3.7% for power supply and clutch respectively) confirm the applicability of a linear model.

A similar linear modelling method for an ER clutch was considered by Xiao[15]. Unfortunately the power supply used by Xiao was unable to provide greater than a ± 100 V modulation on the mean voltage and no significant torque modulation was observed.

The ER actuator described forms part of a closed loop positional controller for a single degree of freedom rotational robot joint. Fig. 9 shows in block diagram form the control strategy adopted. The transfer functions $G_V(s)$ and $G_E(s)$ are those determined in the present work and the arm dynamics are found as Eq. 27.

$$G_R(s) = \frac{1}{J s^2 + k_1 s} \tag{27}$$

All of the necessary information is thus available for the design of the position and torque controllers.

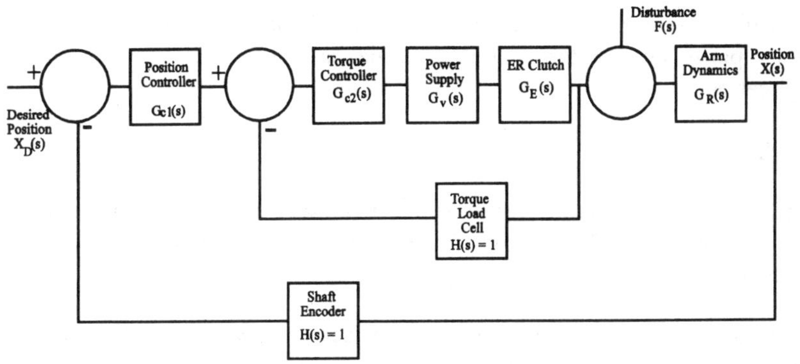

Fig. 9 ER Robot Actuator Control Block Diagram

The ER component of the output torque from the actuator is typically 3 Nm at a field of $E = 4.5 \times 10^6$ V m^{-1} (4.5 kV mm^{-1}). This field represents a reasonable upper limit to avoid breakdown of the ER fluid. In conventional industrial robots the rotational joints are driven by DC motors via harmonic drive speed reducers having a typical speed reduction ratio of 150:1 and thus a corresponding torque multiplication. The present small scale ER actuator is thus capable of providing approximately 450 Nm of joint torque and this is comparable to the maximum joint torques in medium scale industrial robots. Work is in hand on the design of a larger actuator to provide joint torques up to 900 Nm and a bidirectional capability.

Conclusion

An ER clutch driven by a constant speed motor has been proposed as a torque actuator. This actuator has been modelled through a linear transfer function and it has been shown that the transfer function is the product of the transfer functions of the power supply and the ER clutch. Amplitude and phase Bode plots for the power supply and ER clutch have been experimentally determined over a 1000:1 frequency range and the transfer functions obtained from these plots. The transfer functions have been shown to be first order and the actuator to be stable.

The clutch transfer function has been used to determine the time constant of 0.0063 seconds. This very rapid response shows considerable promise for robot control applications, particularly as the available torques are sufficient for many medium scale robot applications..

Acknowledgements

One of the authors (ZBD) has been supported by a scholarship from the National University of Science and Technology, Bulawayo, Zimbabwe, and this support gratefully acknowledged.

An initial study on closed loop control of an ER actuator was undertaken by H.W. Ng as final year project at Liverpool University in 1991 and the authors are grateful for the contribution made by Ng to this work.

References

1. Winslow, W., "Induced Fibration of Suspensions", *J. Applied Physics*, Vol. 20, 1949, pp. 1137-1140.

2. Block, H. and Kelly, J.P, U S Patent 4 687 789, 1987.

3. Filisko, F.E. and Armstrong, W.E., U S Patent 4 744 914, 1988.

4. Stevens, N.G., Sproston, J.L. and Stanway, R., "An Experimental Study of a Simple Electroviscous Damper", *J. Electrostatics,* **15**, (1984), pp. 275-283.

5. Williams, E.W., Rigby, S.G., Sproston, J.L. and Stanway, R., "Electro-Rheological Fluids applied to an Automotive Engine Mount", *J. Non-Newtonian Fluid Mechanics,* Vol. 47, 1993, pp. 221-238.

6. Peel, D.J. and Bullough, W.A., "Prediction of Electro-Rheological Valve Performance in Steady Flow", *J. Mechanical Engineering Science*, Proc. I.Mech.E., Vol. 208, C4, 1994, pp. 253-266.

7. Stevens, N.G., Sproston, J.L. and Stanway, R., "An Experimental Study of Electro-Rheological Torque Transmission", J. Mechanisms, Transmissions and Automation in Design, Trans. ASME, Vol. 110, pp. 182-188, June 1988.

8. Monkman, G.J., "Electro-Rheological Robot Actuators", *Proc. IEE Colloquium on Robot Actuators,* London, 1991, digest 1991/146.

9. Craig, J.J., "Introduction to Robotics: Mechanics and Control", Addison Wesley, Reading, Mass., 1986, p. 320.

10. Tao, R., Woestman, J.T. and Jaggi, N.K., "Electric Field Induced Solidification", *Appl. Phys. Lett.*, Vol. 55 (18), October 1989, pp. 1844-1846.

11. Bode, H.W., "Relations between Attenuation and Phase in Feedback Amplifier Design", Bell Systems Technical J., 1940, pp 421-454, also in Thaler, G.J. (Ed.), "*Automatic Control: Classical Linear Theory*", Dowden, Hutchinson and Ross, Stroudsberg, Pa., 1974, pp. 145-178.

12. Dorf, R.C., "Modern Control Systems", 6th. Ed., Addison Wesley Publishing, Reading, Mass., 1992.

13. Brookfield, D.J., "Identification of Coulomb Friction in Robot Drives and other Mechanical Systems through observation of Third Harmonic Generation", *J. Mechanical Engineering Science*, Proc. I.Mech.E., part C, Vol. 208, 1994, pp. 329-336.

14. Bullough, W.A., Johnson, A.R., Hosseini-Sianaki, A and Firoozian, R., "The Electro-Rheological Clutch: Design, Performance, Characteristics and Operation", J. Systems and Control Engineering, Proc. I.Mech.E., part I, Vol2. 207, 1993, pp. 87-95.

15. Xiao, Shi, "Investigation of Kinetic and Dynamic Characteristics of an Electro-Rheological Clutch, Ph.D. Thesis, University of Sheffield Faculty of Engineering, 1992.

THE EFFECT OF SCALE FACTOR ON THE PERFORMANCE OF ELECTRO-RHEOLOGICAL FLUID DAMPERS

F. Gordaninejad[1] and R. Bindu[2]
Mechanical Engineering Department
University of Nevada, Reno, Nevada 89557, USA

ABSTRACT

Theoretical and experimental scale analyses are conducted to investigate the effects of length of the damper and electric field strength on the damping and stiffness coefficients of electrorheological fluid (ERF) dampers. It is demonstrated that by employing the results obtained from smaller size prototypes, the energy absorbing capacity of larger dampers can be estimated.

1. Introduction

Large civil structures such as bridges undergo predictable (e.g., wind excitations and traffic loads) as well as unpredictable (e.g., severe storms and earthquakes) vibrations. Predictable motions can be suppressed by using passive damping devices such as friction or viscoelastic dampers and elastomeric bearings. Control of unpredictable structural vibration is receiving increasing attention. One type of device that can potentially be used for unpredictable vibration control of structures is an ERF damper system.

Performance of cylindrical ERF dampers to suppress vibration was addressed by Ehrgott and Masri [1]. Bang-bang, and linear vibration control of cylindrical ERF dampers, and the effects of number of electrodes, length, electrode gaps were investigated by Gordaninejad, et al. [2,3]. Performance of hybrid ER/viscous-fluid dampers in controlling vibration, and comparisons with regular ERF dampers was studied by Gordaninejad, et al. [4]. Recently, the behavior of ERF dampers was addressed by Gavin [5]. An extensive summary of the development of ERF vibration dampers was presented by Coulter, et. al. [6].

Full scale studies of ERF dampers, for use in large structures, are time consuming and costly. The objective of this work is to develop a scale model by which the characteristics of large-scale ERF dampers can be accurately estimated via the study of the

---

[1] Professor
[2] Graduate student

performance of small prototypes. In the following, theoretical and experimental methods used to develop the scale relations and results for damping and stiffness coefficients are presented.

2. Theoretical Analysis

A mass-spring-damper system is considered for the analytical study. A simplified model of this system is shown in Figure l. This system consists of two springs, with spring constants of k_1, and a mass, m, which includes the sum of the mass of the arrangement between the shake table and the mass of the piston. The restoring force of the damper, $f(x,\dot{x})$, is assumed to be of the following form:

$$f(x,\dot{x}) = c\dot{x} + kx \tag{1}$$

where x and $\dot{x}$ and the displacement and the velocity of the mass-spring-damper system, respectively. c and k are the damping and stiffness coefficients, respectively. The system can be considered as a single-degree-of-freedom and its equation of motion can be expressed as

$$m\ddot{x} + c\dot{x} + k_2 x = k_2 y \tag{2}$$

where $k_2 = 2k_1 + k$ and y is the displacement of the shake table, $y = x\sin(\omega t - \phi)$. Here, ω is the angular frequency of the excitation and ϕ is phase angle. The solution of Eq. (2) yields

$$c = \frac{1}{\omega}\sqrt{\frac{k_2^{\,2}}{r^2} - (k_2 - m\omega^2)^2} \tag{3}$$

where $r = \ddot{x} / \ddot{y}$.

The variable damping coefficient of ERF dampers is a function of the following variables:

$$c = c(d, D, \omega, L, x, E) = C\pi DLdxE^n \tag{4}$$

where, d is the gap between the electrodes, D is the damper diameter, L is the damper length, , E is the applied field strength, C is the scale factor coefficient and n is a constant which ranges from 1.75 to 2. For steady-state forced vibration, the loss of energy is balanced by the energy due to excitation. In the case of cyclic oscillation the energy loss per cycle, W_d, due to damping force, F_d (which is linearly proportional to the velocity), is:

$$W_d = \pi c \omega x^2 \tag{5}$$

Considering both Eqs. (4) and (5), one has

$$c = C\frac{\pi DLdE^n}{\omega x} \tag{6}$$

To focus on the effects of the length of the damper and the field strength, the gap between the electrodes, the diameter of the cylinder and the frequency of the motion are fixed. Therefore, C can be of the following form:

$$C = \overline{C}L^{\alpha}E^{\beta}x^{\gamma} \tag{7}$$

where $\overline{C}, \alpha, \beta$ and γ are constants that are to be determined experimentally.

Similarly the stiffness coefficient of the damper can be expressed as follows [6]:

$$k = \frac{K\pi DLdxE^n}{x^2} \tag{8}$$

where

$$K = \overline{K}L^{\lambda}E^{\zeta}x^{\xi} \tag{9}$$

and $\overline{K}, \lambda, \zeta$, and ξ are constants that are to be determined experimentally.

3. Experimental Study

The performance of single-electrode dampers with different lengths was experimentally studied under a forced vibration. A spring-mass-damper system was designed, built and set up as illustrated in Figure 2. A schematic of the cross-section of a single-electrode damper is shown in Figure 3. An aluminum rod was placed at the center of the damper to transmit the force directly to the moving cylindrical inner electrode inside the plexiglass casing of the damper. The aluminum rod, which was rigidly fixed to the cylindrical inner electrode was connected to the ground. The rod was constrained by a bearing at the top and a pair of nylon guides located below the inner, moving electrode and can move only in the axial direction. The voltage was applied to the stationary outer electrode which was fixed tightly inside the cylindrical plexiglass casing of the damper. The damper is filled with Versa Flo ER-100. The fluid between the electrodes is electrically stressed by the voltage applied to the damper using a 408B Fluke High Voltage Power Supply. The vibration was measured by using two accelerometers. The signals obtained from the accelerometers were amplified and passed through a band pass filter. The filtered signals were fed into a Keithly Metrabyte DAS 1600 data acquisition board.

Logarithmic Decrement method was employed to obtain the damping coefficient of the dampers. The damping coefficient can be expressed by

$$c = \frac{2m\ln(x_1 / x_2)}{(t_2 - t_1)} \tag{10}$$

where , t_1 and t_2 are the times at which the consecutive amplitudes occur. A typical decaying vibration of the motion using a one-electrode ERF damper is presented in Figure 4. The stable decaying part of the experimental results and Eq. (10) were used to determine the damping coefficients. Stiffness coefficients, *k*, were obtained by combining Eqs. (3) and (10), and using the experimental data of decaying vibration.

4. Results and Discussion

Experiments were conducted on six single-electrode dampers ranging from 0.6 in to 3.0 in. The experiments were performed in three different days to evaluate the consistency of the results. In each day five experiments were conducted on each damper. A maximum relative error of 3% was observed [7]. The gap between the electrodes, the diameter of the cylinder and the frequency of the motion were fixed at, 0.10 in, 1.17 in and 10 Hz, respectively. This was done to focus on the effect of the length of the damper and the field strength on the damping coefficient. The damping coefficient of each prototype was determined experimentally by using the Logarithmic Decrement method.

In order to estimate the constants in Eqs. (6) and (8), experimental results of the two smaller dampers (0.6 in and 0.8 in) at 4 KV and 5 KV, and Eq. (3) were used. The detailed description of the procedure to find the following results were presented in [7]:

$\overline{C} = 6901$	$\overline{K} = 69415$
$\alpha = -0.79$	$\lambda = -1.02$
$\beta = -1.75$	$\zeta = -1.90$
$\gamma = 1.42$	$\xi = 1.95$

These values were substituted in Eqs. (6) and (8), and the damping and stiffness coefficients were computed for dampers of the lengths of 1.2 in, 1.5 in, 2.4 in and 3.0 inches.

The theoretical results were compared with experimental data obtained by using the Logarithmic Decrement method. The comparisons are shown in Figures 4 and 5. As can be seen in Figure 4 the effects of field strength and length of the damper on the damping coefficient is considerable. In addition, it should be noted that the maximum relative error between the experimental and the theoretical results are less than 10 percent. The effects of field strength and length of the damper on the stiffness coefficient are presented in Figure 5. Although, for field strengths less than 1 KV/mm the relative error is higher than those of higher field strengths, the maximum error does not exceed 8 percent. Also, it is interesting to note that the stiffness coefficient is not sensitive to changes in the field strength and the length of the damper.

5. Concluding Remarks

Theoretical and experimental scale studies was performed to assess the damping and stiffness coefficients of large-scale ERF dampers. The focus of the research was on single-electrode cylindrical dampers. It was demonstrated that the damping coefficient of the larger dampers can be predicted by using the data obtained from small prototypes with a relatively low percentage of error. It was also shown that the variation of stiffness coefficient due to changes in the field strength and the length of the damper is minimal.

6. Acknowledgments

This study was supported in part by a grant from the U. S. National Science Foundation. The authors would like to thank Dr. S. C. Liu of the NSF for his continuous encouragement. The authors also are thankful to Dr. A. Ray for his contribution in the experimental set up.

7. References

1. Ehrgott, R. C. and Masri, S. F., "Experimental Characterization of an Electrorheological Material Subject to Oscillatory Shear Strain," ASME *Journal of Vibration and Acoustics*, 1994, pp. 53-60.
2. Gordaninejad, F., Ghazavi, A., Ray, A., Wang, H. and Saiidi, M., "Intelligent Electro-Rheological Dampers for Active Vibration Control: An Experimental Study," *Ceramic Matrix Composites and Other Systems*, Vol. II, Ed. by A. Miravete, Woodhead Publishing Limited, Madrid, Spain, 1993, pp. 473-480.
3. Gordaninejad, F., Ray, A., and Wang, H., "Use of Hybrid Dampers for Vibration Control of Structures," *Proceedings of the Third US/Japan Conference on*

Earthquake Protective Systems for Bridges, Berkeley, California, 1994, pp. 3-15 - 3-29.

4. Gordaninejad, F., Ray, A. and Bindu, R., "Vibration Control of Structures Using Hybrid ER/Viscous Dampers," *Proceedings of the First World Conference on Structural Control*, 1994, pp. TA2-41- TA2-48.
5. Gavin, H. P., "Electrorheological Dampers for Structural Vibration Suppression," *Ph.D. Dissertation*, Department of Civil and Environmental Engineering, University of Michigan, 1995.
6. Coulter, J. P., Weiss, K. D., and Carlson, J. D., "Engineering Applications of Electro-rheological Fluids," J. Intelligent Materials Systems and Structures, Vol. 4, 1993, pp. 248-259.
7. Bindu, R., "Design of Electro-Rheological Fluid Dampers for Hybrid Vibration Control: Experiments and Scale Analysis," *M.S. Thesis*, Department of Mechanical Engineering, University of Nevada, 1995.

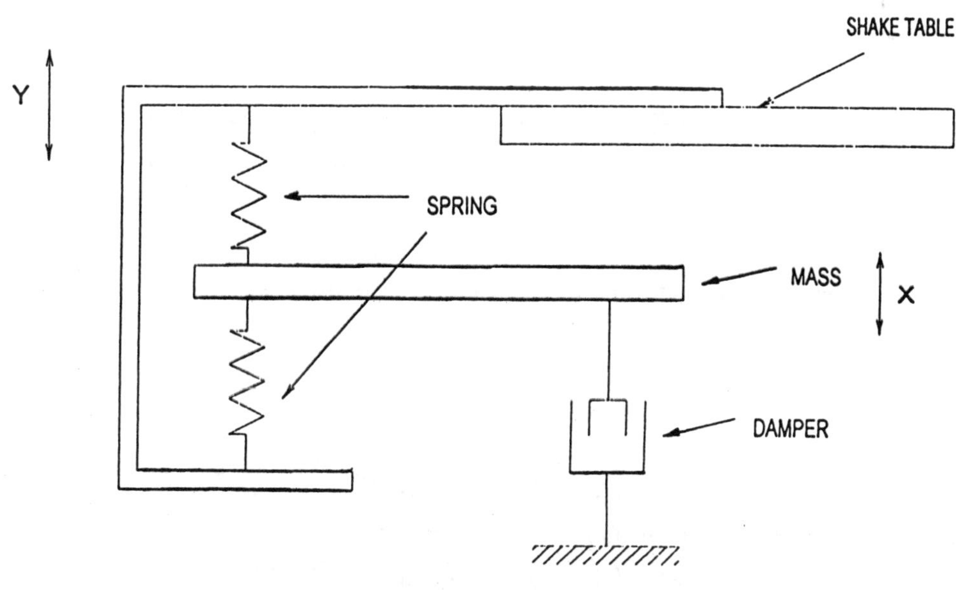

Figure 1. Schematic of the mass-spring-damper system used in the scale analysis.

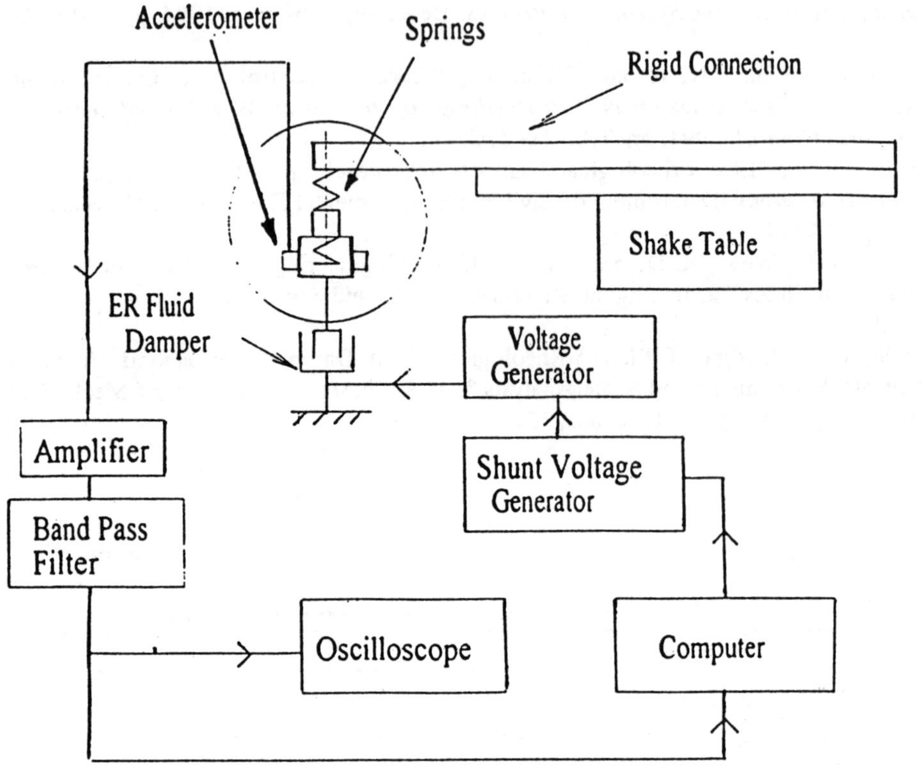

Figure 2. Schematic of the experimental set up for the scale analysis.

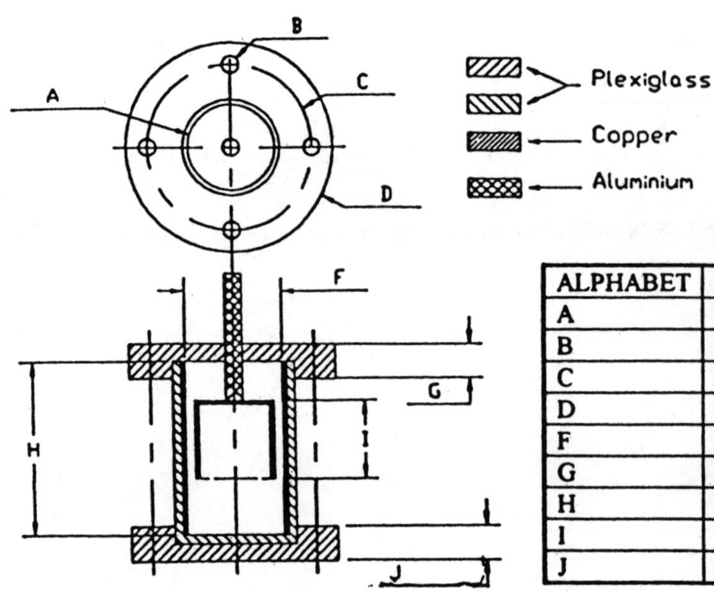

ALPHABET	DIMENSION
A	R1.25"
B	R0.12"
C	R1.11"
D	R1.40"
F	1.29"
G	0.50"
H	2.22"
I	1.40"
J	0.50"

Figure 3. Schematic of the single-electrode ERF dampers.

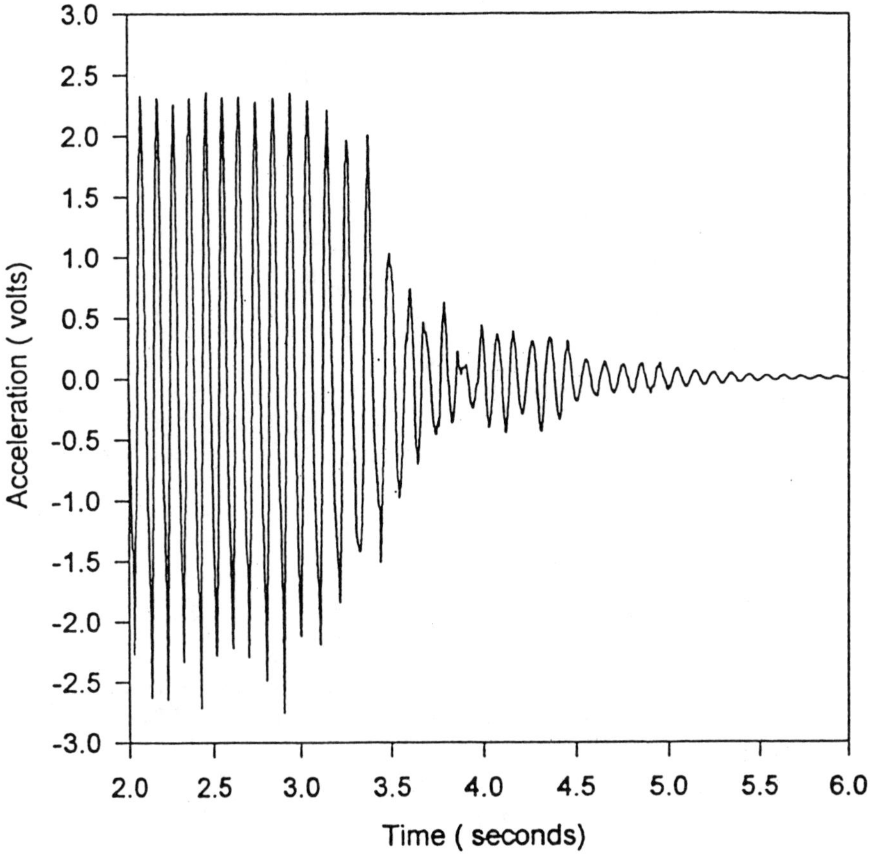

Figure 4. A typical experimental results for decaying vibration of a single-electrode ERF damper.

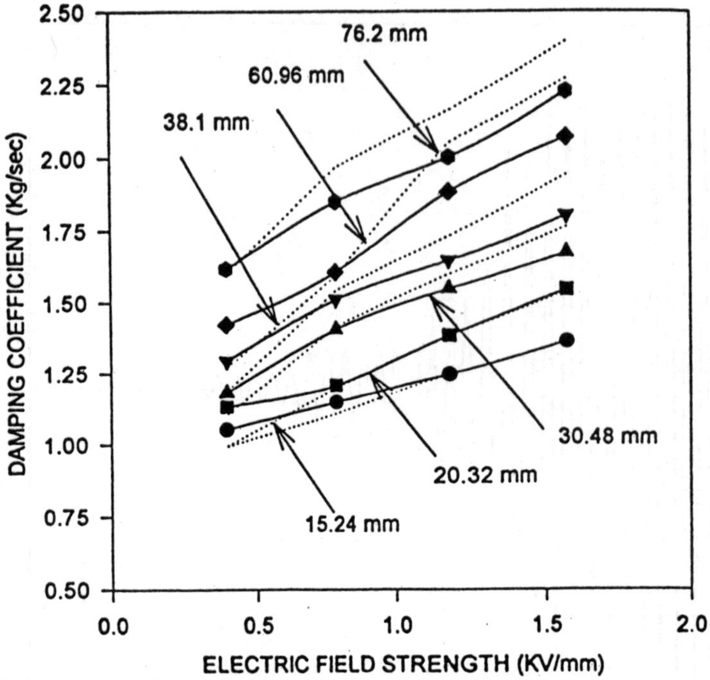

Figure 5. Comparison of the experimental (solid lines) and theoretical (dotted lines) results for damping coefficient of a single-electrode dampers with different lengths.

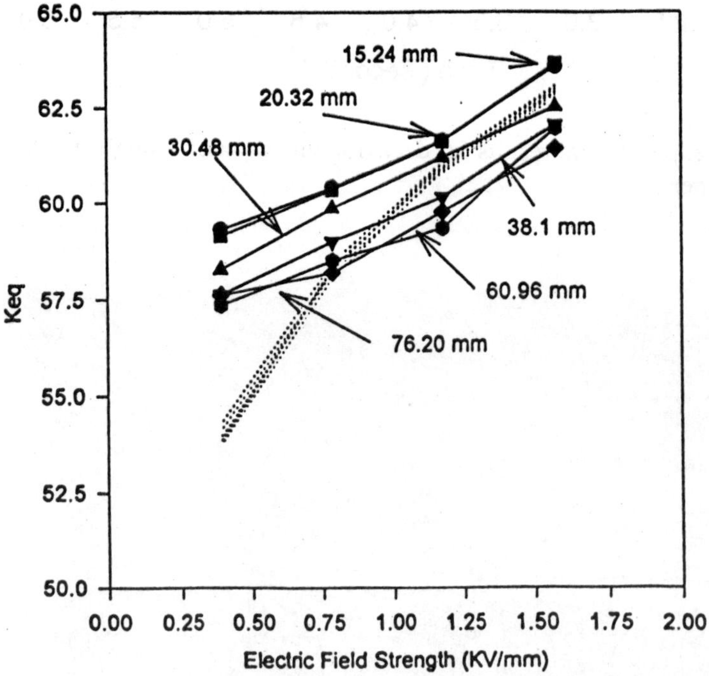

Figure 6. Comparison of the experimental (solid lines) and theoretical (dotted lines) results for stiffness coefficient of a single-electrode dampers with different lengths.

PERFORMANCE ANALYSIS OF AN ENGINE MOUNT FEATURING ER FLUIDS AND PIEZOACTUATORS

S. H. CHOI, Y. T. CHOI, S. B. CHOI AND C. C. CHEONG
Smart Structures and Systems Laboratory
Department of Mechanical Engineering, Inha University
Incheon 402-751, KOREA

ABSTRACT

Conventional rubber mounts and various types of passive or semi-active hydraulic engine mounts for a passenger vehicle have their own functional aims on the limited frequency band in the broad engine operating frequency range. In order to achieve high system performance over all frequency ranges of the engine operation, a new type of engine mount featuring electro-rheological(ER) fluids and piezoactuators is proposed in this study. A mathematical model of the proposed engine mount is derived using the bond graph method which is inherently adequate to model the interconnected hydro-mechanical system. In the low frequency domain, the ER fluid is activated upon imposing an electric field for vibration isolation while the piezoactuator is activated in the high frequency domain. A neuro-control algorithm is utilized to determine control electric field for the ER fluid, and H_∞ control technique is adopted for the piezoactuator. Comparative works between the proposed and single-actuating(ER fluid only or piezoactuator only) engine mounts are undertaken by evaluating force transmissibility over a wide operating frequency range.

1. Introduction

A modern vehicle features its high engine power as well as light body structures. This draws an attention on noise and vibration problems of the vehicle. A vehicle engine is one of the most dominant noise and vibration sources because it rotates itself and transmits the torque through a driveline. Therefore for the purpose of resolving noise and vibration problems due to the engine movement, numerous researches on the design and control of a high-performance engine mount have been conducted with great effort.

The engine mount basically supports the engine and restricts excessive movement of the engine. In addition, it reduces engine excitation force transmitted to vehicle frame and absorbs road-induced vehicle vibrations by means of proper tuning. Performance requirements of the engine mount vary according to the the frequency range of the engine operation. In the low frequency range(5-30 Hz), where the movement of the engine is relatively large and the resonance of the vehicle frame exists, the engine mount needs not only to constrain abrupt movement of the engine with high stiffness but to reduce the force transmitted to the vehicle frame with large damping. On the other hand, in the high frequency range(30-200 Hz), where excitation displacement of the engine is relatively small, more flexible engine mount needs to be installed.

The stiffness and damping characteristics of most conventional rubber engine mounts are invariant with respect to excitation frequency or excitation displacement. Some engine mounts whose dynamic characteristics are changeable according to the excitation displacement using felt or wire mesh elements were proposed. However, the damping force and the stiffness are not simultaneously controllable to meet imposed performance criteria[1]. The hydraulic engine mount was developed to meet the large damping criteria in the low frequency range. Large fluid damping force is induced through the fluid channel in the hydraulic engine mount. By inserting a decoupler between separated fluid chambers, it would meet the criteria without producing fluid damping force in the high frequency range. However, the force transmissibility is increased in the high frequency range despite of the installation of the decoupler. Furthermore, due to the difficulty in predicting operation characteristics of the hydraulic engine mount, exhaustive experiments by trial and error must be performed in advance for appropriate design and tuning[2-3].

In the mean time, researches on the semi-active engine mount have been widely undertaken in order to improve performance characteristics. A solenoid actuator or a step motor is used to change the diameter of the fluid path. The complexities associated with the design and tuning are reduced and the fluid damping force can be exactly controlled by employing semi-active control action[4-5]. Being different from these mechanical actuation devices, a semi-active engine mount using ER fluids, which change rheological characteristics according to the imposition of applied electric fields, was experimentally investigated and theoretically analyzed[6-7]. The ER engine mount features miniaturization of the control devices, fast response time, continuously controllable damping force, and low energy consumption. However, since the main purpose of the semi-active engine mount is to control the damping force in the low frequency range, the noise and vibration problems in the high frequency range are still in question.

Recently an active engine mount using a piezoactuator which has salient fast response time was proposed by Ushijima *et al.*[8]. However since the piezoactuator produces only small displacement, the active engine mount could be not applied to the lower frequency range below 30 Hz in which the engine displacement is large. This implies that the piezoactuator may not assure the effectiveness of the vibration isolation in the low frequency range. Muller *et al.* presented an active engine mount combined with inertia track and electromagnetic actuator[9]. The electromagnetic actuator supplements fluid damping force in the low frequency range, and produces active control force for vibration isolation in the high frequency range. However, this mechanism requires relatively large power consumption to achieve favorable system performance.

In this paper a new prototype of engine mount, which provides superior vibration isolation performance in the whole engine operating frequency range with a small amount of energy consumption, is proposed. This engine mount comprises of two actuators; ER fluid and piezoactuator. These actuators have unique operation characteristics in each specific frequency range. A neuro-control algorithm and H_∞ control method are applied to the ER fluid and piezoactuator, respectively. Computer simulation is undertaken to demonstrate the efficiency of the vibration isolation and the robustness to parameter variations of the proposed engine mount.

2. Model Synthesis and Modeling

Fig. 1 presents a configuration of the proposed engine mount. The upper chamber and the lower chamber separated by the electrodes are filled with the ER fluid. The upper chamber is enclosed with the rubber and steel case. When compression force is applied to the rubber from the engine, there exists a volume change in the upper chamber due to the relative movement of the rubber toward the steel case. Therefore the ER fluid in the upper chamber is forced to flow into the lower chamber. A diaphragm isolates the lower chamber from the air, and the lower chamber accumulates the ER fluid flooded from the upper chamber. The ER fluid plays a role of an internal actuator between the engine and the piezoactuator to complement the damping force of the rubber. The piezoactuator installed below the steel case supplies the engine mount with active control force(hereinafter called 'hybrid engine mount').

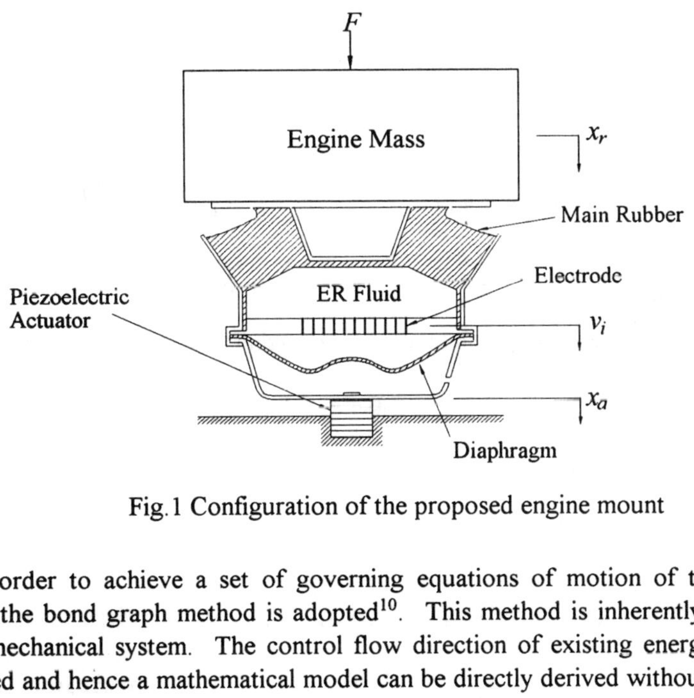

Fig.1 Configuration of the proposed engine mount

In order to achieve a set of governing equations of motion of the hybrid engine mount, the bond graph method is adopted[10]. This method is inherently adequate to the hydro-mechanical system. The control flow direction of existing energies can be easily identified and hence a mathematical model can be directly derived without confusion. The bond graph model of the proposed hybrid engine mount is presented in Fig. 2. In this figure, m_r , m_a and I_i are the mass of the engine and rubber, the mass of the piezoactuator, and the fluid inertance between the electrodes, respectively. And k_r, k_a,

C_u and C_l represent the stiffness of the rubber, the stiffness of the piezoactuator, the compliance of the upper chamber, and the compliance of the lower chamber, respectively. The compliance of each chamber, C_u or C_l is obtained under assumptions that whole edges of circular type rubber plate are clamped and pressure is uniformly distributed upon one face. A_p is the piston area of each chamber, and b_r is the damping coefficient of the rubber. b_i is the fluid resistance between fixed plates given by $b_i = 12\eta_e L / bh^3$. Here η_e is the fluid viscosity and L is the electrode length along the flow direction, b is the electrode width, and h is the gap between the electrodes. $F(t)$, $P_{ER}(t)$ and $F_a(t)$ represent excitation force from the engine, pressure drop due to the field-dependent yield stress of the ER fluid, and control force of the piezoactuator, respectively.

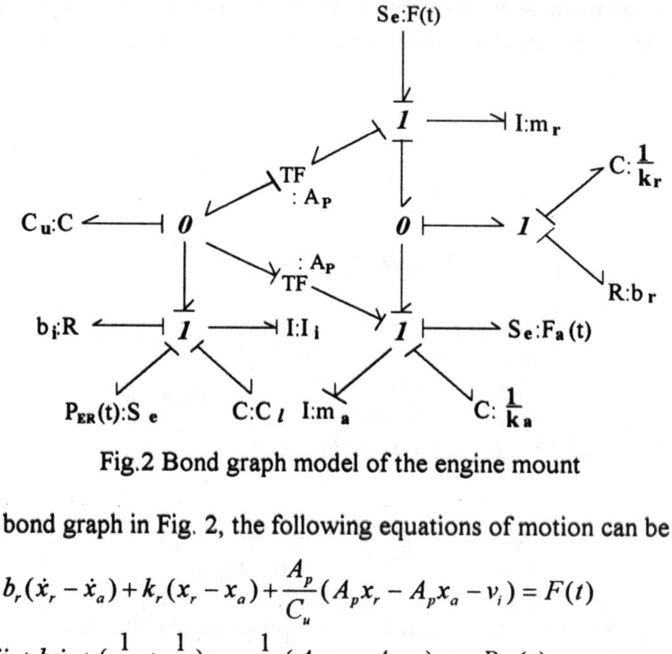

Fig.2 Bond graph model of the engine mount

From the bond graph in Fig. 2, the following equations of motion can be obtained.

$$m_r\ddot{x}_r + b_r(\dot{x}_r - \dot{x}_a) + k_r(x_r - x_a) + \frac{A_p}{C_u}(A_p x_r - A_p x_a - v_i) = F(t) \tag{1}$$

$$I_i\ddot{v}_i + b_i\dot{v}_i + (\frac{1}{C_u} + \frac{1}{C_l})v_i - \frac{1}{C_u}(A_p x_r - A_p x_a) = -P_{ER}(t) \tag{2}$$

$$m_a\ddot{x}_a + b_r(\dot{x}_a - \dot{x}_r) + k_r(x_a - x_r) + k_a x_a - \frac{A_p}{C_u}(A_p x_r - A_p x_a - v_i) = -F_a(t) \tag{3}$$

Here x_r is the engine displacement, x_a is the displacement of the piezoactuator, v_i is the volume of the fluid passing through the electrodes. If the pressure difference between the upper and lower chambers is smaller than $P_{ER}(t)$, the fluid flow passing through the electrodes does not occur, ($b_i = \infty$), and hence the Eq. (2) becomes

$$\dot{v}_i = 0 \tag{4}$$

3. Characteristics of ER Fluids and Piezoactuators

ER fluids are typically suspensions of conductive particles in the suitable non-conductive carrier liquids, which undergo significant instantaneous reversible changes in material characteristics when subjected to applied electric fields. Fig. 3 shows that the ER fluid is changed from Newtonian flow in which particles move freely to Bingham behavior in which particles are aligned in a chain and thus having yield stress upon applying electric field to fluid domain. The strength of the electric field which is needed to undergo phase-change from an isotropic material to an anisotropic material is about 1-2 kV/mm. However, due to small current density below about $10\,\mu A/cm^2$, consuming power is very low. In this paper, maximum control electric field applied to the hybrid engine mount is chosen to be 3 kV/mm, due to the electric breakdown of ER fluids and the difficulty of the miniaturization of high voltage supplier in practical situations.

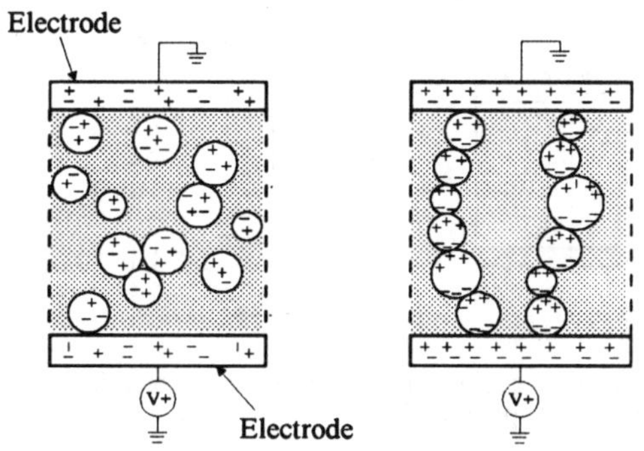

(a) no voltage applied (b) voltage applied

Fig.3 Phenomenological behavior of ER fluids

When the electric field is applied to the ER fluid, the Bingham behavior of the ER fluid is described by

$$\begin{aligned} \tau &= \eta_e \dot{\gamma} + \tau_y(E) \\ &= \eta_e \dot{\gamma} + \alpha E^{\chi} \end{aligned} \tag{5}$$

Here τ is shear stress, η_e is viscosity, $\dot{\gamma}$ is shear rate and $\tau_y(E)$ is yield stress. The parameters of α and χ are the characteristic values for certain ER fluid and can be usually determined by experiments. In this paper, these values are obtained by experiments using couette type electroviscometer.

Piezoelectricity is a phenomenon that when force or pressure is applied to material, an electric charge is produced from the material. It is called a direct piezoelectric effect. Reversibly, when the electric field is supplied to the material, a mechanical stress or strain takes place. It is called a inverse piezoelectric effect. Thus, the electric energy can be changed to the mechanical energy and vice versa.

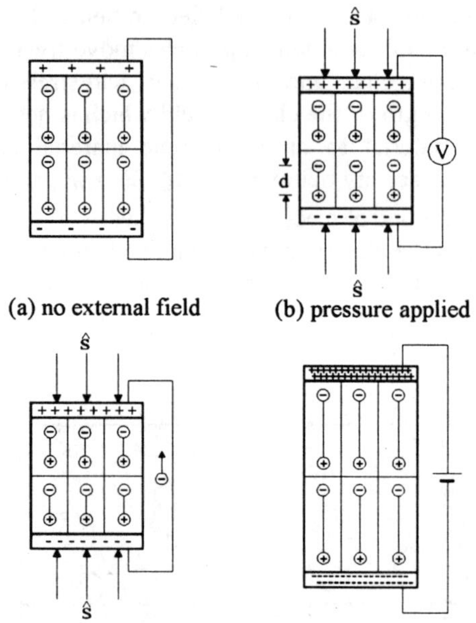

(a) no external field (b) pressure applied

(c) pressure applied and electrical path present (d) external potential applied

Fig.4 Characteristics of piezoelectric materials

Fig. 4(a) shows a distribution of the electric charge in the absence of the electric field(or external force) applied to the piezoelectric material[11]. A polarization occurs according to a set of arrays of the dipole moment. The arrays possess many unit cells in the material. By this polarization, positive charges move to one end of the material, and negative charges move to the other end. If the material is stressed by $\hat{s}$, the material is deformed. This deformation alters the dipole length d and thus causes polarization to be changed. This small polarization due to compression of the material causes current density of both ends of the material to be increased. If the ends are isolated, a voltage drop takes place. And if the ends are electrically connected, electrons move from one end to the other (refer to Fig. 4(c)). On the other hand, if the voltage is supplied to the material as shown in Fig. 4(d), the current density of the ends of the material increases. Since the negative charges in the piezoelectric material are attracted to the positive electrode and vice versa, both the dipole length d and the length of the material are changed.

Without loss of generality, input voltage(or electric field) to the piezoelectric material is to be upper-bounded due to nonlinearity, hysteresis and also electrode breakdown. Normally the magnitude of the input voltage of commercial piezoceramic(PZT) to ensure stable operation is not more than about 100 V.

4. Control Algorithms

4.1 Neuro-Controller for the ER Fluid

A pressure drop due to the yield stress of the ER fluid activated by the electric field always reacts against the fluid flow. Therefore the pressure drop $P_{ER}(t)$ can be represented using the signum function as follows:

$$P_{ER}(t) = 2\frac{L}{h}\tau_y(E)\cdot \mathrm{sgn}(\dot{v}_i) \tag{6}$$

In physical sense, the purpose using the ER fluid as an actuator is to reduce displacement and velocity of the engine, which can be achieved by appropriately controlling the pressure drop $P_{ER}(t)$. Therefore, in order to determine proper magnitude of the pressure drop in a certain time, a neuro-control algorithm is employed in this study.

Fig. 5 presents a neuro-control block diagram and architecture. The neuro-controller is a multi-layer neural network whose neurons in the input and output layers have a linear activation function, $f(x) = x$, while neurons in the hidden layer have a sigmoid activation function, $f(x) = 2/[1+\exp(-x)]-1$, where x is a certain variable. The input-output relationships of neurons units in the hidden layer are given by

$$net_j = \sum_{i=1}^{n} w_{ji}\, O_i + \theta_j \tag{7}$$

$$O_j = f(net_j) = \frac{2}{1+\exp(-net_j)} - 1 \tag{8}$$

where w_{ji} is the interconnection weight between the i-th neuron and the j-th neuron, O_i is the output of the i-th neuron, θ_j is the bias of the j-th neuron, net_j is the sum of the weighted outputs of the i-th neuron, and O_j is the output of the j-th neuron. In the input layer, inputs of neurons are state vectors, while outputs of neurons in the output layer are adopted as control input to the engine mount, *i.e.*,

$$u_n(t) = \sum_{j=1}^{n} w_{kj}\, O_j + k_1 S \tag{9}$$

Here w_{kj} is the interconnection weight between the j-th neuron and the k-th neuron. k_1

is a positive constant and S is a sliding hyperplane represented by $S = c_1 e + \dot{e}$ with a positive constant of c_1[12]. The error between desired displacement x_d and actual displacement of the engine mount x_r is denoted by e.

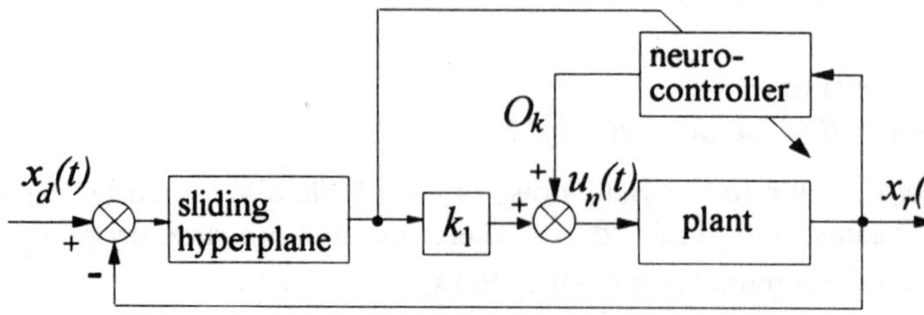

(a) block diagram of control flow

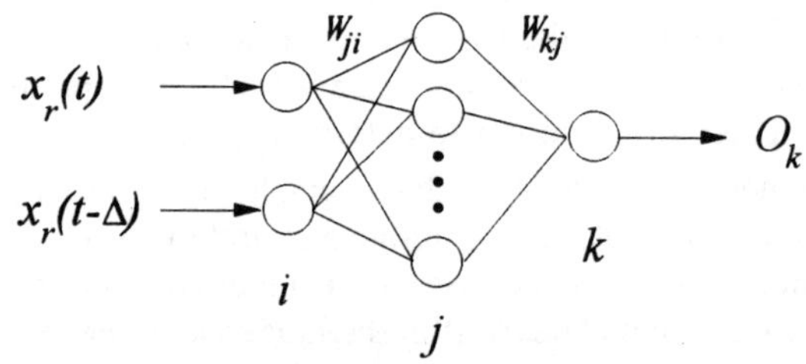

(b) architecture of neuro-controller

Fig.5 Neuro-control scheme

In addition, in order to train the neural network on line, a learning algorithm is adopted to drive the error dynamics S and its derivative $\dot{S}$ asymptotically to zero as soon as possible[13]. By the learning algorithm, the incremental change of the interconnection weight is obtained as follows:

i) between the hidden layer and the output layer;

$$\Delta w_{kj} = \eta \delta_k O_j, \qquad \delta_k = S + k_2 \dot{S} \tag{10}$$

ii) between the input layer and the hidden layer;

$$\Delta w_{ji} = \eta \delta_j O_i, \qquad \delta_j = f'(net_j) \sum_{k=1}^{n} \delta_k w_{kj} \tag{11}$$

In the above equations, η is learning rate, k_2 is a positive constant and $f'()$ is the first derivative of the activation function.

Determining the magnitude of $P_{ER}(t)$ through the neuro-control method, the control electric field applied to the engine mount can be obtained from Eq. (5), Eq. (6) and Eq.

(9) as follow.

$$E = \left[u_n \cdot (\frac{h}{2L\alpha}) \right]^{1/x} \tag{12}$$

In this paper, the neuro-controller is of size 2-5-1 having one hidden layer of 5 neurons. And following numerical values are employed ; $\eta = 0.1$, $\theta_j = 0$, $c_1 = 2, k_1 = 450, k_2 = 0.1$.

4.2 H_∞ Controller for the Piezoactuator

In order to formulate a H_∞ controller for the piezoactuator, the Eq. (1)- Eq. (3) are rewritten in a matrix form as follows.

$$\begin{gathered} \mathbf{M}\ddot{\mathbf{X}} + \mathbf{B}\dot{\mathbf{X}} + \mathbf{K}\mathbf{X} = \mathbf{G_w}F(t) + \mathbf{G_u}F_a(t) \\ \mathbf{X} = \{x_r, v_i, x_a\}^T, \mathbf{G_w} = \{1,0,0\}^T, \mathbf{G_u} = \{0,0,-1\}^T \end{gathered} \tag{13}$$

Since the design of a robust controller using only the piezoactuator is required, $P_{ER}(t)$ is not included in the above equation. $\mathbf{M}, \mathbf{B}$ and $\mathbf{K}$ denotes the mass, damping and stiffness matrix of the engine mount system, respectively. It is noted that the fluid resistance coefficient b_i in $\mathbf{B}$ is derived for laminar flow.

To design a controller which guarantees robust stability and high performance for uncertain system parameter b_i, a modeling uncertainty needs to be considered first. Assuming that the b_i has additive uncertainty of 50%, $\mathbf{B}$ can be decomposed as follow:

$$\mathbf{B} = \mathbf{B}^0 + \Delta\mathbf{B} \tag{14}$$

where $\mathbf{B}^0$ is a nominal portion, $\Delta\mathbf{B}$ is a perturbation portion due to the uncertainty of b_i. $\Delta\mathbf{B}$ can be represented as follows to describe uncertainty structure:

$$\begin{gathered} \Delta\mathbf{B} = \boldsymbol{M}_1\Delta_1\boldsymbol{N}_1 \\ \boldsymbol{M}_1 = \left[0, \sqrt{\frac{b_i}{2}}, 0 \right]^T, \quad \boldsymbol{N}_1 = \left[0, \sqrt{\frac{b_i}{2}}, 0 \right] \\ |\Delta_1| < 1 \end{gathered} \tag{15}$$

where Δ_1 is uncertain perturbation. Defining $\tilde{z}_1 = \boldsymbol{N}_1\dot{\mathbf{X}}$, $\tilde{w}_1 = -\Delta_1\tilde{z}_1$ and using Eq. (13)- Eq. (15), the engine mount system which takes account into the uncertainty of b_i is represented by following equation.

$$\mathbf{M}\ddot{\mathbf{X}} + \mathbf{B}^0\dot{\mathbf{X}} + \mathbf{K}\mathbf{X} = \mathbf{G}_{\tilde{w}_1}\tilde{w}_1 + \mathbf{G_w}F(t) + \mathbf{G_u}F_a(t) \tag{16}$$

where $\mathbf{G}_{\tilde{w}_1} = \{\boldsymbol{M}_1\}$. The fluid resistance due to the uncertainty of b_i is defined as a disturbance $\tilde{w}_1$.

Fig. 6 represents the block diagram of the piezoactuator system P with perturbation feedback Δ_1. The force transmitted to the frame is supplied with controller K as feedback signal(y). The control target is to determine the force transmitted to the frame ($\tilde{z} = F_t(t) = k_a x_a + F_a(t)$) and the perturbation input($\tilde{z}_1$). Disturbances are chosen to be the excitation force of the engine ($F(t)$) and perturbation output($\tilde{w}_1$). The dynamic characteristic of the piezoactuator between input voltage u and output displacement $\Delta\ell$ can be modeled by a first order linear system as follow.

$$\Delta\ell(s) = P_u(s)u(s) = \frac{d_{eff}}{Ts+1}u(s) \tag{17}$$

where $P_u(s)$ is the transfer function of the piezoactuator, s is a Laplace operator and T is time constant of the piezoactuator. And d_{eff} is effective piezoelectric constant. From this, the actuating force of the piezoactuator is obtained by

$$F_a(t) = k_a \cdot \Delta\ell(t) \tag{18}$$

It is generally known that most commercial piezoactuators have a certain level of hysteresis(less than 15% of real strain)[14]. This implies that in practice the linear model $P_u(s)$ in Eq. (17) can not exactly describe operating characteristic of the piezoactuator.

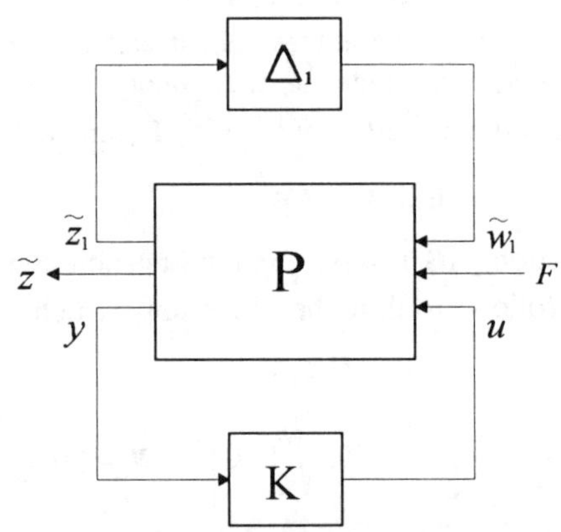

Fig.6 Block diagram with perturbation feedback

Therefore, it needs to be assumed that the piezoactuator dynamic $P_u(s)$ has multiplicative uncertainty of 15%. Then, the perturbation model of the piezoactuator is described as follows:

$$\begin{aligned} P_u &= P_u^0 + \Delta P_u \\ &= P_u^0(1 + \boldsymbol{M}_2\Delta_2\boldsymbol{N}_2) \end{aligned} \tag{19}$$

$$\boldsymbol{M}_2 = \boldsymbol{N}_2 = \sqrt{\frac{3}{20}}, \quad |\Delta_2| < 1$$

where P_u^0 is nominal part and ΔP_u is perturbation part of P_u.

The interconnection structure of the control system containing uncertainties $\mathbf{\Delta B}$ and ΔP_u is given by Fig. 7. The control input due to the perturbation of the effective piezoelectric constant is defined as the disturbance $\widetilde{w}_2$. P is the transfer function of the piezoactuator containing the nominal model $P_u^{\,0}$. It is assumed that measurement noise n is added to the feedback signal y. The signal $\widetilde{z}_2$ and the input voltage u are also chosen as control targets. Therefore, the final system is a multivariable system having 5 inputs($\widetilde{w}_1, \widetilde{w}_2, F, n, u$) and 5 outputs($\widetilde{z}_1, \widetilde{z}_2, \widetilde{z}, u, y$). By multiplying $\widetilde{z}_1, \widetilde{z}_2, u$ by weighting factors $\gamma_1, \gamma_2, \gamma_3$, the design freedom to appropriately control system performance in the frequency domain is obtained. The sub-optimal algorithm of H_∞ and the calculation of μ are designed through μ-analysis and synthesis toolbox of the MATLAB software[15]. By properly tuning the weighting factor γ_i, a 7-th order controller to produce the desired closed-loop response (transmissibility) is achieved.

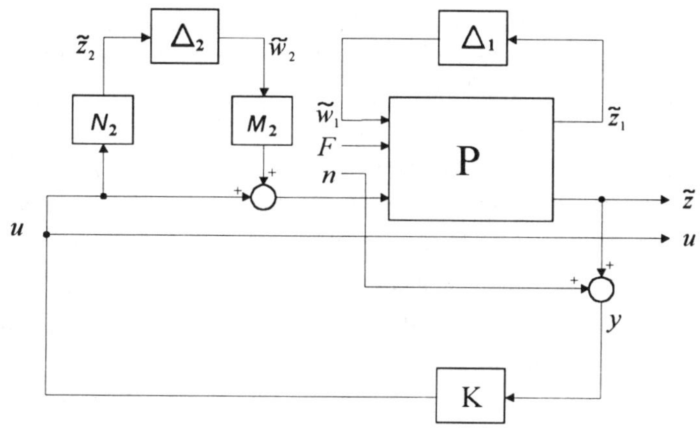

Fig.7 Interconnection structure of the control system

5. Simulation Results and Discussion

In order to demonstrate the effectiveness of the proposed hybrid engine mount, a set of model parameters, which are given in Table 1, are considered. Fig. 8 compares isolation performance in terms of the force transmissibility without and with the neuro-controlled electric field for the ER fluid. It is clearly observed that the isolation effect was substantially improved in the neighborhood of the system natural frequency. However, there is no performance improvement above 10 Hz. This indicates that the activation of the ER fluid is effective only in a low frequency range.

The robust stability of the H_∞ controller for the piezoactuator is to be checked prior to implementation. This can be advocated by μ value shown in Fig. 9. Since μ value is always less than 1 in whole frequency range, the H_∞ controller guarantees the robust stability to parameter uncertainties[16]. Fig. 10 presents the transmissibility performance with the H_∞ controlled input voltage for the piezoactuator. The applied voltage is assumed to be bounded by ±50 V. It is obvious that there is no control effect below 10 Hz. This is attributed to the limit of actuating force produced by the piezoactuator. However, the piezoactuator provides favorable active force above 10 Hz yielding much improved isolation performance.

These results lead to intuition that the isolation performance of the hybrid engine mount will be considerably enhanced by activating the ER fluid in the low frequency range, and the piezoactuator in the high frequency range. However, there is a difficulty in determining a threshold frequency in practice. Therefore, a certain value of the

Table 1. Parameters of the hybrid engine mount

Parameters	Values
effective piston area (A_p)	32×10^{-4} m^2
fluid resistance between electrodes (b_i)	1.780×10^{7} N·s/m^5
damping coefficient of the main rubber (b_r)	116 N·s/m
compliance of the lower chamber (C_l)	3.891×10^{-8} m^5/N
compliance of the upper chamber (C_u)	5.337×10^{-11}m^5/N
fluid inertance between electrodes (I_i)	4555 kg/m^4
stiffness of the piezoactuator (k_a)	2.1×10^{8} N/m
stiffness of the main rubber (k_r)	114×10^{3} N/m
mass of the piezoactuator (m_a)	0.043 kg
mass of the engine and the main rubber (m_r)	58.6 kg
electrode width (b)	0.04 m
electrode gap (h)	0.001 m
electrode length (L)	0.01 m
proportional of the yield stress function (α)	151.47
exponent of the yield stress function (χ)	1.8579
viscosity of the ER fluid (η_e)	0.2 N·s/m^2
number of electrodes	60
effective piezoelectric constant (d_{eff})	3.4×10^{-7} m/V
time constant of the piezoactuator (T)	60×10^{-6}

displacement of the upper rubber could be used as the threshold. In this study, the threshold of the displacement is chosen to be 0.28 mm. Fig. 11 presents the performance improvement achieved by the hybrid actuating control strategy. The isolation effect was enhanced over the whole frequency range, as expected.

In order to explicitly observe the performance improvement, control responses are presented in the time domain as shown in Fig. 12. The system is subjected to time-varying excitation force as given in Fig. 12(a). The closed-loop controlled transmitted force is very low compared with the open-loop response. It is seen from Fig. 12(b) that there is chattering near 0.4 sec and 0.6 sec. This arises because the input voltage applied to the piezoactuator is saturated in the transient frequency range. We observe from the control input histories that the ER fluid is activated only in the low frequency range, while the piezoactuator is activated within the voltage limit over the whole frequency.

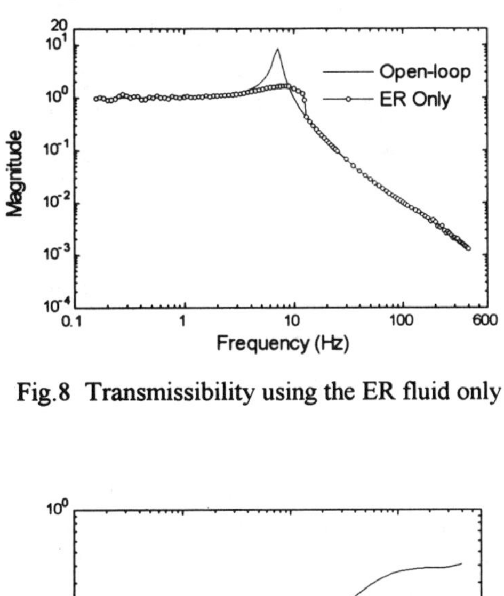

Fig.8 Transmissibility using the ER fluid only

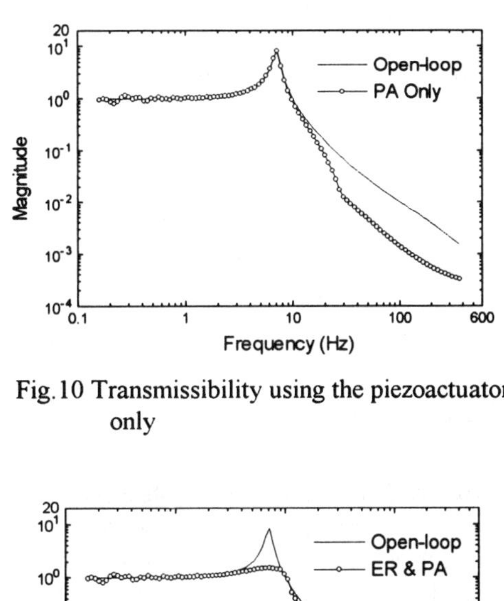

Fig.10 Transmissibility using the piezoactuator only

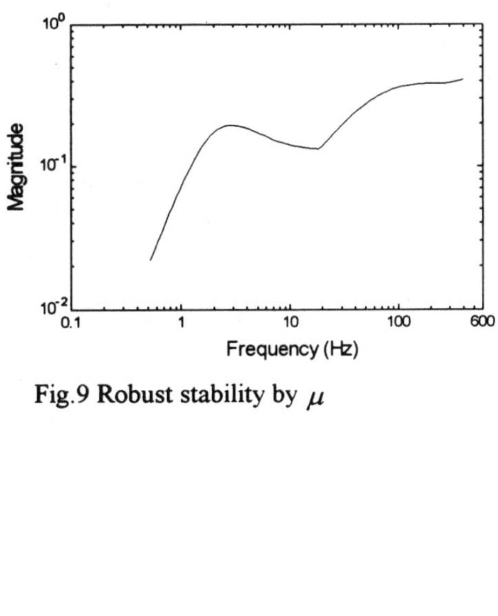

Fig.9 Robust stability by μ

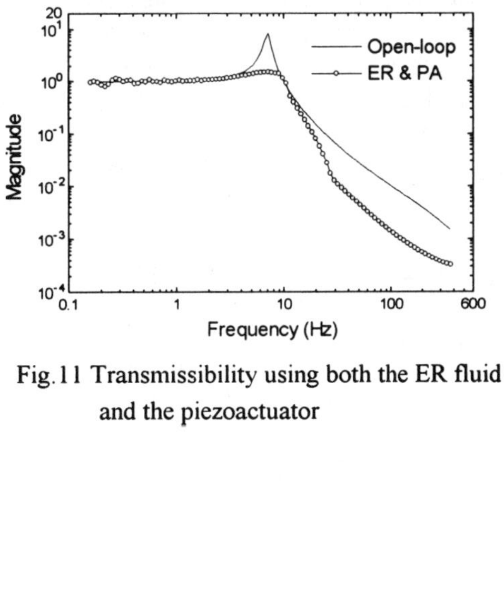

Fig.11 Transmissibility using both the ER fluid and the piezoactuator

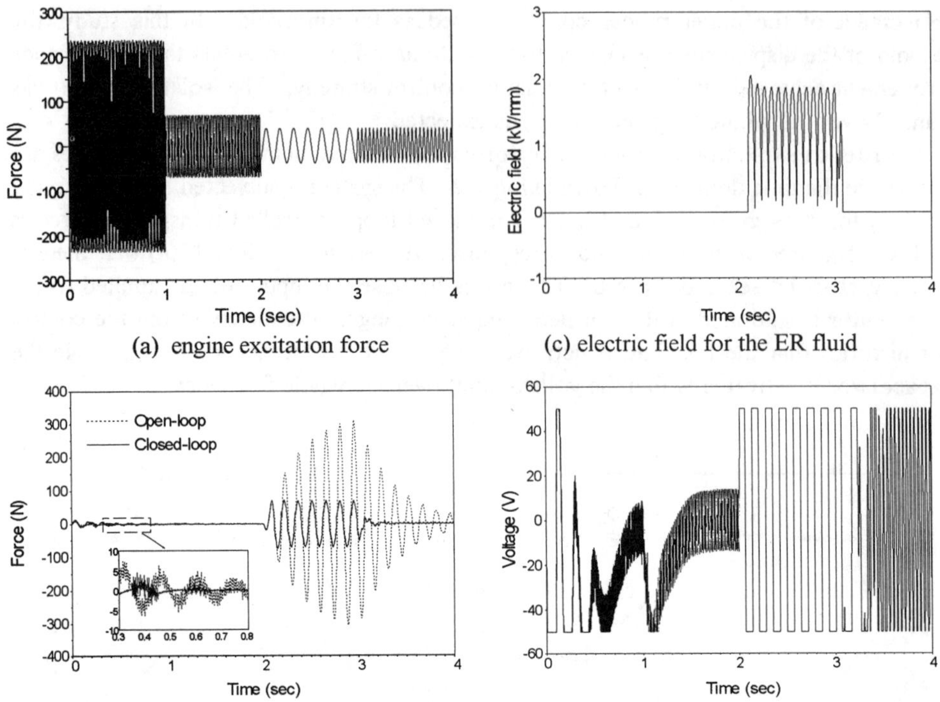

(a) engine excitation force

(c) electric field for the ER fluid

(b) transmitted force

(d) control voltage for the piezoactuator

Fig. 12 Control responses in the time domain

6. Conclusions

A novel type of engine mount using ER fluids and piezoactuators was proposed to improve isolation over the whole operating frequency range. A governing equation of motion was derived from the bond graph model, and it was represented in a state space form to formulate control algorithms. The neuro-control scheme was formulated for the ER fluid and subsequently employed to achieve performance improvement in the low frequency range. On the other hand, the H_∞ controlled system was synthesized for the piezoactuator to achieve robust stability and also isolation effect. It was demonstrated that the performance with the piezoactuator was very effective in the high frequency range. Consequently, two distinct characteristics were adopted to achieve performance improvement over the whole frequency range, and favorably verified in both the frequency and the time domains. The results presented in this work are quite self-explanatory in justifying that the proposed hybrid engine mount is very effective for improving the isolation over a broad frequency range. Experimental implementation of the proposed engine mount is being studied.

References

1. E. I. Rivin, *SAE Technical Paper Series* **850481** (1985).
2. M. Bernuchon, *SAE Technical Paper Series* **840259** (1984).
3. R. Singh, G. Kim and P. V. Ravindra, *Journal of Sound and Vibration* **158** (1992) 219-243.
4. P. L. Graf and R. Shoureshi, *ASME Journal of Dynamic Systems, Measurement, and Control* **110** (1988) 422-429.
5. S. E. Smith and J. F. Hoying, *United States Patent* **4,789,143** (1988).
6. S. Morishita and J. Mitsui, *SAE Technical Paper Series* **922290** (1992).
7. N. K. Petek, R. J. Goudie and F. P. Boyle, *SAE Technical Paper Series* **881785** (1988).
8. T. Ushijima and S. Kumakawa, *SAE Technical Paper Series* **930201** (1993).
9. M. Muller, U. Weltis, D. Law, M. M. Roberts and T. W. Siebler, *SAE Technical Paper Series* **940607** (1994).
10. D. C. Karnopp, D. L. Margolis and R. C. Rosenberg, *System Dynamics : A Unified Approach*, John Wiley & Sons, Inc., (1990).
11. L. H. V. Vlack, *Elements of Materials Science and Engineering-Sixth Edition*, Addison-Wesley Publishing Company, (1989).
12. S. B. Choi, D. W. Park and S. Jayasuriya, *Automatica* **30** (1994) 899-904.
13. Z. M. Yeh, *Proceedings of IEEE Int. Conf. on Neural Networks* (1994) 2586-2591.
14. E. Takahashi, T. Miyoshi, T. Inoi, T. Maeda and H. Takada, *NEC Res. & Develop.* **99** (1990) 39-46.
15. G. J. Balas, J. C. Doyle and K. Glover, A. Packard and R. Smith, *μ-Analysis and Synthesis Toolbox*, The MathWorks, Inc., (1991).
16. A. Packard and J. C. Doyle, *Automatica* **29** (1993) 71-109.

PROPERTIES OF ELECTRORHEOLOGICAL FLUIDS CONTAINING SULFONATED POLY(STYRENE-CO-DIVINYLBENZENE) PARTICLES

YOSHINOBU ASAKO, SATORU ONO, RYUJI AIZAWA,
and TOSHIHIRO KAWAKAMI
Tsukuba Research Laboratory, Nippon Shokubai Co., Ltd.
1-25-12, Kannnondai, Tsukuba, Ibaraki, 305, Japan

ABSTRACT

Two series of sulfonated poly(styrene-*co*-divinylbenzene) particles (SSD), different in average particle diameter and in sulfonic acid substitution level were synthesized. The vertical sectional views of the SSD particles showed that the sulfonic acid groups were uniformly distributed all over the SSD particles. The electrorheological behavior of the suspensions of the SSD particles in silicone oil was investigated. Induced shear stress under an electric field increased with the increase of average particle diameter and sulfonic acid substitution level. The sedimentation velocity of the dispersed particles was monotonous with the average particle diameter and the sulfonic acid substitution level. Another series of sulfonated poly(styrene-*co*-divinylbenzene) particles (SSDH), different in sulfonic acid substitution level, was separately prepared. The vertical sectional views of the SSDH particles showed that the particles had sulfonated surface layer and unsulfonated core. The suspensions of the SSDH particles having the sulfonated layer of not less than 1.4 microns induced similar shear stress at the same electric field strength. The sedimentation velocity of the dispersed particles was lowered with diminishing the thickness of the sulfonated layer.

1. Introduction

It is well-known that suspensions of hydrophilic particles with small amounts of moisture induced relatively large shear stress under an electric field.[1,2,3,4] Sugimoto investigated electrorheological behavior of the suspensions containing ion-exchange resins.[1] This investigation was attractive in the field of electrorheology, but didn't follow up the relation between composition of the ion-exchange resin used and electrorhogical behavior of the corresponding suspension.

We synthesized two series of sulfonated poly(styrene-*co*-divinylbenzene) particles (SSD), different in sulfonic acid substitution level and in average particle diameter, and investigated electrorheological behavior of the suspensions of the SSD particles. The sulfonic acid groups were distributed all over the SSD particles. In this paper, effects of the particle diameter and the sulfonic acid substitution level on electrorheological properties will be discussed. Further, another series of sulfonated poly(styrene-*co*-divinylbenzene) particles (SSDH) was separately prepared. The SSDH particles had sulfonated surface layer and unsulfonated core. In this paper, the electrorheological properties of the suspensions of the SSDH particles will be also reported.

2. Experimental Section

2.1. Syntheses of SSD1-7 and SSDH1-5

Four kinds of spherical particles of styrene-*co*-divinylbenzene (SD1-4), different in average particle diameter (SD1 = 44 microns, SD2 = 24 microns, SD3 = 16 microns, SD4 = 4 microns) were obtained from a mixture of styrene and divinylbenzene according to the manner similar to that described in the literature.[5] The average particle diameters were measured on a particle size analyzer (Shimadzu SALD-1000).

100 g of SD1 was mixed with 500 g of tetrachloroethane and then the mixture was reacted with 1000 g of 98 wt% sulfuric acid at 80 °C for 5 hours. The reaction mixture was poured into large amounts of iced water, then the solid was separated by filtration and washed by water several times. The obtained particles were poured into 700 g of aqueous solution of 10 wt% sodium hydroxide and the mixture was stirred at 20 °C for 10 hours. After separation by filtration, the particles were washed with water several times. Drying the particles at 150 °C for 24 hours gave 173 g of SSD1.

SSD2-7 were synthesized in the manner similar to the synthesis of SSD1. SSDH1-5 were synthesized in the manner similar to the synthesis of SSD1 without using tetracholoroethane. The law material SD particles and synthetic conditions for the preparation of these sulfonated particles were summarized in Table 1.

Table 1. Reaction Conditions to Prepare Sulfonated Polymer Particles

Particle	SD 1)	Solvent 2)	Temp(°C)	Time(h)
SSD1	SD1	TCE	80	5
SSD2	SD2	TCE	80	5
SSD3	SD3	TCE	80	5
SSD4	SD4	TCE	80	5
SSD5	SD2	TCE	80	3
SSD6	SD2	TCE	80	2
SSD7	SD2	TCE	80	1
SSDH1	SD1	non	80	50
SSDH2	SD1	non	80	36
SSDH3	SD1	non	80	24
SSDH4	SD1	non	80	8
SSDH5	SD1	non	80	5

1) Each SD was obtained by suspension polymerization of a mixture of styrene and divinyl benzene.
2) TCE: Tetrachloroethane. Non: without using solvent.

The average particle diameter of the sulfonated particles was measured on the particle size analyzer. The sulfonic acid substitution level of the sulfonated particles was determined based on the data obtained by an elemental analysis. The sulfonic acid substitution level means the average number of sulfonic acid substituents which each aromatic ring has in the sulfonated particles. The specific gravity of the sulfonated particles was determined according to the floatation method.[6] The vertical sectional views of the sulfonated particles were collected on a scanning electron microscope (Shimadzu EMX-SM7). The sulfur elemental analyses on the sectional views were carried out by use of the same SEM.

2.2. Preparation of suspensions of SSD1-7 and SSDH1-5

After SSD2 was divided into four portions, each portion was allowed to stand with the atmosphere for 3, 20, 40 and 60 minutes. The water content of each portion was measured on a Karl-Fisher moisture meter (Kyoto Electric Manufacturing MKS-3P). 30 g of each portion was dispersed in 70 g of dimethyl silicone oil (Shin-Etsu Silicone Oil KF96-20CS).

The water content of SSD1, SSD3-7 and SSDH1-5 was controlled so that the current density of the resultant ER suspension was about 30 micro A/cm^2 under electric field strength of 4 kV/mm (Table 3). Suspensions of the sulfonated particles were obtained likewise.

2.3. Measurements of Electrorheological Properties of the Suspensions

In order to evaluate shear stress property and current property of the suspensions, the cup and bob of a rotary viscometer (Contraves Rheomat115) were modified to facilitate the supply of an electric field across 1 mm clearance between the cup and bob. An AC electric source (Tokyo Seiden ITS-5005-A-AP) was used to supply the electric field. Each of the suspensions was placed between the cup and bob. The values of shear stress and concurrent current density were measured under 60 Hz electric fields at different field strength. The shear rate investigated was 200/s.

Zero-field viscosity of the suspensions was measured on an E type viscometer (Tokyo Keiki Visconic ELD).

Each of the suspensions (10 ml) was placed in a test tube of 1 cm diameter. In order to evaluate dispersion stability of the suspensions, the time to form the supernatant of 5 mm was measured.

3. Results and Discussion

3.1. Characteristics of SSD1-7 and SSDH1-5

Characteristics of SSD1-7 and SSDH1-5 were summarized in Table 2.

Figure 1a shows a photograph of the representative vertical sectional view of an SSD1 particles. The SEM view indicated that the particle had a homogeneous internal structure. Fig. 1b shows a photograph of the sulfur atoms distribution in the same particle. The view indicated that sulfur atoms were uniformly distributed all over the particle. These results suggest that the sulfonic acid groups in the SSD1 particle were uniformly distributed all over the particle (Fig. 1c). The SSD2-7 particles were found to have the homogeneously sulfonated structure similar to the SSD1 particle from the same SEM experiments. Fig. 2a shows a photograph of the representative vertical sectional view of an SSDH3 particle. The SEM view indicated that the particle had a core-shell structure. Fig. 2b shows a photograph of the sulfur atoms distribution in the particle. The view indicates that sulfur atoms were distributed only in the shell layer. These results suggest that the SSDH particle had the sulfonated shell layer and the unsulfonated core (Fig. 2c). The other SSDH particles were found to have the core-shell structure similar to the SSDH3 particle from the same SEM experiments.

The difference of the internal structure between the SSD particles and the SSDH particles can be attributed to the mode of sulfonation to prepare the particles. The

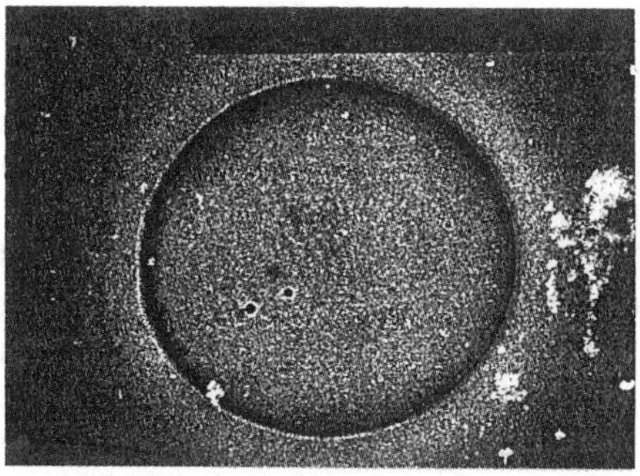

Fig.1a Photograph of the vertical sectional view of an SSD1 particle.

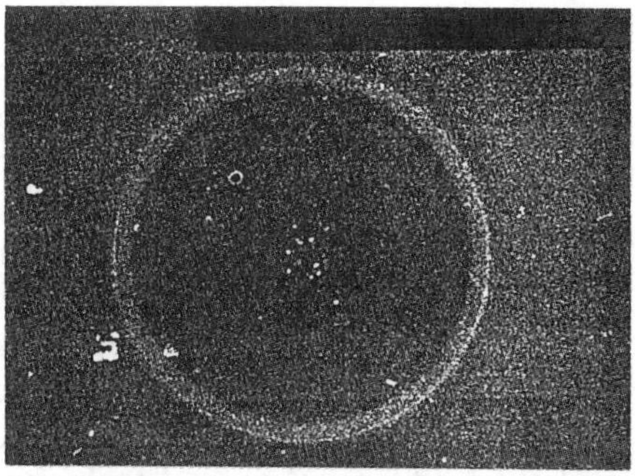

Fig.2a Photograph of the vertical sectional view of an SSDH3 particle.

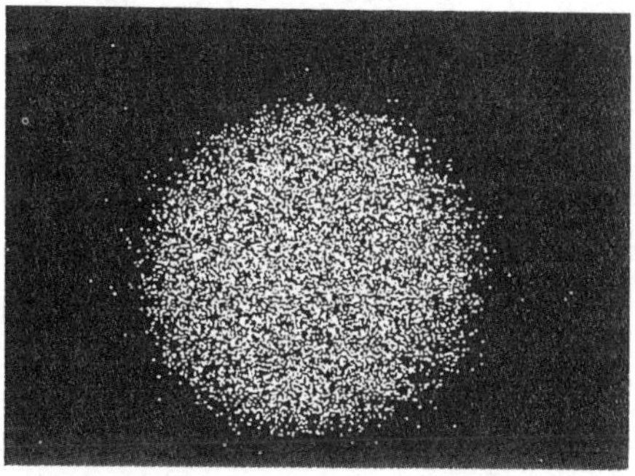

Fig.1b Photograph of the sulfur atoms distribution in the SSD1 particle.

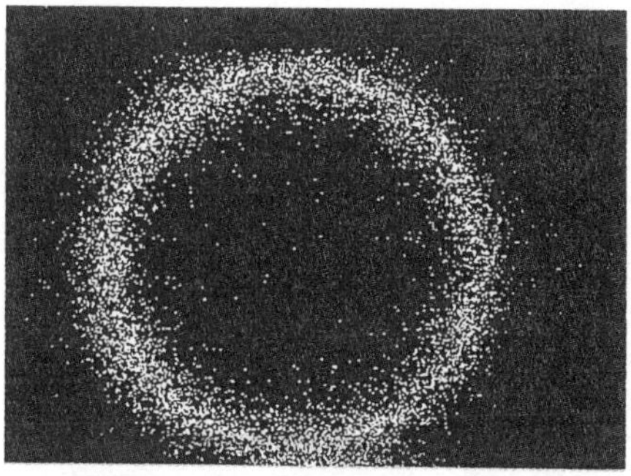

Fig.2b Photograph of the sulfur atoms distribution in the SSDH3 particle.

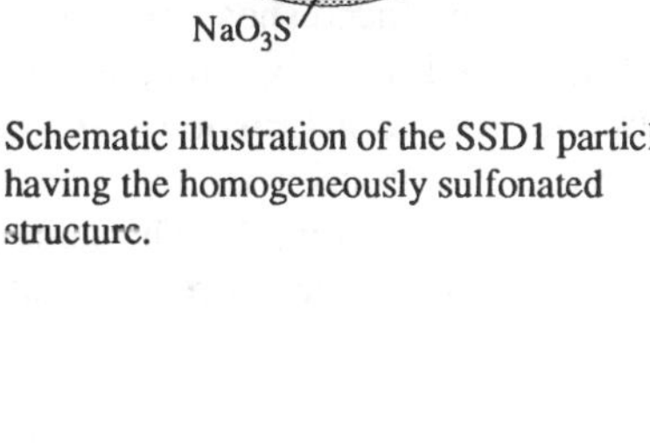

Fig.1c Schematic illustration of the SSD1 particle having the homogeneously sulfonated structure.

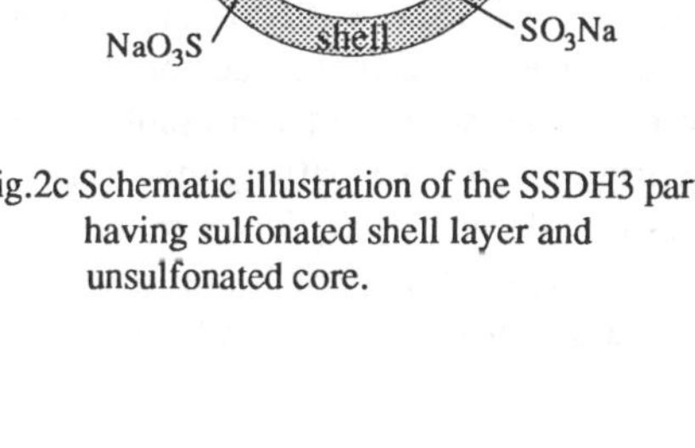

Fig.2c Schematic illustration of the SSDH3 particle having sulfonated shell layer and unsulfonated core.

sulfonation to prepare the SSD particles was carried out by reacting the SD particles with sulfuric acid in the presence of tetrachloroethane. The SD particles swelled in tetrachloroethane. Therefore, the sulfonation is considered to proceed by degrees all over the particles, resulting in the formation of the homogeneously sulfonated structure of the SSD particles. On the other hand, the sulfonation to prepare the SSDH particles carried out by reacting the SD particle with sulfuric acid without using tetrachloroethane. The SD particles could not swell in sulfuric acid. Therefore, the sulfonation is considered to proceed gradually from the particle surface to the center, resulting in the formation of the core-shell structure of the SSDH particles. The thickness of the sulfonated shell layer of SSDH1-5 was determined based on the SEM views and the sulfonic acid substitution level obtained from elemental analysis (Table 2).

Table 2. Characteristics of Sulfonated Polymer Particles

Particle	Average Particle Diameter 1) (μm)	S-level 2)	Thickness of Sulfonated Shell Layer 3) (μm)	Specific Gravity 4) (g/cm^3)	Water Content 5) (wt%)
SSD1	45	0.82	...	1.53	3.3
SSD2	25	0.91	...	1.54	3.1
SSD3	17	0.86	...	1.51	3.2
SSD4	4.5	0.82	...	1.49	3.2
SSD5	25	0.43	...	1.25	3.2
SSD6	25	0.31	...	1.16	3.2
SSD7	25	0.12	...	1.07	2.6
SSDH1	45	0.69	7.3	1.40	3.3
SSDH2	45	0.59	5.8	1.37	3.0
SSDH3	45	0.42	3.7	1.25	3.2
SSDH4	45	0.18	1.4	1.11	1.5
SSDH5	45	0.08	0.7	1.06	0.7

1) Measured on a particle size analyzer.
2) Sulfonic acid substitution level. Determined based on the data obtained by elemental analysis.
3) Determined based on SEM views and S-level.
4) Measured according to floatation method.
5) Measured on a Karl-Fisher moisture meter.

3.2. Electrorheological Properties of Suspensions of SSD1-7 and SSDH1-5

Suspensions of SSD1-7 and SSDH1-5 were prepared by dispersing each of the sulfonated particles in the silicone oil of the kinetic viscosity of 20 cSt. The viscosity of the suspensions was in the range from 48 to 79 mPas (Table 3).

Figure 3a shows plots of shear stress against applied electric field strength for suspensions of SSD2 with different water content. As applied field strength increased, the induced shear stress increased for all the suspensions. The induced shear stress increased with the increase of water content at any field strength applied. Fig. 3b shows plots of concurrent current density against applied field strength for the suspensions. As applied field strength increased, the current density increased for all the suspensions. The current density increased with the increase of water content at any field strength. The same effect of contained water on the induced shear stress and the concurrent current density was observed in all of the suspensions of SSD1, SSD3-7 and SSDH1-5. In this study, the water content in the other sulfonated particles was controlled so that the current density of the corresponding suspension was about 30 micro A/cm^2 under field strength of 4 kV/mm. The water content was listed in Table 2.

Figure 4 shows plots of shear stress against applied electric field strength for suspensions of the SSDs with different average particle diameter. As applied field strength increased, the induced shear stress increased for all the suspensions. The induced shear stress increased with the increase of average particle diameter at any field strength

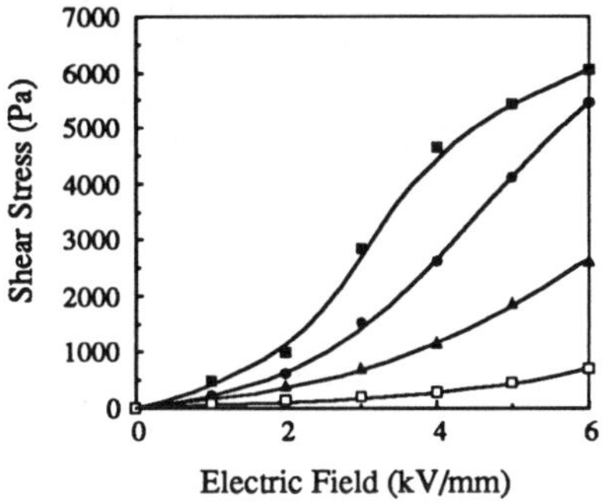

Fig.3a Plots of induced shear stress against applied electric field for suspensions of SSD2 with different water content. Water content; (■) 4.0 wt%, (●) 3.0 wt%, (▲) 2.0 wt%, (□) 1.0 wt%.

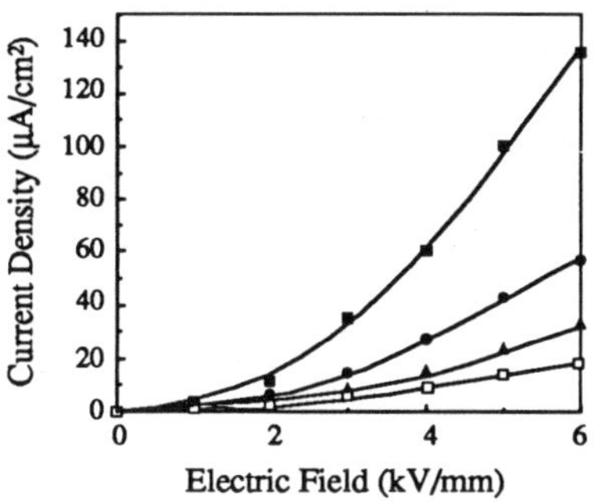

Fig.3b Plots of current density against applied electric field for suspensions of SSD2 with different water content. Symbols are the same as in Fig. 3a.

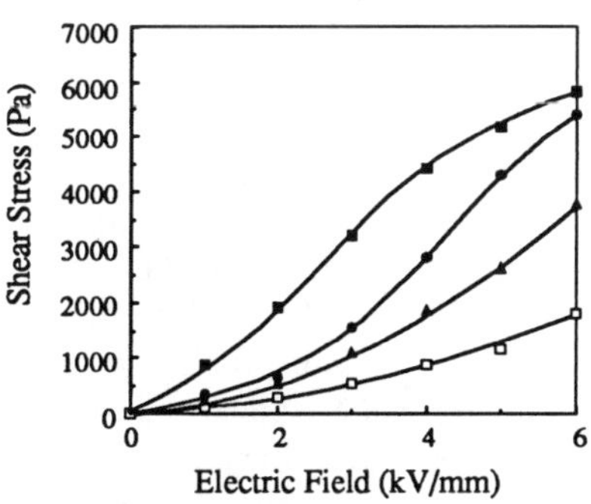

Fig.4 Plots of induced shear stress against applied electric field for suspensions of SSD1-4 with different average particle diameter. (■) SSD1 (average particle diameter: 45 μm), (●) SSD2 (25 μm), (▲) SSD3 (17 μm), (□) SSD4 (4.5 μm).

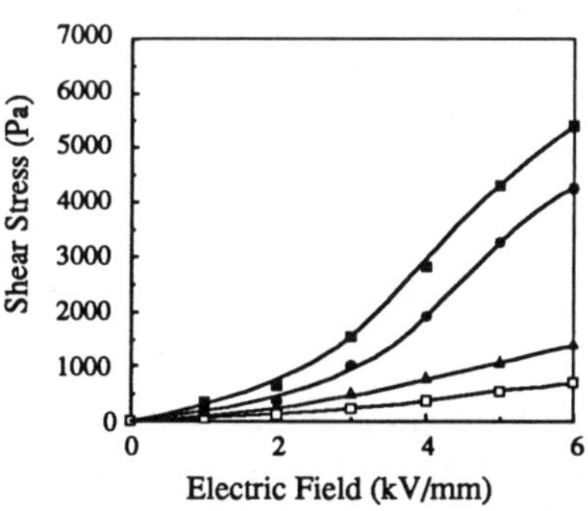

Fig.5 Plots of induced shear stressy against applied electric field for suspension of SSD2 and 5-7SSD with different sulfonic acid substitution level. (■) SSD2 (substitution level: 0.91), (●) SSD5 (0.43), (▲) SSD6 (0.31), (□) SSD7 (0.12).

applied. The induced shear stess of the suspension of SSD1 (average partiucle diameter of 45 microns; 4500 Pa) was about 5 times that of SSD4 (4.5 microns; 900 Pa) at 4 kV/mm. On the other hand, the sedimentation velocity of the dispersed particles became larger as the average particle diameter increased (Table 3).

Figure 5 shows plots of shear stress against applied electric field strength for suspensions of the SSDs with different sulfonic acid substitution level. As applied field strength increased, the induced shear stress increased for all the suspensions. The induced shear stress increased with increasing the substitution level at any field strength applied. The induced shear stress of the suspension of SSD2 (substitution level of 0.91; 2800 Pa) was about 7 times that of SSD7 (0.12; 400 Pa) at 4 kV/mm. On the other hand, the sedimentation velocity of the dispersed particles became larger as the sulfonic acid substitution level increased (Table 3). The increase in substitution level gave rise to the the increse in specific gravity of the particle (Table 2). The sedimentation velocity is considered to depend on the specific gravity of the particle.

Table 3. Zero-Field Viscosity of Suspensions and Sedimentation Velocity of Dispersed Particles

Particle	Zero-Field Viscosity 1) (mPas)	Sedimentation Velocity 2) (min)
SSD1	50	3
SSD2	48	9
SSD3	53	20
SSD4	56	120
SSD5	58	25
SSD6	63	35
SSD7	70	90
SSDH1	54	5
SSDH2	56	6
SSDH3	57	9
SSDH4	64	17
SSDH5	79	60

1) Measured on an E type viscometer.
2) Time to form the supernatant of 5 mm.

Figure 6 shows plots of shear stress against applied electric field strength for suspensions of SSD1 and the SSDHs, different in thickness of sulfonated shell layer. As applied field strength increased, the induced shear stress increased for all the suspensions. The induced shear stress of the suspensions of SSDH1-4 was almost the same as that of SSD2 at any field strength. In the case of the SSDH5 suspension, dielectric breakdown took place under the field strength over 4 kV/mm. The induced shear stress for the suspension of SSDH4 was about half that of SSDH1-4 at any field strength below 4 kV/mm. Thus, the suspensions of the SSDHs having the sulfonated layer of not less than 1.4 microns induced similar shear stress regardless of the thickness of the sulfonated layer at the same field strength. The sedimentation velocity of the dispersed particles was lowered with diminishing the thickness of the sulfonated shell layer. The sedimentation velocity is considered to depend on the specific gravity of the particle, similarly to that of the SSDs with different sulfonic acid substitution level.

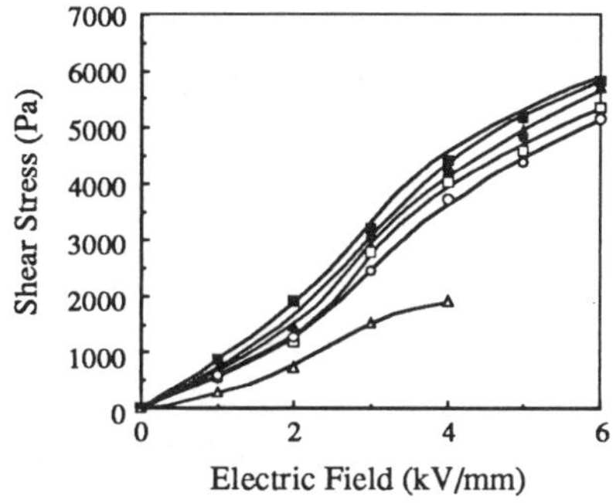

Fig.6 Plots of induced shear stress against applied electric field for suspensions of SSD1 and SSDH1-5 with different thicknesses of sulfonated shell layer. (■) SSD1, (●) SSDH1 (thicknesses of sulfonic shell layer: 7.3 μm), (▲) SSDH2 (5.8 μm), (□) SSDH3 (3.7 μm), (○) SSDH4 (1.4 μm), (△) SSDH5 (0.7 μm).

4. Conclusions

The sulfonation of styrene-*co*-divinylbenzene particles (SD) with sulfuric acid in the presence of tetrachloroethane gave the homogeneously sulfonated polymer particles (SSD). Using the suspensions of SSDs, effects of the particle diameter and the sulfonic acid substitution level on induced shear stress and dispersion stability of the dispersed particles were investigated. The induced shear stress and the dispersion stability were largely affected by the particle diameter and the substitution level. The induced shear stress increased with increasing the average particle diameter and the sulfonic acid substitution level at the same electric field strength, whereas the dispersion stability decreased. The sulfonation of the SDs gave the composite particles having the sulfonated shell layer and the unsulfonated core (SSDH). Upon the sulfonated shell layer of not less than 1.4 microns, the resultant suspensions induced similar shear stress at the same field strength. The dispersion stability improved with diminishing the thickness of the sulfonated shell layer.

5. Reference

1. N. Sugimoto, *Trans. Jpn. Mech. Eng. (Part2)*, **43** (1977) 1075.
2. J. E. Stangroom, **UK Patent 1570284** (1980).
3. J. Goosens, G. Oppermann, W. Padzun and V. Härtel, **US Patent 532689** (1994).
4. Y. Asako, S. Ono, R. Aizawa and T. Kawakami, *Polymer Preprints* **35(2)** (1994) 352.
5. J. Hitner and W.F. Batoe, **US Patent 2264375** (1941).
6. T. Dann, *in Encyclopedia of Experimental Chemistry*, Vol. 17, ed. Chem. Soc. Jpn. (Maruzen, Tokyo, 1978) p. 551.

Composite Structure Analysis of a Hollow Cantilever Beam Filled with Electrorheological Fluid

by

C.D.Berg
Control Systems Centre,
E. & E. E.,
UMIST, PO Box 88,
Sackville Street,
Manchester M60 1QD.
United Kingdom

L.F.Evans and P.R.Kermode
ER Fluid Developments Limited,
Vincent Works,
Brough,
Bradwell,
Sheffield S30 2HG,
United Kingdom

Abstract

Unwanted vibrations within structures can become an ever increasing problem as structural components are made lighter with the use of advanced materials and tighter design tolerances. Increased operating bandwidth further compounds the problem resulting in the build up of unwanted vibrations within optimally designed structures. The alteration of the spectral signature of a structural component will enable the aforementioned design criteria to be satisfied. The use of a hollow structural component filled with Electro-Rheological fluid will enable the spectral signature of the component to be altered electrically.

Experimental data obtained for a hollow cantilever beam filled with Electro-rheological fluid is presented in this paper. Analysis of the data is carried out to determine heuristic values for the beams effective stiffness and damping coefficient. The effects of applying an electric field to the ER Fluid are outlined from the results obtained. Comparison of the results is then made with those predicted by a model of an oscillating composite structure cantilever beam. Differences between the predicted and the heuristic data is then examined with regard to a hypothesis based on the behaviour of the induced ER Fluid's microscopic structure.

1. Introduction

The use of Electrorheological fluids for the construction of smart structural components has been suggested and tried previously by many researchers. This paper attempts to categorise an ERF filled cantilever beam for control purposes and develops a hypothesis based on the observed behaviour of such a structure.

The use of beams as structural components is not new, but the advent of computer aided design has meant that designers can reduce traditional 'over-design' necessary to produce a reliable structure. This is often achieved by reducing design safety factors to a minimum. The efficient design of a system sometimes requires minimal mass for a structure which is subjected to vibration. This is not always achievable due to vibrations building up in the structural components resulting in instability and induced stresses which can cause failure. Thus a structure can be unnecessarily heavy to give the required stiffness and damping to eliminate vibration. The use of so called 'smart' structural components which are able to alter their stiffness or damping coefficients will enable lighter designs to be realised. The use of a 'smart' component within a structure which alters its spectral signature to an optimum for system inputs can increase the system bandwidth. This can be an important factor for system cost, energy efficiency, physical size or mass.

Unwanted vibration within a system can be eliminated by the use of a vibration absorber which is tuned by the design engineer. The vibrational absorption ability of such a device would be a compromise between optimal system performance and all round performance. However the use of a variable damper allows a higher bandwidth of vibration to be absorbed. Numerous vibration absorber designs could be developed, but one of particular interest here is that of a tuned cantilever beam. The tuning of the cantilever beam can be achieved by making the beam up as a sandwich of two dissimilar materials. The outer layers are two thin steel sheets either side of a vibration absorbing material the properties of which are chosen to suit the application and the frequency of the vibrational source. The use of a cantilever vibration absorber with the ability to electrically alter the material properties of the middle layer would significantly enhance its effectiveness.

The ability of fluid to change its rheology under the influence of an electrical field was first discovered by Willis Winslow[1] in 1949. Since then Winslow's work has generated much interest and what was originally known as Winslow fluid is now referred to as electrorheological fluid (ERF). Whilst the early ERFs were sometimes unstable and user unfriendly, modern day ERFs are much improved. ERFs are now practical fluids and the performance of a calibrated ERF is predictable. An ERF consists essentially of a polarizable hydrophillic particulate dispersed within a high dielectric constant carrier fluid. The change in the rheology of an ERF under the influence of an electrical field is considered here to be brought about by the polarisation of the particulate due to the applied electric field. The polarisation causes water molecules within a particle to collect around one end of the particle which aligns itself with another particle. Particles are electrostaticaly attracted to each other and the water molecules form a bridge between them. The forces required to separate the particles comes from the interfacial tension between the water and the oil and the electrostatic attraction forces. If the applied electrical potential is of sufficient strength then the particles align to form chains from one charge plate to the other. The particulate structure which is induced has a yield strength and thus the fluid appears to have an electrically induced resistance to shear and yield strength. The behaviour of an ERF is dependent upon the relative movement of one charge plate to the other and/or the pressure drop normal to the applied field. The modelling of ERF has been considered by many researchers, ERF is considered here to behave as a Bingham plastic with a yield strength, τ_{yield}, which is a function of the applied electrical potential, E_{erf}.

Previous work has been undertaken on the properties of ERF filled cantilever beams and their use as smart structural components[2-5]. The investigations here show that there is a change in the beam's properties dependent upon the applied electric potential, E_{erf}. The beam's structural properties are characterised using a simple second order model to enable the control of vibrations induced in the beam. It is shown that the beam does not behave as a composite structure formed by the use of homogenous layers and a hypothesis is presented to describe its behaviour. The effects of the induced field on the first three harmonic frequencies are shown. The level of the vibrational source, F_m, or more precisely the amplitude of the deflection of the beam and its effect on the natural frequency, ω_n, as well as its spectral signature are shown.

2. Experimental Setup

The cantilever beam, see figure_1, was constructed from two stainless steel strips 225 mm long, 25.25 mm wide and 0.4 mm thick. The two steel strips are separated by a 2.45 mm wide bead of silicone rubber 1.55 mm thick. The bead of silicone rubber has a gap in it along the top edge to allow the void in the centre of the beam to be filled with ERF.

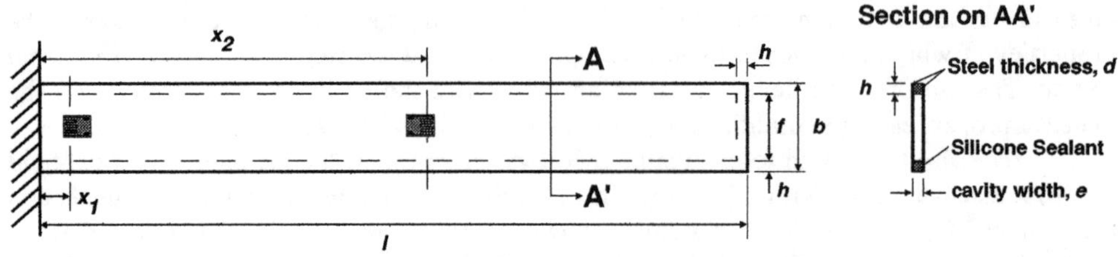

Figure_1, Schematic of ERF Filled Cantilever Beam

Construction of the beam was achieved by placing a strip of PTFE 1.55 mm thick and 20.5 mm wide centrally and 2 mm from one end between the two steel strips. Clamps were then placed over the steel strips and silicone rubber applied around the edges of the beam and allowed to cure. The PTFE strip was then removed and a 20 mm wide and 1.55 mm thick polyurethane spacer placed in the open end of the beam. Silicone rubber was applied around the polyurethane insert leaving a small gap to allow the beam to be filled and left to cure. Once the silicone rubber had cured two strain gauges, G_1 and G_2, were fitted to one side of the beam 192 mm and 90 mm from the free end respectively. The use of two gauges in these positions being sufficient to examine the vibration of the beam for the first three harmonic frequencies. The strain gauges were wired to a switch such that each could be connected separately to the instrumentation. The beam was mounted with the wide face vertical, see figure_2, and clamped on the end fitted with the polyurethane. Electrical connections were made to enable an electrical potential to be applied across the ERF, the ground potential was connected to the side of the beam fitted with the strain gauges.

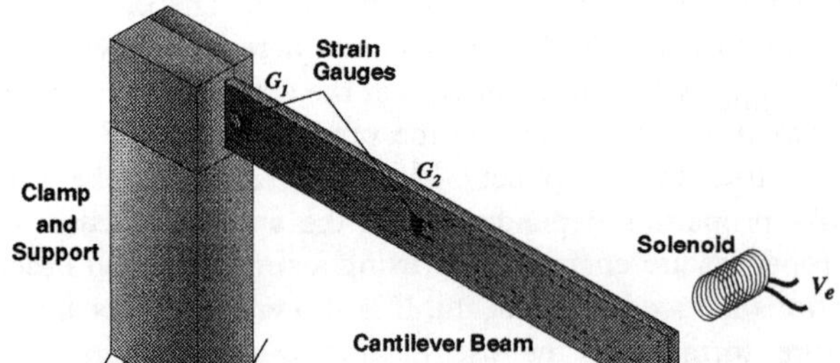

Figure_2, ERF Filled Cantilever Beam Test Setup

Excitation of the beam was achieved by attaching a magnet of mass 3.3 gramme 10 mm from the free end of the beam and mounting a solenoid horizontally 9.5 mm away from the face of the magnet. A voltage, V_e , at the solenoid generates a magnetic force, F_m , which attracts or repels the free end of the beam.

The instrumentation is shown schematically in figure_3. The strain gauges were connected via a 2-pole, 2-way switch to an RS strain gauge amplifier the output of which was connected to a secondary amplifier to increase the gain. The output of the secondary amplifier was connected to a frequency response analyser (hereafter referred to as an FRA). The solenoid was driven from an amplifier designed to generate enough power to deflect the free end of the empty beam sufficiently when fed with a drive signal output from the FRA. A storage oscilloscope was used to monitor the FRA input and output signals and to record the transient response of the beam.

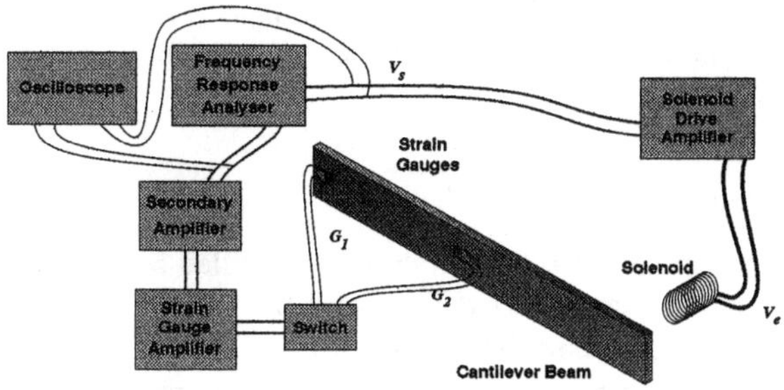

Figure_3, ERF Filled Cantilever Beam Instrumentation

The beam was filled with a 40% volume fraction Reslinol ERF as supplied by ER Fluid Developments. The characteristics of the fluid are that at 800V, 1KV and 1.2KV the yield strength, τ_{yield}, is 3.8 KPa, 4.38 KPa and 4.96 KPa at current densities of 164 μAcm^{-3}, 295 μAcm^{-3} and 463 μAcm^{-3} respectively.

Data Acquisition

A Solartron 1070 FRA was used to obtain the frequency response of the beam. The strain gauge outputs, G_1 and G_2, are compared with the solenoid amplifier input voltage, V_s, generated by the FRA. Response data for 0 to 200 Hz in 10 Hz steps was taken and detail obtained around the resonant peaks. The strain per unit load at the gauge positions was obtained for all test conditions by applying static loads to the free end of the beam and measuring G_1 and G_2. A Gould 1425 storage oscilloscope was used to measure the transient response characteristics of the beam. The beam was deflected a known amount to a stop, released and the strain gauge output, G_1 recorded. The damped natural frequency, ω_d, and the amplitude of the first few oscillations were measured directly from the oscilloscope screen. The damping factor, ζ, was then calculated from the logarithmic decrement[6].

3. Analysis of Data obtained for an ERF Filled Cantilever Beam

Firstly the frequency response of the empty and the filled beam (E_{erf}= 0V) was obtained and the strain gauge outputs G_1 and G_2 examined with respect to the modal shapes of the beam. The effect of the mass of the ERF, m_{erf}, on the damped natural frequency is calculated to verify the validity of the frequency response. The harmonic frequencies are shown to be less than those for a solid beam. A simple second order model is used to categorise the behaviour of the beam by examining the response and the effects of E_{erf} on the damping factor, ζ, and the damped natural frequency, ω_d . This is then expanded to examine the output G_2 which shows that a simple second order is not valid for an ERF filled beam. A common control engineering technique for the examination of system behaviour is to use the root-locus method[7]. This is used to determine the nature of the system model. A hypothesis is then presented which describes the behaviour of the E_{erf} induced microstructure when the beam is subjected to vibration. Data obtained for different driving force amplitudes for a constant charge voltage is shown to support the hypothesis. Strain per unit load data also appears to support the hypothesis as well as strain outputs G_1 and G_2 at the resonant frequencies.

Empty Beam Frequency Response

The frequency responses for the empty and the filled beam (E_{erf}= 0 V) are shown in figure_4 as gain and frequency plots. The solid lines show the strain gauge response G_1 and the dashed line G_2. The data shows that the addition of the ERF causes a reduction

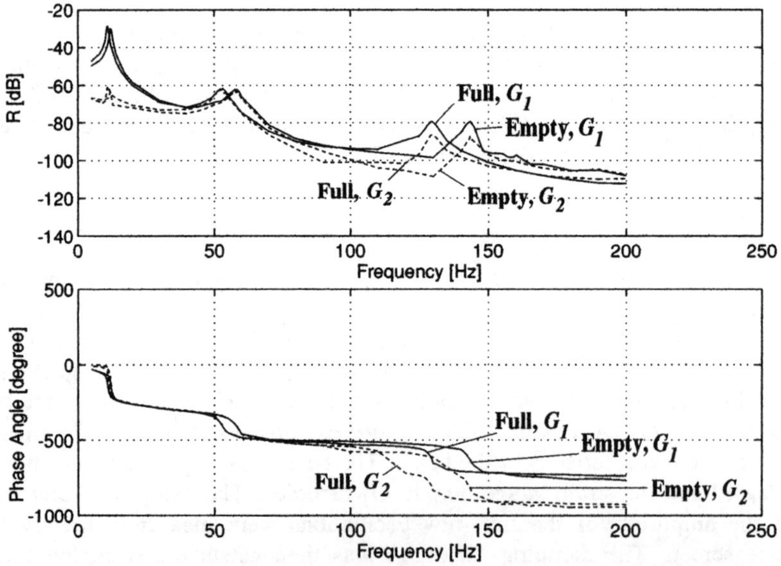

Figure_4, Empty and Zero Field ERF Filled Beam Spectral Signature

in the damped natural frequency, ω_{d1} , and also the first two harmonic frequencies, ω_{d2} and ω_{d3} . Figure_5 shows the modal shapes for the first three resonant frequencies of a solid cantilever beam and the position of the gauges as mounted on the ERF filled beam. With reference to figure_5 and figure_4 it can be seen at the damped natural frequency that the two gauges outputs are in phase but the gain is less for G_2. At the second harmonic frequency, ω_{d2} , G_1 and G_2 are in phase as expected and also have the same gain and damping. The third harmonic frequency, ω_{d3} , shows different gains for G_1 and G_2 as expected from the modal shape and a phase difference of 180°.

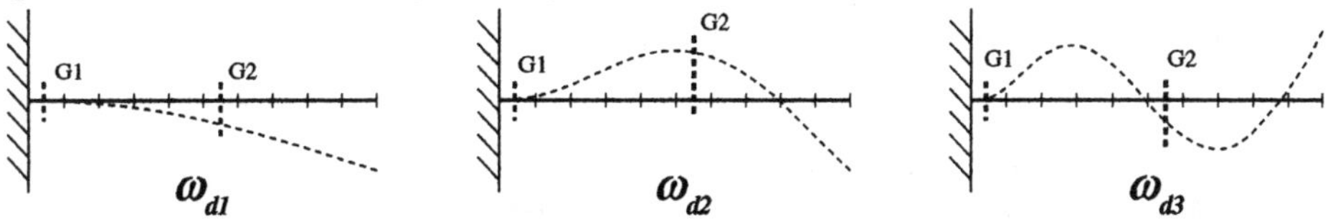

Figure_5, Modal Shapes for a Cantilever Beam Sinusoidally Forced at the Free End

The relation between the harmonic frequencies is governed by the solution to the following general equation for the transverse vibration of a uniform beam

$$\frac{\delta^4 y}{\delta x^4} + \frac{\rho A}{EI}\frac{\delta^2 y}{\delta t^2} = 0 \tag{1}$$

the deflection of the beam is given by,

$$y = X(B_1 \sin\omega t + B_2 \cos\omega t) \tag{2}$$

where X *is a* $f(x)$ which defines the shape of the beam

thus
$$\frac{\delta^4 X}{\delta x^4} = \frac{\rho A}{EI}\omega X = \lambda^4 X \tag{3}$$

The general solution of Eq. 3 is $X = C_1\cos\lambda x + C_2\sin\lambda x + C_3\cosh\lambda x + C_4\sinh\lambda x$ (4)

The solution of Eq. 4 is dependent upon the end conditions, which for the ERF filled beam are clamped-free, thus for a solid beam the frequency equation is,

$$\cos\lambda l + \cosh\lambda l = -1 \tag{5}$$

The natural frequencies are given by,

$$\omega = \frac{\alpha}{l^2}\sqrt{\frac{EI}{\rho A}}\,\text{rad s}^{-1} \tag{6}$$

where $\alpha = 3.52$ for ω_{d1} , 22.4 for ω_{d2} , 61.7 for ω_{d3}

The first three resonant frequencies, ω_{d1}, ω_{d2} and ω_{d3} for the empty and the filled beam as given by the frequency response data are 12 Hz, 58 Hz, 143 Hz and 10.8 Hz, 53.5 Hz, 129 Hz . This is not in the ratio that the solution for a solid cantilever beam would give. This is due to the composite nature of the structure, solution of Eq. 3 for the correct values of α was not undertaken as the data required is heuristic and the effect on the values of α for the different modes of beam excitation and E_{erf} is complex.

However to validate the frequency response data the effects of adding the mass of ERF, m_{erf} to the beam are calculated using a simple second order model as shown schematically by figure_6. The model consists of a mass, m, a spring of stiffness, k , and a damper with a damping coefficient, b.

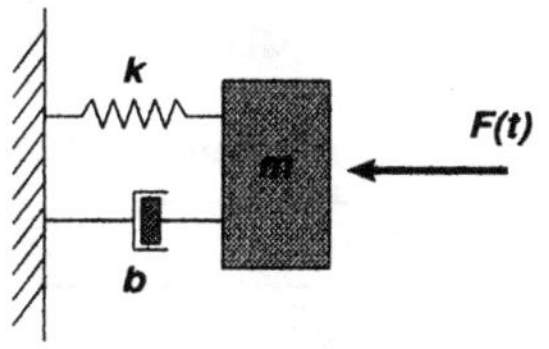

Figure_6, Simple Mass-Spring-Damper Second Order Model

The undamped natural frequency, ω_n , and the critical damping, C_c , are given by

$$\omega_n = \sqrt{\frac{k}{m}} \quad (7) \qquad C_c = 2\sqrt{mk} \quad (8)$$

The damping factor, ζ, and the damped natural frequency, ω_d, from,

$$\zeta = \frac{b}{C_c} \quad (9) \qquad \omega_d = \omega_n\sqrt{1-\zeta^2} \quad (10)$$

From the density and dimensional data the mass of the empty beam, m_e, is calculated to be 33.474 g and m_{erf} to be 7.86 g . The FRA data gives the empty beam's natural frequency to be 12 Hz and the damping factor as 0.03. The damping factor is calculated from the data using the half power points[8]. The stiffness of the empty beam, k_e, is calculated as 190.64 Kgs^{-2}.The addition of the ERF will cause little change to the damping and stiffness as can be seen from the FRA data. Therefore using the stated values of stiffness and damping the damped natural frequency of the filled beam is calculated to be 10.81 Hz and from the FRA data it is found to be 10.8 Hz.

Categorization of System Constants

The FRA was used to obtain ω_n at E_{erf} = 800 V, 1 KV and 1.2 KV from the gauge output G_1 shown as figure_7. Table_1 shows the natural frequency and damping factor determined from the FRA data for each charge potential.

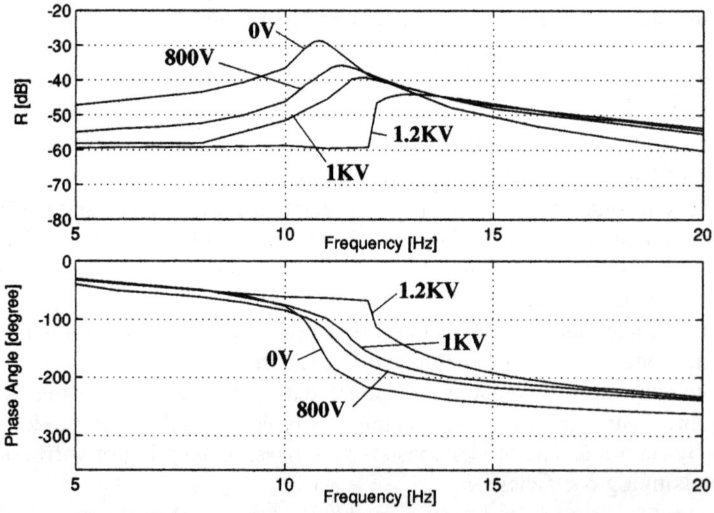

Figure_7, Natural Frequency Data for G_1 at Differing Values of E_{erf}

Table_1, Natural Frequency Data for Differing Values of E_{erf}

E_{erf}	0V	800V	1.0KV	1.2KV
ω_n	10.8	11.35	11.85	13.1
ζ	0.055	0.056	0.076	0.12

The transient response of the ERF filled beam to a step input was recorded and values for ω_n and ζ determined as shown in table_2.

Table_2, Transient Response data for Differing Values of E_{erf}

E_{erf}	0V	800V	1.0KV	1.2KV
ω_n	10.68	11.02	11.11	overdamped t=52ms
ζ	0.04	0.09	0.247	

The transient response data gives significantly different results to those obtained from the FRA. This was investigated further by examining the output G_2, figure_8 shows the FRA data obtained. The data shows different response characteristics for the beam at G_2 which are not those of a simple second order model but those of a higher order model.

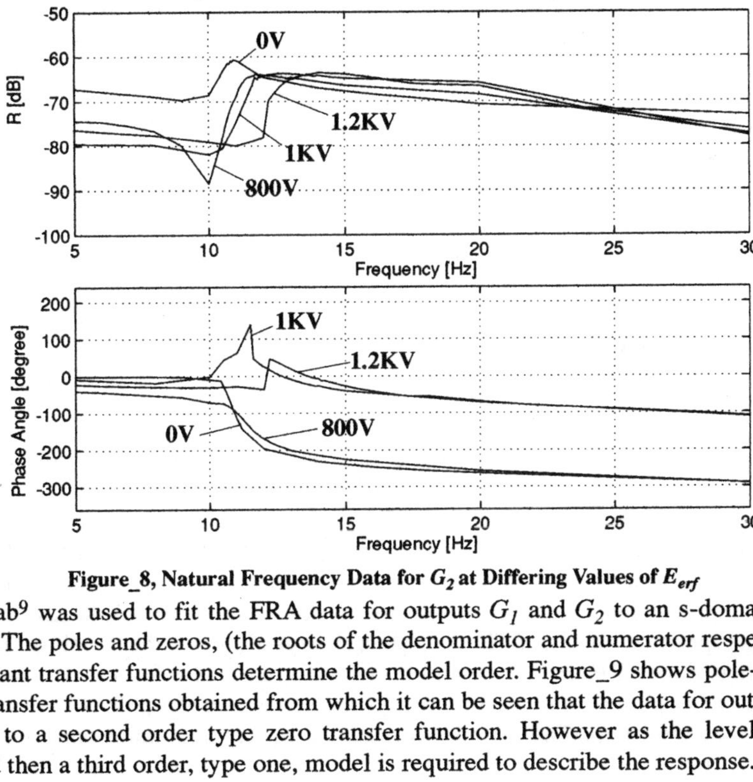

Figure_8, Natural Frequency Data for G_2 at Differing Values of E_{erf}

Matlab[9] was used to fit the FRA data for outputs G_1 and G_2 to an s-domain transfer function. The poles and zeros, (the roots of the denominator and numerator respectively) of the resultant transfer functions determine the model order. Figure_9 shows pole-zero maps for the transfer functions obtained from which it can be seen that the data for output G_1 can be fitted to a second order type zero transfer function. However as the level of E_{erf} is increased then a third order, type one, model is required to describe the response.

ERF filled Beam Induced Microstructure Hypothesis

The anomalies in expected behaviour seem to be related to the level of bending of the beam. Specifically the amplitude of the step for the transient data was larger than the maximum amplitude of vibration of the beam for the acquisition of the FRA data. The difference in the response at G_1 and G_2 also suggests that the behaviour of the beam differs along the length of the beam.

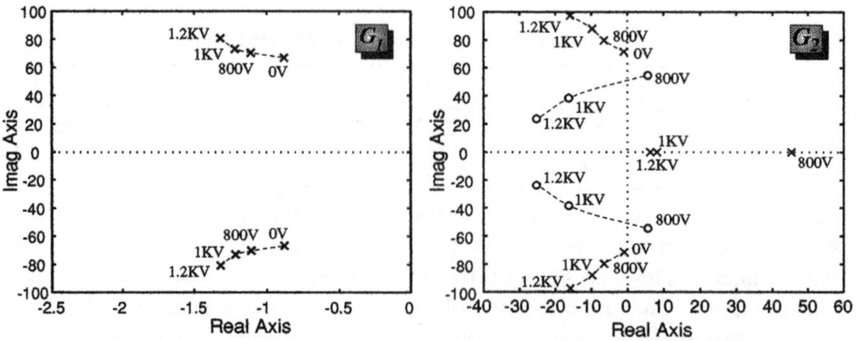

Figure_9, Pole-Zero Maps for G_1 and G_2 for Differing Values of E_{erf}

It is hypothesised that the induced ERF microstructure is destroyed at the free end of the beam if the amplitude of deflection, or more precisely the relative displacement of the outer plates of the beam, is great enough. This is shown schematically in figure_10.

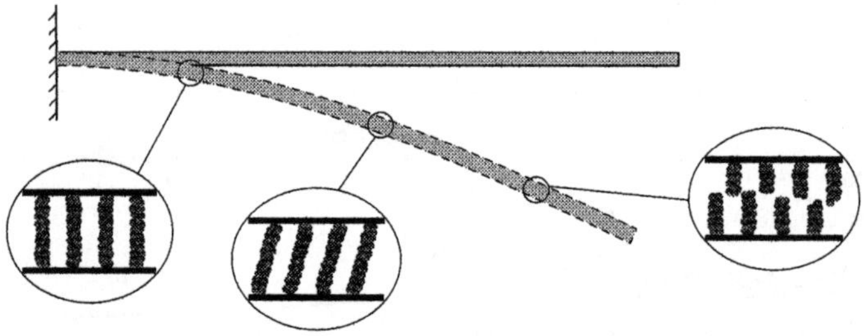

Figure_10, Schematic Diagram of ERF Filled Beam Structure Hypothesis

The structure within the beam will therefore either be deformed within the elastic region (pre-yield) or will be constantly destroyed and reformed as the beam oscillates. The point at which this occurs will be referred to as the structural transition point (STP). The area of the beam from the STP to the free end will behave as a non-linear element as the beam oscillates. The length of beam from the support to the STP will behave in a linear manner. The position of the STP will change according to the level of E_{erf}, the characteristics of the ERF and the amplitude of oscillation. It also follows that for transient response as the amplitude of oscillation decays the position of the STP will move towards the free end of the beam. The hypothesis needs to be verified and a complex structure model developed based on the relative shear between the two outer plates and the modal shape of the beam. Such a model is beyond the scope of this paper but is the subject of current research. To this end data was obtained for differing values of V_s for E_{erf} = 800 V. The response data is given as figure_11 and shows that as the amplitude of oscillation is increased then the response of the beam alters. This data was also analysed to determine the system transfer function and the resultant pole-zero maps for G_1 and G_2 are given as figure_12. The pole-zero maps show that for G_2 as the level of V_s is increased the order of the system model changes and a simple second order model is no longer valid.

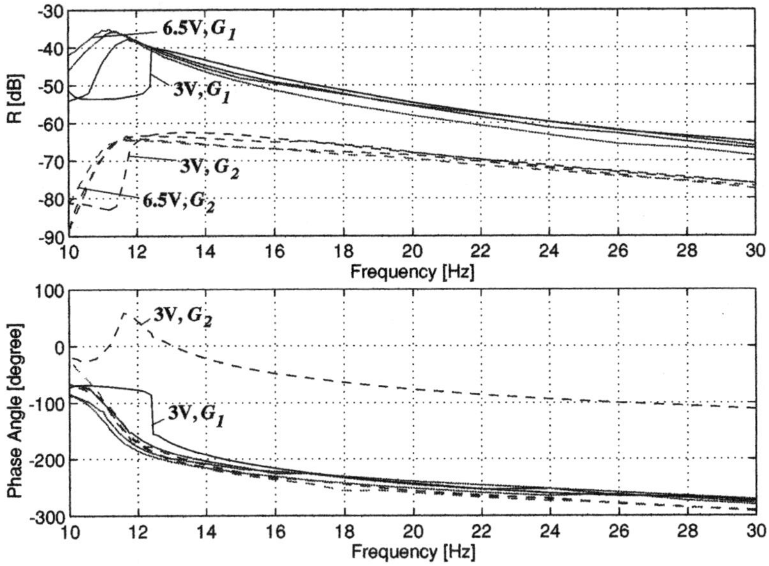

Figure_11, Frequency Response Data for Differing V_s Constant E_{erf}

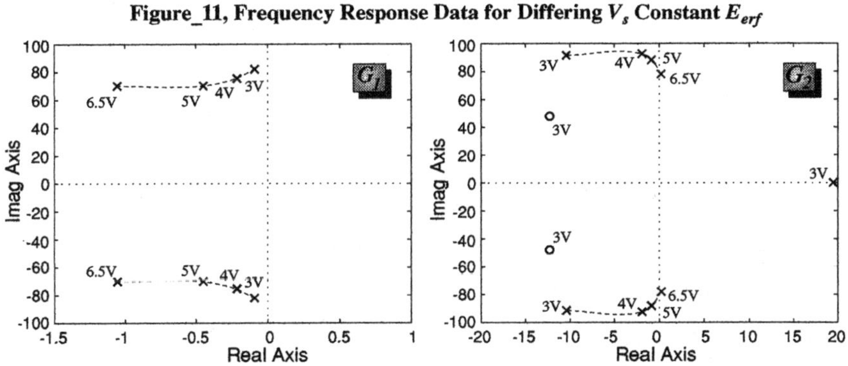

Figure_12, Pole-Zero maps for G_1 and G_2 for differing values of Vs at constant E_{erf}

Data was also obtained for the strain per unit load and tabulated in table-3. This data shows an anomaly at 800V and 1KV as the hypothesis suggests that as the level of E_{erf} is increased then the strain per unit load reduces. This anomaly is shown for both gauges and requires further investigation.

Table_3, Strain/Unit Load for Differing Values of Eerf

E_{erf}	0V	800V	1KV	1.2KV
G_1	1.7875	1.7171	1.828	1.6856
G_2	0.4769	0.392	0.4526	0.3908

Figure_13 shows the full spectral response data for the ERF filled beam from which it can be seen that the second harmonic frequency, ω_{d2} , is less at G_2 than at G_1 for values of E_{erf} = 800 V and 1 KV. This would seem to suggest that the hypothesised STP is nearer to the support than G_2 at these potentials and therefore the beam's apparent stiffness is less at G_2 than at G_1. The data obtained at higher frequencies is given although the gain is very low and the validity of the data uncertain. However the third harmonic frequency, ω_{d3} , appears to increase with E_{erf}. This is as expected because with the application of a higher field and a lower level of excitation the beam will behave more as a solid and less like a composite structure. This is borne out if the values of the first three resonant frequencies, ω_{d1} , ω_{d2} and ω_{d3} are compared with those predicted using the values of α for a solid beam as given earlier.

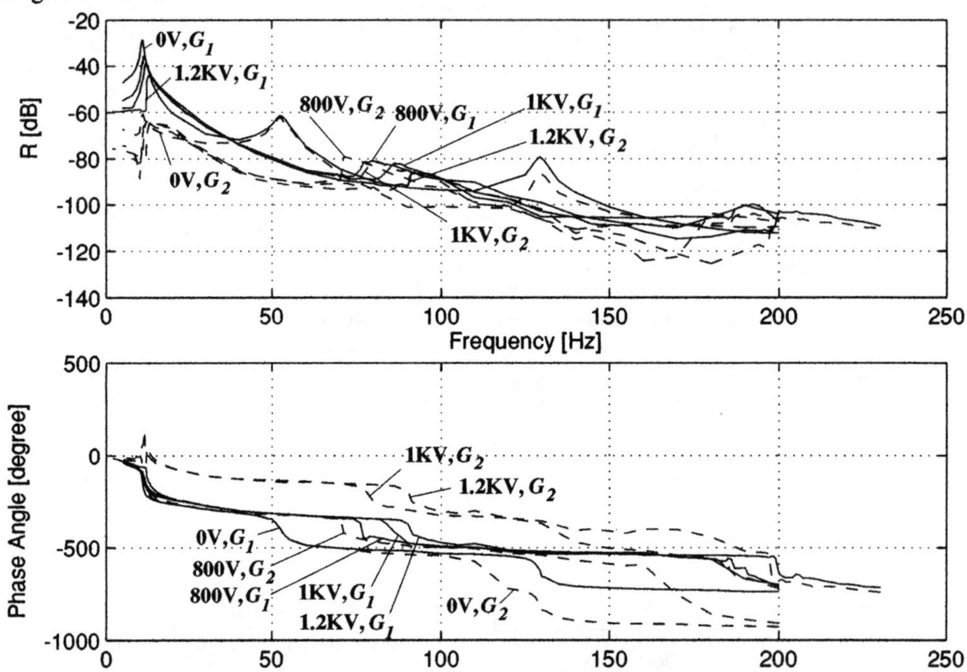

Figure_13, Spectral Signature for ERF filled beam at different values E_{erf}

6. Conclusion

In conclusion tests were carried out to categorise the behaviour of an ERF filled cantilever beam as a variable stiffness and damping control element. It appears that the behaviour of an ERF filled cantilever beam is highly complex and cannot be modelled as a homogeneous structure. A composite structure analysis is not suitable as data indicates that the induced ERF microstructure is not homogenous but dependent upon the relative shear between the two outer layers of the beam. A hypothesis is presented that there is a change in the induced ERF structure along the length of the beam. It is proposed that if the relative shear between the two outer layers of the beam is great enough then the structure will be continuously destructed from that point to the free end of the beam. The development of a model which takes account of the relative shear is required to enable effective control of such a structural component. The data does show however that significant shifts in damping and the resonant frequencies are possible using an ERF filled cantilever beam. The alteration of the frequency response is dependent upon the level of the applied ERF potential, E_{erf}, and the amplitude of the driving force or more precisely the amplitude of vibration. The data suggests that as E_{erf} is increased, or for the same value of E_{erf} but at lower amplitudes of vibration, the apparent damping and stiffness of the beam are increased.

7. Further Work

To develop a complex model based on the induced ERF microstructure behaviour hypothesis and match it to the actual beam behaviour. Investigate the effects of excitation amplitude on the perceived stiffness at different points along the length of the beam. Obtain FRA data for higher values of E_{erf} and V_e. Investigate the effects of varying the amplitude of the step input on the transient response characteristics.

References

1. W.M.Winslow, *J.Appl.Phys.*,**20**,(1949) 1137
2. M.V.Gandhi, B.S.Thompson, S.B.Choi, *J. Composite Materials,* **23**, (1989) 1232
3. S.B.Choi, Y.K.Park, *J. Sound and Vibration,* **172**(3), (1994) 428
4. S.B.Choi, Y.K.Park, *AIAA Journal,* **32**(3), (1994) 438
5. W.I.Kordonsky, A.D.Matsepuro, S.M.Demchuk, Z.A.Novikova, *Electrorheological Fluids proc. of 4th conf Austria,* (World Scientific, 1994) 440
6. C.F.Beard, *Vibration Analysis and Control System Dynamics*, (Ellis Horwood, Chichester, 1981) 31
7. J.Schwarzenbach, K.F.Gill, *System Modelling and Control,* (Edward Arnold, London, 1979) 170
8. C.F.Beard, *Vibration Analysis and Control System Dynamics*, (Ellis Horwood, Chichester, 1981) 45
9. Matlab, *Numeric Computation Software,*The Mathworks Inc., MA USA.

MR FLUIDS WITH NANO-SIZED MAGNETIC PARTICLES

Cl. Kormann, H. M. Laun, H. J. Richter
BASF Aktiengesellschaft, D-67056 Ludwigshafen, Germany

ABSTRACT

Recently magnetorheological fluids with nanosized magnetic ferrite particles have become available. Their composition, rheological and magnetic properties are described. A comparison with conventional MR fluids based on micron-sized particles is given. The yield stress of nano-MR fluids can be increased by a moderate magnetic field (0,2 T) by 4000 Pa. It can be modulated by the magnetic field with a response time of less than 5 ms. Details are given on the long term thermal stability at 150 °C, on flow properties at elevated temperatures and at high shear rates. Design principles for MR fluid actuator design are outlined.

1. Motivation

Magnetorheological (MR) fluids are colloidal suspensions that reversibly stiffen under the influence of a magnetic field. Most of the known MR fluids consist of magnetic carbonyl iron particles, that are coated with a surfactant for stabilization in a carrier solvent[1-3]. In order to overcome some problems associated with the relatively large size of the carbonyl iron particles (diameter ≈ 1-3 μm), namely the tendency to sediment and the abrasivity of the particles, another approach was chosen.

Nano-sized magnetic particles (diameter ≈ 0.01 μm) have been known for a long time as the key component in ferrofluids[4]. And ferrofluids have been known in turn for their high stability and reliability in rotary seals, where they function as (non abrasive) lubricants. While true ferrofluids show a negligible increase of the yield stress in strong magnetic fields[5] for the present purpose they were modified in order to obtain strong magnetorheological effects: particle size was somewhat increased, and the dispersing layer (with a reduced thickness) was adapted to the carrier solvent (cf. Figure 1).

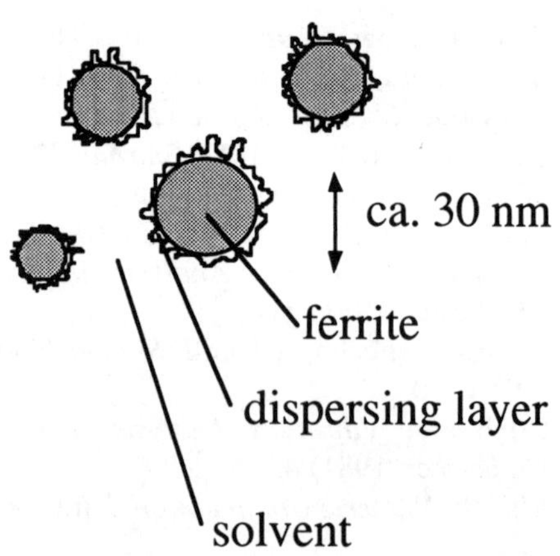

Figure 1

2. Materials

Nano-sized magnetic particles in this study are ferrites prepared by a wet chemical precipitation technique. The dispersing layer consists of molecules carrying on average more than five charges each. The high-boiling polar solvent (similar to the refrigerant in a car radiator) is already familiar to the automobile industry. All components were chosen to be stable at high temperatures in air and to be non-toxic. The total solids content is about 60% by weight resulting in a density of the fluid of about 2 g/cm^3.

3. Magnetorheological measurements

Steady shear viscosities (η) were measured at shear rates $\dot{\gamma} \leq 1000\ s^{-1}$ in concentric cylinder (Couette) and cone plate geometries as described elsewhere[6-7]. Viscosity at high shear rates ($>1000\ s^{-1}$), in addition, was measured in a slit die with height 0.4 mm, breadth 4 mm and length 20 mm.

4. Results and discussion

Figure 2 shows the shear stress τ of a nano-MR fluid as a function of applied magnetic field at operating conditions that are considered typical for magneto- and electrorheological fluids in practical devices: 80 °C and shear rate $\dot{\gamma} = 1000\ s^{-1}$. Shear stress, viscosity η and shear rate $\dot{\gamma}$ are related by $\tau = \eta\ \dot{\gamma}$.

While the shear stress can be further increased using even higher magnetic fields, data are shown only for moderate fields that can be generated easily (e. g.) with permanent magnets or rapidly with low electrical power.

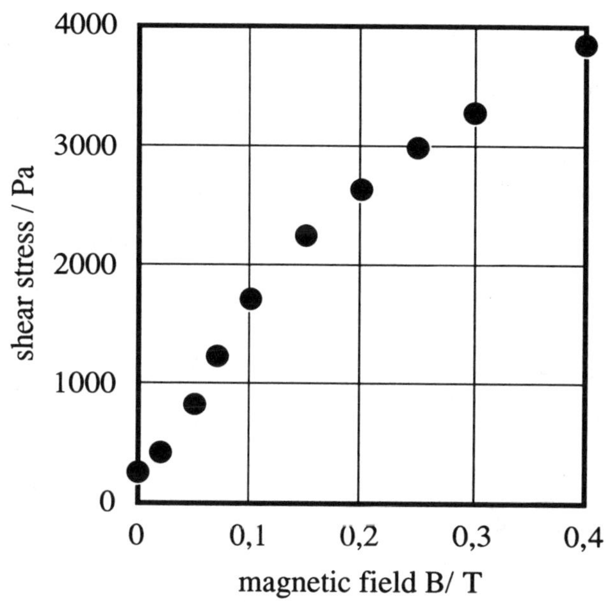

Figure 2

Figure 3 shows how fast the shear stress of a nano-MR fluid responds to an AC magnetic field in the fluid of 0.05 Tesla effective strength. A variation of the shear stress of more than 1000 Pa is observed. For this experiment the cone-plate fixtures of the rheometer were inserted into a transformer. The fluid was sheared at a constant shear rate $\dot{\gamma} = 28\ s^{-1}$ at room temperature while the transformer was powered by a 50 Hz electrical current. In fact, the absolute value of the magnetic field was changing with 100 Hz frequency. Note, that the direction of the magnetic field does not influence the yield stress. From this experiment it can be seen that the response time of the yield stress to the changing field is less than 5 ms.

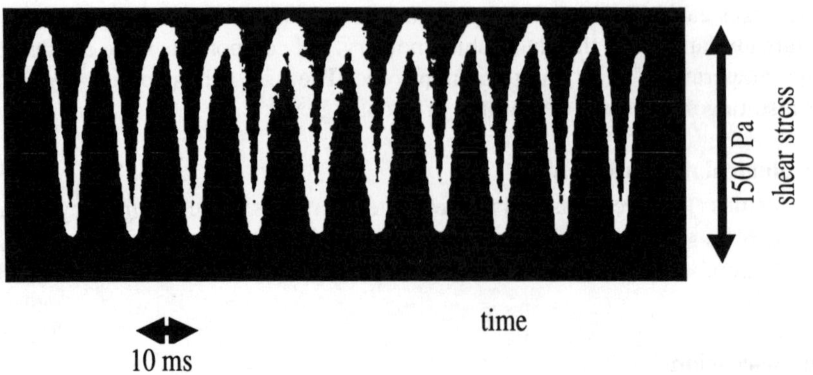

Figure 3

Figure 4 shows the change of the viscosity as a function of temperature (shear rate $\dot{\gamma} = 1000\ s^{-1}$). From these measurements the shear stress is found to decrease with increasing temperature. A correlation with the decreasing viscosity of the carrier liquid at higher temperatures is observed up to 100 °C, where the no-field viscosity reaches a minimum. Above 100 °C a reversible increase of the no-field viscosity is observed which could be due to a conformation change of the dispersing layer. At 0 °C the increase of the yield stress in a 0.2 T magnetic field, viz. the magnetorheological effect, is found to exceed 8000 Pa.

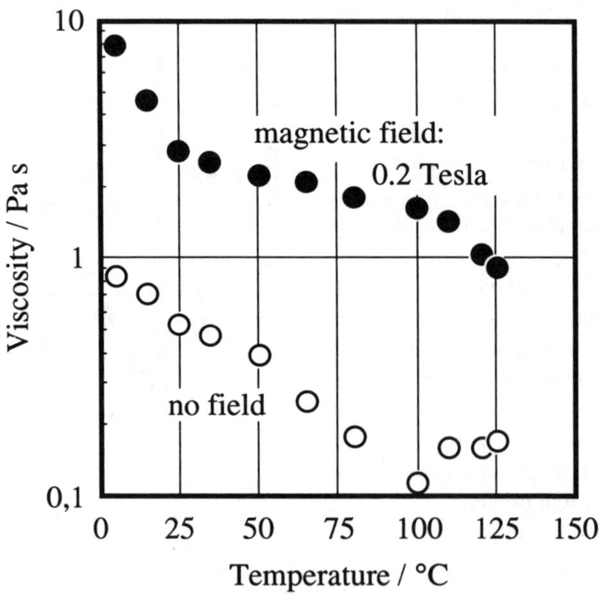

Figure 4

The dependence of the shear stress on the shear rate is given in Figure 5 without and with a 0.2 T applied magnetic field (temperature 80 °C). The data for this figure were obtained in the Couette rheometer for shear rates up to 1000 s^{-1}. For shear rates above 1000 s^{-1} the slit die rheometer was used. At shear rates above 10000 s^{-1} the shear stress change due to the field becomes small with regard to the no-field value.

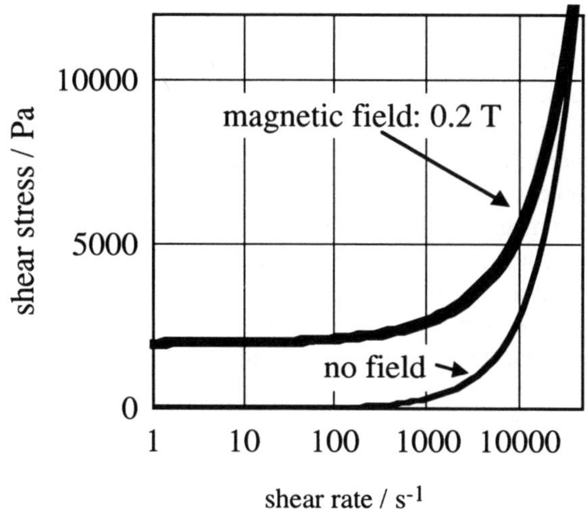

Figure 5

The nano-MR fluid is found to be liquid and stable over a wide temperature range. In fact, storage for 2 weeks at 150 °C under air does not deteriorate its performance significantly: Figure 6 shows the shear stress vs. shear rate behavior before and after the heat treatment (for this sole experiment a slightly different MR fluid composition was chosen, so the data do not compare exactly with the other figures).

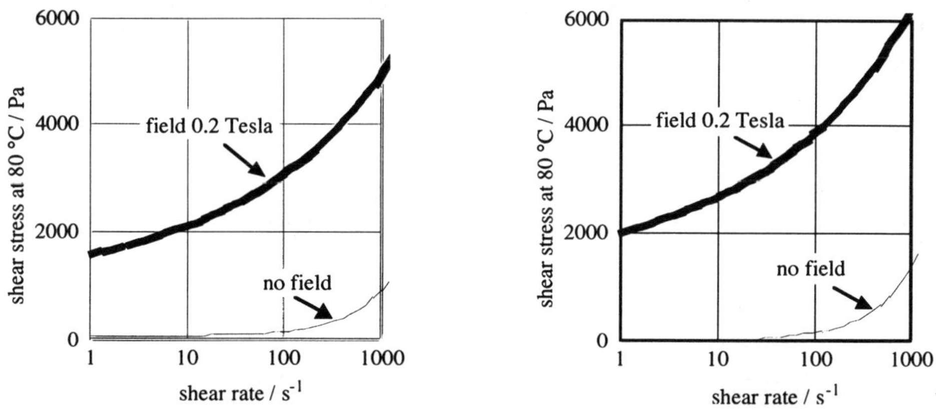

a) virgin sample

b) after 2 weeks at 150 °C

Figure 6

From the perspective of magnetic properties there is a fundamental difference between nano-MR fluids and conventional MR fluids. Conventional MR fluids use micron sized carbonyl iron particles where the particle size is such that these are multidomained. On the other hand, the size of the magnetic particles in nano MR fluids is so small that only single domain particles exist. While single domain particles represent permanent dipoles, multi domain particles need an external field to be magnetized (i. e. there are induced dipoles). For example, consider spherical particles without anisotropy. In the case of a single domain particle the magnetization M will rotate into any field direction imposed externally: only a very small magnetic field H_{ext} is required (cf. Figure 7). On the other hand, the same external field will magnetize a multidomain particle only to $M=3H_{ext}$. (since the effective permeability of spheres is 3) which is much smaller than the magnetization obtained in the previous case. Thus MR fluids with single domain particles require considerably less magnetic field (=energy) to show the same effect. They yield larger values for the "intrinsic" permeability.

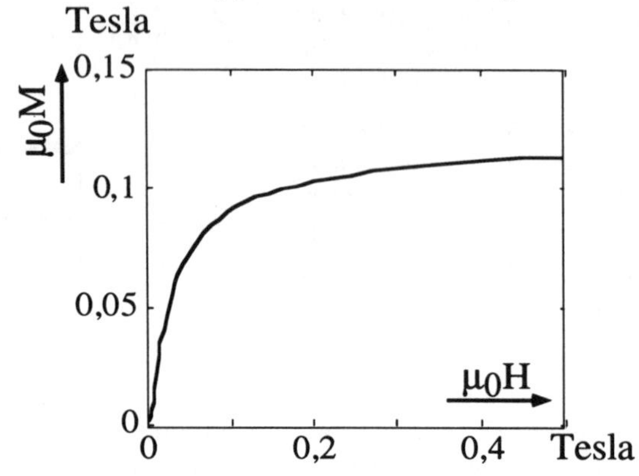

Figure 7

A high intrinsic permeability (of the particles) translates itself to a high permeability of the fluid μ_r. It is important to know μ_r in order to calculate the electrical power N that is needed to generate a magnetic field B in a given fluid volume V. Neglecting losses due to Eddy currents and stray fields the following formula relates the electrical power to the time constant t_k of the magnetic circuit ($\mu_o = 1.26\ 10^{-6}$ Vs/Am):

$$N = V\ B^2 / (\mu_o\ \mu_r\ t_k)$$

Obviously, large values μ_r lead to lower electrical power requirements to operate a MR fluid device.

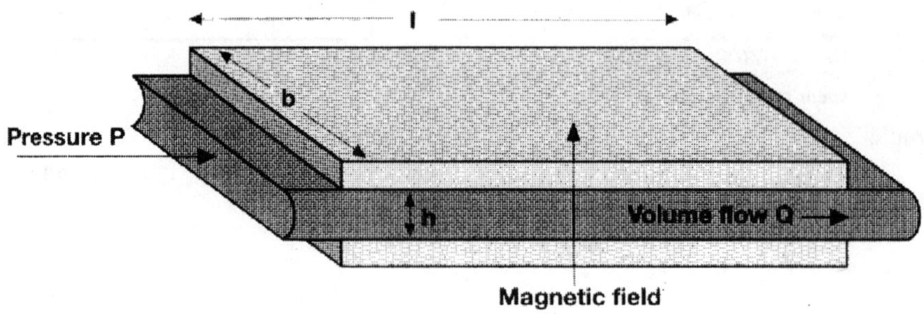

Figure 8

When the magnetorheological fluid is operated in a flow geometry (cf. Figure 8), i. e. when the fluid is squeezed with a flow rate Q through a flow channel of length l, height h, and width b (volume V= l h b) the pressure drop along the flow channel follows as

$$P = 2\,\tau\, l/h.$$

The shear rate of the fluid can be estimated according to

$$\dot{\gamma} \approx 6\, Q/(b\, h^2)$$

Figure 9 depicts the calculated pressure vs. flow rate behavior of a hydraulic valve using the flow geometry (cf. Figure 8) with a nano-MR fluid at 80 °C (data taken from Figure 5). The geometrical dimensions are: l=800 mm, g=4 mm, b=8 mm. Thus an active volume of V≈26 cm^3 needs to be set under the magnetic field (0.2 T).

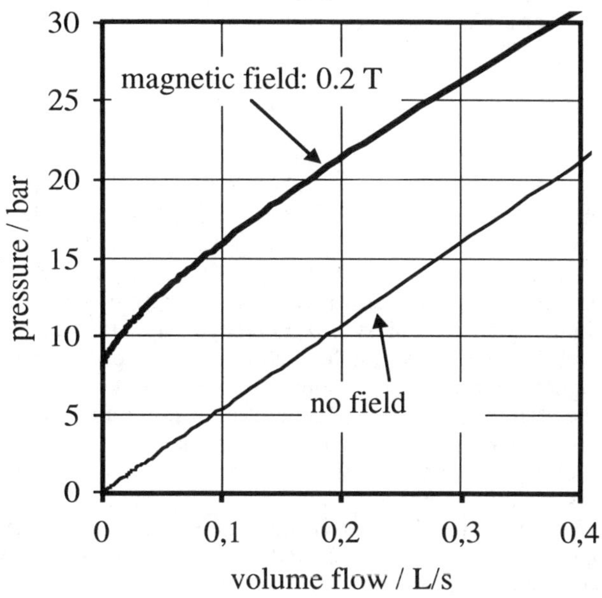

Figure 9

5. References

1. J. Rabinow, *AIEE Transactions* **67** (1948) 1308-1315.
2. D. Carlson, *Proc. Actuator 94* (Bremen, 1994) 266.
3. V. I. Kordonsky et al. *J. Mag. Mag. Mat.* 85 (1990) 114.
4. R. E. Rosensweig, *Ferrohydrodynamics*, Cambridge Univ. Press (1985)
5. R. E. Rosensweig et al. (J. Colloid and Interface Science, 29(4) (1969) 680.
6. Cl. Kormann, H. M. Laun, G. Klett, *Proc. Actuator 94* (Bremen, 1994) 271.
7. H. Janocha, B. Rech, *Rheology* **12** (1994) 198.

DEFORMATION AND RUPTURE MECHANISMS OF ER AND MR FLUIDS

E. Lemaire, G. Bossis, O. Volkova
Laboratoire de Physique de la Matière Condensée
CNRS Université de Nice, 06108 Nice cedex 2, France

ABSTRACT

In this paper, we discuss the different modelisations of the rupture of aggregates in ER or MR fluids. In particular, we show that microstructural models which take into account the separation between the particles inside tha aggregates give predictions very different from those based on a continuum approach. Some experimental results help us to discuss the reliability of each approach.. Creep measurements were made on TiO2 particles dispersed in silicon oil. They give information on the critical deformation at which the aggregates break. An experimental model consisting of millimetric steel spheres in a magnetic field allows us to understand the different possible mechanisms of rupture.

Introduction

Since the discovery of electrorheological and magnetorhelogical fluids many attempts have been done in order to predict the yield stress of these fluids from the knowledge of the respective electric or magnetic properties of the particles and of the fluid. Some models are based on a dipolar approach[1], but it is well known that this approach strongly underestimates the strength of the structure as soon as the ratio of the permittivity (permeability) : $\alpha = \varepsilon_p/\varepsilon_f$ (or μ_p/μ_f) of the particles respectively to the liquid carrier is larger than a few units. A multipolar approach has been developed by Klingenberg which gives more reasonable predictions[2] but still underestimates the value of the static yield stress. A general multipolar approach for periodic networks having N particles per unit cell is now usable[3] and can give reliable predictions in the range $0<\alpha<100$. For higher values of α, the price to pay becomes more and more expensive in terms of the numbers of multipoles we need to introduce in the calculation. This numerical approach is based on a determination of the electrostatic or magnetostatic forces acting on each particle and depends on the precise motion of the particles inside the aggregates. An alternative way to predict the electrorheological stress is to use some mean field theories where only the global shape of the aggregates is taken into account but without any attempt to model the internal stress generated by the relative motions of the particles inside the aggregate. Such a model has been used recently by Rosensweig[4] in order to predict the yield stress for a layered structure. In the first part of this paper we extend this model for any deformation and also to cylindrical structures. In the second paragraph we compare the predictions of the average field model to the more sophisticated multipolar approach. In the third paragraph we discuss these models in view of some experimental results obtained on a macroscopic model of suspension and on an electrorheological fluid.

1. Continuum approach of the yield stress

The elastic shear stress can be directly obtained from the derivative of the electrostatic or magnetostatic energy relatively to the strain γ

$$\tau = \frac{\partial W(\gamma)}{\partial \gamma} \qquad (1)$$

For the case of an isolated sample brought in an external field $\mathbf{E_o}$ we have $W = -(1/2)\mathbf{M}(\gamma) \cdot \mathbf{E_o}$ where $\mathbf{M}(\gamma)$ is the total dipolar moment of the sample. In this paper we are not going to consider finite size effects and we take an infinite medium with an average Maxwell field **E** directed along the z axis. Then the stress is given by [3] :

$$\tau = -\frac{1}{2} E^2 \frac{\partial \varepsilon_{zz}}{\partial \gamma} = -\frac{1}{2} \varepsilon_f E^2 \frac{\partial \varepsilon^*_{zz}}{\partial \gamma} \qquad (2)$$

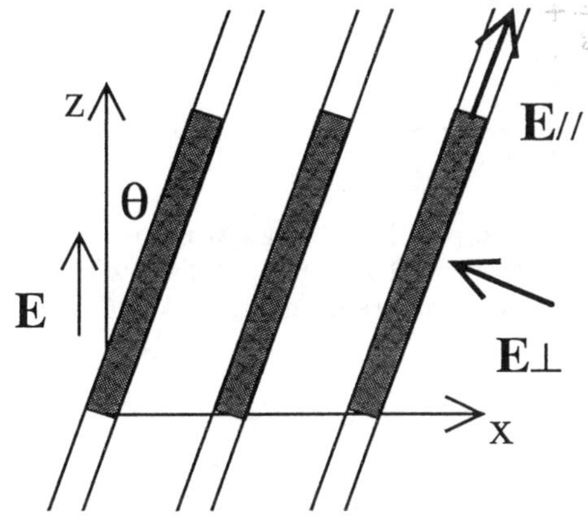

Figure 1 : Shematic view of particles stripes

In this expression, ε_{zz} is the component of the dielectric tensor along the z axis and $\varepsilon^*_{zz} = \varepsilon_{zz}/\varepsilon_f - 1$ is the part of the dielectric constant which comes from the presence of the particles. If we note θ the angle between the principal axis of the aggregates and the z direction we have :

$$\varepsilon^*_{zz} = \varepsilon^*_{//}\cos^2\theta + \varepsilon^*_{\perp}\sin^2\theta \qquad (3)$$

In Eq. 3 the notations // and ⊥ stand for parallel and perpendicular to the direction of the aggregates (cf fig. 1). In order to calculate $\varepsilon^*_{//}$ we have to consider a field $\mathbf{E}_{//}$ parallel to the direction of the aggregates. In this case, the Maxwell average field $\mathbf{E}_{//}$ is identical to the local field inside or between the aggregates since we have no depolarization effects. Then :

$$\varepsilon_{//} = \phi_s\, \varepsilon_s + (1 - \phi_s)\, \varepsilon_f \qquad \text{or } \varepsilon^*_{//} = \phi_s\, \varepsilon_s^* \qquad (4)$$

The quantity ϕ_s represents the volume fraction occupied by the aggregates. It must not be confused with the internal volume fraction of particles inside the aggregates : ϕ_a . If ϕ is the initial volume fraction of particles, we have $\phi_s = \phi/\phi_a$. The permittivity ε_s is the one of a suspension of volume fraction ϕ_a.

When the average field is perpendicular to the main axis of the aggregates the local field depends on depolarization factors related to the shapes of the aggregates. In order to calculate $\varepsilon_\perp$ we have to specify the shape of the aggregates. We shall consider two kinds of structures based on layers and on cylinders. Actually these structures are found in ER and MR fluids[4,5]. The calculus of $\varepsilon_\perp$ in a layered structure is quite simple since the induction **D** must be continuous across the planar boundaries of the layer :

$$\mathbf{D} = \varepsilon_\perp\, \mathbf{E} = \varepsilon_s\, \mathbf{E}_s = \varepsilon_f \mathbf{E}_f \qquad (5)$$

on the other hand the average field $\mathbf{E}_\perp$ is given by :

$$\mathbf{E}_\perp = \phi_s \mathbf{E}_s + (1 - \Phi_s) \mathbf{E}_f \qquad (6)$$

Combining Eq.5 and Eq.6 yields the well known formulae for a layered structure :

$$\varepsilon_\perp = \left[\frac{\phi_s}{\varepsilon_s} + \frac{1-\phi_s}{\varepsilon_f}\right]^{-1}$$

$$\text{or } \varepsilon_\perp^* = \frac{\varepsilon_s^*\phi}{1 + \varepsilon_s^*(1-\phi)} \qquad (7)$$

The case of a field perpendicular to an assembly of cylinders is more involved but if we note that it is equivalent to a distribution of discs with two-dimensional electrostatics we can use the 2d version of mean field theories like Maxwell-Garnet theories. Then we shall obtain[6] :

$$\frac{\frac{\varepsilon_\perp}{\varepsilon_f} - 1}{\frac{\varepsilon_\perp}{\varepsilon_f} + 1} = \phi_s \frac{\frac{\varepsilon_\perp}{\varepsilon_f} - 1}{\frac{\varepsilon_\perp}{\varepsilon_f} + 1}$$

$$\text{or } \varepsilon_\perp^* = \frac{2\varepsilon_s^*\phi_s}{2 + \varepsilon_s^*(1-\phi)} \qquad (8)$$

With Eq. 7 and Eq.8 for the layers and the cylinders and Eq 4 for both of them we can find ε^*_{zz} (cf Eq 3) and noting that $\cos^2\theta = 1/(1+\gamma^2)$ and $\sin^2\theta = \gamma^2/(1+\gamma^2)$ we obtain for the stress τ:

$$\frac{\tau}{\varepsilon_f E^2} = \frac{1}{2}\varepsilon_s^*\phi_s \frac{2\gamma}{(1+\gamma^2)^2} \frac{\varepsilon_s^*(1-\phi_s)}{C_s + \varepsilon_s^*(1-\phi_s)} \qquad (9)$$

The constant C_s is equal to one for a layered structure and to two for the cylindrical structure. The maximum of the stress is obtained for $\gamma = 1/\sqrt{3}$ that is to say for an angle $\theta = 30°$; this maximum will be the value of the yield stress :

$$\frac{\tau^y}{\varepsilon_f E^2} = \frac{3\sqrt{3}}{16}(\varepsilon_s^*)^2\phi_s \frac{\varepsilon_s^*(1-\phi_s)}{C_s + \varepsilon_s^*(1-\phi_s)} \qquad (10)$$

We can see from Eq 10 that the yield stress depends on the structure : it is larger for the layered structure (C_s=1) than for the cylindrical structure (C_s = 2) , as can be seen in table 1. The second line of this table corresponds to the yield stress of layers and the third line to the one of cylinders. We have supposed that the internal structure inside the aggregates was simple cubic so we have taken $\phi_a = \pi/6$ and the corresponding internal permittivity $\varepsilon^*_s = 5.26$ (for a ratio $\alpha = \varepsilon_p/\varepsilon_f = 38.$)has been obtained by the multipolar method in order to carry out the comparison between these two methods.

ϕ	0.1	0.2	0.3	0.343	0.373	0.4	0.45	0.5
$\frac{\tau_y^l}{\varepsilon_f E^2}$	0.26	0.50	0.68	0.73	0.72	0.72	0.62	0.31
$\frac{\tau_y^c}{\varepsilon_f E^2}$	0.22	0.40	0.52	0.52	0.53	0.50	0.40	0.17

Table 1
Yield stress predicted by Eq.10 for a layered and a cylindrical structure. The volume fraction inside the aggregates is $\phi_a = \pi/6$ and their internal permittivity : $\varepsilon^*_s = 5.266$ corresponding to a ratio $\alpha = \varepsilon p / \varepsilon_f = 38$.

2. Microscopic approach of the yield stress

The calculation of the yield stress based on a multipolar development has been exposed elsewhere [3,7] We just want to point out that, once we have given the way the structure deforms we can have a numerical accuracy better than 1% for the value of the yield stress in a range $0 < \alpha < 100$.

We have taken $\alpha = 38$ which corresponds to the average permeability of steel spheres in a magnetic field H = 310 Oersted. The structures we have used are formed by placing a few particles in a unit cell which is replicated on a network. The cell, and then all the network can be deformed on any axis and the resulting yield stress and critical strain is obtained. We have used two basic structures : one formed with infinite simple chains disposed on a simple square lattice and the other formed with infinite double chains (with a shift of half a diameter of one chain relatively to the other). We have also begun to study layered structures but the results are not very different from the chain structure and we shall present these results more extensively elsewhere. In order to compare with the values presented in table 1 and with experiments we have carried out calculations with Φ=0.2 and 0.4 for the single and double chains. We find τ^y =2.0 and 4.0 for the single chains and τ^y= 1.9 and 3.8 . for double chains If we compare these values to these predicted by Eq 7 we see that already for Φ=0.2 , the microscopic model predicts a yield stress which is four times larger than the mean field model. Furthermore the critical strain is always between 0.08 and 0.09 whereas the analytical model predicts $\gamma_c = \sqrt{3}/3$. An other striking difference is that in the analytical model we have a maximum of the yield stress for $\Phi \cong 0.35$ whereas the microscopic model predicts a continuous increase of the yield stress with the volume fraction (up to the simple cubic volume fraction we observe

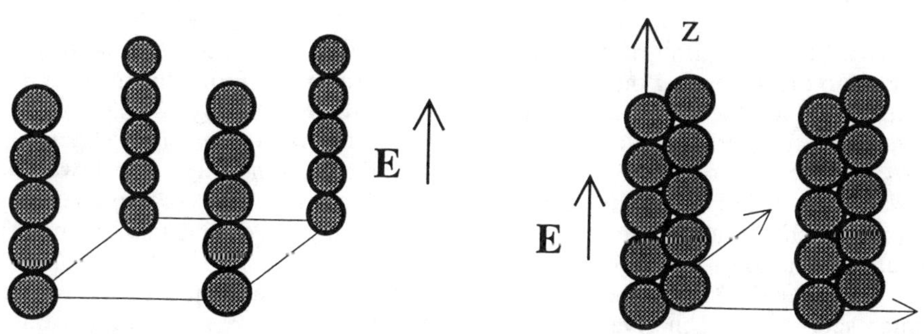

Figure 2 : Basic structures used to modelize the suspension in the microscopic approach

a linear increase with Φ). The results of the multipolar model are exact within a few percent so it is clear that the mean field model fails to predict the value of the yield stress. The reason is easy to understand : in the mean field model we have neglected the local stresses due to the change of the relative positions of the particles inside each aggregate. For instance if we shear a layer in its own plane the mean field model will predict a zero yield stress whereas using the multipolar model we obtain $\tau y/\varepsilon_f E^2 = 1.98$ for $\Phi = 0.2$. This is the proof than a continuum model cannot properly represents the stress coming from the local anisotropy inside the aggregates.

3. Experimental results

We have seen in the preceeding sections that the two approaches (discrete microstructure or continuum) give quite different results as well for the value of the yield stress as for the critical deformation. In order to determine the reliability of these two models, a macroscopic experiment was performed.

The device is represented in fig.3, it consists of two parallel rings whose average diameter is 3,5 cm. The lower one rotates with a controlled velocity and the upper one, is motionless and connected to a torquemeter. The rings are placed inside Helmoltz coils which produce a magnetic field parallel to the rotation axis of the lower ring.

Within the gap between the rings, we introduce steel spheres whose radius is 1 millimetre. They are arranged in such a way that they connect the two rings. In order to prevent the slipping of the structures we have glued on each ring 18 spheres which are embedded of one radius depth inside the ring and postionned face to face. Some different kinds of connecting structures can be built inside the gap which is 14 mm. (6 free particles and 2 fixed one). The magnetic permeability of spheres μ_p was determined by measuring the interaction force between two particles. This force is directly related to the value of the permeability. The measurements were carried out for different values of the magnetic field. As expected, the permeability decreases when the magnetic field increases.
For a field of 310 Oe, we obtain $\mu_p = 38\mu_o$.

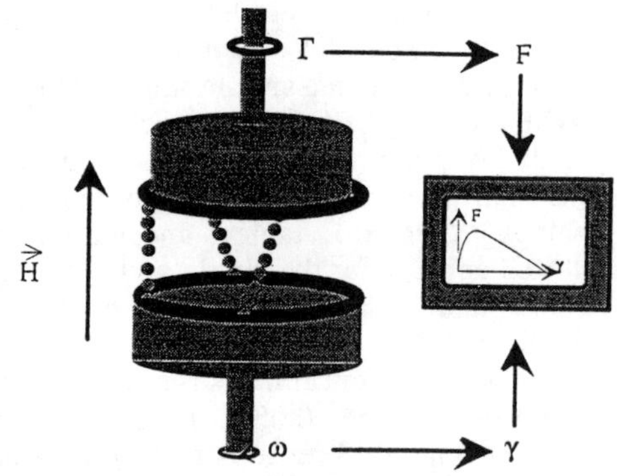

Figure 3 : Experimental device

After the formation of structures, the lower ring begins to rotate and the torque transmitted to the upper ring through the structures is recorded. From this measurement, we are able to deduce the variation of the restoring force exerted by a chain or a column sheared in the magnetic field as a function of the deformation. In a first experiment, we built 18 single chains of 6 free particles plus the 2 fixed on the rings. The lower curve of Fig 4 shows the behavior of one single chain restoring force. This force passes through a maximum which is equal to 6 mN for a magnetic field of 310 0e. This maximum occurs for a critical deformation γ_c=0,12 for which the chains break .

These results have to be compared with the theoretical results of the two models. The microscopic approach gives a normalized yield stress of 1,9 for a volume fraction of 0,2. This corresponds to a restoring force per chain equal to 15 m N for a magnetic field of 310 Oe. So the experimental value is more than two times lower there the predicted value. This difference can be explained by the non linearity of the magnetic material. Indeed, for a field of 310Oe, the permeability is 38 whereas for 190 Oe it is 50. This sensitivity of the magnetic permeability to the field is consistent with a reduction of the magnetic

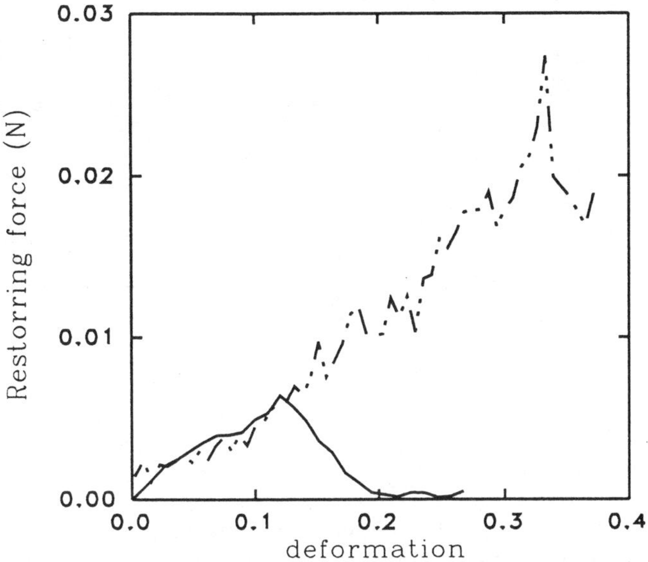

Figure 4 :
Restoring force for a chain of 6 particles (———) and of 7 particles (–··–··–), the gap remaining constant.

interaction since in the areas where the particles are touching each other, the field is higher and so the permeability is less than it would be in a linear medium. This difference can also be explained by the difference of rupture mechanisms. Indeed, in the theoretical approach, the chain deformation is affine: each couple of spheres is separated by the same distance while, in the experimental situation, the separation occurs only between two spheres. This possible mechanism has been treated in dipolar approximation by R.Tao[1] who shows that for a critical angle the absolute value of the energy of affine deformation becomes larger than the rupture energy of two half chains.

On the other hand, the continuum approach gives an equivalent maximum restoring force of 3mN. So the experimental results lie between the predictions of the two models Nevertheless, the experimental critical deformation 0,125 is much closer to the microscopic prediction: 0, 08 than to the continuum one:0,58.

In a second experiment, we have added one sphere to each chain. The position of the additional sphere plays an important role on the values of the critical deformation and of the maximum restoring force. If it has been juxtaposed to the chain forming a triangle with two other particles of the chain, the restoring force is only slightly enhanced by its presence. We obtain a maximum of 6,3mN compared with 6mN for a single chain. There is no significative change in the critical deformation. On the contrary, if the additional sphere is introduced inside the chain without changing the gap width, the deformation increases up to 0,44 before rupture. The maximum restoring force is also much higher : 27 mN. At this stage, if we add some spheres to the chain, the maximum restoring force per chain increases proportionally to the number of particles inside the aggregate (Cf table 2) and the critical deformation seems to be independent of this number.

Nevertheless it should be noted that whatever the number of spheres inside an aggregate the experimental situation corresponds to low volume fractions ; so this experiment does not allow to determine if the variation of the yield stress is or is not proportional to the volume fraction.

N particles	8	9	9	15	21	27
Structure		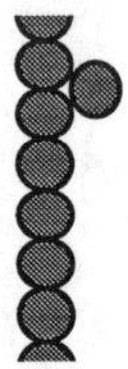		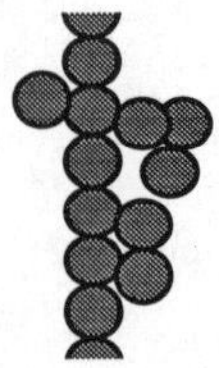		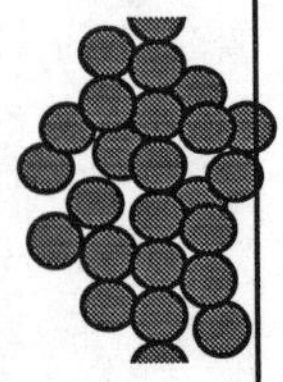
Fmax (mN)	6	6.3	27	41	61	77

Table 2

We have begun to study rupture mechanisms with a suspension of TiO_2 in silicone oil, with a solid volume fraction of 15 %. The creep experiments were carried out on a rheometer carrrimed CSL100 with coaxial cylinders whose radii are 13,8 mm and 15 mm. A high alternative voltage, 920 V rms, is applied to the motionless external cylinder with a frequency of 1 kHz. The inner cylinder is connected to the earth through an electrode dipping in a mercury reservoir. The creep measurements were made for different values of the stress. Then the creep functions were analysed with Inokouchi method.

The viscoelastic behavior can be represented by the generalised Kelvin Voigt model with two basic units since the applied stress is lower than the yield stress

$$J(t,\tau) = \sum_{i=1}^{2} J_i(\tau)\,(1-e^{-t/\theta_i(\tau)}) \qquad (10)$$

The creep function depends on the value of τ since the viscoelastic behavior is linear only for applied stresses lower than 0.8 which corresponds to deformation smaller than 0.2. The equilibrium deformation under the application of a stress is given by :

$$\gamma(\infty,\tau) = \tau \sum_{i=1}^{2} J_i(\tau) \qquad (11)$$

and, in particular, the critical deformation is given by:

$$\gamma_c = \tau_y \sum_{i=1}^{2} J_i(\tau_y) \qquad (12)$$

In our case, we obtain : $J_1(\tau_y)=0.49\ Pa^{-1}$, $\theta_1(\tau_y)=425$ s
$J_2(\tau_y)=0.19\ Pa^{-1}$, $\theta_2(\tau_y)=0.25$ s

Figure 5 shows the variation of the stress with the equilibrium deformation. The critical deformation is about 0.7 which is not so far from the prediction of the continuum approach.

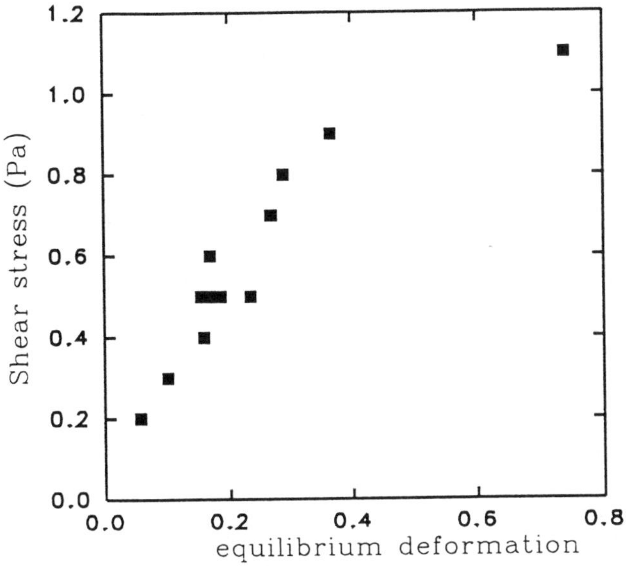

Figure 5 : variation of the stress versus the equilibrium deformation obtained through creep measurements on TiO2 particles in silicone oil.

Conclusion.

Since the ER suspension forms thick aggregates in the presence of a field, a model of structure based on chains which deform in an affine way does not allow to predict the rupture deformation. It seems clear that , before rupture, the aggregates stretch and this mechanism leads to a much higher critical deformation . So, the continuum approach seems to be more reliable in order to predict the critical deformation. Nevertheless, this model underestimates the yield stress because it neglects the internal anisotropy inside the aggregates for example, it predicts no restoring force when a sheet of particles is deformed in its own plane. We think that a discrete model using more realistic initial configuration as for example curved chains should lead to reliable predictions as well for the critical deformation than for the yield stress.

Acknowledments

The realisation of the experiment with steel spheres has been supported by the CNES and the simulations by the computer centre IDRIS. Financial assistance from NATO grant 921196 is equally acknowledged We are also grateful to A. Audoly and R. Clapier for their technical support.

References

1. G.L. Gulley and R.Tao, Electrorheological Fluids, ed R.Tao and G.D.Roy ((World scientific, Singapore, 1994)
2. D.J. Klingenberg and C.F. Zukoski, Langmuir **6** (1990) 15
3. G. Bossis, H. Clercx, Y. Grasselli, E. Lemaire, International Journal of Modern Physics B **8** (1994) 2747
4. R.E. Rosensweig, J. Rheol. **39** (1995) 179
5. G. Bossis, Y. Grasselli, E. Lemaire, J. Persllo, L. Petit, Europhys. Lett. **25** (1994) 335
6. G. Bossis, Mol. Phys. **38** (1979) 2023
7. H. Clercx, G. Bossis, Phys. Rev. E **48** (1993) 2721

Elastohydrodynamic Performance of Unexcited Electro-Rheological Fluids.

R.S. Dwyer-Joyce[1], W.A. Bullough[1], and S. Lingard[2]

1. *Department of Mechanical & Process Engineering, University of Sheffield, Mappin Street, Sheffield, UK*

2. *Department of Mechanical Engineering, The University of Hong Kong, Pokfulam Road, Hong Kong.*

ABSTRACT

Exploratory test results are presented for a series of mixtures of unexcited electro-rheological (ER) fluids under elastohydrodynamic lubrication (ehl) conditions. These were obtained from direct observation of film formation in an optical interferometric apparatus. Results are presented as photographs of the fluid film and plots of film thickness versus speed for a range of ER fluid solid fractions. Adequate film formation is limited by the tendency of the solid particles to evade the contact region. At very low contact speeds particles enter the ehl contact and generate a fluid film. At higher speeds the particulates do not become entrained in the contact; the film formation is then determined by the viscosity of the base fluid.

1. Introduction

ER fluids promise their greatest potential in fast, flexibly operated machines where inertial loading is high. In such operating regimes the fluid must fulfil a number of functions. It must minimise friction at the contact points to reduce heat generation and power loss. It must prevent excess wear to achieve an extended machine service life. And it should carry away any generated heat from the interface sites. The fluid must therefore perform adequately as a lubricant. Depending on the nature of the machinery this lubricity may be required over the boundary, hydrodynamic, and elastohydrodynamic lubricating regimes.

In earlier work[1] a pin-on-disk machine was used to study the lubricating properties of unexcited ER fluids in the boundary regime. The chemical reactivity of the base fluid was considered to be most important in the reduction of friction and wear. The chemically active base fluid (polychlorinated biphenyl) showed lower wear rates than the relatively non-polar base fluid (silicone oil). The active components in the former are believed to be physically or chemically absorbed onto the component surfaces providing a thin 'boundary' film. The wear behaviour in the presence of a suspended solid phase showed anomalous behaviour. The chemically active base with suspended particles showed a higher wear rate than the base fluid alone. The non-polar fluid showed similar wear rates both with and without particulates (although the particulates were noted to reduce friction coefficients). The particles appear therefore to be able to act as either suspended abrasives or solid lubricants. In all cases wear rates were found to be significantly higher than those for a mineral oil (of comparable viscosity).

The film formation in the hydrodynamic regime has also been investigated[2]. The test procedure involved pressure measurements in a Rayleigh step bearing at constant film thickness (higher viscosity fluids generate greater pressures and therefore have the best load carrying capacity). The behaviour of an ER fluid, and ER base fluid was compared to a mineral oil (the viscosity of which was chosen to be in between that of the base fluid and the mixture). The pressures developed were found to be in proportion to their viscosities. It was the viscosity of the particle/fluid mixture that was found to be significant in determining the pressure generation (and hence lubricity) of an ER fluid. These findings were then used in later work[3,4] to predict film formation in unexcited and excited hydrodynamic applications.

The purpose of the present work programme was to examine the performance of an ER fluid in the elastohydrodynamic lubricating regime. The study is limited to an investigation of the basic parameters controlling film formation.

2. Test Method

2.1 Apparatus

In order to study the generation of an elastohydrodynamic lubricant film, it is instructive to replace one of the contacting elements by glass. The motion of the suspended particles and the generation of a separating film can then be directly observed. The apparatus used for these tests is shown in figure 1(a).

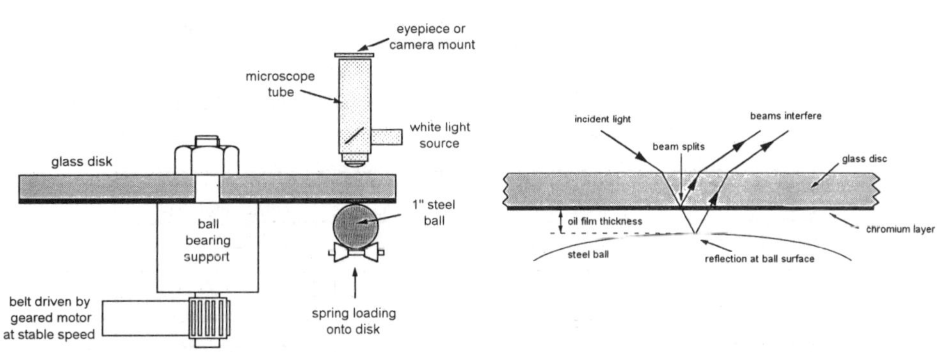

Figure 1. Schematic diagram of (a) the optical interferometric apparatus, (b) interference of an incident and reflected light.

A steel ball is sprung loaded onto the underside of a glass disk. The spring force is maintained constant throughout the test. The load is such that a peak contact pressure is 0.5 GPa and the radius of the contact area is 180 μm. The disk is rotated at constant speed driving the ball. White light is shone through a beam splitter onto the contact. Figure 1(b) shows the mechanism whereby the light rays reflected back at the disk underside and the ball surface interfere to generate a fringe pattern (similar to Newton's rings). This fringe pattern is observed from above through a microscope. The quality of the image is improved by the addition of a thin chromium coating as shown in figure 1(b). The colour of the fringe may then be related to the thickness of the lubricant film. Figure 2 diagramatically shows a typical fringe pattern generated by a mineral oil. The central region is at a greater separation compared to the surrounding 'horseshoe'.

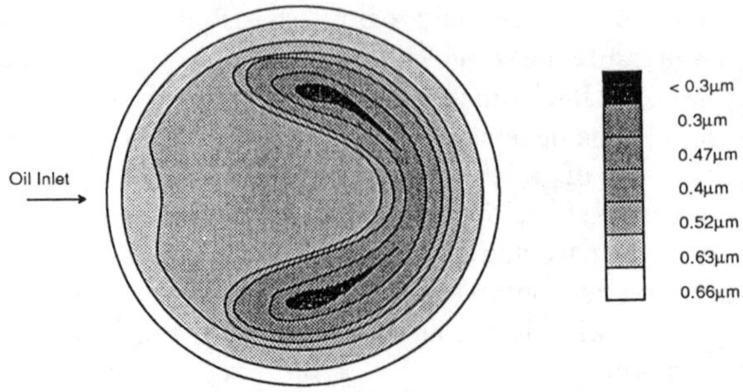

Figure 2. Diagrammatic map of a typical elastohydrodynamic fluid film generated by a mineral oil.

2.2 Test Fluids

The following fluids were used in the experiments. Table 1 summarises the properties of these test fluids relevant to this study.

(i) Electrorheological fluid consisting of a polydimethyl siloxane base oil (silicone oil) with a water free suspension of 6 μm mean sized polyurethane particles. Used in solid particle fractions of 30, 50, and 60%.

(ii) Electrorheological fluid consisting of 5 μm mean sized lithium polymethacrylate (lipol) particles in a dielectric base fluid.

(iii) The silicone and dielectric base fluids alone.

(iv) A standard mineral oil (Shell Tellus 68), with a similar viscosity to the ER fluid mixtures, for comparative purposes.

Base test fluid	Solid particle content	Viscosity, mPas at 22.5°C	Pressure viscosity coefficient, GPa-1
Silicone oil	30%	-	-
Silicone oil	50%	54	-
Silicone oil	60%	188	-
Silicone oil	0%	5	12
Dielectric fluid	30%	187	-
Dielectric fluid	0%	40	58
Shell Tellus 68	-	160	22

Table 1. Properties of the test fluids used in experimentation.

2.3 Test Details

The test fluid was fed onto the glass disk using a constant flow syringe so as to become entrained into the ball on disk contact. The temperature was maintained constant throughout the test (at 22.5°C). A video camera, with a shuttering speed of 1/4000th of a second, and recording at 25 frames per second, was used to record and freeze the moving images.

The fringe order is converted to an optical path difference and then to a fluid film thickness (by dividing by the refractive index). The lowest film thickness measurable by this method is 0.13 μm (corresponding to first order yellow fringe). Measurements are then limited to thicknesses corresponding to successive fringes (first order red, blue, green, then second order yellow and so on). In practice it is convenient to gradually increase the speed and record the point at which the central fringe changes colour. In this way the variation of fluid film thickness with speed can be investigated. Further details of this optical technique for measuring film thickness can be found in the text by Gohar[5].

For the test fluids two sets of tests were performed;

(i) Film generation with speed of rotation. The film thickness was monitored as rolling speeds were increased from zero to 50 mm/s.

(ii) Film generation with time. It was noted that fluid films were not generated immediately on start-up. For each fluid, therefore, film build at low speed up was monitored with time.

All tests were carried out at room temperature. ER fluid film measurements (because of the irrepeatability of the particle entry process) were carried out four times.

3. Results

3.1 ER Fluid Base Oils and Mineral Oil Film Generation

The silicone base oil produced no measurable fluid film in the speed range used[†]. The more viscous dielectric base oil generated a measurable film at around 150 mm/s. The mineral oil produced a measurable film at a speed of 70 mm/s. Both fluid films were stable with time, responded immediately to speed alteration and showed a thickness increase with speed increase. Figure 3(f) and (h) shows the test measurements.

Hamrock and Dowson[6] give a relation for the thickness[‡] of an elastohydrodynamic lubricant film, h_c generated by a Newtonian fluid;

$$\frac{h_c}{R} = 2.69\left(\frac{U\eta}{2E^*R}\right)^{0.67}\left(2\alpha E^*\right)^{0.53}\left(\frac{W}{2E^*R^2}\right)^{-0.067}\left(1-0.61\mathrm{e}^{-0.73k}\right)$$

Where R is the ball radius, W the applied load, U is the mean speed of the surfaces, k is the ellipticity parameter (in this case k=1.03), α the pressure viscosity coefficient, η is the viscosity of the fluid at the inlet temperature, and E^* is the reduced modulus, given by;

$$E^* = \left[\frac{1-\nu_1^2}{E_1} + \frac{1-\nu_2^2}{E_2}\right]^{-1}$$

Where E_1 and ν_1 are the modulus and Poisson's ratio of surface 1. Figures 3(i), (g), and (j) show the fluid film thickness predictions for the mineral oil and the two base fluids using this relation. Agreement between test data and prediction for the mineral and dielectric base fluid is good. The prediction for the silicone base fluid shows film thickness below the measurable limit.

[†] It is possible that at higher speed a film would have been generated. However operating the apparatus at low film thickness and high speed tends to result in costly wear damage to the disk.

[‡] This value corresponds to the film thickness in the central plateau region (see figure 2) and is usually taken as being representative of the average thickness of lubricant film.

3.2 Preliminary Observations on ER Fluid Film Generation

Fluid films were only produced by the silicone based mixtures at low speeds. Video of the contact showed the bulk of the suspended particles being swept around the contact sides. A mass of particles appeared trapped in the inlet region from which some passed into the contact (the particle entry being dominated by contact speed). These became plastically flattened out and produced regions of separating film. The film formation was apparently controlled by this particle entry process. Ball motion appeared slightly jerky as discrete masses of particles entered the contact.

The photograph included as figure 4(a) shows the fluid film. The film is typically patchy, with thicker regions corresponding to a mass of particles. Frequently a central strip of greater separation is observed (corresponding to the region where most entrained particles pass).

Some of the glass disks had some fine surface scratches present after testing with ER fluids. It is not clear whether this surface damage was caused by the particles themselves, some foreign contaminant, or from direct contact between the ball and disk when no film is present.

3.3 Film Generation with Speed

As the speed was increased fewer particles become entrained into the contact and the film thickness falls. At speeds of around 30 - 50 mm/s particles were unable to enter the contact and no measurable film was formed. Figure 3 (a), (b), and (c) show the plot of film variation with speed for three particle concentrations.

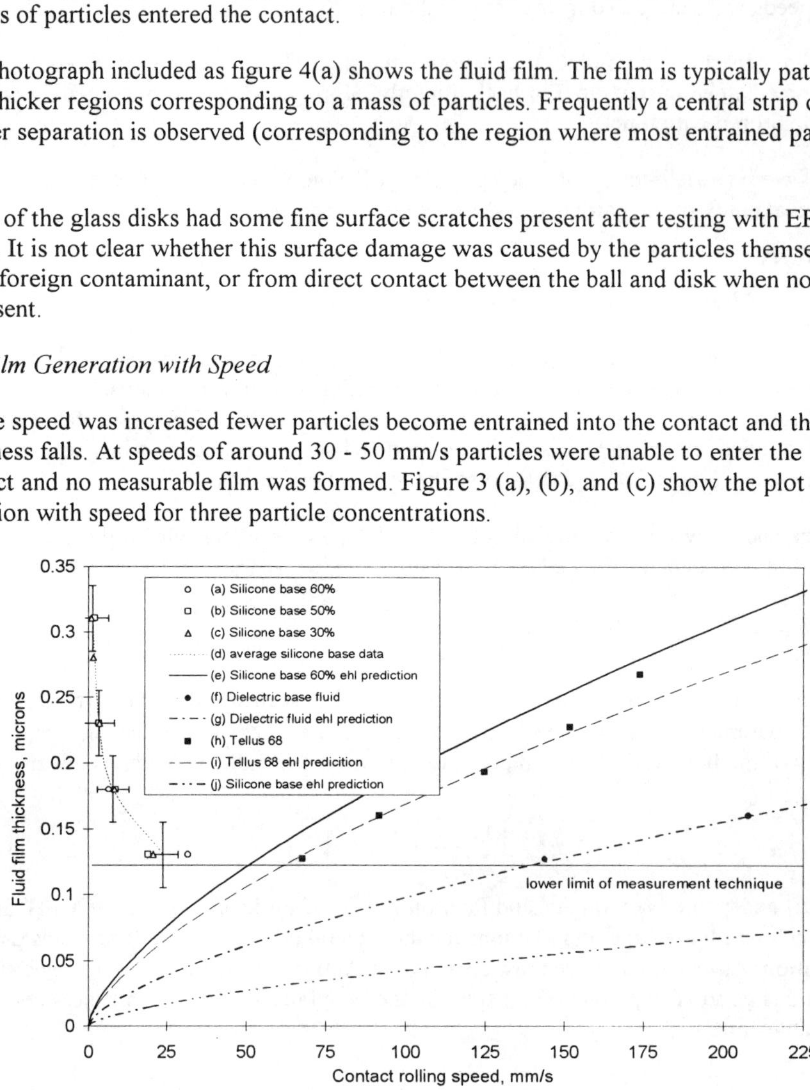

Figure 3. Plot of film thickness against rolling speed. Test data for: three silicone based ER fluid mixtures, a dielctric base fluid, and Shell Tellus mineral oil. Hamrock and Dowson theoretical predictions for: mineral oil, dielectric base fluid alone, silicone base oil, and a 60% mixture ER fluid.

Each data point is the average of four measurements, the scatter is demonstrated by the error bars. The horizontal line represents the lower limit of the measuring technique (the thickness corresponding to the formation of the first light interference fringe). In figure 3 the theoretical predictions of the Hamrock and Dowson equation are also displayed. The mineral oil and dielectric fluid alone shows good agreement with test data. The predictions for the silicone based ER fluid, shown as curve (e), are calculated assuming the fluid is Newtonian with a viscosity equivalent to that of the fluid mixture. Clearly for ER fluids this assumption is not appropriate. Figure 4 shows a sequence of photographs of the film formation behaviour at increasing rolling speed.

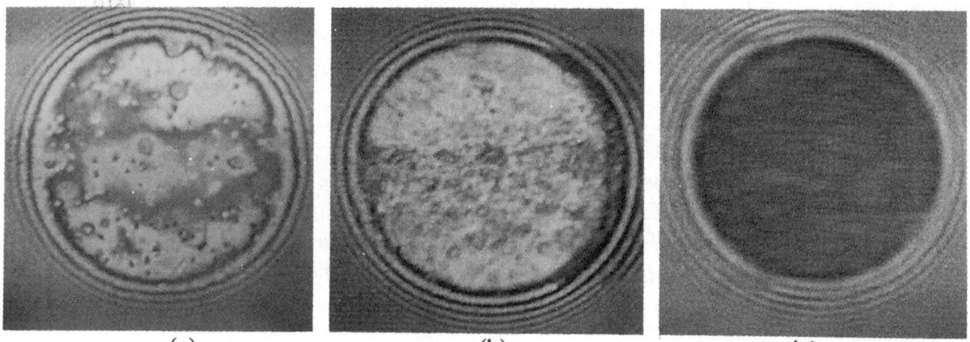

(a) (b) (c)

Figure 4 - Photographs of the fluid film generated by 60% silicone based ER fluid mixture. The sequence shows the effect of the increase in speed (a) 1.7 mm/s, (b) 15 mm/s, and (c) 27 mm/s.

After testing, the glass disk was careful removed (without disturbing the adhered fluid) and examined under an optical microscope. Figure 5 shows the tracks where the ball has passed at (a) high and (b) low rolling speed.

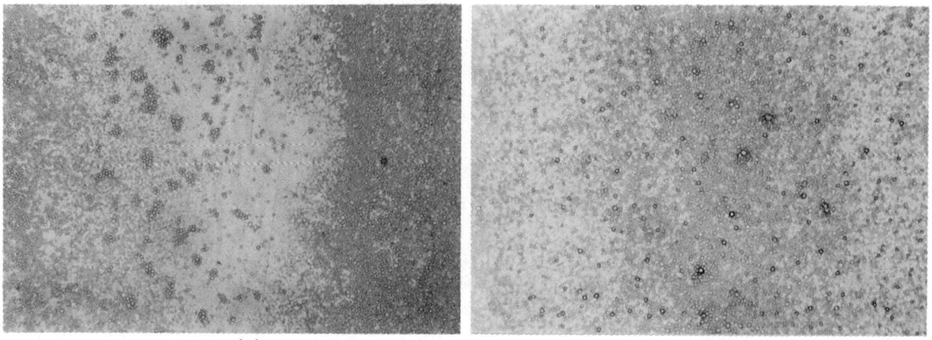

(a) (b)

Figure 5. Tracks on the disk left by the passage of the ball rolling at; (a) high speed - 50 mm/s; and (b) low speed - 3 mm/s.

At low speed the contact has a concentrating effect. Particles are entrained into the contact and assist in the generation of a film. A track of fluid with a higher particle concentration than the bulk is left on the surface. At higher speeds these particles are swept to the sides of the contact. This leaves a depleted region of fluid on the rolling track surface.

It is likely that at still higher speeds the ER silicone base fluid alone will start to generate a measurable film; and the curves 3(a), (b), and (c) will tend to curve 3(j). However, this could not be verified since operating the apparatus at high speed with low film thickness causes costly surface damage to the coated glass disk.

3.4 The Effect of Particle Concentration

Fluids of 30%, 50%, and 60% solid content were tested in the above manner. For these fluids, particle concentration seems to have no substantial effect on film formation within the scatter in the results (see curves a, b and c on figure 3). These three fluid mixtures are of greatly differing viscosity (60% is some three times more viscous than 50%). This has little effect on the film thickness; again we see that it is the entry behaviour of the particles, and not the mixture viscosity, that dominates film behaviour.

Video photography shows a build up of particles in the inlet region which then pass into the contact. The concentration in this area would therefore be of most relevance to the entry process. This 'inlet region concentration' may not be particularly dependent on the 'bulk particle concentration'. However, it is likely that at lower bulk concentrations, the entry of particles (and film formation) will show a greater dependency on concentration.

3.5 Film Generation with Time

The mineral oil and dielectric base oil immediately developed a fluid film on start-up which subsequently remained stable with time. The silicone base oil negligible film condition also did not alter with time. However the silicone based ER fluids did not generate fluid film immediately. Figure 6 shows the variation of film thickness with time for a 60% mixture ER fluid at a rolling speed of 3 mm/s.

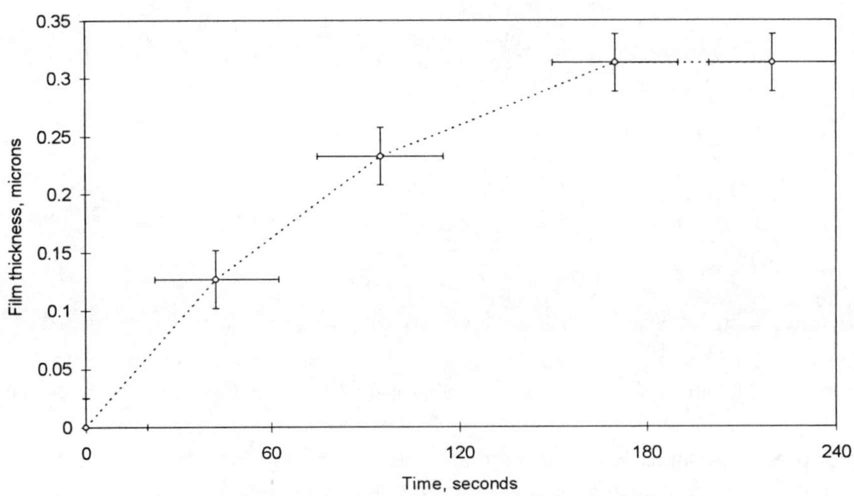

Figure 6. Variation of film thickness with time for a silicone base 60% mixture ER fluid at a rolling speed of 3 mm/s.

Some three minutes was required before a film was generated (this corresponds to several complete revolutions of the ball). It appeared that during this time solid particles were adhering to the ball and disk surface. A few revolutions were required before sufficient particles were adhered and/or entrained into the contact. The film formation therefore seems also to be partly governed by the ability of the solid particles to adhere to the rolling elements. Examination of the rolling tracks did not suggest firmly bonded particles; rather particles weakly adhered or electrostatically attracted to the surfaces.

4. Discussion

The limit of ER fluid machinery is frequently determined by thermal considerations (heat generation and removal) resulting in melt, expansion, or breakdown of continuum. The presence of increasingly rapid accelerations and high inertia has resulted in flow mode designs requiring high system pressures (fulfilling the need to keep the parasitic mass of hydraulic components low). Under these conditions of high temperature, load, and speed, the service life of the system is likely to be governed by failure of the lubricated conjunctions.

Fluid film formation is dependent on the entry of solid particles into the loaded conjunction. In turn this entry process is dependent on fluid forces acting on particles in the contact inlet region. At higher speed these drag forces increase and particles are more likely to be swept around the contact sides. Thus slow speed operation allows particle entry whilst higher speed reduces it. Complimentary research[7] into the entry process for other particle sizes and materials has also the importance of film thickness and particle size. The results presented in this paper represent the case relevant to ER lubricity.

The continuum assumption, which appears to hold for hydrodynamic lubrication processes[2], is clearly not appropriate in the elastohydrodynamic case. The separation of moving components is of the same order as the particle size (in contrast to that in hydrodynamic pad or journal bearings).

These results show the difficulty in achieving adequate lubricating films at realistic contact speeds (typically a rolling bearing might operate at speeds of 0-5 m/s). Lubricating films should ideally be thicker than the combined surface roughness of the contacting elements (typically 0.1 - 1 μm for a ground surface) to minimise surface contact and hence friction and wear. Thus to achieve adequate elastohydrodynamic lubrication with an ER fluid a viscous base is required. This raises a dilemma. The contribution of this base viscosity to the overall ER fluid mixture viscosity would, at high speeds, result in excessive heating and power loss by viscous shearing.

If such a condition cannot be tolerated then components which would usually operate under ehl (counterformal contacts such as bearings, gears, or cams) will experience a boundary lubrication regime. It is then the ‘surface activity’ of the base oil which will be important in the control of friction and wear[1].

5. Conclusion

The generation of elastohydrodynamic films by the ER fluids tested is controlled by the viscosity of the base oil and the entry of particulates into the loaded conjunction. At realistic operating speeds particle entry is restricted and a viscous base oil is required to generate a sufficient separating film.

With the present generation of fluids the average particle sizes are an order of magnitude greater than typical concentrated contact film thickness using good lubricants. The continuum assumption for these ER fluid mixtures is therefore not valid under elastohydrodynamic conditions. However, at low speeds some particles do pass through the contact and assist in film formation. However, there is also evidence that compaction of these particles in the contact zone may cause surface damage.

Based on the results of this study, the following are broad recommendations for machine operation and fluid design to optimise elastohydrodynamic lubricity:

(i) Base oil viscosity should be maximised, within the constraints of fluid stability and viscous heating effects.

(ii) Low operating speeds and periods of start-up under low load, would be beneficial in the formation of 'particle assisted' separating films.

(iii) Where a high viscosity cannot be tolerated a surfactant base should be used to maximise boundary lubricating properties.

(iv) Fluids with much lower particle sizes, may fulfil the continuum assumption, and perform improved elastohydrodynamic lubrication.

Acknowledgements

The authors are grateful to Dr P. Cann of Imperial College, for helpful discussion and the loan of test equipment.

References

1 **Lingard, S., Bullough, W.A.,** (1990), "Tribological Aspects of Electro-Rheological Fluid Behaviour with Respect to Non-Polar Base Fluids", *Proc. 2nd Int. Conf. ER Fluids, Raleigh, NC*, pp 185-195.

2 **Lingard, S., Bullough, W.A., Ho, L.E.,** (1991), "Hydrodynamic Pressure Generation with an Electrorheological Fluid", *Wear*, Vol 142, pp 373-381.

3 **Leek, T.H., Lingard, S., Bullough, W.A., Atkin, R.J.,** (1994), "Hydrodynamic Pressure Generation with an Electrorheological Fluid. Part I - Unexcited Fluid", *Proc. 4th Int. Conf. ER fluids, July 1993, Feldkirk, Austria*, pp 609-624.

4 **Leek, T.H., Lingard, S., Bullough, W.A., Atkin, R.J.,** (1994), "Hydrodynamic Pressure Generation with an Electrorheological Fluid. Part II - Excited Fluid", *Proc. 4th Int. Conf. ER fluids, July 1993, Feldkirk, Austria*, pp 625-642.

5 **Gohar, R.,** (1988), "Elastohydrodynamics, 1st ed.", Ellis Horwood Ltd.

6 **Hamrock, B.J. and Dowson, D.,** (1977), "Isothermal Elastohydrodynamic Lubrication of Point Contacts, Part III, Fully Flooded Results", *ASME J. Lub. Tech.*, Vol. 99, pp 264-276.

7 **Dwyer-Joyce, R.S. and Heymer, J.,** (1995), "The Entrainment of Solid Particles into Rolling Elastohydrodynamic Contacts", *to be published in Proc. 22nd Leeds-Lyon Symp., Elsevier.*

ELECTRORHEOLOGICAL EFFECT OF LIQUID CRYSTALLINE POLYMERS

AKIO INOUE, SYUNJI MANIWA, YOICHIROH IDE, and HIROJI ODA
Central Laboratory, Asahi Chemical Industry Co., Ltd.
2-1 Samejima, Fuji, Shizuoka 416, Japan

ABSTRACT

A series of side chain liquid crystalline polymers (LCPs) containing different spacer length and various terminal structures has been prepared and their electrorheological (ER) effect has been investigated. Although the LCP itself showed small ER effect, a mixture of the LCP with a plain polydimethylsiloxane (PDMS) oil was found to show a large increase in shear stress upon application of an electric field. The shear stresses under DC 2 kV/mm of an electric field were 14 to 4 times those without the application of an electric field over the temperature range of 30 to 160 °C for a 2:1 mixture (by weight) of a LCP containing an oxyethylene spacer and a mesogen based on 4-cyanophenyl benzoate with PDMS. Two mechanisms responsible for the generation of the ER effect was suggested from the investigation using phenyl-substituted PDMS and an analysis by DSC: oriented LC network induction, and phase separation between the LC base polymer and PDMS.

1. Introduction

Electrorheological (ER) fluids have long held promise for use in vibration control and torque transmission devices, based on the characteristic dependence of their viscosity on applied electric field strength. Since their initial discovery by Winslow[1] many particle-dispersion ER fluids, consisting of dielectric particles dispersed in insulating oil, have been reported, but their practical utilization has been limited by their tendencies for particle sedimentation, aggregation, or solidification; particle or electrode abrasion; and poor durability or temperature dependence.[2-5] Polar liquids,[6] low molar mass liquid crystals,[7-9] lyotropic polymeric liquid crystals,[10,11] ferroelectric polymer solutions,[12] and other homogeneous ER fluids[13] have been proposed as a means of avoiding these problems, but in all cases their ER effect has been found too weak for practical use.

We have been particularly interested in liquid crystals, because of their fast response to weak electric fields. As previously reported[10,14] we speculated that the poor ER effect of low molar mass liquid crystals was attributable to the weak interaction between the liquid crystalline domains induced under an electric field, and that a stronger interaction could be obtained if the domains were linked by flexible molecular chains (illustrated in Figure 1). On this assumption we prepared a series of liquid crystalline polymers (LCPs) including side chain and main chain liquid crystalline polysiloxanes, and bimesogenic and monomesogenic polysiloxanes which bear mesogenic groups on both chain ends and only one chain end of the polydimethylsiloxane, respectively, and their ER behaviors were investigated. It was observed that the side chain LCP exhibited the largest increase in shear stress (3,000-4,000 Pa) by the application of an electric field and that the ER effect decreased in the order of the side chain LCP, the main chain LCP and the bimesogenic polysiloxane. The increase of shear stress was only several Pa for the monomesogenic polysiloxane as found in low molar mass liquid crystals. This dependence of the ER effects on the polymer architecture supports our initial assumption; an oriented LC network structure is induced under an electric field, where molecular chains bearing more than two mesogens span and anchor neighboring LC domains to exhibit resistance against an applied shear. The formation of the oriented LC network structure was also evidenced by the ER normal stress measurements.[15]

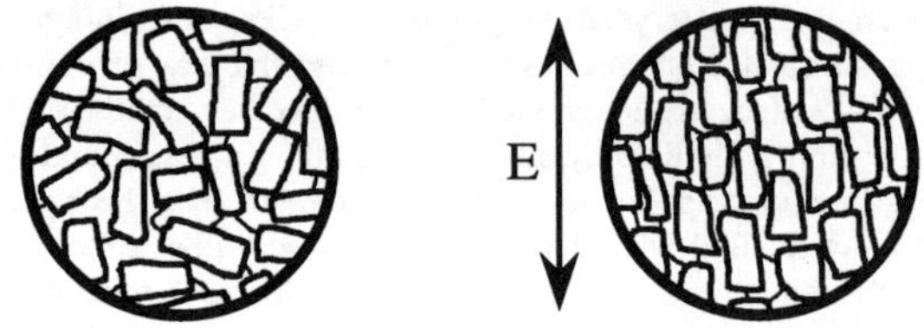

Figure 1. Schematic illustration of LC network structure.

Throughout the previous investigation, determination of ER effects was carried out for mixtures prepared from the LCPs with a plain polydimethylsiloxane (PDMS) oil (viscosity 20 cSt). The dilution by PDMS was originally intended to ensure a low zero-field viscosity of the ER fluid since the inherent viscosity of LCPs is generally very high and has a large temperature dependence. Due to this temperature dependence, the system, even if it is the one diluted with PDMS, loses a sufficient fluidity and does not provide stable operation when subjected to temperatures blow 40°C. On the other hand, it was also found that PDMS had an unexpected effect to dramatically enhance the ER effect rather than simply serving as a dilution oil. The objective of this research work is thus to take an insight into the role of the dilution by PDMS in the generation of ER effect and to design the system that operates over a wide temperature range including room temperature.

2. Experimental

2.1 Materials

The structures of side chain LCPs used in this study are shown in Figure 2. They were prepared by a hydrosilation reaction of poly(hydrogenmethyl-*co*-dimethylsiloxane) (Shin-Etsu Chemical) with terminal alkenyl-functionalized phenyl benzoate derivatives in the presence of $H_2PtCl_6 \cdot 6H_2O$ as a catalyst in tetrahydrofuran (THF) at 60 °C. The details of synthetic procedures will be reported elsewhere. Polydimethylsiloxane (PDMS) and phenyl substituted PDMS (Ph-PDMS) used as a dilution oil have structures shown in

Figure 3 and are purchased from Toray Dow Corning Silicone and Petrarch Systems. Mixtures of the LCP with PDMS or Ph-PDMS were prepared by dissolving them in THF followed by evaporation of the solvent under vacuum.

For comparison a conventional particle dispersed ER fluid was prepared by adding silica particles (Mizusawa Co., AMT 300, *ca.* 3 mm diameter, 8 wt% water content) into PDMS (20 cSt) in a weight ratio of 30:70.

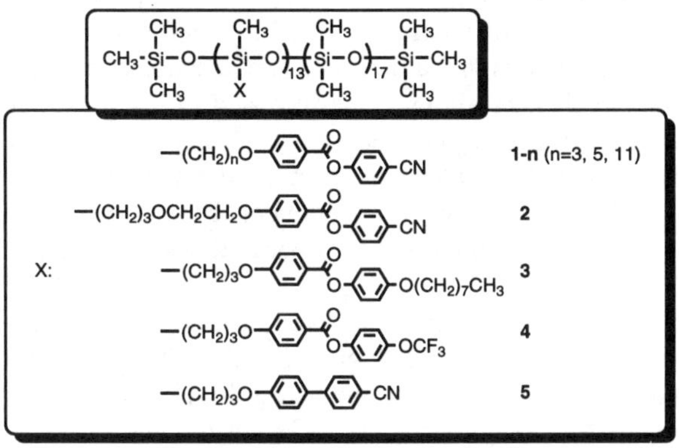

Figure 2. Structures of side chain LCPs.

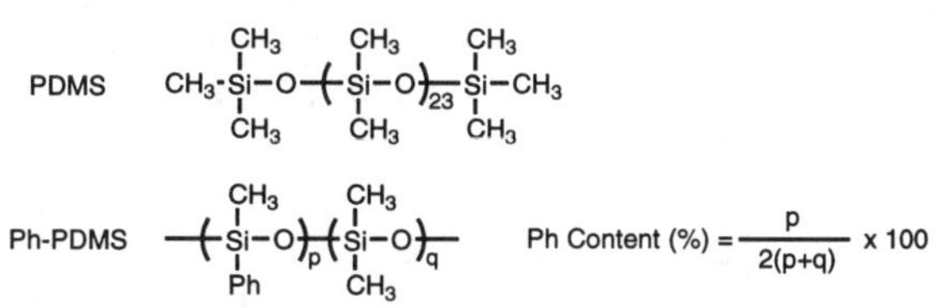

Figure 3. Structures of PDMS and Ph-PDMS.

2.2 *Measurements*

Shear Stress was measured with a parallel-plate rotational viscometer (Model MR 300, Rheology Co., Ltd.) modified to permit the application of an electric field over the range of DC 0-2 kV/mm by a high voltage power supply (Model PT-3503, Toa Electronics Ltd.). The distance of electrodes was 0.5 mm throughout this study and measurements were performed at a constant (100, 200 sec^{-1}) or stepwise increasing (0-200 sec^{-1}) shear rate. A Perkin-Elmer DSC-7 differential scanning calorimeter (DSC) equipped with Intracooler II was used to determine the thermal transition temperatures and enthalpy changes. Heating and cooling rates were 10 °C/min.

3. Results and Discussion

3.1 ER effect of side chain LCPs

The ER effect of plain LCPs with different spacer length (**1-n**, n=3, 5, 11) is shown in Figure 4. The measurement temperatures shown in this figure are those resulting in the maximum shear stress changes for each LCP. In the case of **1-3** the increase in shear stress by the application of an electric field is small and therefore, the two curves which represent the shear stresses in the presence and absence of an electric field (solid line and dotted line, respectively) are almost superimposed in Figure 4. On the contrary, **1-5** and **1-11** exhibit 2.0 to 2.5 times larger shear stress in the presence of an electric field than in its absence. Longer spacer length tends to promote the decoupling of mesogens from the polymer backbone to which the mesogens are connected[16] and therefore, the larger increase in shear stress observed for **1-5** and **1-11** is probably due to the enhanced freedom of the mesogenic motion.

When **1-3** is diluted with PDMS in a weight ratio of 2:1, the mixture showed an unexpectedly large ER effect as shown Figure 5. The ER effect that is observed for a conventional particle-dispersed system composed of hydrous silica in PDMS (30 wt%) is also shown for comparison in Figure 5. The LCP system shows Newtonian flow both in the presence and absence of an electric field. On the contrary the particle dispersed system shows Newtonian flow in the absence of an electric field, and Bingham flow in an applied electric field. Figure 6 displays the temperature dependence of the shear stress collected for a **1-3**/PDMS (2:1 by weight) mixture. The practical operating temperature of this mixture ranges from 50 °C to 90 °C. If the mixture is exposed to temperatures below 40 °C, its zero-field viscosity increases to such an extent that the mixture does not provide a sufficient

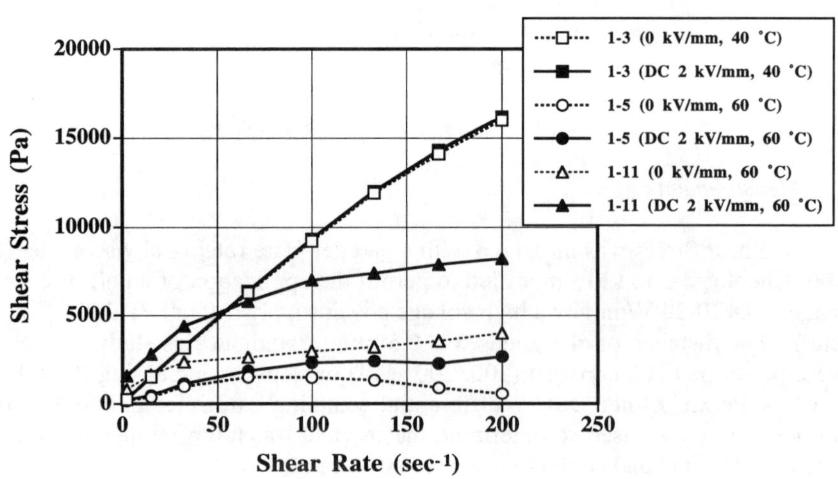

Figure 4. ER effect of side chain LCP **1-n** (n=3, 5, 11).

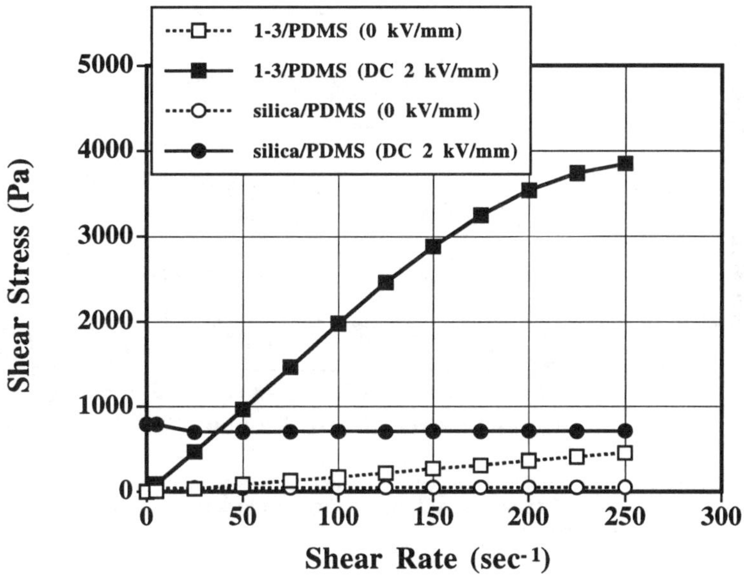

Figure 5. Shear stress of side chain LCP (**1-3**/PDMS=2/1 by weight at 50 °C) and hydrous silica dispersion (silica/PDMS=3/7 by weight at 25 °C).

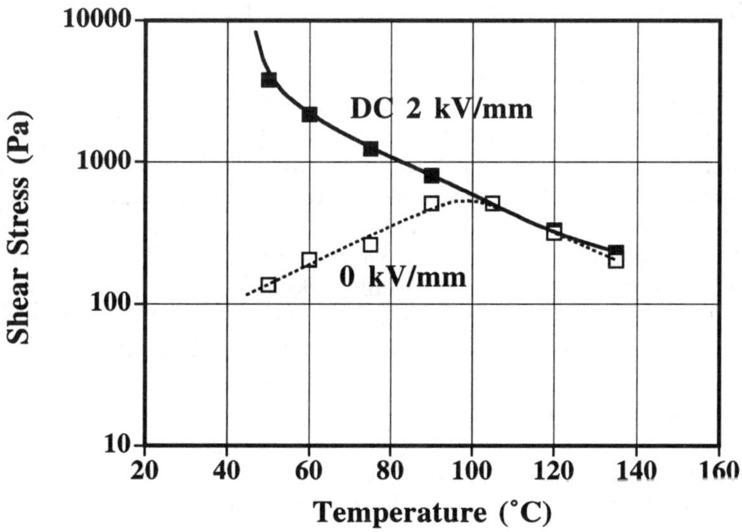

Figure 6. Temperature dependence of the ER effect of **1-3**/PDMS mixture (2/1 by weight) measured at 200 sec^{-1}.

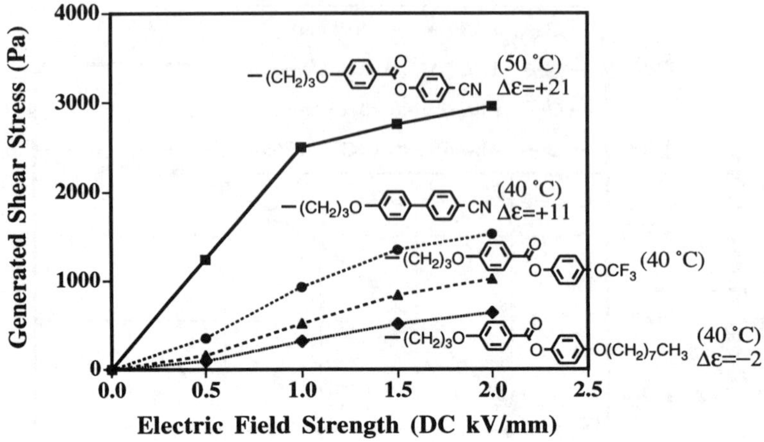

Figure 7. Influence of mesogenic structure on the generated shear stress of LCP/PDMS mixture (2/1 by weight) at various electric field strengths at 200 sec^{-1}.

fluidity necessary as an ER fluid. In an attempt to lower the operating temperature various mesogenic structures were examined. Figure 7 summarizes the maximum generated shear stress obtained for each of the tested mesogen, and the temperatures at which the maximum stress was observed. This figure clearly indicates the difficulty in controlling the operating temperature just by changing the mesogenic structures. Another message drawn from this figure is that the dielectric anisotropy of the mesogens can be directly related to the generation of ER effect. The LCPs having greater dielectric anisotropy tend to exhibit larger ER effects, presumably because of its stronger orientation capability under an electric field. This observation is consistent with the speculation that the oriented network of LC domains is induced in response to an applied electric field.

One of the candidates for lowering the operating temperature is a side chain LCP containing an oxyethylene spacer (**2**) which can be easily synthesized from commercially available ethylene glycol monoallyl ether. A side chain LC polysiloxane with an oligooxyethylene spacer was found to show a lowered and wider LC temperature range compared to that with a paraffinic chain spacer[17] and therefore, improved temperature characteristics desired for a base polymer of an ER fluid can be expected. Figure 8 presents the temperature dependence of the shear stress collected for a 2:1 mixture of **2**/PDMS under the conditions of DC 2 kV/mm, 200 sec^{-1} shear stress. As expected, smooth and large response to an electric field was achieved at 30 °C. Furthermore, ability to exhibit large shear stress changes is maintained at fairly high temperatures up to 160 °C.

3.2 Effect of the dilution by PDMS

As described above, PDMS was first used in order to reduce the high inherent viscosity of the base LCP so that a sufficiently low zero-field viscosity suitable for an ER

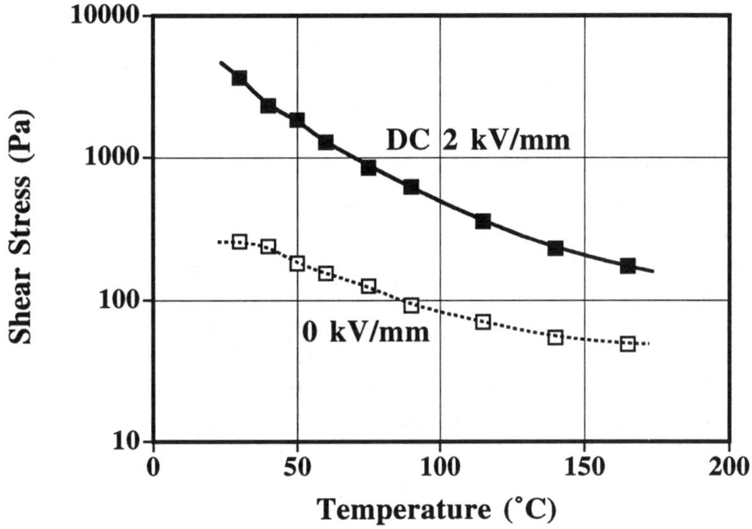

Figure 8. Temperature dependence of the ER effect of **2**/PDMS mixture (2/1 by weight) measured at 200 sec^{-1}.

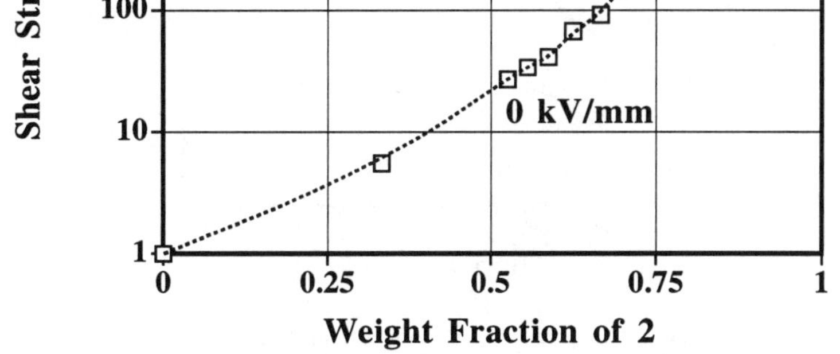

Figure 9. Dependence of shear stress on the composition of **2**/PDMS mixture measured at 60 °C, 100 sec^{-1}.

fluid was obtained. Figure 9 shows the shear stress of the **2**/PDMS mixture in the presence and absence of an applied electric field as a function of base LCP weight fraction. The shear stress of the mixture is nearly constant and the increase in the shear stress by the application of an electric field is very small when the weight fraction of **2** is larger than 0.9. With decreasing the weight fraction of polymer **2** in the mixture, the zero-field shear stress drastically falls to generate a large contrast between the shear stresses with and without an applied electric field. It is remarkable and unexpected to see that a large ER effect is achieved even in such a concentration that the weight fraction of **2** goes below 0.5.

In order to clarify the role of PDMS in the generation of ER effect, phenyl-substituted polydimethylsiloxane (Ph-PDMS) with various contents of the phenyl group was used instead of a PDMS oil. Figure 10 summarizes the normalized DSC thermogram of the base polymer **1-3** and those of the mixtures between **1-3**, and PDMS or Ph-PDMS with a different phenyl content taken during the second heating scan. Transition temperatures and associated enthalpy changes are also given in this figure. The base polymer **1-3** exhibits a glass transition and a smectic A mesophase which melts into an isotropic phase at 43.6 °C. Characterization of the S_A phase was performed with polarized

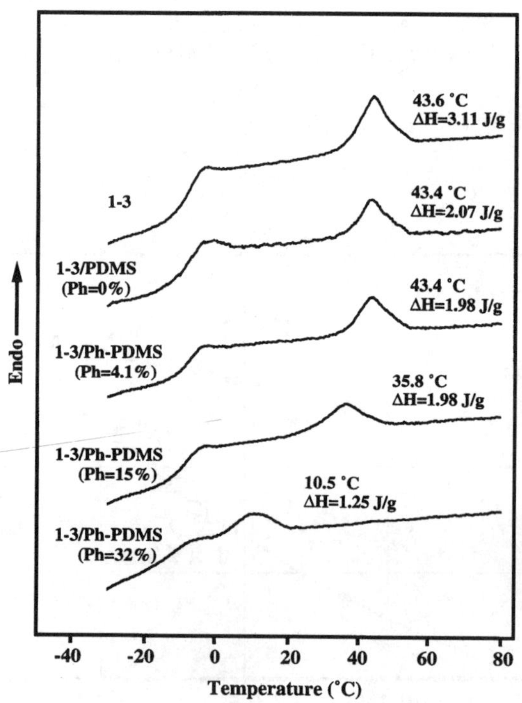

Figure 10. DSC thermograms (10 °C/min, heating) of **1-3**, **1-3**/PDMS, and **1-3**/Ph-PDMS (2/1 by weight) with various Ph contents.

optical microscope and X-ray diffraction measurements. The mixture of **1-3** with PDMS shows the identical DSC trace to that of the base polymer **1-3** except that the enthalpy change associated with the isotropization of **1-3**/PDMS is approximately 66.6% reduced compared to that of **1-3**. This experimental observation clearly indicates that the PDMS is not miscible with **1-3** and the LC structure of the base polymer **1-3** is maintained in the mixture. It also should be noted that the operating temperature of the **1-3**/PDMS mixture as an ER fluid corresponds to the isotropic phase of the base LCP **1-3**. When Ph-PDMS is used instead of PDMS, the isotropization transition is shifted to lower temperatures and the enthalpy change decreases as the phenyl content of Ph-PDMS increases, indicating that the liquid crystallinity of the base polymer is disrupted, most probably because Ph-PDMS is partially miscible with the base polymer **1-3**. The generated shear stress of these mixtures decreases with increasing the miscibility of Ph-PDMS with the base polymer **1-3** as shown in Figure 11.

The results shown in Figures 9-11 can be interpreted in the following two ways; firstly, the ER effect is generated by the induction of the oriented LC network as stated above, and in this network PDMS plays a role to enhance the molecular motion of LC side chains in response to an electric field. When Ph-PDMS is used as a diluent instead of PDMS, miscible Ph-PDMS oil molecules affect the interaction between mesogens to seriously interrupt the LC network formation, resulting in the disappearance of the ER effect. Secondly, the generation of the large ER effect in those mixtures is attributable to the existence of the phase-separated oil which is probably homogeneously dispersed in the base LCP. In this case an applied electric field may cause changes in morphology of the fluid to increase the shear stress. Ph-PDMS does not provide effective morphological changes, since it is not dispersed but dissolved in the LCPs.

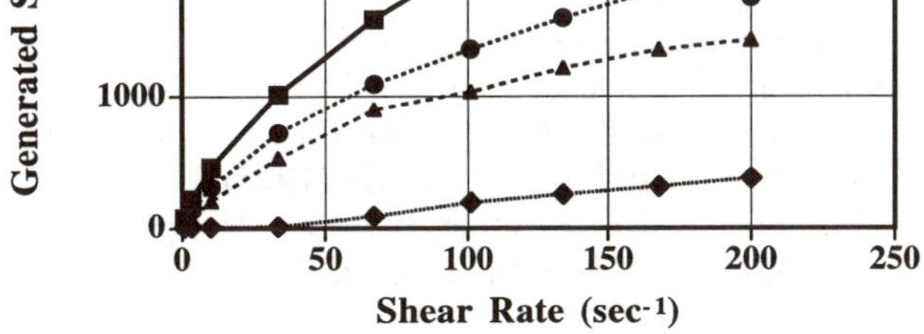

Figure 11. Influence of the Ph content of Ph-PDMS on the generated shear stress of **1-3**/Ph-PDMS mixture (2/1 by weight) measured at 60 °C, DC 2 kV/mm.

4. Conclusions

A new type of ER fluid based on a mixture of side chain liquid crystalline polysiloxane with polydimethylsiloxane has been developed. This fluid is free from the particle sedimentation, aggregation or abrasion problems that are encountered in conventional particle dispersion ER fluids, and has been found to exhibit Newtonian flow and ER effects stronger than those of the particle dispersion ER fluids under an applied electric field.

With respect to the mechanism by which the LC polysiloxane mixture exhibits the ER effect, in addition to the hitherto proposed oriented LC network induction mechanism, another possibility that the phase separation of PDMS oil from the base LC polymer plays a significant role, has been suggested by the investigation using phenyl-substituted PDMS and a DSC analysis. The study is not conclusive and therefore, it has not been evidenced whether either of these two mechanisms is dominant or both are cooperative. Investigation focused on the morphological aspects of the mixture as well as that focused on the molecular orientation would be required for the further elucidation of the mechanism.

6. References

1. W. M. Winslow, *J. Appl. Phys.* **20** (1949) 1137.
2. H. Block and J. P. Kelly, *J. Phys. D, Appl. Phys.* **21** (1988) 1661.
3. T. C. Jordan and M. T. Shaw, *IEEE Transaction on Electrical Insulation* **24** (1989) 849.
4. H. Block, *Chemtech* June (1992) 368.
5. R. Tao and G. D. Roy ed., *Electrorheological Fluids* (World Scientific, Singapore, 1994).
6. T. Honda and T. Sasada, *Jpn. J. Appl. Phys.* **16** (1977) 1775.
7. T. Honda and T. Sasada, *Jpn. J. Appl. Phys.* **17** (1978) 1525.
8. K. Skarp, S. T. Lagerwall, and B. Stebler, *Mol. Cryst. Liq. Cryst.* **60** (1980) 215.
9. M. Fukumasa, K. Yoshida, S. Ohkubo, and A. Yoshizawa, *Ferroelectrics* **147** (1993) 395.
10. A. Inoue and S. Maniwa, EP478034 (1992)
11. I-K. Yang and A. D. Shine, *J. Rheol.* **36** (1992) 1079.
12. K. Tanaka, A. Fujii, and K. Koyama, *Polym. J.* **24** (1992) 995.
13. T. Uemura, K. Minagawa, and K. Koyama, *Chem. Lett.* (1994) 1995.
14. A. Inoue and S. Maniwa, *J. Appl. Polym. Sci.* **55** (1995) 113.
15. H. Kimura, K. Minagawa, and K. Koyama, *Polym. J.* **26** (1994) 1402.
16. V. Percec and C. Pugh, in *Side Chain Liquid Crystal Polymers*, ed. C. B. McArdle (Blackie, Glasgow, 1989).
17. C.-S. Hsu, L.-J. Shih, and G.-H. Hsiue, *Macromolecules* **26** (1993) 3161.

THE DEVELOPMENT OF ELECTRORHEOLOGICAL FLUIDS FOR AN AUTOMOTIVE SEMI-ACTIVE SUSPENSION SYSTEM

THOMAS R. WEYENBERG *and* JOSEPH W. PIALET
The Lubrizol Corporation, 29400 Lakeland Blvd., Wickliffe, Ohio 44092, USA

and

NICHOLAS K. PETEK
RheoActive Inc., 35585 Curtis Blvd., Eastlake, Ohio 44095, USA

ABSTRACT

The feasibility of electrorheological (ER) dampers for an automotive semi-active suspension was evaluated in a three phase program. In the first phase, ER fluid performance targets were derived. The desired ride and handling attributes of the suspension system were translated into damper specifications, which were then translated into the ER fluid performance targets. The damper specifications included dynamic range, bandwidth, power draw, and packaging. The ER fluid performance parameters then included zero-field viscosity, ER stress, response time, and power density. In the second phase, the dampers and the ER fluid were developed to meet the performance targets. Trade-offs were made between damper design and fluid formulation to achieve the desired damper dynamic range and power draw. A state-diagram approach using screen test data was used to select candidate ER fluids. In the third phase of the program, a prototype semi-active suspension system using fast, continuously variable ER dampers was installed on a demonstration vehicle. Heave, pitch, and roll motions of the vehicle were controlled by applying voltages independently to the four dampers as determined by a modified sky-hook algorithm. The system was designed to respond in less than 10 ms with an average power requirement less than 40 W for normal road surfaces and handling. Laboratory data from a pressure driven flow screen test and a damper test are presented that document the ER fluid performance specification and selection process. Vehicle performance data are presented that demonstrate the features of ER technology for the semi-active suspension application. Remaining issues for commercialization of ER fluids are discussed.

1. Program Overview

1.1. Program Scope

Automotive customers are demanding vehicles with both improved ride and improved handling. *Ride* refers to isolating the vehicle from road noise, vibration, and harshness, providing comfort for the passengers. *Handling* refers to the agility of a vehicle while cornering, accelerating, and braking, providing responsiveness for the driver. The challenge in suspension design is that soft springs and dampers (shock absorbers) are generally preferred for ride, while firm springs and dampers are preferred for handling. Implementation of electronically controlled suspensions, such as semi-active systems that adjust damping in response to road inputs, are intended to achieve both ride and handling objectives. Semi-active suspensions using electromechanical valves in the dampers to achieve the variable damping are currently being commercialized by several companies.

Electrorheological (ER) fluids, which change flow properties in the presence of an electrical field, offer the potential for achieving easily controlled, fast acting, continuously variable dampers[1]. In particular, unlike an electromechanical valve, an ER valve achieves flow control without moving parts. To determine if electrorheology can technically offer competitive advantages when used in a semi-active suspension, a three phase program was initiated to install a prototype ER semi-active suspension on a demonstration vehicle in order to identify overall system performance. The scope of the program was to evaluate technical feasibility, and not to optimize a semi-active suspension. Technical merit was evaluated objectively in the laboratory and subjectively using test tracks and public roads.

1.2. Program Responsibilities

The implementation of an ER semi-active suspension system required a multi-disciplinary effort. Expertise was required in ER fluids, ER devices, high voltage amplifiers, sensors, controllers, control algorithms and evaluating the performance of the total system. Three companies formed a partnership for the demonstration program[2]. Monroe Auto Equipment Company was responsible for the sensors, control hardware and software, and ride and handling evaluations of the demonstration vehicle. The Lubrizol Corporation was responsible for providing the ER fluids. RheoActive provided the test vehicle, the ER dampers and the high voltage amplifiers.

2. Performance Targets

2.1. Vehicle

In the first phase, ER fluid performance targets were derived. The desired ride and handling attributes of the suspension system were translated into damper specifications, which were then translated into the ER fluid performance targets. The overall system performance objective was to improve vehicle ride comfort without sacrificing handling. A Ford Thunderbird Super Coupe (SC), known for aggressive handling, was selected because it was non-strut front and rear, and because there was previous experience designing ER dampers for the rear location[1]. To achieve the improved ride, softer springs and anti-roll bars were installed, similar to the baseline Thunderbird LX. The ER semi-active system was relied upon to provide sufficient damping during transient handling events to compensate for the lower roll and pitch stiffness. The ER test vehicle performance was compared to the factory stock Thunderbird SC and LX models.

2.2. Damper and Fluid

The primary design objective was to accomplish a high dynamic range, high bandwidth, low power damper within the original equipment package size. As shown in the ER damper schematic, Figure 1, movement of the piston pushes the ER fluid through a stationary, annular duct. Voltage across the duct creates an electric field perpendicular to the fluid flow, causing an increase in the damping force. The dimensions of the duct and the piston bleed orifice were designed consistent with the damping requirements of the selected control algorithm and the performance of the ER fluid.

The damper dynamic range, from minimum to maximum force, is directly related to the degree of rheological change for the ER fluid. The minimum damping force occurs when the voltage is off and is due entirely to the viscosity η of the fluid and is proportional to its flow rate through the duct. A nominal target of 500 N at 0.4 m/s piston velocity was established for minimum damping in the front damper, corresponding to 60 mPa•s viscosity for the fluid. When voltage is applied, the damping force is modeled as increasing by an amount proportional to the ER fluid yield stress, τ_y, regardless of piston speed. The maximum targeted damper force of 3000 N for the front damper established the maximum ER fluid yield stress, corresponding nominally to 4 kPa at the maximum electrical field available in the damper, based on a 5 kV amplifier. The target forces for the rear damper were slightly lower, and thus achieved with the same ER fluid.

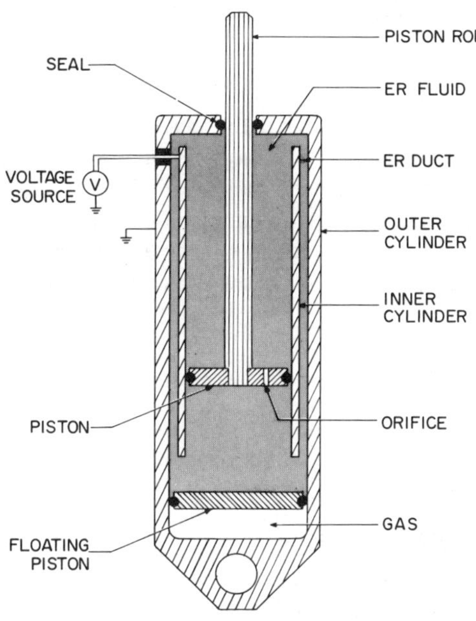

Figure 1: ER damper schematic.

Damper bandwidth is desired to be as high as possible to enable the semi-active system to respond to body heave (2 Hz), body roll (5 Hz) and wheel hop (15 Hz). Time lags in each system component (sensor, controller, amplifier, damper) combine to give an overall response time. A system response time of 10 ms was established as the target for continuously variable semi-active control, including a target of 5 ms for the ER fluid filled damper.

Damper power draw is desired to be kept to a minimum to conserve vehicle power and to minimize the cost and size of the high voltage amplifier. Maximum electrical power to the damper was established to be 20W. Given previous experience with the volume of ER fluid in the annular duct, a fluid power density ceiling of 500 kW/m^3 was established for the front damper. Additional damper and fluid performance targets such as durability, lubricity, settling and cost, which do not directly affect damper performance were considered but will not be discussed herein.

3. Fluid Selection and Performance

3.1. Operating Temperature Range

In the second phase, the dampers and the ER fluid were developed to meet the performance targets. Trade-offs were made between damper design and fluid formulation to achieve the desired damper dynamic range and power draw. The performance of a semi-active suspension is desired to remain stable over a broad temperature range, but ER fluid performance is known to vary with changing temperature[3,4]. As fluid temperature increases, zero-field viscosity and response time decrease, power density increases and yield stress may go through a maximum.

3.2. State-Diagram Approach

A state-diagram approach using screen test data was used to select and optimize candidate ER fluids which met design targets over a temperature range[5]. One complexity in optimizing fluid and damper design is that variables in each can be modified, within packaging constraints, to optimize the overall damper performance. To simplify the classification and description of fluids, a relationship was developed which relates dynamic range to power density. This dimensionless measure is known as the Winslow number[6], W_n:

$$W_n = \frac{(\tau)^2}{\eta \; \phi} \qquad (1)$$

Where

τ = ER fluid stress, stress increase or yield stress (Pa)
η = viscosity without electric field (Pa•s)
ϕ = power density (W/m^3) = electric field (V/m) • current density (A/m^2)

The Winslow number can be thought of as an overall measure of merit for an electrorheological fluid. It is essentially the ratio of ER Effect, a measure of fluid rheology change, to Power Density, a measure of the electrical cost necessary to achieve the rheology change.

$$\text{ER Effect} = \frac{(\tau)^2}{\eta} \qquad (2)$$

Whether a yield stress, steady state shear stress change, or dynamic stress is used in the calculation depends on what is pertinent to the device requirements. For ER semi-active damping, a shear stress measurement taken 5 milliseconds after the application of electric field in an oscillating duct flow screen test was used. Figure 2A illustrates how Winslow number varies over a temperature range for a select ER fluid. If Winslow number were the only criteria, a W_n of 500 would indicate an operational temperature

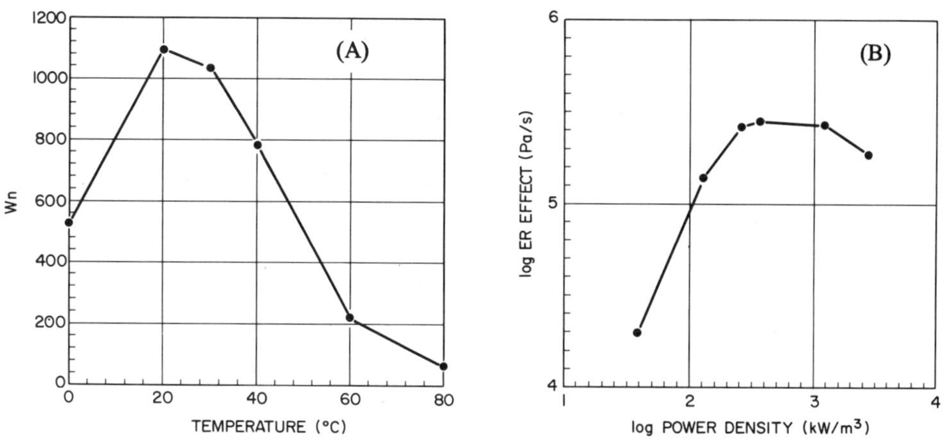

Figure 2: (A) Winslow Number vs. Temperature. Stress measurements made 5 ms after field pulse at 6000 1/s. (B) State-Diagram showing relationship between rheological change and power draw.

range of 0° to 50°C for the semi-active system. Because of damper and high voltage amplifier design constraints, the temperature range is further limited from the Winslow number calculation. The effect of these constraints can best be seen by examining the same data in state-diagram format, Figure 2B, in which ER Effect is plotted against Power Density. The target performance area is toward the upper left hand quadrant. Note that the operational range of this fluid is bounded at high temperatures by high power density and at low temperatures by low ER Effect.

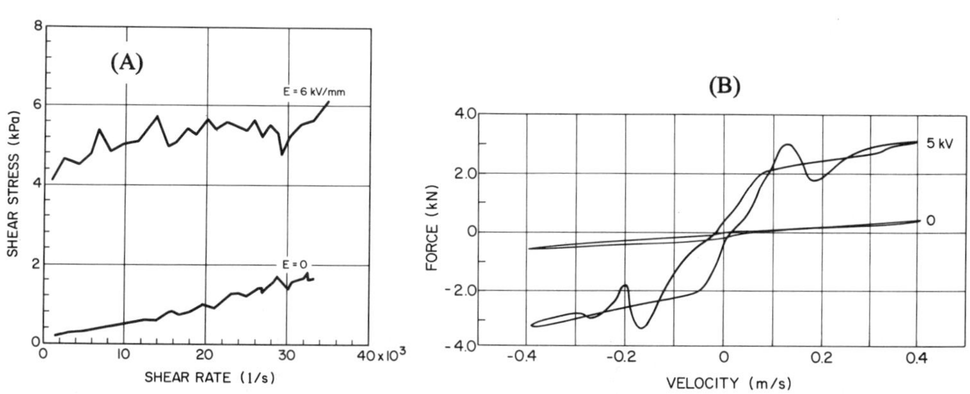

Figure 3: (A) ER fluid performance in an oscillating duct flow screen test device. (B) Steady-state properties of front ER damper.

3.3. Fluid and Damper Performance

The ER fluid selected for the vehicle damper program had a broad dynamic range and high bandwidth over a temperature range broad enough for vehicle demonstrations. The fluid rheology illustrated in Figure 3A from an oscillating duct flow screen test translates into the ER damper performance shown in Figure 3B. The effect of the piston bleed orifice is observed at low piston velocities and serves to reduce harshness associated with piston direction reversal. The damper response time plot, Figure 4, was obtained using one of the prototype high voltage amplifiers with a step input. The ER damper and amplifier combined response time is nominally 5 ms.

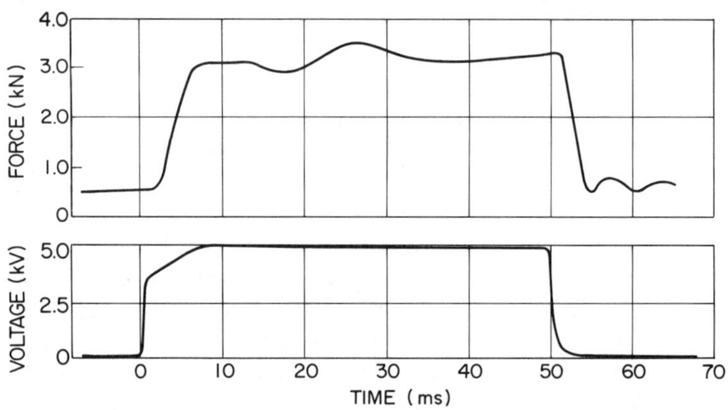

Figure 4: Transient response of ER damper to step voltage.

4. System Performance

4.1. Configuration and Components

The ER semi-active suspension installed on the demonstration vehicle during Phase 3, as shown in Figure 5, included the following components at each of the four vehicle corners: 1) ER damper, 2) high voltage amplifier, and 3) acceleration and displacement sensors. A controller was centrally located with specially designed software that monitored all the sensors and determined damping level independently at each damper. The ER semi-active system components are further described in Table 1.

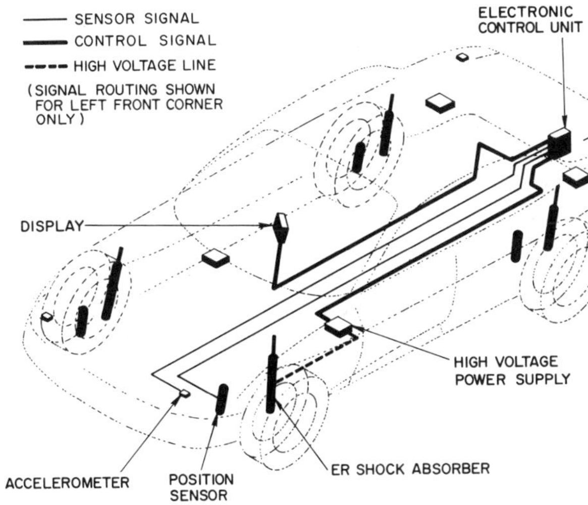

Figure 5: ER semi-active suspension schematic.

Table 1 ER Semi-active System Components

ER Dampers	• No moving parts needed to achieve variable damping • Continuously variable force, independent of piston velocity
ER Fluid	• Experimental material for semi-active suspension • Wide dynamic range and fast response within temperature range • 5 ms time constant for ER damper
High Voltage Amplifiers	• 4 prototype x1000 DC voltage amplifiers, max. output 20W, 4mA • Input requirement: 12V, 1W with zero output; 35 W with full output
Control Hardware	• 4 accelerometers and 4 inductive displacement sensors • Intel 80C196 16 bit microprocessor, 16 Hz, optically isolated • Under 2 ms from sensor input to power supply output
Control Software	• Modified sky-hook algorithm with 4 separate, continuously variable outputs • Independent control of heave, pitch, roll and wheel damping • 2 ms program cycle time
Vehicle	• Ford Thunderbird Super Coupe • Lower rate springs and anti-roll bars • No packaging modifications to vehicle required to accommodate ER units

4.2. Vehicle Evaluations

A single wheel shaker was used to characterize the suspension system. The right front corner of the ER semi-active vehicle and two baseline vehicles were excited vertically with a sinusoidal frequency sweep (1 to 20 Hz) while body acceleration and tire load variation were measured. Passenger ride comfort has been shown to correlate with vertical acceleration of a vehicle body, with lower accelerations corresponding to greater comfort. Figure 6 (top) illustrates that the ER semi-active suspension causes generally lower accelerations than the standard Thunderbird LX and SC suspensions. Road holding and handling have been shown to improve when tire load variation is minimized. Figure 6 (bottom) shows that the ER suspension reduces tire load variation at the body heave resonance, compared to standard suspensions. At the wheel hop resonance, the suspensions exhibit similar magnitudes of tire load variation.

On test tracks, the ER vehicle demonstrated aggressive handling characteristics in lane change and braking maneuvers. The system demonstrated a particular ability to

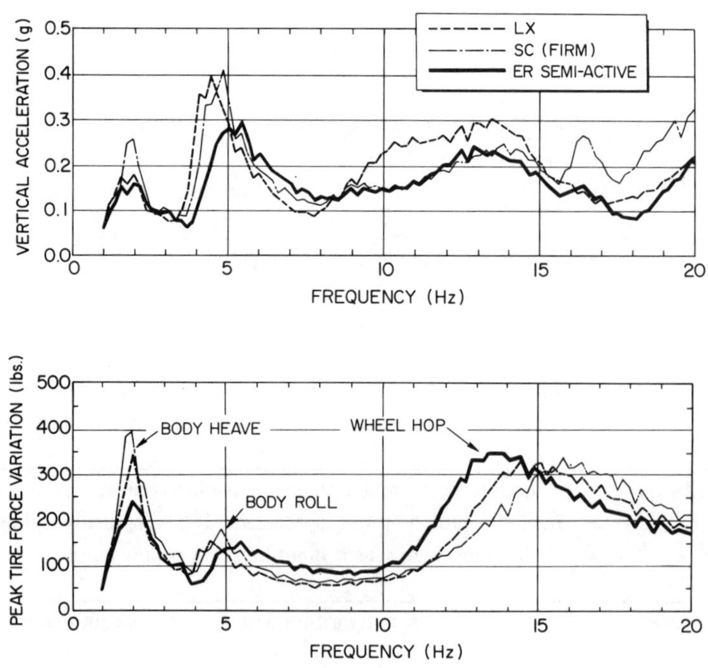

Figure 6: Single wheel shake test data, right front corner.

control body heave over large inputs when compared to the baseline LX and SC vehicles. This is attributable to the high damping forces at low piston velocity achieved by the ER dampers. The system power requirements were dependent on the magnitude and frequency of the road inputs and the temperature of the ER fluid within the dampers. Near the maximum operating temperature of the fluid and under normal road conditions with relatively aggressive driving, power requirements of the individual power supply/damper combinations averaged less than 7 W. Including the controller, the average system power requirement was less than 40 W in normal operation. At lower temperatures within the optimal range, system power requirement was reduced. Transient peak power requirement for the system was 150 W.

5. Conclusions

This vehicle demonstration of a continuously variable semi-active suspension system showed that high bandwidth, high dynamic range ER dampers were constructed within a conventional package size and required no moving parts to achieve variable damping. When operated within the optimal temperature range of the ER fluid, the suspension had a system response time less than 10 ms and required less than 40 W of power under normal driving conditions. Quantitative results showed that the level of body acceleration (an indicator of comfort) and the tire load variation (an indicator of handling) were improved, compared to standard Thunderbird LX and SC suspensions. At present, the application of the system is limited by the temperature range in which the ER fluid meets the fluid design targets.

6. Acknowledgments

The authors wish to acknowledge the considerable contributions of Monroe Auto Equipment Company to the success of this demonstration program.

7. References

1. N. K. Petek, *An Electronically Controlled Shock Absorber Using Electrorheological Fluid*, SAE 920275 (SAE, Warrendale, PA, 1992).
2. N. K. Petek, D. J. Romstadt, M. B. Lizell and T. R. Weyenberg, *Demonstration of an Automotive Semi-Active Suspension Using Electrorheological Fluid*, SAE 950586 (SAE, Warrendale, PA, 1995).
3. H. Conrad, Y. Chen and A. F. Sprecher in *Proc. 2nd Int. Conf. on Electrorheological Fluids*; ed. J. D. Carlson, A. F. Sprecher, and H. Conrad, (Technomic, Lancaster, 1990) p. 252.
4. D. R. Clark and J. W. Pialet in *Progress in Electrorheology: Science and Technology of Electrorheological Materials*; ed. K. Havelka and F. Filisko, (Plenum, in print).
5. K. H. Hagele, *et. al*, *Continuously Adjustable Shock Absorbers for Rapid-Acting Ride Control Systems*, SAE 905125 (SAE, Warrendale, PA, 1990), fig. 5.
6. F. P. Boyle in *Electrorheological Fluids - Mechanisms, Properties, Structure, Technology, and Applications*, ed. R. Tao (World Scientific, 1992).

SIMULATIONS OF ELECTRORHEOLOGICAL FLUIDS WITH HYDRODYNAMIC LUBRICATION

J. R . MELROSE†, S. ITOH† AND R. C. BALL*

Polymers and Colloids† and Theory of Condensed Matter* groups,
Cavendish Laboratory, Madingley Road, Cambridge CB3 0HE

ABSTRACT

We report preliminary results for 3D simulations of a reduced model of electrorheological (ER) fluids. The hydrodynamic interactions are just squeeze flow lubrication and the electrostatic interactions that of point dipoles. In the low Mason number regime the same layer structure seen in earlier non-hydrodynamic simulations is found.

1. Introduction

Computer models can provide insight into structure of ER fluids[1-3]. Simulations in 3d have not to date included hydrodynamic interaction[2,3]. The most complete hydrodynamic model[4] is restricted to small systems in monolayers and therefore of limited use for structure in 3D. This may be serious , as the 3D simulations without hydrodynamics[2,3] and experiment[5] both indicate that the phase of enhanced viscosity will show layer formation normal to the vorticity axis. To facilitate large system studies in 3D, which we consider to be crucial to progress on structure, we are developing models in which hydrodynamic effects are limited to just two body viscous interactions.

Modelling the flow of densely packed particles in a hydrodynamic medium is a challenge and requires approximation. The most complete technique[6] uses low moment expansions of the far field hydrodynamics to generate approximate many body drag terms in combination with two body near field terms - the algorithm requires inversion of non-sparse matrices. These $O(N^3)$ operations limit the algorithm to small numbers ($N < 100$) of particles. This is the algorithm used in Ref. 4. In our reduced model we only include the two-body near field terms. These terms are singular on close approach of particles, with the dominant squeeze mode diverging with the inverse inter particle gap, and are anticipated to dominate structure under flow in concentrated systems. The resulting algorithm requires inversion of sparse matrices which can be solved by $O(N^2)$ iterative techniques.

2. The Model

Hydrodynamic interactions between close particles can be divided into squeeze modes due to relative motion along their line of centres and shear terms arising from shear motion and their relative rotation. The fluid mechanical creeping flow equations can be solved in the narrow gaps between close surfaces by use of the lubrication approximation. The leading order the squeeze flow hydrodynamic force on a particle is given by the well known Reynolds formula.

$$\mathbf{f}_i = - \Sigma_j \, (3 \pi \mu / (8 h_{ij})) \{ (\mathbf{v}_i - \mathbf{v}_j) . \mathbf{n}_{ij} \} \mathbf{n}_{ij} + O(\ln(2/h_{ij})) , \qquad (1)$$

where μ is the viscosity of the solvent; the sum is over nearest neighbour particles j (see below),; h_{ij} is the gap between the neighbour surfaces; $\mathbf{n}_{ij}$ is the unit vector along the line of centres i to j and $\mathbf{v}_i$, $\mathbf{v}_j$ are the particle velocities. We included higher order logarithmic terms using the expressions in Ref. 7. We report here on models with just squeeze terms. Note the viscous interaction diverges as the inverse of the gap between the particle surfaces. This is a consequence of the stick boundary conditions : the fluid and particle velocities match on the surface of the particles. If close approaching particle surfaces have very different velocities very strong hydrodynamic forces occur . In practice close particles have close velocities and the denominators and numerators in (1) are of the same order.

For N particles we can encode the model in terms of the 3N dimensional vector of the components of hydrodynamic force on all the particles, $\mathbf{F}_H$, related linearly to the corresponding 3N velocity vector, $\mathbf{V}$, by

$$\mathbf{F}_H = \mathbf{R} \cdot \mathbf{V} , \qquad (2)$$

where $\mathbf{R}$ is the 3N x 3N dimensional resistance matrix. The motion of the particles in the fluid medium is overdamped - that is the forces balance. The hydrodynamic forces balance the electrostatic and other colloid interactions, $\mathbf{F}_C$, and the equation of motion is given by

$$\mathbf{F}_H + \mathbf{F}_C = \mathbf{0} . \qquad (3)$$

The simulation algorithm for a given configuration of particles at time t is as follows :

1) evaluate the forces $\mathbf{F}_c$ by summing the net force on each particle from the electrostatic and other colloid interactions;

2) evaluate the resistance matrix $\mathbf{R}$ from a sum of the matrix of coefficients in the pair terms of (1) i.. e. $(3 \pi \mu / (8 h_{ij})) \mathbf{n}_{ij}\mathbf{n}_{ij}$. - the matrix is sparse and therefore the elements for just each pair of neighbours is stored;

3) solve equations (2,3) for the velocities $\mathbf{V}$ by an iterative inversion of the matrix $\mathbf{R}$, applying shear boundary conditions to the velocity field;.

4) move the 3N particle co-ordinates $\mathbf{X}$ over a time step dt :

$$\mathbf{X}(t+dt) = \mathbf{X}(t) + \mathbf{V} \, dt \ .$$

In our simulations the viscous interaction (1) is applied between nearest neighbours identified by the construction of Delaunay tetrahedra. The technical details of this will be reported elsewhere[9]. This is not an essential feature of the model, simpler ways of identifying nearest neighbours could be used. The matrix **R** is sparse and the solution of (2,3) is therefore carried out by iterative techniques inverting the matrix **R**. Techniques for the inversion of sparse matrix are well documented[10] - we used were conjugate gradient techniques. Step (3) dominates the computation.

Flow is driven by Lee-Edwards boundary conditions[8]. The simulations are carried out in system of sheared periodic images. That is the N particles sit inside a central box which is surrounded periodically by images of itself. Particles near the edges of the box interact with image particles. If a pair of such nearest neighbours lies across the box edge then a box velocity is added to the relative velocity of the particles when evaluating the viscous interactions. The effect of this is to apply via the velocity dependent viscous interactions a net shear across the system. This application of shear is more physically sensible than the globally imposed background flow used in simulations without hydrodynamics[1-3]. Indeed for simulations of repulsive sphere we have found that such artificially imposed flow fields can lead to ordered arrays of particles under flow which are not observed in corresponding experiment.. Note the inversion of **R** in the solution of the velocities introduces highly correlated collective motions of the particles. For example if the equations were solved for a chain of particles with an applied force to end particle of the chain, then the solution of (2,3) would generate velocities on all particles in the chain.

The divergent squeeze terms alone would prevent any particle overlap, but simulations of smooth hard spheres lead to unphysical small gaps between particles. Indeed we have recently shown that perfectly smooth hard spheres do not in fact reach a steady state. We find that before a strain of 0.5 is reached some gaps between particles have closed to 10^{-10} the diameter of the particles. To avoid this we include short range spring forces as a crude model of a porous surface coat. This controls how close the particles can approach each other. The interaction forces between spheres i and j separated by r_{ij}

$$\begin{aligned} \mathbf{f}^{c}_{ij}(r) &= -\mathbf{n}_{ij}\, F_0(1-r_{ij}/L) \quad \text{if } r < L, \\ \mathbf{f}^{c}_{ij}(r) &= 0 \qquad\qquad\qquad\qquad r > L. \end{aligned} \tag{4}$$

In the present work we set L = 0.01 particle diameters.

We have extended the model to include Brownian forces. These random forces are correlated through the matrix **R**. We can generated these via sums of weighted random forces generated between each pair of neighbours.

The electrostatic interactions are approximated by point dipoles as detailed elsewhere[3] - although this captures the symmetries of the ER electrostatic interaction it is a weak representation of electrostatic interactions in real ER fluids. It is straightforward to include more complex interactions in the model, these are evaluated at step (1) of the computation for each configuration and therefore their complexity does not increase much on the overall computational load.

The reduced units of the simulation for viscosity, length and time are ($\eta = \eta^* \mu$, $L = L^* d$, $t = t^* (1/\dot{\gamma})$) where the units of viscosity, length and time are the solvent viscosity, particle diameter and inverse shear rate. The parameter space of the model is the Mason number[11], Mn, characterising the ratio of the shear to electrostatic forces we utilse the definition from our earlier work[3] and the West number

$$W = (\dot{\gamma}\, d / \mu\, F_0) \tag{5}$$

which characterises the ratio of the shear forces to the surface spring. The stress tensor was calculated by pair wise sums over all pair forces, including the hydrodynamic pair forces Eq. 1.

3. Results

In the simulations we fixed the shear rate at $\gamma = 1$, West number at 10^{-6} and varied Mn - equivalent to an experiment at fixed shear rate and varying applied field. We will report results for systems at a volume fraction of 40 % of 200 particles over a range of Mason numbers and 700 particles at Mn = 0.01 The authors of Ref.11 were able to collapse a series of experimental data at different volume fractions by plotting the viscosity divided by its high shear rate limit and the Mn by volume fraction. We adopt the same procedure in figure (1) to compare our data with previous

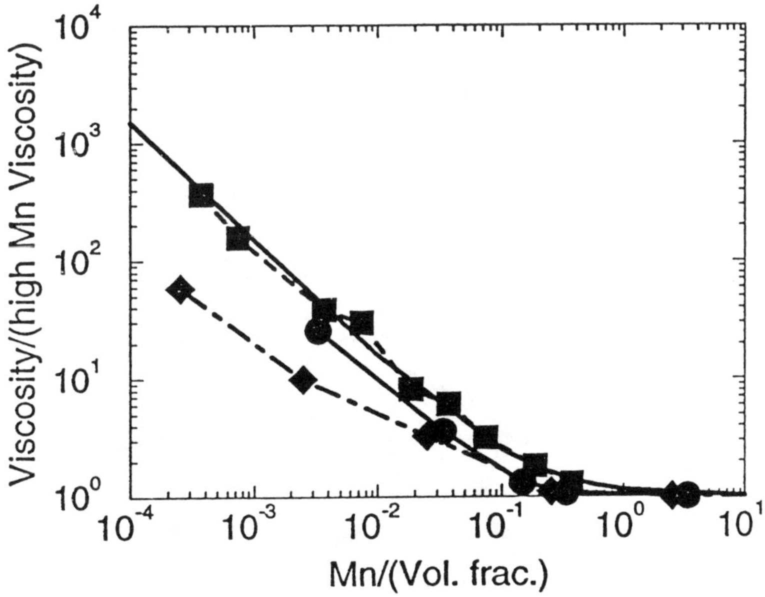

Figure 1. A plot of viscosity normalised by its high shear rate limiting value against the Mason number divided by volume fraction (c.f. Ref. 11) Diamonds : the simulation data of figure 1 ; Squares : the data of Ref. 4 table II with equivalent volume fraction 0.27; Circles : the data of Ref. 3. Solid line : the fit of Ref. 11 to their collapse of experimental data.

simulations[3,4] and experiment[11]. All the simulations roughly agree on the Mason number below which viscosities are enhanced. The data of Ref. 4 fits the experiment but the weak representation of the electrostatic interactions probably explains the low values for viscosity at low Mn here and in Ref. 3.

All three simulations, however, differ in the way they treat the interactions of close particles. In Ref. 3 a steep power law replușion is employed which complicates comparison. In Ref. 4 smooth hard spheres are employed. Our experience with hard sphere systems[12] leads us to suspect that at low Mn numbers hydrodynamic contributions are dominated by unphysically close approaching particles. The hydrodynamic contibution in Ref. 4 grows with decreasing Mn. In contrast to our results whereit remains close to constant. A more serious complication with the data of Ref. 4 is substantial system size effect with a reported halving of the viscositiy on doubling the monolayer size from 25 to 50 particles.

At high Mason numbers the structure is disordered in contrast to the string phases found in simulations without hydrodynamics[2,3]. In the low Mn regime, we find the same layering phenomena seen in the earlier work[2,3]. In the case of the 200 particle system layers first formed at Mn = 1.0. Figures 2 and 3 shows snapshot projections of the N=200,700 systems at Mn = 0.01. Increasing the system size brought about thicker layers; moreover one layer orders whilst the other does not. The viscosity of the two system sizes agree $\eta = 10.3$ (10.4) +/- 0.08 for N=200 (700). The 3D simulations do not show the large size effects of monolayers[4].

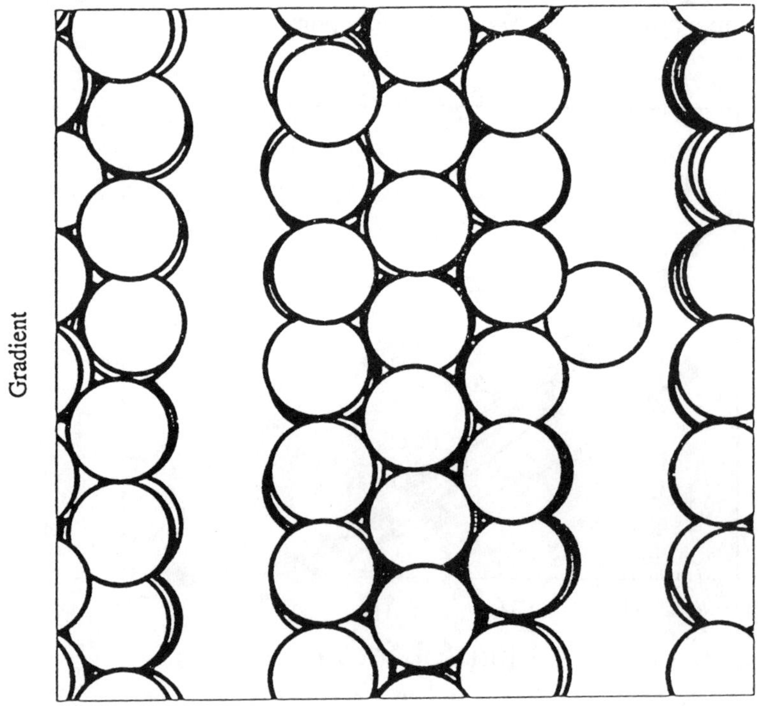

Figure 2. A configuration of 200 particles at Mn=0.01 projected onto the shear gradient-vorticity plane; box side length is 6.4 particle diameters.

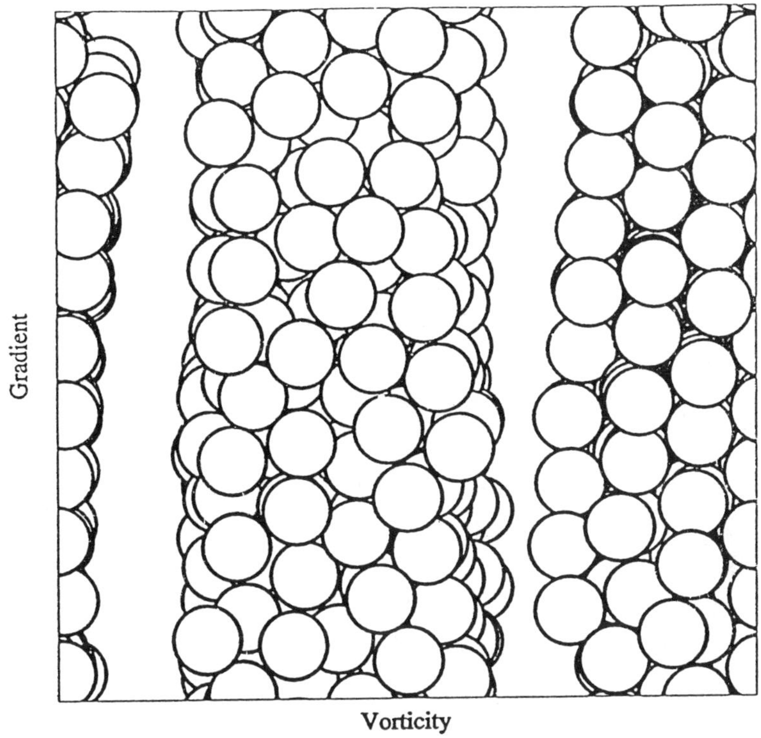

Figure 3 A configuration of 700 particles at Mn=0.01 projected onto the same plane. Box side length is 9.7 particle diameters.

4. Discussion

We wish to study structure in large 3D systems and therefore are pursuing reduced models of hydodynamics. We have presented preliminary results for a model not likely to be applicable much below 40 % volume fraction as many body hydrodynamics is increasingly influential at lower volume fractions. To assess how well the model performs quantitatively the large systems will require a better representation of the electrostatic interactions. We note it is quite feasible on workstations to simulate the current model with several thousand particles and hope to report on such simulations in the future. The hydrodynamic model we have developed[12] also includes particle rotation, shear lubrication terms and Brownian motion. We also hope to report on this more complete model in the future.

5. Acknowledgment

We thank the DTI/colloid technology project co-funded by the DTI, Unilever, Schlumberger and ICI for funding.

6. References

1. D. J. Klingenberg, F. van Swol and C. F. Zukoski, *J. Chem. Phys.* **91** (1989) 7888; *J. Chem. Phys.* **94** (1991) 6160.
2. D. M. Heyes and J. R. Melrose, *Mol. Sim.* **5** (1989) 293 ; *J. Chem. Phys.* **98** (1993) 5873; M. Whittle, *J. Non-Newtonian Fluid Mech.* **37** (1991) 223.
3. J. R. Melrose, *Phys. Rev. A* **44** (1991) R4700; *Mol. Phys.* **76** (1992) 635.
4. R. T. Bonnecaze and J. F. Brady, *J. Chem. Phys.* **96** (1992) 2183.
5. G. Bossis, Y. Grasselli, Lemaire E:, J. Persello, L. Petit, *Europhysics Lett.* **25** (1994) 202.
6. G. Bossis and J. F. Brady, *J. Chem. Phys.,* **80** (1984) 5141; J. F. Brady, R. J. Phillips, J. C. Lester and G. Bossis, *J. Fluid. Mech.* **195** (1988) 257 and references therein.
7. S. Kim and S. Karrila, *Microhydrodynamics* (Butterworths 1992).
8. A. W. Lees and S. F. Edwards, *J. Phys.,* **C5** (1972) 1921.
9. R. C. Ball and J. R. Melrose, Unpublished.
10. W. H. Press, S. A. Teukolsky, W. T. Vettering and B. P. Flannery, *Numerical Recipes* Cambridge Univeristy Press. (1986).
11. L. Marshall, C. F. Zukoski and J. W. Goodwin, J. Chem. Soc. Faraday Trans. 1 **85**, 2785 (1989).
12. J. R. Melrose and R. C. Ball, Unpublished.

POLARIZATION OF SURFACE IONS ON A SPHERICAL PARTICLE IN AN ELECTRIC FIELD

Yue Hu, Vidya Kumar, and Kristen E. Blackmon
Department of Physics, Wellesley College, Wellesley, MA 02181, U.S.A.

ABSTRACT

Polarization of surface ions on a spherical colloidal particle in a transient electric field is simulated using a random walk method. Total polarization increases as ion density increases but saturates when the Coulomb repulsive forces between the ions become dominant. The time constant of polarization is shorter for larger ion densities, and non-linear relations between polarization and the electric field strength are also seen.

1. Introduction

In designing practical electro-rheological (ER) fluids, one of the main objectives is to maximize the strength of the polarization of the suspended colloidal particles in electric fields, because the strong attractive forces between the polarized particles cause the particles to form chains and columns and thereby drastically increase the viscosity of the fluids. Recent experiments have shown that for samples of insulating particles suspended in insulating fluids, the strength of polarization of the particles is dominated by the *conductive* mismatch in DC fields, even though care has been taken to remove possible conducting elements (moisture) from the fluids.[1] The investigators in these experiments have attributed the residual conductivity to conducting elements adsorbed on the surfaces of the particles.

Because adsorbed ions are confined to the interface between the particles and the fluids and are not present in the bulk fluids, standard methods for solving the Poisson equation for homogeneous materials do not apply. In this paper, we use a random walk method to study the polarization of ions confined to the surface of a spherical dielectric particle immersed in a dielectric fluid. We will explore how the polarization depends on the field strength and ion density, and we will also study the time constants of polarization in transient fields.

2. Model

2.1. Electrostatic Forces

We consider a configuration of N ions of charge q confined to the surface of a spherical particle of radius R. The particle has dielectric constant ε_p and is surrounded by a uniform fluid of dielectric constant ε_f. We choose a spherical coordinate system with the origin at the center of the particle, and a uniform external electric field of magnitude E is applied in the z-direction.

Due to the dielectric mismatch between the particle and the fluid, the external electric field is distorted near the particle, and this field is easily calculated by the standard boundary-value method of electromagnetism,[2] yielding a potential on the surface of the particles of the form

$$\varphi_E = -\frac{3\varepsilon_f}{\varepsilon_p + 2\varepsilon_f} ER\cos\theta, \tag{1}$$

where θ is the angle between the external electric field and the position vector of the point of interest. This field exerts a force on each ion on the surface of the particle and causes a biased diffusion of the ions in the direction of the component of the field tangential to the spherical surface of the particle (assuming that the ions are confined to the surface of the particle).

In addition to the force from the external field, each ion also experiences Coulomb repulsive forces from other ions, and the ions also polarize the particle and the surrounding fluid. To calculate the polarization of the media by an ion, we set up a boundary-value problem with a charge q located at the north pole ($\theta = 0$) of the particle. The exact solution[2] for the potential at a point on the surface of the particle is

$$\varphi_q = \frac{q}{4\pi\varepsilon_o\varepsilon_f R}\left(\frac{1}{2\sin(\theta/2)} + \sum_{n=0}^{\infty}\frac{n(\varepsilon_f - \varepsilon_p)}{(n+1)\varepsilon_f + n\varepsilon_p} P_n(\cos\theta)\right), \tag{2}$$

where $P_n(\cos\theta)$ is the Legendre polynomial of order n. Noting that $2R\sin(\theta/2)$ is the distance between the charge and the point of interest, we recognize the first term in Eq. 2 to be the potential of a point charge q. The second term in Eq. 2 is the result of the polarization of the particle and the fluid due to the point charge. Using the Legendre polynomial generating function, the sum in the second term of Eq. 2 can be expressed as

$$\sum_{n=0}^{\infty}\frac{n(\varepsilon_f - \varepsilon_p)}{(n+1)\varepsilon_f + n\varepsilon_p} P_n(\cos\theta) = \frac{\varepsilon_f - \varepsilon_p}{\varepsilon_f + \varepsilon_p}\frac{1}{2\sin(\theta/2)}$$
$$+\sum_{n=0}^{\infty}\left(\frac{n(\varepsilon_f - \varepsilon_p)}{(n+1)\varepsilon_f + n\varepsilon_p} - \frac{\varepsilon_f - \varepsilon_p}{\varepsilon_f + \varepsilon_p}\right) P_n(\cos\theta). \tag{3}$$

The first term in Eq. 3 is proportional to the potential of the image charge of the ion if the interface between the particle and the fluid is flat. The second term is the correction for the spherical interface. Numerical evaluation shows that the second term is about 15% of the first term, and we will only use the first term for simplicity.

Adding the potentials due to the external field and those due to the presence of the other ions, we can write down the electrostatic potential at the position of the ith ion (located at (θ_i, ϕ_i)) as:

$$\varphi_i = -\frac{3\varepsilon_f}{\varepsilon_p + 2\varepsilon_f} ER\cos\theta_i + \frac{q}{4\pi\varepsilon_o(\varepsilon_p + \varepsilon_f)R}\sum_{j\neq i}^{N}\frac{1}{\sin(\gamma_{ij}/2)}, \tag{4}$$

where γ_{ij} is the angle between the position vectors of the ith and the jth ions:

$$\cos\gamma_{ij} = \cos\theta_i\cos\theta_j + \sin\theta_i\sin\theta_j\cos(\phi_i - \phi_j). \tag{5}$$

The electrostatic force on the ith ion is the negative gradient of Eq. 4 multiplied by charge q, and the velocity of this ion is

$$v_i = -\mu \nabla \varphi_i, \tag{6}$$

where μ is the mobility of the ion (velocity per unit field). Because we assume that the motion of the ions is confined to the surface of the particle, we only consider the components of Eq. 6 that are tangential to the spherical surface.

2.2. Random Walk

The electrostatic forces discussed in the previous section cause a net motion of the ions in the electric field, and this motion is effectively a biased diffusion process because of thermal agitation. It has been well established that a diffusion process can be simulated with a random walk.[3] The random walk is a biased one in our case because of the net motion of the ions. If the probability for a walker to make a step of size Δl in a specified direction in time Δt is p_+ and the corresponding probability in the opposite direction is p_-, the average drift velocity of the walker is

$$v_d = (p_+ - p_-)\frac{\Delta l}{\Delta t}. \tag{7}$$

The diffusion coefficient D is related to Δl and Δt by

$$D = \frac{1}{2}\frac{(\Delta l)^2}{\Delta t}. \tag{8}$$

In our simulation the ions move in a curved two-dimensional space on a spherical surface. A two-dimensional random walk is usually carried out by allowing a walker to make steps in four directions along two perpendicular axes, with the total probability along either axis being one half. In a flat geometry, random walk steps can be taken by changing either an x- or y-coordinate in a Cartesian coordinate system. On a spherical surface, however, simply changing either the θ- or ϕ- coordinate results in an eventual build-up of walkers near the poles. To avoid this bias problem, our simulation uses random walk steps taken along two great circles intersecting orthogonally at the current walker's location. We have verified that in the absence of any drift velocity, a randomly distributed set of walkers remains randomly distributed after thousands of random walk steps taken using this method.

2.3. Simulation Method

The goal of the simulation is to study the response of the distribution of ions on a spherical particle surface to a DC electric field that is turned on at time $t = 0$. First, we scale length, electric field strength, and time to be dimensionless by normalizing them relative to the following quantities:

$$r_o = R, \quad E_o = \frac{q}{4\pi\varepsilon_o\varepsilon_f R^2}, \quad t_o = \frac{R}{\mu E_o}. \tag{9}$$

In our simulation, we use $R = 1\ \mu\text{m}$, $q = e$, $\varepsilon_f = 80$, and $\varepsilon_p = 3$, which yield $E_o = 18\ \text{V/m}$. If we take the mobility of ions μ as that in a bulk solvent (typically on the order of $10^{-4} cm^2V^{-1}s^{-1}$), t_o is roughly 6 s. Because the ions in our study are tightly bound to the surface of the particle, we expect the mobility to be much lower, and therefore the value of t_o should be much larger than 6 s.

We start with *N* ions randomly distributed on a spherical surface. The drift velocity of each ion due to electrostatic forces is calculated according to Eqs. 4 and 6, and the probabilities of random walk steps are calculated according to Eq. 7. A random number generated by the computer then determines which step to take in accordance with the calculated probabilities. Polarization of the ions is calculated according to the following equation:

$$p_z = \frac{1}{N}\sum_{i=1}^{N} z_i , \tag{10}$$

where z_i is the coordinate of the *i*th ion along the axis of the electric field.

A practical difficulty arises when the total number of ions *N* in the simulation becomes large, because computation time is essentially proportional to N^2 due to the calculation of pair interactions. We argue that by renormalizing the pair interactions, we can simulate the polarization of a large number of ions *N* by using a smaller set of ions *N'*. First, we examine the forces on each ion when the total number of ions *N* increases. The first term of Eq. 4 represents the influence of the external field on each ion and is determined only by the position of each ion. This term is therefore independent of the number of ions, and its contribution to the polarization per ion (Eq. 10) is unaffected by the number of ions in the simulation. The second term of Eq. 4, however, is sensitive to the total number of ions, because in a fixed surface area, more ions means a smaller average distance between the ions and a correspondingly stronger Coulomb repulsion between the ions. We can renormalize the number of ions by considering *N* ions uniformly distributed in a flat circular area of radius *R*. The charge density is

$$\sigma = \frac{Nq}{\pi R^2}, \tag{11}$$

and the potential energy of the ion at the center of the circle due to the pair-interaction with other ions is

$$U = \frac{1}{4\pi\varepsilon_o}\int_0^R \frac{q\sigma}{r} 2\pi r dr = \frac{1}{2\pi\varepsilon_o}\frac{Nq^2}{R}. \tag{12}$$

If we wish to maintain the same potential energy for the ion while reducing the total number of ions to *N'*, we have to renormalize the pair potential by a factor of *N/N'*, i.e.,

$$U = \frac{N}{N'}\frac{1}{4\pi\varepsilon_o}\int_0^R \frac{q\sigma'}{r} 2\pi r dr = \frac{N}{N'}\frac{1}{2\pi\varepsilon_o}\frac{N'q^2}{R} = \frac{1}{2\pi\varepsilon_o}\frac{Nq^2}{R}. \tag{13}$$

This means that in order to simulate the polarization of *N* ions, we can use a set of *N'* ions and add a factor of *N/N'* to the second term of Eq. 4. To confirm the validity of this approach, we have compared simulation runs with 100, 200, and 500 ions all renormalized to 1000 ions against a run with a true set of 1000 ions without renormalization. Fig. 1 plots the polarization per ion for all cases, and we can see that the renormalization method works

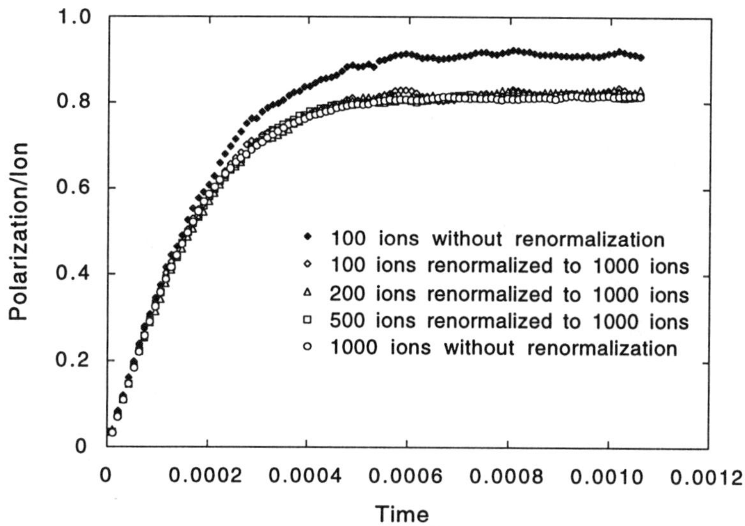

Fig. 1. Simulating polarization of 1000 ions with 100, 200, and 500 ions by renormalization of pair potential. The results agree well with the polarization of a true set of 1000 ions without renormalization.

well. The polarization of a set of 100 ions without renormalization is also plotted in Fig. 1 for comparison.

3. Simulation Results

We first examine the effect of the total number of charges on the polarization of the ions. Even though all ions are drawn to the north pole under the influence of the external electric field, not every ion can reach the north pole. The larger the number of total ions N, the greater the percentage of ions that have to stay away from the pole because of increasing Coulomb repulsive forces between the ions, and because of this, we expect the polarization per ion to decrease as N increases. Fig. 2 confirms this.

The quantity that really matters, however, is the total polarization of the ions. In Fig. 3 we plot the equilibrium value of the total polarization of the ions as a function of N. We can see that even though the polarization per ion decreases with N (Fig. 2), the total polarization *increases* as N increases. Eventually the total polarization saturates when ions are so closely spaced that the Coulomb repulsive forces become dominant in determining ion distribution.

We also expect the polarization to increase when the strength of the applied field increases. However, because the north pole is as far as an ion can migrate (corresponding to unit polarization), we also expect the polarization to saturate at certain field strengths. I.e., we expect non-linearity of polarization as a function of the applied field. Fig. 4 shows polarization as a function of field strength for several values of N. We can see that the non-linear behavior sets in sooner for smaller N. This is not difficult to understand, because for

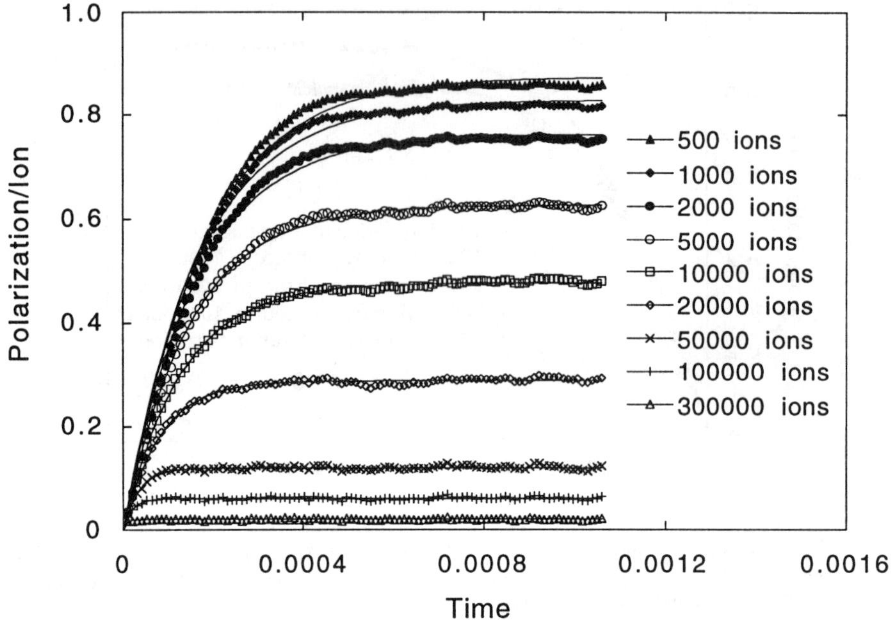

Fig. 2. Polarization per ion as a function of time after a DC field is turned on at $t = 0$. Polarization per ion decreases as the total number of ions increases because of the Coulomb repulsive forces between the ions. The field strength is 4000 in dimensionless units (Eq. 9). The solid lines are fits to single exponential functions.

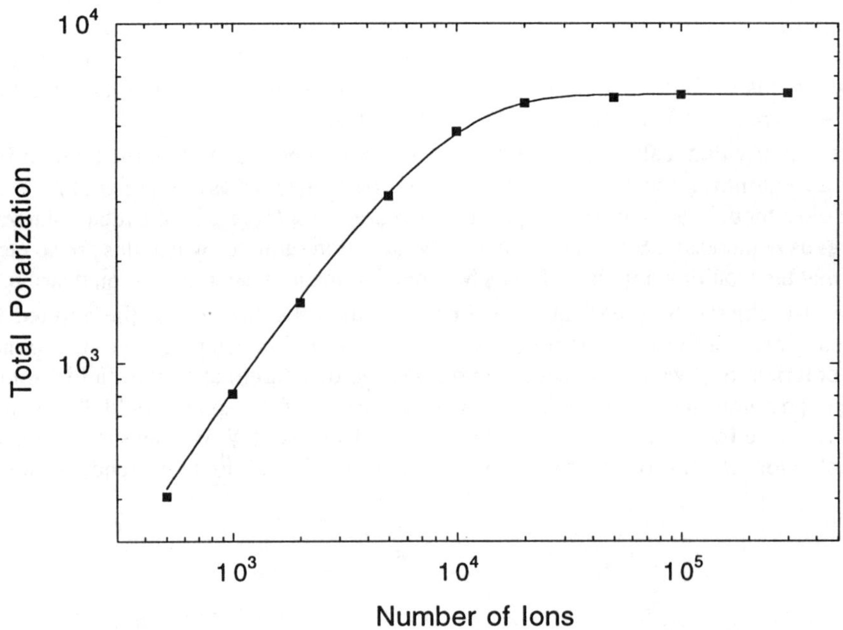

Fig. 3. Total polarization of the ions as a function of the number of ions. The solid line is to guide the eyes. The field strength is 4000 in dimensionless units (Eq. 9).

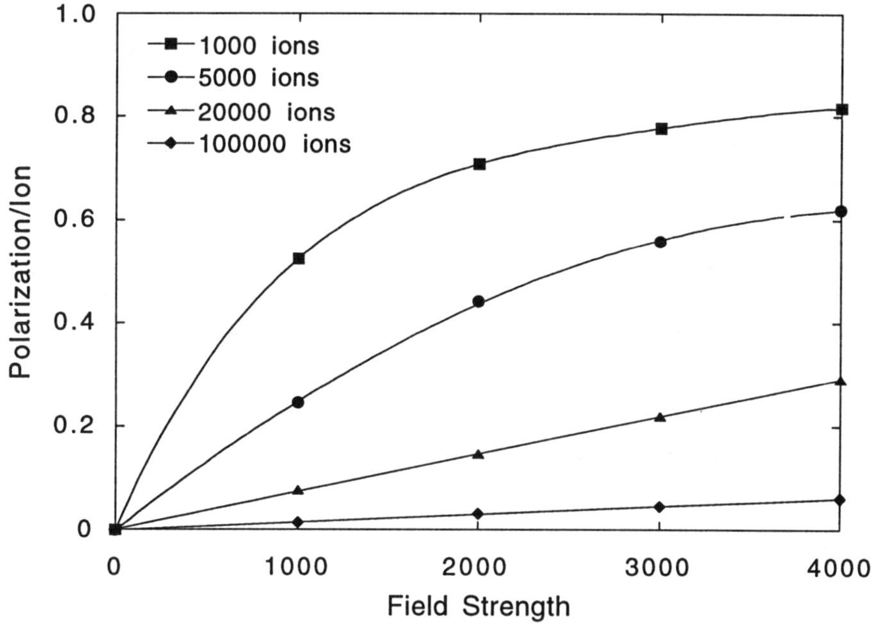

Fig. 4. Polarization as a function of field strength. Non-linear relations between the polarization and the applied field are seen, especially for low ion densities.

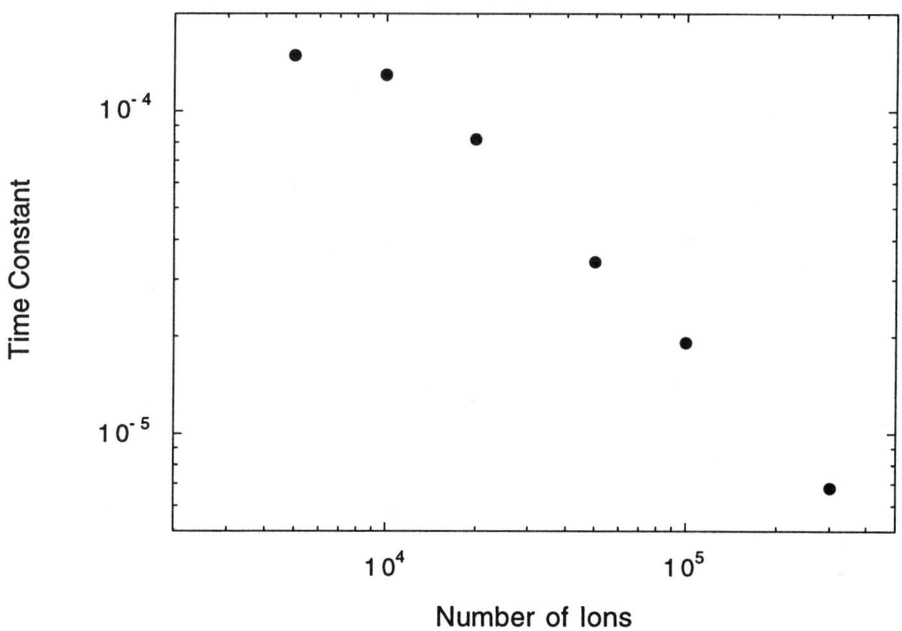

Fig. 5. Time constant of polarization as a function of total number of ions. The field strength is 4000 in dimensionless units (Eq. 9).

small N most ions can migrate to the north pole, while for large N Coulomb repulsive forces between the ions keep a larger percentage of ions away from the pole, and the distribution of the ions in this case represents a small perturbation from a uniform random distribution.

We can also extract the time constant of polarization from the polarization-versus-time data. Fig. 2 shows the fit to single exponential functions. For systems with large N, where polarization increases linearly with the applied field, we can see that using a single exponential for polarization as a function of time works very well. For systems with small N, where polarization is no longer linear with the applied field, however, it is not surprising that exponential functions do not fit well. As we have discussed earlier, when N increases, the ions become more densely packed, and Coulomb repulsive forces between the ions also become stronger. This means that it is more difficult for an ion to migrate to the north pole, and the average displacements of ions from the equilibrium positions for a uniform distribution are smaller for larger N. Because diffusion over smaller distances takes less time, we expect the time constant for polarization to also decrease as N increases. Fig. 5 shows this effect.

4. Conclusions

In conclusion, we are able to carry out molecular dynamic simulation of polarization of surface ions on a spherical colloidal particle in transient DC electric fields using a random walk method. Renormalization of pair interactions allows us to model a large number of ions using a much smaller set of ions in the actual simulation. As the total number of ions N increases, the average distance between ions decreases, and the Coulomb repulsive forces between the ions increase. This means that the average displacement of ions in response to the external electric field becomes smaller as N increases. Because smaller displacements take less time, the time constant of polarization can be significantly smaller than the time required for an ion to diffuse the whole length of the particle. The total polarization of ions, however, increases as N increases and eventually reaches a saturation value. Non-linear effects of polarization as a function of field strength are also seen.

5. Acknowledgments

This work has been supported by grants from the Henry Luce Foundation, the Research Corporation (grant no. C-3591), and the Department of Energy (grant no. DE-SG02-94ER45522.A000).

6. References

1. J. M. Ginder, L. C. Davis, and S. L. Ceccio, to appear in *Proceedings of the Electrorheological Materials and Fluids Symposium*, 208th National ACS Meeting, Ed. by F. E. Filisko and K. O. Havelka (Plenum Publishing, 1994).
2. J. D. Jackson, *Classical Electrodynamics*, Second Edition (John Wiley & Sons, 1975).
3. C. V. Heer, *Statistical Mechanics, Kinetic Theory, and Stochastic Processes*, Sect. 3.8 (Academic Press, 1972).

CHARACTERISATION OF THE COMPLEX SHEAR MODULUS PROPERTIES OF ELECTRO-RHEOLOGICAL FLUIDS BY THE DIRECT STIFFNESS AND MASTER CURVE TECHNIQUES

S O Oyadiji
Dynamics and Control Research Group
Division of Mechanical and Nuclear Engineering
Manchester School of Engineering
University of Manchester
Manchester M13 9PL, UK

ABSTRACT

The direct stiffness technique was employed to characterise the complex modulus properties of a silicone oil-based electrorheological fluid over a frequency range from 30Hz to 300Hz and a temperature range from 0°C to 60°C. The ER fluid device utilised was a set of concentric cylinders possessing a radial gap of 3mm between adjacent cylinders. Electric field strengths of between 0kV/mm and 2kV/mm were applied across the ER fluid. The results show that the shear modulus of the ER fluid decreased monotonically as the temperature was increased from 0°C to 60°C. Overall, the shear modulus decreased by a factor of up to 20. On the other hand, the shear loss factor increased from a low value of about 0.05 at 0°C to a high value of about 1.0 at 60°C. Conversely, as the electric field strength was increased from 0kV/mm to 2kV/mm, the shear modulus increased whereas the loss factor decreased. At all temperatures and electric field strengths of these investigations, both the shear modulus and loss factor increased in value as the excitation frequency was increased. The sets of measured temperature- and frequency-dependent data were converted, using the master curve technique, to master curves of shear modulus and loss factor which vary with frequency over several decades at a constant reference temperature and for varying levels of the electric field strength.

1. Introduction

In the design, specification and performance of ER fluid devices, the dynamic mechanical properties of the ER fluids play a major role. These properties are dependent on imposed and environmental conditions such as applied electric field strength, operating temperature, excitation frequency and dynamic amplitude of the excitation. The last two conditions are usually expressed as strain and strain rate. In service, most ER fluid devices, which have been proposed or are in the stages of development, such as clutches, automotive engine mounts and shock absorbers, dissipate significant amounts of energy as heat which results in a rise in temperature of the ER fluids. Because of the effects of such temperature rise on performance of the devices, because of the variation of the electric field strength characteristics with temperature for most ER fluids, and because of the need for accurate complex modulus data of ER fluids for the purposes of design and analysis, it is important to characterise the dependence of the complex dynamic mechanical properties of ER fluids on these variables.

Measurements of the rheological properties of ER fluids are usually made by means of a rotary rheometer or viscometer which consists essentially of two concentric cylinders with the ER fluid sandwiched between the two cylinders [1-6]. From the magnitude and phase of the excitations applied to the outer cylinder and the torque or rotary motion transmitted to the inner cylinder, the shear stress, shear strain and shear strain rate of the ER fluid under the test conditions are determined. In this work, a different approach is adopted. An ER fluid device consisting of a nest of three fixed and three movable cylinders is used. The resulting annular cylinders of fluid are subjected to axial shear vibrations via the excitations of the movable cylinders along the direction of their common central axis. From the force transmitted to the fixed cylinders and the input excitation applied to the movable cylinders, the complex axial shear stiffness of the ER fluid cylinders is obtained. Application of appropriate geometrical factors to this complex stiffness results in the complex shear modulus of the ER fluid at the operating conditions. The determination of the complex stiffness from the transmitted force and the input excitation is the basis of the direct stiffness method which is one of the principal techniques for the determination of the dynamic mechanical properties of solid viscoelastic materials[7-9].

The master curve methodology is a technique which is widely used for the reduction of the dependence of the viscoelastic properties of polymeric and elastomeric materials on temperature and frequency (or time) to a dependence on a single composite parameter called the reduced temperature or reduced frequency (or reduced time)[7-9]. The technique, which is based on the temperature-frequency superposition principle, was derived originally on an empirical basis by Williams, Landel and Ferry[7]. The WLF Equation named after them relates actual frequency to actual temperature for most thermorheologically simple materials. By means of the master curve technique, the temperature- and frequency-dependent complex shear modulus properties of the ER fluid investigated in this work were reduced to a pair of master curves of dynamic shear modulus and loss factor which depends only on one parameter, namely, the reduced frequency. A polynomial function, which was derived empirically from the measured data was used to relate the actual temperature and frequency. It is shown that when the master curves derived for the ER fluid at different electric field strengths are overlaid, the general trends of increasing shear modulus and decreasing loss factor as electric field strength increases are preserved.

2. Principles and Methods

2.1 Direct Stiffness Method

The direct stiffness test method is a non-resonance, forced vibration technique for determining the complex dynamic stiffness K^* of a resilient element. The element is connected between the top of the ram of an eletrohydraulic actuator or the top of the vibration table of an electrodynamic exciter and a rigid termination of theoretically infinite impedance. The end of the element which is connected to the actuator or exciter is subjected to either sinusoidal or random displacement excitation. In the case of sinusoidal excitation, the input displacement is of the form $x(t) = Xe^{j\omega t}$. The ratio of the corresponding output force $f(t) = F^*e^{j\omega t}$ to the input displacement gives the complex dynamic stiffness K^* at the excitation frequency ω. The dynamic stiffness K, phase (loss) angle θ, and loss factor η are related to K^* by

$$K = |K^*| = |F^*/X| \; ; \qquad \theta = \angle K^* = \angle(F^*/X) \; ; \qquad \eta = \tan\theta \tag{1}$$

Thus, the complex dynamic stiffness K^* of the mount can be related to the real part K' and imaginary part K'' of K^* by

$$K^* = K' + jK'' = K'(1 + j\eta) \; ; \qquad K = |K^*| = K'\sqrt{1 + \eta^2} \tag{2}$$

In the case of the characterisation of the complex shear modulus G^* of an element, the measured complex stiffness is the complex shear stiffness K_s^* which is related to the complex shear modulus by

$$G^* = K_s^* t/A \qquad \text{where} \qquad G^* = G' + jG'' = G'(1 + j\eta) \; ; \qquad G = |G^*| = G'\sqrt{1+\eta^2} \tag{3}$$

The parameters t and A are the shear thickness and cross-sectional area of the element parallel to the shear direction. Thus, the magnitude G and loss factor η of the complex shear modulus are obtained through the application of Eqs. 1 to 3.

2.2 *Derivation of Master Curves*

Master curves of shear modulus and loss factor, which cover many decades of frequency, are obtained from the measured data by means of the master curve methodology which involves the application of the temperature-frequency superposition principle. There are two steps in the application of this principle. The first step is to choose a reference temperature T_r and to convert the complex modulus G obtained at the different temperatures T to their corresponding values G_r at T_r using the following equation derived from kinetic theory

$$G_r = \frac{\rho_r T_r}{\rho T} G \tag{4}$$

where ρ_r and ρ are the mass densities at the absolute temperatures T_r and T respectively.

The loss factor remains invariant to these temperature changes. Hence,

$$\eta_r = \eta \tag{5}$$

The second step is to plot the sets of reduced shear modulus G_r and loss factor η_r on an extended frequency scale. Starting with the data sets closest to the reference temperature data set, these sets of data are shifted, simultaneously, parallel to the frequency axis to produce smooth curves, known as master curves, of the dynamic shear modulus G_r and loss factor η_r. The plot of the log of frequency shifts against the relative temperatures $(T\text{-}T_r)$ results in the so-called shift factor curve which has the general form

$$\log(f_r/f) = \phi(T - T_r) \qquad (6)$$

where ϕ is some empirical function of $T-T_r$, and f_r is the reduced frequency.

3. Experimental Procedure

The ER fluid used in these investigations was made by mixing 50 wt% of Analar starch (analytical reagent grade) with 50 wt% of Dow Corning silicone oil of 100 centistoke viscosity. A measured volume of the fluid was poured inside the fluid compartments of the ER fluid concentric cylinder device which had an inter-electrode gap of 3mm. The device was assembled in a direct stiffness test rig as shown in Figure 1 which also shows the associated test and analysis instrumentation. The operating temperature of the device was maintained by means of a temperature-controlled environmental chamber which employed liquid nitrogen to achieve temperatures below the laboratory ambient temperature and electrical heating for achieving temperatures greater than the ambient temperature. The operating temperature was varied in steps from 0°C to 60°C. At each temperature, the complex axial shear stiffness of the device was determined initially at an electric field potential of 0kV and at increments of 1kV in the electric field potential up to the maximum

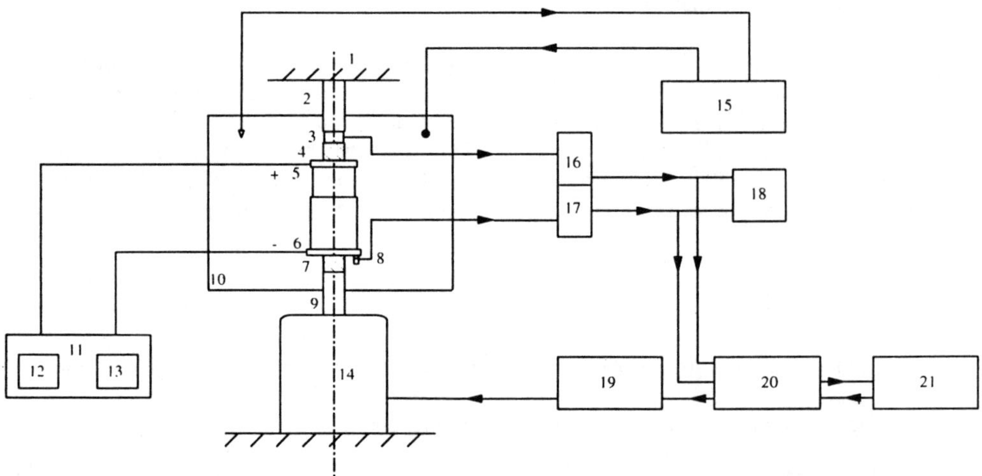

1.	Fixture	8.	Accelerometer	15.	Temperature Controller
2.	Top Connector	9.	Bottom Connector	16.	Charge Amplifier
3.	Force Transducer	10.	Environmental Chamber	17.	Charge Amplifier
4.	Top Nylon Insulator	11.	High Voltage Power Supply	18.	C.R. Oscilloscope
5.	Top Set of Cylinders	12.	Voltmeter	19.	Power Amplifier
6.	Bottom Set of Cylinders	13.	Ammeter	20.	Spectrum Analyser
7.	Bottom Nylon Insulator	14.	Electrodynamic Exciter	21.	Computer

Figure 1 : Experimental Configuration for the Characterisation of the Complex Shear Modulus of Electrorheological Fluids

voltage that the power supply could deliver to the ER fluid at the particular temperature. It was found that the maximum supply voltage varied from 1kV at 0°C to 5kV at 20°C and 13kV at 60°C. This difference was due to the variation in the electrical and mechanical properties of the ER fluid with temperature. The movable cylinders of the ER fluid device were excited using random vibration of 0 - 1000 Hz frequency bandwidth. From the measured complex axial shear stiffness, the complex shear modulus of the ER fluid was computed using Eq. 3.

4. Results and Discussions

For an applied electric field potential difference of 1.0kV, which corresponds to an electric field strength of 0.33kV/mm, Figure 2 shows the variation of the dynamic shear modulus and loss factor of the ER fluid with temperature and frequency. It is seen that increasing the temperature from 0°C to 60°C has a much more pronounced effect on the complex modulus properties than increasing the excitation frequency from 30Hz to 300Hz.

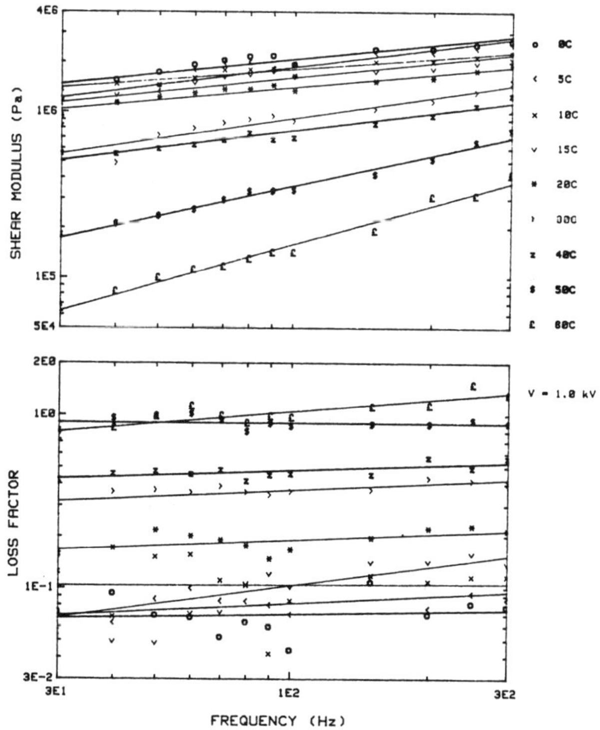

Figure 2 : Complex Shear Modulus versus Frequency for a Silicone Oil-Based ER Fluid subjected to an Electric Field Voltage of 1.0kV at Temperatures of 0°C to 60°C

Specifically, when the temperature is increased from 0°C to 60°C the shear modulus is decreased by a factor of up to 20 while the loss factor is increased from a value of 0.05 to 1.0, that is by a factor of 20. Conversely, increasing the frequency from 30Hz to 300Hz causes an increase in the shear modulus by a maximum factor of 7 but does not significantly affect the loss factor values.

The effects of increasing the applied electric field strength at a constant operating temperature of 20°C is shown in Figure 3. When there is no applied electric field, the shear modulus is relatively low whereas the loss factor is relatively high. But when an electric potential of 1.0kV is applied, the shear modulus increases by a factor of about 1.8 while the loss factor decreases by a factor of about 2. Increasing the applied electric field further causes a less dramatic change to the values of the shear modulus and loss factor. Thus, there is a very significant gap in the characteristics of the ER fluid between applied electric field potentials of 0 and 1kV. However, as shown in Figure 4, the size of this gap decreases as the temperature is increased. Figures 3 and 4 also show that the effects of excitation frequency on the shear modulus increases as the temperature is increased.

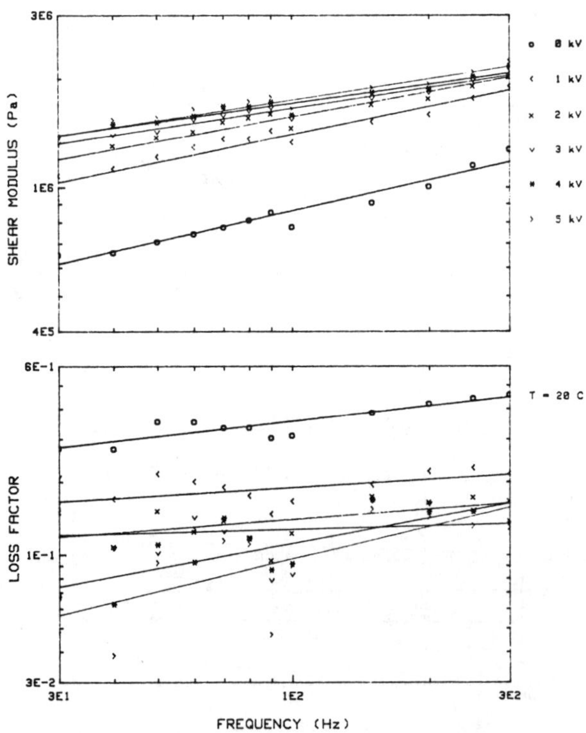

Figure 3 : Complex Shear Modulus versus Frequency for a Silicone Oil-Based ER Fluid at 20°C and for Electric Field Voltages of 0 kV to 5 kV

The global differences in the separate effects of temperature and applied field strength of the ER fluid will now be compared. From Figures 2 and 4 it is seen that as the operating temperature is increased within the range 0 - 60°C, the shear modulus of the ER fluid decreases while its loss factor increases. Conversely, Figures 3 and 4 show that as the applied electric field strength is increased, the shear modulus of the ER fluid increases whereas its loss factor decreases. This effect of increasing field strength on the viscoelastic properties of the ER fluid is similar to the increase in compression modulus of anti-vibration mounts and decrease in the associated loss factor when the applied static pre-load (pre-strain) is increased[10].

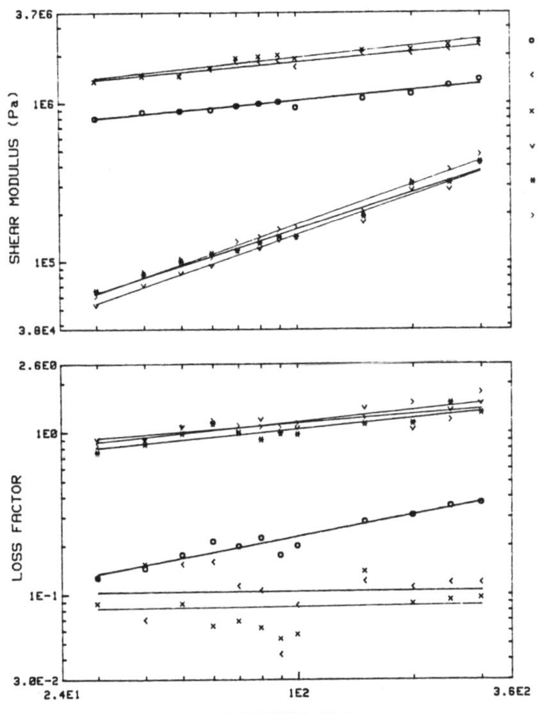

Figure 4 : Complex Shear Modulus versus Frequency for a Silicone Oil-Based ER Fluid at 10°C and 60°C and for Electric Field Voltages of 0, 1 and 2 kV

By means of the master curve methodology outlined previously, the frequency- and temperature-dependent complex shear modulus data shown in Figure 2 was reduced to a pair of master curves of shear modulus and loss factor at a reference temperature of 20°C as shown in international plot format in Figure 5. It is seen that the family of curves in Figure 2 which span the limited frequency range of 30 to 300Hz and extend over the temperature range of 0 - 60°C have been reduced to a pair of curves which cover about 6 decades of frequency at a reference temperature of 20°C. The similarity between the characteristics of these master curves and those of clasical viscoelastic materials suggests that the ER fluid could be considered as an electroviscoelastic material under the conditions in which it was characterised. The master curves which were derived from the complex shear modulus data obtained at electric field potentials of 0kV and 2kV were overlaid with those obtained at 1kV and are shown in Figure 6. It is obvious that the general increase in the shear modulus and the corresponding decrease in the loss factor of the ER fluid as the applied electric field strength is increased is preserved.

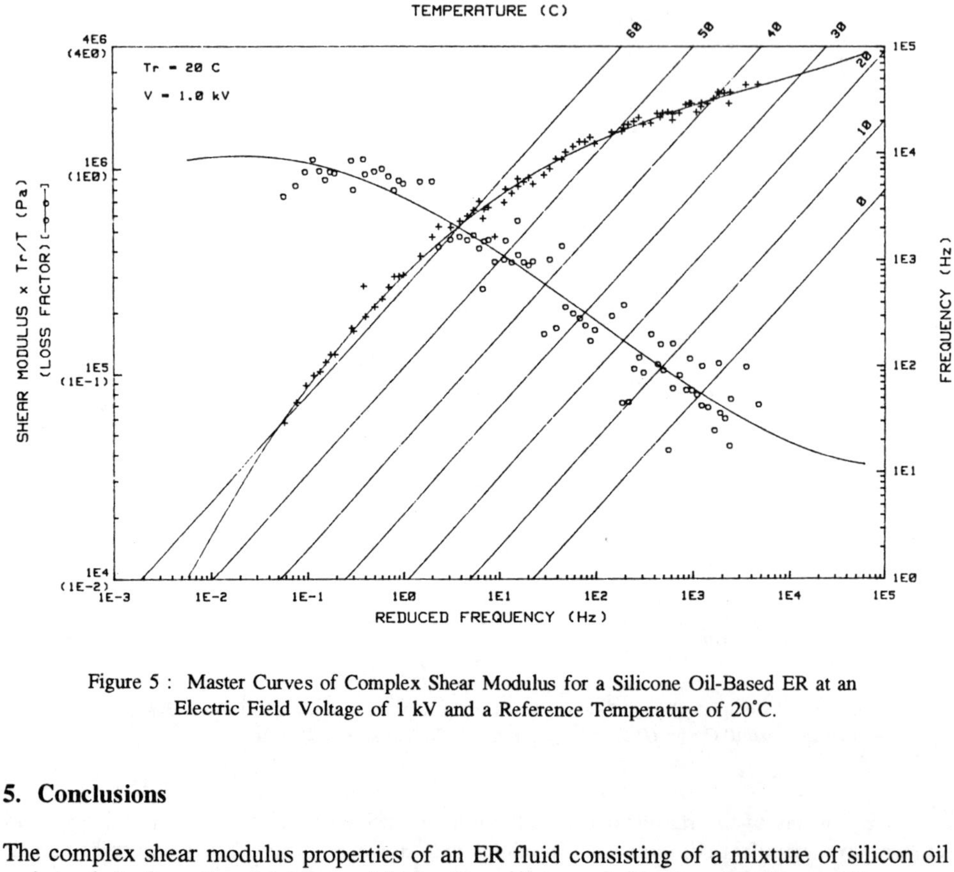

Figure 5 : Master Curves of Complex Shear Modulus for a Silicone Oil-Based ER at an Electric Field Voltage of 1 kV and a Reference Temperature of 20°C.

5. Conclusions

The complex shear modulus properties of an ER fluid consisting of a mixture of silicon oil and starch in the ratio of 1:1 by weight has been characterised using the direct stiffness test method. It has been shown that within the temperature and frequency ranges of the tests, temperature variations have a much more significant effect on the complex modulus properties of the ER fluid than the effect of frequency variations. As the temperature was increased from 0°C to 60°C, the shear modulus decreased in magnitude whereas the loss factor increased. Conversely, as the electric field strength was increased from 0 kV/mm to 2 kV/mm, the shear modulus increased whereas the loss factor decreased. The application of the temperature-frequency superposition principle has enabled the reduction of the temperature- and frequency-dependent complex shear modulus data to master curves of dynamic shear modulus and loss factor which cover about six decades of the reduced frequency scale.

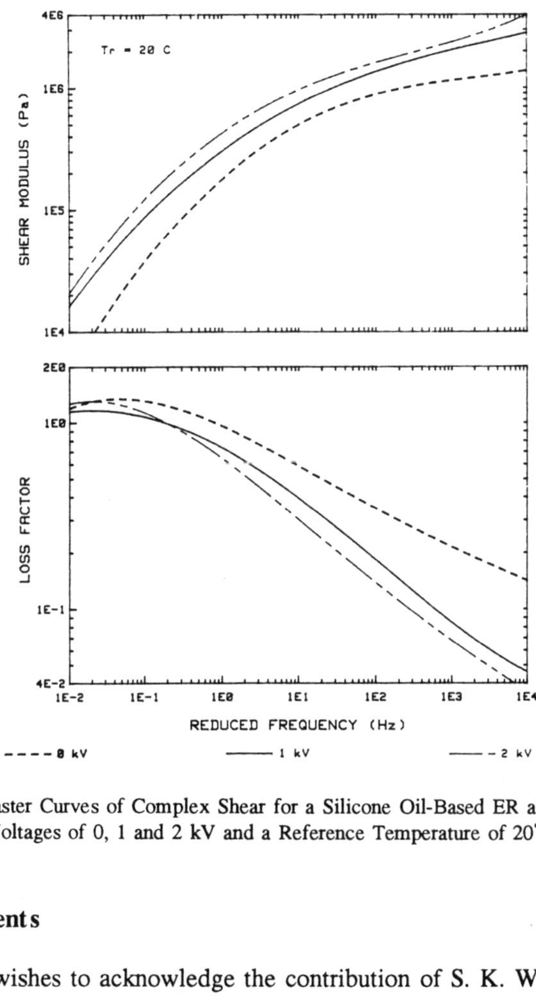

Figure 6 : Master Curves of Complex Shear for a Silicone Oil-Based ER at Electric Field Voltages of 0, 1 and 2 kV and a Reference Temperature of 20°C.

6. Acknowledgements

The author wishes to acknowledge the contribution of S. K. W. Denyer, 1993 final year undergraduate student, to the design, manufacture and development of the concentric cylinders ER fluid device. Thanks are also expressed to Dr R. Stanway of Sheffield University for recommending the ER fluid characterised in the present work.

7. References

1. D. R. Gamota and F. E. Filisko, High Frequency Dynamic Mechanical Study of an Aluminosilicate Electrorheological Material, *J. Rheol* **35**(7) (1991) 1411-1425

2. H. Conrad, A. F. Sprecher, Y. Choi and Y. Chen, The Temperature Dependence of the Electrical Properties and Strength of Electrorheological Fluids, *J. Rheol* **35**(7) (1991) 1393-1410

3. W. S. Yen and P. J. Achorn, A Study of the Dynamic Behaviour of an Electrorheological Fluid, *J. Rheol* **35**(7) (1991) 1375-1384

4. R. Bloodworth, Electrorheological Fluids based on Polyurethane Dispersions, *Proceedings of the 4th International Conference on Electrorheological Fluids*, edited by R. Tao and G. D. Roy (World Scientific, Singapore, 1994), pp. 67-83

5. D. A. Nelson and E. C. Suydam, The Thermal Aspects of the Electrorheological Effect and its Impact on Application Design, *Proceedings of the Symposium on Electrorheological Flows*, edited by D. A. Siginer, J. H. Kim and R. A. Bajura (FED-vol. 164, ASME 1993), pp. 71-84

6. J. A. Powell, The Mechanical Properties of an Electrorheological Fluid under Oscillatory Dynamic Loading, *Smart Mater. Struc.* **2** (1993) 217-231

7. J.D. Ferry, *Viscoelastic Properties of Polymers*, John Wiley & Sons, New York, 3rd Edition (1980).

8. A.D. Nashif, D.I.G. Jones and J.P. Henderson, *Vibration Damping*, John Wiley & Sons, New York (1985).

9. S.O. Oyadiji and G.R. Tomlinson, Characterisation of the Dynamic Properties of Viscoelastic Elements by the Direct Stiffness and Master Curve Methodologies, *J. Sound Vib.* (1995), Part 1(accepted), Parts 2 & 3 (to be submitted).

10. S.O. Oyadiji and G.R. Tomlinson, Relating the Complex Moduli of Viscoelastic Materials to the Complex Stiffness of Anti-Vibration Mounts, *Proceedings of Smart Structures and Materials Conference: Passive Damping,* edited by C. D. Johnson, (vol. 2193, SPIE 1994), pp. 226-237

ELECTRORHEOLOGICAL NORMAL STRESS MEASUREMENTS OF POLYMER SOLUTIONS AND SUSPENSIONS

KEIJI MINAGAWA †, HIROSHI KIMURA, JUN-ICHI TAKIMOTO, and KIYOHITO KOYAMA*

Department of Materials Science and Engineering, Yamagata University, Yonezawa 992, Japan

ABSTRACT

Normal stress measurement under an electric field is demonstrated as a new method of evaluating ER effects. The measurement was applied to liquid crystalline polymers, low molecular-weight liquid crystals, polymer suspension, and aluminum suspension. The normal stress of the liquid crystalline polymers drastically increased under an electric field at high shear rate, which suggests existence of an elastic network structure of the polymer chains. The normal stress of the suspensions also changed in the field, indicating that interaction of particles results in an increase of both shear stress and normal stress. These changes of the normal stress give additional information which is helpful for characterizing ER effects.

1. Introduction

Usual evaluation of ER effects has been based on the measurements of changes in the shear stress and/or dynamic moduli of the fluids under an external electric field. For example, changes of the steady shear stress or shear viscosity are useful for describing the ER effects. Recently we suggested a new method of evaluating the ER effects in different viewpoint from the conventional methods by introducing normal stress measurement under an electric field[1,2]. This method is applicable to evaluation of elasticity of ER fluids subjected to a constant shear flow. We reported that the effects of electric field on the elastic property of polymeric materials such as liquid crystalline polymers[3] can be analyzed with this method. Here we show application of the normal stress measurements to a few kinds of ER solutions and suspensions.

2. Materials and Method

Comb-shaped liquid crystalline polymer solutions[3] (P-type and N-type LCP) and low molecular-weight liquid crystals, Chisso Petrochemical LIXON 2010 (P-type LC) and LIXON EN-37 (N-type LC), were used. The P-type mesogens align along the applied electric field, while the N-type mesogens align perpendicularly to the field. Suspensions of ion-exchange resin and of aluminum[4] particles were also used. The LCPs and aluminum suspension were supplied by Asahi Chemical Industry, Co. Ltd.

---

† Present address: Department of Chemical Science and Technology, The University of Tokushima, Minamijosanjima, Tokushima 770, Japan
*To whom correspondence should be addressed.

A parallel-disc rheometer (MR-300V2, Rheology Co., Ltd.), modified for simultaneous measurements of both shear stress and normal stress [2], was used. The diameter of the discs was 32 mm, and the gap between the disc electrodes was 1mm for the LCPs and 0.5mm for the others. Here the stress to extend the gap was defined as the positive normal stress. The measurements were carried out under 0~3 kV/mm electric field. Normal stress of the samples in a steady shear flow was measured in the presence and absence of electric field [1,2]. The primary normal stress difference Δn_1 was determined with a simplified equation $\Delta n_1 = 2F/\pi R^2$, without correcting for the inhomogeneous shear. Here F and R are the thrust and the radius of parallel discs, respectively. The Δn_1 value obtained here is not precise primary normal stress difference but is a simple estimation for the force normal to the parallel discs.

3. Results and Discussion

3.1. Liquid Crystalline Polymers

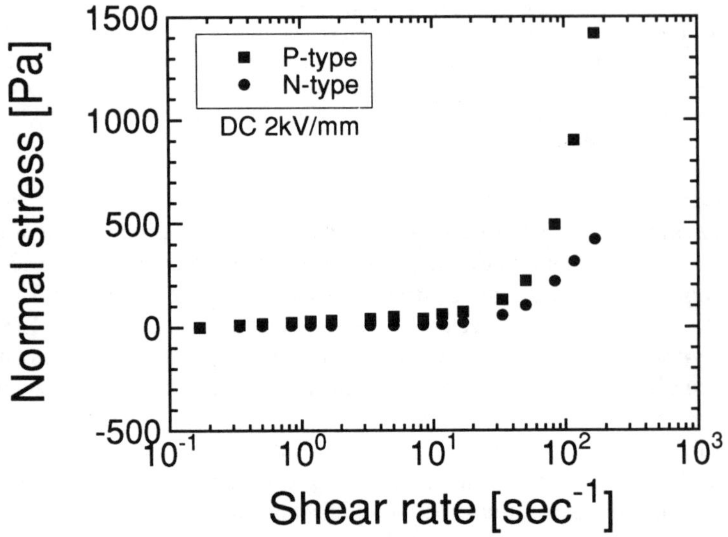

Fig. 1 Induced normal stress vs. shear rate for P-type and N-type LCPs.

We have already described characteristic features of induced normal stress for the P-type LCP in previous papers [1,2]. Under an electric field, the LCP showed apparent negative normal stress at low shear rate and rapid increase at high shear rate. We concluded that the negative stress is due to an electrostatic attractive force between the electrodes and that the rapid stress increase is due to an induced polymer network. Here we compare the P-type and N-type LCPs with the plots of normal stress induced by 2kV/mm d.c. field, against shear rate (Fig. 1) and against shear stress (Fig. 2). In these figures the electrostatic attractive force was subtracted from the original data, and thus the normal stress shown here is just the electrically induced stress increments. As shown in Fig. 1, a large increase in the normal stress is observed at high shear rate region under an electric field for the both fluids. Though the increase of the stress for P-type is larger than that for N-type, the tendency of shear-rate dependence is the same for the both types. This behavior indicates a network structure in the LCP solutions because of the interaction among the oriented mesogens connected to a flexible main chain.

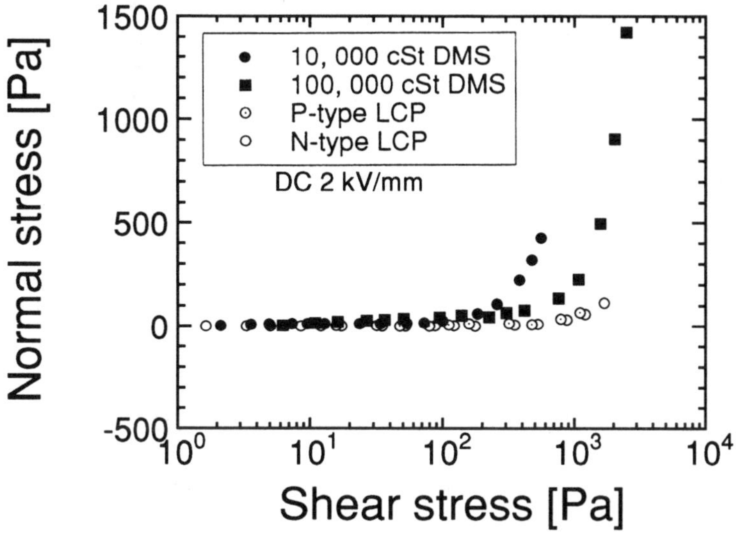

Fig. 2 Relationship between induced normal stress and shear stress for P-type and N-type LCPs.

On the other hand, the normal stress of both LCPs rapidly increases with the increase of shear stress, as shown in Fig. 2. The results of two dimethyl siloxanes without mesogen (10,000 cSt and 100,000 cSt), shown in the same figure, confirm the significance of the mesogens for the ER normal stress effect. It is found that the N-type LCP shows larger normal stress than the P-type one when they are compared at the same shear stress, though the maximum value for the N-type one is not so large. The P-type mesogens oriented perpendicularly to shear flow are much perturbed by the shear force, while N-type mesogens oriented along the shear flow receive little force. Therefore, the shear flow is effective to break the networks of P-type LCP comparing to N-type one. The resistance of network in N-type LCP results in the observed relationship between the shear stress and normal stress.

3.2. Low Molecular-weight Liquid Crystals

Figs. 3 and 4 show the normal stress of P-type and N-type LCs, respectively, plotted against shear rate. Apparently negative normal stress is observed for the both fluids, and the absolute values of the negative stress for P-type is larger than that for N-type. The negative normal stress suggests that the electric field causes electrostatic attractive force between the parallel discs [1, 2] and thus the difference in the stress values is due to the different dielectric properties of these LCs. Since the stress under each field condition is independent of the shear rate, it is concluded that intermolecular interactions of the LCs are too weak to generate any elastic property in these fluids. This fact supports the importance of flexible chains connecting mesogens for ER effects.

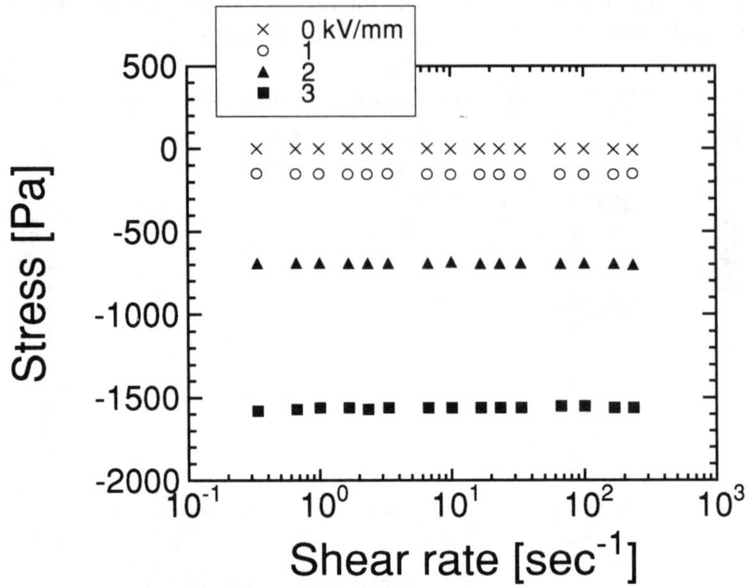

Fig. 3 Shear rate dependence of normal stress for P-type LC.

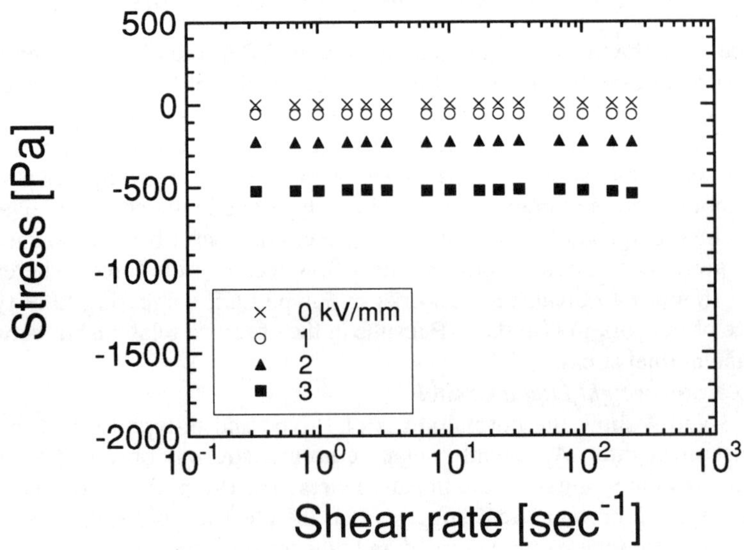

Fig. 4 Shear rate dependence of normal stress for N-type LC.

3.3. Suspensions

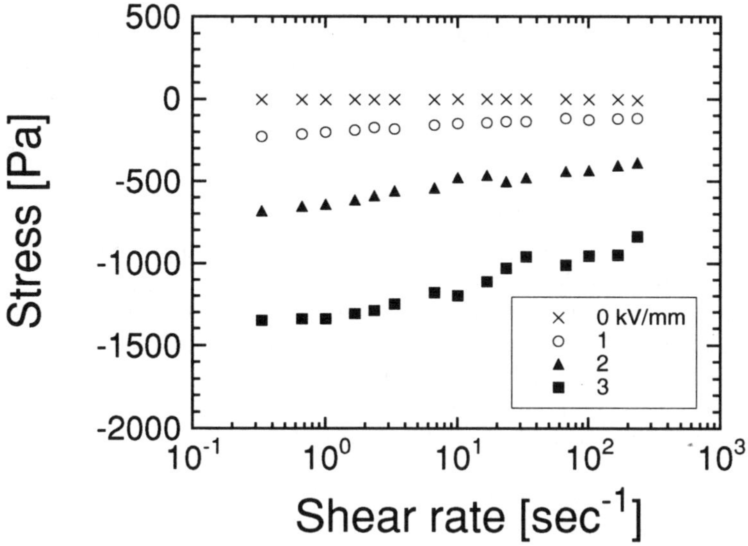

Fig. 5 Shear rate dependence of normal stress for ion-exchange resin suspension.

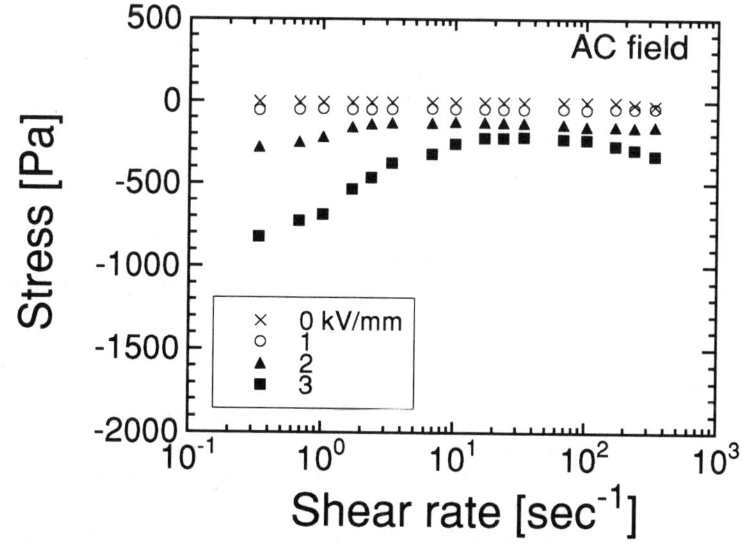

Fig. 6 Shear rate dependence of normal stress for aluminum suspension.

Measurements of normal stress with the present ER apparatus provide information not only on the elastic property of polymeric materials but also on some other changes in the flow behaviors of ER fluids. Here we apply this method to ER suspensions. Figs. 5 and 6 show normal stress for the ion-exchange resin suspension and for the aluminum particle suspension, respectively. The measurements were carried out under d.c. field for the polymer suspension and a.c. field for the aluminum suspension because the aluminum suspension does not show significant ER effect under d.c. field with usual conditions [4].

The stress values, as a whole, are negative similarly to the cases of LCs. This result is also due to the electrostatic attraction between the electrodes. For these fluids, however, there is a tendency that a small normal stress counteracting the attracting stress generates as the shear rate increases. This tendency is more obviously seen in the aluminum suspension under the stronger field condition.

The reason for appearance of normal stress counteracting the electrostatic attraction is not clear, but it can be explained with particle behaviors. For example, if it is assumed that the particle clusters bridging the parallel discs reorient along the shear flow keeping interaction with each other, the particles can form some layers lying between the discs. Since the fluid velocity is faster than that of particles in such two-phase systems, the base fluid, i.e., silicone oil in this case, flows between the particle layers at a high velocity. Accordingly the rapid flow extends the fluid pathway, generating a stress which extends the gap between the discs.

4. Conclusion

The induced normal stress of LCPs suggests existence of an elastic network structure of the polymer chains under an electric field. The normal stress of suspensions also changed in the field though the value was not so large. The induced change in normal stress indicates that interaction of particles results in an increase of the normal stress in addition to the increase of shear stress known as an ER effect. The present method gives additional information on the property of various ER fluids, which could be helpful for evaluating the ER effects.

5. Acknowledgements

The authors are grateful to Asahi Chemical Industry for providing the samples. This work was partly supported by the Grants-in-Aid from the Ministry of Education, Science and Culture of Japan. The JSPS Research Fellowships for Young Scientists for one of the authors (H. K.) is also acknowledged.

6. References

1. H. Kimura, K. Minagawa, and K. Koyama, *Polym. J.* **26** (12) (1994) 1402.
2. K. Minagawa, H. Kimura, and K. Koyama, in *Electrorheological (ER) Fluids/Materials,* eds. K. O. Havelka and F. E. Filisko (Plenum, New York) in press.
3. A. Inoue and S. Maniwa, *J. Appl. Polym. Sci.* **55** (1995) 113.
4. A. Inoue, in *Proceedings of the 2nd International Conference on ER Fluids,* eds. J. D. Carlson, A. F. Sprecher, and H. Conrad (Technomic, Lancaster, PA, 1990) p. 176.

PRACTICE-RELEVANT ASPECTS OF CONSTRUCTING ER FLUID ACTUATORS

H. Janocha, B. Rech, R. Bölter

Laboratory for Process Automation (LPA)
University of the Saarland
P.O. Box 15 11 50
66 041 Saarbrücken, Germany

ABSTRACT

The flow resistance of electrorheological fluids (ER fluids) can be controlled by applying electric fields. Thus, ER fluids are suitable for the application in actuators, using high-voltage sources for the generation of the field. The behaviour of an ER fluid actuator not only depends on the properties of the individual actuator components (ER fluid, energy transducer and energy source) but especially on their combined efforts as a system. Based on a possible scheme for the design of ER fluid actuators, this paper presents important practice-relevant aspects of a systematic actuator construction. Here the behaviour of a commercial ER suspension is examined and compared to a homogeneous ER fluid without yield point using a rotational viscometer and a flow-mode damper realized at the Laboratory of Process Automation (LPA) of the University of Saarland.

1 Introduction

Electrorheological fluids (ER fluids) respond to constant and alternating electric fields with a reversible increase of their flow resistance, which is controlled by the field strength. Thus, they are suited as simple, fast reacting interfaces between control electronics and mechanical system in energy transducers of actuators, as e.g. in clutches, vibration dampers and engine bearings. Here high-voltage sources are used as energy sources for field generation.

The behaviour of ER fluid actuators not only depends on the properties of the used ER fluid but also on the structure of the energy transducer as well as on the high-voltage source for the field generation. Moreover, the combined action of all actuator components are of fundamental importance. Hence, they should not be considered individually when realizing an ER fluid actuator but as a total system. This paper first deals with the subsystems of an ER fluid actuator and then concentrates on practice-relevant aspects of a systematic actuator construction based on a possible scheme for the design of an ER fluid actuator.

2 Subsystems of ER Fluid Actuators

ER fluid actuators are complex systems which, according to the definition of an actuator, can be subdivided into the subsystems ER fluid, energy transducer and energy source with various basic possibilities of realization.

2.1 Electrorheological Fluids

Currently, there are two different types of ER fluids: the mainly used ER suspensions of polarizable solid particles in a non-conducting carrier fluid and the recently developed homogeneous ER fluids on the basis of olegomeric carboxylic acids and their salts in mineral oil[2]. Without a control field, these suspensions have a low viscosity η. When exposed to a field they show an increase of the flow resistance as well as a yield point $\tau_y(E)$ which is dependent of the field strength. Usually, this behaviour is explained by the theory of chain formation. The flow resistance of homogeneous ER fluids can also be controlled by electric fields. Compared to ER suspensions, they show no yield point under field influence. In addition, problems with electrophoresis, particle coagulation, sedimentation and abrasion can be avoided because of the lack of suspended parts. The effective principle of the homogeneous ER fluids is currently investigated.

2.2 Energy Transducers

Depending on the kind of interaction between the electrode arrangement and the ER fluid, energy transducers of ER fluid actuators are divided into three basic principles: the shear mode, the flow mode and the squeeze mode. In case of the shear mode, two oppositely poled electrodes move in relation to each other (clutch principle). The so transferred torque can be controlled by an electric field. In an energy transducer using the flow mode, the ER fluid flows between electrodes which are stationary in relation to each other (valve principle). Here an electric field influences the flow resistance. With the squeeze mode, a distance alteration of the parallel electrode surfaces causes a squeeze flow and the development of a pressure pad between the electrodes. With the help of an electric field, the pressure and thus the normal force required for the movement of the electrodes can be determined.

2.3 Energy Sources

The electrical equivalent circuit diagram of an ER fluid energy transducer basically consists of the parallel connection of a capacitor and of an ohmic resistor. Their values result from the dimensions of the electrode arrangement as well as from the electric ER fluid conductivity γ_{ERF} and the ER fluid permittivity ε_{ERF}[4]. ER fluid energy sources must thus be suited for the control of ohmic-capacitive loads. Here a distinction has to be made between single-quadrant, two-quadrant and four-quadrant operation. The mode of operation and the suitable concept of the high voltage source are determined by the actuator application (quasi-static or high-dynamic drive), the applied ER fluid (ER suspension or homogeneous ER fluid) and the signal form to be generated as well as by the thus resulting build-up time and the decay time.

Energy sources for single-quadrant operation operate as current source, i.e. they can only charge the capacity of the energy transducer. The resistance of the ER fluid causes the discharge. The decay time is thus determined by the electric ER fluid conductivity and consequently depends on the temperature. In two-quadrant operation, ohmic-capacitive loads can be charged and discharged. The energy source here functions as current source and current sink. However, it is not possible to generate signals with an avarage value of zero. If an ER fluid energy transducer is to be charged positively and negatively, an energy source for four-quadrant operation is required to operate as positive and negative current source and as positive and negative current sink. These high-voltage sources are required to drive energy transducers using ER suspensions to avoid the electrophoresis effect[7].

In order to drive ER fluid energy transducers in single-quadrant operation, the principle of the switching power supply is especially suitable. Here a constant voltage, which is transformed into the required output voltage level by using a transformer, is converted into a high-frequent alternating voltage using a switching unit. There a rectification and a filtering is achieved. A more favourite realization of a two-quadrant high-voltage source is given by the combination of a controllable single-quadrant high-voltage source and a controllable high-voltage current sink (cf. Fig. 1). The high-voltage source serves to charge the load and can e.g. be realized by a switching power supply. It is switched off when discharching the load. The discharching is implemented by the current sink R which consists of a series connection of power MOSFETS and forms a switch with adjustable switching speed[7].

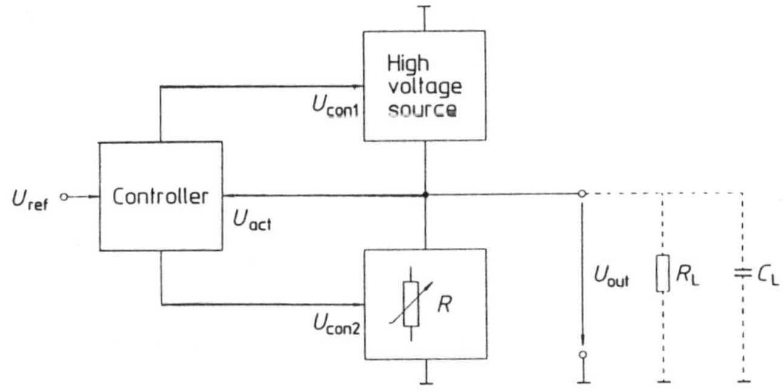

Fig. 1: Principle of a two-quadrant high-voltage source

A simple way of designing high-voltage sources for four-quadrant operation is provided by using transformers. Due to their band-pass behaviour, however, they cannot generate output signals with low frequencies down to constant voltage. For the frequency range of the output voltage from constant voltage up to some hundred Hertz, the concept of a full bridge amplifier is suitable. It essentially consists of two separately adjustable and stabilized single-quadrant high-voltage sources HVS_1 and HVS_2, two controllable high-voltage current sinks R_1 and R_2 as well as of a controller (cf. Fig. 2). A positive output voltage U_{out} is generated when the high-voltage source HVS_1 charges the load capacitor.

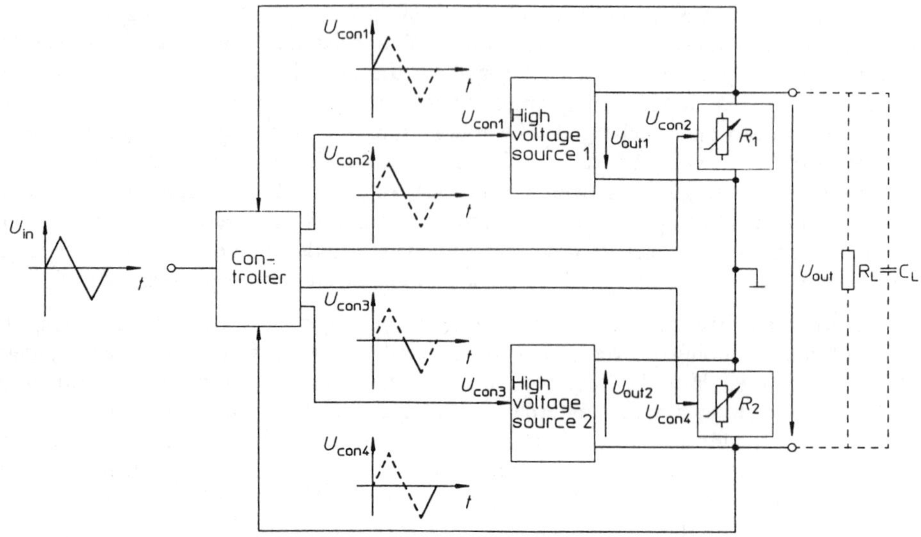

Fig. 2: Functional principle of the full bridge amplifier

In order to decrease the output voltage, HVS_1 is switched off and the transducer is discharged by the current sink R_1. The discharge speed is determined by the resistance value of R_1. Correspondingly, negative output voltages can be generated with HVS_2 and R_2[7].

3 Conceptualization of ER Fluid Actuator

The construction of an ER fluid actuator can be realized in four working phases: clarification of the problem, conceptualization, generation of the draft design and finally of the detailed design. Here a systematic proceeding is advisable to facilitate a possibly good realization of the actuator. This proceeding should be objectiv, planable, reconstructable and justifiable[6]. A possible scheme for the first two phases of a systematic construction of ER fluid actuators is illustrated in Fig. 3.

The efficient realization of an actuator construction requires a precise, possibly complete knowledge of the problem. In case of a damper, this could include the knowledge of the maximum damping amplitude and velocity as well as of the damping force, but also of the housing, the maximum driving power etc. For this reason, the phase "clarification of the problem" arranges and structures all requirements and requests on the ER fluid actuator as working basis for the constructor in a possibly clear form. The requirements are to be met by all means (e.g. certain power data) in order to make the solution acceptable. Requests are to be taken into account, probably with some concessions.

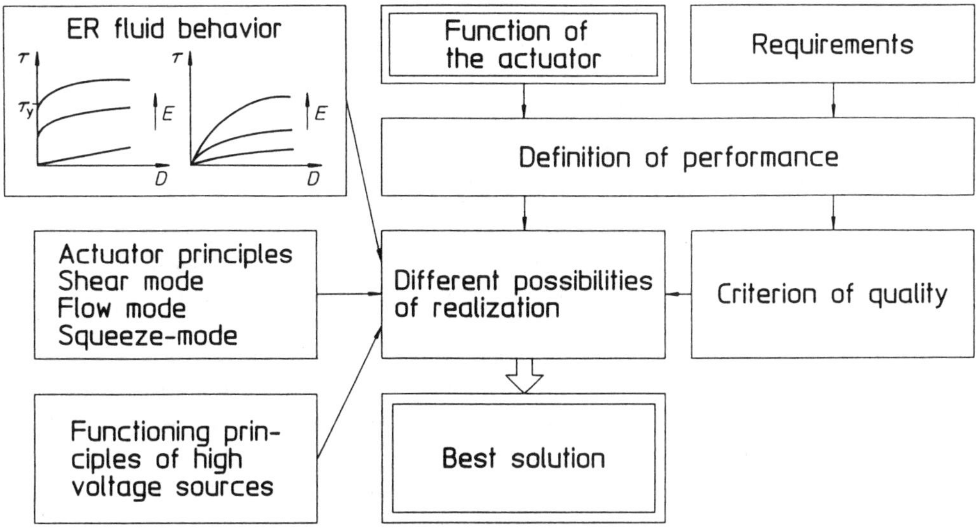

Fig. 3: Scheme for the working phases "clarification of the problem" and "conceptualization" of the construction of an ER fluid actuator

There are a great number of possible realizations of the subsystems of an ER fluid actuator (ER fluid, energy transducer, energy source). Thus, the developer disposes of various ER suspensions as well as of homogeneous ER fluids with different rheological and electrical properties. In addition, energy transducers can operate according to the shear mode, the flow mode, the squeeze mode or a combination of the basic principles. Besides, there are various possibilities for the design of the energy source (single-quadrant source, two-quadrant source, four-quadrant source). The actuator subsystems are again subdivided into system elements, for which different solutions are available. A rectifier in an energy cource can e.g. be realized as half-wave rectifier, full-wave rectifier or four-wave rectifier.

Theoretically, the combination of all possible realizations for the actuator system elements meeting the requirements of the requirement list is the total of all solution variants for the actuator system. The optimum solution for the actuator realization is given by one element of this total. In praxi, it is of course impossible to consider the totality of all solutions of the actuator, and generally no solution can be found to meet all requirements and requests of the requirement list in an optimum way. But a solution quantity consisting of a possibly large number of variants can be developed in the phase "conceptualization" meeting the requirements completely and the requests to some extend. By evaluation, this leads to the solution offering the most favourable system properties.

When developing this solution quantity, it is advisable to subdivide the actuator into its less complex subsystems according to the theoretical approach. Then these subsystems are subdivided again into their system components. First, separate realizations must be found for the system components which are evaluated by allocating points (e.g. between 1 and 10) with the evaluation criteria deriving from the requirement list. Here an evaluation can only be made after a rough estimation and after taking into consideration the technological possibilities. The combination of the system components leads to the possible realizations of the total actuator system. The best solution variant is determined by comparing the allocated quantities of points.

After determining the basic solution, the constructor generally develops the design structure in the design phase according to the technical and economical aspects. In the last phase, the final draft design is generated and the production-technological solution is determined. These working steps require exact dimensioning calculations, which, however, can currently not be implemented for ER fluid transducers since no rheological models can yet meet the requirements given in practical operation in a satisfactory way.

Homogeneous ER fluids have only recently been available in small amounts for laboratory use and have thus only been little tested in ER fluid actuators. For this reason, no calculation models have been developed yet. In order to implement approximation calculations for the design of transducers with ER suspensions, mainly the model of a Bingham-body is used[1], the flow curve of which (shear stress τ as function of the shear rate D with the parameter electric field strength E) can be described by equation

$$\tau(E, D) = \tau_y(E) + \eta D \tag{1}$$

The measurement of ER fluid flow curves can be implemented by using rotational viscometers which allow the application of a voltage to the measuring arrangement. Usually, the shear rate is determined as quotient of the relativ velocity of the shearing surfaces and of the surface distance. This applies for a Newtonian fluid flowing in laminar flow with linear speed progression, but also for the ER fluid measurement. If ER suspensions are exposed to an electric field, this approach offers only a very rough approximation. Due to the ER fluid yield point, the flow field has two areas the width of which depends on the field strength and the shearing velocity: one area in which the fluid layers flow with the same velocity and another area with a very large shear rate.

When examining ER fluids operating in the flow mode, the pressure difference Δp is measured at an ER fluid measuring valve as function of the volume flow q_v. Frequently, however, the flow mode data provided by the ER fluid producers include pressure differences converted into shear stresses and volume flows converted into shear rates. Here the approach valid for Newtonian fluids serves for the determination of D. The shear stress τ can be calculated on the basis of the Newtonian model and an arrangement consisting of two parallel electrodes with the length l and the distance h according to equation

$$\tau = \frac{h}{2\,l}\,\Delta p \tag{2}$$

As the measurement of ER fluids is not standardized yet, the rheological and electrical ER fluid parameters provided by the ER fluid producers are determined under measuring conditions which are generally hard to verify for the user.

Hence, the question arises whether the conversion of flow-mode data into shear-mode data and vice versa is permissible for his calculations and what kind of measuring data he needs. An answer should be provided by the measurements of ER fluids implemented at the Laboratory for Process Automation (LPA) of the University of Saarland using a rotational viscometer and a flow-mode damper realized for this special purpose. An ER suspension by Bayer AG consisting of polyurethane particles in silicone oil[8] and a homogeneous ER fluid by "RWE Gesellschaft für Forschung und Entwicklung"[2] were applied. The measuring values are processed and compared as described above, with the equations of Herschel-Bukley and Ostwald serving for the approximation of the measuring courses. The measurements are presented in the following.

4 ER Fluid Behaviour in the Rotational Viscometer

For the examination of the ER fluid behaviour using the shear mode, a speed-controlled Searl rotational viscometer type HAAKE M10 was available. The measurements were implemented using cylindrical measurement elements with a non-conducting cylinder cup base to avoid measuring errors due to shear stress caused by an electric field between the cylinder base and the base of the cup. Inhomogeneities of the electric field at the edges between the cylindrical surface and the cylinder base or cover cause measuring errors as well. When determining the shear stress τ this errors were eliminated by a difference procedure which is based on torque values measured with cylinders of different lengths[5]. For the measurements inner cylinders with a length $L_l = 60$ mm and $L_s = 48$ mm were used. The gap width was $h = 0{,}5$ mm.

Fig. 4 shows yield stresses of an ER suspension at a temperature of $\vartheta = 23$ °C. In order to avoid the measured values to be influenced by electrophoresis, a bipolar rectangular control field with a signal frequency of $f = 50$ Hz was applied. These curves are typical for ER suspensions. Without an electric field ($E = 0$), the shear stress τ linearly depends on the shear rate D (behaviour of a Newtonian fluid). When exposed to the electric field, an increase of the flow resistance and a yield point $\tau_y(E)$ rising with the field strenth can be observed.

Especially if esposed to high field strengths, the measured curves show strong deviations compared to the Bingham-body. The determinded shear stress courses are not shifted parallelly to the zero-field-curve, but after an initial increase the shear stress values decrease again at higher shear rates. At least for smaller field strengths, the equation of Herschel-Bulkley with coefficients A and B, which are dependent on the field strength, provides a better description of the flow curve of a real ER suspension. Here $\tau_y(E = 0) = 0$ and $B(E = 0) = 1$ applies without a control field, and $\tau_y(E > 0) > 0$ und $B(E > 0) < 1$ applies under field influence.

$$\tau(E, D) = \tau_y(E) + A(E)\, D^{B(E)} \tag{3}$$

Fig. 5 displays flow curves of a homogeneous ER fluid acquired at direct voltage control. It is especially striking that there is no yield stress, and that consequently all shear stress curves start at $\tau = 0$. Moreover, compared to commercial ER suspensions, higher field strengths are required to attain a distinct ER effect.

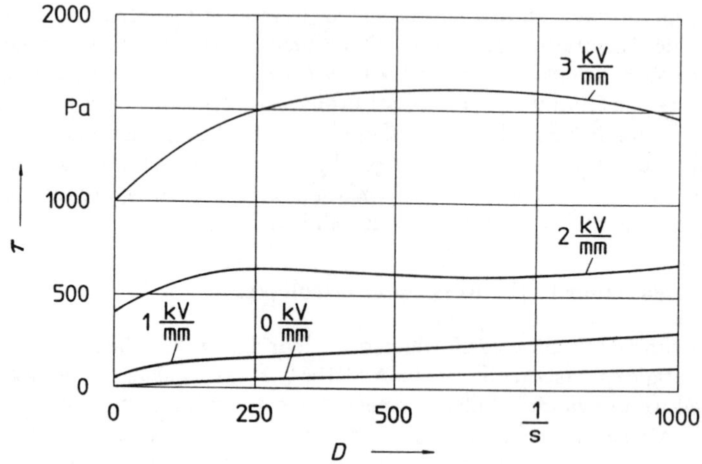

Fig. 4: Flow curves of an ER suspension when controlling with rectangular alternating fields

The behaviour of this homogeneous ER fluid can be described as pseudoplastic. The flow resistance increases with rising control field amplitude. Hence, the behaviour can be presented by applying the Ostwald equation. Just like equation (3), the coefficients A and B are dependent upon the applied electric field strength.

$$\tau(E, D) = \mathrm{A}(E)\ D^{\mathrm{B}(E)} \tag{4}$$

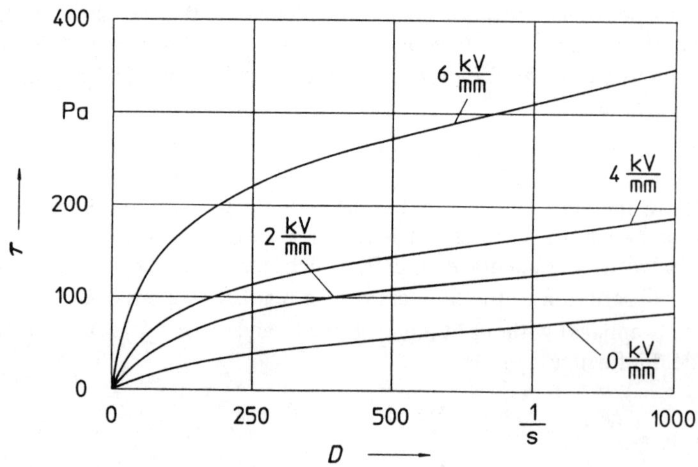

Fig. 5: Flow curves of a homogeneous ER fluid

5 ER Fluid Behaviour in a Flow Mode Damper

5.1 Structure of the Damper

In order to examine the behaviour of ER fluids using the flow mode at an oscillating flow, the ER fluid damper presented in fig. 6 was developed at the Laboratory for Process Automation (LPA) at the University of Saarland. Its bypass is realized as double tube. The modular structure of the damper allows the installation of flow channels with different electrode lenths and gap widths. In the presented design, the piston has a stroke of $x_{pmax} = 20$ mm with an electrode length varying between minimum 30 mm and maximum 60 mm. For the measurements presented in the following, electrodes of the length $l = 40$ mm, a gap width of $h = 0{,}75$ mm and an average channel diameter $d_m = 25{,}75$ mm were applied.

The outer cylinder, which is electrically insulated from the housing, is the high-voltage electrode, while the inner cylinder is the grounded electrode. The piston and the housing are grounded as well. The piston diameter is $d_p = 22$ mm, and the diameter of the rod is $d_{ro} = 8$ mm. In order to keep the displacement volume of the piston constant independent of its position, the piston rod is realized as passing rod. During measurement, the piston moves in an oscillating way causing an oscillating ER flow through cross holes in the inner cylinder and through the flow channel between inner and outer cylinder. A voltage applied to the electrodes generates an electric radial field in the flow channel, which can control the ER fluid flow resistance.

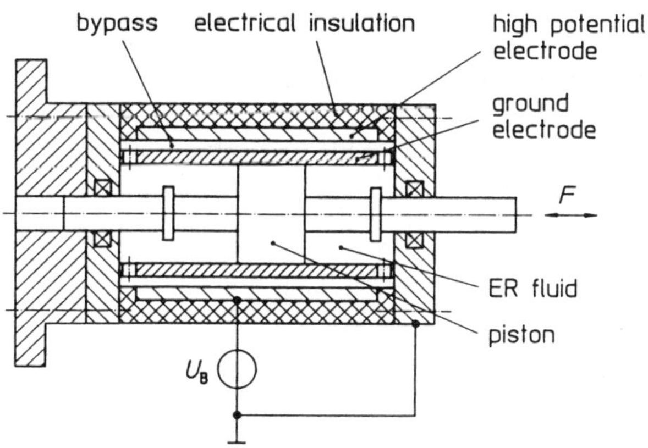

Fig. 6: Structure of an ER fluid damper

5.2 Damper Test Stand

The damper test stand realized at the LPA is illustrated in fig. 7. The damping piston has a linear drive with a recirculation ball screw and a slide able to generate a given path-time curve. The displacement is determined by an ohmic sensor. The damper as well as the linear drive are screwed to a common base.

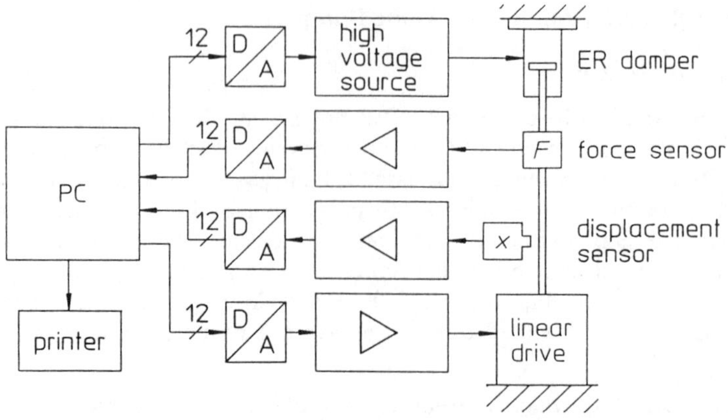

Fig. 7: Principle Structure of a Damping Test Stand

A piezoelectric force sensor for the measurement of the resulting force-time-function can be found between the piston rod of the damper and the slide. Taking into account the geometry of the damper, the volume flow-time-curve of the examined ER fluid can be given, and the pressure-time-curve can be determined. Moreover, the test stand disposes of a computer-controllable high-voltage source to drive the ER fluid damper. The drive motor is controlled by an analog amplifier.

5.3 *Measurement Results*

The flow-mode measurements were implemented at constant voltage control at a temperature of $\vartheta = 23$ °C. Fig. 8 illustrates the volume flow dependence of the pressure difference Δp for an ER suspension over the measuring valve at different control field strengths.

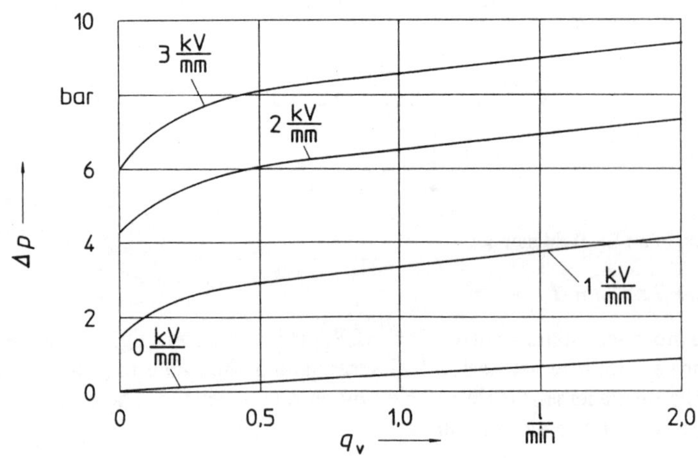

Fig. 8: Volume flow dependence of the pressure difference of an ER suspension

Without being exposed to a field ($E = 0$), the flow characteristic of a Newtonian fluid is shown with a pressure difference which is proportional to the volume flow. If a control field is applied, a field-strength-dependent pressure difference $\Delta p_y(E)$ results due to the yield point $\tau_y(E)$ which cannot be reached by any volume flow ($q_v = 0$). If $\Delta p_y(E)$ is exceeded, the ER fluid begins to flow, and Δp increases with q_v. Unlike the measurements using the shear mode, the curves do not descend at high field strengths. Thus, the pressure curves are well approximated by the equation of Herschel-Bulkley (equation 3).

Fig. 9 illustrates the dependence of the pressure on the volume flow. Like in measurements using the shear mode, a pseudoplastic behaviour is exhibited which can be described by the equation of Ostwald with field-strength-dependent coefficients (equation 4).

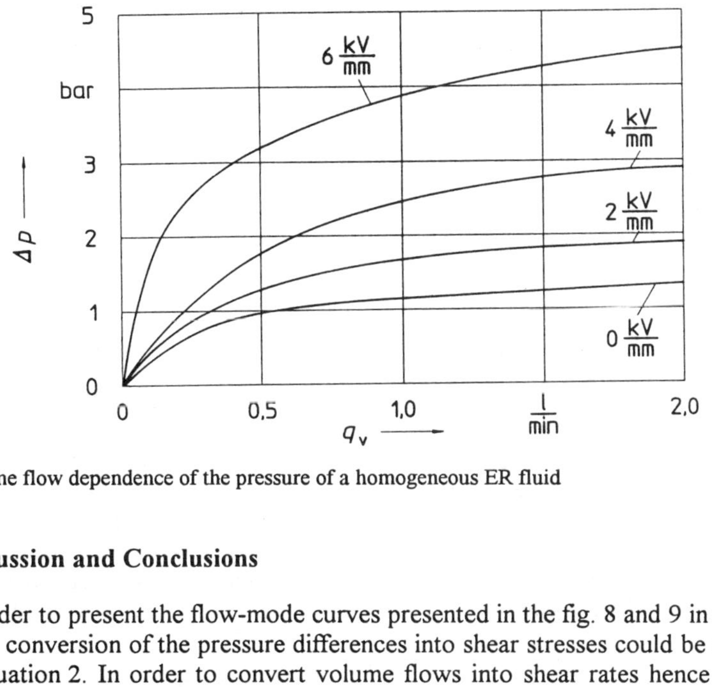

Fig. 9: Volume flow dependence of the pressure of a homogeneous ER fluid

6 Discussion and Conclusions

In order to present the flow-mode curves presented in the fig. 8 and 9 in shear-mode quantities, a conversion of the pressure differences into shear stresses could be achieved by applying equation 2. In order to convert volume flows into shear rates hence follows for the axial flow through an annular gap to be found in the ER fluid damper

$$D = \frac{6}{h^2 \, d_m \, \pi} \, q_v \qquad (5)$$

Hence follows a shear rate of $D = 4400$ l/s with the used damper geometry at a volume flow of $q_v = 2$ l/min. The so processed curves of the measurements using the shear mode and the flow mode show differing curves within the same shear rate ranges and at the same electric field strengths. In addition, these curves exhibit differing slopes (cf. Fig. 10). Here different field strengths cause different deviations.

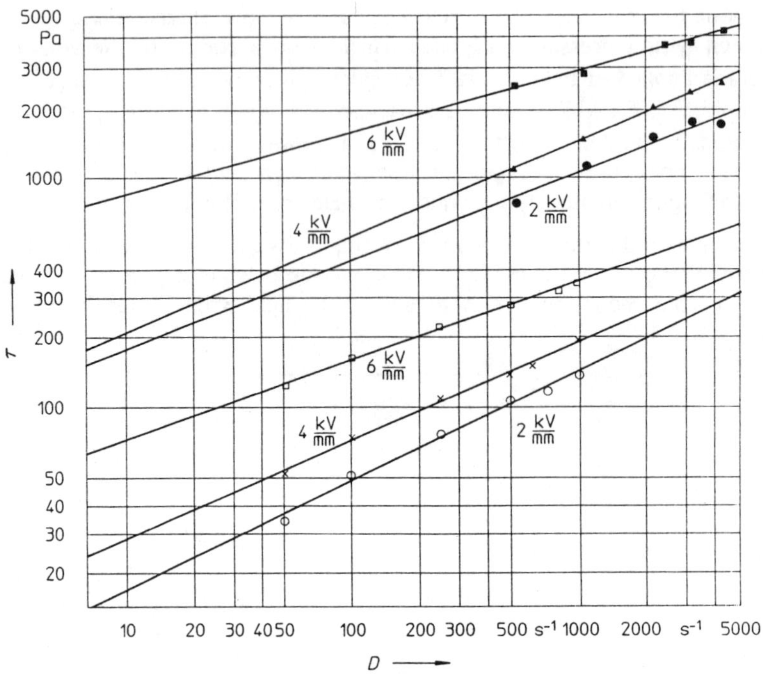

Fig. 10: Flow curves of the shear mode (O,X, □,) and the flow mode (●,▲,■) measurements for the homogeneous ER fluid

In case of the homogeneous ER fluid, the shear stress values of the shear mode measurement are higher by about 30 % to 50 % than the values of the flow mode measurement. The measuring curves of the ER suspension partly show far greater differences. At high control field strengths and high shear rates the shear stress values of the shear mode measurements are higher by up to 100 %. Correspondingly, the determination of the coefficients of the Ostwald equation or the equation by Herschel-Bulkley provides differing coefficients for the shear mode or the flow mode measurements (cf. Table 1).

	ER suspension						Homogeneous ER fluid					
Field	1 kV/mm		2 kV/mm		3 kV/mm		2 kV/mm		4 kV/mm		6 kV/mm	
Mode	Shear	Flow	Shear	Flow	Shear	Flow	Shear	Flow	Shear	Flow	Shear	Flow
A	6,354	39,98	18,12	21,01	11,63	208,3	5,07	70	10,36	77,51	31,6	519
B	0,536	0,504	0,436	0,636	0,648	0,32	0,488	0,388	0,424	0,432	0,352	0,276

Table 1: Coefficients A and B of the equations by Ostwald and Herschel-Bulkley

Based on the present measurements and the described simple rheological ER fluid models, a conversion of the measured values from the shear mode into the flow mode or vice versa is neither possible for the ER suspensions nor for the homogeneous ER fluid. For this reason, a calculation of shear-mode energy transducers can only be implemented by applying shear stress values and shear rate values determined by using the shear mode. Equally, the calculation of flow-mode energy transducers requires values for pressure differences and volume flows acquired by using the flow mode.

Moreover, the implemented examination shows that an extrapolation of measured values to other shear rate and volume flow ranges is hardly possible. Thus, it is important to acquire the measured values based on the calculation of an ER fluid transducer under similar conditions as they might occur later in real energy transducers (e.g. same shear rate range). Apart from the rheological quantities, the electrical ER fluid parameters conductivity and dielectric constant must be measured as well since they determine the concept and the dimensions of the energy source.

7 Literatur

1 Carlson, J. D.: *The Promise of Controllable Fluids.* Proc. Actuator 94 (Bremen, Germany, 15./17.06.94). p. 261-270

2 DE 41 39 065 A1

3 Janocha, H.; Rech, B.: *Determination of Rheological and Electrical Parameters of ER Fluids Using Rotational Viscometers.* Proc. 4th Int. Conf. on ER Fluids (Feldkirch, Austria, 20.-23.07.93). p. 344-357

4 Janocha, H.; Rech, B.; Bölter, R.: *Practice-Oriented Measuring of ER Fluid Parameters.* Proc. Actuator 94 (Bremen, Germany, 15./17.06.94). p. 294-298

5 Janocha, H.; Bölter, R.; Rech, B.; Hellbrück, St.: *The Influence of Cylinder Geometry in the Measurement of ER Fluids with Cylindrical Rotational Viscometers.* Presented at the 5th Int. Conf. on ER Fluids, MR Suspensions and Associated Technology (Sheffield, England, 10.-14.07.95)

6 Pahl, G.; Beitz, W.: *Konstruktionslehre.* Springer Verlag, 1986

7 Rech, B.: *Elektrorheologische Aktoren.* in "Technischer Einsatz Neuer Aktoren", Jendritza, D. J. et al., expert verlag GmbH, 1995

8 Wendt, E.; Büsing, K.: *Properties of a New Generation of Non-Abrasiv and Water-Free Electrorheological Fluids.* Proc. Actuator 94 (Bremen, Germany, 15./17.06.94). p. 309-311

LIGHT SCATTERING STUDIES OF AN ELECTRORHEOLOGICAL FLUID IN OSCILLATORY SHEAR

JAMES E. MARTIN AND JUDY ODINEK
Advanced Materials Physics Division, Sandia National Laboratories, Albuquerque, New Mexico 87185-1421

ABSTRACT

We have conducted a real time, two-dimensional light scattering study of the nonlinear dynamics of field-induced structures in an electrorheological fluid subjected to oscillatory shear. We have developed a kinetic chain model of the observed dynamics by considering the response of a fragmenting/aggregating particle chain to the prevailing hydrodynamic and electrostatic forces. This structural theory is then used to describe the nonlinear rheology of ER fluids.

1. Introduction

The rheology of ER fluids at low to moderate volume fractions is due to the aggregation of particles into volatile chain-like structures whose size adjusts in response to the flow, fragmenting and aggregating as the shear rate increases or decreases. Understanding the size and orientation of particle structures has been the basis of the development of the microscopic droplet model[1] of electrorheology in steady shear, and light scattering[2] has proven useful in establishing many of the predictions of this model. Much less is known about ER fluid structure in oscillatory shear, where the orientational dynamics of chain-like aggregates is the key issue. Since an understanding of electrorheology in nonstationary flows is essential to modeling the behavior of ER devices, we have undertaken a light scattering study of chain dynamics in oscillatory shear.[3]

Light scattering enables us to determine both the orientation and size of chains as functions of time, revealing a class of nonlinear dynamics dominated by fragmentation/aggregation processes that occur throughout each shear cycle. We have developed a simple *kinetic chain model*[3] that is in good agreement with the observed dynamics and that allows us to predict the nonlinear rheology. In fact, direct rheological measurements[4,5] show that the framework of linear viscoelasticity- a phase-shifted stress proportional to an applied strain- is of limited utility for ER fluids, so a nonlinear theory is appropriate.

Previous light scattering and diffraction studies focused on the quiescent fluid and on steady shear. The kinetics of structure formation[6] and the equilibrium state of the quiescent fluid[7] have been reported, as has the manner in which the chain size and orientation increases with shear rate.[2] These latter measurements are in accord with earlier experiments and theory of the power-law shear thinning viscosity.[1]

2. Experimental

2.1. Sample and Apparatus

Our light scattering sample is a transparent model silica fluid developed for light scattering and electrorheology studies.[6] The 700 nm spheres are coated with a silane coupling agent and dispersed in 4-methylcyclohexanol to 7.5 wt % silica. Aggregation is reversible by Brownian motion alone.

The scattering cell consists of an inner 40x2 mm circular disk electrode that is concentric to a 42 mm hole in an outer electrode, thus creating a radial electric field in a 1 mm gap. The outer electrode is sandwiched between plastic, and both are embedded between glass plates, with a fluid-filled 2 mm gap between the electrodes glass plates. The inner electrode is oscillated by a rod connected to a 25 mm lever on the electrode shaft, driven by an eccentric on a 300 oz.-in. microstepping motor.

The electric field is a 1 kHz square wave from a Trek™ power supply driven by Wavetek™ signal generator. Voltages are reported peak-to-peak.

Scattering data are captured by a Pulnix™ video camera, and stored on video tape. Data were then real-time digitized by a Perceptix™ frame grabber on a Macintosh Quadra 950™. Optical corrections were then made to the scattering data.[6]

A problem in this experiment is the phasing of the applied strain with the scattering data. We devoted a small corner of the scattering screen to an optical "strain phase clock" created by rotating a prism in synchrony with the stepping motor. The prism deflects a He-Ne laser beam that thus scribes a circle on the scattering screen, encoding the absolute strain phase on each image.

2.2 Data analysis

A scattering image contains a pair of scattering lobes[6] and the strain phase clock. These scattering lobes are tilted relative to the orientation of the scattering lobes obtained for the quiescent fluid, which are *orthogonal* to the electric field. The orientation of the scattering lobes reflects the orientation of the chains.

The rotation angle θ of the chains was determined by dividing the scattering image into 360 wedges, each subtending 1° of arc, and then integrating to find the average intensity in each wedge.[2] The orientation of the scattering lobes was then obtained by: locating the intensity maximum, θ_{max}; or finding the median, θ_m, that divides the integrated scattered intensity into equal halves, so $\int_{\theta_m}^{90^\circ} I(\theta)d\theta = \int_{-90^\circ}^{\theta_m} I(\theta)d\theta$. The median is useful when the dynamics is strongly nonlinear, but gives an orientation angle about five times smaller than θ_{max}.

3. Measurements

At low strain amplitudes $\gamma_0 << 1$ the response of the orientation angle θ_{max} to the shear strain $\gamma = \gamma_0 \sin(2\pi\omega t)$ was nearly sinusoidal. This sinusoidal response is exemplified in Fig. 1, where clockwise Lissajous plots of $\tan(\theta_{max})$ against γ are shown to be nearly elliptical. At high frequencies the shear rate, $\dot{\gamma} = 2\pi\omega\gamma_0\sin(2\pi\omega t)$, is large, and the chain orientation is nearly in phase with the fluid shear, since the hydrodynamic dominates electrostatic torque. Affine deformation is nearly achieved, as shown by the dashed line. At low shear frequencies the chain orientation leads the strain by 57° and thus lags the strain rate by 33°. In this regime electrostatic torque evidently dominates the hydrodynamic torque.

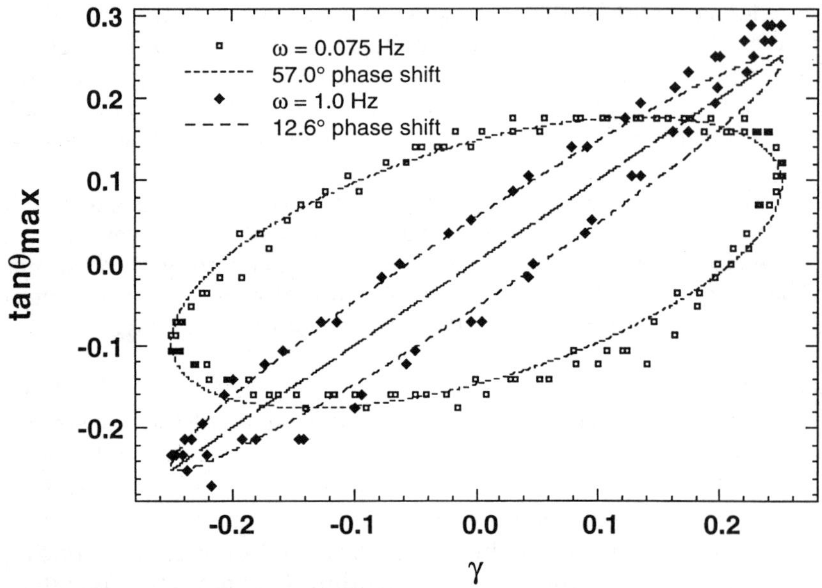

Figure 1: Clockwise Lissajous plots of $\tan(\theta_{max})$ against γ.

At higher strain amplitudes, Fig. 2, the sinusoidal motion becomes "clipped" as the chains fragment and aggregate during a cycle to maintain good field alignment, giving parallelogram Lissajous plots. At lower voltages we observe a highly nonlinear fluid response, as shown in Fig. 3. Starting at maximum positive strain, the chain half cycle can be described as follows. As the strain reverses, the chains co-move with the fluid, tilt to a maximum angle at half maximum strain, whereupon they fragment and undergo retrograde motion to realign with the electric field. Analysis shows that as they co-move with the

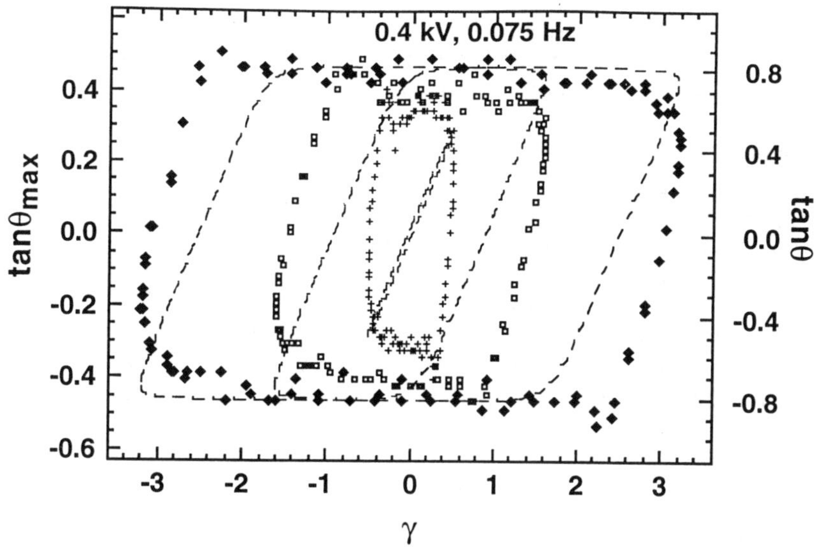

Figure 2: Lissajous plots at γ_0 = 0.5, 1.6, and 3.2 have parallelogram shapes that indicate a "clipping" of the motion. Theoretical curves are computed in the limit of large k.

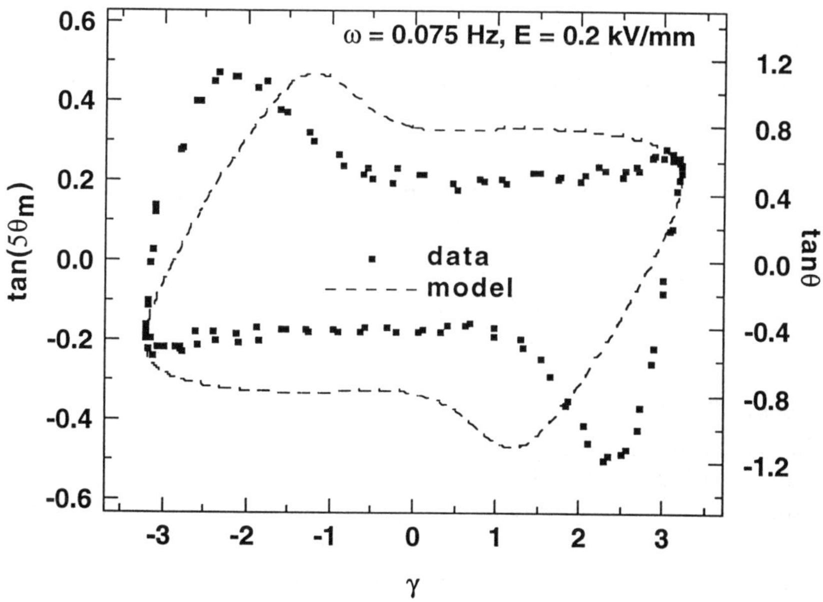

Figure 3: At low voltages a novel retrograde motion of fragmenting chains can be observed (dashed line is theory). The amplitude of motion is sensitive to the dipolar model and the method of data reduction.

shearing fluid, the chains aggregate, and indeed the scattering lobes brighten considerably during this time. These observations of chain motion have led us to propose the following model of ER fluid microstructure and rheology.

4. Chain Model

The salient features of experimental results we have shown can be understood in terms of a kinetic model of the dynamics of volatile chains. In this model we assume that dipolar interactions and hydrodynamic forces dominate thermal forces, which is certainly true in our experiments. This kinetic chain model differs from the elliptical droplet model, which minimizes a free energy and is thus is an equilibrium model. Although it is not appropriate to give a full derivation of the model here, a brief description will aid the reader.

4.1. Equation of Motion

We consider a linear chain of $2N+1$ spheres of radius a in an ocillatory shear field. The chain has a rotational velocity $\dot{\theta}$ and makes an angle θ to the electric field. The hydrodynamic force the fluid exerts can be decomposed into a tangential component that causes chain rotation, and a radial component that causes tension or compression. The tangential force is a maximum at the chain center, where in low Reynolds number flow is balanced by the tangential component of the dipole-dipole interaction force. Balancing the tangential hydrodynamic and electrostatic forces at the chain center[8] gives the damped oscillator equation

$$\dot{\theta} + \omega_d \sin 2\theta = \dot{\gamma}\cos^2\theta, \quad \text{where} \quad \omega_d = \dot{\gamma}\Big/16MnN^2 \tag{1}$$

Here $Mn=\mu_0\dot{\gamma}/2\varepsilon_0\varepsilon_c\beta^2E_0^2$ is the Mason number, which expresses the ratio of hydrodynamic to electrostatic forces, β is the dielectric contrast factor $(\varepsilon_p-\varepsilon_c)/(\varepsilon_p+2\varepsilon_c)$, ε_c and ε_p are the liquid and particle dielectric constants. [9]

4.2. Mechanical Stability

The oscillator frequency ω_d depends strongly on chain size. Physically acceptable values of N must correspond to mechanically stable chains or fragmentation will occur. The radial component of the hydrodynamic force is a maximum at the chain center, puts the chain in tension when $\theta\dot{\gamma} > 0$ and under compression when $\theta\dot{\gamma} < 0$. For the chain to be stable to fracture this hydrodynamically induced force must be smaller than the radial component of the electrostatic interaction.[9] The maximum stable chain number is determined by balancing these forces at the chain center.

$$N_{\max} = \begin{cases} \sqrt{\dfrac{2}{Mn}\dfrac{3\cos^2\theta - 1}{\sin 2\theta}} & \dot{\gamma}\theta \geq 0 \\ \infty & \dot{\gamma}\theta < 0 \end{cases} \tag{2}$$

N_{max} is dependent on chain orientation and strain rate, causing nonlinear response when driven by oscillatory shear $\dot{\gamma}=2\pi\omega\gamma_0\cos(2\pi\omega t)$. The maximum stable chain length diverges when the chain is aligned with the field, when the instantaneous strain rate is zero, and when the chain is under compression.

4.3. Kinetics of aggregation and fragmentation

If a chain finds itself far from its maximally stable size then its size will adjust by aggregation or fragmentation. The kinetics of aggregation and fragmentation can be described by the phenomenological formula

$$\frac{dN(t)}{dt} = \frac{k}{N(t)}\left[1 - \frac{N(t)^2}{N_{\max}^2(t)}\right] \tag{3}$$

where because induced dipolar forces drive aggregation the rate constant $k=k_0(8\varepsilon_0\varepsilon_c\beta^2E_0^2/\mu_0)$ where k_0 is a concentration dependent constant. When the chain is at its maximum length no aggregation or fragmentation occurs since $dN(t)/dt=0$. When the chain is much smaller than its maximum stable length, the chain will aggregate according to $N(t) = \sqrt{N(0)^2 + 2kt}$, in agreement with the root time prediction of See and Doi.[10] Thus aggregation is a slow power law growth process that is independent of the often divergent size N_{max}. If the chain is much larger than its stable length, then it will fragment exponentially quickly according to $N(t) = N(0)\exp(-kt / N_{\max}^2)$. Note that the fragmentation rate k/N_{max}^2 is proportional to the strain rate and is *independent* of the electric field.

Equations 1-3 are a set of coupled nonlinear equations that model the dynamics of chains in shear flow. There are four independent parameters in the system; Mn, ω, γ_0, andk_0. A reduction to a two-parameter model occurs by noting that solutions to the kinetic equation are of the form $N(t) = (\dot{\gamma} / 16\omega Mn)^{1/2} n(\omega t)$. The two parameter ($\gamma_0$, k_0) model is a result of the rate equation we have chosen, and because the strain amplitude is fixed by experiment, this is really a single free parameter model.

4.4. Electrorheology

Before showing the behavior of these equations we would like to discuss the predicted rheology. The stress σ in the sample is the hydrodynamic torque density, which equals the electrostatic torque density since inertia is small. In terms of the volume fraction ϕ of spheres this is

$$\sigma = 6\mu_0 N^2 \phi(\dot{\gamma}\cos^2\theta - \dot{\theta}) = \frac{3}{4}\varepsilon_0\varepsilon_c\beta^2 E_0^2 \phi \sin 2\theta. \qquad (4)$$

This equation demonstrates that the stress is nearly proportional to the angle θ, and that light scattering is an indirect probe of stress.[11] Furthermore, it is apparent that when $\theta = 0$, the chain is comoving with the fluid so $\dot{\gamma}\cos^2\theta = \dot{\theta}$.

Because the dynamics of chain orientation $\theta(s)$ is independent of electric field and shear frequency, we conclude that the stress scales purely as the square of the electric field. In steady shear it is readily shown that $\sin 2\theta = 2\sqrt{6}/5$, so the the *inherent* viscosity is given by the Bingham plastic form

$$\mu_I \equiv \frac{\mu - \mu_\infty}{\phi\mu_0} = \frac{1}{5}\left(\frac{3}{2}\right)^{5/2} Mn^{-1}. \qquad (5)$$

Note that we have subtracted off the infinite shear rate viscosity and divided by the liquid viscosity: at low volume fractions these viscosities are virtually equal and the lhs of Eq. 5 is the conventional definition of the inherent viscosity.

The shear thinning result $\mu \propto \dot{\gamma}^{-1}$ has been obtained in many experiments, including those conducted on our silica fluid.[1] However, for the silica fluid at low fields we find $\mu \propto \dot{\gamma}^{-2/3}$, a result consistent with the droplet model.

4.5. Numerical Results

The nature of the nonlinear response is illustrated in Fig. 4 (small strains give a sinusoidal response). For $k = 0$ the oscillator reorientation rate ω_d is constant, and the remaining nonlinearities in Eq. 1 are small enough that a nearly elliptical response is observed. As k increases, the chains fragment and aggregate to achieve mechanical stability while trying to maximize their length, and Lissajous plots approach a parallelogram, as in Fig. 2. Finally, for intermediate values of k an interesting crossover to retrograde motion is observed. This is observed in our data at low fields, as illustrated in Fig. 3.

4.5. Chain Size

It is clear from the model that chain size variations due to aggregation and fragmentation are the dominant factors driving the nonlinear dynamics. The predicted dynamics of fragmentation, which occurs at twice the oscillation frequency, is shown in Fig. 5. Chain aggregation is very pronounced just as the electrode is about to return to zero angle, with chains reaching a maximum size

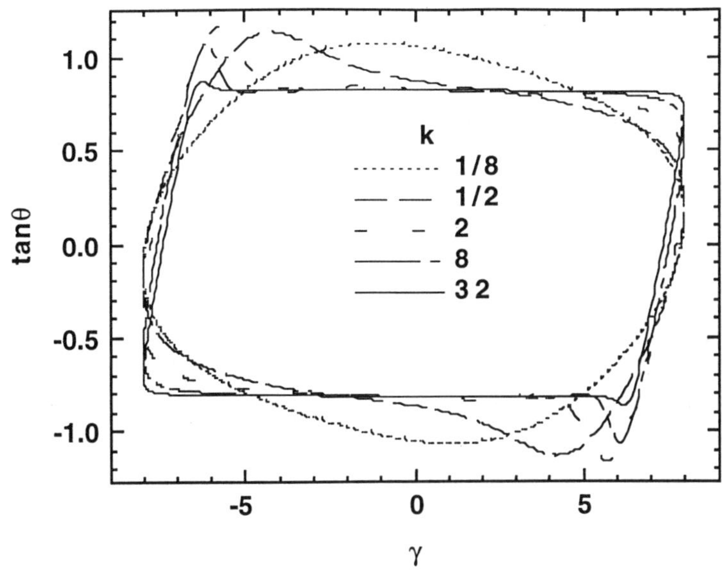

Figure 4: Lissajous plots change unexpectedly as the rate constant k is varied. For $k = 0$ the oscillator reorientation rate ω_d isconstant, and the nonlinearity is small. For large k Lissajous plots approach a parallelogram, whereas for intermediate k values retrograde motion is observed.

just after the maximum strain. During this time interval the chain is more or less comoving with the flow. Fragmentation occurs much before the electrode returns to zero angle because the chain is nearly 90° ahead of the strain and thus is already at its maximum tilt angle in the opposite direction. The power law aggregation and exponential fragmentation leads to a decided asymmetry in the chain size peaks.

Large aggregate structures give rise to intense light scattering, so we decided to revisit the data to see if there is any experimental evidence that would directly support the chain model. At each time we determined the wedge integrated maximum intensity I_{max} (as opposed to the peak position θ_{max}) and plotted this versus time in Fig. 6. We were frankly surprised at the close correspondence between these data and the prediction of the kinetic chain model. The peak position relative to the applied strain is about right, and even the peak asymmetry, due to slow aggregation and fast fragmentation, can be seen. Moreover, this pattern was observed in all of the data sets, except those taken at a strain amplitude of 0.25, despite the tremendous variations in Lissajous plots.

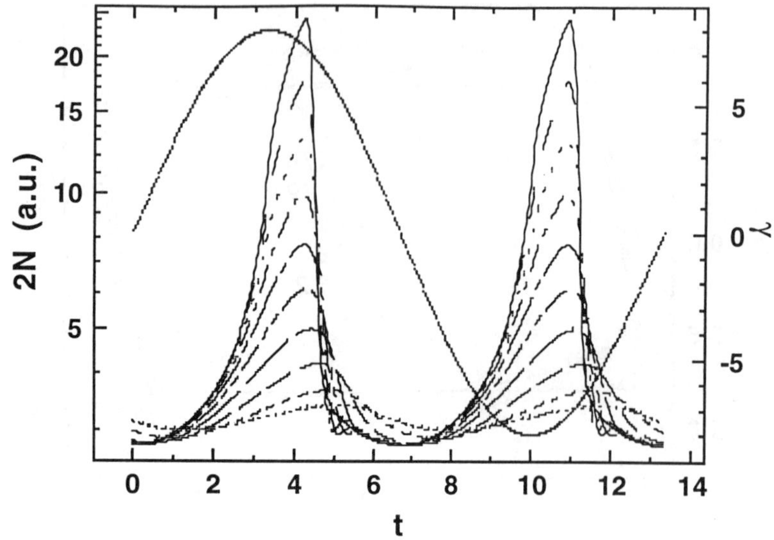

Fig. 5 The kinetic chain model predicts chain size fluctuations during each cycle, the magnitude of these fluctuations increases with the rate constant *k*.

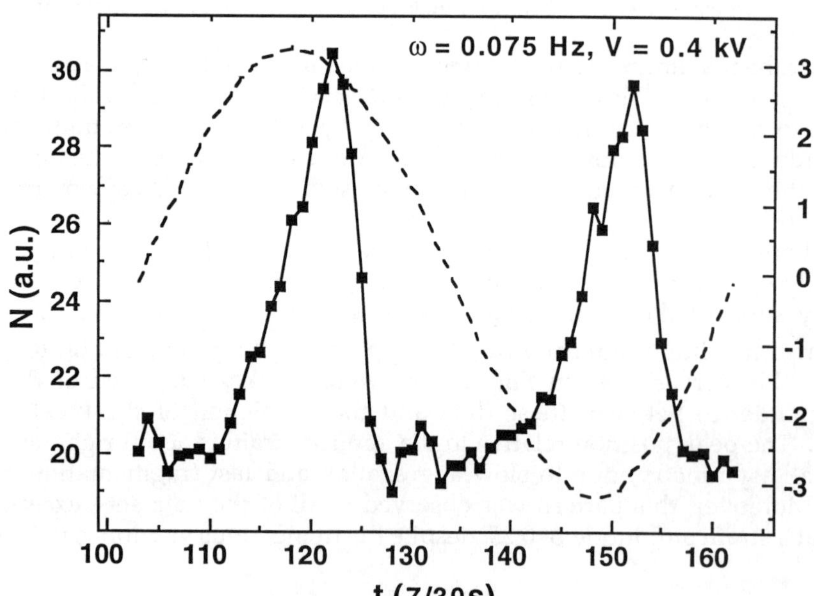

Fig. 6 Light scattering measurements indicate that the predicted chain fluctuations do occur. Note the asymmetry between the aggregation and fragmentation.

5. Conclusions

We have conducted light scattering studies that show that ER fluids can have strongly nonlinear behavior in oscillatory shear. The dipolar kinetic chain model gives a qualitative account of the nonlinearities observed in our data, without generating behaviors that we have not observed. However, it makes a number of quantitative predictions that are not correct, such as predicting chain angles that are too large and not accounting for residual field/frequency trends in the data. It is easy to modify the theory to give much better fits to the data, but only at the expense of introducing arbitrary parameters. What are the problems? Multipolar interactions actually increase the chain angle.[12] However, the distribution of chain sizes has been ignored, yet light scattering samples all chains and thus *understimates* the maximum chain orientation. Nonetheless, we feel that this model provides a framework that captures the essence of the observed dynamics.

6. Acknowledgment

This work was performed at Sandia National Laboratories and was supported by the U. S. Department of Energy under Contract No. DE-AC04-94AL85000. J.E.M acknowledges helpful theory discussions with T.C. Halsey and R. Anderson.

7. References

1. T.C. Halsey, J.E. Martin and D. Adolf, *Phys. Rev. Lett.* **68**, 1519 (1992).
2. J.E. Martin, J. Odinek and T.C. Halsey, *Phys. Rev. E* **50**, 3263 (1994).
3. J.E. Martin and J. Odinek, to appear in *J. Rheol.*
4. J.E. Martin, D. Adolf and T.C. Halsey, *J. Coll. Inter. Sci.* **167**, 437 (1994)
5. T.C.B. McLeish, T. Jordan, M.T. Shaw, *J Rheol.* **35**, 427 (1991); T. Jordan, M. T. Shaw, and T.C.B. Mcleish *J. Rheol.* **36**, 441 (1992).
6. J.E. Martin, J. Odinek and T.C. Halsey, *Phys. Rev. Lett.* **69**, 1524 (1992).
7. T. Chen, R.N. Zitter and R. Tao, *Phys. Rev. Lett.* **68**, 2555 (1992).
8. Balancing the force along the chain gives a chain shape that resembles an arcsine; Linearity avoids a dynamical system of $2N+1$ coupled nonlinear equations.
9. D.J. Klingenberg and C.F. Zukoski, *Langmuir* **6**, 15-24 (1990).
10. H. See and M. Doi, *J. Phys. Soc. Jpn.* **60**, 2778 (1991).
11. Like the Stress-Optical Theorem, which relates birefringence to stress.
12. J.E. Martin and R. Anderson, to be published.

Phenomenological Interpretation of Energy Loss and Storage Mechanisms in Electrorheological Fluids

J. A. Powell

Department of Mechanical Engineering, University of Strathclyde
James Weir Building, Glasgow G1 1XJ

Abstract

The application of a three element model to the oscillatory behaviour of an electrorheological fluid is given. This model consists of an irreversible yielding type element, a non-linear softening elasticity and viscous energy loss terms. The combination of the elastic, viscous and yielding type elements model the form of the stress-strain rate and stress-strain diagrams very well. These diagrams were obtained experimentally from an oscillatory electroviscometer. Application of the model is shown for a variety of strain amplitudes, oscillatory frequencies and electric field strengths. From these data the variation of energy storage (elasticity) and loss (yielding and Newtonian flow) mechanisms are shown to vary in a complicated manner, dependent for a given field strength on strain amplitude, frequency and hence strain rate.

1. Introduction

The material being considered in this investigation is an Electrorheological (ER) Fluid [2]. Rheology is the study of the deformation and flow of matter and so the name "Electrorheological" indicates an electrically induced deformation or change in flow conditions. In fact, an ER fluid is a liquid with suspended particles, which, upon the application of an electric field, changes to a semi-solid state having a yield stress, elasticity and increased viscosity. The liquid is non-conducting and can range from silicon oil to chlorinated hydrocarbon oil. The particles - the active ER ingredient - are semi-conducting and examples of these are starch and lithium polymethacrylate [3].

The ability of the fluid to change its mechanical properties, which can be controlled by variation of the electric field, opens up a wide variety of possible applications. Commercial systems are still rare, due to the practical limits of ER fluids, such as abrasion, sedimentation and low achievable yield stresses, but it would appear that there will be widespread applications in the automotive industry when these problems are overcome. Engine mounts and clutches utilising ER fluids as their main component are also currently being developed.

1.1 Description of ER Effect

The change of state of an ER fluid takes place within a matter of milliseconds upon application of the electric field. The change is also completely reversible - the fluid returns to its quiescent colloidal state within milliseconds of removing the electric field. Several theories have been investigated as to why the fluid displays such characteristics under the effect of an electric field. The opinion of many is that the suspended particles become polarised and form chains which align along the direction of the applied field [3]. The theory that the particles form chains is now widely accepted but what is still not fully understood is why the particles polarise and arrange themselves in this way. A common explanation is that the particles form dipoles. For particles containing water, as many of the particles used in ER fluids do, this is a fair assumption as water droplets do form dipoles under the influence of an electric field. This theory is also attractive because it has a magnetic counterpart in Magnetorheological fluids, which contain dispersed ferromagnetic particles. However, particles which do not contain water have been used in ER fluids and so it would appear that this dipole theory is an over simplification of the actual ER effect. Another common theory for the electrical behaviour within an ER fluid is the induced polarisation of an ionic double layer around each particle [4]. This layer deforms in the electric field and interacts with, or overlaps other layers. The fact that ER response times are so short, makes this a more probable explanation as it is only the ionic layer which rearranges and not the actual particle [5].

2. Applications of ER Fluids

Research into ER fluids is continuing, particularly in the USA, into new ER materials and fluid compositions. In 1990, it was estimated that the 1995 USA and world-wide market figures would run to $3.5 billion and $20 billion respectively [8]. However, the fact remains that there are still no actual mass produced engineering components which utilise ER fluids, and so figures such as these are unlikely to be realised for perhaps another 10 years or more.

There have been three broad areas of application previously identified [9]:-

1. Clutch based devices :- The fluid is constrained between either concentric cylinder or parallel plates, and sheared rotation. The shear stress in the activated fluid transmits a torque from a plate on the driving shaft, to a plate on the driven shaft.
2. Flow based systems (Valve mode) :- The fluid is pumped through a straight flow path and when the fluid is activated, there is a pressure drop across the path.
3. Damping :-The fluid is used to damp rotational and translational vibration in either a passive or semi-active mode.

Probably the first major applications of ER fluids will be in the automotive industry. Many of the prominent American and Japanese car manufacturers have designed and tested prototype devices.

2.1 Automotive Engine/Machinery Mounts

Engine or machinery mounts utilise ER fluids in shear, extensional and compressional modes. The fact that an ER fluid can provide a variable damping force suggests that amplitude of vibration at resonance can be greatly reduced as well as reducing high frequency vibrations. These two conditions cannot be met using an isolator with a fixed damping level. A number of different designs have been discussed, and one which was recently patented was an automotive engine mount designed by Sproston *et al.* [10], and operated in squeeze mode. This mount was tested and the authors results expressed in terms of output/input transmissibility. The results showed a tenfold reduction in transmissibility at resonance, but an increase in transmissibility at higher frequencies. An on-off feedback control system was developed by Petek [11] where the "off " state represented very low damping and the "on" state signified high damping. This achieved control over transmissibility at resonance and anti-resonance.

Another type of engine mount discussed by Morishita and Mitsui [12] uses an electrode with a hole in the centre so that the ER fluid flows through the electrode under a pressure gradient. As well as the viscous damping and stiffness provided by the extensional and compressional damper, this mount provides an element of friction type damping due to the fluid flowing over the electrode surface. These authors identified a level of electric field strength above which the fluid appeared to become clogged in the electrode which reduced the flow, causing a reduction in the friction type damping. Stanway *et al.* [29] presented some results on a vibration damper and showed that transmissibility could be reduced. A variety of other materials and designs for applications can be found in the literature [13-22].

3. Modelling

3.1 Explanation of Model

When mathematically modelling the behaviour of an ER fluid, there are two separate paths which can be followed. Firstly, the fluid can be modelled on a microscopic scale, dealing with molecular and particle interactions, and secondly, discrete elements such as springs and viscous dashpots can be combined to model the materials rheological behaviour.

The first type of modelling has been carried out by authors such as Mason [6], Martin and Odinek [23], and Zukoski [4]. Mason and co-workers analysed the behaviour of a pair of interacting di-electric spheres of relative permittivity, ε_2', dispersed in a liquid of relative permittivity, ε_1', and viscosity, η_1. They modelled the effects in terms of the dimensionless Mason number

$$M_n = \eta\dot{\gamma} / 2\varepsilon_o\varepsilon_1`(\beta E)^2 \quad (1)$$

This is the ratio of the viscous to electrostatic forces set up under a shear rate, $\dot{\gamma}$, and a d.c. electric field, E. ε_o is the permittivity of free space and β is the internal field correction term for spheres. These authors show that the dynamics of an ER fluid can be understood in terms of a balance of viscous forces and polarisation forces.

This paper uses the second type of modelling mentioned and so did not consider further, the microscopic properties of ER fluids. When activated, the ER fluid displays an increased viscosity, elasticity and a yield stress. When it is subjected to a small steady state oscillatory deformation, the particles structure deforms, ruptures and then reforms. The rupturing of the particle formations is responsible for the yielding characteristic [4]. A viscous fluid with a yield stress can be modelled as a Bingham plastic [24]. This is described by the equation

$$\tau = \tau_y + \eta\dot{\gamma} \quad (2)$$

where τ is the shear stress, τ_y is the shear stress at yield, η is the plastic viscosity and $\dot{\gamma}$ is the shear rate. This has been used for steady uni-directional, rotational tests on ER fluids where a given rotational frequency corresponds to a particular shear rate. However, if the fluid is sheared in a bi-directional or oscillatory fashion, the Bingham equation does not take account of the elasticity of the fluid. The variation of the elasticity is non-linear, and is modelled as a non-linear softening spring. The Newtonian losses are modelled as a viscous dashpot, which only leaves the yielding characteristic to be emulated. A friction or Coulomb damping type yielding element lends itself to this purpose quite readily. A combination of a viscous damper, friction damper and non-linear stiffness would appear to represent the response of an activated ER fluid. This model was devised by Powell [1] and was the model used in this investigation.

The three elements are combined in parallel as shown in figure 1.

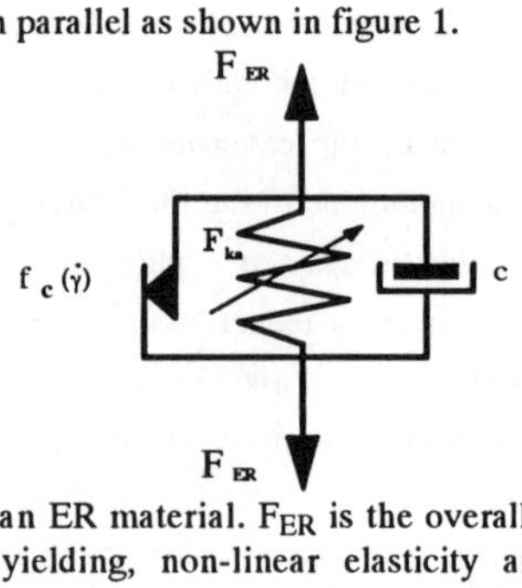

Figure 1. Mechanical analogy of an ER material. F_{ER} is the overall force on the ER material. f_C, F_{ka} and c are the stresses due to yielding, non-linear elasticity and Newtonian flow respectively. (Reproduced from [1]).

Descriptions of the three elements will now be given.

3.1.1 Viscous Damper

The energy dissipated by the Newtonian flow within the activated material was assumed to be linear and was represented mathematically by the expression

$$F_D = c\,\dot{z} \qquad (3)$$

where c is the coefficient of viscous friction (Nsm^{-1}) and $\dot{z}$ is the relative velocity across the damper.

3.1.2 Yielding Damper

The experimental results indicated that a type of yielding stress was evident in the ER material when acted upon by a shear deformation [25]. These yielding stresses were noted to decrease with increasing shear rate, rather like a friction force with differing static and dynamic coefficients of friction, F_{CS} F_{CD}, respectively. When the yielding force or stress changed sign (change in sign of shear rate) the process was noted to be irreversible. To take account of this two exponentially varying functions were utilised.

The equations which describe the upper and lower characteristics are :-

$$F_{CU}(\dot{z}) = F_{CS}\,(1 + (F_{CD}/F_{CS})\,e^{-a|\dot{z}|}) \qquad (4a)$$

and

$$F_{CL}(\dot{z}) = F_{CD}\,(1 - e^{-b|\dot{z}|}) \qquad (4b)$$

The exponential terms model the observed gradual decrease and increase in friction (yielding) force adequately. The shear stress changes at the peak amplitudes were noted not to be abrupt. To take account of this, a smoothing function was included in the equations to produce a finite change in force rather than a sudden impulse. This smoothing function comes in the form of a hyperbolic tangent.

So the equations become :-

$$F_{CU}(\dot{z}) = F_{CS}\,(1 + (F_{CD}/F_{CS})\,e^{-a|\dot{z}|})\tanh(\varepsilon.\dot{z}) \qquad (5a)$$

and

$$F_{CL}(\dot{z}) = F_{CD}\,(1 - e^{-b|\dot{z}|})\tanh(\varepsilon.\dot{z}) \qquad (5b)$$

where $\dot{z}$ is the relative velocity and ε denotes the abruptness of the tanh function and hence, the abruptness of the friction force change.

3.1.3 Non-linear Stiffness

From the experimental results, it was observed that the stiffness was non-linear. It is represented by a hyperbolic tangent with a variable maximum amplitude. The equation is

$$F_{ka} = A' \tanh(d.z) \qquad (6)$$

where z is the relative displacement, A' is the maximum amplitude corresponding to the maximum spring force and d controls the abruptness of the tanh function.

3.1.4 Overall Force Balance Equation

The force equations for each element in the model have now been derived and so the overall force balance equation can be developed from figure 1.

For equilibrium,

$$\text{Force in ER fluid, } F_{ER} = \begin{cases} F_{CU} + F_{ka}(z) + c\dot{z} & \text{when } \ddot{z}.\dot{z} \geq 0 \\ F_{CL} + F_{ka}(z) + c\dot{z} & \text{when } \ddot{z}.\dot{z} < 0 \end{cases} \qquad (7)$$

3.2 Matching Model to Experimental Data

To show the validity of the mechanical analogy described in section 3.1, it was necessary to match it to the experimental results. This would also show the variation of the forces in the viscous damper, spring and friction damper for varying electric field strengths, strain amplitudes and strain rates (oscillatory frequency).

A FORTRAN 77 program was used to solve equation (7). This program plotted graphs of shear stress against strain - and shear stress against strain rate. Also, a shear stress time history for the experimental data was plotted. Then after solving (7), the same graphs were plotted for the theoretical model. By plotting the corresponding experimental and theoretical graphs on the same axes, the parameters of the theoretical equation could be adjusted until there was as close a correlation as possible between the two curves, by using a trial and error approach. This was carried out for each experimental case. The model was matched to all the data found in [25]. The fluid used was 100 cSt silicone in a 30% particle/fluid mass ratio with starch which had 12% by weight absorbed water. Shown in figures 2-5 are a selection of experimental results [25] and the associated results for the model.

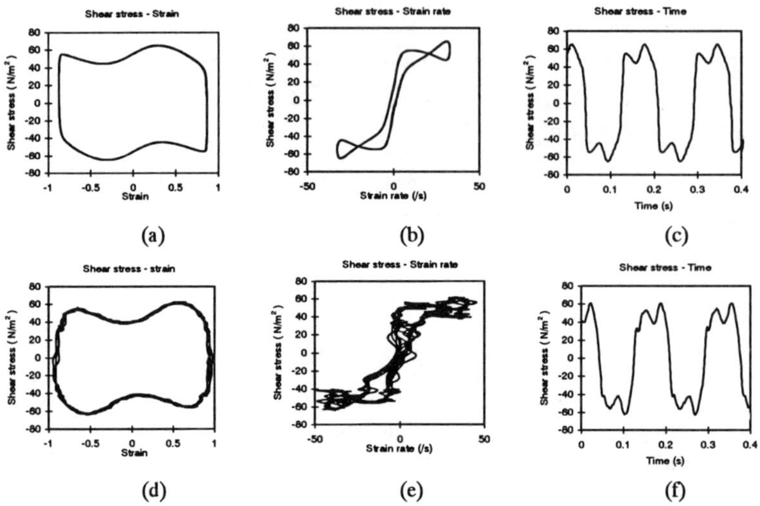

Figure 2. (a) - (c) Theoretical model curves for field strength 0.33 kV/mm, frequency 6 Hz and non-dimensional strain 1.7452. (d) - (f) Equivalent experimental curves.

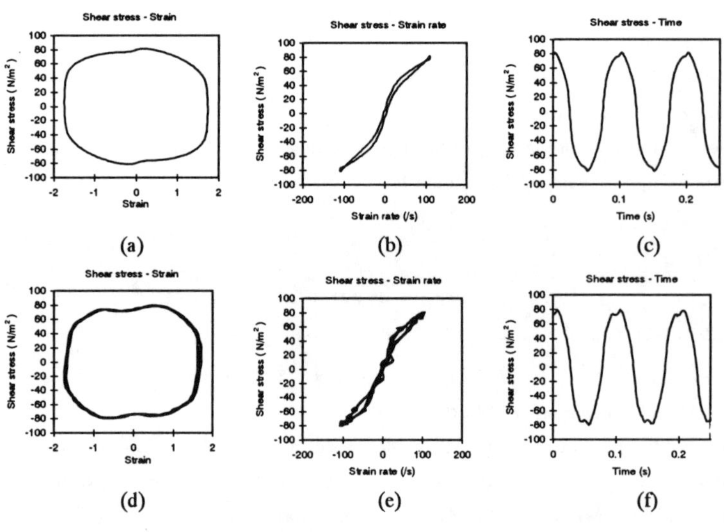

Figure 3. (a) - (c) Theoretical model curves for field strength 0.33 kV/mm, frequency 10 Hz and non-dimensional strain 3.4904. (d) - (f) Equivalent experimental curves.

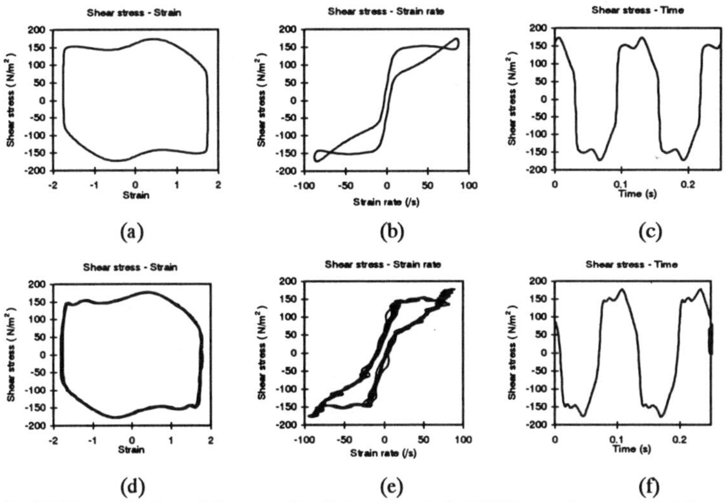

Figure 4. (a) - (c) Theoretical model curves for field strength 0.66 kV/mm, frequency 8 Hz and non-dimensional strain 3.4904. (d) - (f) Equivalent experimental curves.

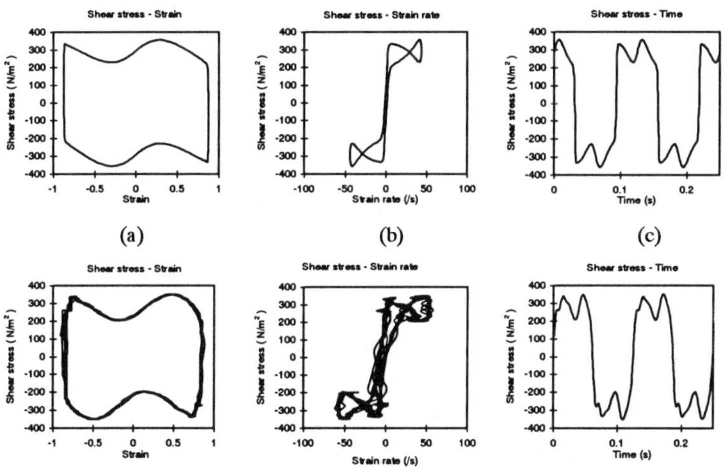

Figure 5. (a) - (c) Theoretical model curves for field strength 1.00 kV/mm, frequency 8 Hz and non-dimensional strain 1.7452. (d) - (f) Equivalent experimental curves.

4. Energy Dissipation and Storage

The primary application of the ER fluid is likely to be as some sort of damper. The effect of damping is to remove energy from a system and so it is desirable to know the energy dissipation and storage capabilities of the damping element. Here, the results gained in the previous section were used to calculate the energy dissipation and storage of the material using the application of the previously described three element model under varying electric field strengths, amplitudes and frequencies of oscillation. The model consists of two mechanisms which dissipate energy - the friction or yielding and viscous damping - and one mechanism which stores energy - the non-linear spring.

4.1 Energy Dissipation

Under oscillatory conditions, the force-displacement curve for the ER fluid encloses an area, referred to as the hysteresis loop. This enclosed area is proportional to the energy dissipated per cycle. It was necessary to consider the friction and viscous damping separately, and to integrate the functions described for each. The energy dissipated per cycle due to a damping force, F_d, is computed from the general equation

$$W_D = \oint F_D \, dz \tag{8}$$

4.2 Energy Storage

For cyclic oscillations, the area between this curve and the displacement axis is proportional to the work done. This area is calculated by integrating the force-displacement equation (6), which was again undertaken using the trapezoidal rule. A separate program was used to calculate the energy dissipated for the upper friction damping characteristic, the lower friction damping characteristic and the energy stored using the non-linear stiffness. The programs calculated the area enclosed between the shear stress-strain curve and the strain axis for each model element. The upper and lower limits of the strain were required for the trapezoidal rule calculation.

The cyclic force-displacement curve for a spring is not a hysteresis loop and, as the spring returns to its original position at the end of the cycle, there is no net work done which could be called the energy stored per cycle, so only 1/4 of a cycle was considered for the calculation.

4.3 Discussion

The results obtained from section 4.2 are presented in the form of graphs showing the variation of energy dissipated and stored per cycle, with cyclic amplitude, oscillatory frequency and electric field strength. It has been realised that the method of modelling the ER fluid behaviour, by matching experimental and theoretical graphs is fairly basic. However, the actual values of energy dissipated and stored are not required to be highly accurate for this investigation. It is the trends which appear in the following graphs which are of importance in attempting to describe the behaviour of an ER fluid. Having said this, it is estimated that the accuracy from initial estimates was never less than 6 - 7%. The method of Tchebycheff polynomial curve fitting [26] could have been used to fit parameters to the curves but this would not have highlighted information on the level of each individual model element's contribution to the overall stresses calculated.

The first thing to consider was the relationship between the energy stored and dissipated and the applied electric field strength. From figure 6, it can be seen that increasing the electric field strength increased the energy dissipation. This increase can be attributed to an increase in the friction damping and viscous losses. The friction damping increased because the yield stress increased with electric field strength. The viscosity also increased. The relationship between energy stored and electric field strength can be seen in figure 7. The energy stored also increased with electric field strength. This is attributed to an increasing elasticity in the ER material.

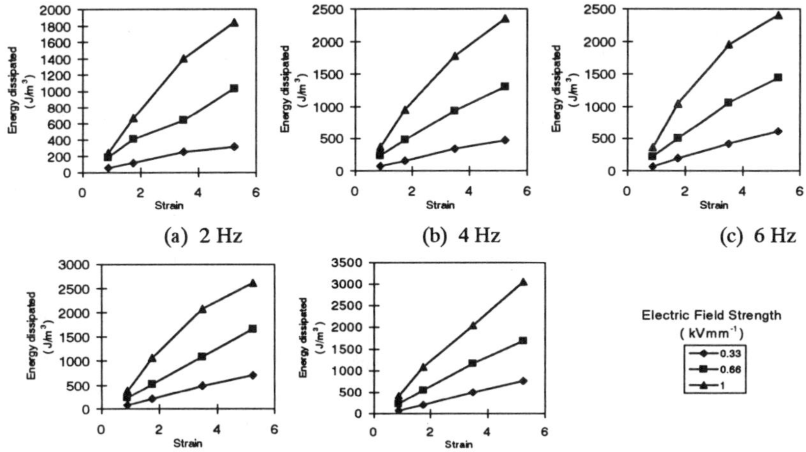

(a) 2 Hz (b) 4 Hz (c) 6 Hz

(d) 8 Hz (e) 10 Hz

Figure 6. Variation of total energy dissipation with strain and field strength for frequencies (a), (b), (c), (d) and (e).

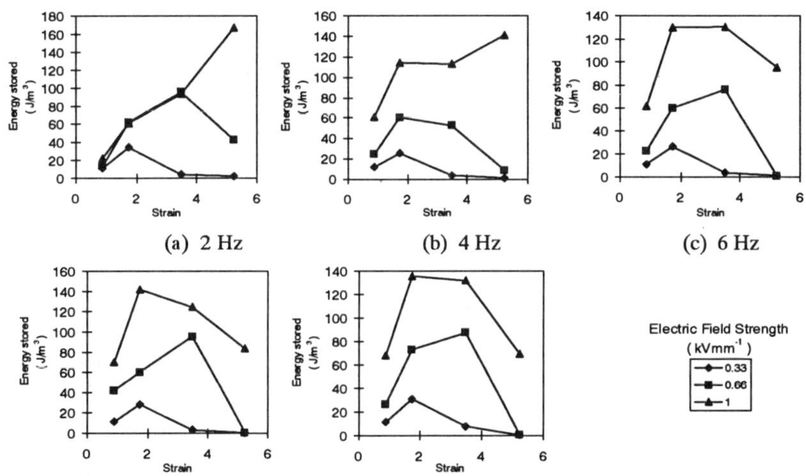

(a) 2 Hz (b) 4 Hz (c) 6 Hz

(d) 8 Hz (e) 10 Hz

Figure 7. Variation of energy storage with strain and field strength at frequencies (a), (b), (c), (d) and (e).

The total energy dissipated and stored also varied with the strain and frequency of oscillation. It was noted that the total energy dissipated increased with strain. At low strains, the total energy dissipated was relatively independent of frequency. That is, it remained reasonably constant as the frequency increased.

4.4.1 Comparison of Friction and Viscous Energy Dissipation, and Energy Storage

The main benefit of having the simple three element model and being able to describe it mathematically, was that each element could be considered individually. This allowed a more detailed analysis of the activated ER material behaviour.

The energy dissipated can be separated into that dissipated by the friction damping and by the viscous damping. Figures 8-10 show how the viscous damping, friction damping and stiffness behaved in relation to one another. Figure 8 is a comparison at the lowest field strength. At low strains, the majority of the energy dissipated was attributed to the friction damper. However, as the strain increased, the viscous damping increased and at the highest amplitudes, it dominated the friction damping. This was because the viscous energy dissipation is proportional to the square of the strain, so it increased more rapidly with strain than the friction energy dissipation. In terms of the rheology of the ER material, as the strain increased the internal particle formations yielded and the viscosity became the main source of damping.

Figures 8-10 show the variation of viscous energy dissipation with strain was parabolic, showing that the energy varied with the square of the strain. From equation (3), this showed that c, the coefficient of viscosity, was constant at all strains for each frequency. However, as the field strength increased (figures 9 and 10), the variation of viscous energy dissipation with strain was no longer a parabolic at low frequencies. This is particularly clear in figure 9 (b) and (c). This is because c was not constant but in fact decreased as the strain increased. This may be attributable to a rheological phenomenon known as 'shear thinning' which is a recoverable decrease in viscosity as the shear rate or frequency increases. This should not be confused with thixotropy which is a time dependent effect. With 'shear thinning', as the shear rate decreases again, the viscosity increases.

Figures 8-10 show how insignificant the energy stored was compared to the energy dissipated. At low amplitudes, the storage and viscous energy dissipation were similar in magnitude, but as the strain increased and yielding took place, the material became more Newtonian, rather than elastic-plastic.

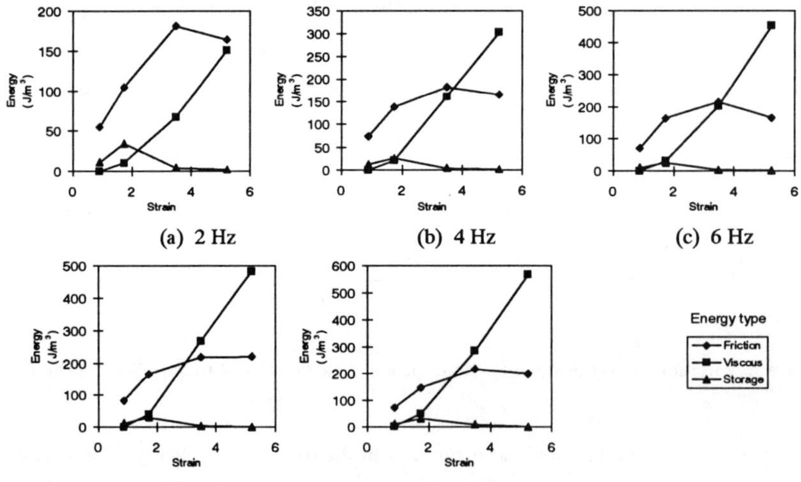

Figure 8. Comparison of friction, viscous energy dissipation, and energy storage against strain for frequencies (a), (b), (c), (d) and (e). Field strength 0.33 kVmm^{-1}.

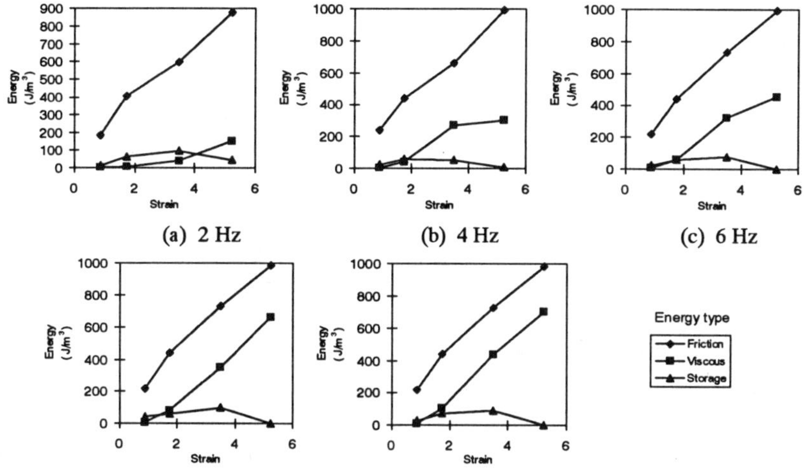

(d) 8 Hz (e) 10 Hz

Figure 9. Comparison of friction, viscous energy dissipation, and energy storage against strain for frequencies (a), (b), (c), (d) and (e). Field strength 0.66 kVmm^{-1}

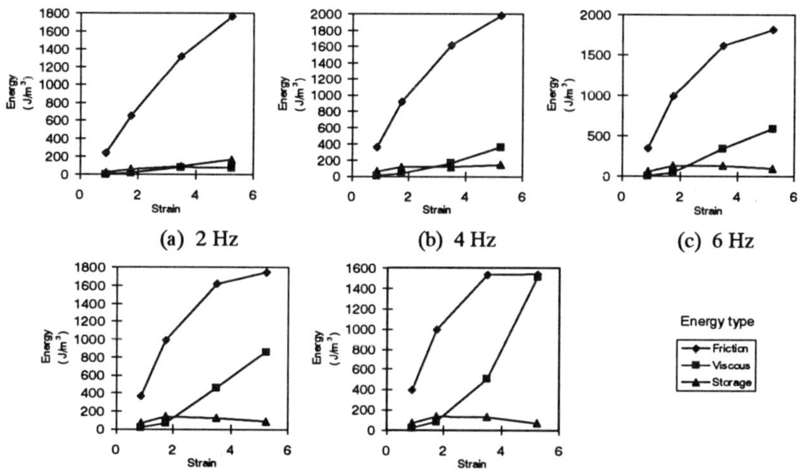

(d) 8 Hz (e) 10 Hz

Figure 10. Comparison of friction and viscous energy dissipation, and energy storage against strain for frequencies (a), (b), (c), (d) and (e). Field strength 1.00 kVmm^{-1}.

4.4.2 Comparison with Previous Investigations

Several authors have previously investigated the energy dissipation and storage capabilities of an ER material under oscillatory loading. So far no literature is available of the type presented in this paper, particularly in terms of quantifying the relationship between the viscous and friction energy dissipation, and the energy storage. One notable study was carried out by Gamota and Filisko [27]. They used an oscillatory electroviscometer to investigate the stress-strain relationship of an ER material for different electric field strengths. They obtained similar stress-strain curves whose area increased as the electric field strength increased, which indicated, as shown in this investigation, that the energy dissipation increases with electric field strength. They did not show shear stress-strain rate curves. Their material required a higher field strength - 2.00 kV/mm - to make the stress-strain curve deviate from an elliptical shape but this was at much lower strains so the material remained in a linear viscoelastic state. This means that at low field strengths - < 1.00 kV/mm - , the material displayed only viscous energy dissipation and that frictional type damping was not evident. However, for the material in this investigation, friction damping energy dissipation was present at all field strengths due to the larger strains used. These authors have been able to calculate the energy dissipated by calculating the area enclosed by the stress-strain curve using some approximate method, but have not been able to separate the viscous and frictional energy dissipations. This could only be undertaken if the ER material behaviour had been effectively modelled using phenomenologically based approaches. A mechanical analogy of an active ER material was proposed in their paper, but it was not described mathematically due to its complexity. It consisted of a stiffness, viscous damping and frictional damping in series.

Studying the actual energy dissipation values obtained in [27], they were found to be much higher than any obtained in this investigation. The energy dissipation was calculated at a strain amplitude of 1.24° peak to peak, electric field strength of 1.00 kV/mm and a frequency of 10 Hz. At this strain, the energy dissipated per cycle per unit volume was 40 kJ/m^3. The smallest strain amplitude considered in this present project was 5° and as energy dissipation increases with strain amplitude, this would be expected to give a higher energy dissipation than an amplitude of 1.24°. However, at 5°, 1.00 kV/mm and 10 Hz, in this present investigation, the energy dissipated was 414.34 J/m^3. So the energy dissipation obtained in [27] is almost 100 times greater than in the present paper, for a strain amplitude four times smaller. This large increase has been achieved by using a different ER material - alumino-silicate particles in paraffin oil. Also, the annular gap in the electroviscometer used was much smaller which could produce a greater damping effect.

Another study worth considering was carried out by Ehrgott and Masri [28]. The experimental apparatus used consisted of two coaxial cylinders with an ER fluid filling the gap between them. The outer cylinder was vertically oscillated and the force transmitted to the inner cylinder was measured. They concluded that the energy dissipated by friction type damping is independent of frequency. The actual values of energy dissipated per cycle per unit volume of ER fluid obtained in [28] are again different from the values obtained in this project. This time, the energy dissipated was lower with the maximum value being 300 J/m^3, and this was at a field strength of 5 kV/mm and a frequency of 35 Hz. These are considerably larger than any field strengths and frequencies used in this project. A comparison can be made using the energy dissipated at 1.00 kV/mm and 10 Hz from [28]. Here, the energy dissipated per unit volume was 30 J/m^3 and in this present project, for the same field strength and frequency, and at the minimum strain amplitude, the energy dissipated was 414.34 J/m^3. Once again, the study by Ehrgott and Masri did not consider the energy dissipated by friction type damping and viscous damping separately, and yet they have still described the variation of friction with frequency which was not shown lucidly. They used equivalent linearisation to approximate the energy dissipated by the friction type damping as the area of an ellipse, making it equivalent to viscous energy dissipation.

5. Conclusions

This paper highlights the use of a simple three element mechanical model to investigate the energy dissipation and storage capabilities of an Electrorheological fluid. It has been accepted that the equations describing this model do not represent the forces in the ER fluid under oscillatory loading exactly. They do provide a fairly simple method of describing the activated ER fluid behaviour and it can be seen from figures 2-5 that the theoretical and experimental shear stress variations can be extremely closely matched.

When the forces associated with each model element were quantified for the experimental parameters of electric field strength, oscillatory frequency and strain amplitude, the energy dissipated and stored in one cycle was calculated. This was undertaken by integrating the shear stress-strain equations for each element. The benefit of having a simple three element model was that the energy dissipated by the friction type damping and the viscous damping could be separated. This allowed a more detailed analysis of the dynamic ER fluid behaviour.

Summarising the results from section 4, some general conclusions can be drawn. Both the energy dissipated and stored increased with increasing electric field strength. The energy dissipated increased with strain amplitude but the energy stored peaked and fell off as the strain amplitude was increased. This was probably due to the elasticity of the ER material decreasing as it yielded.

From the comparison of friction damping and viscous damping energy dissipations, the friction type losses were much greater at low strain amplitudes. As the amplitude increased, the viscous damping increased until it was similar in magnitude to the friction type damping. The viscous energy dissipation varied linearly with frequency and with the square of the strain.

References

1. Powell, J. A. : "Application of a Non-linear Phenomenological Model to the Oscillatory Behaviour of ER Materials", Journal of Rheology, 1995, In press.
2. Askeland, D.R. : "The Science and Engineering of Materials", 2nd Ed., Published by Chapman and Hall, 1990.
3. Block, H., Rattray, P. : "Recent developments in ER fluids", to be published, 1995.
4. Zukoski, C.F. : "Material properties and the Electrorheological response", Annual Review of Materials Science, 1993, 23, 45 - 78.
5. Klass, D.L., Martinek, T.W. : "Electroviscous fluids I and II; Rheological properties; Electrical properties", Journal of Applied Physics, 1967, 38, 67 - 80.
6. Chaffey, C.E., Mason, S.G., Journal of Colloid Interface Science, 1968, 27, 115.
7. Austin, S.A. : "The vibration damping effect of an Electrorheological fluid", Journal of Vibration and Acoustics, 1993, 115, 136 - 140.
8. Coulter, J.P. : "An investigation of Electrorheological material based controllable damping devices", Electrorheological Flows, proceedings of Fluids Engineering Conference, 1993, 164, 115 - 127.
9. Brooks, D. : "Fluids get tough", Physics World, 1989, 2, 8, 35 - 38.
10. Sproston, J.L., Stanway, R., Williams, E.W., Rigby, S. : "The Electrorheological automotive engine mount", Journal of Electrostatics, 1994, 32, 253 - 259.
11. Petek, N.K., Goudie, R.J., Boyle, F.P. : "Actively controlled damping in Electrorheological fluid filled engine mounts", proceedings of The Second International Conference on Electrorheological Fluids, 1989, 409 - 418.
12. Morishita, S., Mitsui, J. : "An electronically controlled engine mount using Electrorheological fluid", SAE Tech Paper, 1992, No. 922290.
13. Powell, J.A. :"Electro-Rheological Fluids as a means of vibration suppression". *KEYNOTE SPEECH:* Presented at "Vibration and Noise '95", Venice, 23rd April 1995.
14. Coulter, J.P., Weiss, K.D., Carlson, J.D. : "Engineering applications of electrorheological materials", Journal of Intelligent Material Systems and Structures, 1993, 4, 248 - 259.
15. Sproston, J.L., Stevens, N.G., Page, I.M., : "An investigation of torque transmission using electrically stressed di-electric fluids", proceedings of Conference on Electrostatics, 1983, 53 - 58.
16. Nelson, D.A., Suydam, E.C. : "The thermal aspects of the Electrorheological effect and its impact on application design", Electrorheological Flows, proceedings of Fluids Engineering Conference, 1993, 164, 71 - 84.
17. Edwards, M. : "Systems benefit from smart fluid advances", Eureka, 1990, 52 - 55.
18. Korobko, E.V., Ragotner, M.M., Gorodkin, R.G., Bukovich, I.V. : "Electrorheological fluid as an element in clutches", Journal of Intelligent Material Systems and Structures, 1991, 2, 215 - 227.
19. Philips, R.W. : "Engineering applications of fluids with a variable yield stress", Ph.D. Thesis, University of California, Berkeley, 1969.
20. Coulter, J.P., Don, D.L., Yalcintas, M., Biermann, P.J. : "An experimental investigation of Electrorheological material based adaptive plates", Adaptive Structures and Material Systems, ASME 1993, 35, 287 - 296.

21. Lee, G.K. Yuan, F., Conrad, H. : "On the use of Electrorheological fluids for smart space materials", Space Exploration Science and Technologies Research, ASME 1992, 31, 107 - 115.
22. Jang, B.Z. : "Intelligent composites and structures - A review", International Conference on Advanced Composite Materials, 1993, 661 - 667.
23. Martin, J.E., Odinek, J. : "A light scattering study of the non-linear dynamics of Electrorheological fluids in oscillatory shear", Journal of Rheology, in press, 1995.
24. Schulz, D.N., Glass, J.E. : "Polymers as Rheology Modifiers", ACS Symposium series 462, Published by American Chemical Society, 1991.
25. Powell, J.A. : "The mechanical properties of an Electrorheological fluid under oscillatory dynamic loading", Smart Materials and Structures, 1993, 2, 217 - 231.
26. Ehrgott, R.C., Masri, S.F. : "Modelling the oscillatory dynamic behaviour of Electrorheological materials in shear", Smart Materials and Structures, 1992, 1, 275 - 285.
27. Gamota, D.R., Filisko, F.E. : "Dynamic mechanical studies of electrorheological materials : Moderate frequencies", Journal of Rheology, 1991, 35(3), 399 - 425.
28. Ehrgott, R.C., Masri, S.F. : "Experimental characterisation of an Electrorheological material subjected to oscillatory shear strains", Journal of Vibration and Acoustics, 1994, 116, 53 - 60.
29 Stanway, R. : "Nonlinear modelling of an electrorheological vibration damper", Journal of Electrostatics, 1987, 20, 167-184.

LIQUID DIELECTRIC WITH DISPERSED PARTICLES AS AN ACTIVE MEDIUM FOR AN "ACOUSTIC" LASER

S.T.Zavtrak
The Institute of Nuclear Problems, Bobruiskaya Str., 11, Minsk 220050, Republic of Belarus

Abstract

The theoretical scheme of an "acoustic" laser is suggested. A liquid dielectric with dispersed particles is considered as an active medium. Pumping is created by an oscillating electric field that deforms dispersed particle volumes or by a piezoelectric emitter of a piston type. Two types of resonators (rectangular and cylindrical) are considered for this laser. Phase bunching of initially incoherent emitters (dispersed particles) is realized by acoustic radiation forces. The generation conditions for this "acoustic" laser are evaluated. It is shown that two types of losses must be overcome in order for generation to begin. The first type results from the energy dissipation in the active medium and the second one is caused by radiation losses at the boundaries of the resonator. The possibility of the creation of an "acoustic" laser with an extraordinary active medium (liquid helium with free electrons) is discussed.

1. INTRODUCTION

At the present time, there are many types of lasers. Their operation is based on the stimulated emission (or scattering) of light by an active medium that is in an optical resonance. Lasers are generators of coherent electromagnetic waves in the range from ultraviolet to submillimeters. But acoustic analogs of such devices have not been created up to now in spite of the progress in laser technology. Meanwhile, a problem of creation of an "acoustic" laser is of great interest because of a variety of potential applications of such devices. They can be used for the direct underwater communication, the impact action at underwater objects, the impact action at armed forces on the ground (an analogy of the shockwave from the supersonic airplane at a low height), medical applications (the destruction of kidney stones, etc), the prospecting, the struggle against the terrorisme.

Recently, a theoretical scheme for an "acoustic" laser was proposed in Refs. [1-3]. A liquid dielectric with dispersed particles was considered as an active medium. Various types of oils or distilled water can serve as a liquid dielectric with gas bubbles as dispersed particles. Gas bubbles can be produced, for example, by electrolyses. The pumping was suggested to create by an oscillating electric field deforming dispersed particle volumes. The phase bunching of the initially incoherent emitters (dispersed particles) was realised by acoustic radiative forces. The suggested scheme is an analog of the free-electron laser (FEL) [4]. It is well known that in FEL the electromagnetic emission is created by electron beams moving through magnetic periodic systems. These systems are called ondulators or vigglers. Ondulators play the role of pumping. Strong laser waves or some electrostatic systems may be used as an alternative to the ondulators. Initially the emission of the electrons is not coherent, but they become grouped due to their interaction with the applied electromagnetic wave. As a result, the emission becomes coherent. It leads to the amplification of the electromagnetic field.

The generation conditions for an "acoustic" laser were evaluated in Ref. [2]. It was shown that two types of losses must be overcome in order for generation to begin. The first type results from the energy dissipation in the active medium and the second one is caused by radiation losses through boundaries of the resonator.

In the case of ordinary piezoelectric emitters only a working surface emits. Thus, these devices are two-dimentional working systems. In the case of the "acoustic" laser the whole volume of an active medium emits. That is why the "acoustic" laser is a three-dimentional working system. It is novel property.

It should be noted that the alternative theoretical schemes of an "acoustic" laser were earlier suggested in a number of papers. For instance, the self-syncronization in a system of noncoherent mechanical oscillators (monopoles) and the amplification of an acoustic field was considered by Kobelev et al. [5]. In this paper nonlinear effects have been taken into account. Sound oscillations in the Helmholtz resonator filled with supercooled vapor were investigated by Kotusov and Nemtsov [6]. However, the schemes of an "acoustic" laser suggested in [5,6] have not been realised in practice because of the weak self-synchronization.

II. SCHEME OF AN "ACOUSTIC" LASER WITH THE PUMPING BY OSCILLATING ELECTRIC FIELD

The scheme of this "acoustic" laser is shown in Fig.1. The active medium is plased between the two semi-infinite walls.

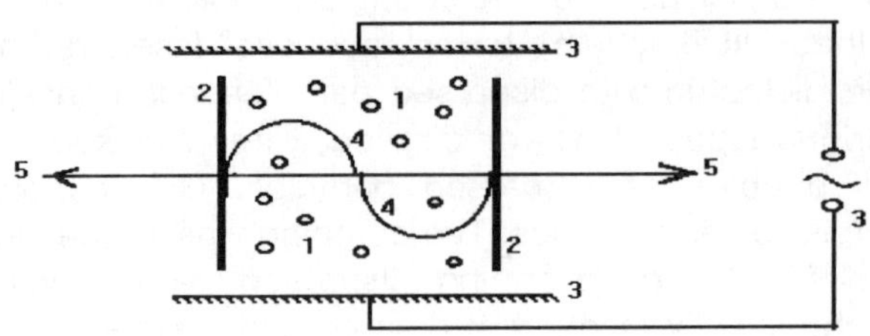

Fig.1. The scheme for an "acoustic" laser with electric pumping. (1) The active medium. (2) Solid walls of the resonator. (3) Electromagnetic system creating the periodic electric field. (4) An acoustic mode. (5) A sound emission.

Electric field acting on such a system results in the deformation (electrostriction) of the dielectric and, hence, changes particle volumes. The value of the effective pressure acting on the particle is equal to [7]

$$\Delta P = \frac{3}{8\pi} \frac{\varepsilon_l E^2(\varepsilon_l - \varepsilon_p)}{(2\varepsilon_l + \varepsilon_p)} \tag{1}$$

Here, ε_l and ε_p are the dielectric constants of the liquid and particle, respectively, E is the electric intensity.

For example, for air bubbles in water ($\varepsilon_p \approx 1$, $\varepsilon_l \approx 81$) and at electric intensity E = 10 kV/cm the value of ΔP is of the order of 0.5 kPa.

Let us suppose that E is a periodic time function: $E = E_0 \cos(\Omega t)$.

The electromagnetic waves propagate through the medium with the velocity of light (for this medium), which is much greater than the velocity of sound. Consequently, the pumping pressure wave may be considered as being independent on the spatial coordinates

$$P(t) = P_E \exp(i\omega t) \tag{2}$$

Here $\omega = 2\Omega$ is the angular frequency. The pressure amplitude of the pumping wave P_E can easily be calculated from (1). The constant term in (2) is omitted.

Under the action of the pumping wave the particles oscillate and emit sound waves. The initial distribution of the particles is spatially homogeneous. As a result, the waves created by particles are added with the different phases. Consequently, the resulting pressure of the useful wave is equal to zero.

However, for the active medium in the resonator, an acoustic mode can be excited. Then the particles can be bunched due to the acoustic radiation forces. Moreover, it is well known that the state of the medium with the spatially homogeneous bubble distribution is unstable not only for a standing but also for a traveling wave [8]. This leads to self-synchronization of the oscillating particles and the amplification of the mode.

The equations describing the interaction between a sound wave and the active medium were obtained in Refs. [1,2]. Two generation conditions for this scheme of an " acoustic" laser were considered in Ref. [2]. The first condition shows that the acoustic mode can be exited only if the length of the resonator *L* contains an integral number of half wavelengths. The second condition gives the starting value of P_E. Two types of losses must be overcome for the beginning of generation. The first type is conditioned by the energy dissipation in the active medium. The second type is caused by the radiation losses through boundaries of the resonator. As usually, this second term is inversely proportional to the volume of the stored energy, i.e. L^{-1}. The starting pressure P_E increases with the increase of the liquid viscosity which prevents from dispersed particles bunching.

At the usual experimental conditions (See Refs. [1,2]) the starting pressure P_E is of the order of a few kPa. It corresponds to the electric intensity E of the order of a few kV/cm.

III. SCHEME OF AN "ACOUSTIC" LASER WITH THE MECHANICAL PUMPING

The scheme of an acoustic laser with the electric pumping has one weak point. The starting electric amplitude, for example, for distilled water with gas bubbles is of the order of a few kV/cm. It is close to the breakdown potential. That is why we shall consider some alternative schemes. Two schemes of an "acoustic" laser with the mechanical pumping are shown in Fig.2,3.

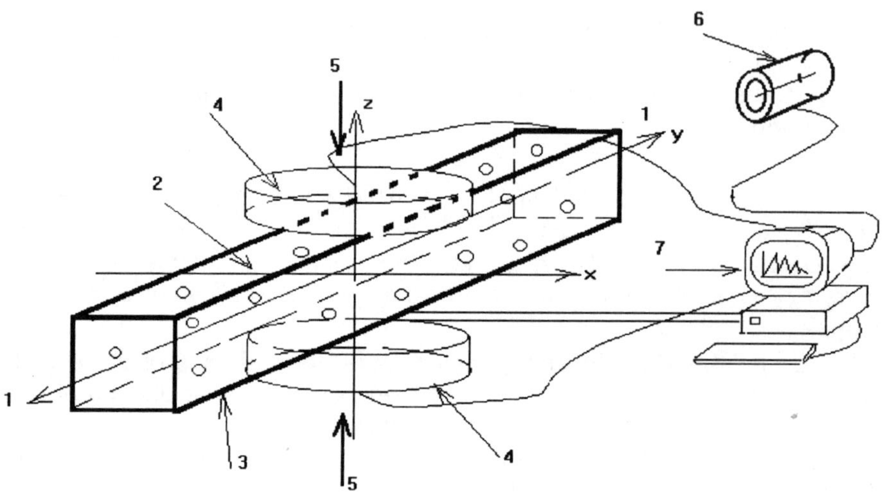

Fig.2. The scheme of an "acoustic" laser with the resonator of the rectangular type. (1) The sound emission. (2) The active medium. (3) The rectangular resonator. (4) The piezoelectric emitters of the pyston type. (5) The mechanical pumping. (6) The monitoring system. (7) The controlling and process computer.

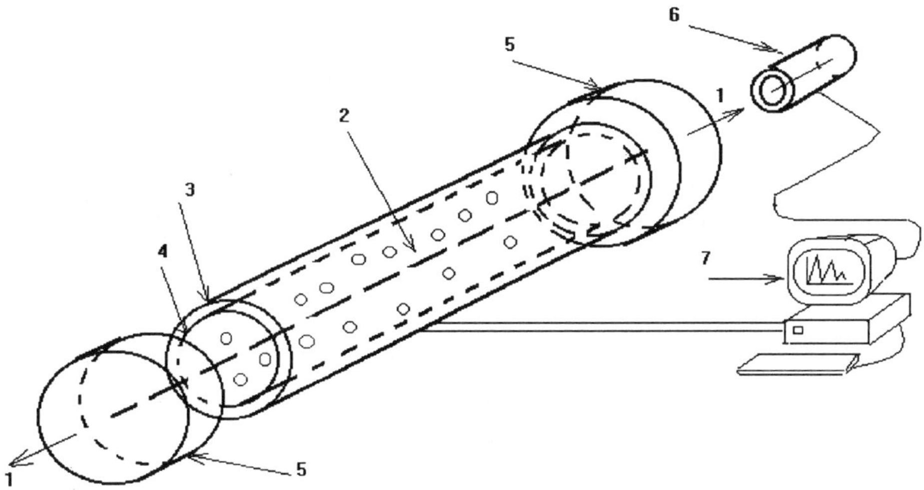

Fig.3. The scheme of an "acoustic" laser with the resonator of the cylindrical type. (1) The sound emission. (2) The active medium. (3) The exterior cylindrical surfase of the resonator. (4) The interior cylindrical surfase of the resonator. (5) The solid walls of the resonator. (6) The monitoring system. (7) The controlling and process computer.

The first scheme (Fig. 2) corresponds to the rectangular resonator and the second scheme (Fig. 3) corresponds to the cylindrical one. In both cases the pumping is created by a piezoelectric emitter of a piston type. In the first case the pumping is created by the oscillating walls and in the second case the pumping is caused by radial mechanical pulsations of the cylinder.

Equations describing the behaviour of the active medium are analogous to those obtained previously (See Refs. [1,2]). Their analysis has shown that the generation conditions for such "acoustic" laser are analogous to those obtained previously for the "acoustic" laser with the electric pumping. For the typical experimental conditions (See Refs. [1,2]) the pumping pressure amplitude is of the order of 1 kPa. It can be seen that this value is easy to obtain.

IY."ACOUSTIC" LASER ON ELECTRON BUBBLES IN LIQUID HELIUM.

The purpose of this paragraph is to analize the possibility of the creation of an "acoustic" laser with an extraordinary active medium: liquid helium with free electrons. It is well known that electron creates the bubble in the liquid helium (see, for example, [10,11]). An electron can penetrate into helium when its kinetic energy exeeds the penetration potential $V_0 \approx 1eV$. The appearance of an electron bubble results from the interaction between electron and helium atoms. The suggested active medium has the next advantages. Firstly, the approximations of Refs.[1,2], where all bubbles were assumed to be spherical and equal in radius hold true for electron bubbles as well. Secondly, it is well known that liquefied gases have a high breakdown potential $(\sim 3 \cdot 10^6 V / cm)$. Thirdly, the helium's viscosity tends to zero if $T \to 0$. As it has been shown in Refs.[1,2], the liquid viscosity prevents from the bunching of bubbles. Fourth, the usual problem of the bubble coagulation is absent for electron bubbles due to their repulsive interaction at small distances. The parameters of an electron bubble were estimated in [10-11]. The electron bubble radius is equal to $R_0 \approx 18 A^0$, the effective electron bubble mass is equal to the apparent mass $M_0 = \frac{2}{3}\pi R_0^3 \rho_l \approx (243 \pm 5) m_e$, the resonance frequency is equal to $f_0 \approx 10^{10} c^{-1}$ (it corresponds to phonon frequencies $\sim 1K^0$).

In the case of an electric pumping at $T = 1{,}3K^0$ with the angular frequency $\omega = 10^4 c^{-1}$ the starting pumping amplitude is equal to $P_E \approx 0{,}08 Pa$ corresponding to the electric field amplitude $E_0 = 5kV / cm$. In the case of a mechanical pumping at the same conditions $P_E \approx 0{,}08 Pa$.

Y. CONCLUSIONS

Thus, there is a strong analogy between FEL and suggested schemes of the "acoustic" laser with dispersed particles. In both cases, we have the same integral parts, such as an active medium, the phase bunching and the starting pressure (or current in FEL). Numerical estimations show that parameters of the generation conditions are reasonable for suggested schemes.

There are many problems which should be consider before an experimental realization of an "acoustic" laser:

- a simulation of the behaviour of the active medium under the action of the pumping (electrical and mechanical) and the useful mode;
- a consideration of the "dansing motion" of bubbles under the pumping of high intensity;
- a simulation of the behavior of this active medium in shock waves;
- a calculation of the radiation pattern of this "acoustic" laser;
- calculations of the influence of bubble size distribution uniformity on the lasing process (the uniformity of bubble size distribution leads to the appearance of integrals over they radii; as preliminary calculations show, in this case the resonanse bubbles play the main role; the calculation of these integrals allows us to obtain the same results for the generation conditions as those obtained before in Refs. [1,2].

ACKNOWLEDGEMENT

This work was partually supported by the grant from International Science Foundation N U9Y000.

REFERENCES

1. S.T.Zavtrak, Phys. Rev. **E51** (1995) 2480.
2. S.T.Zavtrak, Phys. Rev. **E51** (1995) 3767.
3. S.T.Zavtrak, NIM **A358** (1995) 473.
4. C.Marshall, *Free-Electron Lasers* (MacMillan, New York, 1985).
5. Yu.A.Kobelev, L.A.Ostrovsky, and I.A.Soustova, Izv. Vuzov. Radiofiz. **29** (1986) 1129.
6. A.N.Kotusov and B.E.Nemtsov, Akust.Zn. **37** (1991) 123 [Sov.Phys.Acoust. **37** (1991) 62].
7. L.D.Landau and E.M.Lifshitz, *Electrodynamics of Homogeneous Mediums* (Nauka, Moscow, 1985).
8. K.A.Naugol'nykh and L.A.Ostrovsky, *Nonlinear Process in Acoustics* (Nauka, Moscow, 1990).
9. S.T.Zavtrak, JASA (in press).
10. R.A.Ferrel, Phys.Rev. **108** (1937) 167.
11. G.Careri, V.Fasoli and F.Gaeta, Nuovo Cimento **15** (1960) 774.

MODEL OF STRUCTURE AND CONDUCTION IN AN ELECTRORHEOLOGICAL FLUID UNDER FLOW

H. See, R. Sakurai and T. Saito
Research and Development Division, Bridgestone Corporation,
3-1-1 Ogawahigashi-cho, Kodaira-shi, Tokyo 187, Japan

Abstract

A theoretical model of moderately concentrated ERF suspensions under shear flow is proposed which explicitly takes into account the effects of conductivity. Under an electric field a layered structure will form between the electrodes, with the solvent flow restricted to a central free-flowing region, and this gives rise to the field-induced viscosity increase. Using this model, the effect on the ERF properties of inhomogeneities of the particle conductivity can also be estimated. It is predicted that the addition of even small amounts of highly conductive particles to an ERF will dramatically reduce the field-induced viscosity increment, a phenomenon that has been observed in recent experiments.

1. Introduction

An Electro-Rheological Fluid (ERF) is a suspension of solid particles dispersed in an insulating solvent. Application of an electric field induces a dramatic increase in shear viscosity, with the rheological response often modelled by the Bingham fluid equation with a yield stress. The reason for the viscosity increase is generally thought to be that the external electric field induces electric polarisation of the particles due to the mismatch in the electric properties of the fluid and solvent, and the resulting dipole-dipole interactions lead to the formation of aggregates of particles.

However, while it is clear that particle polarisation is the main driving force behind the ERF viscosity increment, many physical details remain unclear. For example, there is still considerable discussion as to whether the mechanism of polarisation is due to dielectric mismatch or conductivity mismatch, although recent calculations show that conductivity effects will dominate at steady state under dc electric fields[1]. Regarding the structures formed by the aggregates which cause the viscosity increase, it has been proposed that chains of particles spanning the electrodes are the main contributor, although there have been observations of a layered structure formation under shear flow[2]. There has also been no model presented for relating the electric current flow through the ERF with the material properties and the structures formed.

In this paper, we propose a model for a moderately concentrated system of conductive ERF particles under shear flow, based on the formation of thick particle layers attached to the electrodes by the electrostatic image forces, which can explain many aspects of ERF behaviour[3]. In Section 2, we present an outline of this model, and calculate the electric field induced viscosity increment of the ERF. In Section 3, we apply this model to the problem of non-uniform conductivity of ERF particles, and we estimate the decrease in shear stress when small amounts of highly conductive particles are added, an effect that has been observed experimentally[4]. A discussion of the results is presented in Section 4.

2. The layered structure model of ERF under shear flow

2.1. Model

The widely used "chain model" of ERF appears to reproduce quite well the behaviour of ERF under very low shear rate conditions, as has been confirmed by direct experimental observation and computer simulation[5]. However, it is doubtful that this picture reflects the situation for systems under moderate shear rates, since we would not expect spanning chains to be able to rearrange fast enough to effectively transmit stress. Microscopic observations of dilute ERF systems under shear flow of order $\dot{\gamma} \sim 5s^{-1}$ and constant electric fields[2,6] reveal that a layered structure is formed, with a thick layer of particles attached to the electrodes and a freely flowing central region. Such observations have led us to propose the layered structure model of ERF, in which the restriction of the solvent flow to a narrow layer is the reason for the viscosity increase[3]. Here we focus on reasonably concentrated systems under dc electric fields and moderate shear rates.

As in Fig.1, we assume that the ERF particles are spheres of radius a with conductivity σ_p that are prevented from direct electrical contact by a polymer layer etc. The surrounding solvent is a Newtonian fluid of viscosity η, conductivity σ_s and dielectric permittivity ϵ_s, and we assume $\sigma_s \ll \sigma_p$. Typical ERF particles are several μm in size, so we can ignore Brownian motion effects. We will also not consider particle sedimentation. When an electric field $\mathbf{E}_{ext}$ is applied, after a time scale $\frac{(\epsilon_p+2\epsilon_s)}{(\sigma_p+2\sigma_s)}$ the current will reach steady state and an electric dipole $\mathbf{p}$ will be induced in each particle due to the difference in conductivities:

$$\mathbf{p} = \epsilon_s \beta_{ps} a^3 \mathbf{E}_{ext} \tag{1}$$

where $\beta_{ps} = \left(\frac{\sigma_p-\sigma_s}{\sigma_p+2\sigma_s}\right)$ is the conductivity mismatch factor.

In the point dipole approximation, the dipole-dipole interaction force between the two particles i and j shown in Fig.1 is given by

$$\mathbf{F}^{ij}_{electro} = 3\frac{p^2}{\epsilon_s |r_j - r_i|^4}\left[\left(3cos^2\theta_{ij} - 1\right)\mathbf{e}^{ij}_r + \left(sin2\theta_{ij}\right)\mathbf{e}^{ij}_\theta\right] \tag{2}$$

For particles in near contact aligned with the electric field, the magnitude of the electrostatic force is $F_{electro} \sim \frac{3p^2}{8\epsilon_s a^4}$.

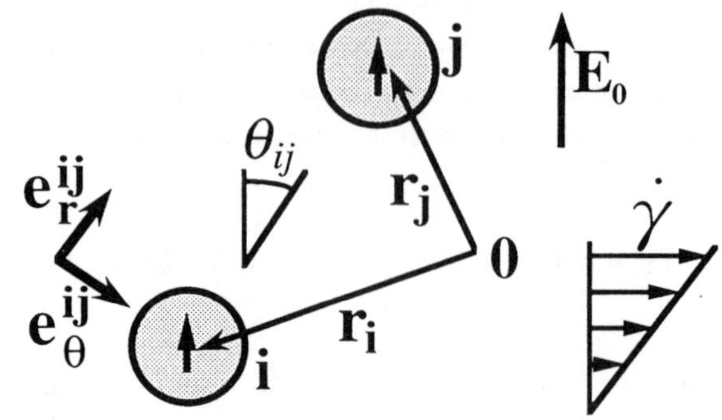

Fig. 1. Sketch showing two ERF particles i and j under a shear flow of rate $\dot{\gamma}$ and an electric field E_{ext}.

Further, since the electrodes are made of highly conducting material, there will be an image dipole force acting on the particles attracting them to the electrode surface. For a particle near the electrode, the magnitude of this force $F_{image} \approx F_{electro}$.

Now let us assume that an electric field is applied to a system under steady shear flow of shear rate $\dot{\gamma}$, which is high enough so that the system is no longer in the "spanning chain" regime (ie $\dot{\gamma} > \dot{\gamma}_{min} \sim 5s^{-1}$). Due to the interparticle force $F_{electro}$ aggregates of particles will form in the fluid. There will also be a movement of single

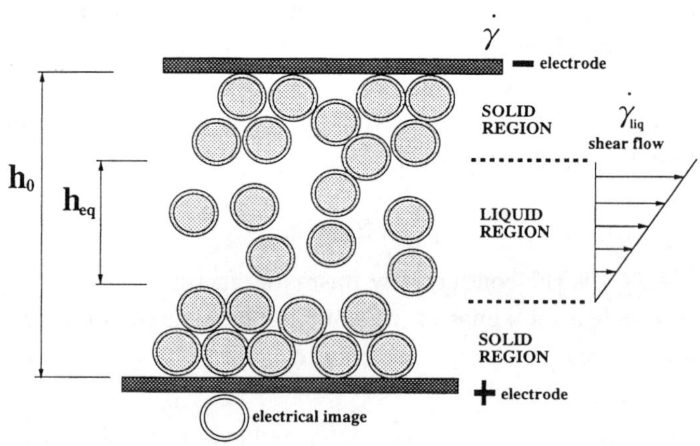

Fig. 2. A schematic diagram showing the layered model of ERF under electric field and shear rate $\dot{\gamma}$. Particle size has been exaggerated for clarity.

particles or aggregates to the electrodes and adhesion due to the force F_{image}. Eventually, we would expect a layered structure as in Fig.2, with layers of attached particles moving together with the electrodes which we will label "solid layers", and a central "liquid layer" where all the flow is concentrated. The restriction of flow to this narrow liquid region causes the viscosity increment.

2.2. Calculation of viscosity increment

To estimate the macroscopically measured viscosity, we need to calculate the characteristic layer thickness h_{eq}. To do this, we will follow the approach of previous workers[6] and assume that the solid region can be replaced by a continuum with a flat interface. Consider a test particle in the liquid layer in the near vicinity of the solid region. This particle will experience two forces: (a) an electrostatic interaction force attracting it to the solid layer which will scale as $F_{electro}$; and (b) a hydrodynamic force F_{flow} due to the solvent shear flow in the liquid layer trying to pull it off the interface, which we approximate by the Stokes equation $F_{flow} \sim \eta a^2 \dot{\gamma}_{liq}$. Due to the restriction of solvent flow to the liquid region, the shear rate in this region will be given by $\dot{\gamma}_{liq} = \dot{\gamma}\frac{h_0}{h_{eq}}$. The equilibrium layer thickness h_{eq} is given by a force balance condition between $F_{electro}$ and F_{flow}, and this in turn enables us to estimate the macroscopically measured viscosity $\eta_{measured}$. The details of the calculations are presented in a previous work[3] but the final result is

$$\eta_{measured} \approx \epsilon_s \beta_{ps}^2 E_{ext}^2 \frac{\eta_0}{\eta} \dot{\gamma}^{-1} \tag{3}$$

Here η_0 is the viscosity of the ERF suspension in the absence of an electric field.

We note that Eq.3 predicts shear thinning behaviour with the exponent of the shear rate equalling -1, which is consistent with the Bingham fluid model. Further, substitution of typical values for the physical quantities shows that the viscosity increment predicted by this model is of the same order of magnitude as experimental measurements[3].

Finally we add that this model can be used to estimate the electric current in an ERF under shear. In particular, due to the variation of solid layer thickness with shear rate, a substantial decrease in electric current with increasing $\dot{\gamma}$ is predicted[3], a phenomenon that has been observed experimentally[7].

3. The effect of non-uniform conductivity on the ER response

We now use the layered model to estimate how the field-induced viscosity increment of an ERF system is changed by the addition of very small amounts of highly conductive particles.

There has recently been reported a set of experiments in which varying quantities

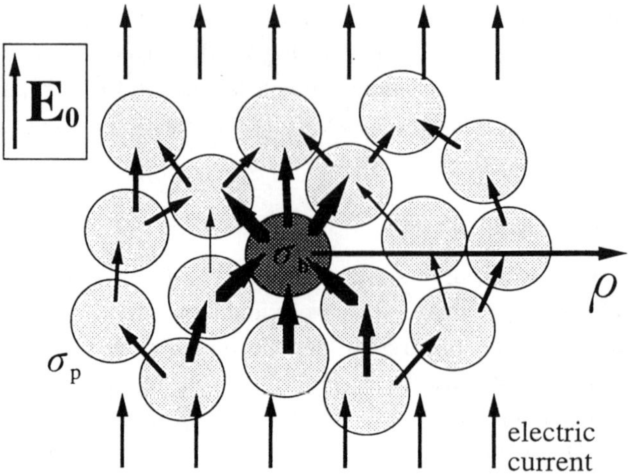

Fig. 3. A sketch of the inside of the solid layer when a high conductivity particle is introduced (dark particle). Concentration of the electric current (magnitude indicated by arrow thickness) through this particle leads to a non-uniform profile of electric current and a reduction in the total cohesive force of this portion of the solid layer.

of highly conducting carbon black powder have been added to a carbonaceous ERF particle system[4]. In this system, the ratio of particle conductivities $\frac{\sigma_h}{\sigma_p}$ is of the order of 10^5. Under an electric field of $2kV/mm$ and shear rate $366s^{-1}$, it is found that addition of 0.07% of carbon black by volume leads to a 5% reduction in shear stress.

Consider a portion of the solid layer where a particle of high conductivity σ_h is surrounded by particles of lower conductivity σ_p, as shown in Fig.3. At steady state the electric current will flow preferentially through the particle of higher conductivity and there will be a reduction in the current flowing through the surrounding particles lying in the plane perpendicular to the electric field. Thus there will be a reduction in the electric polarisation in these particles and the diminished dipole-dipole interactions will cause a weakening of the structure along this plane. We now estimate this effect by a crude model.

As an approximation let us replace the surrounding particles by a continuous medium of conductivity σ_p, and assume that the highly conductive particle is imbedded in this. When the flow of current reaches steady state, the electric field created in the medium by this particle due to the conductivity difference is given by

$$\mathbf{E}(\mathbf{r}) = \mathbf{E}_0 + \beta_{hp}(\frac{a}{r})^3 \mathbf{E}_0 \cdot (3\hat{\mathbf{r}}\hat{\mathbf{r}} - \mathbf{I}) \tag{4}$$

where $\hat{\mathbf{r}}$ is the unit radial vector from the particle centre, $\beta_{hp} = \frac{\sigma_h - \sigma_p}{\sigma_h + 2\sigma_p}$ is the conductivity mismatch factor of the particles and the field $\mathbf{E}_0$ is the average electric field in the system of low conductivity particles if the high conductivity particle were to be removed. We would expect $E_0 \propto E_{ext}$.

We now consider those particles which lie in the plane that contains the high conductivity particle and is perpendicular to the electric field. Let ρ be the radial distance measured in this plane from the centre of the high conductivity particle. We focus attention on those particles with centres that lie in the radius range $\rho_{min} < \rho < \rho_{max}$. We now consider the total force F_{tot} acting through this plane due to the interparticle electrostatic interactions. F_{tot} can be written as

$$F_{tot} = F_{cent} + \tilde{F} \tag{5}$$

where F_{cent} is the interaction force associated with the particle in the centre and $\tilde{F}$ is the sum of the interaction forces of the surrounding spheres due to their dipoles.

To estimate $\tilde{F}$, we note that the strength of these dipoles will be given by Eq.1 under the electric field Eq.4 and will depend on distance ρ. Since the interaction force between two dipoles of magnitude p scales as $\frac{p^2}{\epsilon_s a^4}$, we can estimate $\tilde{F}$ as follows

$$\tilde{F} \sim \sum_{\rho_{min}<\rho<\rho_{max}\ (particles)} \frac{(\epsilon_s \beta_{ps} a^3 E(\rho))^2}{\epsilon_s a^4} \tag{6}$$

Approximating the summation by an area integral, and assuming that $\rho_{min} \sim a$ and $\frac{\rho_{min}}{\rho_{max}} \ll 1$ we find

$$\tilde{F} \sim \epsilon_s \beta_{ps}^2 E_0^2 \left[\rho_{max}^2 - a^2 - a^2(4\beta_{hp} - \frac{1}{2}\beta_{hp}^2) \right] \tag{7}$$

On the other hand, the force F_{cent} can be estimated as $F_{cent} \sim \epsilon_s \beta_{hs}^2 E_0^2 a^2$, since it will essentially act over an area of magnitude $\sim a^2$.

Therefore, from Eq.5 we find the total interaction force across the plane

$$F_{tot} \sim \epsilon_s E_0^2 \left[\rho_{max}^2 - a^2(4\beta_{hp} - \frac{1}{2}\beta_{hp}^2) \right] \tag{8}$$

where we have assumed $\beta_{ps} \sim \beta_{hs} \sim 1$, ie the conductivities of both types of particles are much larger than that of the solvent.

The final step is to estimate the reduction in field-induced shear stress $\Delta\tau$ due to this weakening of the solid layer. To estimate ρ_{max}, we assume that for low volume fractions ϕ_h of the highly conductive particles they will be uniformly distributed in the system. We thus approximate ρ_{max} by the average interparticle distance

$$\rho_{max} \sim \frac{a}{\phi_h^{\frac{1}{3}}} \tag{9}$$

Since the stress will scale as $\frac{F_{tot}}{\rho_{max}^2}$, we find that the reduction in shear stress compared to the original value $\tau_0 \sim \epsilon_s E_0^2$ is

$$\frac{\Delta\tau}{\tau_0} \sim \phi_h^{\frac{2}{3}}(4\beta_{hp} - \frac{1}{2}\beta_{hp}^2) \tag{10}$$

Eq.10 tells us that the decrease in the shear stress will depend on the concentration ϕ_h of highly conducting particles as well as on the degree of conductivity mismatch through the parameter β_{hp}. This behaviour is presented schematically in Fig.4. For example, letting $\phi_h \sim 10^{-3}$ and $\beta_{hp} \sim 1$ (ie addition of very highly conductive particles) we find $\frac{\Delta\tau}{\tau_0} \approx 0.035$. This is of the same order of magnitude as the experimental result discussed above[4]. We thus see how disparity in conductivity leads to an overall decrease in the structural strength of the solid layer, making it easier to be degraded and leading to a reduction in the field-induced viscosity increment.

4. Discussion

In this paper we have presented a layered model of ERF systems under constant electric field and moderately strong flow conditions, which explicitly takes into account the effect of particle and solvent conductivity.

This model is able to reproduce the Bingham fluid nature of the rheological response and can also explain the variation of electric current with shear rate. Application of this concept of layers to ERF suspensions into which small amounts of highly conductive particles have been mixed, leads to the prediction that there will be a reduction in field-induced viscosity as the concentration of high conductivity powder is increased.

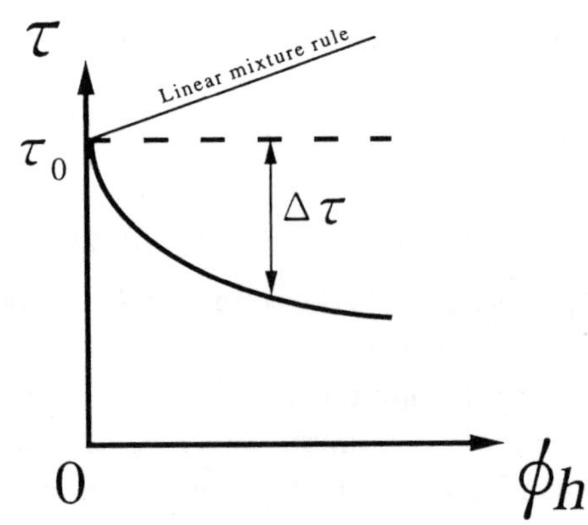

Fig. 4. The reduction of shear stress τ with the volume fraction of high conductivity particle ϕ_h, predicted by Eq.10. A simple "linear mixture rule" would predict an increase in shear stress, which is also plotted for comparison.

Clearly the nature of the conductivities of the ERF particles and solvent are of crucial importance to the performance of the fluid. As is suggested by Eq.10 and Fig.4, it is necessary to have a high degree of uniformity in the particle conductivity: variations in conductivity will lead to non-uniformity of electric current and the overall weakening of the stress-producing structures. Further, when attempting to "blend " ERF powders of different conductivity, the results may not follow a simple "linear mixture rule" formula.

A major assumption we have made is the adhesion of the particles to the electrodes, which may not be completely realisable in some ERF systems. If we were to relax this condition and permit a degree of slippage between the particles and electrode surfaces, the equations of mechanical equilibrium will be altered, but the electrical image forces will still act to induce formation of a layered structure.

The dynamics of structure formation and the details of the transition from a chain-dominated configuration at low shear rates to a layered one as $\dot{\gamma}$ is increased are currently under investigation.

This work has not considered the effects of alternating or time-varying electric fields. We would expect that the essential concept of particle-particle or particle-image forces leading to layers of particles on the electrodes will remain unchanged, but that the interparticle interaction will be governed by the mismatch of complex permittivities.

Acknowledgements

The authors gratefully acknowledge fruitful discussions with L.C. Davis, H. Tamura, T. Maruyama, S. Endo and M. Hashimoto.

References

1. L.C. Davis, *J. Appl. Phys.* **72** (1992) 1334.
2. Y.H. Shih and H. Conrad, in *Electrorheological Fluids: Mechanisms, Properties, Technology and Applications*, ed. R.J. Tao and G.D. Roy (World Scientific, Singapore, 1994), p. 294.
3. H. See and T. Saito, submitted to *Rheol. Acta.* (1995).
4. R. Sakurai and T. Saito, presented at Polymer Society of Japan Meeting (Yokohama, 31 May 1995).
5. D.J. Klingenberg, F. van Swol and C.F. Zukoski, *J. Chem. Phys.* **94** (1991) 6160.
6. D.J Klingenberg and C.F. Zukoski, *Langmuir* **6** (1990) 15.
7. T.Y. Chen and P.F. Luckham, *J. Phys. D Appl. Phys.* **27** (1994) 1556.

TEMPERATURE EFFECTS OF DIELECTRIC PROPERTIES OF ER FLUIDS

Z.Y. QIU, L. HU, M.W. LIU, H.X. BAO, Y.G. JIANG, L.W. ZHOU
T.D. Lee Physics Laboratory & Department of Physics, Fudan University, Shanghai 200433, China

Y. TANG, Z. GAO
Department of Chemistry, Fudan University, Shanghai 200433, China

M. SUN
Department of Polymer Science, Fudan University, Shanghai 200433, China

and

E.V. KOROBKO
A.V. Luikov Heat and Mass Transfer Institute, Minsk 220072, Belarus

ABSTRACT

Under the consideration of the role that energy transfer and dissipation play in ER effect, an improved theory frame for ER effects, polarization-dissipation-structure-rheology, is suggested. The theory frame is substantiated by the basic physical laws and certain critical experimental facts. The dielectric response of a diatomite ER fluid to temperature is measured in the temperature range of 140 K to 400 K. By comparison of the DC conductivity with the AC effective conductivity of the sample, we found that the AC dielectric loss consists of two parts. One part comes from the DC conductivity, the other from the response of the bound charges in scope of particle to AC field. It is suggested that the response of bound charges is very important to ER effects. Besides, the effect of temperature on shear stress is measured, and interpreted based on the dielectric measurements. The source of two loss peaks in the curve of the dielectric loss versus temperature is not clear.

1. Introduction

It is widely accepted that the interfacial or orientational polarization produces the structure in ER fluid under external field and, in turn, causes dramatic rheological response. This theory structure can be identified as "polarization-structure-rheology," in which the role of polarization is emphasized in the electrorheological responses to external electric field. However, some questions still remain unsolved. For example, calculated yield stress of ER fluid is much less than measured one. In addition to polarization, ER fluids under electric field show certain level of dielectric loss, while this theory does not allow for this factor and its possible effect.

The objects of work are as follows: 1. suggesting a new theory frame for ER effect considering the role that energy transfer and dissipation play in ER effect; 2. designing experiments to get insight into how environmental factors and the solid phase structure influence the migration of charge carriers and into how the mobility of charge carriers affects the rheological response of ER fluid.

2. An Improved Theory Frame for ER Effect

Energy transfer and dissipation in the rheological response are important to ER effect. The energy transfer and dissipation take place due to the migration of charge carriers under external electric field. The mobility of charge carriers is closely related to particle structure and liquid environment. In spite of the fact that leaking current arises from other possible sources, there exists a basic loss process that shows loss current. To convey this idea clearly, we propose a theory frame for ER effect, i.e., polarization-dissipation-structure-rheology. Compared with the earlier theory frame, it allows for the energy transfer and dissipation as well as polarization in the rheological process. First, we will show that this theory frame obeys fundamental physical laws and has a solid physics basis. As usual, dielectric property of ER fluid is studied by measuring its complex dielectric spectra, $\varepsilon = \varepsilon' - i\,\varepsilon''$. ε' and ε'' play equal role in describing the dielectric response of ER fluid to external field. ε' and ε'' describe two different aspects of a dielectric process - polarization and dielectric loss, respectively. Both physical variables can be derived from each other. In this sense, both variables are equally important for understanding the dielectric response of ER fluid to external electric field. Secondly, an ER fluid exposed to external field has highly ordering micro-structure, and its entropy is lower than that without field. According to the second law of thermodynamics, the decrease in entropy of a system has to be accompanied with energy dissipation. In other words, the structure ordering of ER fluid occurs at a cost in the energy loss. Thus the energy transfer and dissipation in the rheological process of ER fluid are necessary, even indispensable. Thirdly, when an ER fluid undergoes shearing, maintenance and reconstruction of its structure also have to be accompanied with the energy transfer and dissipation. Therefore, it is not only relevant but also necessary that energy transfer and dissipation are considered in the ER theory frame.

It may be asked, what is the relation between ER effect and energy transfer and dissipation.

The problem related to the leaking current flowing through ER fluid has not received enough attention. In fact, all ER fluids show conduction, and the conduction mechanism is not known completely. It is known only that the conduction of ER fluids is non-ohmic and has features of two-phase system conduction. Hoever, that is not quantitatively, not even qualitatively, understood. It is apparent that fluid's conductivity under the combination of electric and shear field restricts applications. Leaking current was thought to be only related to the external source and was independent of ER effect itself. However, experimental results strongly suggest that leaking current is closely related to ER effect, and is a part of ER effect itself.

It was observed that, a polymer ER fluid initially showed strong ER activity, while its leaking current was quite large; after the fluid was chemically processed, its leaking current was greatly reduced, but its ER activity almost disappeared. Otsubo *et al.*[5] carried out an experiment with a polymer coated with an inorganic material. He found that, when electric field of 1 kV/mm and stress of 70 Pa were applied to this ER fluid, leaking current remained unaffected; while when the externally applied stress increased to 100 Pa, the leaking current was sensitive to external stress. This result clearly shows a relation between leaking current and the structure change of the ER fluid. Block and co-workers[8] verified that yield stress of the polymer ER fluid was closely related to its conductivity: when conductivity of solid phase was 10^{-7} $(\Omega cm)^{-1}$, the yield stress reaches its maximum; for conductivity higher or lower than 10^{-7} $(\Omega cm)^{-1}$, the yield stresses reduced remarkably. The experimental results cited above and many others convincingly substantiate the idea that the energy transfer and dissipation play an important role in ER effect.

How do the energy transfer and dissipation take place? Many experimental results shed light on this question. The migration of charge carriers in ER fluid serves as a medium in the energy transfer and dissipation. There are many factors, such as temperature, liquid viscosity, chemical properties of solid phase, which influence the mobility of charge carriers, and then ER effect.

Makatun *et al.*[2] presented a convincing evidence that sufficient proton mobility was necessary for inducing ER effect. They found that, upon heating a hydrate of aluminum dihydrotripolyphosphate, a buildup of ER activity occurred in conjunction with increased mobility of protons in the hydrate layer. This mobility was measured as a decrease in line width of the broad-line NMR spectra. Filisko *et al.*[6] also found that an electrolyte ER fluid was electrorheologically inactive in room temperature, but very strongly active at 100°C. Under an equal electric field, the increment in its viscosity at 100°C was one hundred times as much as that at room temperature.

Carreira *et al.*[1,3] gave a direct relation between the mobility of charge carriers and ER activity for a photoelectrorheological material. The light activates donor electrons into the conduction band allowing them to migrate under electric field. The following facts may support the idea proposed above in more extensive sense. During the earlier development of ER material, water was thought to be indispensable for activating ER effect. Traditionally, water plays a role in ER effect in two ways. First, by virtue of its relatively high dielectric constant, water may impart a high polarizability to the particles thus increasing electrostatic force between particles. Second, the adsorbed water creates a conductive layer around the surface of particles. Ions in water may drift across the surface in response to an electric field. In a non-conductive hydrophobic environment, the water layer serves as a mechanism by which local conduction of charge carriers can occur. This process was further confirmed by Wang and Shaw[9]. Their experimental results showed that, if water did not constitute an environment that facilitated the migration of charge carriers, it did not make a contribution to ER effect. Here, the role of water in ER effect is explained in the following way. Water in a water-activated ER system facilitates the migration of charge carriers, which serve as media for the energy transfer and dissipation in the dielectric and rheological processes.

The experimental results mentioned above, as well as many others, support the new ER theory frame, which establishes the cause-effect relations among ER activity, energy transfer and dissipation, polarization and the mobility of charge carriers. Understanding the mobility of charge carriers is an objective of this study, which is one of a series researches into the role that the energy transfer and dissipation play in ER effect.

3. Experiments

The sample ER fluid used in this study consists of a diatomite dispersed in a transformer oil. The solid phase is made from natural diatomite using chemical methods, and contains 97% of SiO_2 and some impurities as well as water. This fluid is a water-activated system in which the optimal water content is 7% by weight. When the temperature effect of dielectric properties of the ER fluid was measured below room temperature, the sample cell was put into a metal dewar. The sample cell was cooled down from room temperature to 140 K in 36 hours, and then warmed up to room temperature in another 45 hours. The temperature of the cell was monitored by two thermal-couple temperature sensors, each of which was mounted on the bottom and top of the cell, respectively. Dielectric constant and loss were measured at a step of 1 K by using high voltage bridge mentioned in Ref. 4. DC conductivity of the sample in the range of 140 to 360 K, and the shear stress of the sample fluid in the range of 235 to 360 K.

4. Results and Discussion

The temperature dependence of the dielectric properties of the diatomite ER fluid is shown in figures 1 and 2. As shown in figure 1, the dielectric constant of the fluid increases linearly in the range from 140 K to 280 K; and increases slightly more quickly above 200 K. In the loss-temperature diagram, there are two loss peaks, one of which is at 166 K and the other, 197 K. The loss increases with temperature in an exponential law for temperature higher than 240 K. The underlied fact is that environmental conditions facilitate the mobility of charge carriers with rising temperature. Higher temperature weakens the binding of charge carriers. The migration of charge carriers is confined to the

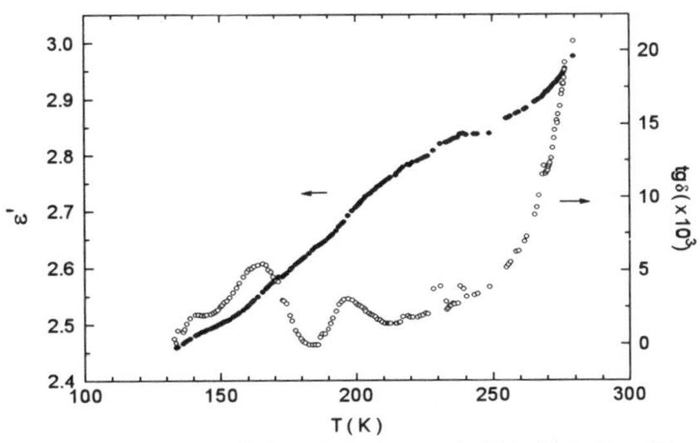

Figure 1 Temperature dependence of the dielectric constant (solid circles) and loss (open circles) of the diatomite ER fluid. Working voltage for the bridge is 1000 V/ 300 Hz.

interface between oil and particle. This leads to interfacial polarization. Above 240 K, the dielectric constant increases more quickly. In this situation, the thermal energy of charge carriers surpasses conduct activation energy, so charge carriers that migrate around the solid powder surface make more contribution to polarization. The two loss peaks may be related to the structure of solid particles, and their origins need further study.

As shown in figure 2, the dielectric constant and loss of the diatomite ER fluid increase in an exponential law in the range from 280 K to 400 K. In this temperature range, the conductivity obeys the mechanism of varying path hopping suggested by Mott[7]. It is recognized that the mobility of charge carriers in this way causes interfacial polarization and creates passage for the energy transfer and dissipation.

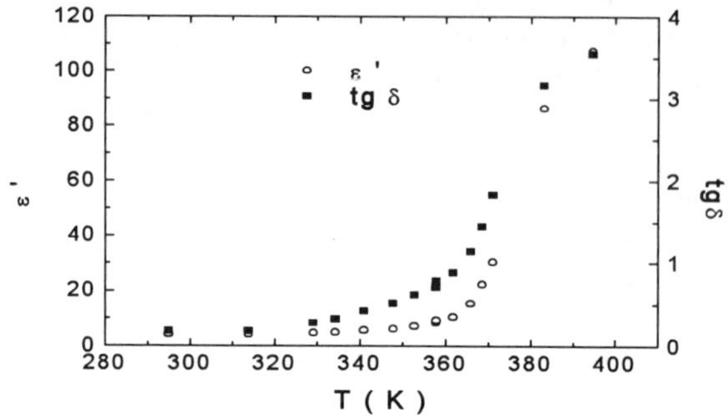

Figure 2 Temperature dependencies of dielectric constant and loss of the diatomite ER fluid from 280 to 400 K. Working voltage for the bridge is 1000 V / 300 Hz.

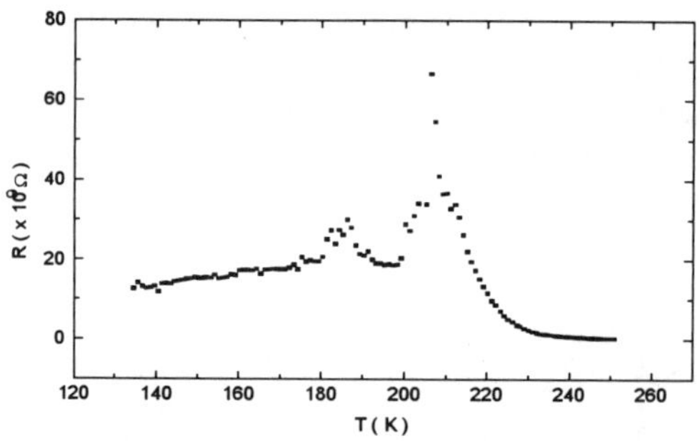

Figure 3 DC resistance of diatomite ER fluid in low temperature. Measurement voltage is 17 V.

In order to further study the dielectric property of the ER fluid in low temperature, we measured its DC resistance using a detecting signal of 17 V from 140 K to 360 K, as shown in figure 3. There is a complex structure in the curve of DC resistance versus temperature, including two peaks at 187 K and 210 K. Above 210 K, the DC resistance of the sample decreases in an exponential law. The two peaks may result from the structure change of the sample. The loss of the transformer oil remains constant in the range from

140K to 260 K, and increases linearly above 260 K. This result verifies that the two peaks are only related to the solid phase of the sample.

In order to establish a direct relation between the rheological and dielectric responses of the sample to external field, we measured the shear stress at difference temperature as shown in figure 4. As shown in figure 4a, shear stress decreases when temperature drops from 290 K to 270 K. After a quick change between 270K to 265 K, shear stress remains constant from 265 to 245 K. The increase of shear stress below 245 K is due to the contribution of the frozen transformer oil at that temperature. This testifies the environment effects on the mobility of charge carriers. This effect is reflected in the dielectric response. As shown in figure 4b, the shear stress remains unchanged between 304 K and 329 K, increases in an exponential law from 329 K to 339K, and finally levels off above 340 K.

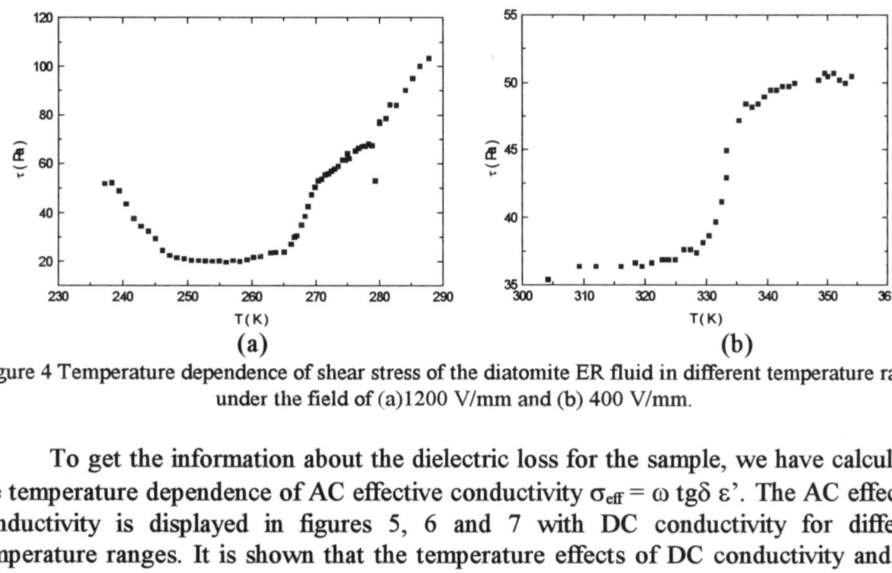

(a) (b)

Figure 4 Temperature dependence of shear stress of the diatomite ER fluid in different temperature range under the field of (a)1200 V/mm and (b) 400 V/mm.

To get the information about the dielectric loss for the sample, we have calculated the temperature dependence of AC effective conductivity $\sigma_{eff} = \omega \, tg\delta \, \varepsilon'$. The AC effective conductivity is displayed in figures 5, 6 and 7 with DC conductivity for different temperature ranges. It is shown that the temperature effects of DC conductivity and AC effective conductivity are different. This difference reflects the different responses of charge carriers to AC and DC electric fields. Considering these facts, the dielectric loss can be divided into two parts. One part originates from the DC conduction, and the other, from the response of the bound charges inside particles to AC field. At the temperature as low as 140 K, the difference of the two parts is quite small, which demonstrates that charge carriers are tightly bounded under such temperature. As temperature increases, more bound charges become movable. When temperature reaches 360 K, the difference become very large, which implies that many bound charges change into free charges to participate DC conduction. One may result in a criterion for a good ER fluid -- a large difference between the two parts. Keeping the difference not changed with temperature would improve the thermal stability of ER fluids.

As shown in figure 5, from 170 K to 220 K, the curves of DC and AC effective conductivity versus temperature are similar to each other. There are two dips in each curve. The dips in one curve take place at about the same temperature as the coincided dips in the other curve. The dip at 185 K in the AC effective conductivity curve is much deeper than the correspondent one in DC curve. In such temperature, the charge carriers are very few, and the charges inside solid phase are tightly bound and not easy to migrate under AC electric field. The facts show interrelations among solid structure, charge carriers, bound charges and environmental temperature. In addition, due to the different detecting signals (DC 17 V, AC 1000 V / 300 Hz) and different water contents of the sample at high and low temperatures, all these factors make contributions to the difference between DC conductivity and AC effective conductivity.

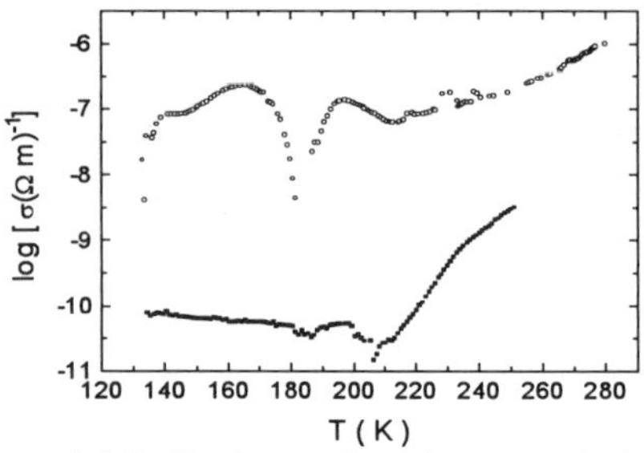

Figure 5 AC effective conductivity (open circles) and DC conductivity (solid boxes) of diatomite ER fluid at low temperature. Solid concentration φ < 20 %.

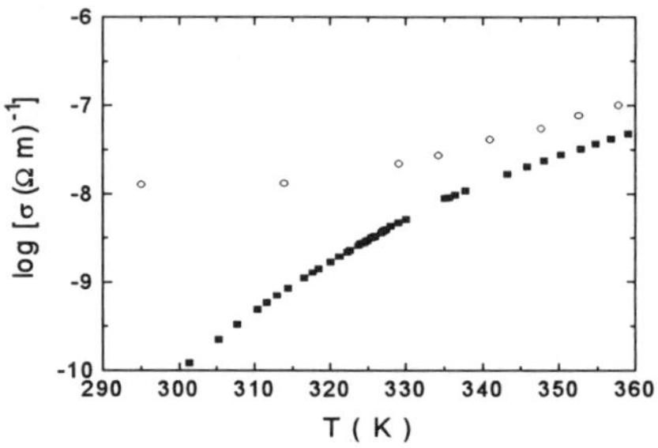

Figure 6 Dependence of the AC effective conductivity (open circles) and DC conductivity (solid boxes) of the diatomite ER fluid on temperature. Solid concentration φ = 36%.

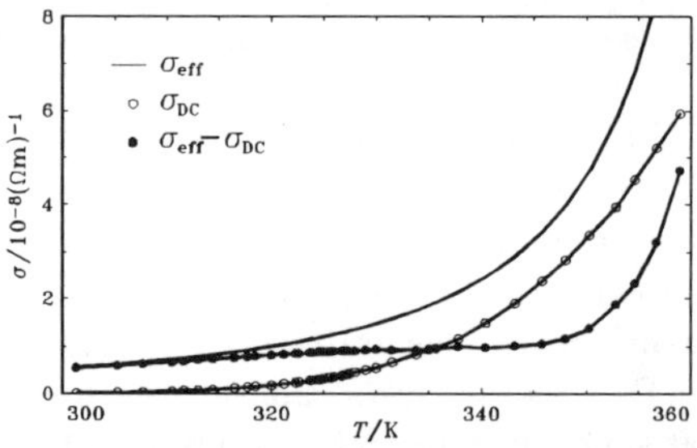

Figure 7. The difference of the temperature dependence of σ_{eff} and σ_{DC}. for temperature higher than room temperature.

5. Conclusions

The improved theory frame for ER effects, i.e. polarization-dissipation-structure-rheology, is proposed. This theory frame allows for the role of the dissipation and energy transfer in ER effect as well as polarization. The theory frame is substantiated by many experimental facts as well as the fundamental physical laws. This theory frame establishes the cause-effect relations among ER activity, polarization, energy transfer and dissipation, and mobility of charge carriers. The energy transfer and dissipation are a part of ER effect, and take place due to the migration of charge carriers. This theory frame is partly confirmed by our experiments.

In the temperature range from 140 to 400 K, the mobility of charge carriers in the sample under AC or DC field are studied experimentally. The DC conductivity and the dielectric spectra of the sample under the AC field of 2000 V/mm of 300 Hz are measured. By comparison of the DC conductivity with the AC effective conductivity, it is found that the dielectric loss can be divided into two parts, one coming from the DC conductivity, the other from the response of bound charges inside the particle to AC field. The difference between the two parts is very small below 250 K, and gradually becomes large from 250 to 360 K. These facts show that the mobility of charge carriers are strongly influenced by the structure of solid phase and temperature. The effect of temperature on shear stress is measured, and explained based on the dielectric measurements. It is suggested, that the loss related to the AC field response of bound charges important to ER effects. this becomes a criterion about ER activity. For good ER materials, it is required that the difference between the two parts of dielectric loss is as large as possible. To improve the thermal stability of ER fluids, the difference should be kept unchanged with temperature through proper molecular designing

The future work is to develop relations between the two parts of dielectric loss and temperature, and structure of particle.

6. Acknowledgment

The research is supported by the National Natural Science Foundation of China under Grant No. 19374018.

7. References

1. L. Carreira and V.S. Mihajlov, *US Patents* No. 3553708 (1971).
2. V.N. Makatun, K.N. Lapko and A.D. Matsepuro, *J. Eng. Phys.* **45** (1983) 1138.
3. F.E. Filisko, in *Electrorheological Fluids*, ed. R. Tao (World Scientific, Singapore, (1991) p. 176.
4. L.W. Zhou, Z.Y. Qiu, Y. Shen, Y. Tang, Z. Gao, M Sun, J.W. Qiu, Z.Z. Yu and L. Hu, *Polymer Preprints* **34** (1994) 323.
5. Y. Otsubo and K. Edamura, *J. Rheol.* **38** (1994) 1721.
6. U.Y. Treasurer, F.E. Filisko and L.H. Radzilowski, *J. Rheol.* **34** (1990) 539.
7. N.F. Mott, *Phil. Mag.* **19** (1969) 835; N.F. Mott, E.A. Davis, *Electronic Processes in Noncrystalline Materials*, (Clarendon Press, Oxford, 1979).
8. H. Block, J.P. Kelly, A. Qin and T. Watson, *Langmuir* **6** (1990) 6.
9. W. Wang and M.T. Shaw, *The Role of Water in ER Effect*, preprint.

FIELD-INDUCED LABYRINTHINE PATTERNS IN FERROFLUID EMULSIONS

GEORGE A. FLORES, MARK L. IVEY AND JING LIU
Department of Physics and Astronomy
California State University, Long Beach, California 90840
And
M. MOHEBI AND N. JAMASBI
Optics Department
Center for Scientific Investigation and Higher Education of Ensenada
Ensenada, Baja California, Mexico

ABSTRACT

A ferrofluid emulsion goes through gas - solid phase transition when an external magnetic field is applied. The solid structures are observed as either column, bent-wall, or labyrinthine patterns. The appearance of these different patterns depend upon the rate of the field applied, thickness of the sample cell along the field direction, and the volume fraction of the emulsion droplets used. Using optical microscopy, formed patterns are recorded and analyzed in which a "phase" diagram of the structural transition from column to bent-wall is measured.

1. Introduction

The key to MR or ER effect is the induced structures by external applied fields. Measuring the induced structures is the first step in understanding the mechanism of interaction and rheological behavior of these system. So far the study of the field-induced structures has been limited to equilibrium state, where the ramping rate of applied field is slow enough that the thermal motion ensures the particles to move freely up to the formation of loosely bounded columns or for coupling constant $\lambda \leq 5$ [1-8]. Field-induced nonequilibrium structures where the ramping rate of applied field is fast, have received almost no attention in either ER or MR fluids.

In this paper we report our initial study of the magnetic-field-induced nonequilibrium structures in a model MR fluid - ferrofluid emulsion. When a strong magnetic field is applied to this fluid, the colloids will go through a structural transition: from a gas of randomly dispersed particles to a solid of aggregates of particles. The structure of the solid phase can be either separated chains, columns, "bent-wall", or "maze" or labyrinthine patterns. We find that not only the strength of the applied field but also the ramping rate of the field changes the formed structures. At a slow rate, an equilibrium columnar structure is observed. At a high rate, a "bent-wall" structure shown from the top as "worms" is induced. We can also control the solid structures by changing the sample confinement along the field direction and particle volume fractions. The higher the rate or the further away the system from the equilibrium state, the more complex the worm pattern becomes. A "phase" diagram of cell thickness L vs. volume fraction Φ is constructed to describe the structural transition from the regime of worm dominated structures to the column dominated structures. A structural transition relation is obtained at the dividing line as: $L = 48\Phi^{-2.53}$ mm .

Labyrinthine patterns have been observed quite frequently in ferrofluid where small permanent dipoles, magnetic grains, are suspended in a liquid [9-11]. When a drop of oil-based ferrofluid is confined between two glass plates filled with water, an applied magnetic field normal to the plate surface induces labyrinthine patterns. The labyrinth formation is found to be sensitively dependent on initial conditions. The higher the rate of the applied field, the more complex the pattern is generated [11]. The behavior is qualitatively explained by their dynamical models based on the competition between surface energy - confining the ferrofluid to minimize surface area, and dipolar energy - pushing the magnetic grains apart within the drop. The same interaction seem to control other systems where labyrinthine patterns are also observed [12-16]. In our ferrofluid emulsion system, without magnetic field, the aggregates disappear and particles re-disperse. Therefore, no intrinsic surface tension exists in the ferrofluid emulsion system. The mechanism here is different from that observed in ferrofluid in the sense that all the interaction energies, attractive or repulsive, are induced by the external applied magnetic field. Therefore, to fully understand the nonequilibrium patterns in ferrofluid emulsion, a new theory has to be developed. The work here provides the experimental data to invite further research especially theoretic models.

2. Experiment

Ferrofluid emulsion is used in this experiment, where droplets of ferrofluid coated with SDS surfactant is suspended in water [6]. The emulsion droplet has an average diameter of 260 nm with 27% standard deviation after extensive use of one year period in the experiments. The droplet volume fraction Φ of the sample used ranges from 2.0 % to 15.0 %. The sample is loaded in a glass capillary cell with rectangular or square cross-section. The capillary cell used with thickness L along the field direction ranges from 20 μm to 1,000 μm.

In this experiment, optical microscopy is used to study the induced structures. An NEC (TI-324A) CCD camera is attached to an Nikon (Diaphot) inverted optical microscope to record the structural images. The images are then sent to a Macintosh IIfx computer through the Perceptics Pixel-Grabber board for analysis. The sample is centered in a magnetic coil which is on the stage of the microscope to provide a magnetic field normal to the stage. The field is applied with a computer-controlled current supply running a predetermined ramping rate. In most of the experiment, we have fixed the maximum field to be 24 mT with the ramping rates ranging from 0.0133 mT/s to 400 mT/s. The highest rate corresponds to a rise time of 60 ms in an exponential increasing function of the field, which is limited by the response of the current supply and the magnetic coil. The field with slow rate rises linearly as a step function. At the highest ramping rate, it takes a fraction of a second to a minute for the structure to reach a steady state depending on L and Φ. The longest time taken to ramp the field from 0 to 24 mT is 1,800 sec., or 30 minutes.

3. Results

3.1. The Effect of Field Ramping Rate

When a magnetic field is applied to the ferrofluid emulsion, randomly dispersed Brownian droplets as shown schematically in Fig. 1a aggregate to form solid structures. At a low ramping rate R = dH/dt, the solid structure is equally spaced columns with equal width as shown in Fig. 1b [7]. However, when *R* is high, the shape of the aggregates become more complex. It can appear as a "folded-sheet" or a "bent-wall" - like from the side view or "worm-like" from the top view as shown in Fig. 1c. If *R* increases further, a maze or labyrinthine pattern from the top view is observed.

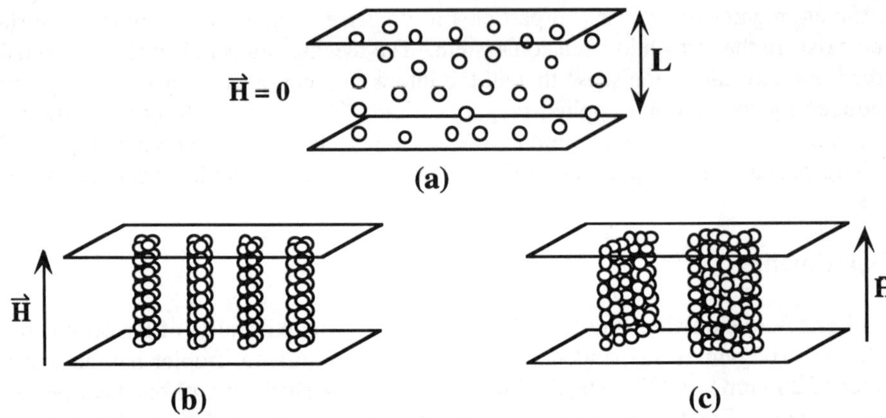

Fig. 1. a) Ferrofluid emulsion droplets inside capillary cell with no magnetic field. b) Columnar solid structure formed with field applied at a slow ramping rate. c) Wall-like solid structure formed with field applied at a high ramping rate.

Figure 2 shows the microscopic images of these patterns in top view for H_{max} = 38 mT (or coupling constant λ = 160), Φ = 12%, L = 50 μm, droplet diameter 2a = 0.33 μm. At the lowest ramping rate 0.0016 mT/s, the saturated equilibrium structure is locally ordered columns with short single chains in between as shown in Fig. 2a. As *R* increases to 0.6 mT/s, Fig. 2b shows the saturated pattern of a mixture of disks and worms reflecting columnar and wall structures in three dimension. At a higher R = 38 mT/s, a maze or labyrinthine pattern is observed as shown in Fig. 2c, where worms and branches dominate with almost no disks left. If *R* increases further to 600 mT/s, all the branches are connected and a random network is observed as shown in Fig. 2d. At this point, it is not clear that the same pattern should go along the field direction; the image is too dense to get a clear view from the side.

The pattern shown in Fig. 2 is very similar to those observed in ferrofluid [10,11] where the intrinsic surface tension between the oil-water interface plays a key role to confine the ferrofluid drop where the dipolar force pushes ferrofluid apart. Competition of these two energies explains the shape evolution. Even though the formation of labyrinthine patterns we observed in the ferrofluid emulsion is very similar to that in ferrofluid, the attractive force to keep aggregates together is not from intrinsic surface energy. Our ferrofluid emulsion system does not have any intrinsic surface tension; droplets coated with surfactant do not aggregate without external magnetic field. To make an analogy to that of ferrofluid, we need to assume that an induced surface energy by the magnetic field [17,18] competes with the induced repulsive energy between dipoles. These two energies along with the geometric constraint sets the many local energy minima which determines the shape of the formed structures.

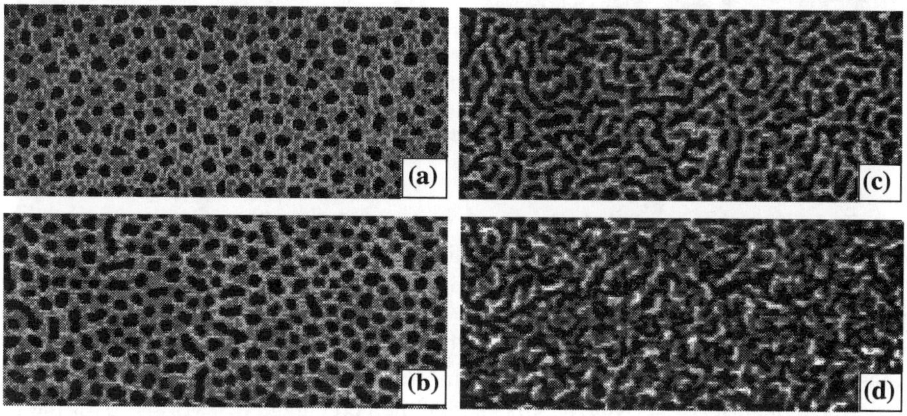

Fig. 2. Microscopic images of formed patterns in top view taken at Hmax = 38.0 mT, Φ = 12%, L = 50 μm for different ramping rates: a) 0.0016 mT/s, b) 0.6 mT/s, c) 38 mT/s, d) 600 mT/s. The horizontal width of each image is 125 μm.

When R is slow, the emulsion droplets are able to settle down to their lowest energy states due to thermal motion where the dipole-dipole interaction energy is comparable to thermal energy or coupling constant $\lambda \leq 5$ [19,6], thereby creating these locally ordered columnar structures. When the ramping rate is high, the emulsion droplets become large dipoles faster than the diffusion time of droplet taken to travel a distance of one droplet diameter. Thermal energy is too weak to balance the dipolar force. Therefore, the droplets cannot find their lowest energy state for column formation but a metastable state or a local energy minimum. The results are these "worm" and maze patterns. Figure 2 indicates that many local energy minima exist induced by the applied magnetic field. These states are sensitive to the initial conditions, the ramping rate may select one particular wavelength in the initial instabilities to grow similar to those observed in ferrofluid [11].

Figure 3 shows more detailed or gradual structural transitions from columns to walls as the ramping rate increases. In obtaining this results, a lower maximum field of 24 mT or $\lambda = 50$ is applied compared to the one in Fig. 2, to a sample at $\Phi = 9.0\%$, $L = 100$ μm.

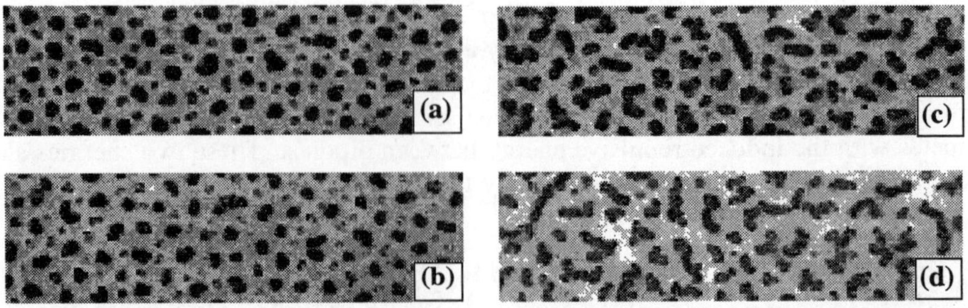

Fig. 3. Microscopic images of solid structures in top view showing transition from columns to walls as R increases: a) R = 0.0133 mT/s, b) R = 0.024 mT/s, c) R = 0.24 mT/s, d) R = 2.4 mT/s. H_{max} = 24 mT, Φ = 9.0%, L = 100 μm and the horizontal width is 80 μm.

To learn quantitatively how the rate of the applied field dictates the formations of the structures, we are going to focus for simplicity on the structural transition only from columns-to-walls, or disks-to-"worms" as those images shown in Fig. 3. A simple way is used to characterize those patterns: counting the number of "worms" N_w in a given area where the total number of aggregates is N_t. The "worm" is defined as that the ratio of the end-to-end length to the width of each "worm", labeled e, is at least a factor of two. We then define a quantity, $P = (N_w/N_t)$, to describe the percentage of "worms" over the total number of aggregate or the complexity of the structure. If $P > 50\%$, a given pattern is "worm-dominated" which might have branches and a partially connected network. If $P < 50\%$, the solid structure is column-dominated. This definition works well for characterizing patterns near $P = 50\%$ transition where most aggregates have $e \sim 2 - 5$, and loses information about details of more complicated patterns for $P >> 50\%$. More work is underway to analyze the shape of aggregates better.

Figure 4 shows the data analyzed based on the above definition where $H_{max} = 24.0$ mT for all the plots. For each set of L and Φ, the complexity of the pattern P increases with the ramping rate R. The solid or dashed lines are logarithmic fitting of the data, $P \propto \log(R)$, which seems describe all the data well to the first order approximation. The slope varies from 1 to 18 depending on the combination of L and Φ. These variation of slope indicates that the rate of increase in P depends strongly on both L and Φ. The increasing field rate changes the structure formation from a locally ordered equilibrium state to that of a random nonequilibrium state.

3.2. Dependence of Structure on L and Φ

As shown in Fig. 4, cell thickness and droplet volume fraction also affect the complexity of the formed structures at a high R and H. In this section, we keep the rate R fixed at 400.0 mT/s (ramping time = 0.06 s) and the final field at 24 mT. We then study the formed structures by varying the cell thickness of the capillary cell, L, and the volume fraction, Φ, of the sample.

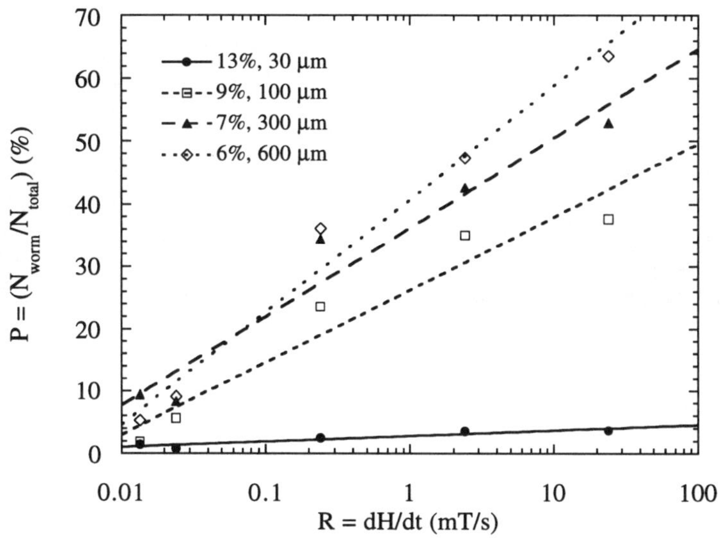

Fig. 4. Dependence of the worm ratio on the ramping rate R for various combinations of Φ and L at R = 400.0 mT/s, and H_{max} = 24 mT. The lines are fits of the relation, $P \propto \log(R)$.

Figure 5a shows microscopic images of the dependence of P on volume fractions at a fixed cell thickness, L = 100 μm. It is clear that the higher the Φ is, the more complex the structure becomes. It is reasonable since the "bent-wall" structure is a metastable state in which the structure stuck in one of the local minimum. At fast R and high H, the dipoles induced in the droplets are very strong so that the attraction between neighboring droplets pull them together without spending time to explore the lowest energy state. The closer the droplets are at higher Φ, the stronger the dipole interaction. The quicker they get together, the more complex the structure is and the farther it goes away from the equilibrium state. At low Φ, not only the separation between droplets are large so that dipolar force is small, but also the viscous force from the fluid will slow down the speed of attraction due to dipolar force. It gives more time for the droplets to explore the lower energy state.

Figure 5b shows microscopic images of the dependence of P, the complexity of the formed patterns, on variable cell thicknesses at a fixed volume fraction, Φ = 6.0%. As the cell thickness increases, the structure contains more "bent-wall"-like aggregates. This indicates that long chains formed first which are limited by the cell thickness. Then chains aggregate to form either columns or "bent-walls". This is understandable since chain formation is the strongest interaction among dipoles. However, it is hard to image all the chains formed with equal length before they aggregate. We suspect that chains form with random length and random position. The longest chains act as nuclei to attract nearby shorter chains or droplets to them [20]. This is because of the anisotropic nature of dipolar interaction between any two dipoles. The detailed aggregating process is still not clear yet.

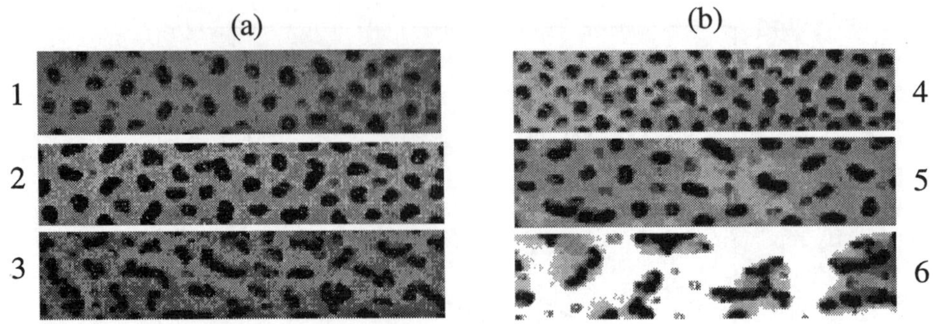

Fig. 5. Images showing the structural transitions at R = 400.0 mT/s, and H_{max} = 24 mT, (a) of volume-fraction dependence at L = 100 μm with Φ = 5.0%, 8.0%, and 11.0% for image 1 to 3; (b) of cell thickness dependence at Φ = 6.0% with L = 50 μm, 200 μm, 500 μm for image 4 to 5.

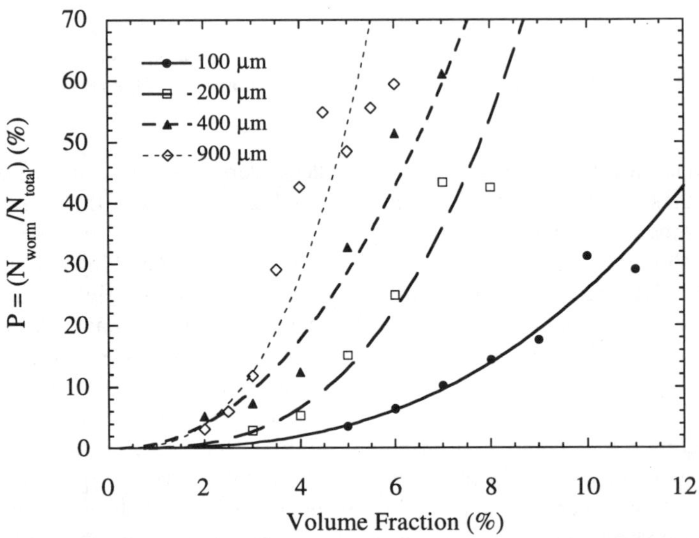

Fig. 6. The dependence of "worm" ratio on the volume fraction for different cell thicknesses at R = 400.0 mT/s, and H_{max} = 24 mT.

The cell thickness dependence of the structure further shows that the interaction between chains depend on their length. The ends of chains having the same monopoles repel each other which plays an important role in the equilibrium structure formation [7,18]. The shorter the chains are, the stronger the effect of repulsion from the ends. As a result, the long chains will attach to each other from the middle of the chain and zip through the rest of the chains. They tend to form aggregates with chains of slightly different length. In equilibrium columnar structures, we found either tapered ends or split ends of columns reflecting the effect of repulsion between ends. The same effect may also exist in the nonequilibrium structure. For short chains, it is hard to bend and therefore smaller aggregates are formed whose shape tends to be less complex.

Figure 6 shows quantitative measurement from analyzing images similar to those in Fig. 5. If we set P = 50% and take a cut in the plot of P - Φ relation as those in Fig. 6, we obtain a set of transition points in (L, Φ) corresponding to equal worms and columns of a structure. By repeating the same measurement for cells with thickness ranging from 20 μm to 1000 μm and droplet volume fractions ranging from 2.0% to 15.0%, we construct a "phase" diagram of column to "bent-wall" structure transition.

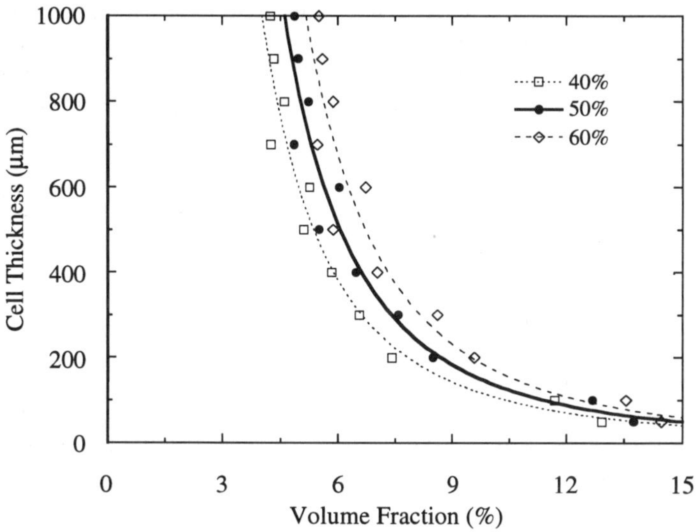

Fig. 7. "Phase" diagram showing the structural transition from columnar solid to "bent-wall"-like solid at R = 400.0 mT/s, and H_{max} = 24 mT. The solid line denotes the equal structure of columns and walls at P = 50%. The dotted and dashed lines are for P = 40%, P = 60%, respectively.

Figure 7 shows the "phase" diagram of the structural transition for P = 50% along with those at P = 40% and 60% for comparison. The P = 40% and P = 60% lines are almost parallel to the P = 50% line. The lines are fits of power-law functions. From the P = 50% transition line, we obtained a power-law relation: $L = 48\Phi^{-2.53}$ mm. These lines divide the space into a "bent-wall"-dominated region above them and a column-dominated region under them. The farther above the transition lines, the more complex and further away the structure is from the equilibrium state. The farther under the transition lines, the closer the structure is to the equilibrium columnar solid. It is not clear how to understand this relation quantitatively. We expect that as R increases, the P = 50% line would move down towards P = 40% line. To reach a "worm" pattern, we can either increase the cell thickness or volume fraction at a given R and final H.

4. Conclusion

By applying an external magnetic field to a ferrofluid emulsion, we can change the formed structures from columns to "bent-walls" or maze pattern. The selection of these structures depends not only on the ramping rate of the applied field for a given strength but also on the cell thickness along the field direction of the capillary cell containing the fluid and the volume fraction of the fluid. The cell thickness dependence of complexity of the patterns indicates that the formation of chains is the first step before forming aggregates. The end-end repulsion among chains is the key for setting the size and shape of the aggregates. Due to the fast field rate, the droplets are not allowed to reach their lower energy states but a metastable state or local minimum state. Therefore, the worm patterns are formed as a result of this nonequilibrium state. The existence of many different patterns indicates a complicated energy map induced in this ferrofluid emulsion system with many local minima in it, which sensitively depend on the set of parameter such as R, H_{max}, L, and Φ. Since the colloidal structure controls the rheological properties of MR fluids, this nonequilibrium-structure study provides a further step toward real situation of application in understanding the MR effects from the study of equilibrium structures.

5. Acknowledgments

We would like to thank K. Javier, G. Hu, Dr. J. Bibette, Prof. R. Sheng, and Prof. E. Wong for technical support, and Dr. R. Rosensweig and Dr. R. Goldenstein for helpful discussion. This work was partially supported by an award from the Research Corporation, and by CSULB University Research and Academic Affairs Office. Acknowledgment is made also to the Donors of The Petroleum Research Fund, administered by the American Chemical Society, for partial support of this research.

6. References

1. T. C. Halsey, and W. Toor, *Phys. Rev. Lett.*, **65**, 2820 (1990).
2. R. Chen, R. N. Zitter, and R. Tao, *Phys. Rev. Lett.*, **68**, 2555 (1992).
3. R. Tao, and J. M. Sun, *Phys. Rev. Lett.*, **67**, 398 (1991).
4. E. Lemaire, Y. Grasselli, and G. Bossis, *J. Phys. II France*, **2**, 359 (1992).
5. E. Lemaire, G. Bossis, and Y. Grasselli, *J. Phys. II France*, **4**, 253 (1994).
6. J. Liu, E. M. Lawrence, M. L. Ivey, G. A. Flores, J. Bibette, and J. Richard, R. Tao (Ed.), *4th Int. Conf. on Electrorheological Fluids* Feldkirch, Austria, July 19-24 1993, (Singapore, World Sci., 1994).
7. J. Liu, E. M. Lawrence, A. Wu, M. L. Ivey, G. A. Flores, K. A. Javier, J. Bibette, and J. Richard, *Phys. Rev. Lett.*, **3**, 2828 (1995).
8. Y. H. Hwang, and X. L. Wu, *Phys. Rev. E*, **49**, 3102 (1994).
9. R. E. Rosensweig, *Scientific American*, **247**, 136 (1982).
10. R. E. Rosensweig, *Ferrohydrodynamics*. (New York: Cambridge Univ. Press, 1985).

11. A. J. Dickstein, S. Erramilli, R. E. Goldstein, D. P. Jackson, and S. A. Langer, *Science*, **261**, 1012 (1993).
12. M. Seul, and M. J. Sammon, *Phys. Rev. Lett.*, **64**, 1903 (1990).
13. P. Molho, J. L. Porteseil, Y. Souche, J. Gouzerh, and C. S. Levy, *J. Appl. Phys.*, **61**, 4188 (1987).
14. H. Mohwald, *Annu. Rev. Phys. Chem.*, **41**, 441 (1990).
15. H. M. McConnell, *Annu. Rev. Phys. Chem.*, **42**, 171 (1991).
16. R. P. Huebner, *Magnetic Flux Structures in Superconductors*. (New York: Springer-Verlag, 1979).
17. W. Toor, and T. C. Halsey, *Phys. Rev. A*, **45**, 8617 (1992).
18. A. Wu, and J. Liu, *"A Cylindrical Model of Field-Induced Structure of Dipolar Colloids",* In preparation.
19. A. P. Gast, and C. F. Zukoski, *Adv. Coll. Int. Sci.*, **30**, 153 (1989).
20. Y. Zhu, and J. Liu, (Ed.), *5th Inter. Conf. on ER and MR Suspensions and Associated Tech.* Univ. of Sheffield, UK, July 10-14 1995.

RHEOLOGY OF MAGNETORHEOLOGICAL FLUIDS: MODELS AND MEASUREMENTS

J. M. GINDER, L. C. DAVIS, and L. D. ELIE
Research Laboratory, Ford Motor Company
20,000 Rotunda Drive, M. D. 3028 SRL
Dearborn, MI 48121-2053 (U. S. A.)

ABSTRACT

Numerical and analytical models of magnetorheological fluid phenomena that account explicitly for the effects of magnetic nonlinearity and saturation are described. Finite-element analysis was used to calculate the field distribution in chains of magnetizable particles. The field-dependent stress required to shear the chains was then obtained using the Maxwell stress tensor. Three regimes are identified: at low applied fields, the stress increase quadratically, as expected from linear magnetostatics. In intermediate fields, the contact or polar regions of each particle saturate, reducing the rate of increase of the stress with increasing field. At high fields, the particles saturate completely, and the stress reaches its limiting value. Approximate analytical expressions for the yield stress and shear modulus in these regimes are also derived. The predictions of these models are compared to magnetorheological experiments in the literature and from our laboratory. These models predict successfully the magnitude of the stresses as well as their field dependence. They also suggest that particles comprised of materials with higher saturation magnetizations will exhibit the largest magnetorheological effects.

1. Introduction

Magnetorheological (MR) fluids were first developed in the late 1940s by Jacob Rabinow.[1] They have recently enjoyed a revival of interest as high-shear-strength, low-voltage counterparts of electrorheological (ER) fluids. Yield stresses of order 100 kPa, some two orders of magnitude larger than in practical ER fluids, have been obtained in MR fluids exposed to applied magnetic fields of 3 kOe, or 240 kA/m,[2,3] which are readily generated by electromagnets or permanent magnets. The substantial field-induced yield stresses exhibited by MR fluids make possible some applications, such as rotary couplings, that are not practical with existing ER fluids.[4] For example, clutches using magnetic powders -- in essence, dry MR fluids -- are used in a number of production applications. Rotary brakes and linear dampers utilizing MR fluids are now in commercial production.[5]

Despite the differences in their composition and properties, the physical phenomena responsible for the MR and ER effects are quite similar. In both these classes of materials, the application of a field induces a dipole moment on each suspended particle; interparticle attractive forces promote the formation of a weakly solid network of particles. Because the electromagnetic energy stored in the network is reduced when it is mechanically deformed, a field-dependent stress must be applied in order to impose a strain.

Using the analogy between linear electrostatics and magnetostatics, the many models developed for electrorheology can be modified to treat MR fluids. However, these models fail to treat the nonlinearity inherent in all magnetic materials: the magnetization M in such materials does not increase indefinitely with increasing applied field, but rather it saturates at the saturation magnetization M_s. We have used finite-element techniques[6] and analytical approximations[7] to construct models of magnetorheological phenomena that explicitly account for magnetic nonlinearity and saturation. These effects have a significant impact on the field dependence of the yield stresses and shear moduli of MR fluids, as well as on their limiting values. Only at the lowest applied fields is a magnetically linear regime identified; as expected, the yield stresses and moduli vary as H_0^2 in this regime, where H_0 is the applied field. For higher applied fields, the polar or contact regions of each particle are saturated; in this regime, the yield stress is predicted to increase as $H_0^{3/2}$ while the modulus varies linearly with H_0. For fields high enough to achieve complete saturation, the particles can be treated rigorously as dipoles; the stresses and moduli are then independent of field and scale as M_s^2.

In this paper, numerical and analytical models of MR fluid phenomena are detailed. In particular, the field dependences and the limiting values of the field-induced yield stresses and moduli are identified. Preliminary rheological measurements on model MR fluids are described. The predictions of the models are compared to these and other measurements.

2. Magnetorheological Models

2.1 Numerical Approaches

In this section, a finite element analysis (FEA) of the stresses in an MR fluid is described.[6] An MR fluid is modeled as a collection of infinite linear chains of spherical magnetizable particles exposed to an average magnetic induction or flux density $\mathbf{B}_{ave}$.

For computational convenience, axial symmetry is assumed;[8] the effects of particle column formation and rotations of the particle moments away from the direction of the applied field are therefore beyond the scope of this model. The axis of symmetry, the z axis, is parallel to $\mathbf{B}_{ave}$. The unit cell is a cylinder of radius W and height 2h, as shown in Fig. 1. The volume of the cylinder is determined by the volume fraction ϕ of the particles: $W=(2R^3/3h\phi)^{1/2}$, where R is the particle radius.

The local magnetic induction **B** is calculated using a commercial two-dimensional finite-element analysis (FEA) solver (Maxwell 2D, Ansoft Corporation) running on an Intel 80486-based personal computer. This program solves Ampere's law $\nabla^2\mathbf{A}=-\mu_0\mathbf{J}$ for the vector potential **A**; here $\mu_0=4\pi\times10^{-7}$ H/m is the permeability of free space and the current density $\mathbf{J}=0$. The magnetic induction is obtained from the vector potential by $\mathbf{B}=\nabla\times\mathbf{A}$. By taking advantage of the symmetry of the model geometry, the complete field distribution can be determined by that in a computational region one-fourth the size of the unit cell, Fig. 1. An average induction B_{ave} is imposed by fixing the angular component of the vector potential at the outer cylindrical boundary: $A_\varphi(\rho=W)=B_{ave}W/2$, where ρ and φ respectively denote the radial and angular coordinates in the cylindrical

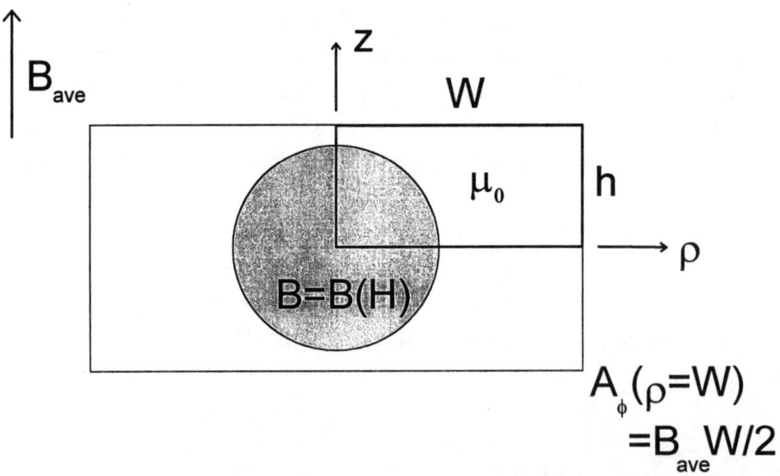

Figure 1. Schematic view of the axisymmetric geometry used to model an MR fluid. Identical magnetizable spherical particles of radius R are aligned in chains along the axis of symmetry, the z axis. An average magnetic induction B_{ave} is imposed along the z direction by the vector potential $A_\varphi(\rho=W)$ on the outer cylindrical boundary of the unit cell. The computational region of radius W and height h is denoted by the small rectangle.

coordinate system. Symmetry boundary conditions on the top and bottom planes are also imposed: $H_\rho=0$ at z=0 and z=h. The constitutive B-H relationship describing the nonlinear magnetic properties of the particulate material was modeled by the empirical Frölich-Kennelly equation[6,9] adjusted to have a low-field relative permeability $\mu_p/\mu_0=10^3$ and a saturation magnetization $\mu_0 M_s=2$ T, properties that approximate those of nearly pure iron. Of course, any B-H relation could be used, including those determined experimentally on bulk magnetic materials.

The chain tension F along z due to the interparticle magnetostatic forces is calculated by integrating the Maxwell stress due to the particles on the lower half of the chain due to those on the upper half:

$$F = \frac{\mu_0}{2}\int_0^W d\rho\, 2\pi\rho\, [H_z(z{=}h) - H_0]^2 . \tag{1}$$

Here H_0 is an average of the magnetic field H over the unit cell. Upon the application of a stress, the chains are assumed to undergo a uniform or affine deformation; nonaffine deformations[10] cannot be treated without discarding the symmetries that simplify the nonlinear magnetostatic calculations. If the chains are sheared at an angle θ with respect to the initial chain axis, the shear strain γ is given by $\gamma=\tan\theta$. The interparticle spacing then varies with strain as $2h=2R \sec\theta=2R\,(1+\gamma^2)^{1/2}$. Within the chain-stretching approximation, the component of the chain tension F acting in the shear direction, normalized by the initial area of the cylinder faces, determines the shear stress τ:

$$\tau = \frac{\gamma}{1+\gamma^2}\,\frac{F}{\pi W^2} . \tag{2}$$

Upon calculating the magnetic force F as a function of interparticle separation at constant volume fraction, as appropriate for shear deformations, Eqs. 1 and 2 enable the prediction of the magnetorheological stress-strain relationship. The static yield stress τ_y is generally identified as the maximum shear stress obtained as the strain is varied.[11] The dependence of the calculated yield stress on average magnetic induction for volume fractions $\phi=0.2$ and 0.5 is shown in Fig. 2.

The dependence of the yield stress and shear modulus on the applied induction consists qualitatively of three regimes: a linear regime is found for very low inductions, in which all field-induced rheological properties vary as B_{ave}^2. For intermediate applied inductions, this familiar linear behavior breaks down: both the yield stress and shear modulus increase slower than B_{ave}^2. In particular, the yield stress increases roughly as $B_{ave}^{3/2}$. These finite-element predictions are in excellent agreement with extant experimental data[12] on the shear stress of a carbonyl-iron-based MR fluid with $\phi \approx 0.5$, as demonstrated in Fig. 2. In high magnetic inductions, the particle magnetization -- as well as the shear moduli and yield stresses -- saturate.

The FEA results can be used to obtain greater physical insight into the consequences of magnetic nonlinearity and saturation in MR fluids. Examination of the calculated magnitude of the magnetization $\mathbf{M}=\mathbf{B}/\mu_0-\mathbf{H}$ reveals that the polar regions are the first to saturate as the applied field is increased, since the magnetic flux is concentrated there. As will be demonstrated below, this local saturation of the particle magnetiza-

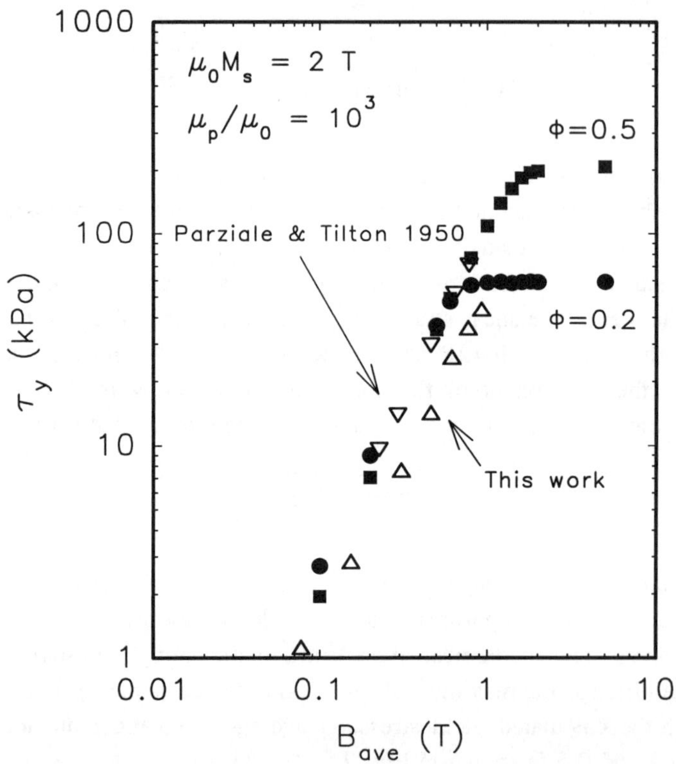

Figure 2. Calculated dependence of the yield stress τ_y on the applied magnetic induction B_{ave} obtained by the finite-element model for two volume fractions, $\phi=0.2$ (●) and $\phi=0.5$ (■). Measurements of carbonyl-iron-based MR fluids with $\phi=0.2$ (△) (present work) and $\phi=0.5$ (▽) (Ref. 12) are also plotted. The FEA results succeed in predicting not only the magnitude of the yield stress but also its roughly $B_{ave}^{3/2}$ dependence on applied induction.

tion leads to a breakdown of the linear approximation and thus to subquadratic field dependences of the forces and shear stresses in MR fluids, as has been suggested previously.[13,14]

FEA calculations have also been used to predict the maximum value of the shear stress, which is obtained when the particles are completely saturated; see Fig. 3. Uniformly saturated particles are rigorously treated as magnetic dipoles whose moment is proportional to M_s; the maximum yield stresses and moduli consequently increase as the square of the dipole moment, i.e., M_s^2. The predictions for ϕ=0.2 and ϕ=0.5 nicely bracket the maximum yield stresses reported for a ferrite-based MR fluid,[15] with $\mu_0 M_s \sim 0.6$ T, and an iron-based MR fluid,[2] with $\mu_0 M_s \sim 2.1$ T. These results underscore that high field-induced yield stresses can be obtained with particulate materials possessing high saturation magnetizations. While iron has the highest saturation magnetization of any element, alloys of iron and cobalt have the highest saturation magnetization of any known material, $\mu_0 M_s \approx 2.4$ T. According to the present model, such a material used in

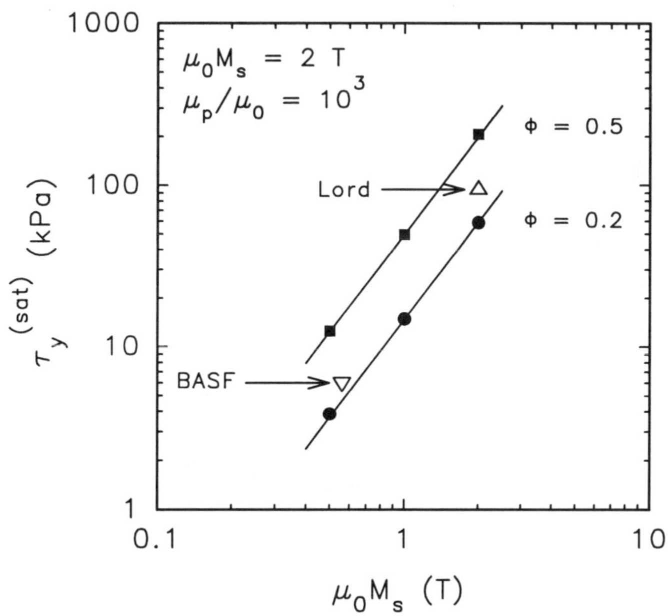

Figure 3. The dependence of the ultimate yield stress $\tau_y^{(sat)}$ on particle saturation magnetization M_s for volume fractions ϕ=0.2 and 0.5, calculated using the finite-element model. The straight lines are fits to a quadratic dependence on saturation magnetization. Also shown are the maximum reported yield stresses for ferrite-based (BASF, Ref. 15) and iron-based (Lord, Ref. 2) MR fluids.

an MR fluid with ϕ=0.5 could possess a yield stress $\tau_y \sim 300$ kPa. These alloys have recently been used to produce an MR fluid with a yield stress substantially higher than that of a comparable carbonyl-iron-based MR fluid.[16]

2.2. Analytical Approaches

The FEA approach is a very powerful and entirely general tool to explore the effects of particle composition, shape, and interparticle interactions on the magnetorheology of MR fluids. In several limiting situations, analytical techniques may also be used to model these materials. For highly permeable particles in the low-field or linear regime, analytical models first derived for ER fluids may be adopted; for examples, see Refs. 17 and 18. We have obtained simple analytical expressions to predict the dependence of the magnetorheological stresses on field and saturation magnetization in the intermediate- and high-field regimes.

In intermediate fields, the magnetic field in the gaps between particles is reduced from the value expected on the basis of linear magnetostatics due to local saturation of the particle magnetization. These saturation effects can be explored using a simple analytical model. This model is reminiscent of that used to treat nonlinear conductivity effects in ER fluids.[17,18] The polar regions of each particle are assumed to be completely saturated for radii inside a saturation radius δ, limiting the magnetic field in the gap in this region to M_s; see Fig. 4. Outside this saturation zone, the field in the gap is assumed to be that for highly permeable linear particles. An approximate expression for the magnetic field in the gap is then given by:

$$H_{gap}(\rho) = \begin{cases} \dfrac{2R}{w+\rho^2/R} H_0, & \rho>\delta; \\ M_s, & \rho<\delta; \end{cases} \tag{3}$$

where w is the interparticle spacing and H_0 is the applied field. Here δ is the radius of the saturation zone:

$$\frac{\delta}{R} = \sqrt{\frac{2H_0}{M_s} - \frac{w}{R}}. \tag{4}$$

The magnetic energy density $\mu_0 H_0^2/2$ can be integrated over the volume of the gap to

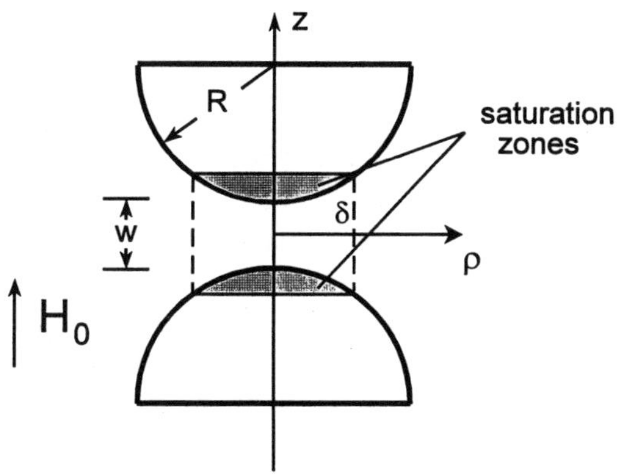

Figure 4: Geometry used in an analytical model of magnetorheology in intermediate fields, in which the polar or contact regions of each particle are saturated. The radius of the saturation zone in each particle is δ; the applied field is H_0.

obtain the effective density of magnetic energy u stored in the chain:

$$u = -\frac{3}{8}\phi\,\mu_0 M_s^2 [\frac{\delta}{R}]^4. \quad (5)$$

By again assuming that a shear strain γ results in a uniform increase in the interparticle spacing, the shear stress τ can be calculated from the first derivative of the energy, $\tau = du/d\gamma$. The static yield stress, again identified as the maximum stress obtained as a function of shear strain, is then given by

$$\tau_y = \sqrt{6}\,\phi\,\mu_0 M_s^{1/2} H_0^{3/2}. \quad (6)$$

This analytical model demonstrates simply the connection between the subquadratic power-law field dependence of the yield stress, observed both in experiments and the FEA results, and the local saturation of the particle magnetization. By taking $\phi=0.3$, $\mu_0 M_s=2$ T, and $H_0=160$ kA/m, $\tau_y \approx 74$ kPa is predicted, comparing favorably to an experimental value of 60 kPa for a proprietary MR fluid of unknown composition and volume fraction reported by Weiss, et al.[2]

In this approach, the shear modulus is predicted to vary linearly with field:

$$G = 3\phi\,\mu_0 M_s H_0. \quad (7)$$

Again taking $\phi=0.3$, $\mu_0 M_s=2$ T, and $H_0=160$ kA/m, the predicted value of the shear modulus is G $\approx$ 290 kPa. By comparison, shear moduli of almost 3 MPa have been reported for comparable applied fields in a proprietary MR fluid.[19] This significant discrepancy is likely due to nonmagnetic colloidal forces between particles, which are beyond the scope of a magnetostatic calculation.

In the high-field regime, the particles are completely saturated. The magnetization of a given particle is not influenced by the local fields produced by its neighbors, and so the particles behave as simple dipoles with fixed magnetic moments **m** of magnitude

$$m = \frac{4}{3}\pi R^3 \mu_0 M_s. \tag{8}$$

The energy of interaction U between pairs of dipoles separated by a position vector **r** making an angle θ with respect to **m** is given by $U=m^2(1 - 3\cos^2\theta)/(4\pi r^3)$. By summing this interaction energy over all pairs, the variation of the stored energy density with shear strain can be determined for any ordered array of particles. For example, the ultimate yield stress for a collection of noninteracting particle chains is given by

$$\tau_y^{(sat)} = \frac{4}{5^{5/2}}\zeta(3)\phi\mu_0 M_s^2, \tag{9}$$

where $\zeta(3)=1.202$. With $\phi=0.3$ and $\mu_0 M_s=2$ T, $\tau_y^{(sat)} \approx 82$ kPa is predicted. Similarly, the ultimate shear modulus obtained at saturation is given by

$$G^{(sat)} = \frac{\zeta(3)}{4}\phi\mu_0 M_s^2, \tag{10}$$

so that $G^{(sat)} \approx 290$ kPa is predicted for the same fluid parameters. Little or no experimental data that is definitely in the saturated regime is available to test these predictions.

3. Magnetorheological Experiments

The rheometer system adapted for preliminary magnetorheological measurements has been described previously.[20] A Couette or concentric-cylinder rheometer fixture constructed of low-carbon steel and incorporating a wire coil was used to deliver a variable magnetic field to the annular sample space. The coil was driven at currents up to 4 A by a current amplifier (Kepco 50-8). A search coil was also embedded in the fixture; the induced electromotive force in the coil was integrated using an integrating fluxmeter (Walker MG-3D) in order to estimate the average magnetic induction B_{ave} in the sample.

Model magnetorheological fluids were prepared using carbonyl iron particles (ISP Technologies, Grade R-1430) dispersed in light mineral oil (Aldrich). Oleic acid (Sigma) was added as a surfactant for the iron particles. These dense particles settled out over time but were redispersable by agitating or mixing the fluid.

The dependence of the shear stress on strain rate at various average inductions was explored for a $\phi=0.2$ carbonyl iron suspension. The dynamic yield stress τ_y was extracted by fitting the stress-strain rate curves to the Bingham plastic model, $\tau=\tau_y+\eta\dot{\gamma}$, where η is the fluid viscosity; the results are plotted in Fig. 2. The yield stress is observed to increase roughly as $B_{ave}^{3/2}$, confirming the predictions of the models described above. Discrepancies between the model predictions and the magnitude of the yield stresses may reflect, e.g., slip at the fixture walls[21], particle roughness, particle shape effects, differences between idealized and real magnetic properties, polydispersity, column formation, or other structural complications. Other interparticle interactions in addition to the magnetic attractive forces treated here are undoubtedly relevant.

4. Summary and Conclusions

Magnetic nonlinearity and saturation play critical roles in the magnetorheology of MR fluids. Local saturation of the magnetization of the particles suspended in these materials determines the field dependence of the yield stress and shear modulus over a wide range of applied fields. In this range, the dependence of the yield stress on applied field H_0 is given by $\tau_y \sim H_0^{3/2}$ while the modulus varies as $G \sim H_0$. The particle magnetization saturates at high fields, giving rise to maximum magnetorheological performance. The ultimate values of the yield stress and modulus are found to increase as M_s^2. Both numerical and analytical techniques can be applied to model the magnetostatics of MR fluids; the predictions of these approaches are in good agreement with magnetorheological measurements from the literature and from our laboratory.

5. References

1. J. Rabinow, *AIEE Trans.* **67** (1948), 1308.
2. K. D. Weiss, T. G. Duclos, J. D. Carlson, M. J. Chrzan, and A. J. Margida, *SAE Paper* No. 932451 (1993); K. D. Weiss and T. G. Duclos, in *Proceedings of the*

Fourth International Conference on ER Fluids, ed. by R. Tao and G. D. Roy (World Scientific, Singapore, 1994), p. 43.
3. A. Pinkos, E. Shtarkman, and T. Fitzgerald, *SAE Paper* No. 930268 (1993).
4. D. L. Hartsock, R. F. Novak, and G. J. Chaundy, *J. Rheol.* **35** (1991) 1305.
5. J. D. Carlson, D. M. Catanzarite, and K. A. St. Clair, these proceedings.
6. J. M. Ginder and L. C. Davis, *Appl. Phys. Lett.* **65** (1994), 3410.
7. J. M. Ginder and L. C. Davis, unpublished.
8. J. M. Ginder and L. C. Davis, in *Proceedings of the Fourth International Conference on ER Fluids*, ed. by R. Tao and G. D. Roy (World Scientific, Singapore, 1994), p. 267.
9. R. M. Bozorth, *Ferromagnetism* (van Nostrand, Toronto, 1951).
10. M. Parthasarathy, K. H. Ahn, B. M. Belongia, and D. J. Klingenberg, in *Proceedings of the Fourth International Conference on ER Fluids*, ed. by R. Tao and G. D. Roy (World Scientific, Singapore, 1994), p. 202.
11. R. T. Bonnecaze and J. F. Brady, *J. Rheol.* **35** (1992), 73.
12. A. J. Parziale and P. D. Tilton, *AIEE Trans.* **69** (1950), 150.
13. P. A. von Guggenberg, A. J. Porter II, and J. R. Melcher, *IEEE Trans. Mag.* **MAG-22** (1986), 614.
14. T. B. Jones and B. Saha, *J. Appl. Phys.* **68** (1990), 404.
15. C. Kormann, M. Laun, and G. Klett, in *Proceedings of the Fourth International Conference on New Actuators*, ed. by H. Borgmann and K. Lenz (Axon Technologie, Bremen, Germany, 1994), p. 271.
16. J. D. Carlson and K. D. Weiss, U. S. Patent 5,382,373 (1995).
17. N. J. Félici, J. N. Foulc, and P. Atten, in *Proceedings of the Fourth International Conference on ER Fluids*, ed. by R. Tao and G. D. Roy (World Scientific, Singapore, 1994), p. 139.
18. L. C. Davis and J. M. Ginder, in *Progress in Electrorheology: Science and Technology of Electrorheological Materials*, Proceedings of a Symposium on Electrorheological Materials and Fluids at the 208th National ACS Meeting, Washington, D. C., August 21-22, 1994, ed. by. K. O. Havelka and F. E. Filisko (Plenum, New York, in press).
19. K. D. Weiss, J. D. Carlson, and D. A. Nixon, in Proceedings of the Second International Conference on Intelligent Materials, ed. by C. Rogers and G. Wallace (Technomic, Lancaster, Pennsylvania, 1994), p. 1300.
20. J. M. Ginder and S. E. Ceccio, *J. Rheol.* **39** (1995), 211.
21. G. Bossis and E. Lemaire, *J. Rheol.* **35** (1991), 1345.

ERF Based Precision Linear Drive System

Hans Kurt Tönshoff
Institute for Production Engineering and Machine Tools,
University of Hannover, Schloßwender Str. 5, D-30159 Hannover,
Germany

and

Axel Stegmann
Institute for Production Engineering and Machine Tools,
Div. Machine Tools and Controls, University of Hannover,
Schloßwender Str. 5, D-30159 Hannover,
Germany

ABSTRACT

The design of clutches using electrorheological fluids (ERF) or magnetorheological suspensions (MRS) offers a new approach to precision linear drive systems based on the 'inch-worm' principle. This principle needs two clutches or controllable clamps and one (or more) elements to vary the distance between these clamps. To perform a caterpillar-like motion the actuators are controlled in a cyclic mode. This paper presents design elements, system structure and first results of motion performance of a prototype drive system.

The presented linear drive combines the high resolution of piezo actuators, the unlimited travel of the 'inch-worm' principle and the clutch effect of ERF without a substantial mechanical excitation. The motion of the two clamps is controlled to move a stage in point to point mode or to follow a given velocity profile. Using two piezo actuators the stage can move continuously. Pre-loaded piezo actuators control the distances between the two clamps and the central slide. Each ERF clamp is designed as a set of thin parallel steel plate electrodes which interlock with a set of ground plates. Drive force limits are given by the maximum yield force of the clamps as well as by the relation of the actuator expansion and the ERF/MRS shear stiffness in "solid" mode.

1. Introduction

The presented development of a linear drive system can be seen as a new approach in precision drive technology. The properties of ERF and MRS provide new design principles of clutches, damping devices and flow based systems concerning integration of electrical and mechanical system properties. Two controllable linear clutches are the main elements of the presented drive system. Other components are two actuators to vary distances, a linear air guide and the motion strategy itself. The basic motion is caterpillar-like: One clutch is clamped, a second clutch is released. After the connecting actuator between these clamps has expanded, the second clutch is clamping. The first clamp releases and a contraction follows. The first clutch starts to clamp again and so forth. If the design consists of two or more elements for varying distances, the motion of a driven slide can be not only discontinuous but also continuous. The drive system described here is being developed to meet feed forces for precision machining and for general precision positioning purposes.

2. Linear Guide

The drive system is guided by an aerostatic pre-loaded linear bearing. A cross section is shown in fig. 1. The ERF drive unit is suspended from the linear slide which provides a friction and stick-slip free motion. The beam is of hardened and ground steel, the twelve air bearing elements are flying at a distance of some micrometers on the beam. The linear guide's load capacity is about 10 kilograms at an air pressure of 6...8 bar. The elements have a sintered porous micro jet surface. The bearing is free of wear and has a high guiding accuracy. Accuracy demands on the slide design are low.

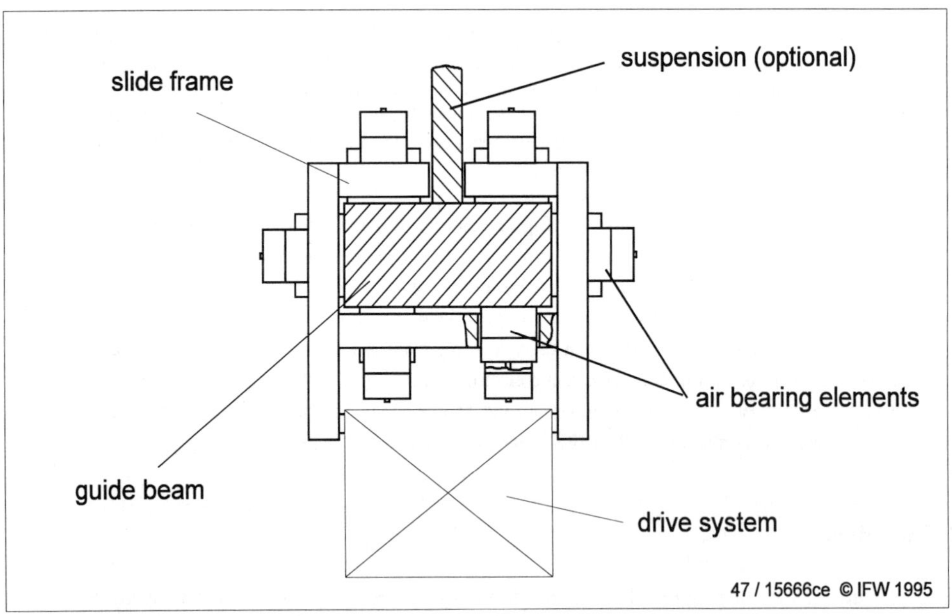

Fig. 1: Cross section of the aerostatic linear guide with suspended drive system

3. The clamping units

Different principles are possible to perform a clamping effect: ERF or MRS clamps can be used as well as clamps based on dry friction force actuated for example by piezo actuators in combination with a flexure hinge design. Mechanisms without substantial mechanical excitation on the structure are clamping devices based on ERF or MRS. This design needs parallel areas to connect the mechanical base and the movable clamp by the fluid's shear stress in a corresponding (electric or magnetic) field.

Using ERF, this area can consist of two sets of parallel and interlocking steel plate electrodes. The lower and upper plates must be guided to keep them constantly parallel. This is necessary to limit the electric field intensity between the plates. To keep them plane and apart at 0.5 mm distance without extra insulation it was necessary to harden and straighten the steel plates which have a thickness of 0.15 mm. The used ERF was RHEOBAY VP AI 3565 of Bayer AG, Leverkusen, Germany.

The plates can be insulated by spacers on the upper plates if necessary (see fig. 2). Some experiments were made using cotton[4] on a self-adhesive layer. First results have shown that a self-adhesive or any other layer of glue suppresses the yield force almost completely. The best results were made using a bare steel surface. The direct contact between the ERF and the steel plates seems to be necessary for a high yield force.

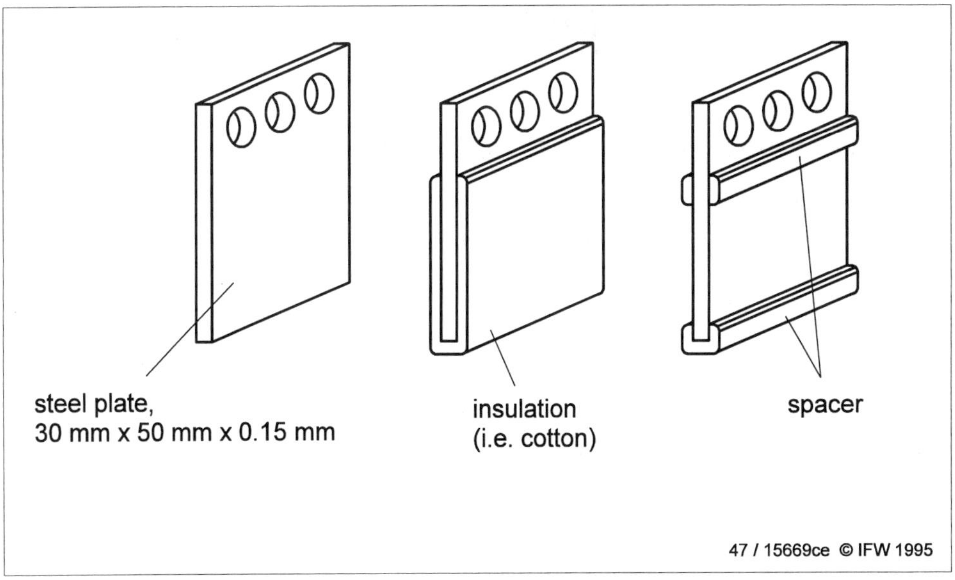

Fig.2: Different designs for upper plates of ERF actuator and insulation

An MRS clamping unit design is shown in fig. 3. A coil is integrated into each plate. The plate thickness is about 0.7 mm. One coil has 300 turns per plate at a wire diameter of 0.05 mm. The distance elements between all plates have to consist of magnetical insulating material for an efficient magnetic flux density through the core and the MRS between the plates. The magnetic flux density is limited mostly by the magnetic resistance of the MRS itself (s. fig. 4). The electrical time constant $\tau = L/R$ of the coil set is calculated to be milliseconds or less.

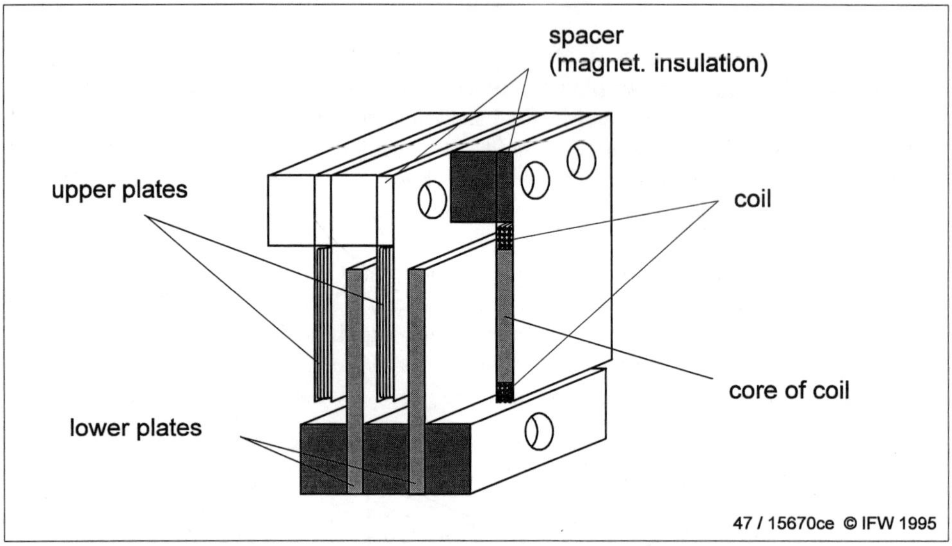

Fig. 3: Design of MRS linear clutch actuator with integrated coils

The above design might be modified for a clutch with electro-magnetorheological suspension to generate a combination of a magnetic and electric field.

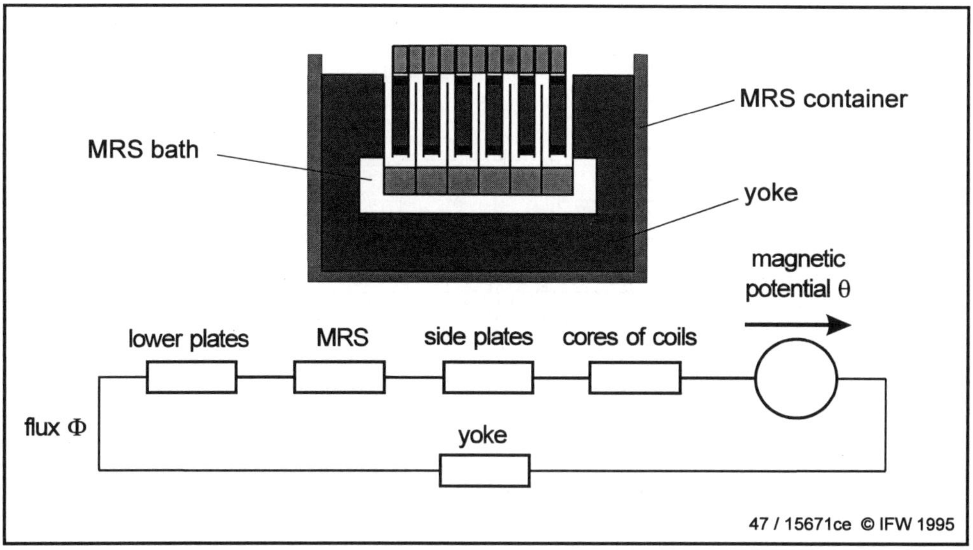

Fig. 4: Cross section of MRS clamping element and scheme of magnetic flux and resistances

4. Mechanical design

The mechanical part is designed as follows: A central slide is mounted on a linear air bearing as described above. Two clamping elements are connected by piezo actuators from both sides of the slide. For a friction and backlash free linear motion between clamps and slide the mechanical link is based on polystyrene flexure hinges which is also an efficient electrical insulation at high voltage. The connection between clamps and slide must be mechanically stiff. The complete experimental design is shown in fig. 5, a simple mechanical model of the system is shown in fig. 6. A symmetric model of the system structure (s. fig. 7) has four control signal inputs (two for the clamps and two for the translators) and the slide and clamp positions as outputs.

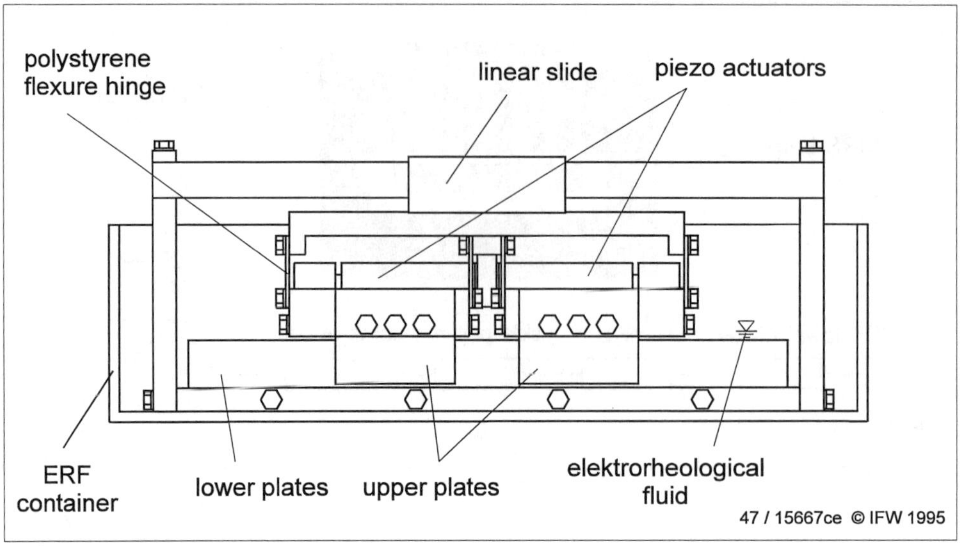

Fig. 5: Linear ERF drive, experimental design

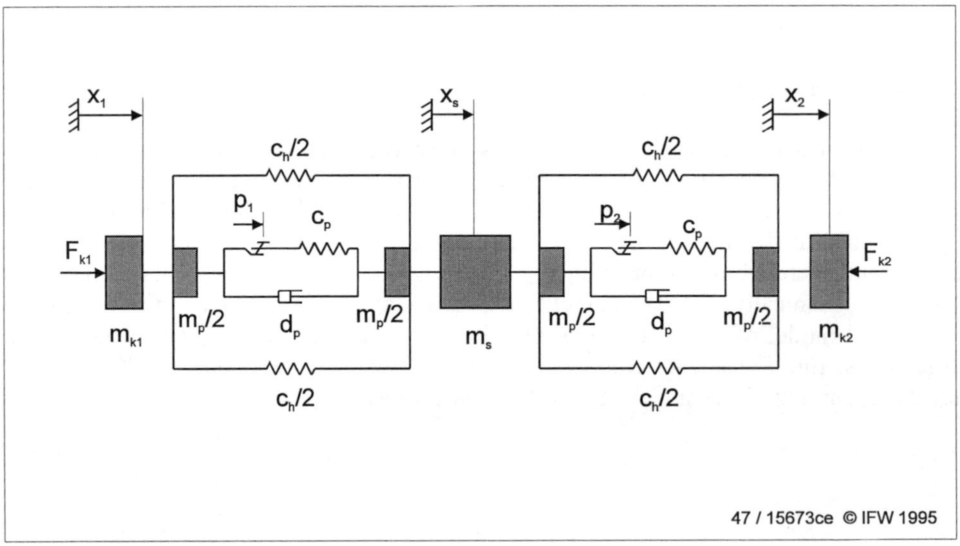

Fig. 6: Mechanical model

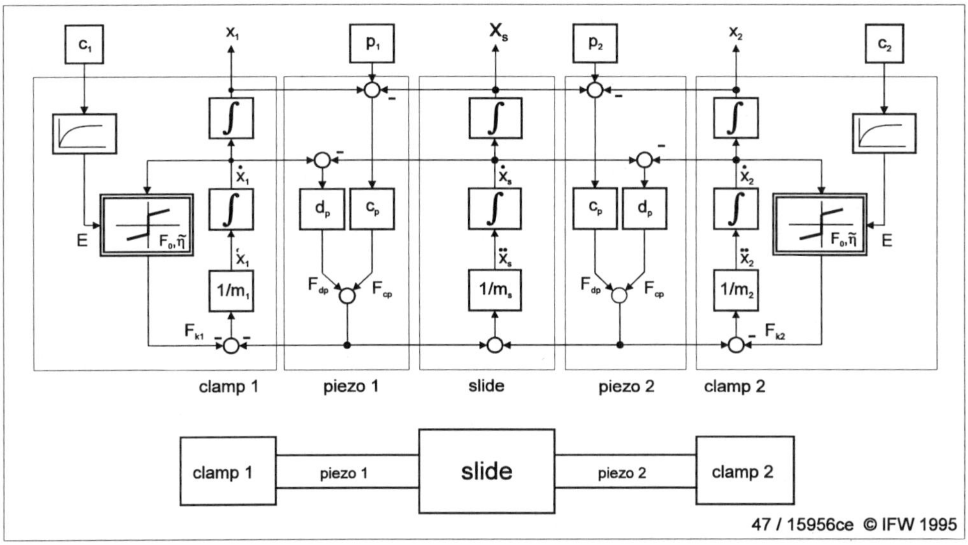

Fig. 7: Structure of the mechanical system model (4 control inputs, 3 motion outputs)

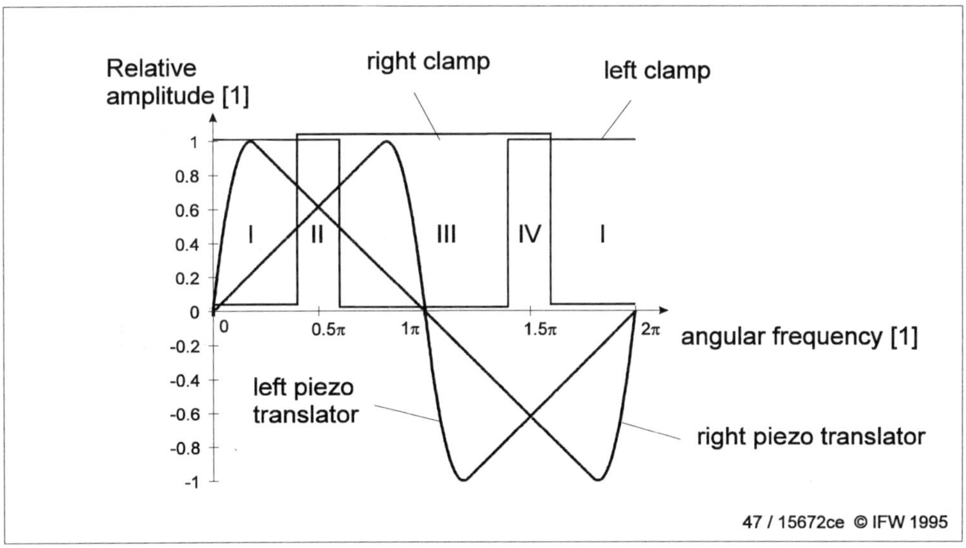

Fig. 8: Possible clamp motion (related to slide) and clamping phases to perform constant slide velocity

5. Strategy of motion and system structure

Fig. 9 shows the experimental control structure of the system. The nominal signal values for all four actuators are computed offline. This data is transferred to analog output and amplified to high voltage. The displacement is measured by a laser interferometer. Using an efficient actuator control curve the velocity is expected to reach 10 mm/s depending on the cycle frequency. Dynamic drive forces are found to be very sensible to the relation of the shear stiffness of the ERF in "solid" mode and the piezo actuator's stroke of about 40 μm.

The basic strategy of motion can be compared with that of a caterpillar[1]. It needs one translator element which gives a more or less discontinuous motion causing high mechanical exitation. The use of two (or more) translators provides a continuous motion with less excitation. This principle has been used in different drive systems[2,3]. Fig. 8 shows one possible way of motion of the two clamps in relation to the central slide. The four clamping phases in one complete cycle are also shown. These signal shapes were used for a first performance test as input signals in open loop mode (s. fig. 10). The drive system control is being optimized.

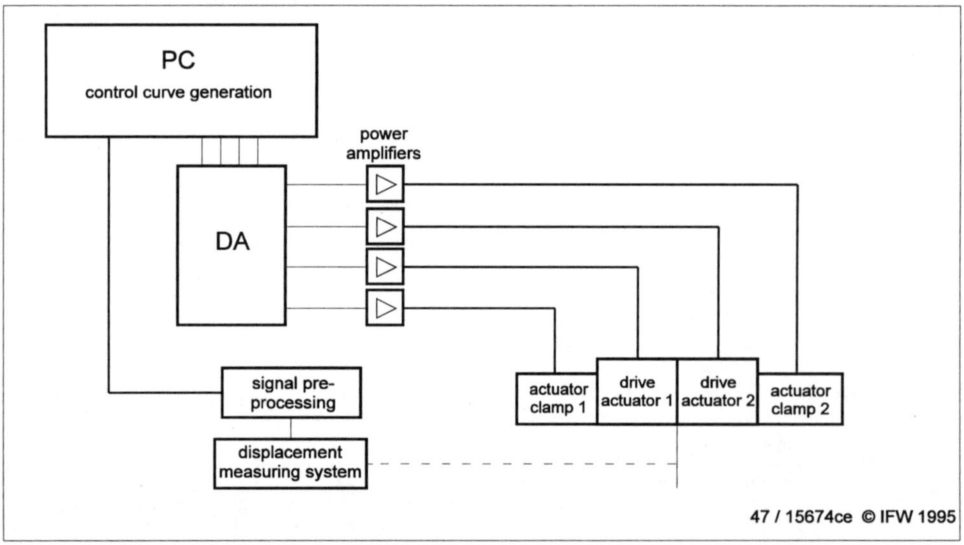

Fig. 9: System structure

To control this drive system at different velocities it is necessary to use frequency (FM) or amplitude modulated (AM) motion or a combination of both. This depends on the motion strategy and its rules. All control signals are to be generated online. In the future adapted algorithms will be used at different velocities and force loads. To meet these options it is also necessary to find an efficient automatic control algorithm.

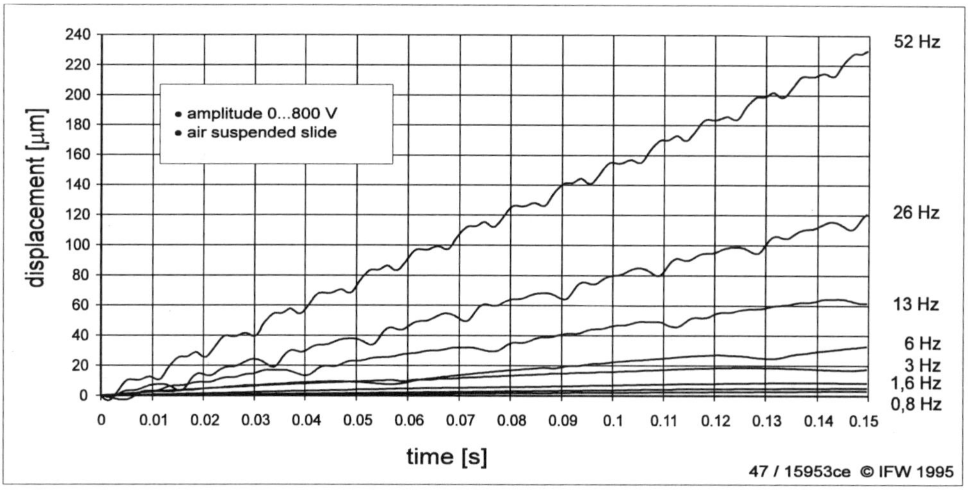

Fig. 10: Examples for first measurements of slide displacement in FM mode control (open loop mode)

6. Summary

The first ideas of a new linear precision drive system are presented. The principle is based on ERF (or MRS) controlled clutches and a caterpillar-like motion strategy which is modified from the known 'inch-worm' principle. Using a linear air guide a friction- and stick-slip-free continuous motion can be created. Further developments focus on optimal motion strategies and online control signal generation as well as closed loop operation(servo mode).

7. Acknowledgements

The authors would like to acknowledge the German Research Foundation (Deutsche Forschungsgemeinschaft, DFG) who supports this project as part of a special program for system integration of electrical drives and the Bayer AG, Leverkusen, Germany, for the supply of electrorheological fluids necessary for this development.

8. References

1. S. Brand, *Proc. ACTUATOR '92*, pp. 59-62, Bremen, Germany, 1992
2. R. Glöß, *Proc. ACTUATOR '94*, pp. 190-192, Bremen, Germany, 1994
3. M. P. Koster, *Proc. ACTUATOR '94*, pp. 144-148, Bremen, Germany, 1994
4. G. J. Monkman, *Mechatronics* **3**, 3, pp. 305-313, 1993

CONDUCTIVITY AND FORCE BETWEEN PARTICLES IN A MODEL ELECTRORHEOLOGICAL FLUID: I CONDUCTIVITY

C. W. WU, Y. CHEN, X . TANG AND H. CONRAD
Materials Science and Engineering Department
North Carolina State University
Raleigh, NC 27695–7907 USA

ABSTRACT

The conductivity of silicone oil with various water contents and that of a single-row chain of glass beads in silicone oil were determined as a function of electric field and spacing between the particles produced by tensile and shear strain. The conductivity of the silicone oil depended sensitively on its water content; that of the chain was three orders of magnitude greater than the silicone oil alone, but decreased rapidly with separation of the particles. The conductivity of the water film on the glass bead surface and the electric field enhancement in the oil gap between the beads were derived from the data. The measured current density along a chain of particles was in accord with that predicted by a conductivity model for electrorheological response. At large strains the separation of the beads in a chain became localized between one set of adjacent beads and they oscillated back-and-forth in the gap, leading ultimately to rupture of the chain with increased strain.

1. Introduction

Although there is now considerable evidence that conductivity is important in ER response[1–7], additional work is needed to more clearly define the conditions and mechanisms that are involved. Desirable are definitive measurements on a single chain of particles spanning the electrode gap. This therefore was the objective of the present study. The ER system chosen was water-activated glass beads in silicone oil, because related measurements had previously been made on single-row and multiple-row chains in this system[8–10]. The first paper of the present two-part series gives the measurements of the conductivity along a single-row chain of glass beads in silicone oil; the second covers the interaction force between the particles and its relationship to the conductivity of the particles and host oil.

2. Experimental

2.1. Materials

The host oil employed was Dow Corning 200 silicone oil with specific gravity 0.973, viscosity 50cs and dielectric constant 2.6 at 25°C. Since the conductivity σ_f of silicone oil depends on its water content W, measurements were made to determine the effect of water content on σ_f. The water content of the oil was varied by first heating the as-received silicone oil for 2 hr at 120°C. The oil was subsequently exposed for increasing time periods at 22°C to the vapors above LiCl–H_2O solutions of various compositions (including 100% H_2O). These treatments gave water contents in the oil up to 0.057 wt.% above that for the heat treatment at 120°C, all determined gravimetrically.

The particles were 222μm soda lime glass beads with specific gravity 2.5 and dielectric constant 7.2 (1 MHz, 20°C) provided by Potter Industries. Prior to mixing with the silicone oil they were held for at least 2 days in an ~ 67% relative humidity chamber containing LiCl/H_2O = 1:4 solution to impart a controlled amount (~ 0.1 wt.%) of adsorbed water, which gives an adsorbed H_2O film whose thickness is estimated to be ~ 200Å[11]. The low voltage, dc dielectric constant of the moisturized glass beads was determined to be K_p = 12.8 using Bruggemans equation[13].

2.2. Conductivity Measurements

2.2.1 Silicone Oil: Two cells were employed to measure the dc conductivity of the silicone oil. One of the cells and the corresponding electric circuit are shown in Fig. 1. The conductivity of the oil was also determined using the cell previously employed to measure the force between glass beads in a chain or cluster[8,10] along with the electric circuit in Fig. 1b. There was good agreement between the results obtained with the

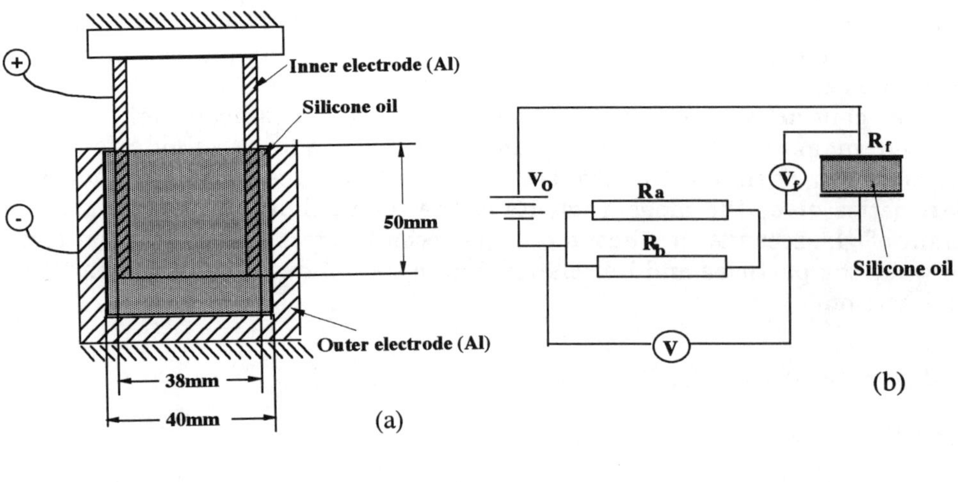

Fig. 1 Schematic of: (a) the large cell used in the conductivity measurements and (b) the electric circuit.

two cells. Two expressions were employed for the conductivity of the silicone oil: (a) $\sigma_f = j_f/E$ where $j_f = I_f/A$ and $E = V_o/h$ and (b) σ_f (= dj_f/dE) given by the slope of a plot of j_f vs E. A is the electrode area in contact with the oil and h the spacing between the electrodes.

2.2.2 Single-Row Chain: The second cell was employed to measure the current density and conductivity of a single-row chain of the glass beads containing from 3 to 6 beads. The current density along a chain was taken to be $j_c = 4A(j - j_f)/\pi d^2$, where j is the total current density flowing through the cell, j_f the current density in the silicone oil surrounding the chain, A (= 0.2375 cm^2) the electrode area of the pendulum bob and d the diameter of the glass beads. Since the total current density j was an order of magnitude larger than j_f, the main contribution to the current in the cell was from the chain of beads. The current density along a chain j_c was determined at zero strain and for increasing separation of the beads produced by tensile and shear displacement of the bob electrode. Straining of the chain was recorded by a VCR tape recorder so that the data could be analyzed at a later time. The electric current (conductivity) measurements on the silicone oil and the single-row chains were made at ambient laboratory conditions, which were typically 22°C and 48% relative humidity.

3. Results

3.1. Silicone Oil

Upon application of an electric field the recorded voltage first increased rapidly, went through a maximum and then quickly decreased again, reaching a steady-state value after ~ 5s. The magnitude of the peak in the measured voltage increased with rate of application of the voltage, similar to the behavior reported by Vermeer[13] during conductivity measurements on glass. The currents reported below are for measurements taken after one minute, which was well into the steady state.

The effect of applied electric field E_o on the current density j_f of the silicone oil with several water contents is presented in Fig. 2. Evident is that j_f (and in turn $\sigma_f = dj_f/dE$) increase with electric field and water content.

Foulc et al[4,5] have derived the following expression for the conductivity of non-polar liquids based on Onsager's theory[14]

$$\sigma_f = \sigma_f(o)\left[(1-A) + A\ \exp(E/E_c)^{1/2}\right] \tag{1}$$

where E is the electric field, $\sigma_f(o)$ the conductivity at low field and A and E_c are constants which depend on the liquid. The effects of water content on the conductivity parameters $\sigma_f(o)$, A and E_c for silicone oil taking $\sigma_f = dj_f/dE$

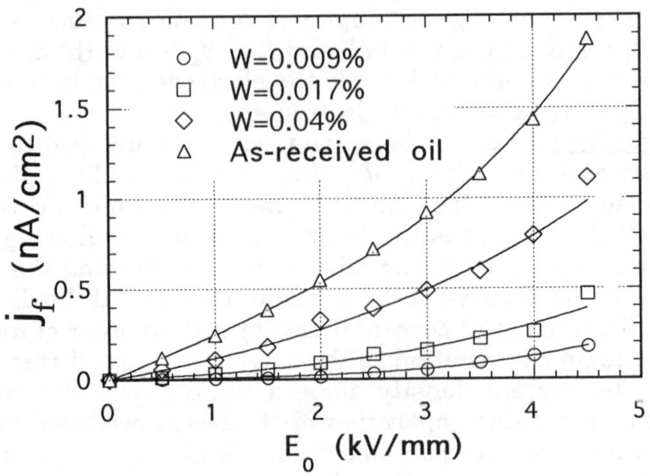

Fig. 2 Effect of applied electric field on the current density j_f for silicone oil with several water contents W.

are presented in Fig. 3. They were obtained by a computer fit of j_f to the equiaton $j_f = \int_0^E \sigma_f dE$. By taking $\sigma_f = j_f/E$ for the conductivity slightly different values of the constants A and E_c are obtained, but their dependence on water content is similar.

3.2. Chain of Particles

Examples of single-row chains of glass beads in silicone oil similar to those produced here are given in Refs. 8, 10. The effect of applied electric field on the current density j_c along chains containing 3, 4 or 5 glass beads in the *as-received* silicone oil is shown in Fig. 4 for no strain. j_c is relatively independent of the number of particles comprising the chain. Behavior similar to that portrayed in Fig. 4 also occurred for chains in silicone oil with the other water contents. The results presented below are for chains of glass beads in as-received silicone oil.

Upon straining a chain in tension or shear, j_c decreased with strain independent of the number of particles along the chain, the rate of decrease being greater the higher the electric field. The normalized *average* separation distance between the glass beads in a strained chain is given by $s = \Delta\ell/Nd$ where $\Delta\ell = (\ell - \ell_0)$ is the increase in axial length of the chain measured directly from the video screen and N the number of particles of diameter d in the chain. The decrease of j_c with *s* for tensile straining is shown in Fig. 5. Initially j_c decreases rapidly with *s*, but the rate of decrease becomes small when s>0.1. The influence of the field on $j_c$was hyperbolic for all separations, and decreased with separation distance.

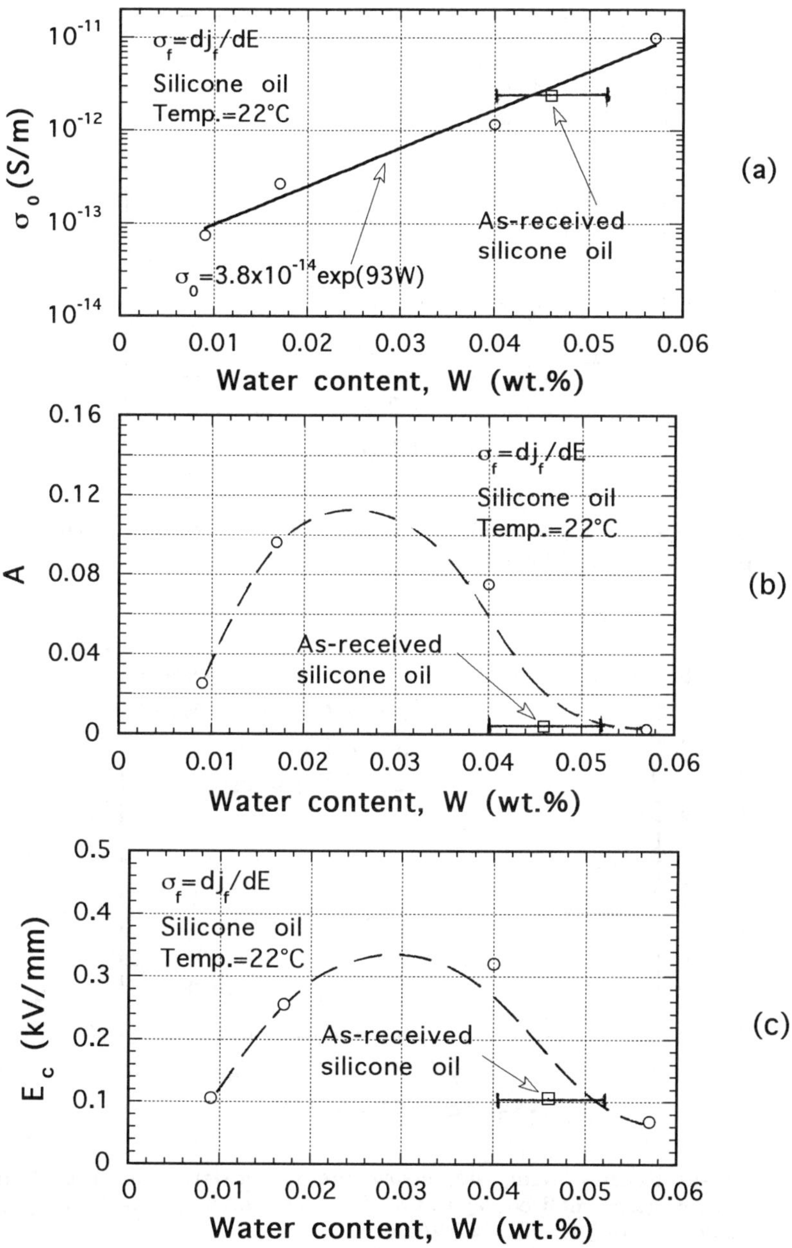

Fig. 3 Effect of water content of the silicone oil on the conductivity parameters $\sigma_f(0)$, A and E_c.

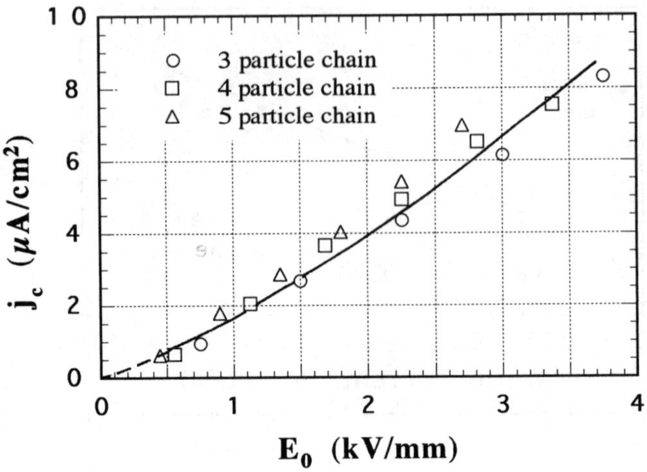

Fig. 4 Current density along single-row chains consisting of 3 to 5 particles vs the applied electric field for zero strain.

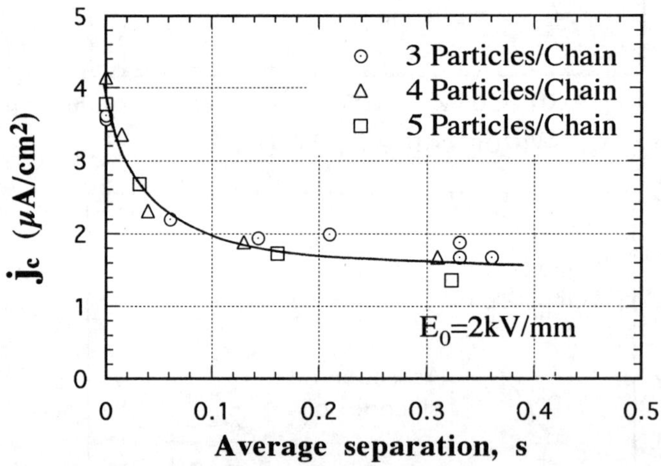

Fig. 5 Current density along single-row chains consisting of 3 to 5 beads vs the average normalized separation distance s between the beads produced by tensile straining for E_0 = 2 kV/mm.

Similar results were obtained for straining in shear as for tension for $s \leq 0.1$; however for larger displacements j_c for shear tended to be smaller than for tension.

It should be mentioned at this point that a uniform separation of the particles along a chain only occurred at small separations ($s < \sim 0.1$). With increase in s the separation usually became localized between two adjacent particles or between the electrode and an adjacent particle. A particle then began to oscillate back-and-forth in the gap between it and its neighbor or the electrode. The frequency of the oscillation was faster than the eye or the video recorder could follow and increased as the localized separation became larger, leading ultimately to rupture of the chain with increased strain.

4. Discussion

In the above it was shown that for a given applied electric field the current density flowing along a single-row chain of glass beads in silicone oil is about three orders of magnitude larger than that for the oil alone, but decreases as the chain is strained. These results will be analyzed in a manner similar to that employed previously[3,15] for water-activated ER suspensions and the outcome compared with predictions based on the conductivity model developed by Tang et al[6].

By considering the resistive components along a chain, it can be shown that[3,15]

$$E_o = \frac{I_c}{d}\left[\frac{\rho_p x_p}{A_p} + \frac{\rho_i x_i}{A_i} + \frac{\rho_f x_f}{A_f}\right] \tag{2}$$

where E_o is the applied electric field, I_c the current along the chain, ρ the resistivity of an element of length x and area A and the subscripts p, f and i refer to the particles, oil and interface, respectively. We will normalize all of the dimensions with respect to the diameter of the particles as follows: $x_p \approx \alpha_1 d$, $A_p \approx \alpha_2 \pi d^2/4$, $x_i \approx \alpha_3 d$, $A_i \approx \alpha_4 \pi d^2/4$, $x_f = \alpha_5 d$ and $A_f = \alpha_6 \pi d^2/4$. Inserting these into Eqn. 2 gives

$$E_o / j_c = \left(\frac{\alpha_1 \rho_p}{\alpha_2} + \frac{\alpha_3}{\alpha_4}\rho_i\right) + \frac{\alpha_5}{\alpha_6}\rho_f \tag{3}$$

To obtain ρ_f (and in turn $\sigma_f = 1/\rho_f$) from the experimental data we first plotted E_o/j_c vs s ($\equiv \alpha_5$). This gave a set of parabolic curves whose level and slope decreased with E_o but had a common intercept of 4×10^{-7} Ω-m. According to Eqn. 3 the slopes of the curves so obtained (i.e. $d(E_o/j)/ds$) give the quantity $[(\rho_f/\alpha_6) + (s/\alpha_6)d\rho_f/ds + (s\rho_f)d(1/\alpha_6)/ds]$. Semilog plots of the slopes vs s for applied field E_o = 1, 2 and 3 kV/mm are presented in Fig. 6.

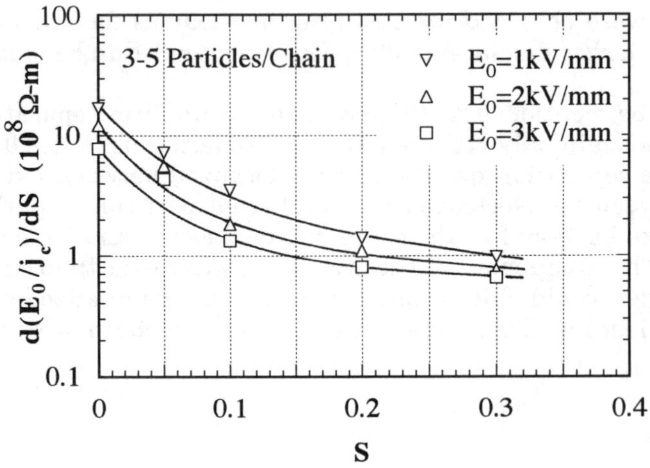

Fig. 6 $d(E_0/j_c)/ds$ vs the separation distance s for 3 to 5 particle chains subjected to applied electric fields E_0 = 1, 2 and 3 kV/mm.

An estimate of the resistivity ρ_f of the oil in the gap between the particles can then be obtained for $s = 0$ from Fig. 6, knowing the value of α_6. For this value we draw on the conductivity model of Tang et al[6] which gives $\alpha_6 \approx 4 \times 10^{-2}$ for humidified glass beads. Using this value for α_6, one obtains $\rho_f = 1.6$–6.0×10^7 Ω-m for E_0 = 3 to 1 kV/mm, respectively, and in turn $\sigma_f = 6 \times 10^{-8}$ and 1.6×10^{-8} S/m. Using Eqn. 2 one can then calculate the local electric field E in the oil between the particles. The resulting local field enhancement E/E_0 ranges from 21 at E_0 = 1 kV/mm to 26 at 3 kV/mm. These values are in accord with those predicted by the conductivity model of Tang et al[6] for a chain of humidified glass beads in silicone oil.

According to Eqn. 3 the ratio E_0/j_c (= 4×10^7 Ω-m) as $s \to 0$ gives the quantity $(\alpha_1\rho_p/\alpha_2 + \alpha_3\rho_i/\alpha_4)$. If we assume as a first order approximation that $\rho_i = \alpha_7\rho_p$, the quantity within the brackets becomes $\rho_p\,(\alpha_1/\alpha_2 + \alpha_3\alpha_7/\alpha_4)$. Furthermore, since the resistivity of the glass (~10^{12} Ω-m)[13] divided by the particle diameter is much larger than that of the water film ($\rho_{H_2O} \approx 10^3$ Ω-m)[11,16] divided by its thickness, the resistivity of the particles $\rho_p \approx \rho_{H_2O}(d/2t)$, so that almost all of the electric current flows in the water film. This then gives for the water film: $\alpha_1 = \pi/2$, $\alpha_2 = \pi t/d$, $\alpha_3 = x_i/d$ and $\alpha_4 = \alpha_6 \approx 4 \times 10^{-2}$. Therefore, one obtains from the measured ratio E_0/j_c (s = 0) the following value for the resistivity term in the brackets on the R.H.S. of Eqn. 3: $\rho_{H_2O}(3.8 \times 10^4 + 4\times10^{-2}\,\alpha_7(x_i/d)) = 4 \times 10^7$ Ω-m. If we make the reasonable assumption that the second quantity is << 3.8×10^4, we obtain $\rho_{H_2O} \approx 1 \times 10^3$ Ω-m. This derived value for the resistivity of the surface water film is in accord with that reported for water films on silica or glass[11,16].

Finally, we wish to compare the current density along a chain of particles j_c with that predicted by the conductivity model of Tang et al[6].

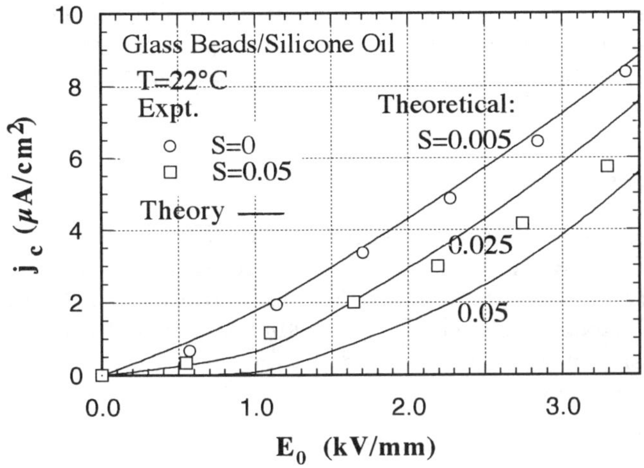

Fig. 7 Comparison of the measured current density j_c along a 4-particle chain with that predicted by the conductivity model of Tang et al[6].

This comparison is given as a function of the separation of the particles in Fig. 7. Good quantitative agreement exists for $s \approx 0$. However, the experimental values of j_c at $s = 0.05$ are slightly higher than those predicted for this separation, being in better agreement with the prediction for $s = 0.025$. This could result from the localized gap which occurs between two adjacent particles as the shear strain is increased, thereby giving a smaller spacing between the remaining particles. The back-and-forth motion of the particles adjacent to the gap could perhaps transfer the charge across the gap. The lower j_c in shear compared to tension for $s > 0.1$ could result from the fact that in shear the field along the chain is at an angle to the applied field.

5. Summary and Conclusions

The following is a summary of the results obtained and the conclusions derived therefrom.

1. The conductivity of a single-row chain of glass beads was three orders of magnitude larger than that of the host silicone oil, but decreased rapidly with separation of the beads produced by tensile or shear straining.
2. The derived conductivity of the water film on the humidified glass beads and the enhanced electric field in the oil between the glass beads were in reasonable accord with those reported in the literature. The measured current density along the chain of beads was in accord with that predicted by the conductivity model.

3. At average strains > ~ 0.1 the separation of the beads in a chain became localized between one set of adjacent beads and they oscillated back-and-forth in the resulting gap, leading ultimately to rupture of the chain with increased strain.

6. Acknowledgements

This research was jointly sponsored by the NSF Engineering Division under Grant CTS–9313897 with Dr. M. C. Roco as technical monitor and the NCSU/Industry Consortium on ER Fluids including Bridgestone, Ford, Paar-Physica and Texaco. Dr. C. W. Wu wishes to express his appreciation to Dalian University of Technology, P. R. C. for their support of his stay at NCSU as a visiting scholar.

References

1. R. A. Anderson, Electrorheological Fluids, R. Tao, ed., World Scientific, Singapore (1992) p. 81–92.
2. L. C. Davis, J. Appl. Phys. 72 (1992) 1334–1340.
3. H. Conrad and Y. Chen, "Electrical Properties and the Strength of Electrorheological (ER) Fluids", Electrorheological (ER) Fluids/Materials, K. O. Havelka, and F. E. Filisko, eds., Plenum Press, New York, in print.
4. J.-N. Foulc, N. Felici and P. Atten, C. R. Acad. Sci. Paris 314 II (1992) 1279–1283.
5. N. Felici, J.-N. Foulc and P. Atten, Electrorheological Fluids, R. Tao and G. D. Roy, eds., World Scientific, Singapore (1994) p. 139–152.
6. X. Tang, C. Wu and H. Conrad, "On the Conductivity Model for Electrorheological Response of Dielectric Particles with a Conducting Film", to be published J. Appl. Phys.
7. X. Tang, C. Wu and H. Conrad, "On the Conductivity Model for the Electrorheological Effect", to be published J. Rheol.
8. A. F. Sprecher, Y. Chen and H. Conrad, Proc. 2nd Int. Conf. Electrorheological Fluids, Technomic, Lancaster, PA (1990) p. 82–92.
9. J. Hill, N. Yaz, D. Klingenberg and C. Zukoski, Electrorheological Fluids, R. Tao, eds., World Scientific, Singapore (1992) p. 280–288.
10. Y. Chen and H. Conrad, Electrorheological Fluids, R. Tao and G. D. Roy, eds., World Scientific (1994) p. 412–420; J. Mod. Phys. B 8 (1994) 2895–2902.
11. L. Holland, The Properties of Glass Surfaces, John Wiley, N. Y. (1964).
12. D. A. G. Bruggeman, Ann. Phys. Lpz. 24 (1935) 636 X referenced in L. K. H. van Beek, Progress in Dielectrics Vol. 7, J. B. Birks, ed., CRC Press, Chemical Rubber Co., Cleveland, OH (1967) p. 69–114.
13. J. Vermeer, Physica XXII (1956) 1257–1268.

14. L. Onsager, J. Chem. Phys. 2 (1934) 599–615.
15. H. Conrad, Y. Li and Y. Chen, "The Temperature Dependence of the Electrorheology and Related Electrical Properties of Corn Starch/Corn Oil Suspensions", submitted to Soc. Rheology Symp. Electrorheology, Ferrofluids and Particulate Flows, Philadelphia, PA Oct. 2–6, 1994.
16. J. B. Hasted, Aqueous Dielectrics, Chapman and Hall, London (1973).

CONDUCTIVITY AND FORCE BETWEEN PARTICLES IN A MODEL ELECTRORHEOLOGICAL FLUID: II INTERACTION FORCE

C. W. WU, Y. CHEN, X . TANG AND H. CONRAD
Materials Science and Engineering Department
North Carolina State University
Raleigh, NC 27695–7907 USA

ABSTRACT

The interaction force between humidified glass spheres forming a single-row chain in silicone oil was determined in shear as a function of the number of particles in the chain, the electric field and the shear strain (separation of the particles). The measured values of the interaction force vs the average distance of separation of the particles and the electric field exponent were in accord with the conductivity model for ER response. With increase in shear strain the separation of the particles soon became localized and one or both of the particles next to the gap began oscillating back-and-forth leading ultimately to rupture of the chain with increased shear.

1. Introduction

In the first paper of this series[1] it was mentioned that the conductivity of the particles and the host fluid play important roles in the response of electrorheological (ER) suspensions at low frequencies of the applied electric field. Recognizing this, the dc current density along a single-row chain in a model ER fluid consisting of humidified glass beads in silicone oil was determined as a function of the number of particles in a chain, the average applied electric field and the separation of the particles produced by straining. In the present paper the interaction force between the particles in a single-row chain of glass beads in silicone oil is determined. The results are compared with predictions of the polarization[2] and conductivity models[3].

2. Experimental

The humidified soda line glass beads (222 μm dia.) and the as-received silicone oil (Dow Corning 200) used in the present experiments were the same as those described in Part I[1]. The cell used here to measure the interaction force between particles in a single-row chain was the same as that employed to measure the current density along a chain. The procedure for measuring the force was similar to that described by Sprecher et al[4]. The shear force F_s acting on a chain of particles exposed to

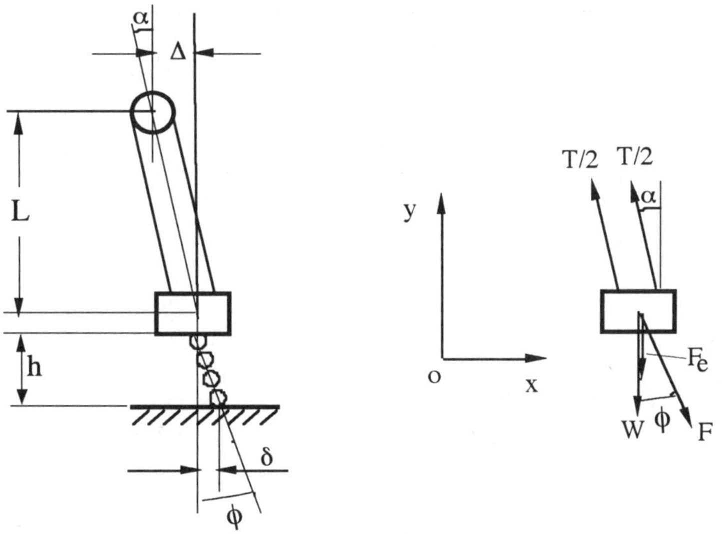

Fig. 1 Schematic of pendulum bob with chain of beads and forces involved.

an electric field is given by (refer to Fig. 1)

$$F_s=(W+F_e)\tan\alpha/(1-\tan\alpha/\gamma) \tag{1}$$

where W is the weight force of the electrode bob and F_e the electrostatic force acting on the bob. In the experiments, the movement distance of the bob Δ was measured, giving $\tan\alpha = \Delta/L$ where L is the length of the suspending wires, and $\tan\phi = \gamma = \delta/h$ where h is the electrode spacing. A brass cantilever beam to which two resistance strain gages had been attached was used to measure the electrostrictive force F_e. All experiments were conducted at ambient conditions, namely temperature T = 22°C and relative humidity RH = 51–53%.

3. Results

The electrostrictive force is given by $F_e = A\varepsilon_0 K_f E_0^2/2$ where A is the electrode area, ε_0 the permittivity of free space, K_f the relative permittivity of the medium (oil) and E_0 the applied electric field[5]. The measured electrostatic force was proportional to E_0^2 as predicted but was ~ 1.5 times larger than predicted. The difference probably results from the fact that the bob is smaller than the cross-sectional area of the cell and therefore the *effective* area is larger than the electrode area A given by the bottom of the bob.

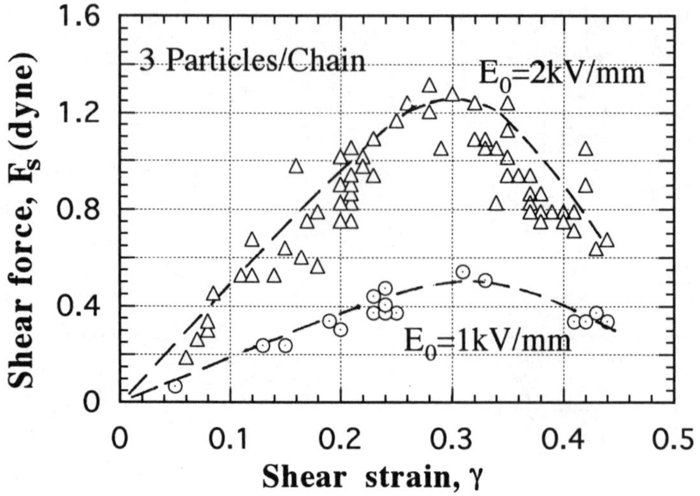

Fig. 2 Shear force vs shear strain for a 3-particle chain with applied electric field E_0 = 1 and 2 kV/mm.

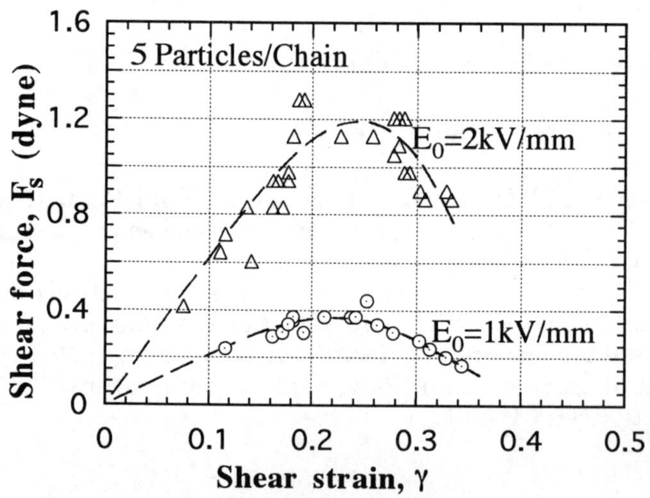

Fig. 3 Shear force vs shear strain for a 5-particle chain with E_0 = 1 and 2 kV/mm.

An example of the shear force F_s vs the shear strain γ at E = 1 and 2 kV/mm is presented in Fig. 2 for a 3-particle chain. The force first increases, reaches a maximum and then decreases. The maximum shear force is about two orders of magnitude smaller than the electrostrictive force. Further, it was relatively independent of the number of particles in a chain (see Fig. 3); however, the shear strain at which the maximum occurred tended to decrease slightly with increase in number of particles. The magnitude of the forces observed here are similar to those reported previously[4]. Since the force F_s to shear a chain was so much smaller than the electrostrictive force F_e meaningful results on the interaction force between particles could not be obtained in tension with the present apparatus. Hence, all measurements of the interaction force between particles were made in shear.

Important to mention is that upon shearing a chain, uniform separation of the particles only occurred at small shear strains ($\gamma < \sim 0.1$). With increase in shear the separation became localized between two adjacent particles located randomly along the chain. When this occurred, one of the particles adjacent to the gap began to oscillate back-and-forth within the gap; see for example Fig. 4. This oscillation was faster than the eye or the video recorder could follow.

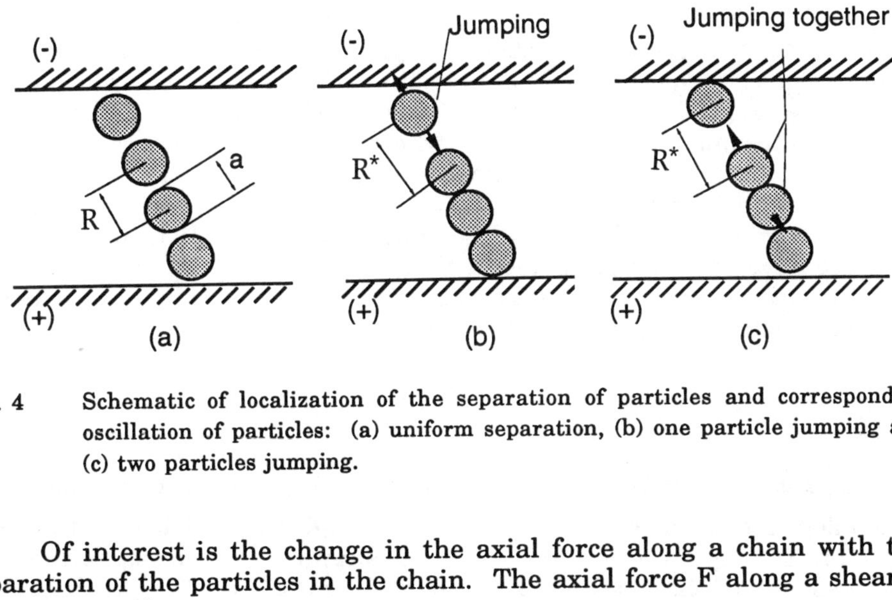

Fig. 4 Schematic of localization of the separation of particles and corresponding oscillation of particles: (a) uniform separation, (b) one particle jumping and (c) two particles jumping.

Of interest is the change in the axial force along a chain with the separation of the particles in the chain. The axial force F along a sheared chain is given by

$$F = F_s / \sin \phi = F_s\left(1+\gamma^2\right)^{1/2} / \gamma \tag{2}$$

where F_s is the shear force, ϕ the angle of the chain from the vertical and γ the shear strain. The *normalized average* separation of the particles is given by

$$s = \Delta\ell / Nd = [(R/a)-2]/2 \tag{3}$$

where $\Delta\ell$ $(= \ell - \ell_0)$ is the measured increase in the length of a chain resulting from shearing, N the number of particles in a chain with diameter $d = 2a$ and R their average center-to-center spacing. s is related to the shear strain γ through $s = (1 + \gamma^2)^{1/2} - 1$. A plot of F vs s for a 4-particle chain is given in Fig. 5 and for a 3-particle chain in Fig. 6. Similar curves were obtained for 5-particle chains. Included in Figs. 5 and 6 are predictions by the conductivity and polarization models to be discussed below.

4. Discussion

The polarization model of Chen et al[2] gives for the interaction force

$$F = 37.44\ \pi a^2 \varepsilon_o K_f (\beta E_o)^2 \exp\{[14.84 - 6.16(R/a)]\beta^2\}/(R/a)^4 \tag{4}$$

where R is the center-to-center spacing of the particles, $a = d/2$ the radius of the particles, ε_0 the permittivity of free space, K_f the relative permittivity of the host oil and E_0 the average applied electric field. $\beta = (K_p - K_f)/(K_p + 2K_f)$ is the dielectric mismatch parameter, where K_p is the dielectric constant of the particles and K_f that of the host oil. A comparison of the measured axial force with that predicted for $K_f = 2.6$ and $K_p = 12.8$[1], for a 4-particle chain is presented in Fig. 5. Reasonable agreement exists between the measured force and that predicted by the polarization model.

The conductivity model of Tang et al[3] gives for the axial force between dielectric particles with a conducting film

$$F = \frac{1}{2}\pi a^2 \varepsilon_o K_f \int_o^{\pi/2} E^2_{f(\theta)} \sin 2\theta d\theta \tag{5}$$

where E is the enhanced electric field in the oil surrounding the particles (which depends on the conductivity of the oil and the particles) and θ is the angle on the sphere surface. Fig. 6 shows that the experimental measurements of the axial force are in good accord with the predictions of the conductivity model with the conductivity ratio Γ $(= \sigma_p/\sigma_f(0)) = 10$, which value was derived from experimental measurements[1]. σ_p is the conductivity of the particles excluding the water film and $\sigma_f(0)$ the conductivity of the oil at low field.

The above indicates that the measured interaction force between the particles in a single-row chain of humidified glass beads in silicone oil is described reasonably well by both the polarization and the conductivity models. However, the measured electric field dependence of the maximum force is closer to the 1.5 power than the quadratic dependence predicted by the polarization model. A value of 1.5 is predicted by the conductivity model

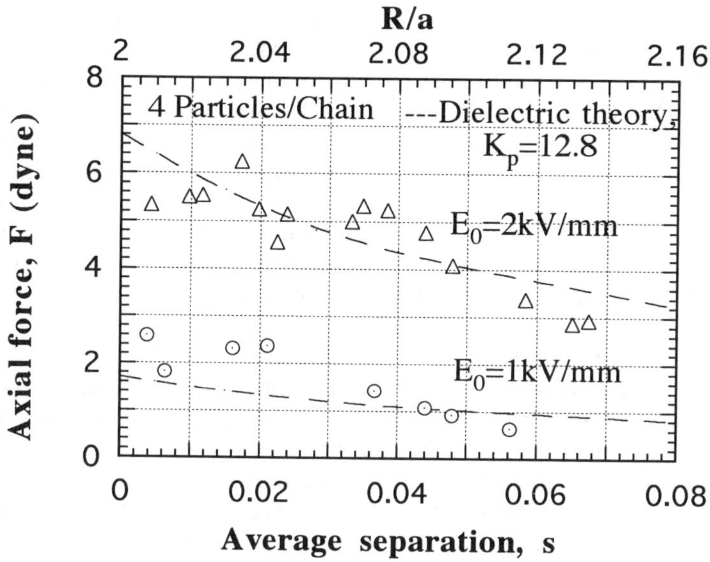

Fig. 5 Comparison of the measured axial interaction force in a 4-particle chain vs R/a with that predicted by the polarization model.

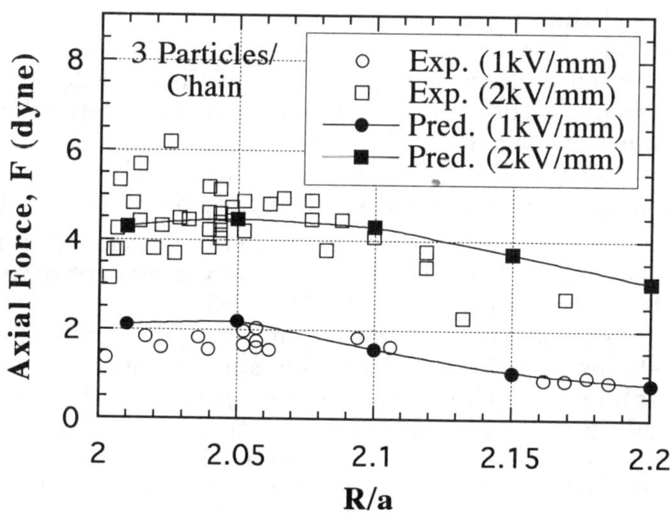

Fig. 6 Comparison of the measured (Exp.) axial interaction force vs R/a for a 3-particle chain with that predicted (Pred.) by the conductivity model.

for the present conditions[3]. Also, the current density predicted by the conductivity model is in accord with that measured[1]. Therefore, it is concluded that the ER response of a chain of humidified glass beads in silicone oil is in better overall accord with the conductivity model than the polarization model. The same conclusion was reached for the water-activated ER suspensions of zeolite particles in silicone oil[7] and corn starch in corn oil[8].

The measured values of the axial force tended to fall below those predicted by the conductivity model at larger separations ($s > \sim 0.1$). This is probably due to the oscillation of the particle in the localized separation illustrated in Fig. 4. With axial loading and zero shear the local electric field and the applied field act on the particles in the same direction. However, when the chain is sheared, the local and applied fields are no longer in the same direction. The larger the shear strain γ the greater will be the angle ϕ ($= \tan^{-1} \gamma$) between the local field E_{loc} and the applied field E_o. Further, an increase in γ causes a larger separation of the particles, giving a smaller E_{loc}[1]. Hence, when the shear strain exceeds some limiting value, the combined electric field causes the particle to move out of the chain with the end result that the chain is broken. The rupture of a chain generally occurred at $\gamma = 0.25 - 0.35$.

6. Summary and Conclusions

The following is a summary of the results obtained and the conclusions derived therefrom.

1. The axial interaction force was proportional to $E^{1.5}$, and decreased with average separation of the particles produced by shearing. The results were relatively independent of the number of particles in a chain.
2. The measured axial interaction force vs separation of the particles was in accord with predictions by both polarization and conductivity models. However, the electric field dependence of the force and the observed current density are in better accord with the conductivity model.
3. At shear strains in excess of ~0.1 the separation of the particles became localized and one or both of the particles neighboring the gap began jumping back-and-forth (oscillating) in the gap. The fact that the interaction force at large separations was smaller than predicted by the conductivity model was attributed to this oscillation, which ultimately led to the rupture of the chain.

7. Acknowledgements

This research was jointly sponsored by the NCSU/Industry Consortium on ER Fluids (including Bridgestone, Ford, Paar Physica and Texaco) and the NSF Engineering Division under Grant CTS–9313897 with Dr. M. C. Roco as technical monitor. Dr. C. W. Wu expresses his appreciation to Dalian University of Technology, PRC for support as a visiting scholar at NCSU.

8. References

1. C. W. Wu, Y. Chen, X. Tang and H. Conrad, "Conductivity and Force Between Particles in a Model Electrorheological Fluid: I Conductivity", this volume.
2. Y. Chen, A. F. Sprecher and H. Conrad, J. Appl. Phys. **70** (1991) 6796–6803.
3. X. Tang, C. Wu and H. Conrad, "On the Conductivity Model for the Electrorheological Response of Dielectric Particles with a Conducting Film", to be published J. Appl. Phys.
4. A. F. Sprecher, Y. Chen and H. Conrad, Proc. 2nd Int. Conf. Electrorheological Fluids, J. D. Carlson, A. F. Sprecher and H. Conrad, eds., Technonic, Lancaster, PA, (1990) p. 82–92.
5. R. L. Reiman, Physics, Barnes and Noble Inc., New York (1973) p. 533–557.
6. X. Tang, C. Wu and H. Conrad, "On the Conductivity Model for the Electrorheological Effect", to be published J. Rheol.
7. H. Conrad and Y. Chen, "Electrical Properties and Strength of Electrorheological (ER) Fluids, Electrorheological (ER) Fluids/Materials, K. O. Havelka and F. E. Filisko, eds., Plenum Press, New York, in print.
8. H. Conrad, Y. Li and Y. Chen, "The Temperature Dependence of the Electrorheology and Related Electrical Properties of Corn Starch/Corn Oil Suspensions", to be published J. Rheol.

MAGNETORHEOLOGICAL MATERIALS BASED ON IRON ALLOY PARTICLES

Anthony J. Margida
Lord Corporation
405 Gregson Dr., Cary, North Carolina 27511 U.S.A.

Keith D. Weiss
Graco Inc.
88 - 11th Avenue N. E., Minneapolis, Minnesota 55440 U.S.A.

J. David Carlson
Lord Corporation
405 Gregson Dr., Cary, North Carolina 27511 U.S.A.

ABSTRACT

A magnetorheological material containing iron alloy particles demonstrates magnetorheological strength dependent upon the elements of the alloy and relative concentration of the alloy elements. Selected iron/cobalt alloys demonstrate improved yield strength over traditional carbonyl iron based MR materials when the iron-cobalt alloy has an iron-cobalt ratio ranging from about 30:70 to 95:5. The iron-nickel alloys which have an iron-nickel ratio ranging from about 90:10 to 99:1 maintains superior strength over iron-nickel alloys outside that range.

1. Introduction

Fluid compositions which undergo a change in apparent viscosity in the presence of a magnetic field are commonly referred to as magnetorheological (MR) fluids. Magnetorheological materials normally are comprised of ferromagnetic or paramagnetic particles, typically greater than 0.1 micrometers in diameter, dispersed within a carrier fluid. In the presence of a magnetic field, the particles become polarized and are thereby organized into chains of particles within the fluid. The chains of particles act to increase the apparent viscosity or flow resistance of the overall material and in the absence of a magnetic field, the particles return to an unorganized or free state and the apparent viscosity or flow resistance of the overall material is correspondingly reduced. These MR fluid compositions exhibit controllable behavior similar to that commonly observed for electrorheological materials, which are responsive to an electric field instead of a magnetic field.

Both electrorheological and magnetorheological materials are useful in providing varying damping forces within devices, such as dampers, shock absorbers and elastomeric mounts, as well as in controlling torque and or pressure levels in various clutch, brake and valve devices. Magnetorheological materials inherently offer several advantages over electrorheological materials in these applications. Magnetorheological fluids exhibit higher yield strengths than electrorheological materials and are, therefore, capable of generating greater damping forces. Furthermore, magnetorheological materials are activated by magnetic fields which are easily produced by simple, low voltage electromagnetic coils as compared to the expensive high voltage power supplies required to effectively operate electrorheological materials. A more specific description of the type of devices in which magnetorheological materials can be effectively utilized is available.[1-3]

Magnetorheological or Bingham magnetic fluids are distinguishable from colloidal magnetic fluids or ferrofluids. In colloidal magnetic fluids the particles are typically 5 to 10 nanometers in diameter. Upon the application of a magnetic field, a colloidal ferrofluid does not exhibit particle structuring or the development of a resistance to flow. Instead, colloidal magnetic fluids experience a body force on the entire material that is proportional to the magnetic field gradient. This force causes the entire colloidal ferrofluid to be attracted to regions of high magnetic field strength.

Magnetorheological fluids and corresponding devices have been discussed in various patents and publications.[4-13]

In many demanding applications for magnetorheological materials, such as dampers or brakes for automobiles or trucks, it is desirable for the magnetorheological material to exhibit a high yield stress so as to be capable of tolerating the large forces experienced in these types of applications. It has been found that only a nominal increase in yield stress of a given magnetorheological material can be obtained by selecting among the different iron particles traditionally utilized in magnetorheological materials. In order to increase the yield stress of a given magnetorheological material, it is typically necessary to increase the volume fraction of magnetorheological particles or to increase the strength of the applied magnetic field. Neither of these techniques is desirable since a high volume fraction of the particle component can add significant weight to a magnetorheological device, as well as increase the overall off-state viscosity of the material, thereby restricting the size and geometry of a magnetorheological device capable of utilizing that material, and high magnetic fields significantly increase the power requirements of a magnetorheological device.

A need therefore exists for a magnetorheological particle component that will independently increase the yield stress of a magnetorheological material without the need for an increased particle volume fraction or increased magnetic field.

Reported herein are magnetorheological materials that utilizes iron-alloy particle components upon which the overall strength of the fluid is dependent. One example is capable of independently increasing the yield stress of the overall magnetorheological material relative to traditional magnetorheological fluids based upon carbonyl iron.

2. Experimental and Materials

2.1 *Example 1: MR Fluid prepared with iron-cobalt alloy*

A magnetorheological material is prepared by initially mixing together 112.00 grams of an iron-cobalt alloy powder consisting of [48%]Fe/[50%]Co/[2%]V obtained from UltraFine Powder Technologies, 2.24 grams of stearic acid (Aldrich Chemical Company) as a dispersant and 30.00 grams of 200 centistoke silicone oil (L-45, Union Carbide Chemicals & Plastics Company, Inc.). The weight amount of iron-cobalt alloy particles in this magnetorheological material corresponds to a volume fraction of 0.30. The magnetorheological material is made homogeneous by a high speed dispersion process. The magnetorheological material is stored in a polyethylene container until utilized.

2.2 *Example 2: MR Fluid "control" prepared with carbonyl iron*

A magnetorheological material control is prepared according to the procedure described in Example 2.1. In this case the particle component consists of 117.90 grams of an insulated reduced carbonyl iron powder (MICROPOWDER R-2521, GAF Chemical Corporation). An appropriate amount of stearic acid and silicone oil is utilized in order to maintain the volume fraction of the particle component at 0.30. This magnetorheological material is stored in a polyethylene container until utilized.

2.3 *Example 3: MR Fluid prepared with iron-nickel alloy*

A magnetorheological material is prepared by initially mixing together 94.25 grams of an iron-nickel alloy powder consisting of [92%]Fe/[8%]Ni obtained from UltraFine Powder Technologies. The fluid can be fully formulated to a particle component fraction of 0.25 with varying amounts of stearic acid, silica, in either silicone or mineral oil. The magnetorheological material is made homogeneous by a high speed dispersion process. The magnetorheological material is stored in a polyethylene container until utilized.

2.4 *Example 4: MR Fluid prepared with iron-nickel alloy*

A magnetorheological material is prepared by initially mixing together 103.38 grams of an iron-nickel alloy powder consisting of [38%]Fe/[42%]Ni obtained from UltraFine Powder Technologies. The fluid can be fully formulated to a particle component fraction of 0.25 with varying amounts of stearic acid, silica, in either silicone or mineral oil. The magnetorheological material is made homogeneous by a high speed dispersion process. The magnetorheological material is stored in a polyethylene container until utilized.

2.5 *Example 5: MR Fluid prepared with iron-cobalt alloy*

A magnetorheological material is prepared by initially mixing together 107.63 grams of an iron-cobalt alloy powder consisting of [48%]Fe/[50%]Co obtained from Micro Materials Technology. The fluid can be fully formulated to a particle component fraction of 0.25 with varying amounts of stearic acid, silica, in either silicone or mineral oil. The magnetorheological material is made homogeneous by a high speed dispersion process. The magnetorheological material is stored in a polyethylene container until utilized.

2.6 *Example 6: MR Fluid prepared with iron-cobalt alloy*

A magnetorheological material is prepared by initially mixing together 101.25 grams of an iron-cobalt alloy powder consisting of [25%]Fe/[75%]Co obtained from Micro Materials Technology. The fluid can be fully formulated to a particle component fraction of 0.25 with varying amounts of stearic acid, silica, in either silicone or mineral oil. The magnetorheological material is made homogeneous by a high speed dispersion process. The magnetorheological material is stored in a polyethylene container until utilized.

2.7 *Magnetorheological Activity*

The magnetorheological materials prepared in the Examples above were evaluated through the use of parallel plate rheometry on a Rheometrics RMS-7200 mechanical spectrometer. The parallel plate fixture was fitted with magnetic circuitry which applied a variable field to the perimeter of the plates. The apparatus was equipped with a 10k gram-centimeter/2k gram transducer.

3. Results and Discussion

As can be seen from the data in Figure 1, the iron alloy particles provide for magnetorheological materials which exhibit substantially higher yield stresses than magnetorheological materials based on traditional iron particles. A summary of the dynamic yield stress values obtained for Example 1 and Example 2 magnetorheological materials at 25°C is provided in Figure 1 as a function of magnetic field. Higher yield stress values are obtained for the magnetorheological material utilizing the iron-cobalt alloy particles (Example 1) as compared to the insulated reduced carbonyl iron powder (Example 2) which is used most commonly in magnetorheological materials. At a magnetic field strength of 6000 Oersted the yield stress exhibited by the magnetorheological material containing the iron-cobalt alloy particles is about 70% greater than that exhibited by the carbonyl iron-based magnetorheological material.

The strength of magnetorheological materials is highly dependent upon the choice of iron alloy and the relative elemental composition. For example, in Table 1 it can be seen that a magnetorheological material based on an iron-nickel alloy with a composition of 92% iron and 8% nickel possesses a higher dynamic yield stress (40.0 kPa) than a material based upon a 38% iron and 42% nickel alloy (26.0 kPa). Similarly, it can be seen that a magnetorheological material based on an iron-cobalt alloy with a composition of 48% iron and 50% cobalt possesses a higher dynamic yield stress (48.0 kPa) than a material based upon a 25% iron and 75% cobalt alloy (38.2 kPa).

Table 1 The dynamic yield stress at 25°C for magnetorheological materials Examples 3-6.

Sample	Alloy Composition	Yield Stress @ 7000 Oe
Example 3	92% Fe / 8% Ni	40.0 kPa
Example 4	38% Fe / 42% Ni	26.0 kPa
Example 5	48% Fe / 50% Co	48.0 kPa
Example 6	25% Fe / 75% Co	38.2 kPa

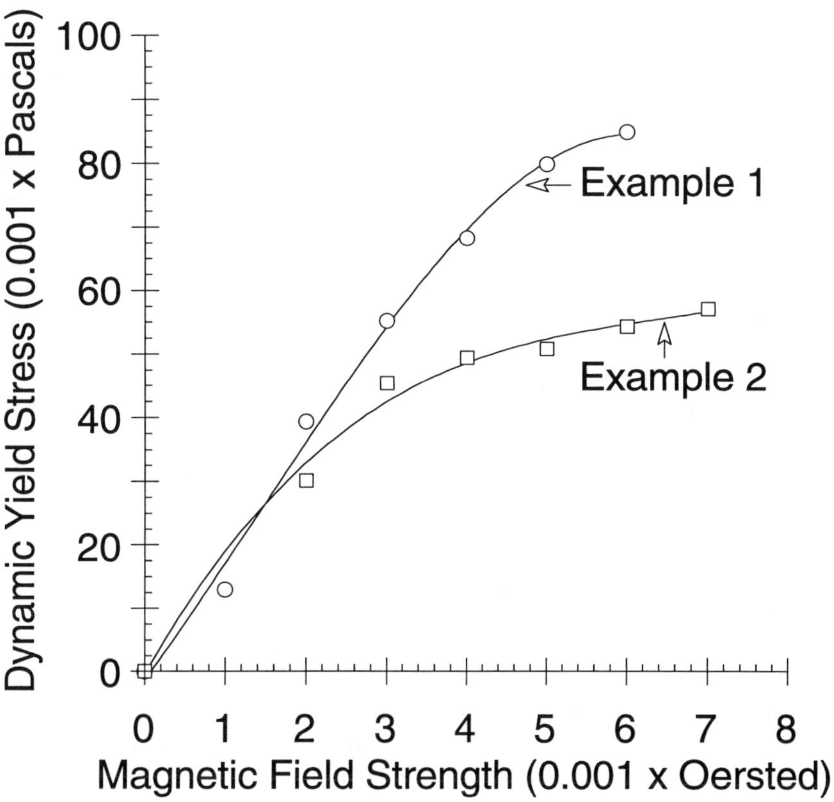

Figure 1 Plot of dynamic yield stress at 25°C as a function of magnetic field strength for magnetorheological materials prepared in accordance with Example 1 and Example 2.

4. Conclusions

A magnetorheological material containing iron alloy particles demonstrates magnetorheological strength dependent upon the elements of the alloy and relative concentration of the alloy elements. Selected iron/cobalt alloys demonstrate improved yield strength over traditional carbonyl iron based MR materials when the iron-cobalt alloy has an iron-cobalt ratio ranging from about 30:70 to 95:5. The iron-nickel alloys which have an iron-nickel ratio ranging from about 90:10 to 99:1 maintain superior strength over iron-nickel alloys outside that range.

5. References

1. J. D. Carlson and K. D. Weiss, US Patent 5,382,373, *Magnetorheological Materials based on Alloy Particles,* (1995).
2. J. D. Carlson and M. J. Chrzan, US Patent 5,277,282, *Magnetorheological Fluid Dampers,* (1994).
3. J. D. Carlson, M. J. Chrzan and F. O. James, US Patent 5,284,330, *Magneto-rheological Fluid Devices,* (1994).
4. Kirk-Othmer Encyclopedia of Chemical Technology, vol. 14, pp. 662-664, (1981).
5. J. Rabinow, "Technical News Bulletin," vol. 32, No. 5, pp. 54-60, issued by U.S. Dept. of Commerce, May, 1948 describing a magnetic fluid clutch.
6. B. A. Bondi, US Patent 2,751,352, *Magnetic Fluids*, June 19, 1955.
7. R. Kaiser, US Patent 3,700,595, *Ferrofluid Composition,* October 24, 1972.
8. F. Bergmann, US Patent 2,847,101, *Overload Releasing Magnetic Power-Clutch,* August 12, 1958.
9. M. S. Chagnon, US Patent 5,147,573, *Superparamagnetic Liquid Colliods,* September 15, 1992.
10. K. Furumura and S. Matsunaga, US Patent 32,573,*Process for producing a ferrofuid, and a composition thereof,* January 5, 1988.
11. K. Ogawa, US Patent 5,013,471, *Magnetic Fluid, Method for Producing it and Magnetic Seal Means Using the Same*, May 7, 1991.
12. E. M. Shtarkman, US Patent 4,992,190, *Fluid Responsive to a Magnetic Field*, Febuary 12, 1991.
13. R. E. Rosensweig, US Patent 3,917,538, *Ferrofluid compositions and process of Making Same*, November 4, 1975.

THE TIME RESPONSE OF AN ELECTRO-RHEOLOGICAL FLUID IN A HYDRODYNAMIC FILM

T. H. LEEK, S. LINGARD, W. A. BULLOUGH, R. J. ATKIN

The work is a practical investigation of an electro-rheological fluid in a hydrodynamic film using a Rayleigh step bearing in which a step electric field is applied across the fluid film. A number of different bearings are used, and all data presented is for the repeatable condition of zero net flow rate. Earlier work on steady state conditions is extended to high speed data logging of the pressure generation response to DC step excitations of various magnitudes.

1. Introduction

There is a considerable amount of literature available concerning the behaviour of electro-rheological fluids in Couette flow and in Poiseuille flow. The current work explores another possibility, being an examination of the practical behaviour of an ER fluid in combined Couette and Poiseuille flow, that is in a hydrodynamic film, a concept that has seen little theoretical analysis[1-3] and experimental investigation[4-6]. The most obvious practical application for such a study is in active ER fluid bearings.

In this work the practical investigation of an ER fluid in a hydrodynamic film is extended from the steady state performance previously examined to the time response characteristics of the ER effect when subjected to DC step excitation.

2. Experimental apparatus - physical arrangement

The required flow condition was achieved using a Rayleigh step bearing. The arrangement is illustrated in Fig. 1 and is similar to that used previously in steady state tests[4-6], although improvements have been made to the bearing design and to temperature stability and monitoring.

A part circular bearing with fixed side walls was loaded against a rotating disc which draws fluid out of a reservoir and into the bearing. The bearing consisted of a piece of conducting material (stainless steel) sandwiched between two pieces of perspex, the latter providing the side walls of the fluid film and electrical insulation. The electric field was applied across the hydrodynamic film by the application of a step D.C. voltage to the metallic portion of the bearing body.

The load and seal fit were such as to approximate the limiting condition of zero net flow rate through the bearing whereby the seal and side walls of the bearing are pressed firmly against the rotating disc; such a condition was chosen as it is easily reproduced. Thus in all tests, with reference to Fig. 2, h_2 is zero and the film thickness is given by the measurement of bearing dimensions h_1 and h_3. Pressure in the bearing is measured by a pressure tapping as shown in Fig. 2 and a semiconductor strain gauge pressure transducer. A capacitive proximity transducer, positioned above the arm carrying the bearing, was used to monitor any relative vertical displacement between the bearing and the rotating disc, *i.e.* any change in fluid film thickness, and another transducer recorded the speed of the rotating disc.

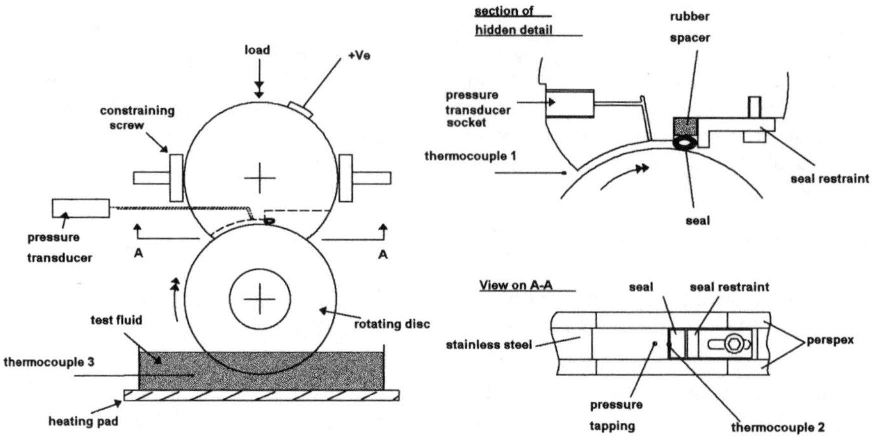

Fig. 1. Experimental arrangement.

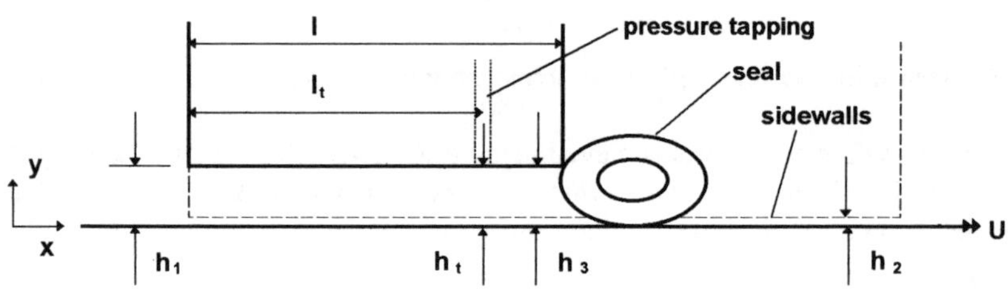

Fig. 2. Nomenclature

3. Equipment specifications and fluid properties

Rotating disc	∅ 58.5 mm, width = 15 mm. Constant 207 rpm, $\Rightarrow U = 0.634\ ms^{-1}$.
Pressure transducer	Kyowa PGF10KC, semiconductor strain gauge, ∅ 6.5 mm pressure face, capacity = 10 kgf/cm^2 = 981 kNm^{-2}, pressure (kNm^{-2}) = 508.7 × voltage on data logger, natural frequency = 30 kHz.
Pressure tapping	originally ∅ 0.5 mm.

High voltage	$V = 112.6 \times$ signal volts, maximum used = 700 V.
Speed transducer	58 pulses per revolution at 207 rpm = 200.1 Hz
Displacement transducer	1 volt $\approx$ 0.15 mm (calibrated each time used)
Data logger speed	minimum of 2 kHz (maximum 0.5 ms intervals)
High voltage rise time:	average 6 MVs^{-1}.
High voltage fall time:	average 3 MVs^{-1}.
Power supply :	0 - 5 kV, 0 - 5 mA, d.c.
Thermocouples :	type K, Chromel-Alumel.

Table 1. Bearing dimensions; l and l_t are arc lengths.

Bearing	w (mm)	l (mm)	l_t (mm)	h_1 (mm)	h_3 (mm)
11	8.00	16.45	14.03	0.156	0.180
12	8.00	16.41	14.02	0.138	0.184
13	8.04	16.69	14.05	0.322	0.347
14	8.00	16.46	13.92	0.267	0.284

Two fluids were used in the tests, ERF1 and ERF2. ERF1 is used in all tests unless indicated otherwise. Fluid properties are as follows:

ERF1 : Dielectric liquid, lithium polymethacrylate particles with sizes of between 2 and 20 μm, a mean size of 5 μm, and a volume fraction of 30%. Water content of particles = 15 - 18 % by weight. Nominally, for the unexcited fluid

$$\eta \ (\text{Pa s}) = 1.43 \times 10^{-10} \ e^{6203/\theta} \ . \tag{1}$$

ERF2 : Cereclor carrier liquid 50 LV with lithium polymethacrylate particles diluted to volume fraction 30 %. Water content of particles 14.5 %.

4. Integrity of equipment

The integrity of the general experimental arrangement for steady state conditions has been established previously[4-6]. All equipment was thoroughly checked and/or calibrated to ensure that all measurements were true to acceptable accuracy.

The manufacturers specify a natural frequency for the pressure transducer of 30 kHz, giving a frequency response in the range of 6 kHz. This is more than adequate for the measurement of events lasting milliseconds. The circuitry of the strain gauge amplifier displayed an exponential output with a time constant of 0.17 ms in response to a step input of 2 volts ($\equiv$ 1017.4 kNm^{-2}). This time constant is negligible in our tests.

5. Experimental results

The pressure response data for all bearings collected in initial tests, at both switch-on and switch-off, is typified by the data shown in Fig. 3 where pressure for each curve is normalised by the pressure at zero volts. The temperatures indicated on the figures in this section are averages for the data shown, variation is of the order of ± 0.8 °C and as such is not important to the time response data. The average pressure at zero volts, $\overline{p}_{V=0}$, is given with each figure.

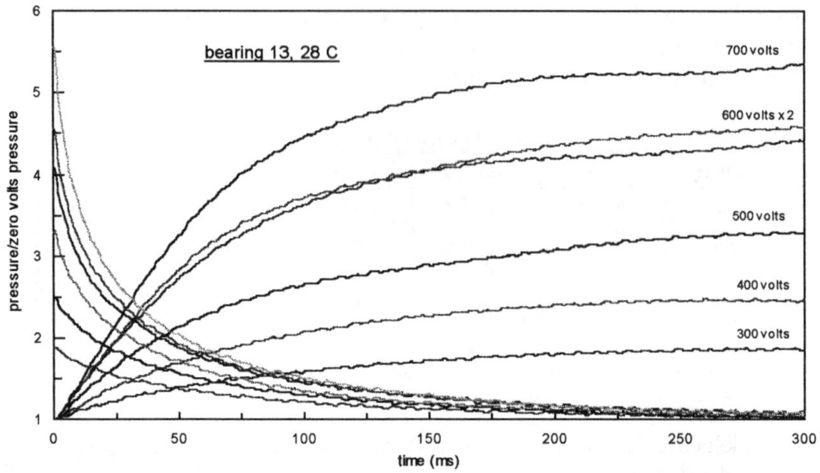

Fig. 3. Example of pressure response data at switch-on and switch-off using ∅ 0.5 mm pressure tapping, voltage switching times < 0.5 milliseconds, $\overline{p}_{V=0} = 63$ kNm^{-2}.

Although not apparent in Fig. 3 the response was preceeded by an initial period of constant pressure immediately following voltage switch-on or off which was approximately 3 ms in duration. This initial delay was not specifically studied, furthermore the absence of perfectly steady signals and the finite data sampling rate used means that only approximations for the duration of this period are given.

Since the time response profile shown in Fig. 3 was considerably slower than expected the experiment was examined for possible reasons as follows:

- No lifting off of the bearing from the disc was detected by the displacement transducer.
- No change in speed was detected by the speed transducer.
- The effect of any compliance in the rubber seal was examined by testing a bearing made with a solid PTFE seal, response results remained similar to Fig. 3.
- The use of a different pressure transducer made no difference.

The initial pressure tappings were ∅ 0.5 mm and to investigate the effect of pressure tapping size all tappings were increased to ∅ 1.0 mm and then ∅ 2.0 mm. The increase from 0.5 mm to 1.0 mm substantially reduced the response time. Increasing the pressure tapping from ∅ 1.0 mm to ∅ 2.0 mm further increased the speed of pressure response, but to a far lesser degree. Examples of the data, which were repeatable, are given in Figs. 4

and 5. The initial no response period was now 1 ± 0.5 ms and is not fixed to greater accuracy because much of the data was collected at a sampling interval of 0.5 ms.

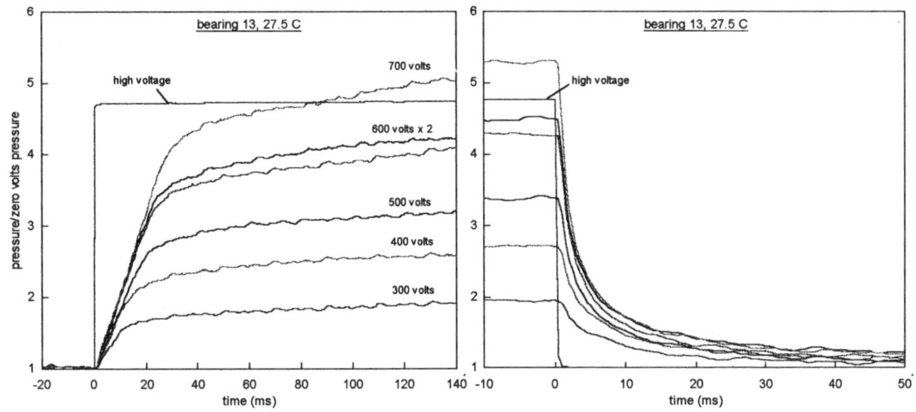

Fig. 4. Pressure response for bearing 13 with ∅ 2.0 mm pressure tapping, $\overline{p}_{V=0} = 63$ kNm^{-2}.

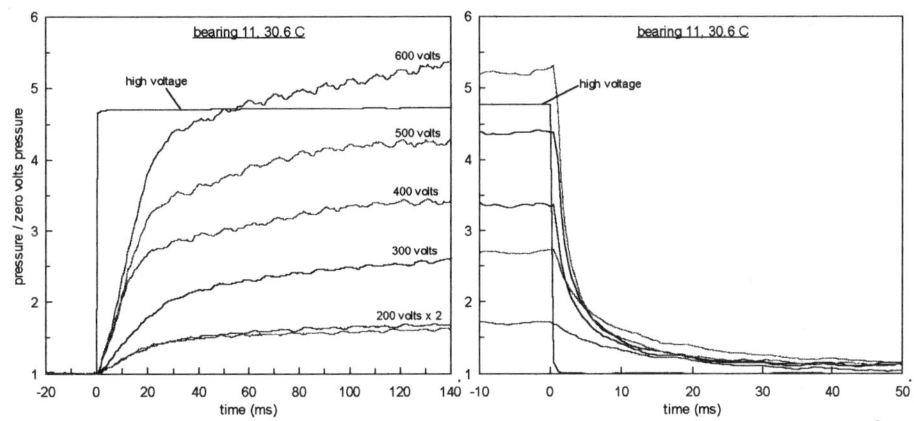

Fig. 5 Pressure response for bearing 11 with ∅ 2.0 mm pressure tapping, $\overline{p}_{V=0} = 125$ kNm^{-2}.

With the new (∅ 2.0 mm) pressure tapping size the bearings all provided a similar pressure response. At switch-on, and after the brief no response period, data showed a rapid increase in pressure towards a knee in the curve at around 20 to 30 ms, followed by a slower rise lasting between 150 and 200 ms towards the final pressure value. At switch-off, again after an initial no response period, a smooth exponential decay in pressure occurs with a time constant of approximately 7 ms, approximately twice as fast as the pressure change at switch-on.

Since the pressure changes along the bearing length, most likely in a non-linear fashion, an increase of pressure tapping size may not result in the same steady state values of pressure being recorded. This explains the small differences that did in fact arise.

Consideration of experimental results indicated that re-doubling the pressure tapping size to ∅ 4.0 mm would have only a marginal, if any, effect upon the pressure response time. The pressure tapping effect upon measured response times is not a linear relationship. Given this, together with the knowledge that the larger the pressure tapping the less the likelihood of measuring hydrodynamic pressures, no further changes were made to the pressure tappings.

Apart from pressure tapping effects the significantly faster pressure response shown in Figs. 4 and 5 over that shown in Fig. 3 requires that other factors in the experiment be investigated once again. Evidence exists, in the form of distinctly different pressure rise and fall profiles, of there being no significant compliance in the experimental arrangement. Furthermore there is no obvious proportionality between the widely different pressures and pressure changes, seen in the different bearings at different applied voltages, and the time response profile. Nevertheless the system was thoroughly investigated.

Examples of all data signals at switch-on and switch-off are shown in Fig. 6. The study of such data indicated no movement of the bearing or change in speed of the rotating disc in response to the rapid pressure change. The use of a compliant rubber seal in the bearings is perhaps the most obvious potential source of errors. Data was collected using a bearing with a solid PTFE seal (otherwise similar to bearings 11 to 14) which indicated that the rubber seal had no undesirable influence. Any unwanted effects of the pressure transducer itself were also discounted by tests conducted using a different pressure transducer which showed no difference in the measured pressure response.

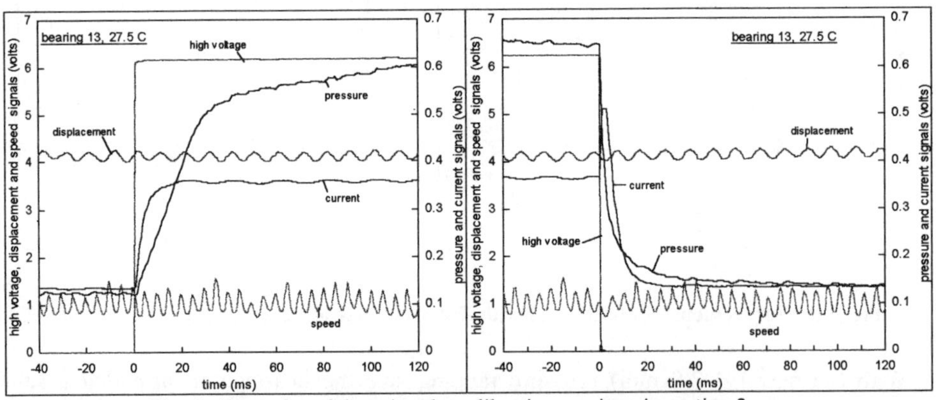

Fig. 6. Examples of data signals, calibration as given in section 3.

The effect of an impulsive disturbance applied to the bearing/disc loading arrangement on the measured pressures is illustrated in Fig. 7. The displacement signal is that provided by the proximity transducer described earlier. No perturbation to the speed or applied voltage signals was detected in response to the disturbance.

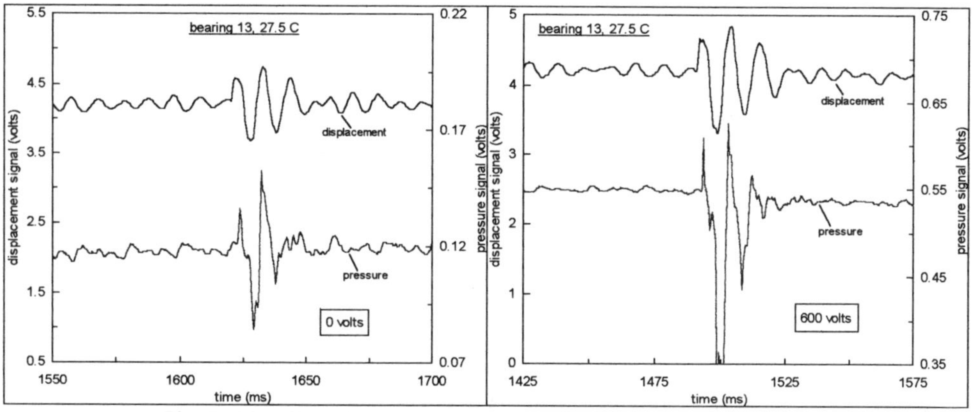

Fig. 7. The effect of a disturbance using pressure tapping of ∅2.0 mm.

The relationship between pressure and disturbance is complex and unknown. However, any disturbance can affect both the condition of net zero flow rate and the film thickness (and hence electric field strength) so a corresponding, and in some way proportional, perturbation in the measured pressures is expected. Fig. 7 shows that in both the unexcited and excited conditions the pressure signal follows the displacement signal, occurring at a frequency of 87 Hz (*i.e.* rise times of approximately 6 ms). This is evidence that the measured pressure response to step voltage, being of the order of 20 ms, is not subject to undesirable influences.

To provide an indication of the importance of which ER fluid is used data was obtained using ERF2 and is shown in Fig. 8. In general ERF2 behaved in a similar fashion to ERF1, showing a period of rapid response of approximately 20 ms followed by a period of slow rise to a final value. However, looked at in more detail two characteristics are apparent in Fig. 8. The knee in the pressure curve is sharper than occurs using ERF1 and the period of slow rise is longer, being approximately 300 to 400 ms as compared to the previous 150 to 200 ms; these are attributed to fluid differences.

6. Analysis and discussion of pressure response results

At present there are no standardised procedures defining ER fluid response times and it is likely to be some time before any are formulated. In this work the time dependent variable is the pressure generated in the hydrodynamic film, whilst in an ER fluid clutch it is the torque response to excitation[7], and in a more fundamental examination it is the formation of the first fibril in a static ER cell[8]. Interestingly, in application based studies there is no understanding of whether the time response depends upon the application itself, *i.e.* does the same fluid have different response behaviour in different flow regimes? This question together with the present absence, to the authors' knowledge, of response times for the hydrodynamic film regime provides the principal thrust behind the current work.

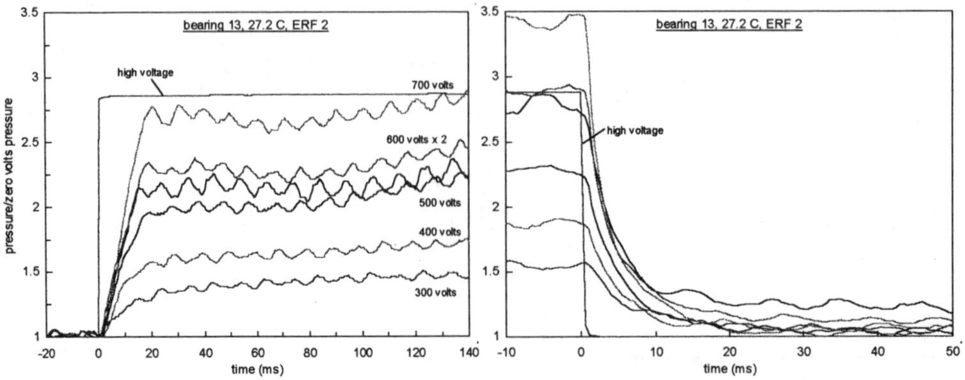

Fig. 8. Data collected using ERF2, $\bar{p}_{V=0} = 99$ kNm^{-2}.

Time response, in its various definitions, is reported to vary with the following factors: fluid composition, shear rate, temperature, solid phase volume fraction, and electric field strength. The variety of methodologies and studies undertaken have led to reported values for time response ranging from less than one millisecond to tens of seconds. In application based studies such as an ER clutch or valve[7,9,10-14] the switch-on response of an ER fluid to a step excitation is characterised by three distinct stages:

- Stage 1 : No discernible response.
- Stage 2 : A period of rapid rise- sometimes accompanied by a small overshoot.
- Stage 3 : A period of slow asymptotic rise to a final value.

The switch-off regime is not as often examined as switch-on. In ER fluid valve tests[12,14] switch-off was a number of times faster than stage 2 of the switch-on response. This is perhaps intuitively expected in that at switch-on both attractive forces between particles and structuring of the particles is predicted whereas switch-off requires only the relaxation of the attractive forces. As shown in Figs. 4 and 5 the pressure response at switch-on measured in this work corresponds to the above three stage behaviour, with no overshoot at the end of stage 2. In addition switch-off response was significantly faster than that at switch-on.

The slower switch-on than switch-off performance concentrates attention on the former, indeed the switch-off can be accurately characterised by a straightforward exponential decay with time constant of approximately 7 ms. For comparison purposes the pressure response profile at excitation is approximated by three straight lines corresponding to the three stage response listed above. This allows the use of the parameters t_2, r_p, r_t, $\dot{p}$.

Since in stage 3 the pressure rises asymptotically to its steady state value and it is subject to small variations at the frequency of rotation of the disc, accurate values of t_3, the time at which pressure reaches its steady state value, have not been achieved. Similarly the duration of stage 1, t_1, is also only approximated as already explained. The approximations for t_1 and t_3 are :

$$t_1 \approx 1 \pm 0.5 \text{ ms for both ERF1 and ERF2.}$$
$$t_3 \approx 170 \text{ to } 220 \text{ ms for ERF1, } 320 \text{ to } 420 \text{ ms for ERF2.} \quad (2)$$

Values for t_2, r_p, and $\dot{p}$ are shown in Fig. 9. Although data for bearing 14 is included in the figures this has been identified as being unreliable since the pressure tapping was not as required.

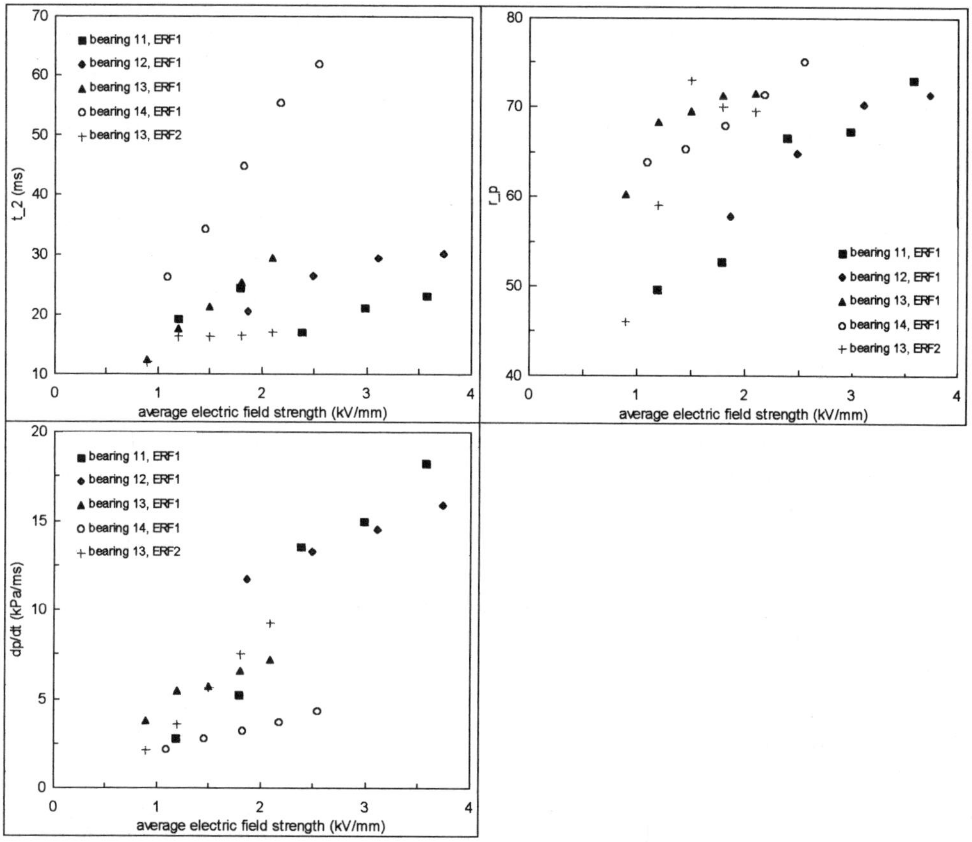

Fig. 9 t_2, r_p, and $\dot{p}$ against the average electric field strength, $\overline{E}$, for the four bearings. The temperature range of the data shown is 27.2 to 33.4 °C. $\overline{E}$ is defined as $= 2V/(h_1 + h_3)$.

The trends in the data are for all of t_2, r_p, and $\dot{p}$ to increase with the level of the applied electric field and thus with Δp. This variation in t_2 is unfortunate since in practical applications relying upon a fast time response anything other than the smallest t_2 is detrimental to performance. However this is assuaged somewhat by the increase of r_p and $\dot{p}$ with E which, although complicating control procedures, indicate that improved time response can be achieved using a two stage application of electric field. This would involve the application of a very high electric field for approximately 20 ms followed by a reduction to the actual electric field required.

Comparison of bearing 13 data for the two ER fluids suggests that t_2 and r_p are, like t_3 fluid dependent. This is not unexpected in the light of the influences, reported in the literature, of fluid constituents.

The different film thicknesses in each bearing provide different nominal shear rates and this is one of the factors which has influenced time response in other work. As for fluid dependence, an exhaustive examination of shear rates has not been performed, nevertheless Fig. 9 does suggest that r_p tends to be smaller the larger the shear rate (*i.e.* the larger the film thickness). This does not bode well for any practical hydrodynamic film application of ER fluids which would most likely involve much thinner films and hence much larger shear rates than used in this work.

Eq. 2 and Fig. 9 provide the following approximations for r_t (neglecting bearing 14 data) :

$$r_t \approx 5 \text{ to } 17\,\% \quad \text{for ERF1}$$
$$r_t \approx 3 \text{ to } 5\,\% \quad \text{for ERF2} \qquad (3)$$

The larger the value of r_t the better, but only if accompanied by a small value of t_2. The opposite occurs in the data presented, that is a larger r_t has a larger t_2.

Comparison of the results in Fig. 9 with other flow regimes would ideally involve a dedicated experimental programme. In the absence of such, some data for the same and similar (same solid phase) ER fluids as used in this work is taken from the literature[7,10-12] and shown in Table 3 where r_p, $\dot{p}$, and Δp refer to pressure drop across the ER valve and r_n refers to torque in the ER clutch. It should be noted that the clutch results are not direct experimental recordings but are experimental data after theoretical manipulation to account for compliance in the test rig.

Table 3. Comparison of time response data in different flow regimes.

parameter:	t_1 (ms)	t_2 (ms)	t_3 (ms)	r_n	$\dot{p}$ (kPa/ms)	Δp (kPa)
ER valve:	≈ 0.5	5 to 10	≤ 1000	≤ 75 %	300	2500
ER clutch:	0 to 0.5	≤ 6.4	≤ 800	≈ 65 %		
bearings:	≈ 1	11 to 30	≤ 420	≤ 75 %	3 to 18	< 700

The data from the literature presented in Table 3 varies with shear rate, electric field strength and temperature, but the examples chosen cover ranges similar to those used in this work. There is general agreement between the t_1 and r_p values shown in Table 3. However t_2 in the hydrodynamic film regime is consistently the largest, and t_3 is consistently the smallest. These may well be flow effects.

$\dot{p}$ for a valve is an order of magnitude larger than in the bearings but is accompanied by a large Δp. Since dependence of $\dot{p}$ upon Δp has been identified in this work a better comparison is the ratio $\dot{p}/\Delta p$:

$$\begin{array}{lrcl} \text{valve} & \dot{p}/\Delta p & = & 0.12 \\ \text{Bearing tests} & \dot{p}/\Delta p & = & 0.02 \text{ to } 0.1 \end{array} \tag{4}$$

Two potential influences upon pressure response have not yet been considered, being the compressibility of the solid phase and the presence of air bubbles in the fluid. In both cases the circumstances of the experiment are such as to rule out these possibilities. The data for the ER valve presented in Table 3 is for a fluid with the same solid phase as used in the bearing tests and involves both a significantly faster rate of change of pressure and a significantly greater change in the pressure itself, yet compressibility has not been identified as an influence. Furthermore the compression of particles would be proportional to pressure, an effect not seen in the experimental results. If air bubbles were present they would introduce considerably more variability into the data than was seen, and significantly longer response times; as an example small proportions of air, 0.5% by volume, have been thought to result in t_2 larger than 100 ms in an ER valve[15].

7. Conclusions

The time response of the pressure in an ER fluid hydrodynamic film to a DC step excitation has been investigated to the extent discussed in the previous section. Considerable efforts have been made to ensure that the pressure responses measured do not include significant compliance errors. The authors cannot conceive of any further unexamined aspect of the experiment which may cast doubt upon the validity of the pressure response data obtained and would be pleased to hear any comments concerning any such possibility.

The pressure response of an already excited ER fluid copes well with small transients (Fig. 7), and thus possibly also with larger transients. This is potentially a useful feature, and one which has not been fully examined as yet. The idea is that time response behaviour may be improved when the fluid is already excited, and thus is structured to some degree, and the change made to effect a response is from low to high electric field strength rather than simply on-off. In the example of a journal bearing such an arrangement is practical and, in fact, advantageous. Without any control circuitry other than a constant level of excitation (constant volts) any sudden eccentricity in the bearing is countered not only by the normal hydrodynamic behaviour but also by the fact that eccentricity reduces film thickness and thus increases electric field strength which in turn increases the yield shear stress of the ER fluid. Thus the correcting forces to a sudden eccentricity are automatically enhanced if the electric field strength-yield shear stress-pressure effect is fast enough. Note that the use of constant excitation, even if small, in ER fluid clutch and valve applications would disadvantage performance, whilst with appropriate design addressing fluid properties and heating effects etc. this is not true in the journal bearing application (other than power consumption). It is also interesting to note that the voltage levels required to activate any practical lubricating film would be of the order of tens rather than hundreds of volts due to the small film thickness. This would greatly simplify the electronic circuitry required.

There are facets of this work which require the illumination of further investigation. As a result authoritative conclusions on some aspects of the present study are reserved, nevertheless some useful data has been obtained and lessons learned, our observations are as follows :

1. The ER effect enhances hydrodynamic pressures, and this is likely to be greater in the more realistic thinner films that practical use would demand.
2. The measurement of pressures within an excited ER fluid flow requires careful attention.
3. The general profile of time response of an excited ER fluid in combined Couette and Poiseuille flow is similar to that seen in pure Couette flow and pure Poiseuille flow.
4. The fall in pressure at switch-off of the electric field is faster than the rise at switch-on. Where time response is important it is restricted by switch-on performance.
5. The stage 2 pressure response in the bearings is of the order of 20 to 30 ms and a significant change of pressure can be achieved in this time. It appears to be slightly slower than occurs in an ER clutch and in an ER valve and is apparently a feature of the flow regime.
6. The steady state condition in the bearing is reached quicker than that in a clutch or valve.
7. A higher rate of increase in pressure is achieved by the imposition of a larger electric field. The most rapid method of achieving a particular pressure is to use a two stage voltage control. This would involve the brief application of as high an electric field as possible followed by a reduction to the actual electric field required.
8. Shear rate and the fluid used have influences upon the pressure response behaviour.

Areas of further interest include temperature effects, speed/shear rate effects (the data in this work is for a constant speed of the rotating disc), alternative methods of pressure measurement, and the behaviour when changing from one electric field strength to another, *i.e.* the response of an already 'structured' fluid, rather than straightforward on-off. Analyses of the steady state pressures generated in the bearings in this work are to be published elsewhere.

Acknowledgements
The authors would like to acknowledge the financial support of The Croucher Foundation, The University of Hong Kong, and The British Council.

Nomenclature

$\overline{E}$	average electric field strength
h_1, h_t, h_3	film thickness at bearing entrance, pressure tapping, and step respectively
l	total curved length of bearing
l_t	curved length of bearing to pressure tapping

p_t	pressure at the pressure tapping
$\bar{p}_{V=0}$	average zero volts pressure
$\dot{p}$	rate of change of pressure with time during stage 2, (= dp/dt)
Δp	change in pressure on the application of electric field
r_p	pressure at end of stage 2 as a percentage of final pressure
r_t	t_2 as a percentage of the time for pressure to reach a steady value
t_n	the time in milliseconds at the end of stage n response, (n=1 to 3).
U	linear velocity of moving surface
V	applied voltage
w	bearing width
η	Newtonian viscosity
θ	absolute temperature

References

1. J. A. Tichy, *J. Rheol.* **35**(4), 1991, pp. 477-496.
2. A. D. Dimarogonas and A. Kollias, *J. of Tribology,* Trans. ASME, June 1992.
3. T. H. Leek, R. J. Atkin, S. Lingard, and W. A. Bullough, (to be published).
4. T. H. Leek, S. Lingard, W. A. Bullough, and R. J. Atkin, *J. Phys. D: Appl. Phys.* **26** (1993) pp. 1592-1600.
5. T. H. Leek, S. Lingard, W. A. Bullough, and R. J. Atkin, *Electrorheological Fluids - Proc. 4th Int. Conf. ER fluids,* Feldkirch, Austria, 1993 (World Scientific Pub. Co., 1994) pp. 608-624.
6. T. H. Leek, S. Lingard, W. A. Bullough, and R. J. Atkin, ibid., pp. 625-642.
7. W. A. Bullough, J. Makin, R. Firoozian, A. R. Johnson, and A. Hosseini-Sianaki, *Int. J. of Mod. Phys. B*, **8,** 20 & 21, (1994), pp. 2855-2876.
8. J. C. Hill, T. H. Van Steenkiste, *J. Appl. Phys.*, **70** (3), (1991), pp. 1207-1211.
9. M. Whittle, R. Firoozian, W. A. Bullough, and D. J. Peel, as ref. 5, pp. 421-439.
10. W. A. Bullough, R. Firoozian, A. R. Johnson, A. Hosseini-Sianaki, J. Makin, and X. Shi, *Mechatronic Systems Engineering*, **1**, (1992), pp. 315-327.
11. A. Hosseini-Sianaki, W. A. Bullough, R. Firoozian, J. Makin, and R. C. Tozer, *Proc. Int. Conf. ER fluids*, S. Illinois Univ. USA, 1991, (World Scientific Pub. Co., 1992), pp. 219-238.
12. D. J. Peel, W. A. Bullough, *J. Intell. Mat. Sys. & Struct.*, **4**, (1993), pp. 54-64, & associated publications.
13. K. Weiss, D. Carlson, as ref. 11.
14. M. Whittle, R. Firoozian, D. J. Peel., and W. A. Bullough, *Proc. conf.: Recent advances in adaptive and sensory materials and their application*, Blacksburg Va. USA, 1992, (Technomic), pp. 495-505, & associated publications.
15. M. Nakano, T. Yonekawa, as ref. 5, pp. 477-489.

BOUNDARY EFFECTS AT ERF SPREADING IN ELECTRIC FIELD

E.V.KOROBKO, R.G.GORODKIN and V.V.MELNICHENKO
A.V.Luikov Heat & Mass Transfer Institute, Belarussian Academy of Sciences
15 P.Brovka St., Minsk 220072, Belarus

and

L.W.ZHOU
Fudan University, Shanghai 200433, People's Republic of China

ABSTRACT

Results of the experimental investigation of the electrorheological fluid flow with a free boundary, in particular, drop spreading over a substrate in an electric field are reported. Residual thixotropic effects and a change in the drop shape have been revealed. A dependence of the characteristic parameters of the drop (its height, radius and wetting angle) on an electric field intensity and observation time have been established.

1. Introduction and Formulation of the Problem

Investigation of liquid spreading over horizontal surfaces is important for many industrial applications, for instance, painting, lubrication, surface cleaning, as well as for examination of adhesion effects and processes that occur in lithographic and xerographic printing. A great number of experimental[1,2] and theoretical works[3,4] have been published on this problem since the pioneering research[5] performed by Hardy in 1919. Recently, the development of new types of fluids with abnormal rheological properties, which can be controlled by external action, for example, by an electric field, have opened up new areas for research.

It is well-known that at equilibrium conditions a fluid spreads due to interfacial tensions determined by physicochemical processes that occur at the phase boundaries. For a slow flow, the Young's equation is valid for an any instant of time

$$\sigma_{sg} - \sigma_{sl} = \sigma_{lg} \cos\theta_o \tag{1}$$

where σ_{sg}, σ_{sl} and σ_{lg} is the surface tension between a solid a gas, a solid and a liquid, and liquid and gas phases, respectively, θ_o is the equilibrium wetting angle.

The contact surface area between liquid and solid phases under non-equilibrium spreading conditions changes quickly and therefore the surface tensions may be treated as constants, the wetting angle, θ, changes and differs from the equilibrium value, θ_o. The motive force for spreading is determined mainly by a decrease of the surface free energy of the system. A force acting on the unit length of the contour of the liquid on the contact line is expressed as

$$\Delta\sigma = k(\sigma_{sg} - \sigma_{sl}) - \sigma_{lg}\cos\theta \qquad (2)$$

where k is the coefficient of the surface roughness.

The flow regimes in the case of non-equilibrium spreading are classified into two groups[9]: (a) the kinetic regime, when the process is determined by the transfer of fluid molecules to a solid across the interface, and (b) the hydrodynamic regime, when the inertial and viscous drag forces predominate.

The kinetic regime, as a rule, takes no more than 10^{-3} s and is characterized by the velocity of a contact spot, v, which, according to the thermodynamics of irreversible processes, is proportional to the motive force

$$v \cong \text{const}\cdot\sigma_{lg}(\cos\theta_o - \cos\theta) \qquad (3)$$

The hydrodynamic regime of spreading involves two stages:

(a) the inertial flow, when the Reynolds number, Re, is great and the rate of change of the wetted area, v=ds/dt, is constant. In this regime the following expressions are obtained for the radius, R, and wetting rate, v:

$$R^2 \cong t, \qquad v = W/(2\pi R) \qquad (4)$$

where $W=\sigma_{lg}(\cos\theta_0-\cos\theta)$ is the work done to displace the wetting front by unit length, and R depends on $\Delta\sigma$ as[6]

$$R^2 \cong (\Delta\sigma R_o/\rho)^{1/2}t \qquad (5)$$

where R_0 is the initial drop radius, ρ is the liquid density;

(b) the viscous flow at incomplete wetting, which is characterized by the relationship[6]

$$R^2 \cong (\Delta\sigma R_o/\eta)t \qquad (6)$$

where η is the dynamic viscosity of the liquid.

When we are concerned with spreading in an electric field, it is necessary to consider, at first, a change in the surface energy of solids. This effect may be estimated using the oscillatory model developed by Khatazhukov (Russia)[7]. It is revealed that in an electric field with a strength of $E\sim10^6$ V/mm the surface energy decreases only by 10^{-15} J/m^2. However, when a conductive or semiconductive layer is polarized near the metal surface (the Johnson-Rabeck effect), a field with a strength of $E\sim10^8$ V/m may be generated. In this case the surface energy estimation the by the Lippman-Helmholtz equation gives the value 100 mJ/m^2. This means that the surface tensions σ_{sg} and σ_{sl} in Eq. 2 may change substantially. Next, it is found that an electric field influences fluid spreading over a solid surface due to structural rearrangements. Almost in all fluids, in an electric field the molecules dissociate, polarize, start moving to electrodes and form anisotropic structures. As has been demonstrated experimentally[8], such rearrangements are insignificant in pure liquids and produce only a slight increase of elecroviscosity (several percents). In this aspect, of the great practical interest are electrorheological suspensions (ERS) characterized by substantial rearrangement of molecules in an electric field, which results in a change of rheological parameters by orders of magnitude.

However, despite the 50 year history of ERS, the electrorheological fluid flows with an open boundary have not been examined yet. By now, the Poiselle and Couette shear flow in closed channels have been all-round studied. In our opinion, this is due to the fact that the scientists and engineers have concentrated their attempts on the optimal implementation of the Winslow effect by increasing the electrosensitivity of working media and improving the design of hydraulic devices in which such media are used. Another types of flows, in particular film flows, single stream flows over solid surfaces or in channels, are still not understood completely. Recent development of novel ERSes for film coatings[9] has necessitated consideration of such problems. Among the constituents of such new ERSes are a fast-drying solvent for the dispersion media and fine-grained powders of oxides that are used as fillers for inks and paints, *e.g.* TiO_2, Cr_2O_3, Fe_2O_3 and others.

The present work is the first in a series of planned studies on the ERS flow with an open boundary in an electric field. Our aim is to estimate qualitatively the electric field effect on fluid spreading by investigating experimentally the changes in drop shape and sizes from the instant of the impact until it attains an equilibrium state after the termination of spreading. We intend to determine a dependence of the drop radius, R, and height, h, on the time and electric field strength in order to use them in future calculations of the electric effect on surface tension.

2. Experimental Studies

As a model ERS, we used a suspension of water-activated diatomite (one of the forms of silica) in a mineral oil with a minor addition of a surfactant (the oleic acid). The solid phase content was varied from 5 to 60 wt%. Figure 1 shows schematic of the experimental setup used for studying the kinetics of drop spreading in an electric field after its fall to a horizontal plane. The setup is assembled of (1) a coordinate guide that ensures displacement of a microdispenser, fixed on it, in two directions; (2) a microdispenser containing a cylinder with a piston (and a microscrew connected with the rod of the piston); (3) a table made of a dielectric material with embedded electrodes; (4) a stand with a coordinate table, onto which the table with electrodes 3 is placed; (5) a video camera, the optical axis of the camera is parallel to the table 3; (6) an impulse generator to set a photographing frequency; (7) a power supply unit for the video camera; (8) a high voltage power supply unit for the table 3 with electrodes.

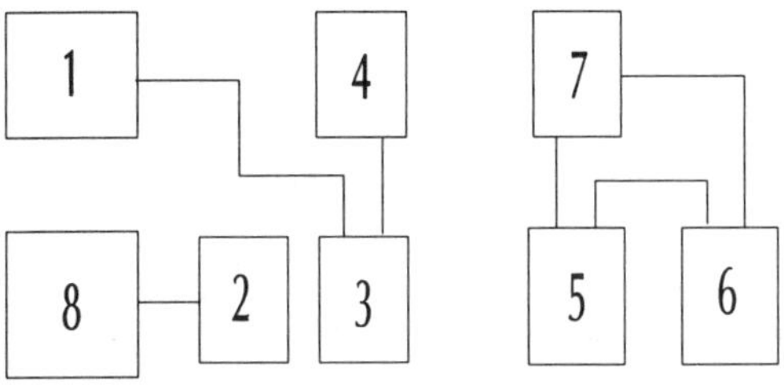

Fig. 1. Block diagram of experimental setup: 1, coordinate guide, 2, microdispenser, 3, table with electrodes, 4, stand, 5, movie camera, 6, impulse generator, 7, power supply, 8, high voltage supply.

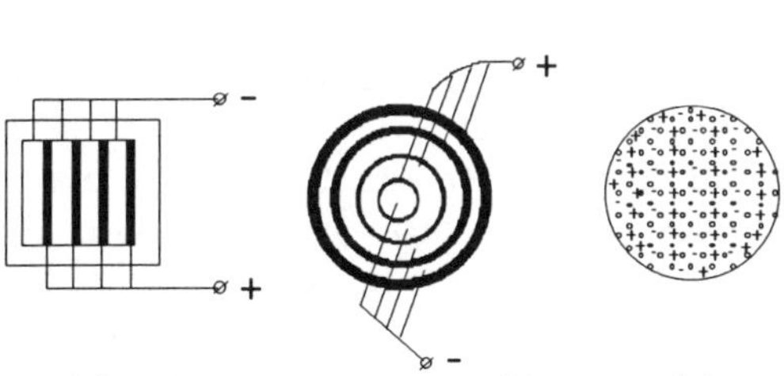

Fig. 2. Types of electrodes: (a), planar, (b), circular, (c), dot-like.

An electric field is generated by plane-parallel electrodes with a thickness of 4 mm and a gap of 1.5 mm between them, which were embedded in the dielectric table flush with the table surface (Fig. 2,a). Up to 10 electrodes were imbedded; the positive potential was applied to the even and the negative potential to the odd electrodes. This design has allowed us to study the spreading kinetics of liquid drops for the following cases:

- spreading between two neighboring electrodes when a drop falls onto the center of an intermediate dielectric strip;
- spreading of a drop over the surface when it falls onto the center of an electrode;
- spreading of a coarse drop that covers several electrodes.

In these cases, the electric field in the drop may be calculated using conformal Schwartz-Christoffelle transform[10]

$$z = w + \exp(z) \tag{7}$$

Here the complex potential of a two-dimensional field on the plane z=x+iy is translated onto the image plane w=u+iv. The linear coefficient of the transform is

$$\frac{du}{dz} = \bar{E}\exp(i\theta) = \frac{\partial u}{\partial x} + i\frac{dv}{dx} \tag{8}$$

where the module $\bar{E}$ is the scalar field gradient on the plane z, θ is the angle of clockwise rotation of a vector when it passes from plane z to plane w,

$$\mathrm{tg}\theta = -\frac{du}{dy} / \frac{dv}{dy} \tag{9}$$

Upon differentiating Eq. 7 and inserting the derivatives into Eqs. 8 and 9, we obtain

$$\bar{E} = \sqrt{1 + \exp(2u) + 2\exp(2u)\cos v}, \quad \theta = \mathrm{arctg}\frac{\exp(u)\sin v}{\cos v \exp(v)} \tag{10}$$

The external static electric field, which is generated in a drop by several pairs of electrodes, can be calculated using the superposition method. The magnitude and direction of the resultant field is determined by the vector sum of the strengths of the fields generated by each pair of electrodes. Figure 3 shows the electric field strength in the drop cross-section for the simplest case when the drop is between two electrodes.

In our experiments, the electrorheological fluid (ERF) was carefully stirred and charged into the cylinder of the microdispenser in the amount of 2-3 ml. The microdispenser was positioned at a required height over the table with planar electrodes. A fluid drop was put onto the table in order to focus the camera lens. For this, a ruler with a millimeter scale was placed near the drop. The ruler also served for estimating the drop size on the film. After the adjustment of the setup all movable parts are fixed by clamping screws. in order to ensure drop falling onto the same place of the table.

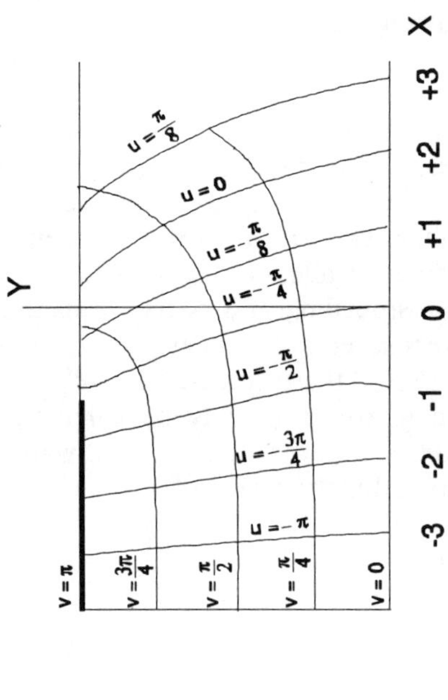

Fig. 3. Electric field in a drop: v=const, equipotential lines; U=const, force lines.

In order to estimate the electric field effect on the shape of a drop fallen onto the table and to examine the flow kinetics, we performed experiments in two ways: with and without applying an electric field. The sequence of operations is the following:

(a) the movie camera was switched on with the photographing frequency of 1 f.p.s. After 1-2 s a fluid drop was forced out onto the table. The camera photographed the collision and further spreading of the drop for 90 s. The readings of the image counter were taken for further analysis of the film;

(b) after all the parameters were chosen (height of falling and the drop mass), a required voltage was applied to the electrodes. The camera was switched on and a drop was forced out onto the table. 30 s later after the collision the electric field was removed. Filming lasted for 120 s since the instant of the collision. The experiments were performed at 250, 500, 1000 and 1500 V.

First of all we'd like to note that the site of collision in relation to the electrodes plays an important role. When a drop falls onto the table between the electrodes, it spreads uniformly in all directions, and the drop cross-section parallel to the table surface retains it round shape. If the drop center coincides with the middle of an electrode, the drop deforms so that the contact site acquires the shape of a figure eight. This is due to a strong non-uniformity of the symmetric electric field. The sizes of the figure eight depend on the drop mass, height of falling and the contact surface area between the drop and the electrode. To ensure a more uniform spreading, another scheme of the electrode arrangement was used (Fig. 2, b and c). The results presented in this paper were obtained with planar electrodes.

Figure 4 shows the photos representing spreading of a ERF drop with a mass of 0.046 g for several sequential instants of time. In the absence of an electric field (Fig. 4, a and b), continuous spreading occurs until an equilibrium state is attained, which is characterized by the drop shape parameters R_c, h_c and θ_c (for t=90 s see Fig. 4,a).

The shape of the drop profile is identical to that obtained for spreading of a media with the wetting angle $\theta>90°$ (low wettability). The application of an electric field changes drastically the flow behavior. Even at a low voltage, U=250 V, the drop parameters, R, h and θ, decrease for fixed instants of time. For a medium-range voltage, U=1000 V (Fig. 4,c and d), a thin film layer is formed on the table surface along the periphery of the drop while inside the drop we observe bulky spreading. The drop profile is usually characterized by two wetting angles, θ' and θ'_1 (see Fig. 4,d). For strong electric fields, U>1500V (Fig.4,e and f), a drop rests on the table surface without any spreading for a certain time, t_n. During this stage the wetting angle is $\theta<90°$. When the electric field is removed, the drop starts fast spreading. However, the drop radius attains the R_c value only after a long time. Thus, an ERF has the ability to retain the elements of its internal structure, which were generated by an external electric field, *i.e.* it possesses memory.

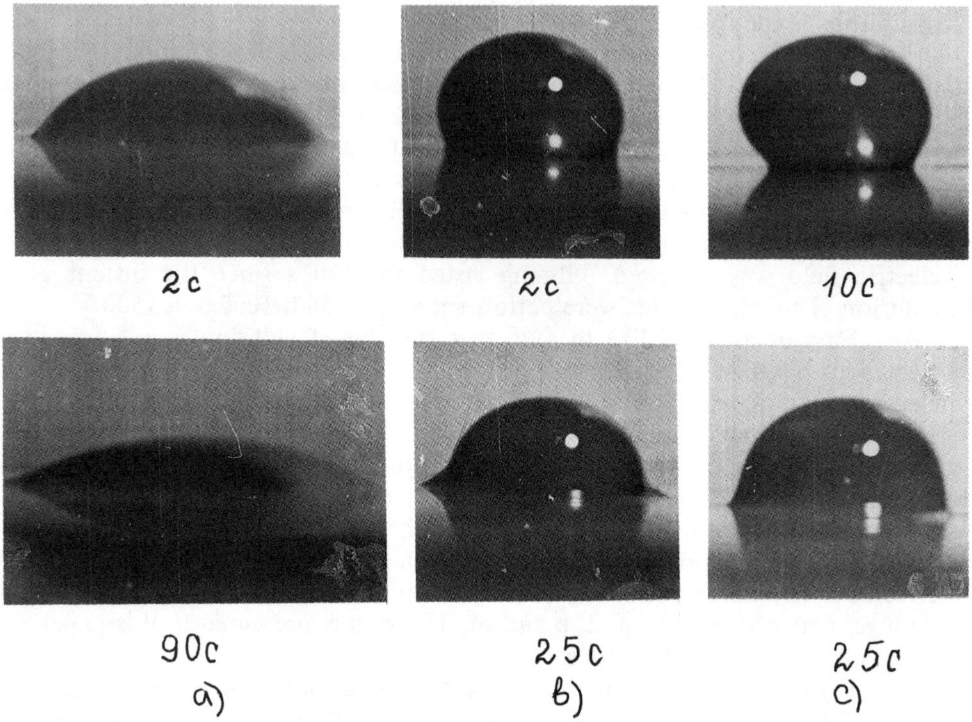

Fig. 4. Profiles of an ERF drop at different instants of time for the voltage U=0 (a and b), U=1000 V (c and d), U=1500 V (e and f); for (a) and (c) t=2 s, for (e) t=10 s, for (d) and (f) t=25 s, for (b) t=90 s.

Typical dependencies of the drop shape parameters on time and voltage are shown in Fig. 5. The reported spreading modes do not change with a filler concentration. Relative changes in the drop radius, $\Delta R=R-R_o$, and height, $\Delta h=h-h_o$, with time, where R_o and h_o are the values at $t<t_n$, may be described by power-law relationships $\Delta h=At^{-n}$ and $\Delta R=Bt^{m}$, which are close to dependencies known for abnormal fluids. The calculated values of A, B, n and m for the obtained experimental results are presented in Table 1.

The above parameters linearly depend on the voltage: $n=c+dU$, $m=c_1+d_1U$, $A=a+bU$, $B=a_1+b_1U$. The estimation of d and d_1 reveals that they are two orders of magnitude lower than c and c_1. Therefore, the dependence of the drop shape parameters on time and electric voltage may be represented in a general form as $\Delta h(t,U)=(a+bU)t^{-c}$

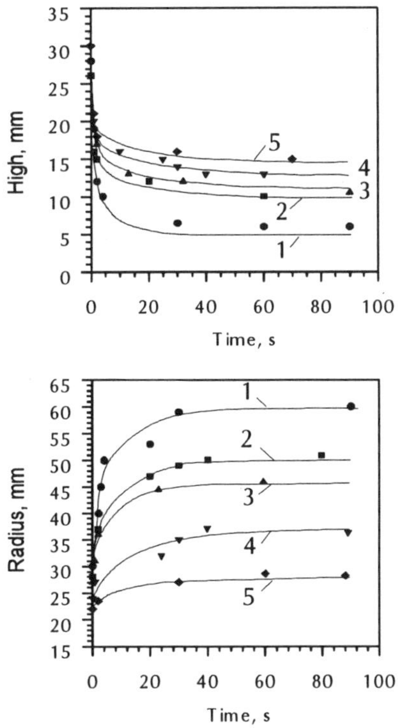

and $\Delta R(t,U) = (a_1 + b_1 U)t^{c_1}$, which agrees well with similar relationships obtained for spreading of abnormal fluid media.

3. Conclusion

The effect of an electric field on spreading of a drop of an ERF consists in a change of the drop shape and flow rate, which depend on the voltage applied. The observed latent period and a period of residual effect are typical of the mechanical behavior of thyxotropic media.

Fig. 5. Change of the contact spot radius, ΔR, and drop height, Δh, versus time: 1, U=0, 2,U=250 V, 3, U=500 V, 4, U=1000 V, 5, U=1500 V.

Table 1. Characteristics of ERF drop spreading

No	Voltage, V	n	A	m	B
1	0	0.226	15.137	0.118	39.316
2	250	0.141	17.451	0.089	35.542
3	500	0.138	18.854	0.104	32.527
4	1000	0.128	24.476	0.066	27.520
5	1500	0.078	20.168	0.036	24.030

4. References

1. Ya.I.Frenkel, *Zhurn. Exper. i Teor. Fiz.*, **16** (1949) 29.
2. Yu.V.Goryunov, *Dokl. AN SSSR*, **137** (1961) 1413.
3. T.D.Blake and J.H.Haynes, *J. Col. Inter. Sci.*, **30** (1969) 421.
4. B.D.Summ and Yu.V.Goryunov, *Physicochemical Foundations of Spreading and Wetting* (Khimiya, Moscow, 1974).
5. W.B.Hardy, *Phil. Mag.*, **38** (1919) 49.
6. E.D.Shukin, *Kolloidn. Zhurn.*, **25** (1963) 108.

7. A.S.Khatazhukov, *Study on the Kinetics of Wetting and Spreading of Conductive Fluids in Magnetic and Electric Fields*, Ph.D. thesis (Kalinin, 1979).
8. T.Sasada, K.Kamijo and J.Kishi, *Proc. 7th Jap. Congr. Mater. Res.* (Tokyo, 1977) p.288.
9. I.N.Ermolenko, T.G.Lazareva and E.V.Korobko, *USSR Patent # 1730100* (1990)
10. E.Janke, F.Emde and F.Losch, *Tabeln hoherer funktionen, Seschte Auflage* (B.G.Teubner-Verlag., Stuttgart, 1960).

THE ANOMALOUS DISPERSION AND RHEOLOGICAL RESPONSE OF ER MATERIALS

Frank E. Filisko
Department of Materials Science & Engineering
The University of Michigan
Ann Arbor, Michigan 48109-2136, USA

and

Adam Schubring
Delco Electronics Corp.
2033 East Boulevard, MS 6205
Kokomo, Indiana 46904-9005, USA

ABSTRACT

A large low frequency dielectric dispersion, termed anomalous because it's source is not clear, is observed in all ER active materials. It follows a classical Arrhenius behavior and as such, characteristc frequencies can be obtained for various temperatures. However it has not been directly associated with the rheological behavior of ER materials. In this study, rheological response times were obtained for an ER material by recording the response to a step increase in field. The response times of 41 msec. obtained dielectrically and 30 msec. obtained rheologically strongly indicate that the anomalous dispersion is associated with the mechanical ER phenomenon.

1. Introduction

Within the last few years, major discoveries have been made in the field of electrorheology, the most significant being that of intrinsic or "anhydrous " systems, i.e. those that work without the need for adsorbed water or other activator but due to their basic chemistry[1,2]. This meant that the numerous insurmountable problems associated with water could now be overcome by choosing the appropriate " intrinsic" systems or more likely custom tailoring the chemistry of the components so as to result in improved ER materials. This seems reasonable because of the very different classes of particles which have been shown to form intrinsic ER fluids such as polyelectrolytes, semiconductors, and inorganic materials. The result has been a few commercially available ER fluids consisting of very different types of particles developed at various industries throughout the world. The approach taken by most of these companies in developing their materials is to focus on molecular mechanisms which seem to be most accepted in terms of being associated with ER activity. While it is clear that molecular characteristics responsive to an electric field must be involved (polarization and/or charge transport) it has become increasingly evident that the phenomenon is extremely complex considering the interactions of mechanical and electrical parameters. A popular hypothesis for the ER phenomenon involves interfacial or Maxwell-Wagner-Sillars polarization which results from mobile charge accumulation at the interfaces between the dispersed particles and matrix phase. Such an effect can be observed through techniques of dielectric spectroscopy and historically it was just these type of measurements which resulted in the original use of the term " anomalous" dispersion, i.e. large low frequency dispersions were observed in materials believed to be homogenous and these dispersions were termed anomalous before they were first explained by Maxwell[3]. We are using the term anomalous because it does not readily fall under

Maxwell's explanation and in fact many other possible explanations have been proposed[4]. What is important however is that such dispersions always seem to be present in ER active fluids and qualitatively related to the rate-temperature performance of the fluids. These dispersions are clearly relaxational and demonstrate an Arrhenius behavior with activation energies in the range of 12-16 kcal/mole and characteristic frequencies at 25 °C for between 1-100,000/sec. for different systems. A clear tie-in however between the "anomalous" dispersion as determined dielectrically and the corresponding rheological performance has not been made nor is it clear how such a relationship can be made.

Two obvious questions arise regarding this dispersion. First; is it directly associated with the materials ER activity, and second, what is it's source? If a positive answer to the first is demonstrated, then the answer to the second will allow us to intelligently approach the choice or synthesis of ER fluid components.

In this paper we will not discuss the source of this anomalous dispersion but instead concentrate on quantitatively relating it with the ER activity. This will be done by relating characteristic frequencies of the dispersion with rheological response times of the ER fluid. How the latter is obtained is still up to debate but the methods we use are similar to those of Ginder and Cessio[5] and are detailed in the text of the paper.

2. Experimental

For this study, the ER fluid consisted of amorphous sodium aluminosilicates dispersed in paraffin oil. The aluminosilicates were synthesized according to procedures similar to those of Kansal and Laine[6]. The method used to prepare dried samples was described by Filisko and Radzilowski[2]. The samples were slowly heated to 600°C then placed under vacuum with a liquid nitrogen trap for four hours and allowed to cool. When the samples cooled to room temperature, zeolitically-dried mineral oil was added without breaking the vacuum seal. All the ER fluids used in this study contained 14g of powder to 30 ml mineral oil.

Measurement of the dissipation factor (tan δ) as a function of temperature (-150°C to 150°C) and frequency (13 Hz to 200 kHz) was obtained with a GenRad 1693 RLC Digibridge connected to a Macintosh computer where the data was compiled using LabVIEW 2 software.

Electrorheological measurements were obtained at room temperature with a Couette cell (bob = 19.5mm dia. x 50mm length, gap = 0.5 mm) using a continuous shear viscometer (Hercules High-Shear Viscometer #TS-8HV) at shear rates of 1 sec^{-1} to 200 sec^{-1}. The electric field of 2 to 6 kV/mm was obtained by a Trek Hi-Voltage Supply #663 and Trek 10 kV amplifier #662, which was attached to a signal generator (Kikusui Electronics Function Generator #455) that applied a 0.1 Hz square wave. The stress applied to the rheometer's cup was recorded with a digital oscilloscope (Nicolet #310), from which the data was compiled using data collection software (LabVIEW 2) driven by a Macintosh computer.

3. Results

A typical dielectric scan of loss factor vs. frequency at various temperatures is shown in Figure 1. The data is usually taken from -150 °C to +150 °C in 5 °C increments, and at 80 different frequencies over a range of 0.013 to 200 kHz. However only the frequency range of interest is displayed and at only sufficiently different temperatures to allow an

Arrhenius plot to be made. Such a plot for this data is shown in Figure 2 which,

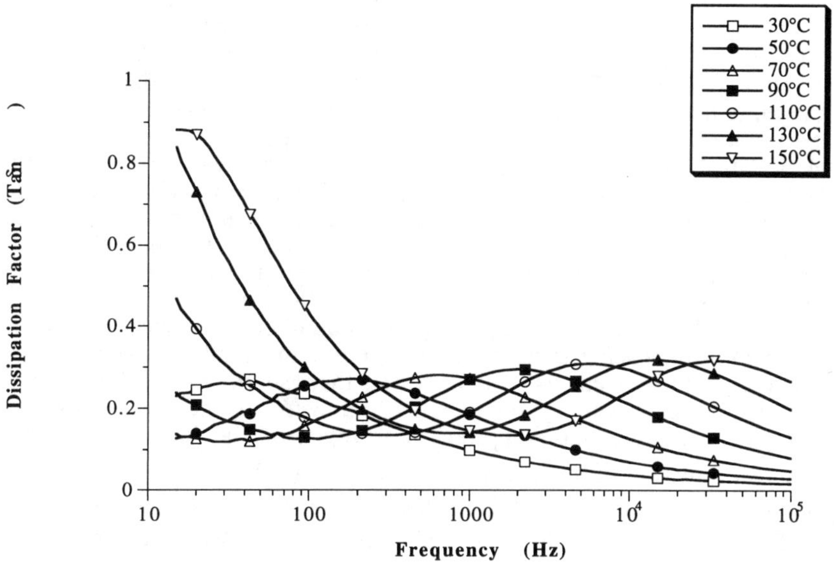

Figure 1: Dielectric study of $NaAlSiO_4$, showing Tan δ as a function of frequency.

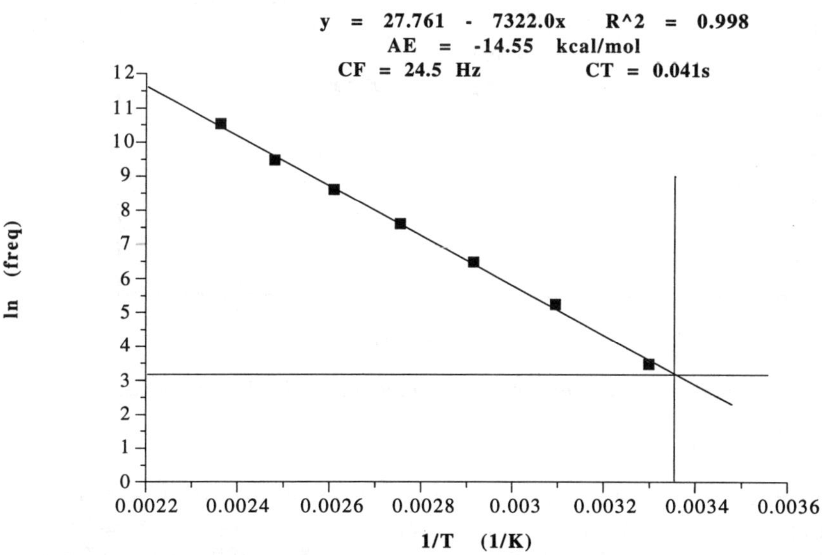

Figure 2: Activation energy plot of the amorphous aluminosilicate suspension.

being linear, clearly shows that the "anomalous" dispersion seems to follow a classical Arrhenius type activation behavior. The activation energy obtained from the slope is -14.55 kcal/mole and the characteristic frequency can be obtained at any temperature directly from the plot. For instance at 25 °C or 1/T(°K) = 0.00336/°K, the characteristic frequency is about 24.5 Hz with a characteristic time of 0.041 sec. At 30 °C the characteristic time would be about 0.027 sec. Thus the results are fairly temperature dependent which is important because the temperature of the rheometer was not controlled.

The rheological data was obtained while shearing the fluids continuously at rates of 1.02, 2.04, 10.21, 20.42, 102.1, and 204.2/sec. Electric fields of 2, 4, and 6 kV were then applied in a 0.1 Hz square wave which rose from 0 to maximum voltage in less than a microsecond, much faster then the effects we were attempting to measure. A typical set of data of shear stress rise vs. time for signals of amplitude 6 kV is shown in Figure 3. Data for field amplitudes of 2 and 4 kV show similar trends except that the maximum (extrapolated) asymptotic stresses at increasing times are longer.

At first the data in Figure 3 appears to be contradictory to what is expected, i.e. stresses obtained at increasing shear rates to be increasingly longer. However, what is being measured in an idealized case is the stress as the result of an angular distortion and the data reflects the fact that at a given time, the strain and thus the stress is less for lower angular velocities. It is thus more appropriate to represent the data not against time as recorded but as a function of the angular displacement (Θ) simply obtained by $\Theta = \omega t$, where ω = angular velocity and t = time. When presented in this fashion the data appears as shown in Figure 4 for 6 kV. The data trend is similar at 2 and 4 kV. Now this data

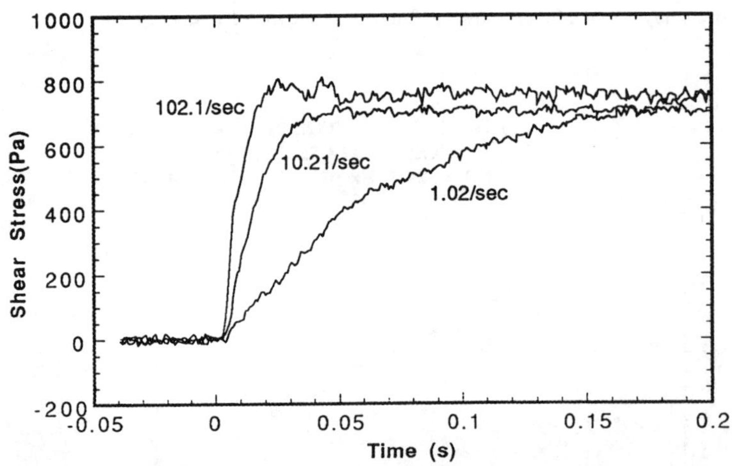

Figure 3: Stress response to a 6 kV step voltage increase for the ER fluid under various continuous shear rates.

may be analyzed by plotting stress vs. time at given Θ from vertical lines drawn on the graph, or by plotting strain vs. time at given stress from horizontal lines drawn on Figure 4. It is more appropriate to represent the data as either shear modulus vs. time at constant strain or shear modulus vs. time at constant stress. The latter was more appropriate consistent with the appearance of the data. For instance, a horizontal line drawn at 550 Pa on Figure 4 gives 6 points which when plotted as shear modulus vs. time [$G(t)=\tau/\Theta(t)$] give curves similar to Figure 5, which is such a curve generated at 6 kV. The modulii

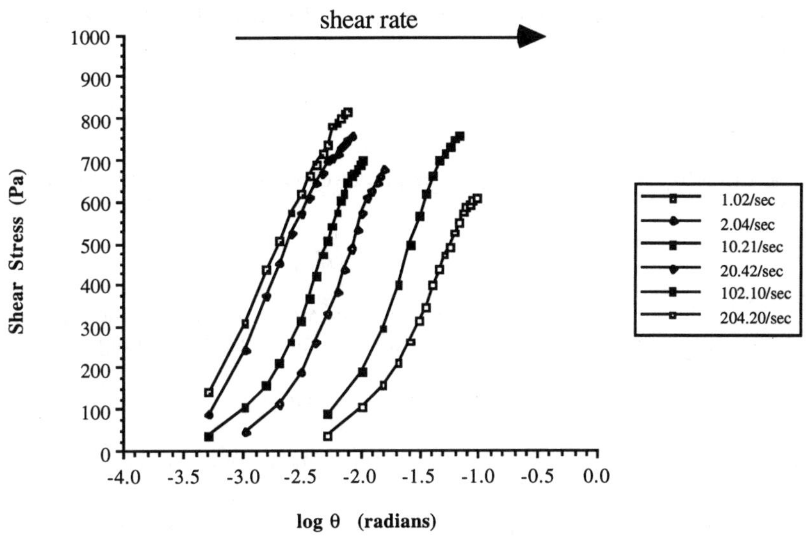

Figure 4: Shear stress versus angle of rotation, θ, at all shear rates at a field of 6 kV/mm.

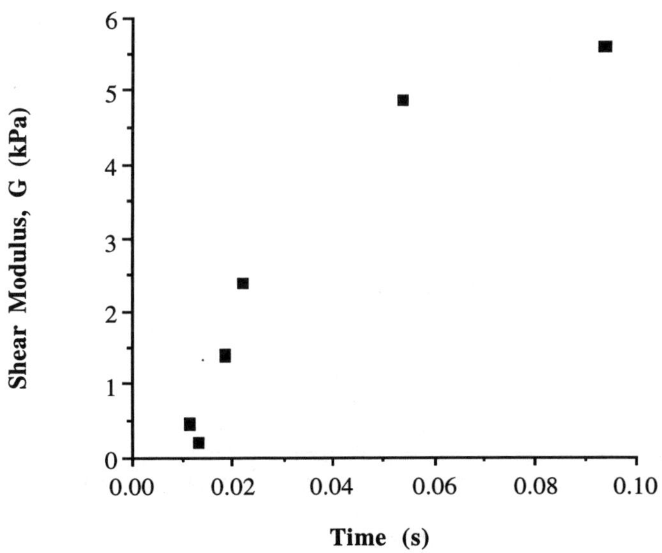

Figure 5: Shear modulus versus time at 6 kV/mm. Stress value taken at 550 Pa. Similar curves for 2 and 4 kV did not approach as near to the asymptotic limit as the 6 kV and therefore the asymptotic values were not readily obtainable.

increases with time, clearly demonstrating a time dependent behavior which approaches an asymptote of about 6 kPa.

4. Discussion

Shear modulus vs. time in Figure 5 can be modeled empirically by an exponential relationship

$$G(t) = G_{\infty}[1 - \exp(-t/T)]$$

where $G(t)$ is the time dependent shear modulus. In order to solve for the material time constant "T" for the electric field of 6 kV/mm, this equation can be rearranged as

$$\ln[G_{\infty} - G(t)] = \ln G_{\infty} - (t/T)$$

such that it produces a linear plot with the slope -1/T. Such a plot is illustrated in Figure 6 which produces a time constant of 0.0302 seconds for G_{∞} = 6.0 kPa. This time constant compares almost identically (considering temperature differences) with the dielectrically obtained characteristic time of Figure 2 (0.041 seconds).

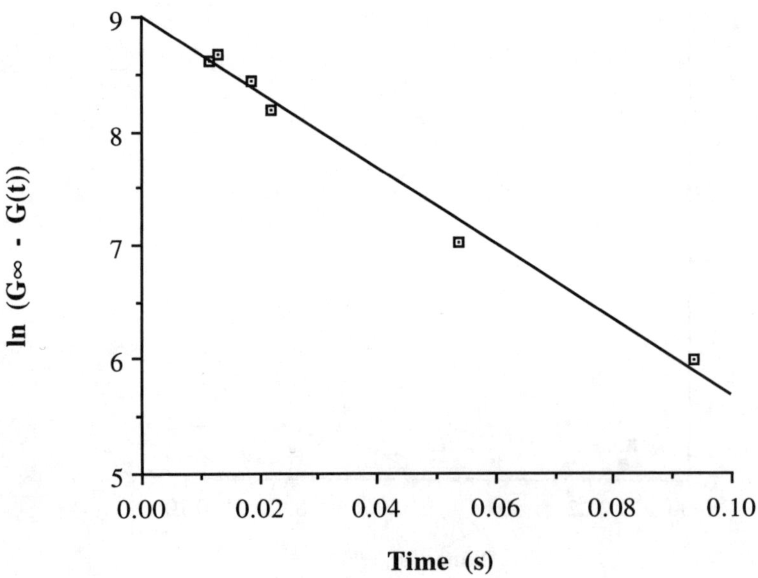

Figure 6: Time constant evaluation, with G_{∞} = 6.0 kPa

The time constants for data at 2 and 4 kV were not as readily obtainable because the G vs. time data, similar to Figure 5, did not extend far enough into the asymptotic range.

However, if these asymptotes are obtained with reasonable approximations, it appears that the time constants increase as the electric fields decrease. This would seem reasonable since the greater electrical forces imposed by the higher fields would cause any interactions to develop more quickly.

We should add that the amorphous alumino-silicate based ER fluids are characteristically "slow" as determined by numerous studies we have performed over the years, i.e. their strength drops off rapidly with increasing shear rates. This is reflected in the time constants presented here. The crystalline alumino-silicate, or Zeolite, based ER fluids are much " faster " in our extensive experience and this is reflected by much faster time constants obtained on these materials in ongoing studies similar to that presented in this paper.

5. Conclusions

From dielectric measurements, it was found that the amorphous aluminosilicate based electrorheological (ER) materials could be represented by an Arrhenius behavior. Furthermore, a characteristic time inherent to the particles was calculated as 41 ms. From the electrorheological measurements, it was found that curves of different shear rates were superimposable along the theta axis. It was also found that the shear modulus was a function of time and that the behavior could be characterized by an exponential function:

$$G(t) = G_{\infty}[\,1 - \exp(-t/T)]$$

The time constant associated with the function was found to be related to the electric field and at 6 kV/mm, a time constant of 30.1 ms was derived. However, as the electric field decreases, the time constant increases.

6. Acknowledgements

This work was supported by NSF grant #BCS-9201787.

7. References

1. H. Block and J. P. Kelly, *J. Phys. D: Appl. Phys.* **21** (1988) 1661-1677.
2. F. E. Filisko and L. H. Radzilowski, *J. Rheol.* **34**(4)(1990) 539-552.
3. J. C. Maxwell, in *Electricity and Magnetism,* (Claredon Press, Oxford , 1892) p.452.
4. M. Fixman, *J. Chem. Phys.* **78**(1983)1483.
5. J. M. Ginder and S. L. Cessio, *J. Rheol.* **39**(1) (1995)211 .
6. P.Kansal and R. M. Laine, *J. Am. Ceramic Soc.* **77**(1994) 875-82 .

Anisotropic Hydrodynamic Screening in Electrorheological Fluids

T.C.B.McLeish

IRC in Polymer Science and Technology, Department of Physics, University of Leeds, Leeds LS2 9JT, UK

Synopsis

We present an approximate calculation of the effective hydrodynamic interaction in a complex fluid of aligned strings in a solvent, an appropriate model for semi-dilute electrorheological fluids in which the string formation has not proceeded to "clumping". Earlier results indicated that the presence of weak screening could be detected in the frequency-dependent modulus of ER fluids with field applied. This calculation supports the conclusion that weak (logarithmic) modification of the imposed fluid flow is effective for relaxation modes which couple to the transverse deformations of the strings, but that shear waves parallel to the aligned strings are much less effectively screened.

I. INTRODUCTION

Until recently, most studies, both experimental and theoretical, of electrorheological (ER) fluids have concentrated on aspects of the high-shear response of the materials such as yield, stress and shear-thinning (see e.g. Klingenberg and Zukoski 1991). While important for many envisaged applications, the case of oscillatory flow is equally important for two reasons: (1) the linear frequency dependent rheology has a strong structure-dependence in many complex fluids (and has been linked in the case of polymers, for example, to molecular weight, branching structure and polydispersity) and is therefore an important analytical tool; (2) several applications require an oscillatory flow rather than a steady one (engine mounts, shock absorbers, etc.). The frequency-dependence of the dissipation in the fluid is therefore a vital property to understand.

The simplest structure that is a candidate model for an ER fluid with imposed field considers that the suspended particles have aggregated into strings parallel to the field, just one particle wide. These may connect the two charged plates bounding the fluid volume ("bound strings"), or may not in the case of "free strings". Such a model was used by McLeish *et al.* (1991) to calculate the frequency-dependent modulus, $G^*(\omega)$, for this structure. We compared our predictions with experiments on ER fluids that displayed a string-structure when examined visually. The calculation was repeated for the cases in which a simple long-range "Oseen" hydrodynamic interaction between the particles was present, and again under the assumption that it was completely "screened" (only a local friction persisting). Although the appropriate choice between these cases was left open at the time, the experimental data clearly favoured the *absence* of a long-ranged interaction, reflecting this by the interesting "signature" of the string model: a frequency plateau in the *dissipative* modulus, $G''(\omega)$, (in contrast to the much more common plateaux observed in $G'(\omega)$ in viscoelastic fluids). It was possible to identify the source of the dissipation within this model as the population of "free strings" (which include strings attached to only one of the bounding plates).

More recently a number of simulations have been directed at the microstructural dynamics of ER fluids and the rheological response. In particular Ahn and Klingenberg (1994) have considered the dissipation in oscillatory shear flow of more complex aggregates, and find that relaxation processes with rather narrow dispersion exist within clusters of particles with internal structure, independently of their connectivity to the plates. However, full treatment of the long-range hydrodynamics of the solvent was not attempted (typically the compputation time with full hydrodynamics becomes prohibitive for simulations large enough to be interesting from the structural point of view), and the assumption of local dissipation was made. This is equivalent to a hydrodynamic interaction which is completely screened. That this may be an oversimplification is indicated by the case of semidilute polymer solutions, for which the dynamics are now relatively well-understood (see Doi and Edwards [1986]). In this case there emerges a length-scale which characterises the range beyond which molecules are overlapped. This acts also as a "screening length": at larger lengthscales long-range hydrodynamic interactions are strongly suppressed (and reflect the composite viscosity of the complex fluid) while at shorter lengths the dilute-solution Oseen interaction dominates, controlled by the viscosity of the solvent alone. Therefore although low-frequency dynamics can be typical of a locally confined dissipation (or "Rouse"-like behaviour), high frequency response can remain similar to that of dilute solution in which long-range hydrodynamics are important (or "Zimm"-like behaviour). Additional evidence for the importance of accounting correctly for solvent hydrodynamics arises from computationally-intensive simulations which have incorporated them in models of limited size (see Bonnecaze and Brady 1992).

The central difference between the polymer and string-like ER fluid cases lies in the anisotropy of the structure. The directed nature of the strings, as opposed to the random walk structure of polymers in solutions, completely alters the relaxation spectrum (McLeish *et al.* 1991, Jordan *et al.* 1992). It would therefore be unsurprising if the hydrodynamic screening, if present, were also anisotropic. Though ferociously difficult to calculate exactly, or to extract from many-body simulations, certain simplifications which give semi-quantitative results in the case of polymers may be expected to work better in the case of directed strings. It is the aim of this work to exploit these analytic approximations to achieve insight into how hydrodynamic screening works in stringlike ER structures.

II. CALCULATION

The route taken here is to consider the complex fluid as an "effective medium" from the point of view of the hydrodynamic interaction. We begin from a "bare" interaction described by the Oseen tensor, $\boldsymbol{H_0(r)}$, which relates a local force $\boldsymbol{f}$ at the origin to the "backflow" velocity field $\boldsymbol{v(r)}$ which it sets up:

$$v(r) = H_0(r).f \equiv \frac{1}{8\pi\eta_s r}(I + \hat{r}\hat{r}).f \tag{1}$$

where η_s is the solvent viscosity. The result of the calculation will be the modified hydrodynamic interation $\boldsymbol{H(r)}$, defined in the same way but in the presence of a uniform distributionof ER strings (modelled as directed polymers). We proceed, as in the case for polymers (see Doi and Edwards 1986) via an auxiliary interaction, $\Xi\ \boldsymbol{(r)}$, which describes the velocity perturbation $\delta\ \mathrm{v}\ \boldsymbol{(r)}$ on an existing flow-field $\mathrm{v}\ \boldsymbol{(r)}$ when inserting into the flow a "test-string" whose particles are at $\boldsymbol{R_m}$ and are exerting a force $\boldsymbol{F_m}$ on the fluid:

$$\delta v(\boldsymbol{r}) = \sum_m \langle \boldsymbol{H}(\boldsymbol{r}) - \boldsymbol{R}_m . \boldsymbol{F}_m \rangle \equiv -\frac{1}{V}\int d^3\boldsymbol{r}'\, \Xi(\boldsymbol{r}-\boldsymbol{r}').v(\boldsymbol{r}') \tag{2}$$

Solving (1) and (2) by Fourier transform produces a useful self-consistent relation between the Fourier transforms of the two response functions $\boldsymbol{H(k)}$ and $\Xi\ \boldsymbol{(k)}$:

$$\tilde{\boldsymbol{H}}(\boldsymbol{k}) = \frac{\boldsymbol{I} - \hat{\boldsymbol{k}}\hat{\boldsymbol{k}}}{\eta_s \boldsymbol{k}^2[1 + \boldsymbol{c}\tilde{\Xi}(\boldsymbol{k})]} \tag{3}$$

Here $\boldsymbol{c}$ is the concentration of ER strings measured as a dimensionless volume-fraction. To solve the self-consistent equations we make the following assumptions (1) the test-string response is *pre-averaged* over its equilibrium structures, which in this case are elongated strings in the z direction so $R_m = an\hat{z}$ where $\boldsymbol{a}$ is the diameter of a single aggregating particle; (2) the strings are taken to move as rigid bodies for calculation of the modified hydrodynamics (not for the dissipation, which relies on internal structural relaxations); (3) the interaction Ξ *(k)* is assumed to be diagonal, but *anisotropic* with respect to its argument. These assumptions are not expected to be critical for the following reasons: (1) and (2) are successful in the case of polymers, but are much more physically appropriate in the case of ER strings, where the conformational dynamics really are perturbations away from a well-defined state (in this case the stretched string in the z-direction). In the polymer case the fluctuations in the relative positions of two monomers are as large as their root-mean-square separations. (3) is an argument from symmetry.

The calculation proceeds via the diagonalising of the interactions in terms of "Rouse modes" on the strings of the form cos($p\pi n/N$) where N is a parameter representing the polymerisation number of the strings, and is unimportant when large. So, for example, we can write the interaction between particles of n and m of a test string as

$$\boldsymbol{H}(\boldsymbol{R}_n - \boldsymbol{R}_m) \equiv h_{nm}\boldsymbol{I} = \frac{1}{N^2}\sum_p h(p)\cos(\frac{p\pi n}{N})\cos\frac{p\pi m}{N} \tag{4}$$

Now all information on $\boldsymbol{H}$ is contained in the set of amplitudes *h(p)*. Applying the approximations (1)-(3) above and assuming the equilibrium configuration of a stretched string for the test chain in the effective medium leads to the following coupled equations for Ξ *(k)* and ***H(k)***:

$$h(p) = \frac{1}{(2\pi)^3 a\eta_s}\int dk_x dk_y \left(\frac{1}{k_x^2 + k_y^2 + \dfrac{p^2\pi^2}{a^2N^2}}\right)\frac{1}{1 + c\Xi\left(k_x,\ k_y,\ \dfrac{p\pi}{aN}\right)} \tag{5}$$

$$\Xi(\boldsymbol{k}) = \frac{3}{2\pi\eta_s k^2}\frac{1}{h(p)\big|_{p=\frac{aNk_z}{\pi}}}$$

in which the Fourier integrals over k_z are trivial as the pth mode on the strings couples to a single shear wave only with the identical spatial wavenumber. Finally the integrals over the remaining Fourier space and over the mode index may be done, the self-consistency simplifying in those integrals dominated by the high-p or high-k cut-offs, which are of the order of the inverse particle size $\boldsymbol{a}^{-1}$.

The final result for the screened hydrodynamic interaction under these approximations is

$$H(k) = \frac{I - kk}{\eta_s \left[k^2 + \frac{3c^2}{4\pi^2 a^2 \ln(\pi / k_z a)} \right]} \tag{6}$$

Where the logarithmic dependence on wavenumber parallel to the strings is common in the hydrodynamics of slender bodies, and typically arises from the integration of the r^{-1} Oseen interaction along the long dimension of the embedded object in the fluid.

III. DISCUSSION

The spatial consequences of the previous section's result can be appreciated by recalling that the Fourier transform of an isotropically screened Coulomb potential (of the form $r^{-1}\exp\{-r/\xi\}$) has a Lorenzian form:

$$\frac{1}{k^2 + \xi^{-2}} \tag{7}$$

ξ is the "screening length", beyond which the interaction is strongly suppressed in real space. The most helpful way of understanding (6) is therefore to consider the logarithmic term as a weakly wavenumber-dependent screening length. It is those shear modes with a component parallel to the strings (and so able to couple to their own relaxation modes) which are damped. The screening lengths themselves are of the order of the typical string spacing except for very long wavelengths. So expectations that some form of anisotropy affects the hydrodynamic screening in such a structurally anisotropic system are confirmed. An unanticipated feature is the very weak screening at long wavelengths.

In the usual ER fluids cell, the same cell boundaries which impose the electric field (and so are perpendicular to the strings) are also used to set up the shear flow. Therefore the natural orientation of the shear waves set up by relaxing Rouse modes of the ER strings are oriented in the direction of strong screening. The calculation of the linear viscoelastic spectrum is detailed in McLeish *et al.* (1991). It proceeds via expressions for the stress (in terms of the averaged string-configurations) and for the dynamics of the configurations (in terms of local viscous drag per particle ζ and any backflow due to hydrodynamics). Specifically, the perturbations are written as

$X_i(n,t)$, the x-displacement of the nth particle in the ith string from its equilibrium position. If there is a tension T in the string the dynamical equation becomes:

$$\zeta \frac{\partial X(n)}{\partial t} = T[X(n+1) + X(n-1) - 2X(n)] + \zeta v_x(X(n)) \tag{8}$$

where v_x is the local velocity of the fluid at the bead site. This term contains all the information on hydrodynamic screening in terms of the string density (e.g. it is zero in the case of total screening). In figure 1 we compare data from McLeish *et al.* (1991) with predicted forms for the dissipative modulus with and without screening, where we see that it has the iportant consequence of increasing the dispersion (in this case set by the finite size of free strings of 30 particles). We suggest that the contribution of orientation and wavenumber-dependent screening to dispersion might also be important in the cluster-induced dissipation of Ahn and Klingenberg, although a treatment of that case will require going beyond the structural simplifications employed here.

It would be interesting to test the prediction that the screening is much reduced for shear-modes parallel to the strings. Though a difficult experiment in ER fluids (the rheometer plates can no longer be used as electrodes), the experiment might be done on a magnetorheological fluid in which string-direction and shear-orientation may be independently controlled. In that case the signature of this result is that the width of the dispersion-peak in G"(ω) becomes dependent on the angle between the imposed magnetic and shear fields.

References

Ahn, K.H. and D.J. Klingenberg, "Relaxation of polydisperse electrorheological suspensions", J.Rheol. **38**, 713-741 (1994)

Bonnecaze, R.T. and J.F. Brady, "Dynamic simulation of an electrorheological fluid", J.Chem.Phys. **96**, 2183-2202 (1992)

Doi, M. and S.F. Edwards, *The Theory of Polymer Dynamics* (Oxford, 1986)

Jordan, T.C., M.T. Shaw and T.C.B. McLeish, "Viscoelastic response of electrorheological fluids. II. Field strength and strain dependence", J.Rheol. 36, 441-463 (1992)

Klingenberg, D.J. and C.F. Zukoski, "Studies on the steady-shear behaviour of electrorheological suspensions", Langmuir **6**, 15-24 (1990)

McLeish, T.C.B., T. Jordan and M.T. Shaw, "Viscoelastic response of electrohreological fluids. I. Frequency dependence", J.Rheol. **35**, 427-448 (1991)

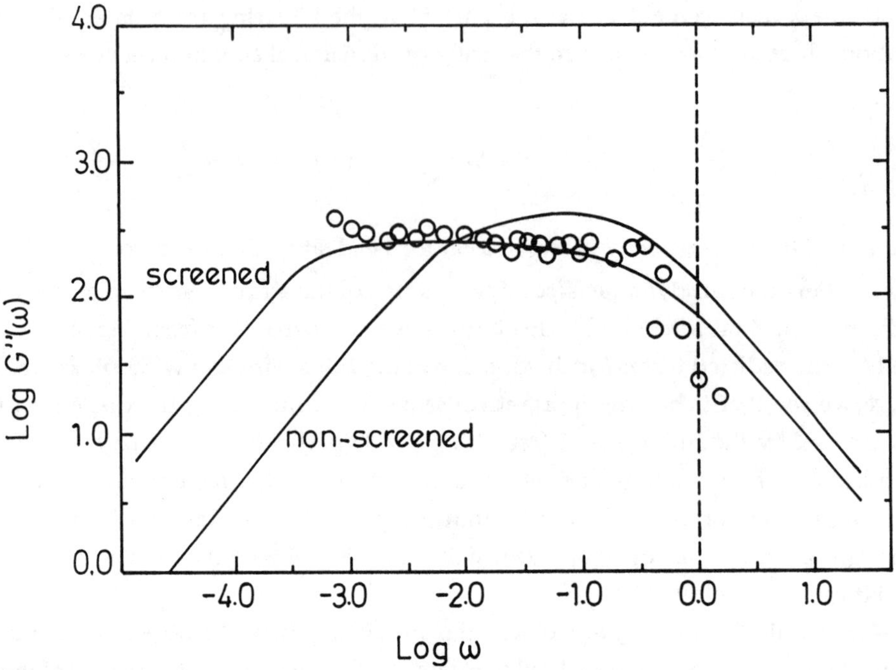

Fig. 1
The dissipative frequency-dependent modulus, G''(ω) predicted in the cases of hydrodynamic interaction present and screened. The system is an ensemble of free-floating strings of 30 concatenated spheres. Data on a silica suspension (Lord Corporation ERX-02) are superimposed. The absolute values of dissipation from all free strings and the relaxation time of the shortest string have been taken from experiment.

EFFECT OF THE ELECTRIC FIELD ON WAVE MOTION IN A THIN SUSPENSION LAYER FLOWING OVER AN INCLINED PLANE

I. SELEZOV
Department of Wave Processes, Inst. of Hydromechanics, Nat. Acad. Sci., 8/4 Sheliabov Str., 252057 Kiev, Ukraine

ABSTRACT

Stability of surface gravity waves in an inclined layer of rheological fluid in the presence of an arbitrarily oriented electric field is considered to investigate the effect of electric fields on wave stability. Rheological equations used for a dilute suspension of the rigid dielectric ellipsoids take into account the influence of electric fields on a particle orientation. The evolution equation is derived as a long-wavelength approximation which includes a set of nondimensional parameters responsible for the behaviour of the suspension. Solution of the evolution equation is obtained in the form of a nonlinear wave with the amplitude and phase slowly varying in time in such a way as to allow determination of the curves of neutral stability, bifurcations, sub- and supercritical instabilities. The results of calculations show that the electric field can both stabilise or destabilise waves.

1. Introduction

The problem of stability of surface gravity waves on a viscous film flowing down an inclined plane has been investigated by Benney[1], Lin[2, 3], Nakay[4], Selezov[5] and others. In the present paper this problem is extended to the case of electrorheological fluids.

In rheological equations for dilute suspensions of rigid dielectric ellipsoids with an effective radius $r_e < 10^{-6}$ m, approximate relationships are used for the moments of the angle distribution function which characterises the angle position of the particle symmetry axis in space. Symmetric and asymmetric components of the suspension stress tensor are composed of a linear part including components corresponding to Newtonian fluids and a nonlinear part including rheological terms of the moments of higher order. The problem is considered in a Cartesian co-ordinate system (x_1, x_2, x_3) with axis $0x_1$ along an inclined plane and $0x_2$ normal to this plane. In the undisturbed state the layer has thickness h_o, and in a disturbed state the free surface is governed by the equation $x_2 = h(x_1, t)$, where $h(x_1, t) = h_o + \eta(x_1, t)$. The geometry of the problem and main notations are shown at Fig. 1.

The evolution equation for the free surface elevation has been derived as a long wave approximation assuming $\varepsilon = (h_o/\ell_o) \ll 1$, where $\ell_o = \ell/2\pi$, ℓ is the wavelength. It includes Reynolds numbers for rheological constants, criterion of relaxation homochronism, angle of γ of the direction of electric field $\vec{E}$, and a Weber number $\varepsilon^2 W \sim 0(1)$. The curves of neutral stability are presented and analysed for a 1% suspension of the rigid dielectric ellipsoids at different values of γ and æ $= q_2 E^2 f_r^{-1} D_r^{-1}$, where q_2 depends on dielectric susceptibilities along the axis of particle symmetry and perpendicular to this axis, f_r is the coefficient of rotational friction, and D_r is the coefficient of rotational diffusion. When $\gamma = 0^o$ the effect is destabilising, while when $0^o < \gamma < 90^o$ stabilising. These effects are amplified with increasing parameter æ.

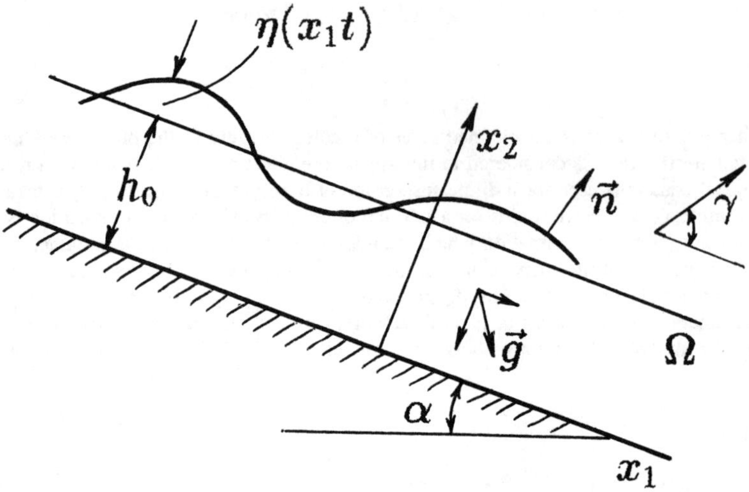

Fig. 1 Geometry of the problem

2. Statement

A statement of the problem for $v_i(x_1, x_2, t)$ includes the equations

$$\rho\dot{v}_i = \overset{*}{t}_{ij,j} + \rho f_{i,} \qquad v_{i,i} = 0 \qquad i,j = 1, 2 \text{ in } \Omega \tag{1}$$

the boundary conditions on the bed

$$v_i = 0 \qquad \text{at } x_2 = 0 \tag{2}$$

the matching conditions on the free surface

$$\left(t^o_{ij} - \overset{*}{t}_{ij}\right)n_j = -skn_i \qquad \text{at } x_2 = h(x_1,t), \tag{3}$$

$$h_t + v_i h_{,i} = v_2 \qquad \text{at } x_2 = h(x_1,t), \tag{4}$$

where $f_1 = g \sin\alpha$, $f_2 = -g \cos\alpha$, α is the angle of inclination of the plane relative to the horizontal, v_i is the component of the velocity vector, $t^o_{ij} = -p_o\delta_{ij}$ is the stress tensor for the external medium, $\overset{*}{t}_{ih}$ is the stress tensor of the suspension, n_i is the i-th component of the normal to the free surface, k is the curvature of the free surface, and s is the coefficient of the surface tension.

The dilute suspension of dielectrical ellipsoids with effective particle radius $r_{ef} < 10^{-6}$ m is considered as an electrorheological fluid. The model of suspension is based on approximate relationships for the moments of the angle distribution function of the particle symmetry axis in space. The equations for the moments of the first and second orders[6] have been closed by using the method of imaginary field[7]. Assuming above mentioned the constitutive equations are presented in the form[8]

$$\overset{*}{t}_{(ij)} = -p\delta_{ij} + 2\mu_o m_o d_{ij} - \frac{\lambda^2\bar{\mu}_4}{15D_r}\frac{dd_{ij}}{dt} + \frac{\lambda^2\bar{\mu}_4}{90D_r^2}\frac{d^2d_{ij}}{dt^2} + \left(\bar{\mu}_s + \frac{2}{7}\bar{\mu}_2\right)$$

$$\times\left[\frac{æ}{15E^2}\left(d_{ik}E_kE_j + d_{jk}E_kE_i\right) + \frac{2\lambda}{15D_r}d_{ik}d_{kj} - \frac{\lambda}{90D_r^2}\left(d_{ik}\frac{dd_{kj}}{dt} + d_{jk}\frac{dd_{ki}}{dt}\right)\right] +$$

$$\lambda\bar{\mu}_4\left[\frac{æ}{15E^2}\left(w_{ik}E_kE_j + w_{jk}E_kE_i\right) + \frac{\lambda}{15D_r}\left(w_{ik}d_{kj} + w_{jk}d_{ki}\right) - \frac{\lambda}{90D_r^2}\left(w_{ik}\frac{dd_{kj}}{dt} + w_{jk}\frac{dd_{ki}}{dt}\right)\right] \tag{5}$$

$$t^s_{[ij]} = \lambda\bar{\mu}_4\left[\frac{æ}{15E^2}\left(E_jE_kd_{ki} - E_iE_kd_{kj}\right) - \frac{æ}{90D_rE^2}\left(E_jE_k\frac{dd_{ki}}{dt} - E_iE_k\frac{dd_{kj}}{dt}\right)\right.$$

$$\left. + \frac{æ}{540D_r^2E^2}\left(E_jE_k\frac{d^2d_{ki}}{dt^2} - E_iE_k\frac{d^2d_{kj}}{dt^2}\right)\right] \tag{6}$$

Here $m_o = 1 + \frac{\bar{\mu}_2}{15\mu_o}\left(1 - \frac{2æ}{21}\right) + \frac{\bar{\mu}_3}{3\mu_o}\left(1 - \frac{æ}{15}\right)$, $t^s_{(ij)}$ and $t^s_{[ij]}$ are the symmetric and asymmetric parts of the stress tensor, d_{hi} and w_{ij} are the symmetric and asymmetric parts of the velocity gradient tensor, $\bar{\mu}_i = \mu f_i(\Phi, p)$ $(i = 1,2,3,4)$ are rheological constants with the same dimensions as the viscosity coefficient μ which depend on the volume concentration Φ and the particle elongation p, $\lambda = (p^2-1)/(p^2+1)$, D_r is the coefficient of rotational diffusion, E_i is the i-th component of the electric field, E^2 = const, $æ = q_2E^2f_r^{-1}D_r^{-1}$ is a dimensionless parameter, $q_2 = \chi_1 - \chi_2$, where χ_1 and χ_2 are the principal values of dielectric susceptibility parallel and perpendicular to the axis of particle symmetry, f_r is the coefficient of rotational friction. The convention of summing over repeated indices is observed here.

3. Solutions and Analysis

The equation of the free surface motion is obtained similar to references[1, 3, 4] with the approximation of long wavelength $\varepsilon = (h_o / \ell_o) << 1$, where , ℓ is the wavelength

$$h_t + L(h)h_{,1} + \varepsilon\left[M(h)h_{,1} + N(h)h_{,111}\right]_{,1} + O(\varepsilon) = 0 \qquad (7)$$

where

$$L(h) = Ah^2, \quad M(h) = Bh^6 + Ch^4 + Dh^2, \quad N(h) = Eh^3$$

$$A = \frac{2}{t_o(m_o + m_1)}$$

$$B = \frac{8R_o}{15f_o^2(m_o + m_1)^3}$$

$$C = \frac{\lambda R_o A_1}{45R_4 f_o^2(m_o + m_1)^2}\left(2\lambda + \frac{æ}{3}\cos 2\gamma\right)$$

$$D = -\frac{2\,\mathrm{ctg}\alpha}{3f_o(m_o + m_1)} + \frac{\lambda R_o æ}{45R_4 f_o(m_o + m_1)}\sin 2\gamma$$

$$E = \frac{2\varepsilon^2 W \operatorname{cosec}\alpha}{3f_o(m_o + m_1)}, \qquad m_1 = \left(\frac{R_o}{R_3} + \frac{2R_o}{7R_2}\right)\frac{æ}{30} - \frac{\lambda R_o æ}{15R_4}\cos 2\gamma$$

In the above expressions R_i (i=1, 2, 3, 4) is the Reynolds number, $R_i = u_o h_o \overline{v}_i^{-1}$. The value R_i corresponds to the rheological constant $\overline{\mu}_i$, $W = s\rho^{-1}g^{-1}h_o^{-2}$ is the criterion of the relaxation homochronism. It is assumed that $\varepsilon^2 W \sim 0(1)$.

The solution to Eq. (7) is presented in the form of nonlinear wave, the amplitude and frequency of which are slowly varying functions of time[9, 2, 4, 10]

$$h = 1 + a\left(e^{i\varphi} + e^{-i\varphi}\right) + a^2\left(h_{2,2}e^{2i\varphi} + h_{2,o} + h_{2,-2}e^{-2i\varphi}\right)$$

$$+ a^3\left(h_{3,3}e^{3i\varphi} + h_{3,1}e^{i\varphi} + h_{3,-1}e^{-i\varphi} + h_{3,-3}e^{-3i\varphi}\right) + \ldots \qquad (8)$$

where a is the amplitude, φ is the phase, $\varphi = x_1 - wt$, $\varphi_{,1} = 1$. In addition to above, the expansions for a_t and φ_t are introduced[9,10]

$$a_t = s_n a^{n+1}, \ \varphi_t = w_n a^n, \ n = 0, 1, 2 \tag{9}$$

The constants $h_{m,-n}$ are the complex conjugates of $h_{m,n}$. Substituting (8) and (9) into (7) yields

$$s_o = \varepsilon(B + C + D - E) \tag{10}$$

$$w_o = -A \tag{11}$$

$$s_2 = \frac{2A^2}{\varepsilon(B+C+D-7E)} + \varepsilon(15B+6C+3D-3E+ \frac{(6B+4C+3D-3E)(6B+4C+3D-21E)}{7E-B-C-D} \tag{12}$$

$$w_2 = A - \frac{A(18B-12C+9D-27E}{7E-B-C-D} \tag{13}$$

$$s_1 = w_1 = 0, \ h_{s,o} = h_{3,1} = 0 \tag{14}$$

The expressions for $h_{2,2}$ and $h_{3,3}$ are very complicated and not presented here. Keeping only the first order terms in a from (9) results in a linear theory with amplitude s_o and frequency w_o. In this case the linear stability or instability of waves is determined by the sign of s_o; the stable waves have $s_o < 0$, and the unstable waves have $s_o > 0$. The equation $s_o(\varepsilon, R_o, W, \Phi, p, æ, \gamma) = 0$ determines the curve of the neutral stability which bifurcates to two branches:
the lower branch has

$$\varepsilon = 0 \tag{15}$$

and the upper branch has

$$\frac{8R_o}{15f_o^2(m_o+m_1)^3} - \frac{2\text{ctg}\alpha}{3f_o(m_o+m_1)} - \frac{2\varepsilon^2 W \text{cosec}\alpha}{3f_o(m_o+m_1)} + \frac{\lambda R_o}{45R_4 f_o^2(m_o+m_1)^3} \left[2\lambda A_1 + \frac{A_1 æ}{3}\cos 2\gamma + æ f_o(m_o+m_1)\sin 2\gamma\right] = 0 \tag{16}$$

The bifurcation point of a neutral curve yields the critical Reynolds number

$$Re_{cr} = \frac{5f_o(m_o+m_1)^2}{4}ctg\alpha - \frac{\lambda R_o}{24R_4}\left[2\lambda A_1 + \frac{A_1 æ}{3}\cos 2\gamma + æfo(m_o+m_1)\sin 2\gamma\right] \tag{17}$$

It follows from Eq. (10) that above the upper branch of the neutral curve (16) $s_o < 0$ and below it $s_o > 0$.

To consider nonlinear effects it is necessary to take into account in (9) higher order terms in ε. The case of special interest is when the order of nonlinear terms equals or exceeds the order of linear terms. In the case when $s_o < 0$ and the second Landau constant $s_2 > 0$ the "supercritical" instability generated by nonlinear terms takes place. For $s_o < 0$ and $s_2 > 0$, the instability originated by the linear term is suppressed by nonlinear terms, and the waves will be "supercritically" stable. In the particular case $s_o = 0$, decreasing or increasing waves are characterised by the sign of s_2. The coefficient w_2 in the (9) characterises the frequency change due to nonlinearity. It follows from (12) that the leading term in s_2 is the first term, which is why the sign of s_2 is determined by the sign of the first term. s_2 has opposite signs above and below the curve described by the following equation:

$$\frac{8R_o}{15f_o^2(m_o+m_1)^2} - \frac{2ctg\alpha}{3f_o(m_o+m_1)} - \frac{2\varepsilon^2 W \operatorname{cosec}\alpha}{3f_o(m_o+m_1)} + \frac{\lambda R_o}{45R_4 f_o^2(m_o+m_1)^2} \left[2\lambda A_1 + \frac{A_1 æ}{3}\cos 2\gamma + æf_o(m_o+m_1)\sin 2\gamma\right] = 0 \tag{18}$$

The second Landau constant s_2 is negative above the curve corresponding to (18) and positive below one. Therefore, in the domain where $s_o < 0$, the value s_2 will be negative. "Subcritical" instability should not appear, and consequently the waves are stable.

Figure 2 presents the curves of neutral stability (15) and (16), as well as the curves of changing sign of s_2 (18) for a one-percent suspension of rigid dielectric ellipsoids for different values of æ and γ.

It follows from Fig. 2 that the electric field can both stabilise or destabilise the waves. When $\gamma = 0^o$, the electric field destabilises the waves, while when $0^o < \gamma < 90^o$, the field stabilises the waves. These effects are amplified as the parameter æ increases.

4. Acknowledgements

This research was supported by the International Soros Science Education Program of the International Renaissance Foundation, grant N059.

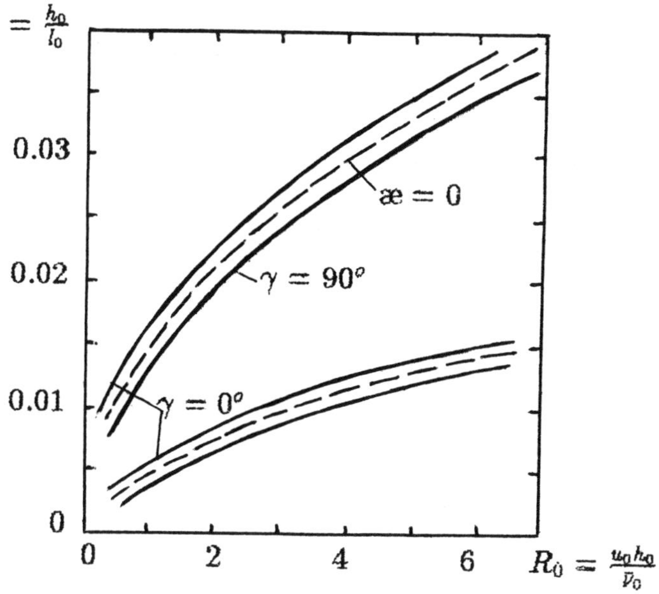

Fig. 2 Curves of neutral stability (upper) and curves of the sign changing of s_2 (lower): $\alpha = 90^o$, W = 470, $\Phi = 0.01$, $A_1 = 0.03$, p = 50, æ = 0.25

5. References

1. D. J. Benney, *J. Math. Phys.* 45 (1966) 150.
2. S. P. Lin, *J. Fluid Mech.* 36 (1969) 113.
3. S. P. Lin, *J. Fluid Mech.* 63 (1974) 417.
4. S. Nakay, *Phys. Fluids* 18 (1975) 1407.
5. I. T. Selezov, *Modelling of wave and diffraction processes in continuious media* (in Russian) (Kiev, Naukova Dumka, 1989).
6. V. N. Pokrovsky, *Statistical mechanics of dilute suspensions* (in Russian) (Moscow, Nauka, 1978).
7. M. A. Leontovitch, *Statistical physics* (in Russian) (Moscow, Gostekhizdat) 1944.
8. I. L. Novak, *Transactions Acad. Sci.*, Ukr. SSR N2 (1980) 48.
9. L. D. Landau and E. M. Lifshitz, *Mechanics of continuous media* (in Russian) (Moscow) 1954).
10. W. C. Reynolds and T. C. Potter, *J. Fluid Mech.* 27 (1967) 465.

BOUNDARY LUBRICATION CHARACTERISTICS OF SILICA-BASED ELECTRO-RHEOLOGICAL FLUIDS

W. C. Park, S. B. Choi and C. C. Cheong
Smart Structures and Systems Laboratory, Department of Mechanical Engineering, Inha University, Incheon 402-751, Korea

and

M.S. Suh and M.S. Yeo
Dynamic Analysis Division, Agency for Defense Development, Taejon 305-600, Korea

ABSTRACT

This experimental study presents friction and wear characteristics under boundary lubrication associated with silica-based ER(Electro-Rheological) fluids. Various types of base liquids(silicone oil, olive oil, transformer oil) and corresponding silica-based ER fluids are adopted as lubricants. As a first step, yield shear stresses of the ER fluids are measured as a function of imposed electric fields. Pin-on-disc wear testings are then performed with the lubricants by evaluating wear rates of the pin specimens. The governing relationship between the wear rate and the field-dependent yield stress is established for the application guideline of the ER fluids. Microscopic observations of worn surfaces of the pin specimens are also analyzed in order to validate the measured wear characteristics. In addition, friction coefficients of the ER fluids are experimentally evaluated with respect to normal loads

1. Introduction

Electro-Rheological(ER) fluids belong to a class of colloidal suspensions whose rheological behavior can be controlled by the imposition of an external electric field upon the fluid domain[1,2]. Under nonzero electric fields, the micro-sized hydrophilic particles suspended in suitable hydrophobic carrier liquids(base liquids) orient themselves in relatively regular chain-patterns to form a mixture. These columnar chain-patterns restrict flow motion and hence increase the flow resistance. From the fluid mechanics point of view, this property can be expressed as a Bingham plastic model. The ER fluid behaves as a Newtonian fluid in the absence of the electric field. However, upon applying the electric field the ER fluid exhibits a yield shear stress which should be overcome prior to initiating fluid flow.

Most previous researches on the ER fluids can be divided into three sectors ; the development of new ER fluids[3], the verification and improvement of the electro-effect[4,5,6], and the application to engineering devices[7,8,9]. Recently, the development of the devices incorporating the ER fluids has been the subject of intensive research activities throughout the world. Especially numerous applications to automotive devices such as shock absorbers, engine mounts, clutches, brakes and valves attract many researchers. Thereby, it is highly hoped that commercial automotive devices will be appeared within this century. However, researches on the incompressibility, specific heat transfer property, wear and lubrication characteristics, and reliability are relatively rare.

Tribological aspects of the ER fluids treating wear performance and lubrication property were studied by only Lingard *et al.*[10,11] When we apply the ER fluids to hydraulic power machines as lubricating oils, microcutting and wear of moving components such as blade of pump arise in practice due to conductive particles. Long period of operating the machine results in the alteration of material characteristics of the component itself and the ER fluid also. This finally causes the lubricant leakage and hence leads to poor performance of the system. Furthermore, the wear debris may change the conductivity property of the ER fluid and as well as that of the electrodes.

This study presents experimental results on the wear and boundary lubrication characteristics of silica-based electro-rheological fluids. Three different types of base liquids; silicone oil, olive oil and transformer oil are employed to make the ER fluids, and these are adopted as lubricants. As a first step, field-dependent yield shear stresses of the ER fluids are measured with respect to different base liquids and different particle concentration by weight. Pin-on-disc wear testings are then performed with the lubricants by evaluating wear rates of the pin specimens. Comparative works on wear characteristics between the base liquids only and the ER fluids are undertaken. And scanning electron microscope observations of worn surfaces of the specimens are also analyzed in order to validate the measured wear characteristics. In addition, the governing relationship between the wear rate and the field-dependent yield stress is established, and friction coefficients of the ER fluids are experimentally evaluated with respect to normal loads.

2. Bingham Characteristics of ER Fluids

Field-dependent rheological behavior of the ER fluid is very complex, especially in the low shear rate. However, to date many experiments verified that the ER fluid has Newtonian characteristics in the absence of the electric field which become Bingham behavior when the electric field is imposed upon the fluid domain. The Bingham-body model for typical ER fluid can be expressed by

$$\tau = \tau_y(E) + \eta_p \dot{\gamma} \qquad (1)$$

Here τ and η_p represent shear stress and viscosity of the suspensions, respectively, and $\dot{\gamma}$ denotes shear rate. On the other hand, the yield shear stress of the ER fluid, $\tau_y(E)$, is a function an electric field E. For most ER fluids, it has been tested that the yield shear stress exponentially increases as the electric field increases[12]. From this sense, Eq. (1) can be rewritten as an explicit function of the electric field as follow :

$$\tau = \alpha E^x + \eta_p \dot{\gamma} \qquad (2)$$

In the above equation, proportional coefficient α and exponent x are intrinsic values of considering ER fluids. These parameters, of course, depend upon the type of ER fluids characterized by viscosity of base liquids, particle size and shape, temperature, water contents, particle concentration and so forth.

The ER fluids used in the present tests were made from silica particle averaging 3.612 μm in size (refer to Fig. 1 for particle size distribution). Three different viscous oils; silicone oil, olive oil and transformer oil were employed as carrier base liquids. In order to investigate the effects of the type of base liquids and particle concentrations on wear and lubrication characteristics, six combinations of the ER fluids were made and designated as given in Table 1. In the table, (cs) represents the viscosity of the base liquid in centi stoke at room temperature, and (%) denotes particle concentration of the mixture by weight. The rheological characteristics of these ER fluids were measured by couette type of electro-viscometer(HAAKE, model VT500). The temperature of the ER fluid was maintained to be room temperature.

Table 1 Specifications of the base liquids and ER fluid lubricants

Lubricant designation	*Base liquids and ER fluids*
IXSX-11	silicone oil (200cs)
ILSC-11C	silicone oil (200cs) + silica (15%)
IXSX-9	silicone oil (50cs)
ILSC-9C	silicone oil (50cs) + silica (15%)
IXOX-10	olive oil (98cs)
ILOC-10C	olive oil (98cs) + silica (15%)
IXTX-6	transformer oil (35cs)
ILTC-6C	transformer oil (35cs) + silica (15%)
ILTF-6C	transformer oil (35cs) + silica (30%)
ILTH-6C	transformer oil (35cs) + silica (40%)

Fig.2 presents field-dependent dynamic yield shear stress of the ER fluids. These data were obtained from the shear stress versus shear rate curve by linear interpolating the curve and adopting the value which intercepts the zero shear rate. Fig. 2(a) compares the dynamic yield stress of the 15% ER fluid with respect to the base liquid. The ILSC-11C ER fluid shows the highest dynamic yield stress and its value at 2kV/mm is 1.2 and 1.6 times higher than those of the ILOC-10C and the ILTC-6C ER fluid, respectively. This is attributed to larger viscosity of the carrier liquid as well as active polarization phenomena of the silicone-oil based ER fluid. Fig.2(b) compares the yield stress of the transformer oil-based ER fluids with respect to the particle concentration. As expected, the yield stress increases as the particle concentration increases. The yield stress of the 40% ER fluid at 2

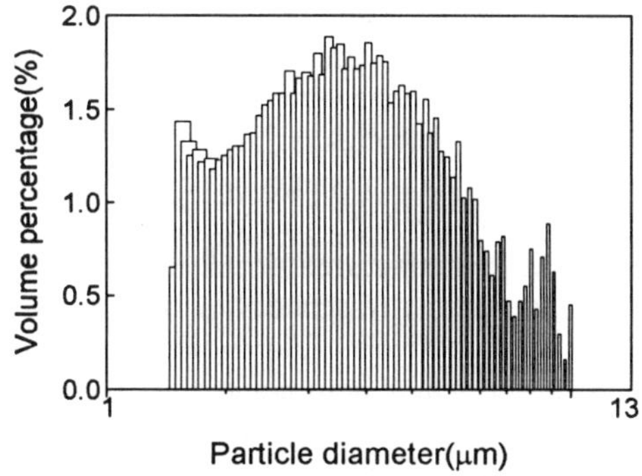

Fig.1 Distribution of the particle size

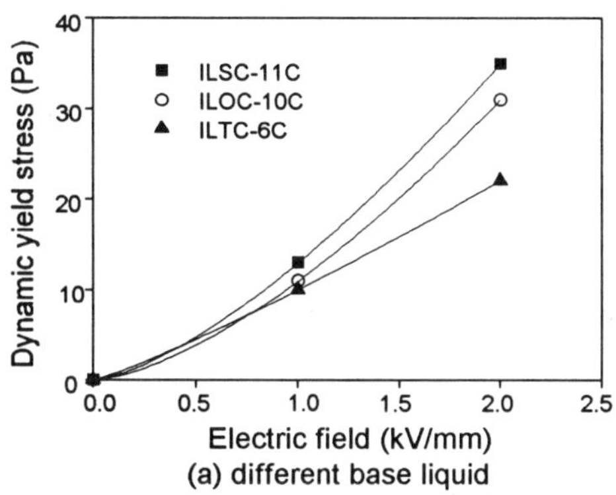

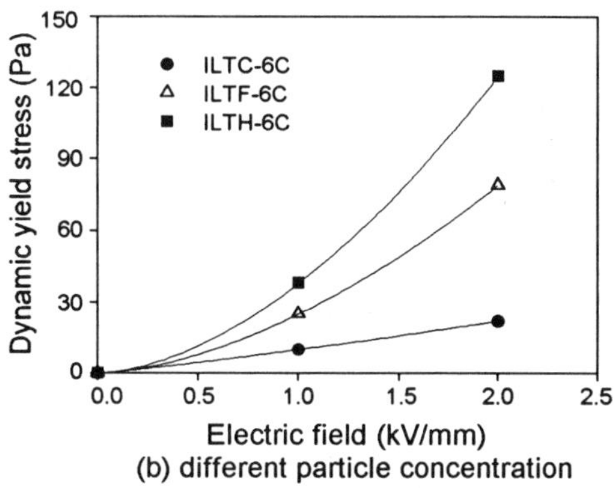

Fig.2 Dynamic yield stress of the ER fluids

kV/mm is 1.58 and 5.68 times higher than those of the 30% and 15% ER fluid, respectively. The distilled yield stress can be expressed as a function of the electric field as follows; $10E^{1.14}$ for the ILTC-6C, $25E^{1.68}$ for the ILTF-6C and $38E^{1.72}$ for the ILTH-6C.

For the application point of view, the higher yield stress, the better system performance we have in general. However, this may not be true in terms of tribology aspects. Later on in the study, the relationship between the field-dependent yield stress and the wear rate will be established.

3. Experimental Apparatus and Methods

The experiments were undertaken using a standard pin-on-disc wear tester as shown in Fig. 3. Boundary lubrication is achieved by rotating a disk dipped in a small reservoir and simultaneously applying a normal load with the air compressor consisting of regulator, lubricator and analog load indicator. The preparation of the uniformity of the pin and disc specimens is very important since the wear characteristics heavily depend upon experimental conditions. The specimens were ground and polished to keep up the surface uniformity, and cleansed by ultrasonic cleaner in order to prevent them being oxidized. A 6mm diameter and 30mm length pin specimen was end-loaded on to the periphery of a 58.5mm rotating disc. The pin specimen was manufactured from mild steel of hardness 180Hv, while the disc was manufactured from high-frequency heat treated steel of hardness 540Hv. And the surface roughness of the specimen is R_{max} 44μm and R_{max} 1.2μm for the disk and pin, respectively. The measurements of the hardness and surface roughness of the specimens were made before and after each experiment.

The wear tests were run for a certain period that depends on the wear rate varying from two minutes to one hour. The wear rate of the pin specimen is obtained by measuring weight loss and dividing it by sliding distance as follow:

$$wear\ rate = \frac{weight\ loss}{sliding\ distance} \tag{3}$$

In other words, the wear rate means the weight variation of the pin specimen per sliding distance, and hence its unit is g/mm. Normal load in the range of 0 ~ 350 N was applied by increasing every 50 N and the rotational speed of the disc was fixed by 200 RPM. This rotational speed was appropriately chosen so that the lubricant on the reservoir could follow up to the rotating disc.

Fig. 4 shows experimental configuration to measure tangential friction coefficient. The strain gage was directly mounted on the pin specimen to measure strain(ε). Then, the friction coefficient (μ) can be obtained by

$$\mu = \frac{(\varepsilon EI / cx)}{P} = \frac{N}{P} \tag{4}$$

Here E and I are Young's modulus and area inertia moment of the specimen, repectively.

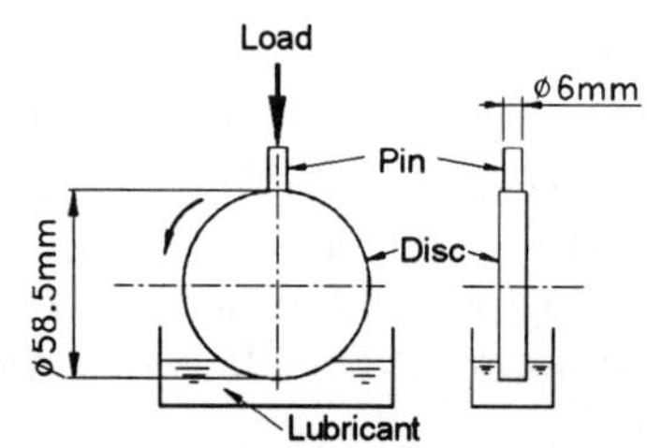

Fig. 3 Pin-on-disc test apparatus

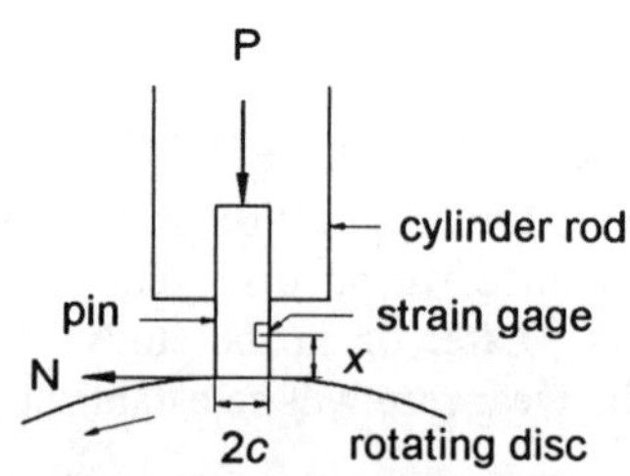

Fig.4 Configuration of friction measurement

The coefficient was measured for 10 seconds after rotating the disc by 200 RPM. A set of linear bearings were employed to minimize the loss of the applying normal loads due to friction between cylinder rod and guide rails.

The worn surfaces of the pin specimens were examined using scanning electron microscope(SEM) to validate the measured wear characteristics. Before the examination, all of the specimens were cleansed using ultrasonic cleaner and by alcohol. The specimens were run by 1000 m of the sliding distance under normal load of 185 N before the examination.

4. Results and Discussion

Fig. 5 shows wear rates of the pin specimens with respect to the normal load. Fig. 5(a) compares the wear rates among the base liquids. The largest wear rate was occurred under the silicone oil lubricant IXSX-11. As given in Table 1, the viscosity of the IXSX-11 is greater than that of the IXSX-9. However, the wear characteristics over 200N are very similar each other. The case of the transformer oil lubricant IXTX-6 produces the smallest wear rate. This is attributed to superior adhesion property of the transformer oil to the metallic surface. Fig. 5(b) presents the wear rates of the specimens which were run under the silica-particle ER fluids made of different base liquids. The particle concentration by weight was fixed to be 15%. It is clearly observed that the wear phenomenon under the ER fluids is very similar to that under the base liquids only shown in Fig. 5(a). For the case of the silicone-oil based ER fluid, the wear magnitude is basically same as that for the case of the silicone oil only. However, for the cases of the transformer oil and olive-oil based ER fluids, the wear magnitudes become much smaller than those under the corresponding base liquids only. This is due to that the particles of the mixtures play a role of anti-wear materials between contact films by reducing metal-metal contact area. Furthermore, from Figs. 5(a) and 5(b), we know that the effect of the base liquid property on the wear is much more significant than the effect of the existence of the particle. Fig. 5(c) compares the wear rates under the transformer-oil based ER fluids with different particle concentrations. As expected, the wear rate increases as the particle concentration does. For the case of the 40%, ILTH-6C, the wear rate is much greater than the case of the base liquid only. If the particle concentration is beyond a certain level, the quantity of the lubricant oil is relatively reduced and hence the wear between two specimens is accelerated. However, for the cases of the 15%(ILTC-6C) and 30%(ILTF-6C), we have smaller wear rates than the case of the corresponding base liquid(IXTX-6). As early mentioned, this arises from the fact that the particles of the contact film play a role of sliding bearing to reduce abrasive wear.

Fig. 6 presents friction coefficients under the base liquids only and the ER fluids. Fig.6(a) compares the coefficients among three different base liquids. As known from Amontons' law, the friction coefficients do not vary with respect to the normal load for three cases. The coefficients were evaluated by 0.47, 0.11 and 0.29 for the silicone oil, olive oil and transformer oil, respectively. The coefficient of the olive oil is smaller than that of the transformer oil. This phenomenon contradicts to the wear characteristics shown in Fig 5(a). This arises from the fact that contribution factors to wear characteristics involve applied loads, molecular structure and adhesive property, while the friction is only function of the loads [13,14]. We see from the Fig. 6(b) that the friction coefficient is much more influenced by the type of the base liquid than the existence of the particle. And it is observed from Fig. 6(c) that the coefficient increases as the particle concentration increases.

Fig. 7 presents worn surfaces of the pin specimens which were taken by the SEM. The surface under the lubricant of the transformer oil IXTX-6 is finer than that under the lubricant of the silicone oil IXSX-9. From the surface of the IXSX-9, granular craters

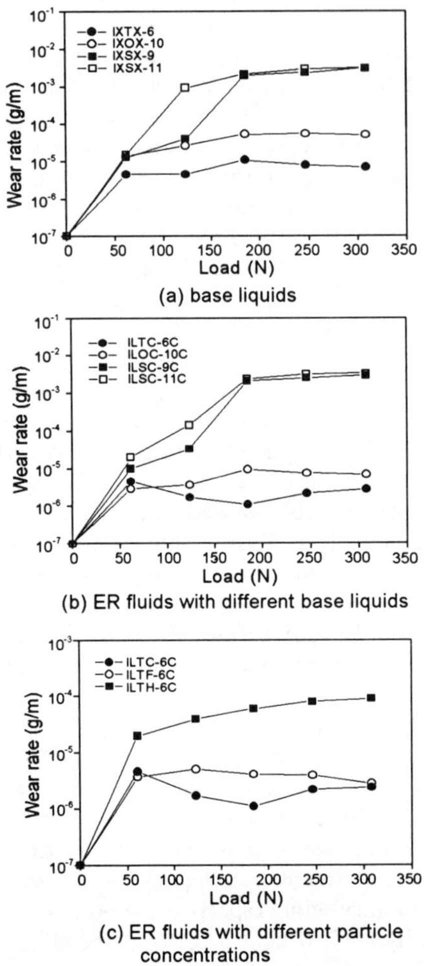

(a) base liquids

(b) ER fluids with different base liquids

(c) ER fluids with different particle concentrations

Fig.5 Wear rates of the pin specimens

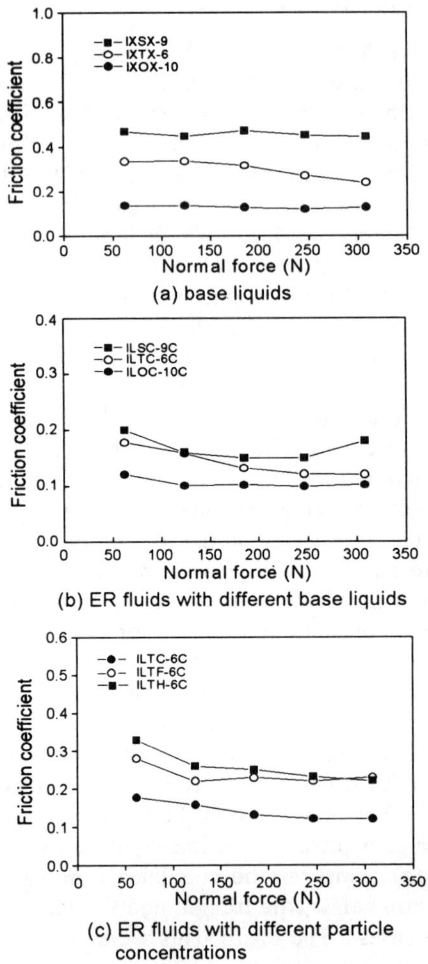

(a) base liquids

(b) ER fluids with different base liquids

(c) ER fluids with different particle concentrations

Fig.6 Friction coefficients under the base liquids and the ER fluids

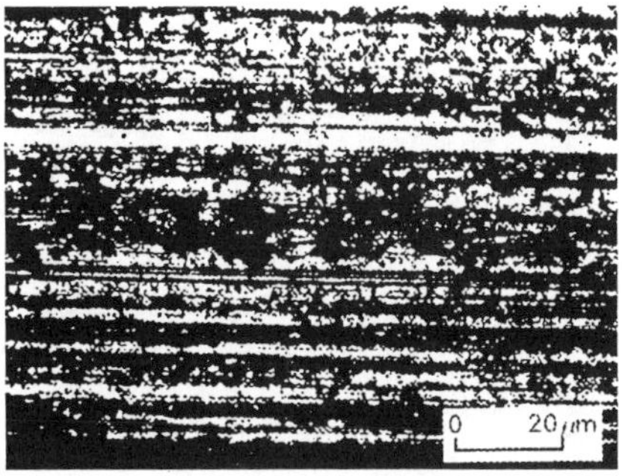

(a) IXSX-9, ×100

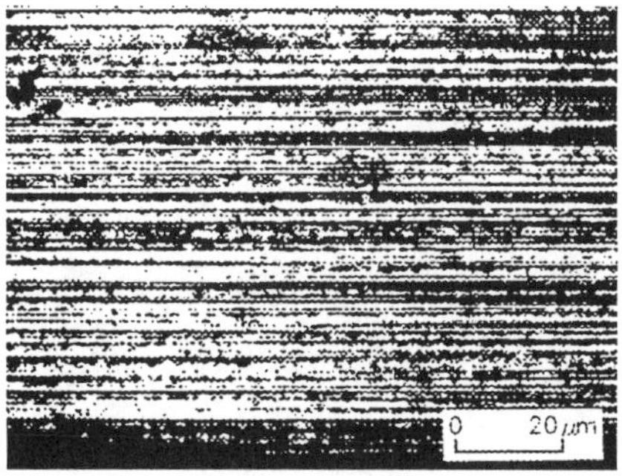

(b) IXTX-6, ×100

(c) ILTC-6C, ×100

(d) ILTH-6C, ×100

Fig. 7 Worn surfaces of the pin specimens

associated with the microcutting due to asperities were occasionally observed. Fig. 7(c) shows the worn surface under the ER fluid with the particle concentration of 15%. We clearly see very fine groves and they are very regular. On the other hand, the worn surface under the 40% of the ER fluid ILTH-6C exhibits large scale delamination. And it is obviously seen that adhesive wear has been occurred.

The objective of this experimental study is to investigate the wear and lubrication characteristics of the ER fluid. However, the comparison between the yield stress and wear rate will be very useful for the guideline of ER applications. Fig. 8 compares these two properties. From this figure, we see that the higher yield stress, the greater wear rate. Therefore, from the practical point of view an optimal choice of a certain ER fluid should be considered for appropriate ER devices.

5. Conclusions

Wear and boundary lubrication characteristics of the silica-based ER fluids were experimentally investigated. It has been shown that the wear and friction properties were much more influenced by the base liquid property than the existence of the particles. Both the wear rate and the friction coefficients increase as the particle concentration increases. However, below certain level of the particle concentration, the wear rate of the ER fluids was smaller than that of the base liquid only. These are due to the fact that the ER

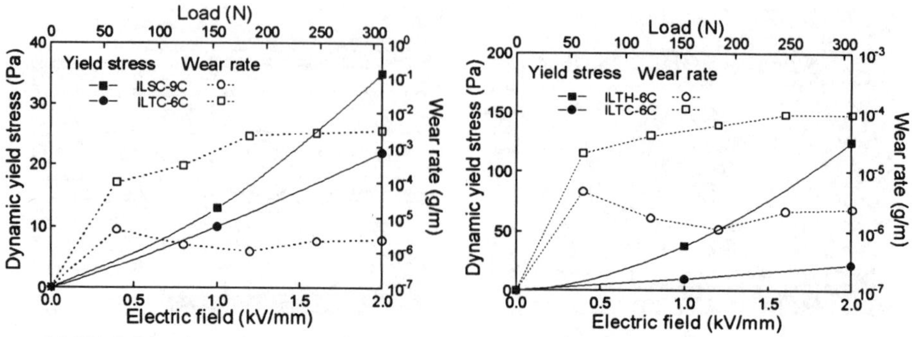

(a) ER fluids with different base liquids

(b) ER fluids with different particle concentrations

Fig.8 Comparison between wear rate and dynamic yield stress

particles play a role like sliding bearings at the contact surfaces. The experimental results presented in this study indicate that an optimal choice of the ER fluid for specific application should be taken into consideration. Tribology characteristics associated with different particles and different materials of the specimens are to be further studied.

References

1. W.M.Winslow, *J. Appl. Physics*, **20**, (1949) 1137~1140.
2. M. V. Gandhi and B. S. Thompson, *Smart Materials and Structures*, (Chapman & Hall, 1992) 147~148.
3. H. Block and J. P. Kelly, *J. Phys.D : Appl. Phys.*, **21**, (1988) 1661~1677.
4. J. E. Stangroom, *Electro-Rheological Fluids*, (ER Fluid Developments Limits, 1992) 22~23.
5. R. Tao, *Phys. Rev. E*, **47**, (1993) 423~456.
6. T. B. Jones, *Proc. 2nd Int. Conf. on Electro-rheological Fluids*, (1990) 14~25.
7. S. B. Choi, Y. K. Park and M.S. Suh, *AIAA Journal*, **32**, (1994) 438~440.
8. D. A. Brooks, *Proc. 4th Int. Conf. on Electro-rheological Fluids*, (1993) 643~656.
9. N. K. Petek, R. J. Goudie and F. P. Boyle, (1988) SAE 881785.
10. S. Lingard and W. A. Bullough, *J. Phys. D: Appl. Phys*, **22**, (1989) 1639~1645.
11. S. Lingard, W. A. Bullough and W. M. Shek, *Wear*, **142**, (1991) 373~381.
12. T. C. Jordan and M. T. Shaw, *IEEE Transaction on Electrical Insulation*, **24**, (1989) 849~878.
13. A. Glaeser, *Wear*, **73**, (1962) 371~386.
14. C.Y. Tung and S. K. Hsieh, *Lubricating Engineering*, **44**, (1987) 856~865.

STRUCTURE AND INTERACTION FORCE IN A MODEL MAGNETORHEOLOGICAL SYSTEM

X. Tang, Y. Chen and H. Conrad
Materials Science and Engineering Department
North Carolina State University
Raleigh, N. C. 27695–7517 USA

ABSTRACT

The force-displacement behavior of specific arrangements of 1.6 mm dia. steel spheres in a magnetic field of 4.2 kGs was determined in tension and shear. The maximum attractive stress between the particles in a single-row chain in *tension* was 411 kPa, independent of the number of spheres in the chain beyond four. The *shear* stress τ-strain γ curve of a single-row chain exhibited a single peak with τ_{max} = 248 kPa at a shear strain γ = 0.32, providing the spheres in contact with the magnet walls were glued in position. Without gluing slippage of a chain or cluster occurred during shearing, giving a maximum stress only about 1/6 that without slippage. The τ vs γ curves for clusters of particles exhibited multiple peaks which corresponded to the movement of the spheres in breaking and reforming bonds. The maximum tensile and shear stresses tended to be smaller for clusters compared to equivalent single-row chains when the number of particles in a row is small.

1. Introduction

Magnetorheological (MR) fluids consist of fine (~ μm) magnetic particles in a host liquid. Upon application of a magnetic field the particles interact and form a chain-like structure aligned in the field direction[1,2]. These fluids exhibit yield stresses of 10 to 100 kPa and have found commercial application. As a result, they have received considerable scientific and technological attention[1–8].

Our understanding of MR behavior is still incomplete. At present, the theories are mostly based on a two-particle model or a single-row chain model. The nonlinearity of the magnetization of the particles increases the difficulty of a theoretical analysis[1,6]. Hence, it is not surprising that the measured static yield stress of an MR suspension was 40% lower than the theoretical prediction, even though the nonlinearity was taken into account[2]. There is thus need for additional definitive measurements on MR suspensions. Especially important is the influence of the arrangement or structure of the magnetic particles on the strength of a suspension subjected to various stress states.

In the present paper we present the results obtained on the influence of the arrangement of the particles on the quasi-static stress-strain behavior of a model MR fluid consisting of 1.6 mm dia. chrome steel balls in air. To evaluate the effect of stress state, tests were performed in both tension and shear.

Considering the interaction force, a sphere of radius a and magnetic permeability μ_p acquires under a uniform magnetic field H_o a magnetic moment[9]

$$M_p = 4\pi\mu_o\mu_f a^3 \beta H_o \tag{1}$$

where $\beta = (\mu_p - \mu_f)/(\mu_p + 2\mu_f)$ and μ_o is the permeability of free space. The subscript p and f refer to the particle and host medium respectively. The point-dipole approximation gives for the interaction force between two magnetized spheres separated by a distance R and aligned in the field direction

$$F = 24\pi\mu_o\mu_f a^2 (\beta H_o)^2 (a/R)^4 \tag{2}$$

Moreover, a magnetic dipole of strength M_p in a medium of permeability μ_1, facing a magnetic wall of permeability μ_2 will generate an image dipole M_p' at its Kelvin image point[1,10], which is given by

$$M_p' = (\mu_2 - \mu_1) M_p / (\mu_2 + \mu_1) \tag{3}$$

Because there exists an interaction force between the magnetic source and the image dipoles, a permeable sphere will sense an attractive force in front of the pole of a magnet. Hence, if we measure the force between two spheres near a magnet pole, the results are actually from the interaction of four magnetic dipoles, two from source dipoles and two from their images. Therefore, to compare the theoretical analysis for a two-sphere interaction with measurements, one must have the test spheres as far as possible from the magnet pole.

As the field is increased, magnetic saturation may occur and affect the interaction between particles. Ginder and Davis[6] have developed a numerical method to estimate this effect for an infinite single-row chain. However, for more complex arrangements such as a double-row chain and other close-packed lattices, it is more difficult to use the numerical method because of the nonsymmetric field distribution and the complicated geometry.

The present experiments were performed to check the above theoretical considerations regarding some specific arrangements of the spheres and their response in both tension and shear.

2. Experimental

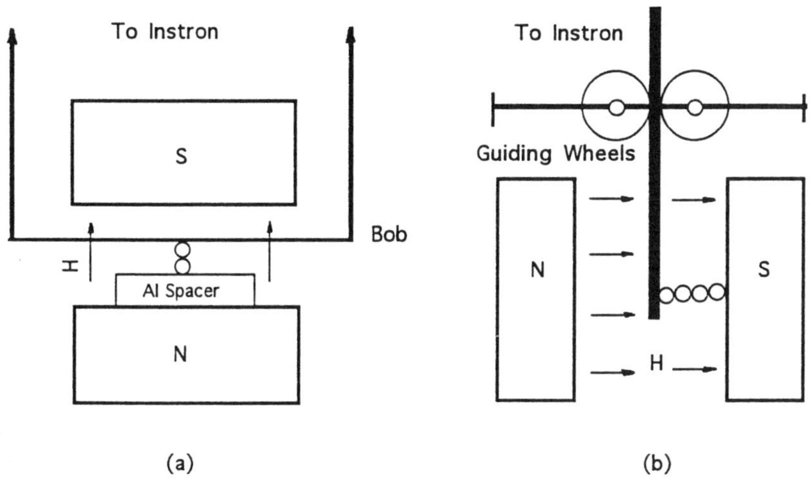

Fig. 1 Schematic of test apparatus: (a) tension and (b) shear.

The magnetic field source was a permanent magnet with a magnetic induction B in the air gap of 4.2 kGs measured by a Model 4048, F. W. Bell gauss meter. Chrome steel balls 1.6 mm dia. were used for the magnetic particles. The tensile and shear forces were measured with an Instron Model 1123 testing machine with a load sensitivity of ~ 1g. Schematics of the test arrangements for tension and shear are shown in Fig. 1. The steel balls were first arranged into single-, double- or triple-row chains and then subjected to loading at a constant crosshead spread of 8.5×10^{-3} mm s^{-1} in tension and 8.5×10^{-2} mm s^{-1} in shear, i.e. in the quasi-static regime.

3. Results and Discussion

3.1 Tensile Loading

For the initial tests, a glass bar was used for the bob in Fig. 1a to reduce the image force. Single-row chains containing two, three and four balls were tensile loaded and their response observed by eye. It was noted that the separation of the spheres did not occur uniformly, but rather was localized in the manner shown in Fig. 2. Plots of the measured force vs the relative separation of the spheres $S/a = (l - l_0)/a$ are presented in Fig. 3. l is the instantaneous length of a chain, l_0 the initial length and a the sphere radius. The solid line depicted is calculated using Eqn. 2, i.e., the point-dipole approximation. From Fig. 3 we note the following: (a) the attractive force decreases markedly with separation of the spheres, (b) the

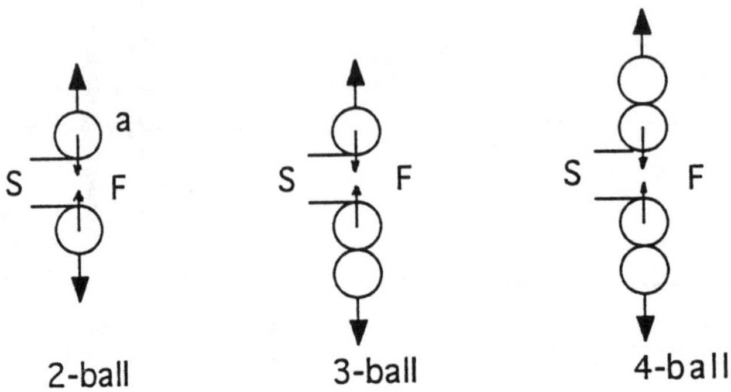

Fig. 2 Illustration of non-uniform separation of the spheres.

total (localized) separation seems to be the dominant factor, rather than a uniform separation of the spheres usually considered in most theories, (c) the number of balls in a chain have only a small effect on the force and (d) at large separations reasonable agreement occurs between the measured force and that predicted by the point-dipole approximation, but a significant difference occurs at small separations where the force seems to drop with separation exponentially.

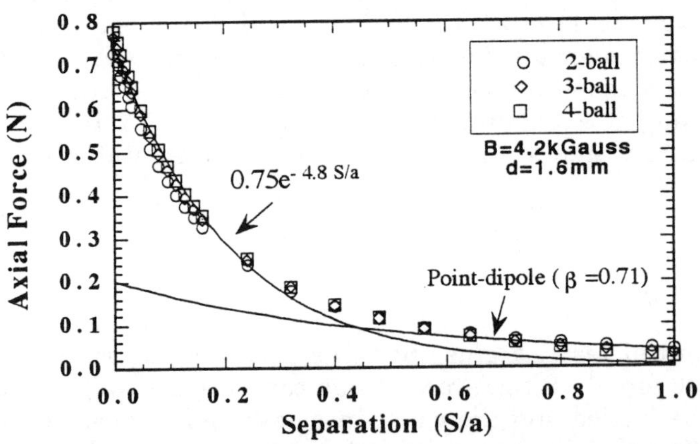

Fig. 3 Measured tensile force vs normalized separation for single-row chains containing 2, 3, and 4 spheres. Also shown is the curve for point dipole approximation with $\beta = 0.71$.

The maximum force F_{max} to break 2-, 3- and 4-ball single-row chains was 74.5, 79.6 and 83.5g, respectively. Upon removing the Al apacer and allowing the balls to be in contact with the magnet pole piece, F_{max} was 83g for all three chains. This agreement results from the fact that due to image forces the measured forces represent the interaction of 4, 6 and 8 balls. The results thus indicate that the force increases with number of spheres in a chain up to a total of four and thereafter remains relatively constant. By taking the maximum tensile strength of a single-row chain to be $\sigma_{max} = F_{max}/\pi a^2$ we obtain for the present tests σ_{max} = 411 kPa. Although the particles used here (1.6 mm dia.) are much larger than those (of the order of microns) normally employed in MR fluids, it is expected according to Eqn. 2 that for a given value of R/a, σ_{max} will be relatively independent of the size of the particles. *This suggests that it may be possible to develop MR fluids with a maximum tensile strength of several hundred kPa.*

Additional arrangements of the spheres that were investigated in tension are shown in Fig. 4. For these tests the Al spacer was replaced by an iron sheet. The filled circles in Fig. 4 indicate that these balls were glued onto the bob (glass) or the pole of the magnet. To ascertain the influence of structure on the tensile strength, we will define a structure factor SF to be the ratio of F_{max} for a given cluster to that for an equivalent number of single-row chains, i.e.

$$SF = \frac{F^c_{max}}{F^s_{max}} \times \frac{N^s}{N^c} \qquad (4)$$

where N is the number of balls in the arrangement and the superscripts s refers to a single-row chain and c to a cluster. The maximum tensile strength of a cluster is then $\sigma_{max}{}^c = SF \times \sigma_{max}{}^s$.

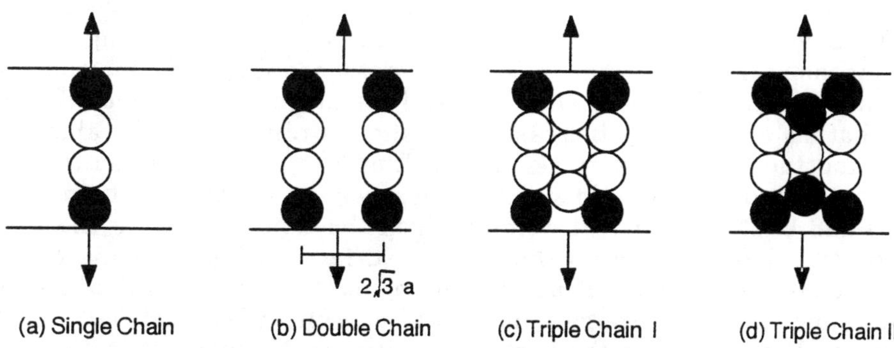

Fig. 4 Clusters of spheres tested in tension. Filled circles are spheres glued to the bob or magnet.

The values of the structure factor obtained for the arrangements shown in Fig. 4 are listed in Table 1. To be noted regarding the double-row chain in Fig. 4b is that the structure factor approaches unity as the number of balls is increased. However, for the triple-row chains (Figs. 4c and 4d) the structure factor remains relatively independent of the number of balls, being lower than that of the equivalent single-row chains. Further, the arrangement in Fig. 4c has a lower strength than that in Fig. 4d. These results suggest that clustering of the particles in MR fluids may lead to a reduction in tensile strength of up to ~ 40%, when the number of particles in a row is less than 10 (including the images)

Table 1. The structure factor for the arrangements shown in Fig.4.

No. of Spheres/Chain		Structure Factor(SF)			
	Arrangement	a	b	c	d
2		1.0	0.89	0.56	—
3		1.0	0.96	0.54	0.82
4		1.0	0.98	0.56	0.83
5		1.0	—	0.57	0.82

3.2 Shear Loading

Presented in Fig. 5 is the shear stress τ (= $F_s/\pi a^2$) vs shear strain γ (= δ/h) curve obtained for a single-row chain (Fig. 4a) using the test arrangement shown in Fig. 1b. δ is the crosshead displacement and h the spacing between the center sheet (Al) and the magnet pole. Evident in Fig. 5 is that the shear stress increases with shear strain to a peak value at γ = 0.32 and then decreases markedly. Taking the peak stress as the yield stress, the quasi-static yield stress in shear for the single-row chain in Fig. 4a is 248 kPa. However, if we only glued the contact balls to the center sheet, the chain slipped on the magnet surface during shearing and the yield stress was 43 kPa. Thus, the interaction force between the balls in a single-row chain is stronger than that between the contact ball and the magnet. It follows that the surface condition of the magnet walls can be important in the strength of MR fluids. Based on the point dipole approximation and image theory, we calculated the attractive force at a rough magnet wall and found that when the roughness of the wall surface is 2–3 times the radius of the particle, the attractive force between the particle and the wall will be ~ 2 times stronger than the particle-particle interaction[11]. Thus, well-defined roughness of the magnet walls can be important in the strength of MR fluids.

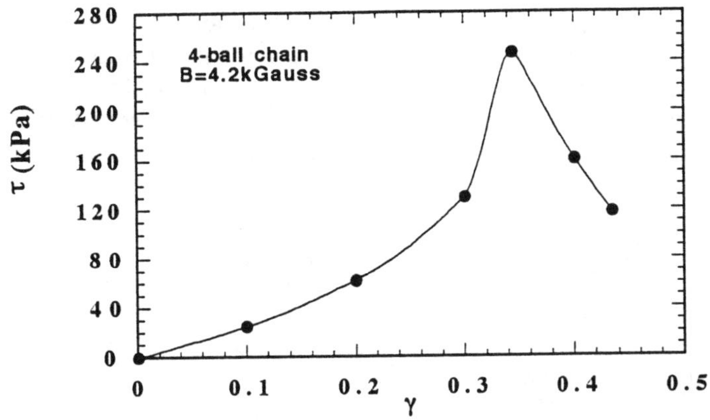

Fig. 5 Shear stress vs shear strain for a single-row chain.

Fig. 6 shows the shear response of some particle arrangements which might occur in MR fluids. Unlike the curve in Fig. 5 for a single-row chain, there are now a number of peaks in the τ vs γ curve. Each peak corresponds to the breaking and reformation of specific bonds, with sliding of an adjacent ball into an opened space. Of special note is the anisotropic behavior of the two double-row chains in C and D. Although they have a mirror-image configuration, their response depends on the position of the second row with respect to the direction of shear.

The results in Fig. 6 indicate that the break-and-reform process does not necessarily lead to an enhanced yield (maximum) stress. On the contrary, in some cases a reduction in the yield stress occurs. For a triple-row chain with a BCT structure, it is expected that there should occur even more frequent break-and-reform processes leading to more peaks, and moreover may behave in an anisotropic manner with respect to shear direction. The present results thus suggest that the single-row chain structure gives the highest yield stress in shear.

4. Summary and Conclusions

4.1. Tension

a. The maximum attractive stress between steel spheres in a single-row chain was 411 kPa, independent of the number of particles in the chain beyond four. This stress was about four times that predicted by the point dipole approximation.

b. Although the maximum stress for a double-row chain with the specific separation distance d = $2\sqrt{3}a$ was approximately the same as that for an equivalent number of single-row chains, arrangement of the spheres in close-packed, triple-row clusters gave a reduction of up to 40% in the maximum stress.

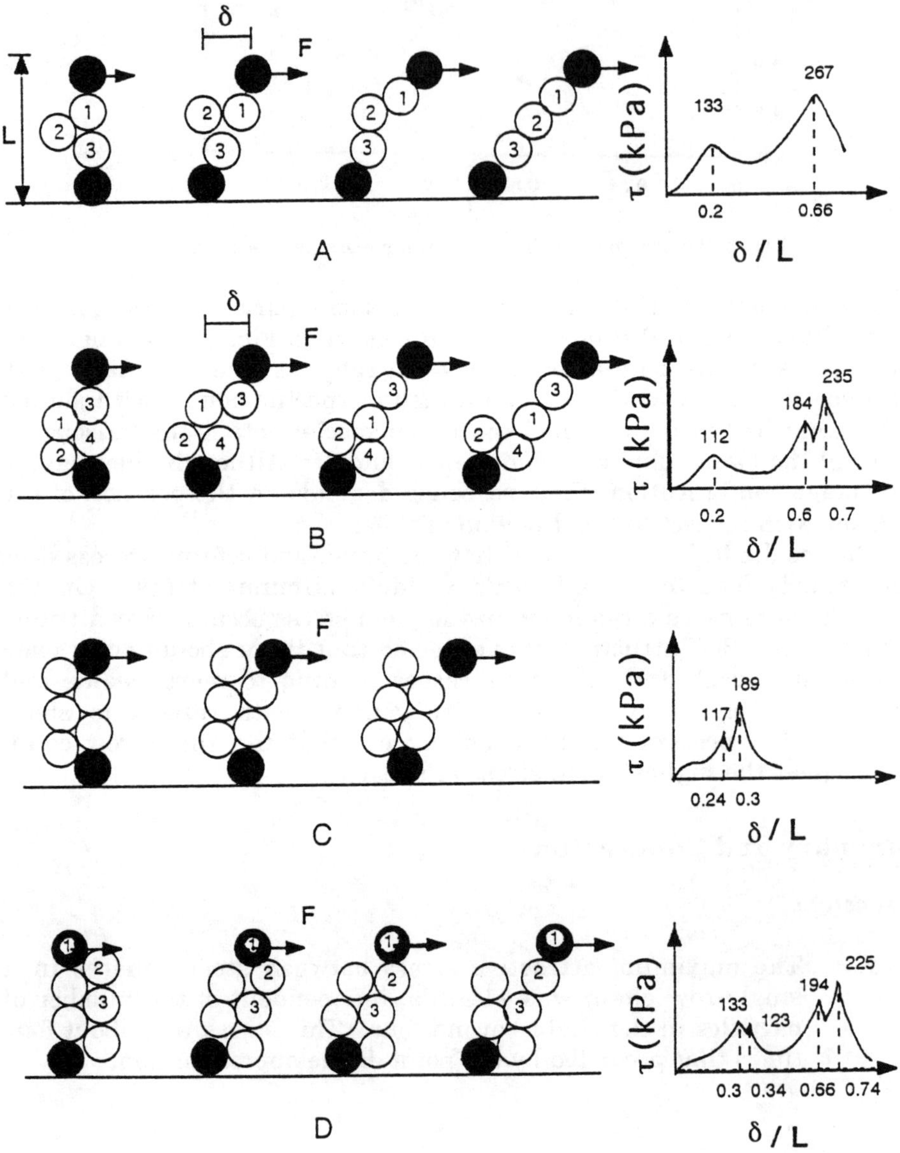

Fig. 6 Schematic of clusters tested in shear and the resulting stress-strain curves.

c. Although calculation based on the point-dipole assumption have shown that double-chain and triple-chain structures have an energy lower than that of a single-chain structure only for more than 7 particles in a row[12], no significant dependence of the strength on the number of particles in a row was found in our experiments with up to 10 particles in a row (including the images). The discrepancy between theory and experimental results may come from the following: (a) magnetization saturation often occurs in MR fluids, which may reduce the interaction among the close-packed particles. (b) the structure deforms in a complex manner up to the point where breaking and reforming bonds occurs rather than in the uniform manner assumed in most theories, and (c) the point-dipole approach may not be strictly valid for a close-packed structure. However, further experiments with larger number of particles in a row are needed to evaluate in more detail the effect of structure on the interaction force.

4.2. Shear

a. The shear stress-strain curve of a single-row chain exhibited a single shear stress peak $\tau_{max} = 248$ kPa at a shear strain $\gamma = 0.32$, providing that the spheres in contact with the magnet walls were glued in position to prevent sliding.

b. The shear stress-strain curves for clusters of particles exhibited multiple peaks which corresponded to the movement of the spheres in breaking and reforming bonds. The magnitude of the highest peak depended on the initial arrangement of the particles, tending to decrease with increase in number of particles in the cluster.

c. The attractive force between the spheres was greater than that between the end spheres and the magnetic walls so that slippage of a chain or cluster occurred during shearing, giving a maximum stress only about 1/6 that when slippage was prevented.

5. Acknowledgements

This research was jointly supported by the NSF Engineering Division under Grant CTS–931387 with Dr. M. C. Roco as technical monitor and the NCSU/Industry Consortium on ER and MR Fluids including Bridgestone, Ford, Paar-Physica and Texaco.

6. References

1. E. Lemaire and G. Bossis, J. Phys. D: Appl. Phys. 4(1991)1473.
2. G. Bossis, H. Clerex, Y. Graselli and E. Lemaire, Proc. of the 4th Int. Conference on ER fluids: Electrorheological Fluids: Mechanisms, Properties, Technology and Applications, edited by R. Tao & G. D. Roy, World Scientific, Singapore, (1994) p. 153.
3. Z. P. Shulman, Magnetorheological Suspensions, Int. J. Multiphase Flow, 12(1986)935.
4. W. I. Kordonsky, J. Intelligent Material Systems and Structures, 4 (1993)65.
5. J. D. Carlson, Actuator 94(1994) p. 266.
6. J. M. Ginder and L. C. Davis, Appl. Phys. Lett. 65(1994)3410.
7. K. D. Weiss, SAE Technical Paper Series 932451, 1993 International Off- Highway & Powerplant Congress & Exposition, Milwaukee, Wisconsin, Sept. 13–15, 1993.
8. V. I. Kordonsky, J. Magnetism and Magnetic Materials, 85(1990)114.
9. M. Zahn, Electromagnetic Field Theory: A Problem Solving Approach, 1979.
10. I. V. Lindell and E. A. Lehtola, IEEE Trans. Magnetics, 28(1992)1930.
11. X. Tang and H. Conrad, "Interaction Between a Permeable Sphere and a Rough Magnet Pole", North Carolina State Univ. (1994) Raleigh, N. C., to be submitted for publication.
12. G. L. Gulley and R. Tao, Proc. of the 4th Int. Conference on ER fluids: Electrorheological Fluids: Mechanisms, Properties, Technology and Applications, edited by R. Tao & G. D. Roy, World Scientific, Singapore, (1994)p.314.

ADDITIONAL MAGNETIC DISPERSED PHASE IMPROVES THE MR-FLUID PROPERTIES

W.I.KORDONSKY,S.A.DEMCHUK

Academic Scientific Complex "A.V.Luikov Heat and Mass Transfer Institute"
Academy of Sciences, Republic of Belarus and Byelocorp Scientific Inc.,
New York, USA

ABSTRACT

The main characteristics of the MR-fluid are the stability and sensitivity of its effective viscosity to an applied magnetic field. Such systems must be well fluid outside the applied field, as the viscosity increases many times in the magnetic field. The method is proposed for improving the MR-fluid properties by adding an additional high-dispersed magnetosolid powder to it. A relatively small content of magnetosolid particles causes a more than 9-fold increase of the MR-fluid viscosity in the magnetic field.

INTRODUCTION

A magnetorheological fluid (MR-fluid) is a new material with controlled rheological properties. It is a stable mixture of noncolloid magnetic suspended particles (about 1 μm) in the fluid which are not subjected to Brownian motion and forms rather a stable structure when acted on by the applied magnetic field. Such a system induced by the applied field manifests itself in the nonlinear rheological effects, i.e. in a sharp and controlled increase of the effective viscosity, plasticity, and viscoelasticity in the field. Thus, the MR-fluid allows direct electric signal control of the flow, heat and mass transfer, and the electric and magnetic characteristics of a fluid medium which, in its turn, provides new processes and equipment[1,2].

The main characteristics of the MR-fluid are the stability and sensitivity of its effective viscosity to an applied magnetic field. By the stability is understood the aggregation and sedimentation stability of such systems which, depending on the type and concentration of the dispersed phase, are provided both by their direct structuring due to the interaction of the molecular forces of particles and by the injection of protective colloid-type substances (such as silica gel,silicon oxide, etc.) into a dispersed medium that form structural-mechanical barriers between particles. The strength of these barriers should be such as to prevent sedimentation and spontaneous irreversible coagulation of magnetic particles.

In the case of sufficiently concentrated MR-fluids, this stability may be achieved by introduction of surfactants and formation of a spatial structure just from ferromagnetic particles. Some peculiarities in the mechanical properties of MR-fluids are brought about by the nature of the dispersed-phase material, viz. magnetosoft or magnetorigid. However, the provision for the stability is a necessary but not yet sufficient condition for a high-effective MR-fluid to exist. Its capability to reversible thixotropic conversions outside a field is still important. The MR-fluid viscosity must decrease sharply under shear flow conditions. The last situation completely specified by the physical and chemical properties of protective colloid is especially of importance in using MR-fluids as a working medium of hydraulic systems.

Usually the classical MR-fluid contains a magnetic dispersed phase of the type, either magnetosoft or magnetorigid. In the case of magnetosoft many-domain particles, for example, carbonyl iron the MR-fluid structure is distorted even at every small shear rate after the field has been cut off. MR-fluids of single-domain particles "store" the viscosity change due to the applied magnetic field even after the field has been cut off. That is why, in this paper, the method is proposed for improving the classical MR-fluid properties by adding an additional high-dispersed magnetosolid powder to it. Such particles form a spatial structure in the carrying medium due to the intrinsic magnetic moment in the absence of the magnetic field, i.e. they serve as an additional stabilizer.

On the other hand, unlike classical nonmagnetic thickening agents, such particles when acted on by the magnetic field play an active role, thereby interacting with more coarse particles of the main magnetic dispersed phase.

As a result, a more stable structure is formed, and this makes an additional contribution to the magnetorheological effect, i.e. the MR-fluid efficiency increases. When the field is cut off, the system reverts to its original stable state.

The above-said is supported by the results of the experiments on determing the type of the magnetosolid dispersed phase and the quantitative relations between the magnetic dispersed phases. As usual, a study of the dynamic structure of MR-fluids encounters some difficulties. The present paper is an attempt to implement MR-fluid microstructure diagnostics by simultaneous measurement of their macroscopic, i.e. magnetic and rheological, characteristics.

EXPERIMENTS AND DISCUSSION OF RESULTS

It is known that in a constant magnetic field the MR-fluid shows a pronounced non-Newtonian pattern of flow. Starting with some shear rate (in this case ,$\dot{\gamma}$=100 s^{-1}), the shear increment induced by a field practically stops independing on the shear rate. The effective viscosity increases with the field and at the same time decreases with the shear rate as shown in Fig.1. Rheological measurements are made over the range of the field strength H= 0 to 84 kA/m; shear rates $\dot{\gamma}$=0.3 to 444 s^{-1}; temperatures T=293 to 323^0K; carbonyl iron P-10. The magnitude of the increase of the effective MR-fluid viscosity strongly depends on the orientation of the force lines of the field relative to the flow direction. The magnetorheological effect is the most pronounced at the mutual perpendicular orientation of the field force lines.

For a specific composition the magnetic field is shown to increase the effective viscosity by more than two orders, Fig.1[3].The rheological properties of MR-fluid in a magnetic field at a temperature from the Curie point of a dispersed ferromagnetic phase are specified by the temperature dependence of the medium viscosity alone. At the same time, the simultaneous effect of the magnetic field and MR-fluid heating up to temperatures close to the Curie point offers ample scope for a considerable increase of the thermosensitivity of the working medium viscosity up to a complete degeneration of the magnetorheological effect. A correlation is established between the temperature-dependent relative increase of viscosity in the field and squared relative magnetization[4].

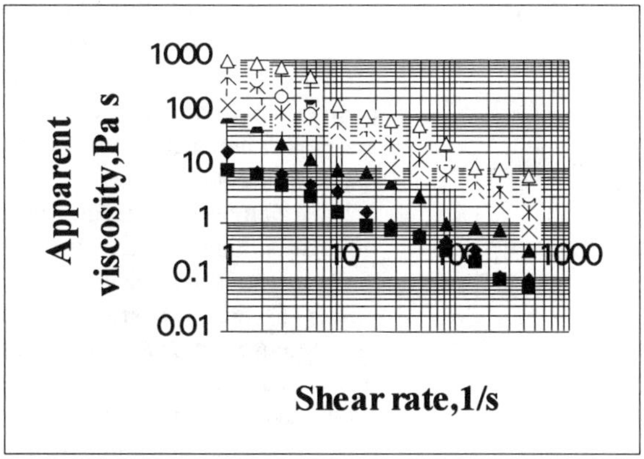

Fig.1. Flow curves for carbonyl iron P-10-based MR-fluid (1,H=0;2,12.7;3,24.2;4,35.0;5,43.6;6,48.0;7,62.0;8,77.0; 9,84.0 kA/m)

Other specific mechanical behavior manifest themselves in the MR-fluid of a magnetosolid material, e.g. of single-domain particles used as the dispersed phase. At low shear rates, the flow curve has a hysteresis, as shown in Fig.2, points 1; at high shear rates, there is no hysteresis, as shown in Fig.2, points 2. Moreover, the suspensions of the single-domain particles when acted on by the magnetic field show the elastic properties which were revealed by the method of free oscillation damping.

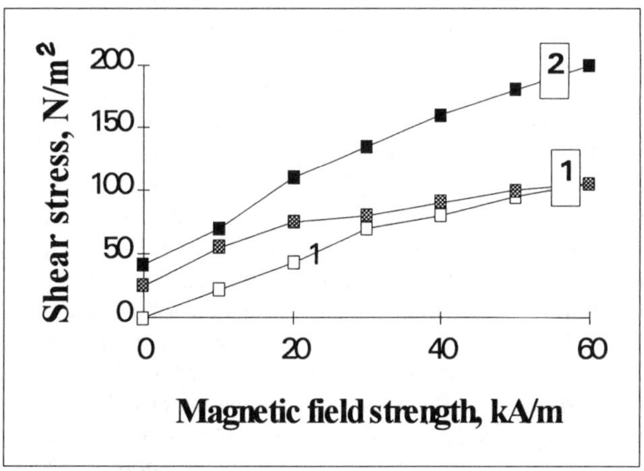

Fig.2. Shear stress as a function of applied magnetic field (φ=25 mass.%).

Suspensions of high dispersed single-domain needle-

-shaped iron gamma-oxide particles in the hydraulic fluid AMG-10 were the subject of study in the first set of experiments. The concentration of the ferromagnetic varied from 9 to 36 mass.%. The curves of flow and magnetization under shear deformation of the MR-fluid were measured on the co-axial cylindrical bell-type magnetorheometer. Qualitatively, the experiments displayed the same behavior of the rheological and magnetic properties of such systems when acted on by the magnetic field.

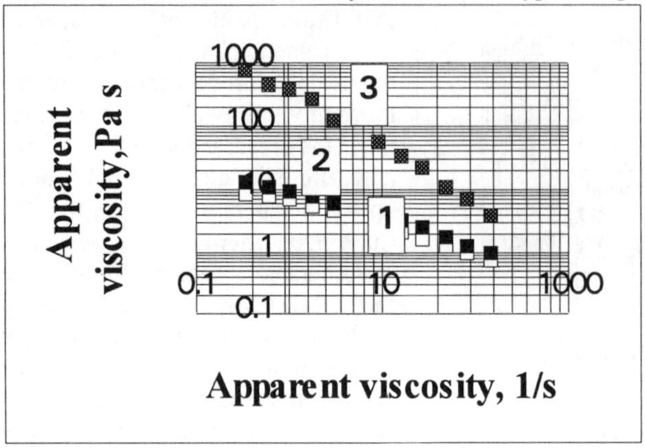

Fig.3.Flow curves for classical MR-fluid (1,H=0;3,H=52 kA/m) and "modernized" MR-fluid 2,H=0.

Namely, for weak shear flow the magnetization curve as well as the flow curve (Fig.2, points 1) has a hysteresis; for high-intense deformation the magnetization of such a MR-fluid occurs having no hysteresis (Fig.2,points 2). This means that a close relation exists between the magnetization of such systems and the state of their micro-structure because such systems are greatly magnetized only due to the orientation (ordering) of their structure. This also means that each level of magnetization is consistent with a completely determined magnetic structure of the particles.

Fig.3 plots flow curves for the classical carbonyl iron-based MR-fluid (spherical particles 3.5μm in size) thickened with a gel-forming addition that provides a sufficiently reliable spatial structure inside the MR-fluid keeping the ferromagnetic particles of the dispersed phase from depositing and sticking together.In addition, such a structure can be reversibly restored after shear deformation of the MR-fluid. This allows one to state that the gel-forming and thixotropic properties of such a system are stable. In the given MR-fluid, it is clearly seen that its magnetic properties and the particle microstructure are related.

But in contrast to the above-described MR-fluid having a magnetosolid dispersed phase, the magnetization curves for this MR-fluid have no hysteresis under any deformation conditions.Considering the above-said and the urgency of the problem on preparing effective and reliable MR-fluids, the following investigation was carried out. The magnetorigid powder was added to the classical MR-fluid having a magnetosoft dispersed phase (carbonyl iron powder). It served as the thickening and, at the same time, stabilizing addition. The rheological and magnetic properties of such a "modernized" MR-fluid were examined on the magnetorheometer under the Couette flow conditions ($\dot{\gamma}$=1.015 to 444.5 s^{-1};H=0 to 80 kA/m) at room temperatures. The iron gamma-oxide powder T-6 prepared by chemical oxidation of the magnetit was used as the high-dispersed magnetorigid addition. The powder T-6 particles were shaped as needles, the major axis length was 0.3 to 0.7 μm, the major-to-minor axis ratio was 10:1. These particles represented miniature permanent magnets which interacted in between and with coarse carbonyl iron particles, adhering to them with one of their ends. In the external magnetic field, the entire ferromagnetic dispersed phase of the MR-fluid was magnetized, and in doing so the iron gamma-oxide particles tended to align with their major axis along the field, formed as if a spatial structure, on the one side, serving in this case as an additional thickening agent, and, on the other, making specific brushes over the surface of the carbonyl iron particles. To our mind, the interaction of such rough formations in the shear flow of the MR-fluid gives rise to additional energy dissipation, i.e. an additional increase of the effective viscosity.

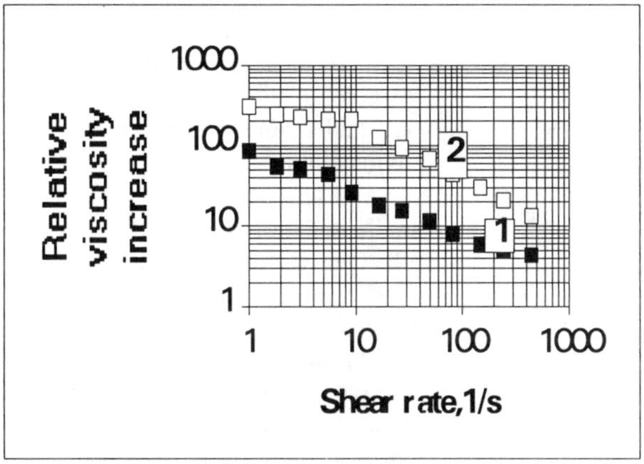

Fig.4.Relative increase of the apparent viscosity of classical carbonyl iron P-10-based MR-fluid (1) and "modernized" iron gamma-oxide powder-added MR-fluid (2);H=62kA/m.

Fig.4 plots curves for the effective viscosity increase of classical carbonyl iron-based MR-fluid (1) and iron gamma-oxide powder-added MR-fluid (2). As is seen, when the MR-fluid stability is preserved, the relative increase of the effective viscosity "modernized MR-fluid (2) with optimal concentration of the magnetorigid iron gamma-oxide powder exeeds almost 4 times that of classical MR-fluid (1). It should be mentioned that, as the conducted experiments have shown, the number of magnetorigid particles must be severely optimal at a given concentration of the magnetosoft particles. When their concentration is too small, the efficiency of the spatial structure formed by them, and of the roughnesses on the surface of the magnetosoft particles is not high; when their concentration is large, the initial viscosity of the MR-fluid increases while its efficiency decreases, because the particles begin to "hinder" each other to stick together, thus lowering the number of particles participating in "useful" formations. Moreover, since the magnetic permeability of the iron gamma-oxide particles is much lower than that of the carbonyl iron ones, in the last case the total permeability of the MR-fluid decreases, which involves a deterioration in the magnetorheological effect.

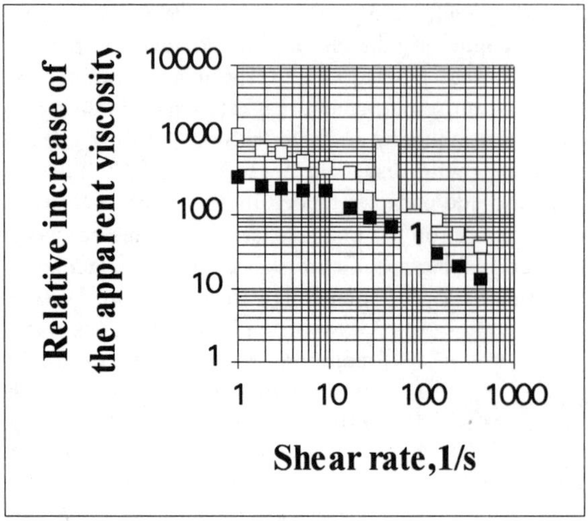

Fig.5.Relative increase of the apparent viscosity of iron gamma-oxide powder-added MR-fluid (1),and of chromium dioxide powder-added MR-fluid (2).

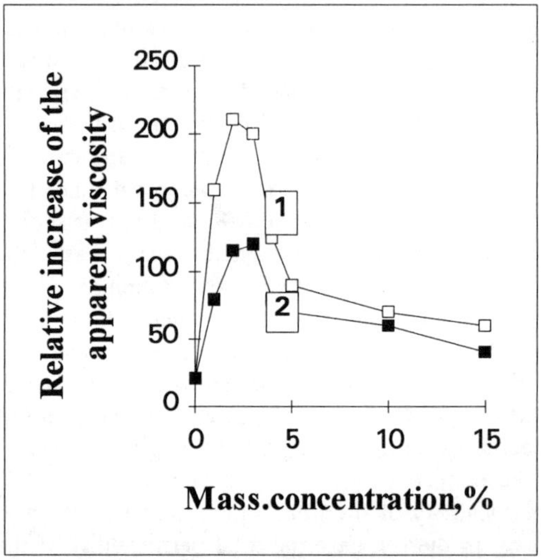

Fig.6.Concentration curves MR-fluids:of chromium dioxide powder-added (2) and of iron gamma-oxide powder-added (1).

In the next set of experiments, the magnetorigid chromium dioxide powder was added to the classical MR-fluid instead of the iron gamma-oxide powder. The particles of this powder were also shaped as needles, but the major-to-minor axis ratio constituted 15:1, the mean particle size was equal to 0.5μ m. The clearly pronounced needle shape of the particles,is, to a considerable extent,responsible for the rectangular outline of the hysteresis loop. This improves residual magnetization and brings it to maximum.

The conducted rheological studies of the relative increase of the effective viscosity of the chromium dioxide powder-added MR-fluid when acted on by the magnetic field having an intensity of 62 kA/m (Fig.5,points 2) have revealed a more than 3-fold growth, as compared to the iron gamma-oxide powder-added MR-fluid(Fig.5,points 1), under all other things being equal.

The improvement of the viscosity increase of the chromium dioxide powder-added MR-fluid is apparently attributed both to the monocrystal structure of the particles and to a more pronounced needle shape. As compared to the iron gamma-oxide particles, this specifies their maximal and residual magnetization and, as a result, thefficiency of the spatial structure formed by them and the "rigidity" of the rough formations on the surface of the carbonyl iron particles.

Besides, here as in the case of the iron gamma-oxide addition the optimal relation exists between the dispersed phase particles when a maximal increase of the effective viscosity is seen, under all other things being equal, Fig.6. However, in this case the magnitude of the optimal concentration of chromium dioxide particles decreases (Fig.6,points 1)as against that of iron gamma-oxide ones (Fig.6,points 2). This is explained by higher magnetization of the chromium dioxide particles and their more elongated shape. The magnetization curves for the investigated MR-fluids remained practically unchanged, as compared to the classical MR-fluid.

CONCLUSIONS

Thus, the performed set of experiments is evident of the fact that a relatively small addition (3-10 mas.%) of high-dispersed magnetorigid particles to the classical MR-fluid causes a more than 9-fold increase of its effective viscosity in the magnetic field, when its main characteristics are preserved outside the field.

REFERENCES

1.Z.P.Shulman and V.I.Kordonsky, *Magnetorheological Effect* (Nauka i Tekhnika, Minsk, 1982) p.184.
2.J.David Carlson and Keith D. Weiss, *Machine Design*, **8** (1994) 61.
3.V.I.Kordonsky,Z.P.Shuiman,S.r.Gorodkin,S.A.Demchuk at al,*Journal of Magnetism and Magnetic Materials,* **85** (1990) 114.
4.Z.P.Shulman, V.I.Kordonsky, E.A.Zaltsgendler, I.V.Prokhorov, B.M.Khusid and S.A.Demchuk, *Magnetohydrodynamics*, **3** (1984) 3.

A VARIABLE STIFFNESS ELEMENT UTILISING ELECTRORHEOLOGICAL FLUID ENABLING TACTILE FEEDBACK IN VIRTUAL ENVIRONMENTS

P. M. Taylor, K. R. McKay, A. Hosseini-Sianaki and C. J. Varley
Department of Electrical Engineering
The University of Hull
HULL, HU6 7RX
U. K.

Abstract

This paper deals with the examination of a variable stiffness structure to enable tactile feedback within a virtual environment. Two methods are conceived, the first utilising the solid like structure of an ElectroRheological (ER) fluid under an applied electric field, the other, using the more conventional shearing electrode configuration.

Experimental results are presented for differing mechanical loading conditions and applied electric field for the two principles described. The discussion comments on the experimentation and the systems applicability to glove technology. The paper is concluded with the future experimental plans highlighted.

Keywords
ER fluids, Virtual Environments, Gloves, Tactile Technology

1. Introduction

At present, Virtual Environment (VE) systems are predominantly vision based and generally lack tactile sensory feedback (though such systems exist[1]) from the features of virtual objects being modelled. Glove like devices should be a particularly convenient and operator-friendly form of interface, particularly compared to active exoskeletons. They can be used to track hand motion, and to provide a platform from which tactile stimuli can be received from and applied to the fingers and joints. The addition of tactile stimuli provides a greater illusion of presence removing some of the 'ghostlike' attributes that users perceive with tracking gloves[2], such as that shown in figure 1. However, in practice, gloves are very difficult to implement due to the volume, weight and safety limitations placed upon their design[3,4].

This paper describes an investigation into electrorheological fluid-based elements for incorporation into gloves or sleeves. If a finger or elbow joint were connected to such an element the user would feel a resistance to joint motion. The controllable compliance of the element could then be set under computer control in order to mimic the resistance which the joint would feel in the virtual environment. Two types of element are described followed by experimental data on their performance.

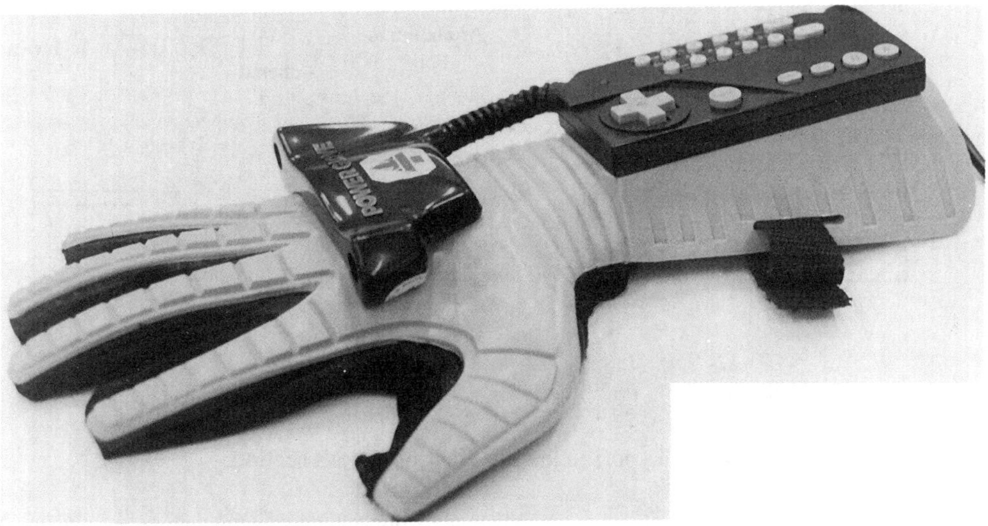

Figure 1. Power Glove capable of tracking hand motion for use in VEs

2. The Electrorheological effect

Electrorheological materials exhibit changes in rheological behaviour with the presence of an electric field, the effect is well documented[5]. ER fluids contain fine particles dispersed within an insulating fluid medium, which align themselves along the field lines between two electrodes. For a dilute suspension these particles form single chains or fibrils in the fluid gap, between the electrodes, whilst for a concentrated suspension the particles aggregate around the field lines forming thick chains many particles wide, with some interconnecting chains or fibrillation's between these main columns[6].

ER fluid behaviour is most commonly related to a Bingham plastic. Ideal Bingham behaviour breaks the fluid response into two distinct states, these being the pre and post yield regimes. Pre-yield characterises the fluid as a solid up to a critical shear stress τ_C after which, the post-yield region, a linear relationship between shear rate and shear stress exists. This τ_C rises as the applied field rises. The failure of this model lies in the inability to predict the fluid behaviour below τ_C. Models exist that show an asymptotic behaviour within the pre-yield region[7].

3. Construction of the compliant elements

Two types of element have been constructed. The first, seen in figure 2, is designed to be used in the shear mode which is the more common mode used for ER applications. The electrodes are made from aluminium foil and are thus incapable of expanding elastically with a similar strain to the insulating rubber. Consequently the

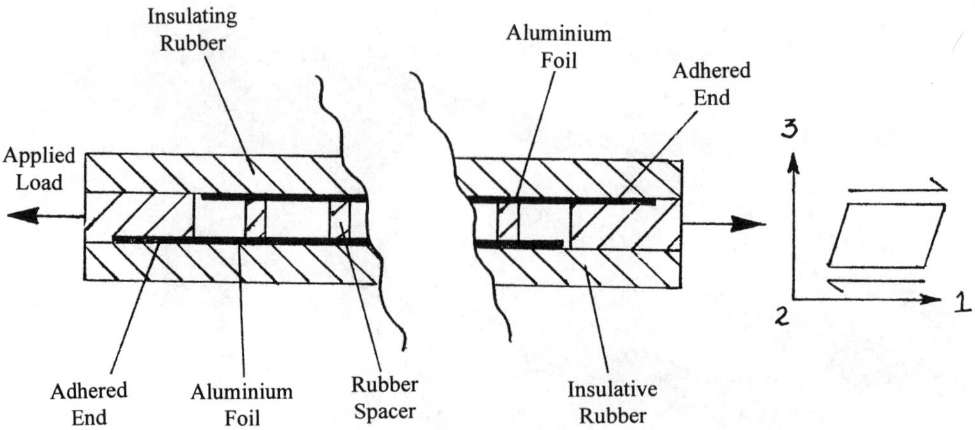

Figure 2. Type 1, shearing electrode cross section

foil electrodes are adhered only at opposing ends, thus when the specimens extend as the external forces are applied the electrodes move with their adhered end, causing the shearing of the fluid to occur. Figure 2 shows an element of fluid under such loading, with shear stress being the principal loading along the 1-2 plane with Poisson's ratio effects occurring in the 2 & 3 directions.

Spacers are necessary between the aluminium foil electrodes to prevent the fluid gap reducing to the state where electrical shoreing occurs between the electrodes. Figure 2 shows thin rubber strips 1mm thick being used for this purpose. Previous work had shown the value of cloth for this purpose, and indeed if a fluid gap remains between the cloth and electrode it has been observed that the ER effect is actually increased[8]. Consequently a second element was constructed identical to that shown in figure 2 but with the addition of a piece of cloth between one electrode and the rubber spacer, which ensured that a distinct fluid gap was maintained. The cloth used in the tests below is a velveted man-made fibre densely punctured with sixteen small holes per square centimetre.

The second type of element (single gap specimen) to be used is shown in figure 3 where the electrodes are now made of conductive rubber thereby allowing strain along the line of action of the forces applied along the principal axis shown as the 1 direction. By treating the fluid as a solid below τ_c any stiffness increase would be due to a rise in the Young's modulus of the specimen as the applied voltage is varied. Insulative rubber spacers are used to ensure that the minimum fluid gap is maintained over the whole length of the sample. A double gap specimen was also produced with the sandwich now comprising outer layers of earthed conductive rubber with insulating rubber spacers and a fluid gap to a centroidal conductive rubber piece that was raised to the required voltages.

From a selection of commercially available fluids, two fluids were chosen for their low corrosive activity on the rubbers and silicone sealant, despite their relatively low shear strength. These fluids are designated as fluid (A) and fluid (B) hereafter.

Both fluids are based on silicone oil for their continuous phase and contain inorganic material ($Na_{12}Al_{12}Si_{12}O_{48}$) for their particulate phase, with an average size of 3 microns. Fluid (A) has 22.5% by volume solid containing 2% water whereas fluid (B) has 24% by volume solid containing 3.5% water. The particles in both fluids were chemically treated for steric stabilisation against sedimentation. Further details of these fluids can be found in reference[9].

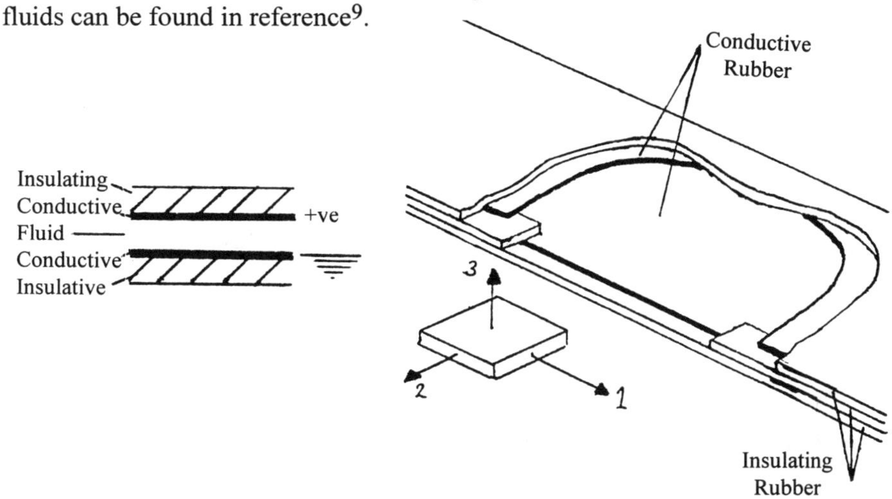

Figure 3. Cutaway section of the type 2 (single gap) specimen containing rubber electrodes

4. Test Apparatus and Procedure

The force displacement curves for the various specimens were taken using a Hounsfield tensometer W-system. The tensometer is manually loaded by turning the loading handle, mechanical linkages both pull the specimen and rotate the drum to provide displacement data. This system was used for convenience in order to gain initial results, despite its mechanical deficiencies. Force is measured from a mercury line via linkages by the bending of interchangeable loading plates, allowing varying maximum loads of 16lbs to 2 tonnes over the full mercury rise.

4.1. Tensile loading

The specimens were placed in the tensometer configured to give tensile loading shown schematically in figure 4. Each specimen was clamped between insulating plates and given a small initial strain in order to take up any slack within the linkages, the force extension curves were then measured, using a strain rate of $10^{-3}s^{-1}$ for various applied voltages.

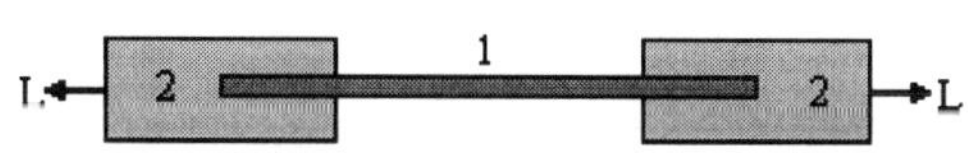

Figure 4. Schematic of specimen in tension.

4.2. Bending

The tensometer and specimens were set up to the configuration shown in figure 5, both ends are encastrely gripped and loaded from the centre, producing a central deflection rate of 10^{-1}mm/s, again curves being obtained over a range of voltages.

Figure 5. Schematic of element in bending.

Key for Figures 4 and 5

1 - Specimen
2 - Clamps
L - Applied Loading

5. Experimental results and discussion

5.1 Aluminium foil shearing electrode with cloth spacer

Results were taken for 0V, 1kV and 1.5kV using fluid (A) with a gap of 1mm. Figure 6 shows the force extension curves for the indicated voltages under a tensile load. Figure 7 depicts the difference between the extensions of the specimens with the applied fields of 1kV/mm and 1.5kV/mm referenced to no applied field. As can be

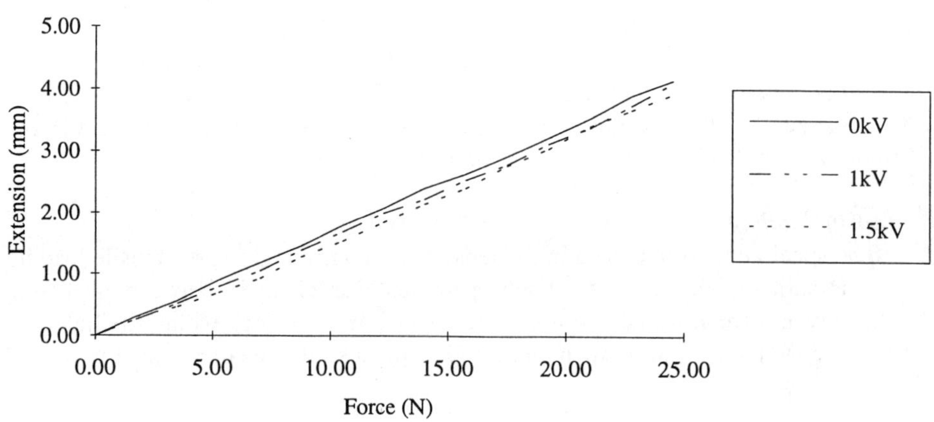

Figure 6. Aluminium foil specimen with cloth spacer using fluid (A) in tension.

seen, true Bingham behaviour does not occur, the curves showing an asymptotic behaviour to a maximum difference in extension under an applied force. The average of this maximum difference has been superimposed, with values of 0.11mm for the 1kV/mm and 0.23mm for the 1.5kV/mm, this was over a total range of 5mm extension. The results presented here demonstrate the feasibility of using ER fluid in this type of structure. Some of the main reasons for the apparent low response are; the specimen's high self stiffness, the low shear strength of the fluid used and the limitation of the power supply to enable higher voltages.

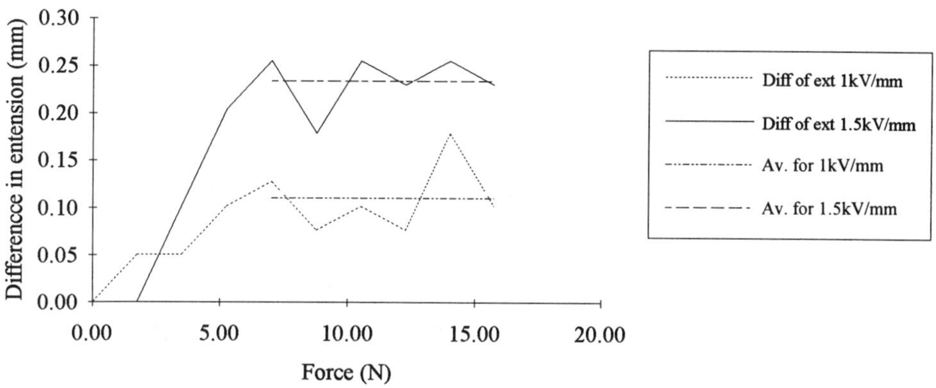

Figure 7. Difference in extensions and averages for data displayed in fig. 6.

5.2 Aluminium foil shearing electrodes with rubber spacers

Fluid (A) was again used at voltages of 0, 1kV and a repeated 1.5kV which was done to ensure that this response was not erroneous, Figure 8 shows the specimen's response to a tensile loading for the aforementioned voltages. The results show a rise to a maximum shear stress with applied voltage after some strain . The behaviour of the 1kV curve after the maximum shear stress follows the gradient of the 0V curve indicating that the maximum fluid shear stress is maintained hereafter. There are two possible reasons for the response of the 1.5kV curves. If the curves maintains its gradient after the maximum fluid shear stress has occurred then this indicates that the Youngs modulus of the element has decreased, alternatively if the curve approaches that of the 0V then the fluid has tended to zero shear stress. In either case experiments recording higher strains are required to determine the correct explanation.

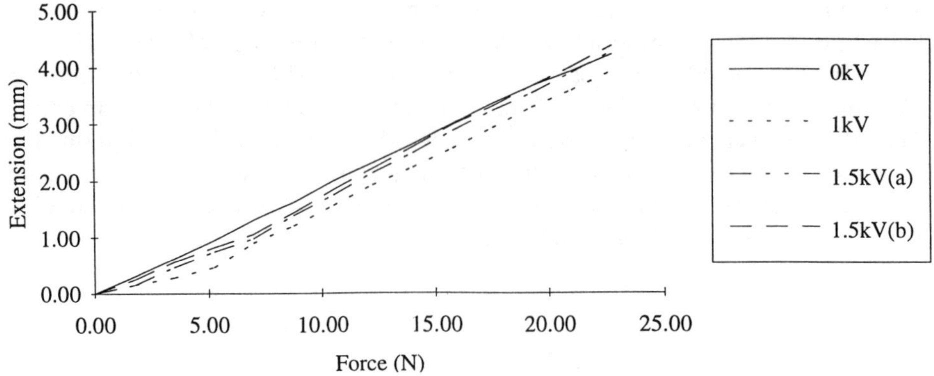

Figure 8: Al foil spacer with rubber spacer

5.3 Type 2 Conductive rubber specimens with rubber spacers

The single and double gap designs were constructed to investigate the effects of fluid cross-sectional area and the movement of fluid away from the neutral axis. The fluid medium in both cases was fluid (A) and the results of tensile tests performed at the indicated voltages are given in Figures 9 and 10 for the single and double gap specimens respectively.

In these tests the tensile effect of any cross linkages between the chains are investigated and this is how the fluid is viewed as a solid like structure. It can be seen that the response is lower than that for the shear case discussed so far, this is even more evident when considering the double gap specimen.

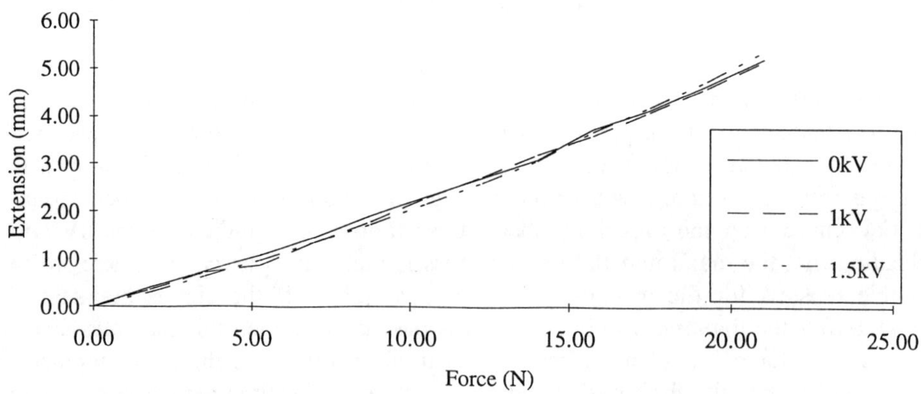

Fig 9. Single gap conductive rubber with rubber spacers using fluid (A) in tension.

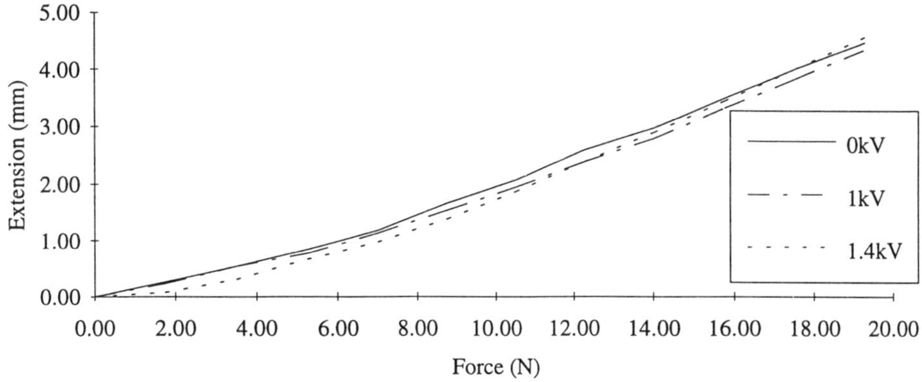

Fig 10. Double gap conductive rubber with rubber spacers using fluid (A) in tension.

5.4 Aluminium foil shearing electrode using fluid (B)

It has been shown that the best response was achieved from the aluminium foil specimen with the fluid in shear mode. For this reason further investigation was carried out using fluid (B). In the first instance the usual tensile tests were conducted for comparison purposes. This was further followed by the encastre grip loading tests the results of which are given in Figures 11 and 12 respectively.

The results shown in Figure 11 indicate that fluid (B) performs considerably better than fluid (A) as the identical tests in section 5.1 show. A further advantage of this fluid is the smaller current demand thus enabling a larger voltage range to be used.

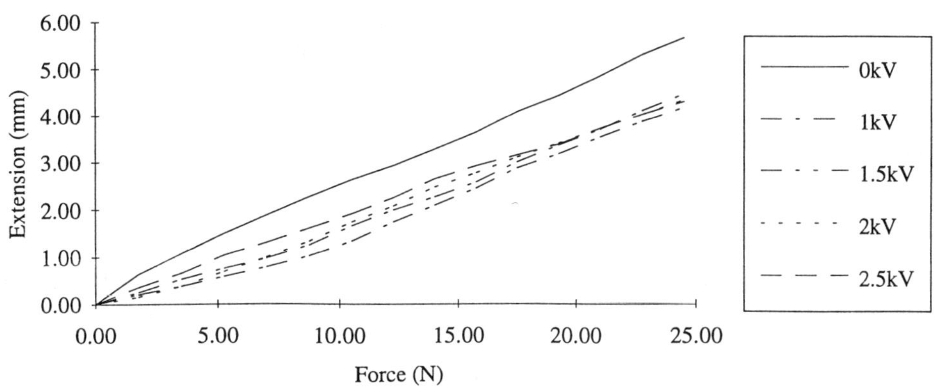

Figure 11. Aluminium foil specimen with cloth using fluid (B) in tension.

In a real application the fluid would be acted upon by various, and probably, simultaneous types of loading. For this reason the encastre grip arrangement explained in section 4.2 was used to apply a bending moment to the specimen. The results presented in Figure 12 clearly indicate a reduced deflection of the specimen with applied voltage. It should be noted that the force does not reflect the tension within the specimen itself but that of the loading mechanism that is normal to it.

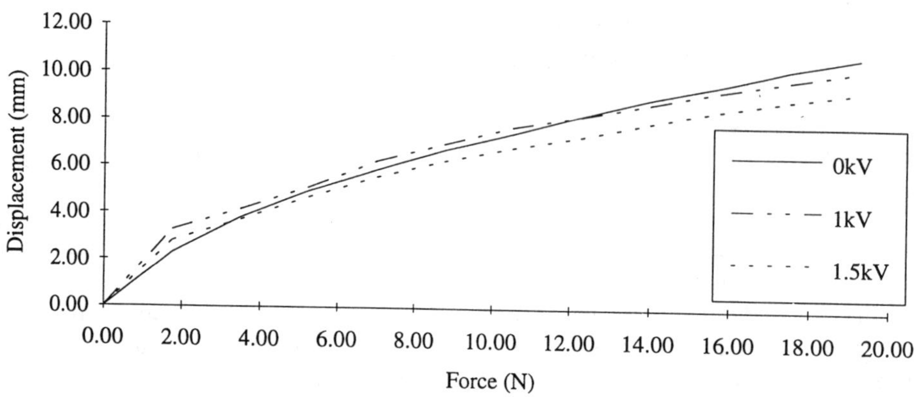

Fig. 12. Aluminium foil specimen with cloth using fluid (B) under encastre loading.

Generally for the tensile load, the typical behaviour of the aluminium foil specimens consists of two domains. The first domain of the response shows that the gradient of the force extension curve is smaller with the presence of an electric field than without. This is followed by an increase in gradient that matches the zero voltage curve. These results indicate that at the boundary of these two domains the shear stress has risen to a maximum and must remain constant thereafter for the gradients to be the same.

6. Conclusions

The work conducted so far has shown, especially for fluid (B), that a variable stiffness element using this method is feasible, however it is still unknown whether this system could provide adequate feedback to a user within a virtual environment. The method is passive in the sense that it can apply only resistive forces to the user's joints, no force can be applied if no movement is attempted. It remains to be seen what psychological effect this has on the impairment of the naturalness of the control and other benefits that such a system could provide. It should be obvious that much more work is required before these very fundamental questions can be answered.

In order to help resolve these unknowns the following future experimental plans have been made:

- The construction of a more automated and accurate testing process,
- more tests on the current specimens especially on the encastre grip are required in order to clarify advantages of specific designs and obtain the bending stiffness for the fluid,
- materials selection process to both reduce the zero voltage stiffness of the specimen and make it more reliable in the experimental process,
- a study of the effect of the second moment of area on the specimens bending stiffness,
- an investigation of specimens undergoing torsion,
- and an investigation of the fluid undergoing lateral compression.

7. References

1. P. M. Taylor, K. G. Swift, D. R. Smith and R. S. Kalawsky, *Proc. 27th ISATA*, Aachen, October, 1994, pp639-646.
2. K. Pimental and K. Teixeira, *Virtual Reality; through the new looking glass* 2nd revised edition, McGraw -Hill, 1995.
3. R. Stone, *Robotica*, (1992) **v10**, pp461-467.
4. M. Bergamasco, D. M. De Micheli, G. Parrini, F. Salsedo and S. Scattareggia Marchese, *Proc. International Conference on Advanced Robotics,* Pisa, 1991, p162-7.
5. E. V. Korobko, *Advances in intelligent materials, systems & structures*, 1994,**v2**, p16.
6. Y. Shih, A. F. Sprecher and H. Conrad, *Proc. Int. Conf. on Electrorheological Fluids*,
7. D. R. Gamota and F. E. Filisko, *Proc. Int. Conf. on Electrorheological Fluids*, 1991, p393.
8. G. J. Monkman, *Journal of Rheology,* **v35** (1991) pp1385-92.
9. P. M. Taylor, A. Hosseini-Sianaki and C. J. Varley, in these proceedings.

INVESTIGATION INTO THE USE OF LIQUID CRYSTALLINE MATERIALS IN ELECTRORHEOLOGICAL FLUIDS

D. LACEY and C. MALINS
School of Chemistry, The University of Hull, Hull, HU6 7RX, UK

P. M. TAYLOR, A. HOSSEINI-SIANAKI and C. J. VARLEY
Department of Electronic Engineering, School of Engineering and Computing, The University of Hull, Hull, HU6 7RX, UK

ABSTRACT

The ER activity of a number of ER fluids incorporating liquid crystalline (LC) and non-LC materials have been measured using a laminar-flow shear cell. The results from this static study are presented and clearly show that the unique self-organising and anisotropic physical properties of the LC materials contribute greatly to the ER activity of LC based ER fluids. A more in-depth evaluation was then carried out on a LC based ER fluid which involved measurements of the steady state behaviour, under various operational conditions, and the results compared with those from two conventional silicone based fluids. From this dynamic study it was found that within the shear rates tested, for an electric field strength above 1kV mm^{-1} there existed a linear relationship, culminating in a shear stress of 12 kPa at an electric field strength of 4 kV mm^{-1}. The unique property of the shear stress increasing with increase in the shear rate makes LC based ER fluids an exciting area for further research and this could be of considerable importance in their use in a variety of control systems.

Introduction

For many years, the influence of electric fields on the deformation and flow properties of materials has been a subject of much interest. Most recently, the electrorheological (ER) effect, in particular, has attracted renewed attention. Interest in these 'smart materials', especially in Japan, the USA and Germany, owes much to the diversity of ER fluid applications, including active-energy dissipation devices such as engine mountings and isolation systems for power generation, ER motion controllers for mechanical components and ER power control devices to control automatic machine operations[1]. Traditionally, an electrorheological (ER) fluid is identified as a suspension of a dielectric solid or polymeric particles (the dispersed phase) in an insulating oil (the continuous phase), which behaves as a Newtonian fluid under normal conditions[2]. On the application of an electric field to this fluid suspension, non-Newtonian or Bingham behaviour is observed and the rheology of the fluid changes to resemble a jelly-like solid.

Thermotropic liquid crystals (LC) are excellent examples of materials that will form a variety of thermally-induced self-organising macrostructures called *mesophases*. These mesophases can vary both in their fluid properties and macrostructure, ranging from the very viscous and highly structured crystal and smectic phases to the more liquid-like and less ordered nematic and chiral nematic phases (see Figure 1). One of the essential features of these mesophases is that their properties are anisotropic. If a LC material is well aligned, then it will show two dielectric permittivities, parallel and perpendicular to the director[3]. The director can be influenced by the application of an electric field and many LC display devices are based on this field-induced reorientation of the director[4]. Electric field strengths in the range 0.5-3 V μm^{-1} are sufficient to reorientate the director and these field strengths are very similar to those used in many ER fluids to produce useful shear stresses in the region of 7-8 kPa[5].

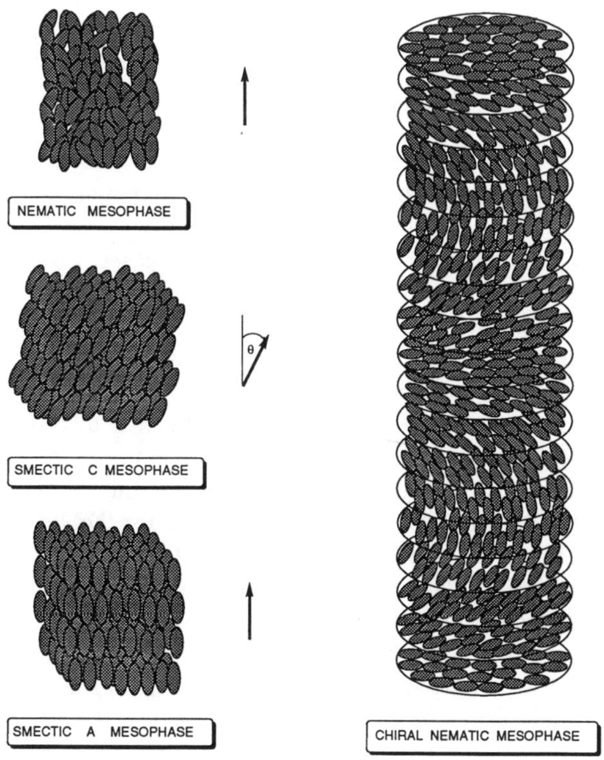

Figure 1 Structure of the nematic, smectic C, smectic A and chiral nematic mesophases

The ER effect is generally thought to be the result of the organisation of electrostatically induced dipolar particles of the dispersed phase of the ER fluid into field-induced microstructures (chains or clusters) between the two electrodes[6]. It was envisaged that the formation of these microstructures would be enhanced by utilising LC materials as the continuous phase. The LC material, rather than simply being a medium that solely provides a dielectric potential between the dispersed and continuous phases, could also act as either a field- or thermally-induced controllable structured medium which would enhance the shear stress of an ER fluid.

In this paper, an initial (static) study is reported which confirmed the advantageous properties of liquid crystals when used in ER fluids. This discovery is followed up with a more comprehensive (dynamic) investigation in to the engineering potential of these fluids.

The Static Study

This static study was undertaken to establish whether LC materials, when used as the continuous phase, are capable of forming ER fluids. The measurement of the shear stress of an ER fluid was carried out using a laminar-flow shear cell, the details of which are given in a previous paper[7]. It is important to emphasise that the method used in this part of the study to measure the shear stress of the LC based ER fluids does not give an absolute value since a zero shear rate is assumed using this static method. However, the method is a useful technique for comparing the shear stresses of the ER fluids.

In this study four commonly used dispersed phases, *i.e.* soda-glass[6], Zeolite Linde 3A[8], granular PAQR[5] and finely divided Lipma[9], were used in conjunction with Cereclor 50LV (a chlorinated hydrocarbon material that is widely used as the continuous phase in many ER fluids) and nematic LC mixture E7 as the continuous phase. All suspensions were made up at room temperature to compositions that achieved a similar flow consistency prior to the application of an electric field. Using the laminar-flow shear cell, the ER activity of the above suspensions were measured at 20 °C over a range of field strengths from 0.2 to 2.2 kV mm^{-1}. All of the suspensions showed ER activity. A comparison of the ER activity for the suspensions containing the various dispersed phases with Cereclor 50LV and E7 at a field strength of 2.2 kV mm^{-1} is given in Figure 2. The effectiveness of the LC mixture E7 in ER fluids is comparable to Cereclor 50LV, which is extensively used in many ER fluids . Although the results shown in Figure 2 clearly show that suspensions comprising of Lipma and E7 give very good ER activity, it was still necessary to establish whether the highly organised nature of the nematic phase had contributed to the ER activity of the LC based ER fluids, and that the LC material had not just simply acted as a very good dielectric medium.

To investigate whether the highly organised nature of the nematic phase had contributed to the ER activity of LC based ER fluids, three further ER fluids were

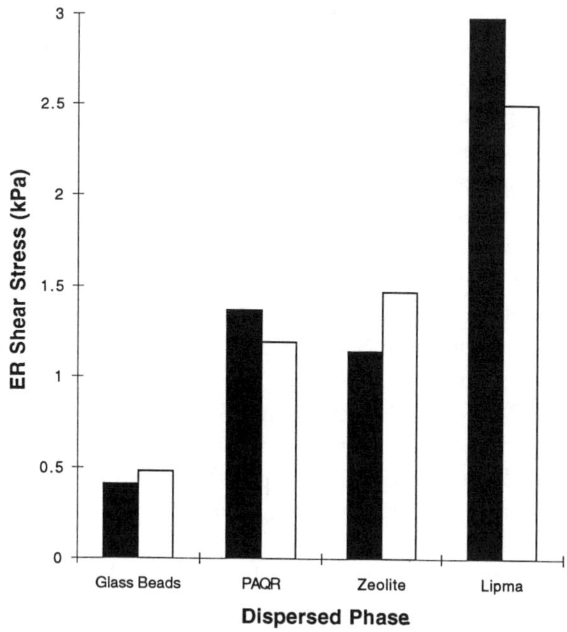

Figure 2 Comparison of yield stresses for ER fluids containing E7 ■ and Cereclor 50LV □ as the continuous phase at a field strength of 2.2 $kVmm^{-1}$

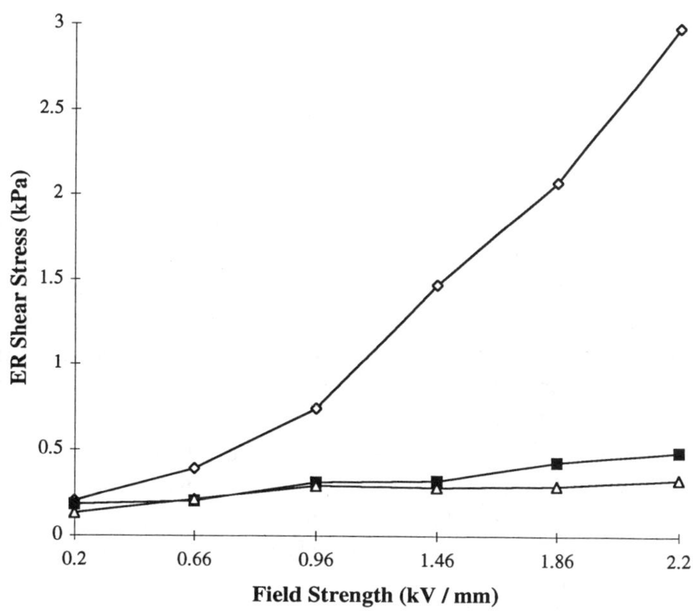

Figure 3 A plot of yield stress against field strength for ER fluids containing 30% lipma and, ◊ 70% E7; ■ 70% LC mixture 17325; and Δ 70% n-pentadecylnitrile

made in which the continuous phase was one of the following: E7, LC mixture 17325 and *n*-pentadecylnitrile (similar in molecular length to the *n*-alkylcyanobiphenylyl compounds used in E7 and LC mixture 17325, but does not exhibit LC behaviour). The three continuous phases, all highly polar due to the terminally positioned cyano-group, were used in dispersions with 30% Lipma. All the tests were carried out at 20 °C. The results from this part of the study are given in Figure 3.

The most interesting result to come out of this part of the study is the comparison in ER activity between E7 and LC mixture 17325. The LC mixture 17325 contains a number of *n*-alkylcyanobiphenylyl compounds, similar to those used to make E7, and so the dielectric anisotropy of the two continuous phases will be very similar. However, at the temperature at which the experiment was carried out (20 °C), only E7 would have exhibited a mesophase (nematic phase) since the T_{N-I} value for the LC mixture 17325 is 9.4 °C, whereas that for E7 is 60 °C. It can be clearly seen from the data presented in Figure 3 that the ER fluid comprising of Lipma and E7 has a far higher yield stress than the ER fluid Lipma/LC mixture 17325. This suggests that the long-range orientational ordering encountered in nematic materials (the director aligning along the direction of the applied electric field), serves to enhance the molecular organisations encountered in ER suspensions, thus promoting and stabilising the formation of molecular aggregates of dispersed particles within the fluid.

One simple explanation for this observation could be that in the E7/Lipma suspension (no applied electric field) there is an apparent quasi-microstructure, mainly due to the self-organising ability of E7, but the system does not have the required energy or inertia to complete a microstructure involving Lipma and E7 which would be far superior to that formed by E7 alone, *i.e.* we have a *frustrated* suspension. The application of the electric field provides the missing inertia.

The Dynamic Study

Due to the success of the LC based ER fluid incorporating E7 as the continuous phase (70% E7 / 30% Lipma), a further set of tests have been undertaken with the E7/Lipma fluid reported in the static study to examine more closely the engineering potential of LC materials in electrorheological (ER) technology. For comparison purposes, two other commercial fluids were tested concurrently under identical conditions. The first of these was a fluid based on silicone oil and had the same volume fraction and particulate phase (Lipma) as the E7/Lipma fluid. The other was also based on silicone oil but had an inorganic (zeolite) particulate phase. The volume fraction was 22.5% and the particles had an average diameter of 3 microns. This fluid was also treated for steric stabilisation in order to prevent sedimentation.

A detailed description of the instrumentation used for these tests is given in reference 10. In short, the test rig comprises a concentric rheometer, encompassed by a fluid jacket to allow accurate temperature control. The voltage, supplied by a TREK

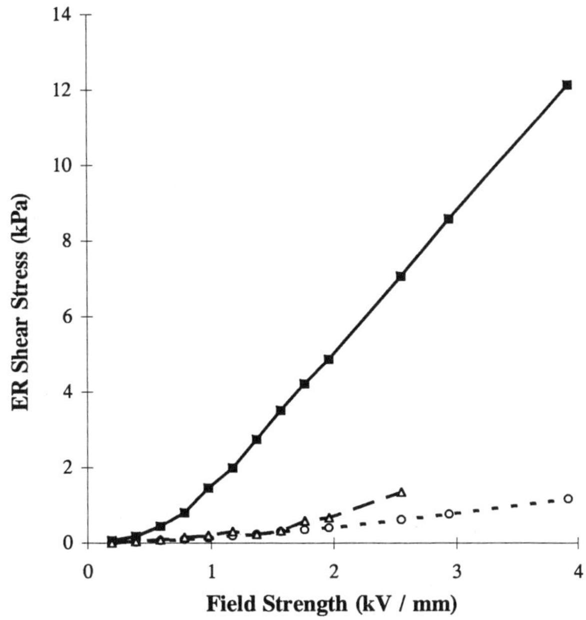

Figure 4 The effect of electric field strength on the associated steady state ER shear stress, –■– E7/Lipma, –Δ– Silicone/Zeolite, --O-- Silicone/Lipma

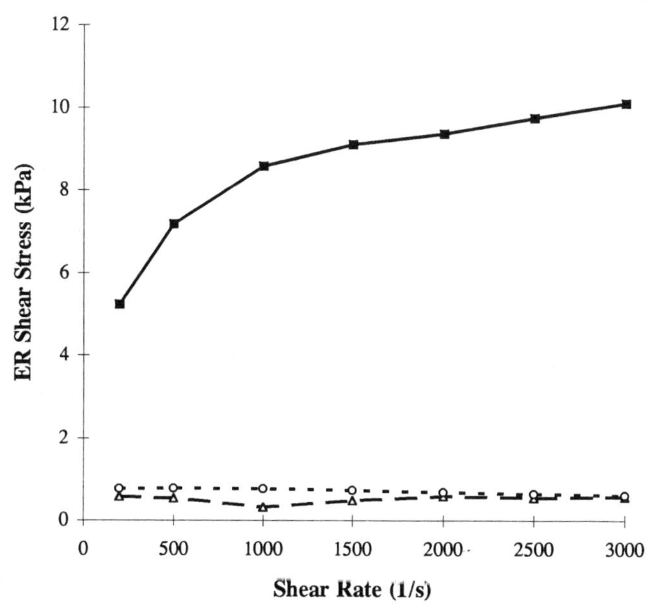

Figure 5 The effect of shear rate on the steady state shear stress, –■– E7/Lipma, –Δ– Silicone/Zeolite, --O-- Silicone/Lipma

unit capable of delivering 10 kV at 20 mA, together with the current and torque response data was captured using a 200 kHz capability data acquisition system. The three fluids were tested using a standard electric field strength of 2 kV mm^{-1} for shear rates of 200, 500, 1000, 1500, 2500 and 3000 s^{-1} and also for a range of electric field strengths (0.2-3 kV mm^{-1}) at a control shear rate of 1000 s^{-1}. The tests were all undertaken with the temperature controller set to 27 °C.

Figure 4 shows the effect of electric field strength on the associated steady state ER shear stress. It is shown for the given shear rate that the liquid crystal exhibits a much greater ER response than either of the other two silicone based ER fluids, *i.e.* an order of magnitude higher at high field strengths. This is of particular importance for applications which require large torque from a small volume of fluid.

By far the most interesting quality of the LC fluid is illustrated in Figure 5. Here, the effect of shear rate on the steady state shear stress is depicted. The LC based ER fluid exhibits a unique characteristic in that unlike the other two fluids and indeed, every other typical ER fluid, the measured torque is actually increasing with shear rate. This result is of extreme importance for applications in which large shear rates are desirable, for example, in high speed clutch devices.

The torque response to a step voltage for the Silicone based fluids are alike[10]; in both cases a maximum shear stress is achieved in the order of 1ms. The E7/Lipma fluid has a very similar characteristic over this time period but exhibits an extra subsequent phase in which the shear stress increases approximately two fold over a 1 second period. This second phase may be attributed to the LC material itself forming an organised lattice around the ER crystalline structure, hence increasing the overall stability of the structure.

The current response to the applied step voltage behaved in a similar manner, exhibiting the peculiar phases for the E7/Lipma fluid, with the Silicone/Lipma fluid demanding the least current. For a shear stress of 1kPa at a shear rate of 1000s^{-1}, the E7/Lipma fluid drew an approximate current of 0.4mA, compared with 12mA for the Silicone/Zeolite fluid and 0.3mA for the Silicone/Lipma fluid.

Conclusions

A static study involving the use of a laminar-flow shear cell has clearly shown that the unique self-organising and anisotropic physical properties associated with LC materials contributed greatly to the ER activity of the LC based ER fluid.

From a more in-depth comparative evaluation of a LC based ER fluid with two silicone based ER fluids, the following conclusions can be made:

1. The E7/Lipma fluid presents an exceptionally larger shear stress than any of the fluids tested.
2. The shear stress of the E7/Lipma fluid shows an increase for an increasing shear rate - a characteristic that is as yet unique to ER fluids. These peculiarities may

result from the several 'Liquid Crystal' related, and substantially slower effects that have been identified with these fluids.

3. A shear stress of 12 kPa for an electric field strength of 4 kV mm^{-1} at a shear rate of 1000 s^{-1} is easily obtainable and could be of considerable importance in their use in a number of control systems.
4. The power demand from the E7/Lipma is much greater than that of the weak Silicone/Lipma fluid but is much more economic than the Silicone/Zeolite fluid. This factor along with other characterisation procedures are currently being addressed and the prospect of low power LCER fluids is not too distant a proposal.

The exceptional ER shear stresses obtainable from the LC fluids indicate great potential for future ER devices. More in-depth chemical and physical tests are required in order to fully understand the nature of the LC behaviour within ER fluids and to minimise the current consumption.

Acknowledgements

The authors wish to thank the EPSRC and Merck Ltd (Poole) for the Case Award to C. Malins. The authors are also grateful to Advanced Fluid Systems Ltd. for allowing use of their rheometer test rig and also provision of the organic fluid. They also wish to express thanks to Metzeler Gimetall AG for supplying the inorganic fluid. Part of this work was funded by the European Union under Brite Euram project number BE-5072.

References

1. W. A. Bullough, *Adv. Mater. Opt. Electron.*, **1** (1992) 159.
2. D. L. Klass and T. W. Martinck, *J. Appl. Phys.*, **38** (1967) 71.
3. L. Pohl, in *Liquid Crystals: Applications and Uses*, ed. B. Bahadur (World Scientific, London, 1990).
4. D. Lacey, in *Introduction to Molecular Electronics*, ed. M. C. Petty, M. R. Bryce and D. Bloor (Edward Arnold, London, 1995).
5. H. Block, J. P. Kelly, A. Qin and T. Watson, *Langmuir*, **6** (1990) 6.
6. H. Conrad and A. F. Sprecher, *J. Stat. Phys.*, **64** (1991) 1073.
7. C. Malins and D. Lacey, *J. Mater. Chem.*, **4** (1994) 1029.
8. F. E. Filisko and L. H. Radzilowski, *J. Rheol.*, **34** (1990) 539.
9. H. Block, J. P. Kelly and T. Watson, in *High Value Polymers*, ed. A. H. Fawcett (The Society of Chemistry, Cambridge, 1991).
10. A. Hosseini-Sianaki, C. J. Varley, P. M. Taylor, D. Lacey and C. Malins, in press.

THE PECULIARITIES OF RELAXATIONAL PROCESSES IN ACTIVE CELLULOSE FILLERS FOR ELECTRORHEOLOGICAL MEDIA

T. G. LAZAREVA

Institute of General & Inorganic Chemistry, Belorussian Academy of Sciences
9 Surganov St., Minsk 220072, Belarus

ABSTRACT

The peculiar features of relaxation processes in active cellulose fillers are elucidated using the dipole relaxation method, the measurements being made in a frequency range from tens to 10^3 Hz at temperatures from -100 to +100 °C. A hypothetical model of structural changes in a cellulose matrix due to the introduction of ionogenic acidic groups is proposed. A possibility is shown of controlling polarization processes in such structures by applying electric fields of variable strength at a room temperature.

1. Introduction

As is known, cellulose compounds with ionogenic groups (carboxyl, amino and phosphate groups) are used as active fillers for electrorheological media[1]. Polymers free of ionogenic groups, namely, methyl, ethyl and oxypropyl cellulose as well as unmodified cellulose are characterized by lower or no electrosensitivity.

To evaluate the contribution of polarization mechanisms to the electrorheological effect (ERE), we have investigated the processes of dipole polarization of cellulose ionites, *viz.* the dipole-group polarization associated with the orientation of small fragments of a side or main chain (the β-relaxation process), and dipole-segment polarization that involves the main chain of a macromolecule (the α-relaxation process). Monocarboxyl cellulose has been used as a model compound with ionogenic carboxyl groups. Monocarboxyl cellulose (MCC) containing a COOH-group in the basic state in a macromolecular unit was obtained by selective oxidation:

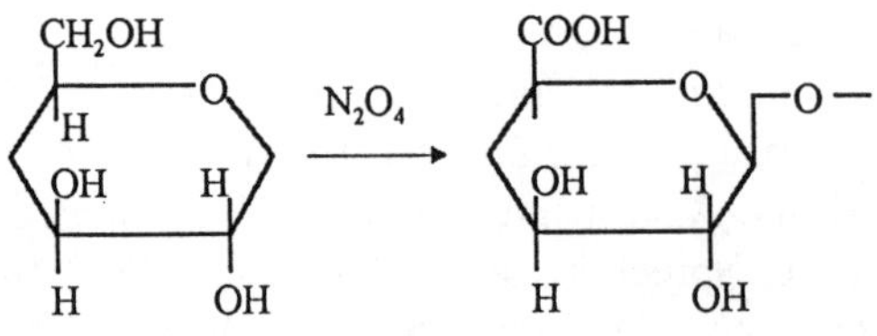

2. Results and Discussion

Figure 1 shows the temperature dependencies of tgδ (at the frequency f=1 kHz) for the predried MCC samples in the H-form with a different content of COOH-groups and cellophane. Oxidation of the cellulose involving the substitution of COOH-groups for OH-groups results at first in a decrease of tgδ values. Moreover, there is observed a characteristic shoulder, *i.e.* broadening of the temperature interval of polarization towards a higher-temperature region. At the content of COOH-groups of about 4.0-5.8% a shift of the polarization maximum is observed: *e.g.*, for MCC samples (5.8% COOH-groups) to the temperature of -41 °C. For the content of carboxyl groups exceeding 6% in the polymer, the dipole-group polarization at the frequency of 1 kHz occurs in the temperature range -30-0 °C.

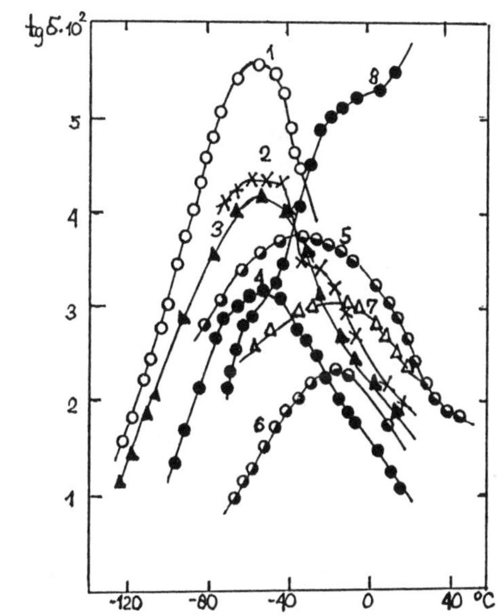

Fig. 1. Temperature dependence of tgδ at a frequency of 10^3 Hz for cellophane (1) and MCC (2-8) on its basis with the following content of COOH-groups: 1.0 (2), 1.5 (3), 2.2 (4), 5.8 (5), 6.8 (6), 12.7 (7) and 17.9% (8).

The dipole-group polarization of cellulose in the temperature range -60 to -70 °C is associated with the orientation of primary hydroxyl groups. At a higher concentration of COOH-groups (over 6%), mainly occurs the polarization of COOH-groups only; this is evidenced by an increase of tgδ in the temperature range -30 to 0 °C with increasing the content of COOH-groups. The broadening of the temperature interval of polarization in the region of small contents of COOH-groups may be caused by the polarization of a complex consisting of OH- and COOH-groups. Such an interpretation is consistent with literature data[2] according to which a shift of the polarization maximum, upon cellulose moistening, is associated with increasing the sizes of polarizable structures by the absorbed water molecules. However, it should not be also ruled out that the broadening of the temperature interval of polarization is connected with overlapping of polarization processes of OH- and COOH-groups, whose polarization maximums appear in different temperature intervals.

A study of the temperature dependence of tgδ in a wide frequency range allows the determination of the activation energy (U) of the dipole-group polarization process to be made by a graphical method combined with simple mathematical operations. A shift of the regions with maximums along the temperature axis upon changing the electric field frequency (f) is described by a relationship between lgf_{max} and $-1/T$, where f_{max} is the frequency and T is the temperature at which the tangent of dielectric loss is maximal. For the dipole-group losses, the lgf_{max} versus $-1/T$ dependencies are linear and can be described by the relationship

$$f_{max} = A \exp[-U/(RT)] = f_o \exp[-U/(RT)]$$

where $f_o = 10^{12}$-10^4 Hz, R is the universal gas constant, U is the activation energy.

As is seen in Fig. 2, the activation energy of the β-relaxation process depends on the content of COOH-groups in MCC: its lowest values pertain to the region with the 1.0-4.0% content of COOH-groups.

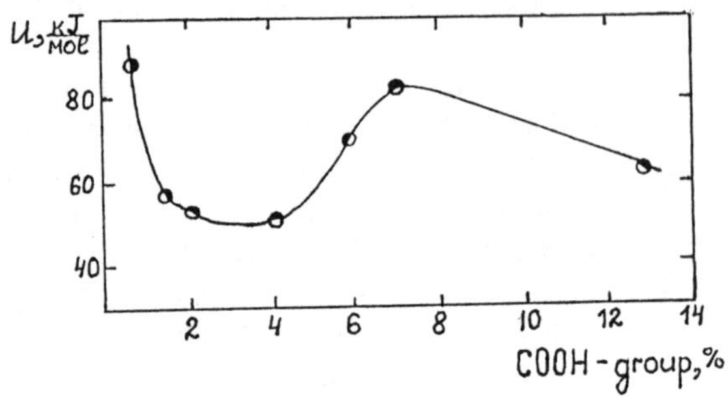

Fig. 2. Activation energy of the β-relaxation process versus the content of COOH-groups in monocarboxyl cellulose.

It is worth noting that we measured the activation energy for different functional groups in dependence on the content of COOH-groups in MCC. At high contents of COOH-groups (over 6%) the maximum of dielectric losses is apparently associated, as noted above, with the polarization of acidic carboxyl groups and the calculated activation energy pertains to these groups, while in the region of low contents of COOH-groups it corresponds to the polarization of primary OH-groups.

The processes of dipole polarization also reflect the α-relaxation process associated with the transformation of glassy cellulose into a highly elastic form. It should be noted that the processes of both β- and α-relaxation may be induced by increasing either temperature or moisture content, or both at once. The vitrification temperature[3], t_c, of the dry regenerated cellulose is 220-230 °C. It is known that t_c substantially decreases with increasing the moisture content and this decrease depends, for instance, on a degree of the polymer crystallinity. With an increase of the moisture content up to 10%, the α-relaxational transition of the cellulose modified by the introduced acidic groups may occur even at a room temperature[2]. The introduction of acidic carboxyl groups into a cellulose matrix, which essentially changes the system of intermolecular interactions, is expected to change markedly the relaxation processes, in particular, those occurring under the action of absorbed water.

Figure 3 shows the frequency dependencies of tgδ for unmodified cellulose in the form of paper as well as for methyl cellulose (MC) on its basis with the COOH-group content amounting to 0.8; 1.7; 2.6; 4.0 and 6.0 wt.%, which have been measured at different relative values of water vapour pressure (from 10 to 93%). As evidenced from the experimental data obtained, an increase of the amount of COOH-groups in the polymer gives rise, even at small water vapour pressures, to a polarization maximum observed at 1.0 kHz. Similar results are obtained in studying the α-relaxation process by the so-called immersion method when the examined polymers in a dispersed form are introduced into an inert medium, *e.g.* oil. In the case of materials with ionogenic groups the α-relaxation occurs at lower temperature and moisture content (see Fig. 4).

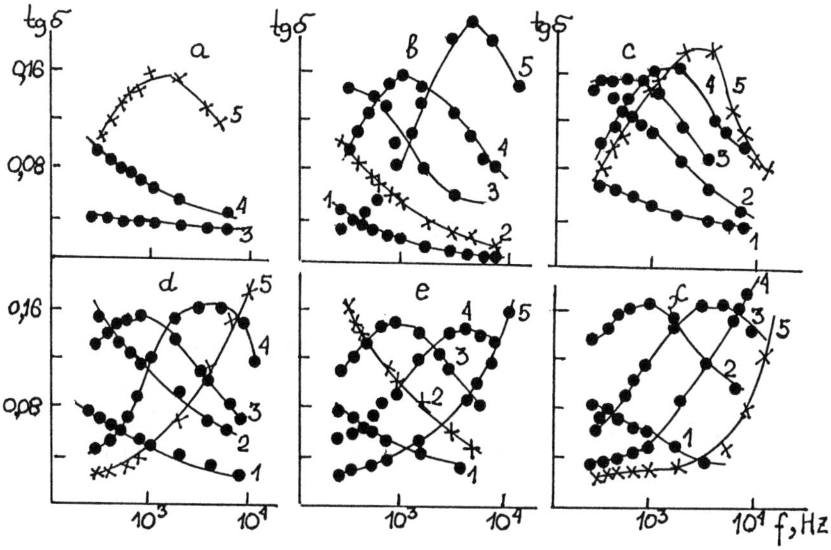

Fig. 3. Frequency dependence of tgδ for cellulose in the form of paper (a) and MCC on its basis (b-f) with the different content of COOH-groups: 0.8 (b), 1.7 (c), 2.6 (d), 4.0 (e), and 6.0 wt.% (f) at the following relative humidity of air: 10 (1), 32 (2), 56 (3), 75 (4) and 93% (5).

To interpret the obtained results in application to the oxidized cellulose, let's consider in more detail the changes in the system of intermolecular interactions caused by the introduction of COOH-groups into cellulose. As judged from literature data[4], the presence of acidic centers in the unmodified cellulose, which possesses, as is known, a small amount of COOH-groups, is responsible for a substantial increase of the reactivity, and in particular, absorptive ability of hydroxyl groups in their nearest environment. This phenomenon is explained by the formation of a specific ensemble around a COOH-group, which contains approximately 120 hydroxyl groups incorporated in about 40 macromolecular units. We consider that such an ensemble does not imply a direct interaction of a COOH-groups with each of the OH-groups but is indicative of the fact that the introduced COOH-group interacts with in its nearest environment to form hydrogen bonds. This results in a rearrangement of the system of intermolecular bonds between hydroxyl groups in a substantially larger volume: it embraces the polymer residues containing hundreds of OH-groups. In fact, the carboxyl group promotes the formation of structural defects in a cellulose matrix. As the content of COOH-groups increases (*e.g.* due to oxidation), such structures may be grouped into several types:

(1) the spheres of interaction including a COOH-group are at a sufficiently large distance from each other. This is the case when MCC or the unmodified cellulose possesses a small amount of COOH-groups.

(2) the spheres join each other.

(3) the spheres become overlapping.

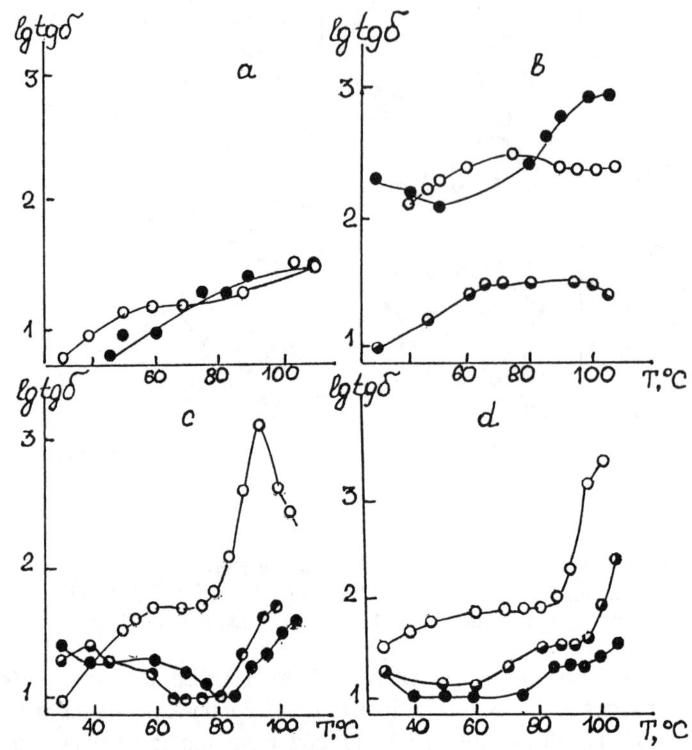

Fig. 4. Temperature dependencies of *l*gtgδ for the dispersions of cellulose (a), diethylaminocellulose (b), carboxylmethyl cellulose (c), and cellulose phosphate (e) dried above the saturated solution of MgO (1) and moistened at the relative air humidity of 31 (2) and 67% (3).

As for the intermolecular interaction, the third type is similar to the second type since in the second case the entire system of intermolecular interactions, involving the participation of OH-groups, is already completely modified. Of course, the latter is valid until a higher concentration of COOH-groups is attained when their own contribution to the system of intermolecular interactions becomes especially pronounced and the concentration of OH-groups is low.

We distinguish the short-range and long-range order of action of COOH-groups on the matrix of cellulose derivatives. The short-range order implies a peculiar polymer "cluster" consisting of a COOH-group which is hydrogen-bonded with OH-groups. The long-range order means a certain polymer volume (the sphere of action of a COOH group) where the intermolecular interaction involving hydroxyl groups has been changed, which brings about an increase of water absorption in these spheres as well as other changes, in particular, under the action of an electric field, mechanical loading and, as a consequence, exerts an influence on the matrix ("cluster") isolation of acidic centers because of the increased molecular mobility of the polymer. The above considered structures may be the constituents of a kinetic segment that changes its the mobility at α-relaxation, or directly represent one of the types of kinetic segments in the matrix of cellulose derivatives. According to data available in literature[5], the segment consists of 2-4, 10-20 or 36 monomeric units and the above described structures may correspond, with respect to their volume, to such kinetic segments whose mobility changes upon α-relaxation. Our hypothetical model of structural changes in the cellulose matrix due to the introduction of

COOH-groups somehow explains an influence of the introduction of COOH-groups on the α-relaxation process and its shift along the temperature and frequency axes.

In experiments we have revealed that the dipole polarization processes in cellulose compounds containing ionogenic groups essentially depend on the electric field intensity. Also, it has been found that a constant electric field affects the dipole polarization processes that are detected in an alternating field.

In order to differentiate, in measurements, the electromigration and polarization processes, we determined electrophysical properties of the air-dried MCC in dependence on the electric field intensity. Electrodes were insulated with a fluoroplastic film in order to decrease the conduction losses.

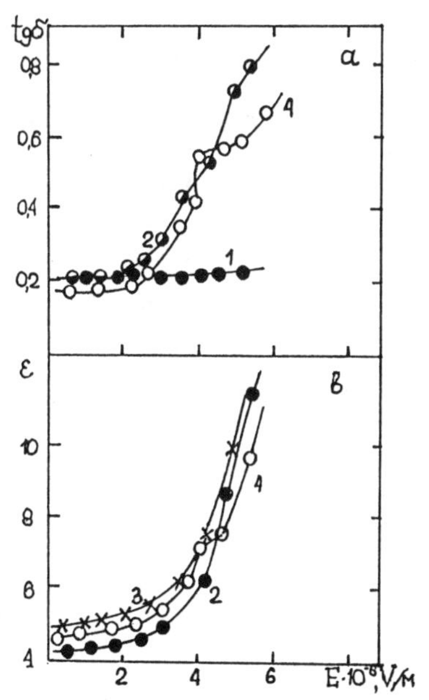

Fig. 5. The effect of electric intensity on tgδ (a) and ε (b) of air-dried cellulose samples in the form of paper (1) and MCC (2-4) on its basis with the following content of COOH-groups: 0.8, 1.4 and 2.6% (curves 2 to 4, correspondingly).

The oxidation of cellulose (Fig. 5) substantially enhances the polarization processes, which results in an increase of dielectric characteristics in the range of higher electric intensities. For unmodified paper, no such pronounced increase of tgδ and ε has been observed. Moreover, as is seen in Fig. 5, the increase of tgδ and ε with the electric field intensity is not monotonic. An electric field begins to exert a polarizing influence on MCC at the intensities from $3.0 \cdot 10^6$ to $4.0 \cdot 10^6$ V/m. However, at first tgδ and ε increase slightly. For instance, for MCC (1.3% of COOH-groups) upon changing an electric intensity from $0.7 \cdot 10^6$ V/m to $2.4 \cdot 10^6$ V/m tgδ increases approximately by 0.0490, while in the range $2.4 \cdot 10^6$ to $4.7 \cdot 10^6$ V/m - by 0.5010, *i.e.*, by the order of magnitude larger. This is indicative, first of all, of the presence of a potential barrier, after overcoming of which the polarization processes in an electric field become much stronger.

3. Concluding Remarks

In conclusion, the reported experimental results point to the increase of the polarization activity upon introducing acidic groups, in particular, the carboxyl groups into a polymer, to the increase of the moisture content or temperature of the tested materials, which is associated with the enhancement of the molecular mobility due to the mentioned factors. We consider that along with the known external factors (moistening and/or heating), an electric field, starting from certain threshold intensities, also weakens the intermolecular interaction in the cellulose structure modified by the introduced acidic groups and enhances the molecular mobility, which results in the enhancement of polarization processes.

To sum up, the polarization processes in cellulose compounds with acidic groups essentially depend on the intensity of a polarizing electric field due both to the presence of an electronegative dipole, *i.e.* an acidic group, and to the formation of a peculiar defective structure, as well as to higher molecular mobility of the kinetic segment. Apparently, this favors the formation of active electrorheological suspensions (ERS) on the basis of such compounds, the polarization properties of which may be controlled by an electric-field. Furthermore, it is concluded that cellulose-based materials, wherein the polarization processes depend on the electric field intensity, may be used as active fillers fo ERSes.

4. References

1. T. G. Lazareva and E. V. Korobko, in *Proc. of the 2nd International Conference of Intelligent Materials* (Williamsburg, USA, 1994).
2. N. I. Naimark, A. G. Zatsepin, V. S. Domkin and M. E. Sha, in *Polysaccharides: Crystallization and Their Interaction with Water* (Riga 1979).
3. V. A. Kargin and N. V. Kozlov, *Dokl. Akad. Nauk SSSR* **130** (1960) 356.
4. M. Rebekin, E. Jakopic, R. Wickenhause and B. Jurkowitsch, Papier (BDR) **29** (1975) 482.
5. S. T. Barsamyan, *Armyan. Khim. Zh.* **31** (1978) 214.

A NOVEL SEMI-ACTIVE CONTROL OF AUTOMOTIVE SUSPENSION USING AN ELECTRORHEOLOGICAL SHOCK ABSORBER

Masami NAKANO

Department of Mechanical Systems Engineering, YamagataUniversity
4-3-16, Jonan,Yonezawa, Yamagata, 992, JAPAN

ABSTRACT

The analytical modelling, the damping characteristics, and the control strategies of a shock absorber filled with an electrorheological fluid have been investigated numerically, focusing on the application of the ER shock absorber to an automotive suspension system. The analytical model of the electrically variable damping force produced in the ER shock absorber was derived, based upon an experimentally obtained post-yield plastic model of the ER fluid passing through the channel between two electrodes. The ER shock absorber under constant input voltage behaves like a coulomb friction damper, and the active shock absorber can be constructed, by electrically varying the friction-like forces in proportion to the relative velocity of the piston, to be a linear viscous damper with variable damping. New conceptual semi-active suspension control strategies using the ER shock absorber were proposed and numerically demonstrated the significant effectiveness in a two degree-of-freedom model suspension system using the ER shock absorber.

1. Introduction

Electrorheological(ER) fluids have intriguing properties of yielding solids in the presence of an electric field and of quick response to the change of electric field[1],[2]. One of the most prospective applications in engineering of the ER fluid was the field of vibration controls, and so far, several dampers and shock absorbers using the ER fluids have been developed[3]-[5]. These kinds of ER damping devices whose damping force can be controlled continuously with electric signals will be used to actively control various kinds of mechanical vibration. In practice, several applications [4]-[6] of the ER shock absorber to the automotive suspension controls have been tried, but unfortunately almost all cases of these suspension controls have been based on the control concept that the damping coefficient would be varied with the magnitude of input voltage to the ER shock absorber. As a result, the control performances were not so good, and exhibited remarkable nonlinearities against the vibration amplitude and frequency due to the significant nonlinearity of the relation between the damping force and the applied voltage. Therefore, it is hoped to establish a novel concept of the suspension control using the ER shock absorber which should be derived reasonably from the ER characteristics of damping force in the ER shock absorber.

The author has investigated the steady characteristics of yield stress caused by ER effects in the flow mode of an ER fluid [7], and has proposed the concept of the ER active damper and effective semi-active control strategies of vibration isolator using the ER active damper, according to the numerical investigation based on the ER characteristics obtained experimentally [8]. In this paper, focusing on the ER shock absorber and its application to the automotive suspension control, the analytical modelling and the damping characteristics of the ER shock absorber are addressed, as well as the control concept of the ER shock absorber in the automotive suspension. And new conceptual semi-active suspension control strategies using the ER shock absorber are proposed and numerically demonstrated the significant effectiveness in a two-degrees-of-freedom model suspension system.

2. Analytical Modelling of Automotive Suspension using ER Shock Absorber

2.1 Two-Degrees-of-Freedom Model Suspension System

The effective semi-active control strategies of automotive suspension using the ER shock absorber are discussed and demonstrated for a two-degrees-of-freedom suspension system modelled after an actual car as shown in Fig.1. If the damping force f_d is produced in the ER shock absorber which will be discussed in the next section, the equations of motion for this suspension system are given as follows:

$$M\frac{d^2x_m}{dt^2}+f_d+k(x_m-x_T)=0 \qquad \cdots (1)$$

$$M_T\frac{d^2x_T}{dt^2}-f_d-k(x_m-x_T)+k_T(x_T-x_{in})=0 \qquad \cdots (2)$$

where, M (=450 kg) is the unsprung mass, M_T (=30 kg) is the sprung mass, k(=19.6 kN/m) is the suspension spring stiffness, k_T(=196 kN/m) is the equivalent spring stiffness of the tire, and x_m, x_T, x_{in} are the vertical displacement of the unsprung mass, the sprung mass and the tire contact point with the road surface, respectively.

2.2 Analysis on Damping Force f_d in ER Shock Absorber

Figure 2 shows the schematic representation of the analytical model of ER shock absorber. Two cylinder chambers divided by a piston are connected to each other by an ER valve with several pairs of electrodes, and are filled with an ER fluid. The ER valve works to change the damping force f_d by controlling the pressure difference P_1-P_2 between both chambers with an electric signal. Assuming that the ER fluid is incompressible and the ER valve response to applied electric signals is quasi-steady, and considering the inertia of ER

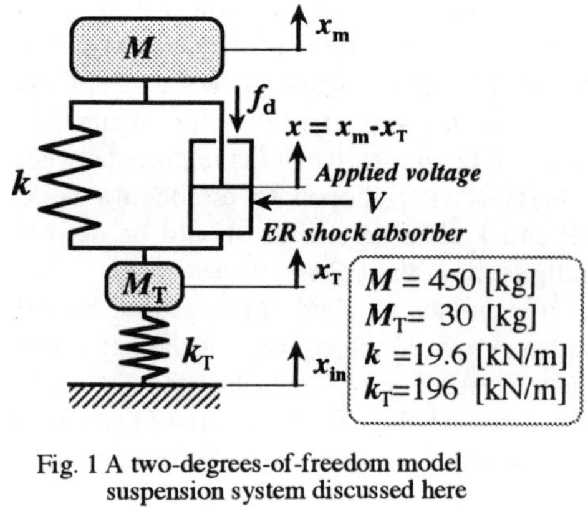

Fig. 1 A two-degrees-of-freedom model suspension system discussed here

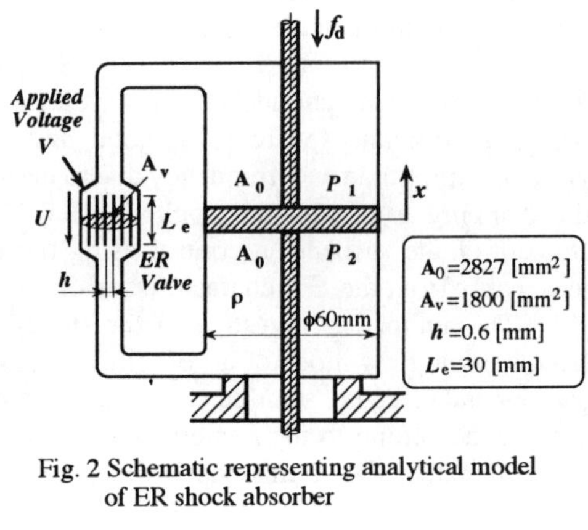

Fig. 2 Schematic representing analytical model of ER shock absorber

fluid within the ER valve channel and the steady pressure drop ΔP_T across the ER valve, the electrically controllable damping force f_d is given as follows:

$$f_d = A_0(P_1 - P_2) = A_0\left\{\frac{m}{A_v}\frac{dU}{dt} + \Delta P_T\right\} \quad \cdots\cdots (3)$$

where, A_0 is the area of the piston, m (= $\rho L_e A_V$) is the total mass of ER fluid within the ER valve channel, A_V and L_e are the total contraction area and the channel length of ER valve respectively, ρ is the density of ER fluid, and U is the mean flow velocity between the electrodes. In this paper, an ER shock absorber was assumed to have a piston(A_0=2827mm^2) of ϕ60 mm in diameter and an ER valve of A_V=1800mm^2, h=0.6mm, L_e=30mm.

And also, considering the mass conservation of ER fluid in the ER shock absorber, the following relationships between the mean flow velocity U and the displacement x (=x_m-x_T) of the piston relative to the cylinder can be given as follows:

$$U = \frac{A_0}{A_V}\frac{dx}{dt}, \qquad \frac{dU}{dt} = \frac{A_0}{A_V}\frac{d^2x}{dt^2} \quad \cdots\cdots (4)$$

2.3 Steady Pressure Drop ΔP_T Across ER Valve

To obtain the characteristics of the steady pressure drop ΔP_T across the ER valve, experiments have been performed for a simple ER valve consisting of two parallel plate electrodes [7]. The ER valve has a channel of h=0.6 mm in height between the two electrodes of L_e=10 mm in length and B=10 mm in width. The steady pressure drops ΔP_T were measured in terms of the mean flow velocity U and the electric field strength $E=V/h$ [kV/mm] (V: the applied voltage to ER valve), using a piston cylinder-ER valve experimental apparatus [7]. The results for an ER suspension of the strong acid ion exchange resin particles of about 5 μm in diameter dispersed into a 20 cSt silicone oil are shown in Fig.3.

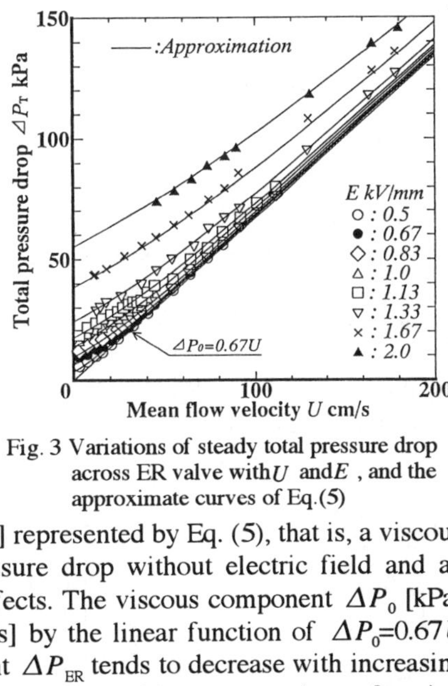

Fig. 3 Variations of steady total pressure drop across ER valve with U and E, and the approximate curves of Eq.(5)

The ΔP_T is divided into two components [7] represented by Eq. (5), that is, a viscous component ΔP_0 corresponding to the total pressure drop without electric field and an electrorheological component ΔP_{ER} due to ER effects. The viscous component ΔP_0 [kPa] can be related to the mean flow velocity U [cm/s] by the linear function of ΔP_0=0.67U obtained experimentally. And also, the component ΔP_{ER} tends to decrease with increasing U, and is a linear function of yield stress τ_{ER} [kPa] represented by an approximate function (6) obtained through a curve fitting process of experimental data[7].

$$\Delta P_T(E, U) = \Delta P_0 + \Delta P_{ER} \cdot \mathrm{sgn}(U)$$
$$\Delta P_0 = 0.67U, \qquad \Delta P_{ER} = (3L_e/h)\,\tau_{ER} \quad \cdots\cdots (5)$$

$$\tau_{ER}(E,U) = \alpha_\tau E^2 - (\alpha_\tau E^2 - \gamma_\tau E^2)(1 - e^{-\beta(E)|U|})$$
$$\alpha_\tau = 0.28\ [\mathrm{kPa \cdot mm^2/kV^2}], \quad \gamma_\tau = 0.07\ [\mathrm{kPa \cdot mm^2/kV^2}],$$
$$\beta(E) = 0.0265/E^2\ [\mathrm{s/cm}] \quad \cdots\cdots (6)$$

The approximate curves of ΔP_T drawn in Fig.3, have a fairly good agreement with the experimental results. Thus, it is possible to estimate the steady total pressure drop ΔP_T across the ER valve with a fairly good accuracy. And also, Fig.3 represents the relationship between a piston speed *dx/dt* and a damping force controlled by the ER valve in the ER shock absorber. So, it is expected that a high level of the damping force due to ER effects is caused in a range of relatively low piston speed, but as the piston speed becomes higher the damping force decreases.

3. Damping Characteristics of ER Shock Absorber

3.1 Evaluation of Damping Characteristics

The damping characteristics of the ER shock absorber are examined from the hysteresis loop of the damping force f_d to the displacement x of the piston moving sinusoidally. The damping forces are explicitly calculated from Eq.(3), (4), (5) and (6). Figures 4 and 6 show the damping characteristics for the two control strategies of the ER shock absorber. The figure (a) illustrates the changes of the hysteresis loops of damping force with electric signals. Since an area within the hysteresis loop corresponds to the energy W dissipated in the ER shock absorber per one cycle of the piston displacement, we can get an equivalent damping coefficient C_{eq} from the following equation:

$$C_{eq} = W / 2\pi^2 f x_a^2 \quad \cdots\cdots (7)$$

where, f and x_a are the frequency and the amplitude of the piston displacement, respectively. The changes of the equivalent damping ratio ζ $(=C_{eq}/2\sqrt{Mk})$ with the frequency f and the amplitude x_a are shown in graphs (b) and (c) of each figure.

3.2 Application of Constant Voltage [Fig. 4]

As obvious from Fig.4(a), the magnitude of the damping force is almost constant through one cycle of the piston displacement x, depending on the electric field strength E, but is just inverted in the direction at the turning points of the piston. These hysteresis loops are similar to those of a coulomb friction damper. The damping ratio ζ increases with the increase of electric field strength E, but dramatically decreases as the frequency f becomes

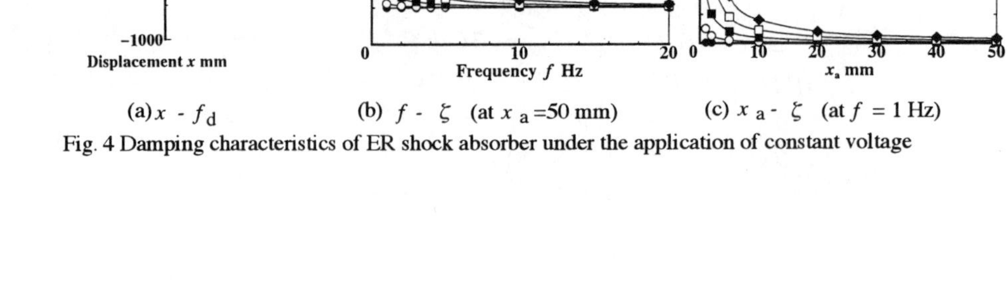

(a) x - f_d (b) f - ζ (at x_a=50 mm) (c) x_a - ζ (at f = 1 Hz)

Fig. 4 Damping characteristics of ER shock absorber under the application of constant voltage

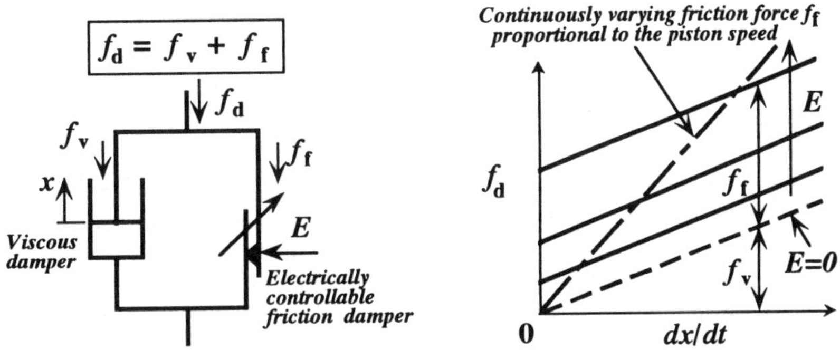

(a) Analogous hybrid damper model (b) Characteristic of damping force

Fig. 5 Analogous hybrid damper model of ER shock absorber and generated damping force

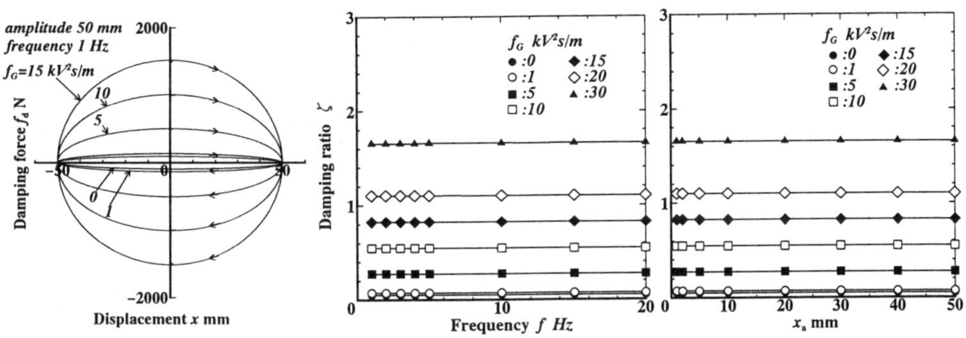

(a) x - f_d (b) f - ζ (at x_a=50 mm) (c) x_a - ζ (at f = 1 Hz)

Fig. 6 Damping characteristics of ER shock absorber under the application of square root of voltage proportional to absolute piston speed

higher or the amplitude x_a increases. Thus, the damping characteristics of the ER shock absorber under the application of constant voltage exhibit significant nonlinearities.

3.3 Application of Square Root of Voltage Proportional to Piston Speed [Fig.6]

As shown in Fig. 5(b), it is easily supposed from Eq.(3) and (5), and Fig.3 that the damping force f_d is divided into two components; i. e., a linear viscous damping force f_v at E=0 and an electrically variable yield damping force f_f . Therefore, the ER shock absorber is analogous to a hybrid one consisting of a viscous damper and an electrically controllable friction damper, as shown in Fig. 5(a). If the yield damping force

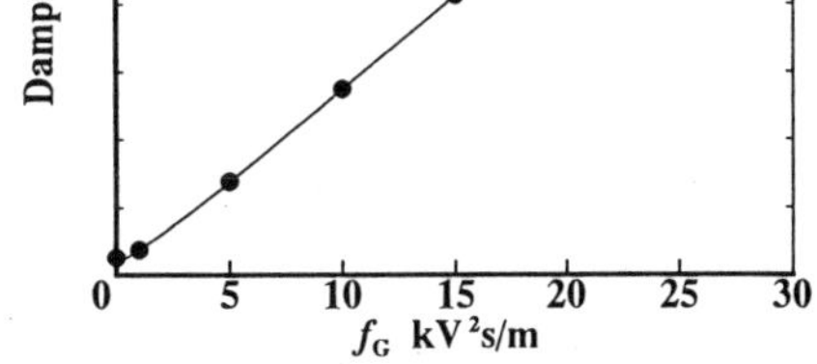

Fig. 7 Variation of damping ratio with feedback gain for the same control strategy as that of Fig.6

f_v is electrically varied in proportion to the piston speed dx/dt as indicated by a broken line in Fig.5(b), the ER shock absorber can be used as an viscous shock absorber with variable damping. Since the electrically variable damping force f_f, corresponding to the component of $A_0 \cdot \Delta P_{ER} \cdot \mathrm{sgn}(U)$ in Eq. (5), are proportional to the square of the electric field strength E^2 as obvious in Eq. (6), the generation of the ER damping forces proportional to the piston speed will be made by the linearization of the feedback control loop in which the square root of voltage proportional to the absolute value of the piston speed $V=\sqrt{|f_G\dot{x}|}$ is applied to the ER shock absorber. As shown in Fig. 6, the hysteresis loops are elliptical in shape, and the damping ratio ζ is crucially determined by the feedback gain f_G, fortunately independent of the frequency f and the amplitude x_a. Figure 7 shows that the damping ratio can be varied in almost proportion to the feedback gain. Thus, this nonlinear feedback control strategy most probably allows one to produce a purely viscous damper with a variable damping ratio controlled by the feedback gain f_G.

4. Vibration Isolation for Several Control Strategies of ER Suspension System

The responses of the unsprung mass to the harmonic excitation of the ground [$x_{in} = x_{ina}\sin(2\pi ft)$] are numerically simulated for the two-degrees-of-freedom ER suspension system shown in Fig.1, over the excitation frequency range from 0.1 to 15Hz. Figure 8 and 9 show the tansmissibility, which is the ratio of the displacement amplitude of the unsprung mass to that of the tire-ground contact point, for three kinds of the control strategies of the ER suspension system.

4.1 Application of Constant Voltage [Fig. 8]

As described in section 3.2, the ER shock absorber under a constant input voltage behaves like as a coulomb friction damper, so that the vibration isolation characteristics for

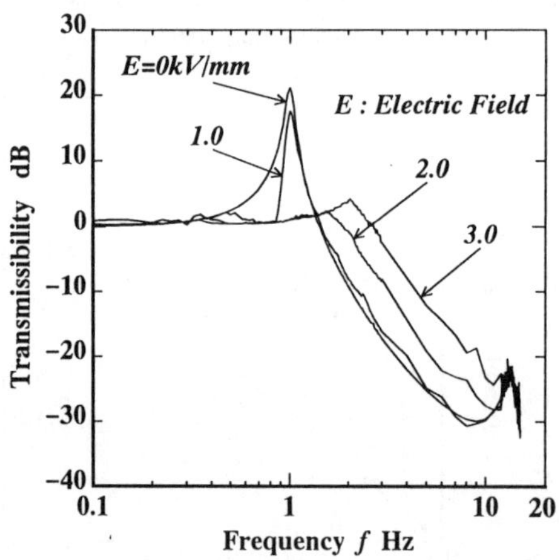

Fig. 8 Changes of the transmissibility with applied electric field strength E in the application of constant voltage. (x_{ina} =10mm)

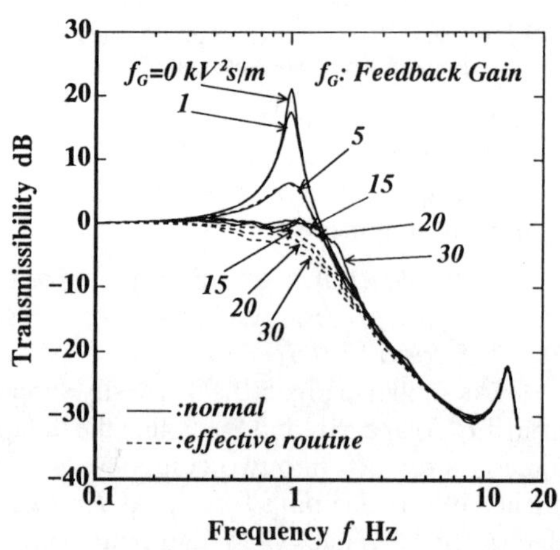

Fig. 9 Changes of the transmissibility with feedback gain in the application of the square root of voltage proportional to absolute unsprung mass velocity. (x_{ina} =10mm)

the suspension system using the ER shock absorber under a constant voltage are very similar to those for that using a coulomb friction damper. That is, the increase of the electric field strength E significantly suppresses the transmissibility down to just one around the natural frequency of the unsprung mass, but inversely increases it at frequencies between the natural frequencies of the unsprung mass and the sprung mass. And it is unfortunately noted that these characteristics of vibration isolation significantly depend on the excitation amplitude, which is easily seen from the damping characteristics of the ER shock absorber under a constant voltage discussed in the section 3.2.

4 . 2 Application of Square Root of Voltage Proportional to Absolute Velocity of unsprung mass [Fig. 9]

What is required for the control strategy to significantly improve the transmissibility of the unsprung mass is a shock absorber producing a damping force proportional to the absolute velocity of the unsprung mass. In the control strategy proposed here, the ER shock absorber is used as an active shock absorber whose electrically variable damping force is tried to be varied in proportion to the velocity $\dot{x}_m$ of the unsprung mass by sensing the velocity $\dot{x}_m$ and then applying the voltage $V = \sqrt{|f_G \dot{x}_m|}$ to the ER shock absorber. The distinguished vibration isolation results for the proposed control strategy named "normal control strategy" are indicated by solid lines in Fig.9. The transmissibility can be effectively decreased to below one at the frequencies around the natural frequency of the unsprung mass by the increase of the feedback gain f_G, but too high feedback gains tend to lead to the increase of the transmissibility by contraries. And also, it should be mentioned that the control results are almost independent of the excitation amplitude, due to the nonlinear feedback control of the square root of the absolute unsprung mass velocity, which is not shown here.

In this normal control strategy of the ER shock absorber, however, the ER shock absorber will only extract energy from the unsprung mass during part of the vibration cycle, spending the remainder of the cycle accelerating the unsprung mass. Then, an effective control routine is added to improve the transmissibility, in which the damping force is reduced to some minimum level by applying zero voltage to the ER shock absorber during that part of the cycle when the damping force tends to accelerate the unsprung mass. That is, the applied voltage V in this effective control strategy is given as follows:

$$\left.\begin{array}{ll} V = \sqrt{|f_G \dot{x}_m|} & \dot{x}_m \cdot (\dot{x}_m - \dot{x}_T) > 0 \\ V = 0 & \dot{x}_m \cdot (\dot{x}_m - \dot{x}_T) \leq 0 \end{array}\right\} \quad \cdots\cdots (8)$$

As indicated by dotted lines in Fig. 9, a significant improvement in the transmissibility is performed around the natural frequency of the unsprung mass at high feedback gains, compared with the above normal control strategy.

5. Transient Responses of ER Suspension System

Figure 10 shows the transient responses of the ER suspension system shown in Fig.1 to the 10 mm step excitation of the tire-ground contact point, when the square root of voltage $V = \sqrt{|f_G \dot{x}_m|}$ proportional to the absolute velocity of the unsprung mass is applied to the ER shock absorber at all times. The responses only depend on the feedback gain f_G in this control strategy. As the feedback gain increases, the step response of the unsprung

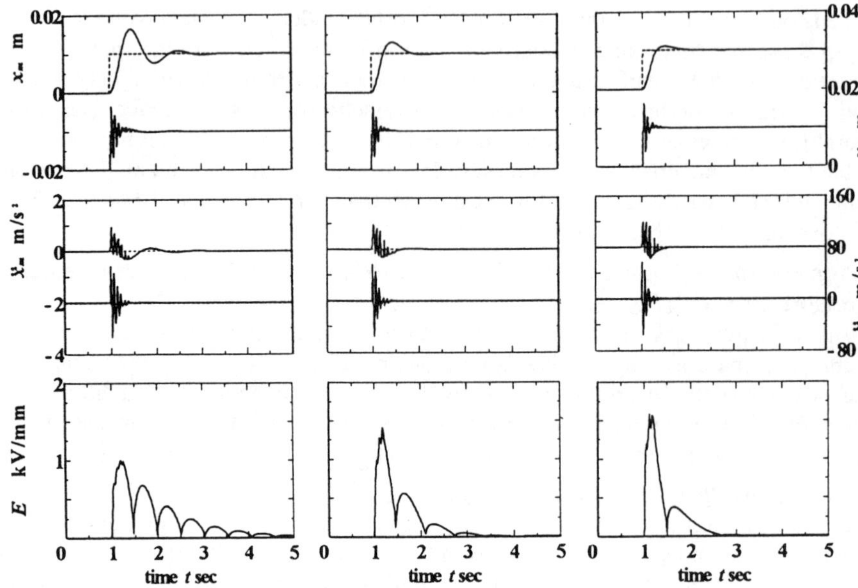

(a) $f_G = 6\ \mathrm{kV^2 s/m}$ (b) $f_G = 12\ \mathrm{kV^2 s/m}$ (c) $f_G = 18\ \mathrm{kV^2 s/m}$

Fig. 10 Changes of the step responses of the ER suspension system with feedback gain in the application of the square root of voltage proportional to absolute unsprung mass velocity

mass displacement x_m becomes to be quickly damped decreasing the overshoot, without any changes of the maximum acceleration $\ddot{x}_m$ level. This fact suggests that the increase of the feedback gain improves the ride comfort favourably. On the other hand, the maximum levels of the displacement x_T and acceleration $\ddot{x}_T$ of the sprung mass remain unchanged for the increase of the feedback gain.

6. Concluding Remarks

The analytical model of the electrically variable damping force produced in the ER shock absorber was proposed, based upon an experimentally obtained post-yield model. Simulation results on the damping characteristics of the ER shock absorber show that the ER shock absorber under constant input voltage behaves like a coulomb friction damper, and under the continuous feedback control of the square root of absolute piston speed exhibits the damping characteristics similar to those of a viscous damper.

Three semi-active control strategies of automotive suspension using ER shock absorber were examined for a two-degrees-of-freedom model suspension system. Among these strategies, the nonlinear continuous feedback control of the square root of the absolute unsprung mass velocity was found to be most effective to reduce the vibration transmissibility of the unsprung mass, and to enable control of transmissibility approaching that which can be achieved with a full active suspension system using an actuator. In addition, it should be possible to reduce component costs, compared with the full active system, by employing the above novel semi-active control strategy using ER shock absorber.

7. References

1. D. L. Klass and T. W. Martinek, *Electroviscous fluids. I: Rheological properties, J. Appl. Phys.*, 38 (1967) 67.
2. H. Uejima, *Dielectric mechanism and rheological properties of electro-fluids, Jpn. J. Appl. Phys.*, 11-3 (1972)319.
3. R. Stanway, J. L. Sproston, and N. G. Stevens, *Non-linear modelling of an electrorheological vibration damper, J. of Electrostatics,* 20 (1987), 167.
4. R. Stanway, J. L. Sproston, and X. Wu, *Variable suspension damping using electro-rheological fluids, Insti. Mech. Eng.*, No. C382/03 (1989), 547.
5. N. K. Petek, *An electronically controlled shock absorber using electrorheological fluid, SAE paper*, No. 920275 (1992), 67.
6. X. M. Wu, J. Y. Wong, M. Sturk, and D.L. Russell, *Simulation and experimental study of a semi-active suspension with an electrorheological damper, Proc. 4th Int. Conf. on ER Fluids, ed. R. Tao & G. D. Roy* (World Scientific Publishing, 1994) 568.
7. M. Nakano and T. Yonekawa, *Pressure responses of ER fluid in a piston cylinder- ER valve system, Proc. 4th Int. Conf. on ER Fluids, ed. R. Tao & G. D. Roy* (World Scientific Publishing, 1994) 477.
8. M. Nakano, *A novel semi-active control of vibration isolator using an electrorheological fluid damper, Proc. of 2nd. Int. Conf. on Motion and Vibration Control*, Yokohama, (1994), 800.

QUASI–STATIC EFFICIENCY OF A CONCENTRIC ER CLUTCH

VLADIMÍR ORAVSKÝ
Institute of Materials and Machine Mechanics,
Slovak Academy of Sciences
Račianska 75, P.O.Box 75, 830 08 Bratislava 38, Slovakia

ABSTRACT

In this contribution an application of electrorheological (ER) fluid in ER clutch is considered. The clutch is embedded in a broader system with a driving machine on one side and a loading machine (brake) on the other one. The quasistatic model of the system at a constant load and speed is described and solved in a dimensionless form.

Efficiency of ER clutch is determined as the ratio of the output and input powers and consecutively analyzed in the space of model parameters. A special sensitivity analysis for many thousands of combinations of parameters (in their possible ranges) is performed and statistically evaluated.

It is pointed out, that for some combinations of parameters and some arrangement of the system, efficiency can be greater than 1 due to additional energy originated from ER fluid. In such cases more refined energy balance and definition of efficiency are needed.

1. Introduction

Modern electrorheological fluids (**ERF**) have rising importance and great future. They are suspensions of electrically conducting particles in non conducting liquids. When an electric potential is applied to these fluids, their rheology and apparent viscosity change in a wide range: they become stiffer and eventually hard [1, 2]. This phenomenon can be utilized in many technical applications.

In the Institute an application of ERF towards concentric electrorheological clutch (**ERC**) has been modelled and solved [5, 6]. It differs from others, e. g. [1, 3, 4] in two features:

1. ERC is embedded into a broader system, so called electro–hydro–aggregate (**EHA**), consisting moreover of an electrodrive (**ED**) and brake (**B**) Fig. 2. It is important that ED has a finite output with its dynamic moment characteristic $M_D(\omega_D)$ and B has a harmonic loading moment M_L. But here only quasistatic (**qs**) regimes and characteristics for a constant load and speed will be considered.
2. nondimensional (**ndim**) form that offers more general, advantageous and simple both formulation and solution.

ERC itself consists of inner and outer concentric rotating discs with ERF between them (Fig. 1). There are two basic arrangements of EHA (Fig. 2):

EHAI when the outer disc is connected with ED and the inner one with B:

$$\omega_o = \omega_D \,, \; M_o = M_C \,, \; \omega_i = \omega_B \,, \; M_i = M_B \qquad (1)$$

and vice versa for EHAII: $\omega_i = \omega_D \,, \; M_i = M_C \,, \; \omega_o = \omega_B \,, \; M_o = M_B \qquad (2)$

More detailed description of EHA and ERC will be given in the next paragraph.

The aim of thispaper is to describe, solve and analyze EHA in qs conditions, nominally its efficiency and to point out an anomalous peculiar situation for EHAII.

2. Quasistatic (qs) model of EHA

For purpose of modelling, the following assumptions have been adopted:

Π 1. discs and adjacent shafts are totally rigid and with unidirectional rotation,

Π 2. on input and output shafts, friction moments are supposed consisting of dry component of Coulomb type and viscous linear component,

Π 3. ERF is a non–newtonian incompressible fluid approx. modelled by a Bingham plastic[2]

Π 4. parameters of ERF: density ρ, basic dynamic viscosity $\mu_o = 1$ and additional shear stress in ERF τ_E due to intensity E of electrostatic field, are supposed to be constant all over volume of ERF and independent of temperature,

Π5. a gap between the outer and inner discs $h = 1 - r_i$ is supposed to be sufficiently narrow for giving rise to a laminar flow of ERF which is moreover assumed to be symmetric regarding axis of rotation and only in tangential direction (influences of gravitation and disc edges are neglected).

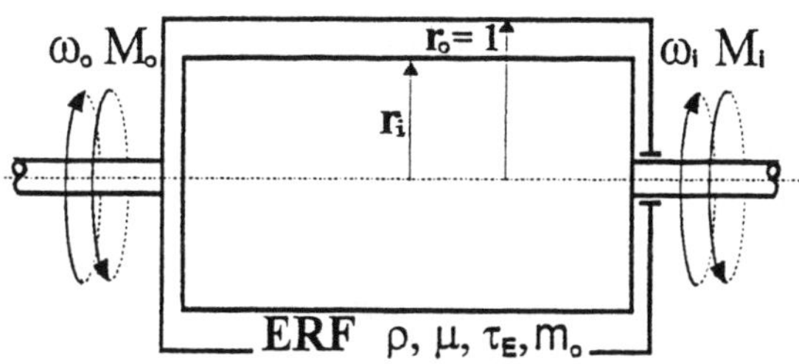

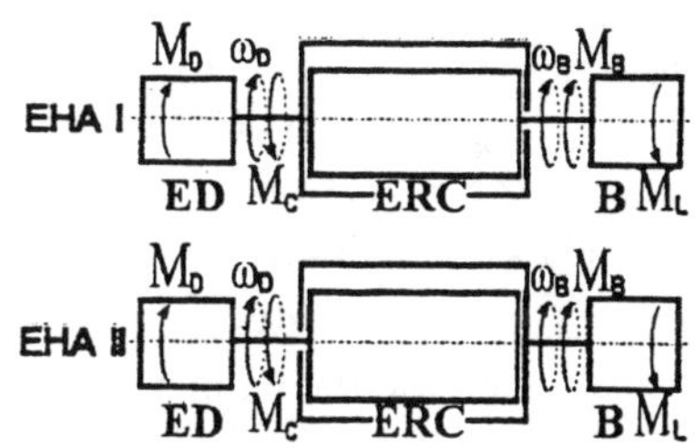

Fig. 1 Scheme of ERC

Fig. 2 Scheme of EHA with ERC

Model of EHA based on the above assumptions is presented in Fig.3 directly in ndim form.

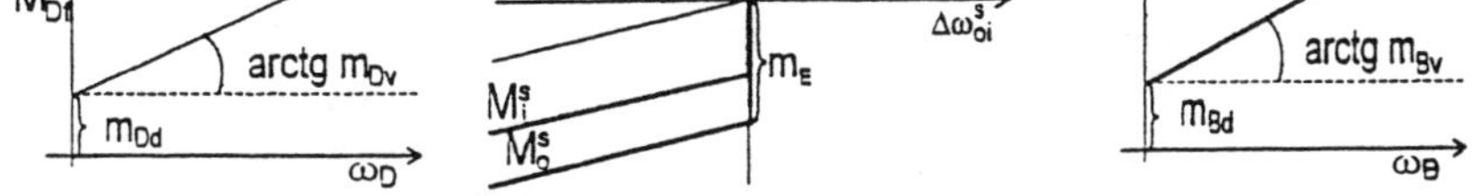

Fig. 3 Structural and moment scheme of EHA

Characteristics (**CH**) in Fig. 3 are described by the following Eqs[7]:

driving moment	$M_D = (1 - \omega_D)/s_K$	(3)
mech. friction moment of ED	$M_{Df} = m_{Dd} + m_{Dv}\,\omega_D$	(4)
mech. friction moment of B	$M_{Bf} = m_{Bd} + m_{Bv}\,\omega_B$	(5)
loading moment	$M_L = m_{\Phi}$	(6)

friction moment on outer disc $M_o = -m\,\Delta\omega_{oi} - q_{oi}\, m_E$ (7)

friction moment on inner disc $M_i = m\,\Delta\omega_{oi} + q_{oi}\, m_E r_i^2$ (8)

where $\Delta\omega_{oi}$ is the difference between angular velocities (**angV**) of outer and inner discs

$$\Delta\omega_{oi} = \omega_o - \omega_i \quad (9)$$

In the model Eq. (3÷9) there are following 9 non–dimensional parameters (**ndimPm**):

s_K is the nominal slip of ED (Fig. 3 left up),

m_{Dd} and m_{Dv} or m_{Bd} and m_{Bv} are ndim dry and viscous components of mechanical friction on ED or B respectively (Fig.3 down, left and right),

m_Φ is ndimPm of mean (constant) load of B (Fig.3 right up),

m is relative ndimPm of size of ERC with respect to size of ED,

m_E is ndimPm of additional moment of ERF due to intensity of electrostatic field,

r_i is ndim radius of the inner disc relative to outer one,

$$q_{oi} = \begin{cases} 1 \\ -1 \\ (-1;1) \end{cases} \quad \text{for } \Delta\omega_{oi} \begin{cases} >0 \\ <0 \\ =0 \end{cases} \quad \text{is a special sign function of difference of angV-s} \quad (10)$$

Note that analytical CH (Eq. 7, 8, 10) and corresponding graphical CH in Fig. 3 (middle, down) is already the result of qs solution of the following

boundary problem for ERF [5]:	$\rho\cdot\partial u/\partial t = \mu\left(\partial^2 u/\partial r^2 + \Omega/r\right)$	(11)
with boundary conditions:	$u(r_i) = w_i\, r_i\,;\quad u(1) = w_o$	(12)
where dim shear rate:	$\Omega = \dot{\bar{\gamma}}/\bar{\omega}_S = \partial u/\partial r - u/r$	(13)
ndim friction moment:	$M(r) = \bar{M}/(2\bar{M}_K) = m_o\,\mu\,\Omega r^2$	(14)
ndim dynamic viscosity:	$\mu = \bar{\mu}/\bar{\mu}_0 = 1 + \tau_E/\lvert\Omega\rvert$	(15)

in which $u = \omega\, r$ is a circumferential velocity on a radius $r = \bar{r}/\bar{r}_o \quad (r_i \le r \le 1)$ (16)

τ_E is ndimPm of electrostatic field, m_o is ndimPm of size of the outer disc.

Moreover binding Eqs for boundary friction moments are $M_i = M(r_i)$, $M_o = -M(1)$ (17)

Manipulation with Eqs (11÷12) for qs state gives dif Eq.

$$\partial^2 u/\partial r^2 + \partial u/(r\,\partial r) - u/r^2 = 0 \quad \text{with solution } u = Ar + B/r, \quad (18)$$

where integration constants A and B are determinable from boundary conditions Eq. (12). After their applications, one obtains

$$u = r\left[\omega_o + \Delta\omega_{oi}/(2R_i)\right] - \Delta\omega_{oi}/(2rR_i), \quad \Omega = \Delta\omega_{oi}/(r^2 R_i) \quad (19)$$

$$M(r) = m\,\Delta\omega_{oi} + q_{oi}\, m_E\, r^2 \quad (20)$$

where $\Delta\omega_{oi}$ or q_{oi} is given by Eqs (9 or 10) and auxiliary ndimPm-s are

$$R_i = \left(1/r_i^2 - 1\right)/2, \quad m = m_o/R_i\,, \quad m_E = m_o\,\tau_E \quad (21)$$

From Eqs (20) and (17), one immediately gets (7) and (8).

Note also that non–dimensionalization:

of all angV is made relative to the synchronous angV of ED ω_S,

of all moments is made relative to the double maximum qs moment of ED $2M_K$,

of all radii is made relative to the radius of outer disc r_o,

of dynamic viscosity is made relative to the base viscosity μ_0 (without ER effect).

In keeping with this one gets also ndim qs moment CH of ED according to Eq. (3) and Fig. 3 (left, up) and ndim friction moments (4, 5).

Further one needs to express the different structural arrangements of EHA mentioned in introduction (Fig. 2). Then for input and output friction moments of EHA one gets:

$$M_C = m\,\Delta\omega + m_C r_C; \quad M_B = m\,\Delta\omega + m_C r_B \tag{22}$$

where now $\Delta\omega$ is a difference between angV of ED and B: $\Delta\omega = \omega_D - \omega_B$ (23)

$$r_C = \begin{cases} 1 \\ r_i^2 \end{cases}, \quad r_B = \begin{cases} r_i^2 & \text{for EHAI} \\ 1 & \text{for EHAII} \end{cases} \tag{24}$$

$m_C = q\,m_E$ is an auxiliary ndimPm of ERC, (25)

$q = \Delta\omega / |\Delta\omega|$ analogically to Eq. (9, 10). (26)

Now one can complete the qs model of EHA with Eqs of equilibrium of moments acting on the input and output side of ERC is follows:

$$M_D - M_{Df} - M_C = 0; \quad M_B - M_{Bf} - M_L = 0 \tag{27}$$

The given ndim qs model of EHA (22÷26, 3÷6) has altogether 9 ndimPm:

3 for ED: s_K, m_{Dd}, m_{Dv}

3 for ERC: r_i, m_o, τ_E or (21): r_i, m, m_E (28)

3 for B: m_Φ, m_{Bd}, m_{Bv}

But, using the second equilibrium Eq. (26) and Eq. (5, 6) it is possible to unify ndimPm m_Φ and m_{Bd} into the common loading ndimPm $m_L = m_\Phi + m_{Bd}$, (29) so that for purpose of the following analysis it will be enough to work only with 8 ndimPm-s. The most important from them, as we shall see, will be the ndimPm of ERC m_C (25) and of load m_L (29).

3. Determination and analysis of qs solution

After insertion of Eqs (3÷6, 22, 23) into Eqs (27) and arrangement one gets the following coupled linear system:

$$(1-\omega_D)/s_K - m_{Dd} - m_{Dv}\omega_D - m(\omega_D - \omega_B) - m_C r_C = 0 \tag{30}$$

$$m(\omega_D - \omega_B) - m_C r_B - m_{Bd} - m_{Bv}\omega_B - m_L = 0 \tag{31}$$

Solution with respect to the unknown angV ω_D, ω_B gives results:

$$\omega_D = \omega_0 - s_0 m_L - s_1 m_C = \omega_{D0} - s_1 m_C, \quad \omega_B = (m\omega_0 - p_0 m_L + p_1 m_C)/m_0 \tag{32}$$

$$\Delta\omega = \omega_D - \omega_B = (m_{Bv}\omega_0 + q_0 m_L - q_1 m_C)/m_0 = (\Delta\omega_0 - q_1 m_C)/m_0 \tag{33}$$

where the following auxiliary ndimPm-s have been introduced:

$$s_0 = s_{K0}\, m_1, \quad s_1 = s_{K0}\, r_{C1}, \quad \omega_0 = (1 - s_K m_{Dd})/(1 + s_K m_{v1}) \tag{34}$$

$$p_0 = 1 + m s_0, \quad p_1 = r_0 - m s_1, \quad \omega_{D0} = \omega_0 - s_0\, m_L \tag{35}$$

$$q_0 = 1 - m_{Bv}\, s_0, \quad q_1 = r_B + m_{Bv}\, s_1, \quad \Delta\omega_0 = m_{Bv}\,\omega_0 + q_0\, m_L \tag{36}$$

in which: $m_0 = m + m_{Bv}$, $m_1 = m/m_0$, $m_{B1} = m_{Bv}/m_0$, $m_1 + m_{B1} = 1$ (37)

$s_{K0} = s_K/(1 + s_K m_{v1})$, $m_{v1} = m_{Dv} + m_1 m_{Bv}$, $r_{C1} = r_C - m_1 r_B$ (38)

After use of Eqs (24, 37) and manipulation with r_{C1}, q_1 from Eqs (36, 38) one gets:

$$r_{C1} = \begin{cases} 1-r_i^2+m_{B1}r_i^2 \\ -1+r_i^2+m_{B1} \end{cases} \quad q_1 = \begin{cases} r_i^2+m_{Bv}\,s_{K0}\left(1-m_1 r_i^2\right) & \text{for EHAI} \\ r_i^2+m_{Bv}\,s_{K0}\left(1-m_1 r_i^2\right) & \text{for EHAI} \end{cases} \tag{39}$$

Using Eq. (33) it is easy to express friction moments M_D, M_B from Eq. (22):

$$M_W = m\,\Delta\omega + m_C\, r_W = m_1\,\Delta\omega_0 + (r_W - m_1 q_1)m_C\,, \qquad W = B, C \tag{40}$$

Note that due to Eq. (31, 29) it also holds: $M_B = m_{Bv}\,\omega_B + m_L$ (41)

Now all qs quantities necessary for further needs are determined. Before using them for determination of efficiency, their concise analysis is presented.

Due to Eq. (25, 26) one can conclude from Eq. (33) that only non-negative solution exists for $\Delta\omega$: $\Delta\omega \geq 0 \;\Rightarrow\; \omega_D \geq \omega_B > 0$ (42)

The limiting case with equal angV of both discs of ERC corresponds to the so called frozen state of ERC: $\omega_D^f = \omega_B^f = \omega^f \;\Rightarrow\; \Delta\omega^f = 0 \;\Rightarrow\; m_C^f = \Delta\omega_0/q_1 = (m_{B0}+m_L)/q_2$ (43)

(corresponding quantities and Pm-s are distinguished by the upper index f)

where:

$$q_2 = \frac{q_1}{q_0} = r_B + m_{Bv}\,s_{K1}\,r_C = \begin{cases} r_i^2 + m_{Bv}\,s_{K1} & \text{for EHAI} \\ 1+m_{Bv}\,s_{K1}\,r_i^2 & \text{for EHAII} \end{cases} \tag{44}$$

$$s_{K1} = \frac{s_K}{1+s_K\,m_{Dv}}, \quad m_{B0} = m_{Bv}\,\omega_1, \quad \omega_1 = \frac{\omega_0}{q_0} = \frac{1-s_K\,m_{Bd}}{1+s_K\,m_{Dv}} \tag{45}$$

It is important to clear up mutual relations between Pm-s: m_C, m_E, q and m_L. According to Eqs (25, 26) in an unfrozen state it is:

$$\Delta\omega > 0 \quad\Rightarrow\quad q = 1 \quad\Rightarrow\quad m_C = m_E\,, \tag{46}$$

so that Pm m_C is dependent on Pm m_E (21), but is independent on Pm m_L (29). On the contrary, in the frozen state Pm m_C^f (43) is dependent on Pm m_L, but is independent on m_E:

$$\Delta\omega^f = 0 \quad\Rightarrow\quad q^f = m_C^f/m_E^f \;\Rightarrow\quad 0 \leq q^f \leq 1, \tag{47}$$

so that from general extent (10) only extent $0 \leq q^f \leq 1$ holds.

The situation is clearly seen in the Fig. 4, where upper bound m_E^f corresponds to the disruptive (breakdown) intensity E^d of D. C. field. At the beginning (point z in the Fig. 4) there is unique relation between all Pm-s m_C, m_E, m_L: $m_C^z = m_E^z = (m_{B0} + m_L)/q_2$. (48)

The corresponding frozen friction moments are easy determined from Eqs (22) and (43):

$$M_C^f = m_C^f r_C, \qquad M_B^f = m_C^f r_B \tag{49}$$

But the frozen angV ω^f must be determined from the original system Eqs (30, 31):

$$\omega^f = \omega_0^f - s_0^f\, m_L, \quad \text{where } \omega_0^f = r_B\,\omega_1/q_2, \quad s_0^f = r_C\, s_{K1}/q_2 \tag{50}$$

It is possible to derive that the following inequalities hold:

$$m^f_{CI} > m^f_{CII}, \qquad \omega^f_I < \omega^f_{II}, \qquad M^f_{CI} > M^f_{BI}, \quad M^f_{CII} < M^f_{BII} \tag{51}$$

Note that the frozen state (or ERC) has been named in Ref. [8] as a blocked state (or ERC) and analysed there in detail. Here only a partial analysis for a basic case is presented, when the viscous friction on the side of B is neglected: $m_{Bv} = 0$. (52)

The structure of Eqs (33÷50) enables one fairly to simplify relations:

$$m_{B0} = m_{B1} = 0, \quad m_0 = m, \quad m_1 = 1, \quad m_{v1} = m_{Dv}\,, \quad r_{C1} = r_C - r_B \tag{53}$$

$$s_0 = s_{K0} = s_{K1}\,, \quad s_1 = s_{K1}/r_{C1}\,, \quad \omega_0 = \omega_1\,, \quad \Delta\omega_0 = m_L\,, \quad q_0 = 1, \quad q_1 = q_2 = r_B \tag{54}$$

so that from Eqs (32, 33, 40, 41) for an unfrozen state ($m_C = m_E$):

$$\omega_D = \omega_1 - s_{K1}(m_L + r_{C1}\, m_E)\,, \qquad \omega_B = \omega_D - \Delta\omega\,, \qquad \Delta\omega = (m_L - r_B\, m_E)/m \tag{55}$$

$$M_C = m_L + r_{C1}\, m_E\,, \qquad M_B = m_L \tag{56}$$

and from Eqs (43, 49, 50) for the frozen state ($m_C = m_C^f$):

$$m_C^f = m_L/r_B\,, \qquad \omega^f = \omega_1 - s_{K1}\, r_C/r_B\,, \qquad M_C^f = m_L + r_C/r_B\,, \qquad M_B^f = m_L. \tag{57}$$

Results (54÷56) for m_L = const are plotted in the Fig. 5 a , b.

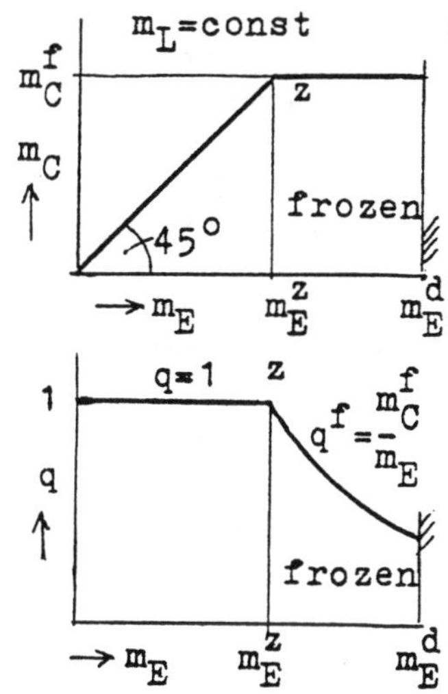

Fig. 4 Connection of parameters m_C, m_E, q

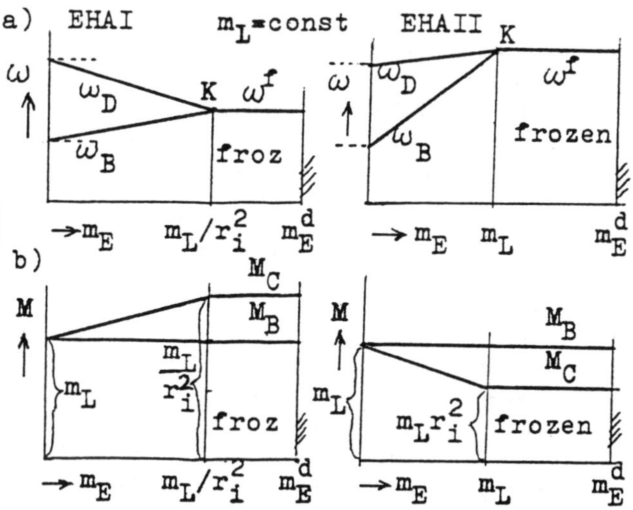

Fig. 5 Courses of angular velocities (a) and moments (b) in ERC for constant load

4. Determination and analysis of efficiency

Efficiency (**Ef**) of ERC is defined as the ratio of output and input powers P_B and P_C given by the products of corresponding moments and angV:

$$\eta_C = P_B / P_C\,, \quad \text{where} \quad P_B = M_B\, \omega_B\,, \quad P_C = M_C\, \omega_D \tag{58}$$

and analogically for EHA:

$$\eta_A = P_L / P_D\,, \quad \text{where} \quad P_L = M_L\, \omega_B\,, \quad P_D = M_D\, \omega_D \tag{59}$$

After use of corresponding power balance instead of moment balance (27):

$$P_D - P_{Df} - P_C = 0, \quad P_B - P_{Bf} - P_L = 0, \quad \text{where} \quad P_{Df} = M_{Df}\, \omega_D,\ P_{Bf} = M_{Bf}\, \omega_B \tag{60}$$

one gets an equivalent definition of Ef of ERC:

$$\eta_C = \left(P_L + P_{Bf}\right) / \left(P_D - P_{Df}\right) \tag{61}$$

The comparison of Eqs (61) and (59) gives inequality:

$$\eta_A < \eta_C \tag{62}$$

From Eq. (58) for frozen ERC (43), (49), with use of Eq. (24), it follows:

$$\eta_C^f = \frac{M_B^f\, \omega_B^f}{M_C^f\, \omega_D^f} = \frac{M_B^f}{M_C^f} = \frac{r_B}{r_C} = \begin{cases} r_i^2 < 1 & \text{for EHAI} \\ 1/r_i^2 > 1 & \text{for EHAI} \end{cases} \tag{63}$$

So for frozen EHAII one gets Ef greater than 1! This is due to fact that for EHAII output moment M_B^f is greater than the input one M_C^f (Fig. 5). Physically it means that for EHAII an additional energy must come into ERC from ERF. This phenomenon calls for detailed both qualitative and quantitative analyses. But due to limited extent of the paper, only results of quantitative analysis can be presented. It is based on special numerical sensitivity analysis, determining and statistically elaborating the partial derivative of Ef with respect to all ndimPm-s. The mean of absolute value of this derivative: $|\partial\eta_C / \partial Pm|_\phi$ all over space of ndim Pm-s, can serve as a measure of influence of individual Pm-s upon Ef. According to it, the ndim Pm-s can be arranged into the so called sequence of dominance.

Realization of the above idea has been performed for the following extent of Pm-s (28) (lower bound, upper bound; step):

$$s_K\ (0, 0.2; 0.05),\ r_i\ (0.85, 0.95; 0.05),\ m_o\ (0, 0.3; 0.05),\ \tau_E\ (0, 2; 0.5) \qquad (64)$$
$$m_{Dd}\ (0, 0.2; 0.05),\ m_{Dv}\ (0, 0.2; 0.05),\ m_{Bv}\ (0, 0.2; 0.05),\ m_L\ (0, 0.4; 0.05)$$

All possible discrete values (64) give altogether 826 875 combinations (points) in the 8-dim space of ndimPm-s and the following sequences of dominance for unfrozen EHA-s:

System	Order	1	2	3	4	5	6	7	8
EHAI	ndimPm	r_i	m_o	m_L	m_{Bv}	s_K	τ_E	m_{Dd}	m_{Dv}
	part. deriv.	2.801	2.337	0.677	0.403	0.072	0.043	0.013	0.012
EHAII	ndimPm	m_o	r_i	m_L	m_{Bv}	τ_E	s_K	m_{Dd}	m_{Dv}
	part. deriv.	3.548	2.185	1.158	0.787	0.209	0.079	0.016	0.014

(65)

So the 3 most important Pm-s influencing Ef of both EHA-s are:
radius of inner disc r_i, size of outer disc m_o and load m_L,
whereby: an increase of m_o always increases Ef,
an increase of m_L always decreases Ef,
but increase of r_i can increase or decrease Ef.

For frozen EHA-s, according to Eq. (63), there is only one important Pm: r_i.

The corresponding mean Ef and its standard deviation as well as number of occurrence for unfrozen, frozen and inadmissible cases (when $\omega_B < 0$) are given in the following table:

System	EHAI			EHAII		
	unfrozen	frozen	inadmissible	unfrozen	frozen	inadmissible
Mean Ef	0.703	0.821	–	0.823	1.242	–
Stand. dev.	0.208	0.873	–	0.270	0.113	–
number	590 002	119 679	117 194	547 702	163 197	115 976
Occur. %	71.35	14.48	14.17	66.24	19.74	14.02

(66)

Important remark is that for EHAI it is always $\eta_{CI} < 1$, while for EHAII $\eta_{CII} > 1$ holds not only for the frozen state (63) but also for some unfrozen cases. Searching for such cases is beyond the given extent of this paper.

5. Conclusion

In the paper one has shown that:

- in operation of ERC (Fig. 1) a frozen regime can occur with equal speed of both discs but different transmitted moments (Fig. 5),
- in arrangement EHAII (Fig. 2) efficiency η_C (58) can be greater than 1, revealing that an additional source of energy is present within ERC,
- presented results of wide numerical analysis (65), (66) can deepen understanding of quasi static properties of ERC and its efficiency.

Finally it is worthwhile to note that efficiency can be refined and redefined in such way to give also for all EHAII efficiency less than 1. Such definition involves all feeding components including power of D. C. field in ERF. This, as well as investigation of efficiency in dynamic regimes will be tackled in further publications.

Acknowledgement

This work has been partially supported by the Grant Agency of the Slovak Academy of Sciences (under the Project 2/1135/94) which is hereby gratefully acknowledged. The author also wishes to thank his colleague O. Šlížek, M. Sc. (Eng.) for executing numerical calculations.

References

1. G. Oppermann, *Grundlagen zur praktischen Anwendung Elektroviskoser Flüssigkeiten* (BAYER AG, 1990).
2. *Electrorheological (ER) Fluids. A Research Need Assessment - Final Report.* (U. S. Department of Energy, Washington, D. C., May 1993).
3. R. Firoosian and al., *Dynamic torque response of an electrorheological clutch,* DE–Vol. 35 Rotating Machinery and Vehicle Dynamics ASME 1991, pp. 295÷301.
4. A. R. Johnson, J. Makin and W. A. Bullough, *Electrorheological catch/clutch: Inertial simulation.* In: Proc. of the 4th Int. Conf. ER Fluids, Feldkirch, Austria. Ed. R. Tao and G. D. Roy (World Scientific, Singapore, 1994) pp. 493÷515.
5. V. Oravský and V. Krč–Jediný, *Analysis of friction moment of a disc rotating in electroviscous fluid* (in Slovak). In: Proc. of 14th Int. Conf. Fluid Mechanisms, Olomouc 1992, pp. 66÷71.
6. V. Oravský and V. Krč–Jediný, *Dynamic model of hydroclutch with electroviscous fluid and its analysis at constant speed* (in Slovak). In: Proc. of 2nd Conf. Hydraulic and Pneumatic Mechanisms, Žilina 1993, pp. 113÷118.
7. V. Oravský and V. Krč–Jediný, *Analysis of constant angular velocities of an electrorheological clutch* (in Slovak). In: Proc. of 9th Scientific Conf. VŠDS, Žilina 1993, pp. 159÷164.
8. V. Oravský, *Analysis of blocked quasi–static regimes of an ER clutch.* In: Abstract and Program of the 4th Int. Conf. on ER Fluids, Feldkirch, 1993, p. 45.

APPLICATIONS OF ELECTRO-RHEOLOGICAL FLUIDS FOR CONSTRAINED LAYER DAMPING TREATMENT OF STRUCTURES

S O Oyadiji
Dynamics and Control Research Group
Division of Mechanical and Nuclear Engineering
Manchester School of Engineering
University of Manchester
Manchester M13 9PL, UK

ABSTRACT

Constrained layer damping (CLD) technique employs solid viscoelastic materials for controlling the vibration characteristics of structures. The effectiveness of the CLD treatment depends on the dynamic mechanical properties and geometrical dimensions of the viscoelastic and constraining layers. It also depends on the percentage coverage of the structure by the viscoelastic and constraining layers. Since the complex dynamic properties of electro-rheological (ER) fluids vary between those of viscoelastic liquids (at low levels of electric field strengths) and those of viscoelastic solids (at high levels of electric field strengths), ER fluids can be used in an adaptive form for CLD applications. The results of experimental and analytical studies on the vibration characteristics of a clamped-clamped plate which is partially and totally filled with an ER fluid are presented. The ER fluid employed is a mixture of 50 wt% of an analytical reagent grade of starch dissolved in 50 wt% of a silicone oil. It is shown that for the particular ER fluid utilised, the variations of the frequency response functions and the modal parameters of the composite plate are more strongly dependent on the location and size of the ER fluid CLD treatment than on the electric field strength.

1. Introduction

The use of solid viscoelastic materials for constrained layer damping (CLD) treatment of vibrating structures is well established[1]. The methodology is used as a relatively inexpensive and convenient means of controlling the vibration characteristics of structures instead of the more expensive alternative of local stiffening of parts of the structure or of complete re-design of the structure. However, the technique sometimes suffers a setback in terms of the unavailability of viscoelastic materials with dynamic properties to match those adjudged to be optimal on the basis of theoretical analysis of the viscoelastically-damped structure. Because the complex dynamic properties of electro-rheological (ER) fluids vary between those of viscoelastic liquids (at low levels of electric field strengths) and those of viscoelastic solids (at high levels of electric field strengths), ER fluids can be used in an adaptive form for CLD applications.

Several investigators have carried out experimental studies on the use of ER fluids for CLD vibration control of structures[2-9]. These investigations have been largely on small cantilever and simply supported composite beams comprising of steel, aluminium, plastic and graphite/epoxy composite beams of thicknesses ranging from 0.4mm to 2.5mm, lengths of between 150mm and 400mm, and widths of between 15mm and 30mm. The thicknesses of the ER fluid constrained damping layers have ranged from 0.1mm to 2.0mm. The composite beams have been 100% filled with the ER fluids. They have shown the potential applications of ER fluids for the constrained layer damping control of the vibrations of these structures.

In this paper, the CLD technique is employed in conjunction with an ER fluid for controlling the vibration characteristics of a clamped-clamped aluminium plate. The dimensions of the free portions of the plate are 500mm long by 100mm wide by 3.3mm thick. A constraining plate of aluminium of length 496mm, width 100mm and thickness 2.2mm is used. The ER fluid is a mixture of a reagent grade starch and a silicone oil. The effects of partial and total ER fluid-based CLD treatment on the vibration characteristics of the composite aluminium plate is investigated. It is shown that for the particular ER fluid utilised, the variations of the frequency response functions and the modal parameters of the composite plate are more strongly dependent on the location and size of the ER fluid CLD treatment than on the electric field strength.

2. Theoretical Equations

The modal frequencies f_n and modal loss factors η_n of a three-layer sandwich beam structure are given by

$$f_n = \frac{\lambda_n^2}{2\pi L^2}\left[\frac{|E^*I|}{\rho_1 H_1 + \rho_2 H_2 + \rho_3 H_3}\right]^{1/2} \quad ; \qquad \eta_n = \frac{\Im(E^*I)}{\Re(E^*I)} \tag{1, 2}$$

where λ_n is a characteristic or eigen value, E^*I is the complex flexural stiffness per unit width of the composite beam, ρ_i ,H_i , L_i are mass density and thickness of layer i, L is length of composite beam, and $\Im$, $\Re$ denote imaginary and real parts. Also, layer 1 is the base structure, layer 2 is the viscoelastic material, and layer 3 is the constraining beam. For the case of negligible damping in layers 1 and 3, and for $E_1 \gg |E^*_2| \ll E_3$, the complex flexural stiffness per unit width of the composite beam is given by

$$E^*I = E_1H_1^3/12 + E_3H_3^3/12 + E_1H_1D^{*2} - E_3H_3(H_{31}-D^*)^2 g^*/(1+g^*) \tag{3}$$

in which E_i is the Young's modulus of layer i, and the complex displacement D^* and complex shear parameter g^* are defined as

$$D^* = \frac{g^*E_3H_3H_{31}}{E_1H_1 + g^*(E_1H_1 + E_3H_3)} \quad ; \qquad g^* = \frac{G_2^*L^2}{E_3H_3H_2\lambda_n^2} \tag{4, 5}$$

where the inter-layer distance H_{31} and the complex shear modulus G_2^* are given by

$$H_{31} = (H_1 + H_3)/2 + H_2 \; ; \qquad G_2^* = G_2'(1 + j\eta_2) \tag{6, 7}$$

These equations are employed in the results section for computing the theoretical values of the modal frequencies and loss factors of the composite plate. Although Eq. 1 applies to a sandwich beam, it is applied for computing the natural frequencies of the first three bending modes of the composite plate structure investigated in the present study for the following reasons. The natural frequencies of a clamped-clamped beam of length L, width b, thickness h, cross-sectional area A, and density ρ are given by[11]

$$f_i = \frac{\lambda_i^2}{2\pi L^2}\left[\frac{EI}{\rho A}\right]^{1/2} = \frac{\lambda_i^2}{2\pi L^2}\left[\frac{Eh^2}{12\rho}\right]^{1/2} \tag{8}$$

The values of λ_i^2 for the first three bending modes of the beam are[11] 21.4, 61.7 and 120.9. Similarly, the natural frequencies of a clamped-clamped plate of length L, width b, and thickness h are given by

$$f_{ij} = \frac{\lambda_{ij}^2}{2\pi L^2}\left[\frac{Eh^3}{12\rho h(1-\nu^2)}\right]^{1/2} = \frac{\Lambda_{ij}^2}{2\pi L^2}\left[\frac{Eh^2}{12\rho}\right]^{1/2} \tag{9}$$

$$\text{where} \quad \Lambda_{ij}^2 = \frac{\lambda_{ij}^2}{(1-\nu^2)^{1/2}} = 1.06\,\lambda_{ij}^2 \quad \text{for } \nu=0.33 \tag{10}$$

As the ratio b/L for the plate approaches zero, the plate becomes a beam, and the eigenvalues λ_{ij}^2 of the plate tend to the eigenvalues λ_i^2 of the beam. Thus, by interpolating between these limiting values of λ_{ij}^2 and those given in standard text[10,11], the values of λ_{ij}^2 for the aluminium plates investigated, for which b/L=0.2 and ν=0.33, are 21.6, 59.3 and 112.0. The corresponding values of Λ_{ij}^2 are 22.9, 62.9 and 118.7. Since these values of Λ_{ij}^2 are approximately the same as the values of λ_i^2 for a clamped-clamped beam given earlier, it is obvious that Eq. 8, the natural frequency equation for the beam, will predict approximately the same natural frequencies as Eq. 9 for the first three bending modes of the aluminium plate used in these investigations. Hence, the use of Eq. 1 for the computations of the modal frequencies of the first three bending modes of the composite plate structure investigated is justified.

3. Experimental Methods

3.1 Details of Composite Plate and ER Fluid

The ER composite plate consisted of an aluminium plate primary structure of total dimensions 600 x 100 x 3.3 mm thick and an aluminium plate secondary structure (i.e. the constraining layer) of total dimensions 496 x 100 x 2.2 mm thick. A length of 50 mm was marked off the two ends of the primary structure and reserved for clamping. The remaining length of 500 mm of the primary structure was divided into five bays using silicone rubber strips of 3 mm thickness as the dividers. The strips were also lined up along the edges of the primary structure plate. The secondary structure or constraining layer plate was bonded to the top of the silicone rubber strips which also served as the electrical insulators between the primary and secondary plates. Thus the cavity of the composite plate was divided into five bays or cavities for the ER fluid. These cavities were numbered sequentially from 1 to 5 with cavity no. 1 at one end and cavity no. 5 at the other end of the composite plate. The ER Fluid employed was made by mixing an analytical reagent grade of starch with a Dow

Corning 200/100 cS silicone oil. The complex shear modulus properties of this ER fluid are shown in a companion paper[12].

3.2 Frequency Response Tests on the Primary Plate Structure

Before assembling the composite plate, the primary plate was tested on its own. The 50 mm lengths from the free ends of the composite plate, which were already marked off, were clamped to the top of two vertical supports as shown in Figure 1. An electromagnetic exciter was connected to the central point of the primary plate structure via a force transducer. The geometrical coordinates of the central point were x=L/2 and y=b/2 where x is coordinate position from the left clamp, L is the distance between the two clamped supports, y is coordinate position along width of plate, and b is the width of the plate. The exciter was connected to the central point of the plate in order to excite only the bending modes of the plate. The plate was then subjected to random excitation over a frequency band of 0 to 500 Hz using a Spectrum Analyser and the associated instrumentation shown in Figure 1. The spectrum of the input point force applied was monitored by the force transducer, which was connected between the exciter and the primary plate. The responses of the primary plate at positions 1 and 2 whose coordinates were (x, y) = (L/2, b/2) and (x, y) = (3L/8, b/2), respectively, were monitored by means of an accelerometer. The charge outputs of both of the piezoelectric force transducer and accelerometer were integrated to voltage and amplified using the charge amplifiers. The resultant voltage outputs of these sensors were connected to the inputs of the two-channel Spectrum Analyser. The ratio of the output response acceleration of the plate to the input force, which was computed by the analyser, resulted in the inertance frequency response function for the primary plate structure.

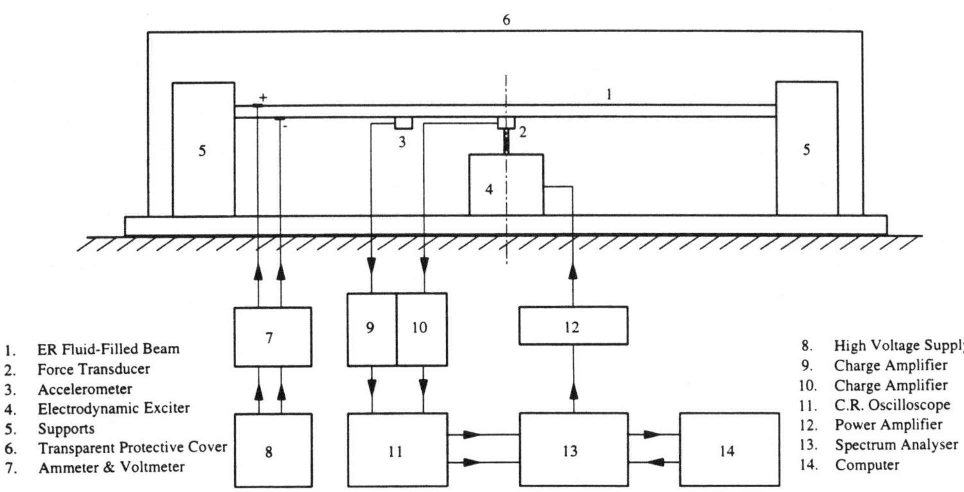

Figure 1 : Experimental Configuration for the Measurement of the Frequency Response Functions of an ER Fluid-Filled Composite Plate

3.3 Frequency Response Tests on the Unfilled Composite Plate

The composite plate was assembled as described in section 3.1. It was clamped at both ends to the top of the two vertical supports as shown by Figure 1. The top or secondary plate of the composite plate was connected to the positive potential, while the bottom or primary plate was connected to the negative potential of a high voltage electric power supply. The electromagnetic exciter was connected to the central point of the primary plate structure. Prior to the application of the ER fluid and of the electric field potential, the composite plate was subjected to random excitation over a frequency band of 0 to 500 Hz using the Spectrum Analyser and the associated instrumentation shown in Figure 1. The inertance frequency response functions of the composite plate assembly were determined in an analogous manner to those of the primary plate described in section 3.2 using the input force spectrum and the acceleration responses measured at positions 1 and 2 on the composite plate. The definitions of these positions is identical to those stated in section 3.2.

3.4 Frequency Response Tests on the Partially and Completely Filled Composite Plate

Following the frequency response tests on the empty composite plate, cavity no. 1 of the composite plate, which was nearest to the left clamped end of the plate, was filled with the ER fluid. This resulted in a partial coverage or filling of 20% of the total cavity size of the composite plate and corresponded to a constrained layer damping (CLD) treatment of 20% of the free (that is, unclamped) length of the primary plate structure by the ER fluid. The inertance frequency response tests were repeated with electric field potentials of 0, 1 and 2 kV applied across the two plates of the composite plate which were separated by a gap of 3 mm. For each test, the random vibration drive voltage supplied to the electromagnetic exciter by the Spectrum Analyser and the Power Amplifier chain was set at a level of 20 mV (RMS). Next, cavity no. 2 of the composite beam was filled with the ER fluid. This resulted in a total partial coverage or filling of 40% of the composite plate by the ER fluid. The inertance frequency response tests were repeated with electric field potentials of 0, 1, 2, 3 and 4 kV. These procedures were repeated until all the 5 cavities of the composite plate were filled and resulted in 100% coverage or filling of the composite plate with the ER fluid. It was found that as the number of cavities of the composite plate filled with the ER fluid was increased from 1 to 5, corresponding to a coverage of 20% to 100%, the maximum electric field potential achievable increased in steps from about 2.0 kV to about 13.5 kV respectively. This was due to the decrease in the equivalent electrical resistance of the ER fluid as the number of cavities filled was increased.

4. Results and Discussions

4.1 Inertance Frequency Response Functions

The inertance frequency response functions (FRF) of the clamped-clamped primary plate structure and of the empty (i.e. unfilled with ER fluid) clamped-clamped composite plate are shown by Figure 2. In this figure, curve 1 denotes the response of the primary plate at accelerometer position 1 whose coordinates are x=L/2 and y=b/2. Curve 2 is the corresponding response at position 2 whose coordinates are x=3L/8 and y=b/2. Similarly, curves 3 and 4 are the inertance FRF's of the empty composite plate which relate to accelerometer positions 1 and 2 respectively. From curves 1 and 2, the modal frequencies of the first three bending modes of vibration of the primary plate are approximately 63, 171 and 329 Hz. Similarly, curves 3 and 4 give the first four bending modes of vibration of the composite plate as 57, 137, 240 and 369 Hz approximately. Comparing curves 3 and 4, it is seen that curve 3 only shows the odd modes of bending vibration. This is because the accelerometer position 1 coincides with the nodal positions of the even modes of vibration.

Figure 3a shows the inertance frequency response functions (FRF) for accelerometer position 1 of the composite plate when it is unfilled (curve 1) and when filled 20% (curve 2), 40% (curve 3), 60% (curve 4), 80% (curve 5) and 100% (curve 6) with the ER fluid. The method for achieving these various degrees of coverage or filling of the composite plate by the ER fluid has been explained in section 3.4. Because accelerometer position 1 coincides with the nodal positions of the even modes of vibration, Figure 3a shows clearly only the odd modes of bending vibration of the composite plate. For all cases of partial or total ER fluid coverage of the composite beam shown in Figure 3, an electric field potential of 2 kV was applied across the 3 mm gap between the top and bottom plates resulting in electric field strength of 0.7 kV/mm. Comparing curve 1 (0% coverage) with curve 2 (20% coverage), it is seen that the filling of the first cavity of the composite plate with the ER fluid resulted in a slight increase in frequency of the inertance FRF characteristics of the composite plate. This indicates that for this condition, the stiffening effect of the ER fluid outweighs it mass loading effects. However, when the ER fluid coverage is increased to 40% and above, the mass loading effects become dominant with the consequent reduction in the frequencies of the FRF characteristics of the composite plate. These facts are shown more clearly in Figures

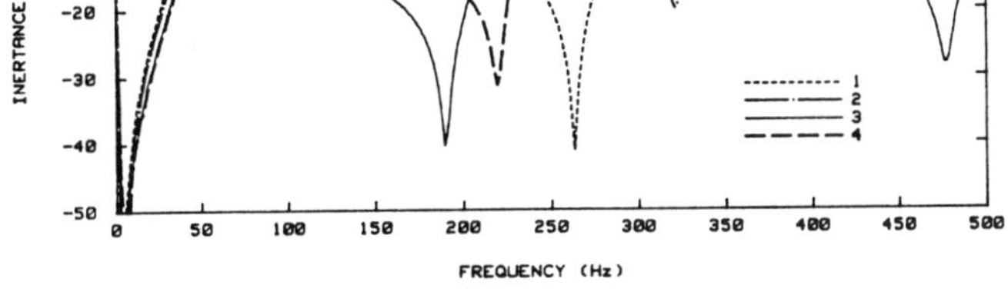

Figure 2 : Inertance Frequency Response Function of Primary Plate (curves 1 and 2) and Unfilled Composite Plate (curves 3 and 4) for Accelerometer Locations 1 (curves 1 and 3) and 2 (curves 2 and 4)

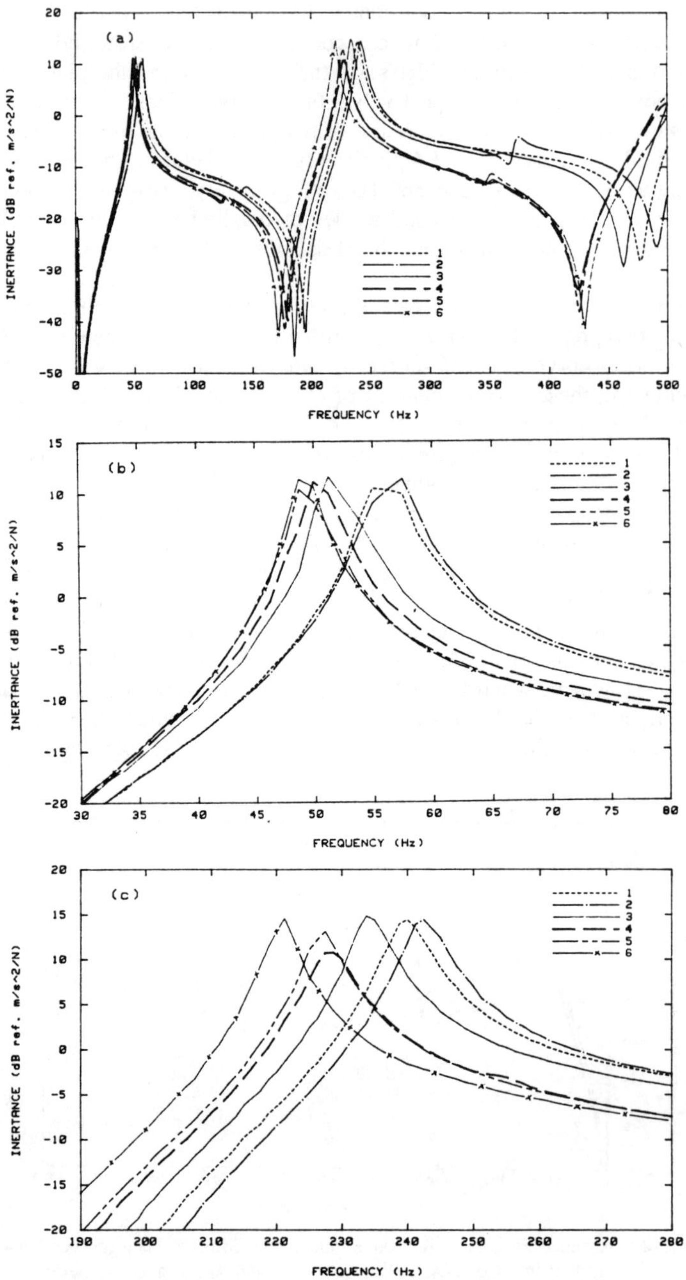

Figure 3 : Inertance Frequency Response Function of ER Fluid-Filled Composite Plate for Accelerometer Location 1; (a) Total FRF, (b) Mode 1 FRF, (c) Mode 3 FRF; ER Fluid Coverage of 1. 0%, 2. 20%, 3. 40%, 4. 60%, 5. 80%, and 6. 100%; Electric Field Potential of 2 kV.

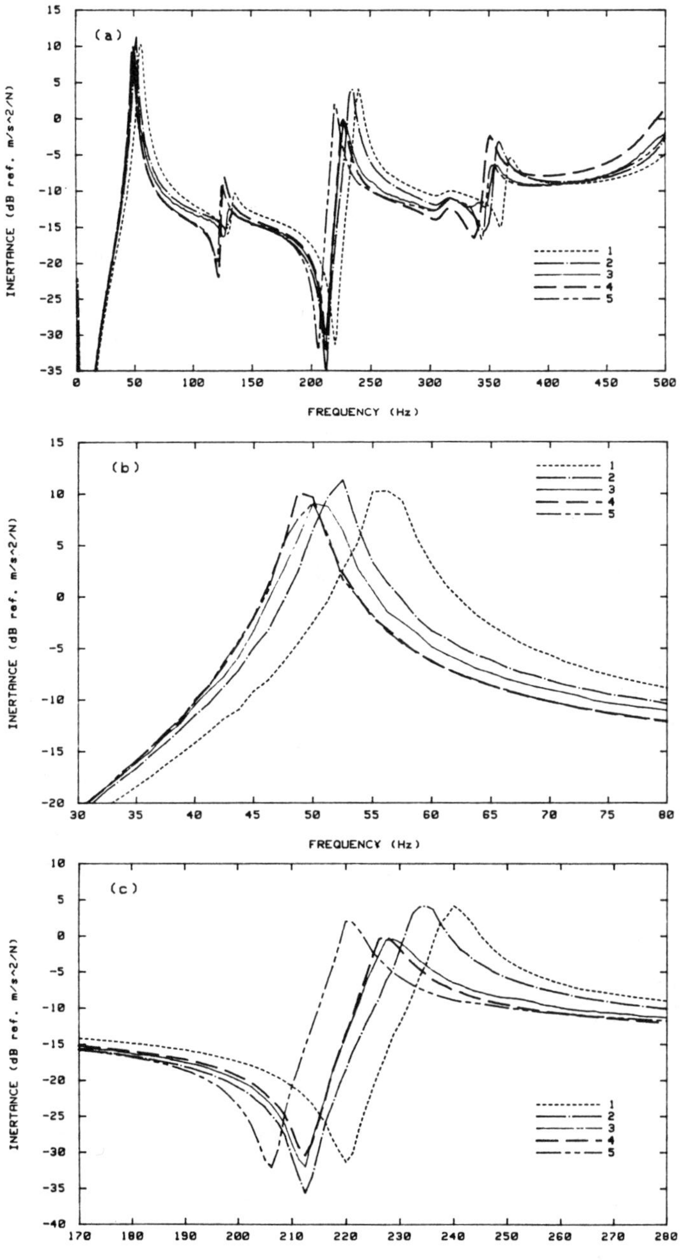

Figure 4 : Inertance Frequency Response Function of ER Fluid-Filled Composite Plate for Accelerometer Location 2; (a) Total FRF, (b) Mode 1 FRF, (c) Mode 3 FRF; ER Fluid Coverage of 1. 0%, 2. 40%, 3. 60%, 4. 80% and 5. 100%; Electric Field Potential of 2 kV

3b and 3c for modes 1 and 3 of the bending vibrations of the composite plate. Another important effect, which is more clearly shown by Figure 3c, is the decrease in the peak amplitude of the inertance FRF (and the corresponding increase in the composite loss factor) observed as the coverage increases to the median value of 60%. At higher levels of coverage, the peak amplitude of the FRF again increases.

The measured inertance FRFs for accelerometer position 2 of the composite plate are shown in Figure 4a. The various curves apply to different Coverage of 0% (curve 1), 40% (curve 2), 60% (curve 3), 80% (curve 4) and 100% (curve 5). The same electric field strength of 0.7 kV/mm was applied in all cases. It is seen from this figure and from the enlarged plots of the inertance characteristics of modes 1 and 3 shown in Figures 4b and 4c that the coverage of 60% gives the inertance FRF with the lowest peak amplitudes. This implies that a partial coverage of 60% gives the highest values of composite loss factor. When the coverage is increased beyond this value, it is evident from these figures that the peak amplitudes of the inertance FRFs increase in magnitude with a corresponding decrease in the associated composite loss factors.

In Figure 5a is shown a comparison of the effects of percentage ER fluid coverage and those of electric field strength applied across the ER fluid on the inertance FRF characteristics of the composite plate. Curves 1, 3, 5 are for ER fluid coverage of 20%, 60% and 100%, respectively, at an electric field potential of 0 kV. On the other hand, curves 2, 4, 6 are for ER fluid coverage of 20%, 60% and 100%, respectively, at an electric field potential of 2 kV. From the enlarged plot of the mode 3 of the composite plate shown in Figure 5b, it is evident that when the ER fluid coverage is increased from 20% to 100%, the peak amplitude of the third bending mode decreases to a minimum at 60% coverage then increases again. Also, this trend in characteristics is maintained both in the presence and absence of the electric field potential. Furthermore, both Figures 5a and 5b show that increasing the electric field potential from 0 kV to 2 kV has a slight effect on the inertance FRF characteristics which is that the curves are very slightly shifted to a higher frequency. This indicates that the application of the electric field causes a very slight increase in the stiffness of the composite plate. Such increase in stiffness is expected but the magnitude of the increase will depend on the strength of the ER fluid and on the electric field strength employed. The figures also show that for the silicone oil-based ER fluid used, and for the maximum electric field potential of 2 kV employed in these investigations, the percentage ER fluid coverage has a more significant effect on the FRF characteristics of the composite plate than the electric field strength.

4.2 Modal Properties of the Composite Plate

The theoretical equations presented in section 2 are now used for predicting the modal frequencies and loss factors of the composite plate. However, to enable predictions for different levels of percentage ER fluid fill of the composite plate, Eq.1 is modified as follows

$$f_n = \frac{\lambda_n^2}{2\pi}\left[\frac{|E^*I|}{\rho_1 H_1 L_1^4 + \rho_2 H_2 L_2^4 + \rho_3 H_3 L_3^4}\right]^{1/2} \tag{11}$$

where L_1 is the length of the primary (base) plate, L_2 is the length of the ER fluid coverage, L_2 is the length of the secondary (constraining) plate. The values of L_2 varied from 99 mm when cavity 1 was filled to 496 mm when all the 5 cavities of the composite plate were filled.

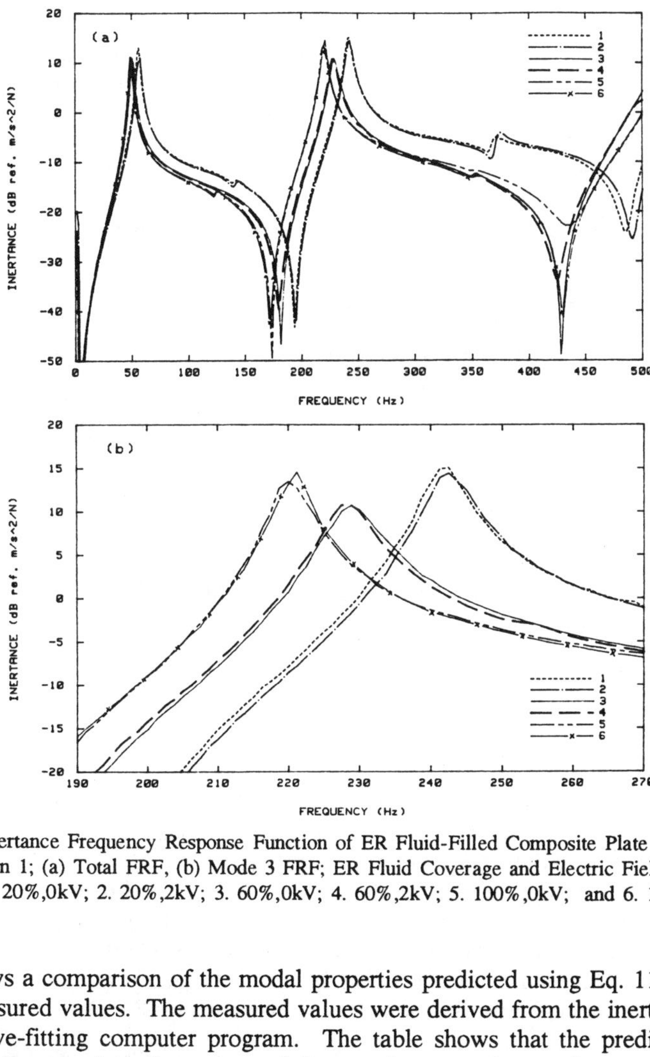

Figure 5 : Inertance Frequency Response Function of ER Fluid-Filled Composite Plate for Accelerometer Location 1; (a) Total FRF, (b) Mode 3 FRF; ER Fluid Coverage and Electric Field Potentials of 1. 20%,0kV; 2. 20%,2kV; 3. 60%,0kV; 4. 60%,2kV; 5. 100%,0kV; and 6. 100%,2kV

Table 1 shows a comparison of the modal properties predicted using Eq. 11 and Eqs. 2 to 7 with the measured values. The measured values were derived from the inertance FRFs using a modal curve-fitting computer program. The table shows that the predicted frequencies follow a similar trend to the measured frequencies, namely: as the percentage coverage increases, the modal frequency decreases. In general, the measured modal frequencies are slightly less than the predicted modal frequencies. This is partly due to the mass-loading effects of the accelerometer which had a mass of 5 gramme. However, the discrepancies between the measured and predicted modal frequencies are less than 5%. Because the present form of the complex flexural stiffness E^*I is invariant to the percentage ER fluid coverage, Eq. 2 predicts a constant value of loss factor for all levels of ER fluid coverage. The measured values for mode 1, on the other hand, show that the loss factor of the composite plate increases as the percentage coverage is increased. For mode 3, the results show that the measured loss factors increase as the percentage coverage is increased from 20% to 60% at which it attains a maximum value. As the coverage increases from 60% to 100% the loss factor decreases. Overall, it is seen that the predicted loss factor for mode 1 is about 3/4 of the measured values whereas the predicted loss factor for mode 3 is only about 1/2 of the measured values. Thus the predicted modal loss factors are less accurate compared to the predicted modal frequencies.

Table 1 : Measured and Predicted Modal Properties of the Composite Plate at an Electric Field Potential of 2 kV and for various Percentage ER Fluid Coverage

ER Fluid Coverage (%)	Mode 1				Mode 3			
	Measured		Predicted		Measured		Predicted	
	f_1 (Hz)	η_1	f_1 (Hz)	η_1	f_3 (Hz)	η_3	f_3 (Hz)	η_3
20	56.7	0.066	54.2	0.052	242.2	0.046	241.3	0.024
40	51.8	0.066	54.0	0.052	234.0	0.046	240.8	0.024
60	50.5	0.072	53.5	0.052	227.9	0.060	238.3	0.024
80	49.4	0.068	52.1	0.052	227.0	0.048	232.2	0.024
100	49.3	0.076	49.4	0.052	220.6	0.040	220.7	0.024

5. Conclusions

The results show that the percentage of ER fluid coverage has a more significant effect on the inertance FRF characteristics of the composite plate than the magnitude of the applied electric field potential. However, this is only strictly true for the silicone oil-based ER fluid and for the maximum electric field potential of 2 kV employed in the present investigations. Before a more general deduction can be made, it would be necessary to employ other ER fluids as well as other gap sizes which will enable the application of higher levels of electric field potential. Furthermore, the results suggest that an ER fluid coverage or fill of 60% gives the highest modal loss factors of the composite plate.

6. Acknowledgements

The author wishes to acknowledge the contributions of T. Overrington and P.

Sarafianos, 1994 and 1995 final year undergraduate students, to the design, manufacture and development of the ER fluid-filled composite plate test system.

7. References

1. A.D. Nashif, D.I.G. Jones and J.P. Henderson, *Vibration Damping*, John Wiley & Sons, New York (1985).
2. J. P. Coulter, T. G. Duclos and D. N. Acker, The Usage of Electrorheological Materials in Viscoelastic Layer Damping Applications, *Proceedings of the Damping '89 Conference*, Palm Beach, Florida (1989), pp. CAA1-CAA17
3. S. B. Choi, B. S. Thompson and M. V. Gandhi, An Experimental Investigation of the Active-Damping Characteristics of a Class of Ultra-Advanced Intelligent Composite Materials Featuring Electro-Rheological Fluids, *Proceedings of the Damping '89 Conference*, Palm Beach, Florida (1989), pp. CAC1-CA14
4. J. P. Coulter and T. G. Duclos, Applications of Electrorheological Materials in Vibration Control, *Proceedings of the 2nd International Conference on Eletrorheological Fluids*, Raleigh, New Conneticut (1989), pp. 300-325
5. Y. S. Kim, K. W. Wang and H. S. Lee, Feedback Control of ER-Fluid-Based Structures for Vibration Suppression, *Smart Mater. Struc.* **1** (1992) 139-145
6. G. Haiqing, L. M. King and T. B. Cher, Influence of a Locally Applied Electrorheological Fluid Layer on Vibration of a Simple Cantilever Beam, *J. Intell. Mater. Syst. Struc.* **4** (1993) 379-384
7. W. I. Kordonsky, A. D. Matsepuro, S. A. Demchuk, Z. A. Novikova, Properties of Electrorheological Fluids under Periodic Deformation, *Proceedings of the 4th International Conference on Eletrorheological Fluids*, edited by R. Tao and G. D. Roy (World Scientific, Singapore, 1994), pp. 440-452
8. Y. Chen, F.-G. Yuan and H. Conrad, Damping Characteristics of Laminate Composites Containing Electrorheological (ER) Fluids, *Proceedings of the 2nd International Conference on Intelligent Materials*, edited by C. A. Rogers and G. G. Wallace (Technomic Publishing Co. Inc, Lancaster, Pennsylvania, 1994), pp. 328-339
9. A. Shiang and J. Coulter, Controllability and Reliability Issues Related to Electrorheological Material Adaptive Structures, *Proceedings of the 2nd International Conference on Intelligent Materials*, edited by C. A. Rogers and G. G. Wallace (Technomic Publishing Co. Inc, Lancaster, Pennsylvania, 1994), pp. 1117-1130
10. R. D. Blevins, *Formulas for Natural Frequency and Mode Shape*, Van Nostrand Reinhold Company, New York (1979).
11. D. J. Gorman, *Free Vibration Analysis of Rectangular Plates*, Elsevier Science Publishers, New York (1982).
12. S. O. Oyadiji, Characterisation of the Complex Shear Modulus Properties of Electro-Rheological Fluids by the Direct Stiffness and Master Curve Techniques, *Proceedings of the 5th International Conference on Eletrorheological Fluids, Magnetorheological Suspensions and Related Technologies*, Sheffield, UK (1995)

DYNAMIC MAGNETO-OPTICAL RESPONSE OF FERRONEMATIC LIQUID CRYSTALS

YURI L. RAIKHER and VICTOR I. STEPANOV

Institute of Continuous Media Mechanics
Urals Branch of the Russian Academy of Sciences, Perm, 614061, Russia
E-mail: ylr@icmm.perm.su

ABSTRACT

We study the time-dependent orientational response of a nematic liquid crystal doped with single-domain ferromagnetic particles. A set of coupled macroscopic equations describing evolution of the director texture, magnetization distribution and concentration of the particles under the change of the external field is derived. For the case of a plane layer, expressions for the main relaxation rates are obtained and analyzed.

1. Introduction

Ferronematics (FN) is the term designating suspensions of single-domain ferroparticles in nematic liquid carriers. One may call them the second generation of complex fluids, since those systems are mixtures of the media each of which is already complex, viz. nematic liquid crystals and magnetic fluids. This crossbreed produces an interesting class of disperse systems, now existing in both lyotropic [1] and thermotropic [2] forms.

The goal of this paper is to analyze field-induced macroscopic magneto-optical properties of FN in view of recently published experimental observations [3].

We begin with recalling the expression for the free energy density of FN with finite anchoring of nematic molecules on the particle surfaces. It has been derived in [4] and under the usual notations for the director $\boldsymbol{n}$, Frank moduli K_i and magnetic field $\boldsymbol{H}$ reads

$$F = \tfrac{1}{2}\left[K_1(\operatorname{div}\boldsymbol{n})^2 + K_2(\boldsymbol{n}\cdot\operatorname{curl}\boldsymbol{n})^2 + K_3(\boldsymbol{n}\times\operatorname{curl}\boldsymbol{n})^2\right] - \tfrac{1}{2}\chi_a(\boldsymbol{nH})^2$$
$$-I_s f\,(\boldsymbol{mH}) + (f k_B T/V)\ln f + (qWf/d)(\boldsymbol{nm})^2\,, \qquad (1)$$

where $\boldsymbol{m}$ is a unit vector along the FN magnetization direction. Here the material parameters describing the liquid-crystalline matrix are: χ_a the anisotropic part of the diamagnetic susceptibility and W the surface density of the anchoring energy; and those referring to the particles: d the mean transverse diameter, V the volume, I_s the saturation magnetization of the particle material and f the volume content of ferromagnetic phase in FN. Orientational interaction between the carrier liquid crystal and ferroparticles on a macroscopic scale is determined by a single dimensionless coefficient q the value of which depends on the actual type of the anchoring condition.

From the structural point of view, the system described by Eq.(1) is a rather dilute ($f \leq 10^{-4}$) suspension of rod-like ferrite grains with major axes about 10^2 nm and aspect ratios around 10. Given that, the particles are single-domain and uniformly magnetized along their long axes because of the magnetostatic shape anisotropy. The boundary conditions for anchoring of nematic molecules on the particle surfaces may vary depending on the treatment. These differences are described in formula (1) with the aid of function $q = -2P_2(\cos\alpha)$, where P_2 is the second Legendre polynomial and α is the angle between the easy-orientation direction imposed on the nematic by the particle surface and the particle major axis. The main mechanism providing the magneto-orientational response of FN, is the particle–matrix interaction, i.e., the ability of the components to influence each other's orientation. In the continuum model (1) the corresponding conjugated parameters are spatial distributions of the magnetization direction $\boldsymbol{m}(\boldsymbol{r})$ of the particle assembly and the liquid crystal director $\boldsymbol{n}(\boldsymbol{r})$. Their coupling is rendered by the last term of the above formula.

2. Equations of motion

We take the general form of equation governing the orientational motion (director rotations) of the FN carrier from the conventional hydrodynamics of nematic liquid crystals [5], and write it down as

$$\gamma_1(\boldsymbol{n} \times \boldsymbol{N}) + \gamma_2(\boldsymbol{n} \times (\boldsymbol{\mathcal{A}} \cdot \boldsymbol{n})) = \boldsymbol{n} \times \boldsymbol{h}\,. \tag{2}$$

Here γ_1 and γ_2 are the usual rotary viscosity coefficients of the nematic, and vector

$$\boldsymbol{N} = \dot{\boldsymbol{n}} - (\boldsymbol{\Omega} \times \boldsymbol{n})\,,$$

accounts for the deviation of the director angular velocity from that of the background fluid, $\boldsymbol{\Omega} = \frac{1}{2}\mathrm{curl}\,\boldsymbol{u}$ being the vorticity of the velocity field $\boldsymbol{u}(\boldsymbol{r})$, and $\mathcal{A}$ a diadic whose matrix form

$$\mathcal{A}_{ik} = \tfrac{1}{2}\left(\frac{\partial u_k}{\partial x_i} + \frac{\partial u_i}{\partial x_k}\right),$$

is the symmetrical part of the deformation rate tensor. The "molecular" field $\boldsymbol{h}$ entering the right-hand side of Eq. (2) is defined (see [5], for example) as a variational derivative of the free energy with respect to director orientation:

$$h_i = -\frac{\partial F}{\partial n_i} + \frac{\partial}{\partial x_k}\left(\frac{\partial F}{\partial\left(\frac{\partial n_i}{\partial x_k}\right)}\right). \tag{3}$$

In the considered case, besides the standard contributions, the product $(\boldsymbol{n} \times \boldsymbol{h})$ contains an additional term

$$-(2qWf/d)\,(\boldsymbol{n} \times \boldsymbol{m})\,(\boldsymbol{n}\boldsymbol{m})\,,$$

which is a torque imposed by the dispersed magnetic particles on the nematic carrier. The structure of this term indicates that the nematic–particle interaction results in a specific uniaxial magnetic anisotropy of FN.

The general linear form of the equation governing the rotational motion of a particle subjected to both a hydrodynamic flow of a suspending fluid and to an external torque $\boldsymbol{L}$ is[6]

$$\omega_i - \Omega_i = B_{ij} L_j + \Lambda_{ijk} \mathcal{A}_{jk} \,, \tag{4}$$

where $\boldsymbol{\omega}$ is the angular velocity of a suspended particle, and B_{ij} and Λ_{ijk} are transport coefficient tensors, whose actual representations depend upon the particle shape as well as the anisotropy of the carrier liquid. For an acircular particle in an isotropic fluid, these tensors may be written as[7]

$$B_{ij} = B\delta_{ij} + \Delta B\, m_i m_j \,, \qquad \Lambda_{ijk} = \tfrac{1}{2}\Lambda \left(e_{ilk} m_j + e_{ilj} m_k\right) m_l \,, \tag{5}$$

with e_{ijk} being the Levi-Civita tensor and

$$B = 1/6\eta V \delta \,, \qquad \Lambda = (\beta^2 - 1)/(\beta^2 + 1) \,; \tag{6}$$

where η is the viscosity of the carrier liquid and β stands for the aspect ratio of the particle. Formfactor δ is a function of β and assumes the minimal value $\delta(1) = 1$ for spherical particles. Note also that in Eqs. (1) and Eq. (5) we use the same notation, $\boldsymbol{m}$, thus identifying the directions of magnetization and the particle major axis. That is possible[4] as long as the rotational thermal motion of the particles is weak: $WV/d \gg k_B T$.

For a liquid-crystalline, i.e., anisotropic, carrier Eqs. (5) must be replaced by more complicated ones whose tensorial structure would incorporate components of $\boldsymbol{n}$. However, up to now the pertinent relations of such a kind have not been found. Because of that hereafter we shall use Eqs. (5) and (6) treating η there as an effective parameter.

Evaluating $\boldsymbol{L}$ from Eq. (1) as a macroscopic torque per single particle, one gets

$$\boldsymbol{L} = -V\boldsymbol{m} \times (\partial F/\partial \boldsymbol{m}) = I_s V\,(\boldsymbol{m} \times \boldsymbol{H}) + (2qWV/d)\,(\boldsymbol{n} \times \boldsymbol{m})\,(\boldsymbol{nm}) \,. \tag{7}$$

Substituting it into (4), with allowance for Eqs. (5) and (6), yields the equation of motion for magnetization. After vectorial multiplication by $\boldsymbol{m}$ (thus retaining only the rotational degrees of freedom) it takes the form similar to Eq. (2):

$$\gamma_{1p}\,(\boldsymbol{m} \times \boldsymbol{M}) + \gamma_{2p}(\boldsymbol{m} \times (\boldsymbol{\mathcal{A}} \cdot \boldsymbol{m})) = I_s(\boldsymbol{m} \times \boldsymbol{H}) + 2\,(qW/d)\,(\boldsymbol{n} \times \boldsymbol{m})\,(\boldsymbol{nm}) \,, \tag{8}$$

where vector

$$\boldsymbol{M} = \dot{\boldsymbol{m}} - (\boldsymbol{\Omega} \times \boldsymbol{m}) \,,$$

has the same meaning for the motion of the magnetic particles as vector $\boldsymbol{N}$ in Eq. (2) for director rotations. The rotary viscosity coefficients in Eq. (8) are defined through parameters (6) as[7]

$$\gamma_{1p} = 6\eta\delta \,, \qquad \gamma_{2p} = -\Lambda \gamma_{1p} \,. \tag{9}$$

In the framework of the continuum model (1), one is dealing with an assembly of particles capable for translational motion, both free (diffusion) and forced. Since the paper[8] which has in fact founded the theory of liquid-crystalline magnetic suspensions, the forced concentrational changes in FN are called the *segregation* effect.

Allowing for diffusion and segregation, the velocity of a suspended particle emerges as a sum of regular and stochastic parts which in a macroscopically quiescent fluid we write down as

$$v_i = -B_t \frac{\partial}{\partial x_i}(VF + k_B T \ln f) = -B_t \frac{\partial}{\partial x_i}\left[-I_s V(\boldsymbol{mH}) + \frac{qWV}{d}(\boldsymbol{nm})^2 + k_B T \ln f\right], \tag{10}$$

with B_t being the translational mobility coefficient. On substituting Eq.(10) into the particle-number conservation law

$$\frac{\partial f}{\partial t} + \frac{\partial}{\partial x_i}(v_i f) = 0,$$

one gets the equation governing the spatial distribution of concentration:

$$\frac{\partial f}{\partial t} - \frac{\partial}{\partial x_i} f B_t \frac{\partial}{\partial x_i}\left[-I_s V(\boldsymbol{mH}) + \frac{qWV}{d}(\boldsymbol{nm})^2 + k_B T \ln f\right] = 0, \tag{11}$$

which is to be solved under appropriate boundary conditions. It may be, for example, impermeability of the layer walls for the particles.

As a final remark, we note that the presence of particles must change the viscous stress tensor, as it happens for any suspension. However, due to the assumed very low concentration of the solid phase, this effect may hardly be of any importance for FN.

3. Dynamics of perturbations in a plane layer

Consider a FN sample confined inside a plane horizontal layer of thickness D bounded by solid walls. The layer is infinite along the directions of the coordinate axes Ox and Oy, and planar boundary conditions along Ox at both layer walls are imposed. The nematic anchoring on the particle surfaces is supposed to favor their equilibrium alignment with the local director, that yields $\alpha = 0$, and thus $q = -2$. The external magnetic field $\boldsymbol{H}_0$ is applied along the Ox axis. In result, in the initial (unperturbed) state vectors $\boldsymbol{n}$, $\boldsymbol{m}$ and $\boldsymbol{H}_0$ make a mutually parallel triad and are uniform over the sample.

This equilibrium texture is perturbed by a magnetic field $\boldsymbol{H}_1$ directed in the plane of the layer perpendicular to $\boldsymbol{H}_0$, i.e., along Oy. Assuming that the induced orientational configuration is also two-dimensional, one may at any time moment t presented it as

$$\boldsymbol{n} = \Big(\cos\varphi(z,t), \sin\varphi(z,t), 0\Big), \qquad \boldsymbol{m} = \Big(\cos\psi(z,t), \sin\psi(z,t), 0\Big), \tag{12}$$

thus describing each of the unit vectors by a single angle with respect to the Ox axis.

Equations (2), (8) and (11) constitute a closed set describing the coupled motion of vectors $\boldsymbol{n}$ and $\boldsymbol{m}$ together with the pertinent response of the concentration distribution. On substitution of formulas (12), one gets

$$\begin{aligned}
\gamma_1 \frac{\partial \varphi}{\partial t} &= K_2 \frac{\partial^2 \varphi}{\partial z^2} + \chi_a \left[H_0 H_1 \cos 2\varphi + \tfrac{1}{2}\left(H_1^2 - H_0^2\right) \sin 2\varphi \right] - \frac{2Wf}{d} \sin\left[2(\varphi - \psi)\right] , \\
\gamma_{1p} \frac{\partial \psi}{\partial t} &= -I_s H_0 \sin\psi + I_s H_1 \cos\psi - \frac{2W}{d} \sin\left[2(\psi - \varphi)\right] , \qquad (13) \\
\frac{\partial f}{\partial t} &= -B_t \frac{\partial}{\partial z} f \frac{\partial}{\partial z} \left[I_s V (H_0 \cos\psi + H_1 \sin\psi) + \frac{2WV}{d} \cos^2(\varphi - \psi) - k_B T \ln f \right] .
\end{aligned}$$

Let us restrict ourselves by the case of small perturbations, setting $H_1/H_0 \ll 1$, and neglecting the diamagnetic term, since the processes in question must take place in the field-range where diamagnetism does not matter. To deal with the spatial derivative in the first of Eqs. (13), we make a standard assumption[5] that the main contribution to the orientational distortion is caused by the longest-wavelength mode, i.e., $\varphi \propto \sin(\pi z/D)$. Given that, the set (13) reduces to a pair of equations

$$\tau_n \frac{\partial \varphi}{\partial t} + \left(1 + \frac{f}{f_*}\right) \varphi - \frac{f}{f_*} \psi = 0 , \qquad \tau_m \frac{\partial \psi}{\partial t} + \left(1 + \frac{H_0}{H_a}\right) \psi - \varphi = \frac{H_1}{H_a} , \qquad (14)$$

since in the linear approximation concentration has no effect. Because of that, further on we set f constant, uniform and equal to its initial value. In Eqs. (14) the following reference quantities have been introduced:

$$\tau_n = \gamma_1 D^2/\pi^2 K_2 , \quad \tau_p = \gamma_{1p} d/4W , \quad H_a = 4W/I_s d , \quad f_* = \pi^2 K_2 d/4WD^2 . \quad (15)$$

Here τ_n is the orientation response time of the nematic matrix, τ_p is that of the particle subsystem and the effective magnetic field H_a measures the uniaxial anisotropy experienced by the FN magnetization. The characteristic value f_* is the lower bound for the concentration range within which FN displays the so-called *collective behavior*—see[8,4]—and may be treated in a continuum approach. That is why in what follows we assume that condition $f > f_*$ always holds. Moreover, given $f_* \ll 1$, as is the experimental case[3,10], the ratio

$$\tau_p/\tau_n = \gamma_{1p} f_*/\gamma_1 \ll 1 , \qquad (16)$$

yields a small parameter for the problem under consideration.

Substituting into Eqs. (14) the time-dependence of perturbations as $\propto \exp(-\lambda t)$, we arrive at the characteristic equation

$$\left(-\lambda\tau_n + 1 + \frac{f}{f_*}\right)\left(-\lambda\tau_p + 1 + \frac{H_0}{H_a}\right) = \frac{f}{f_*} ,$$

Solving it to the first order in the small parameter (16), one gets the roots in the form

$$\lambda_1 = \tau_n^{-1}\left[1 + \frac{fH_0}{f_*(H_a + H_0)}\right], \qquad \lambda_2 = \tau_p^{-1}\left[1 + \frac{H_0}{H_a}\right]. \tag{17}$$

These eigenvalues determine the rates of relaxation to the equilibrium state in a FN layer.

4. Discussion

Expressions (17) show that the rates of the studied transient process depend upon the applied field strength. For $H_0 < H_a$ any of them may be schematized by

$$\tau^{-1}(H_0) = \tau_0^{-1} + \alpha H_0 , \tag{18}$$

with an appropriate τ_0 and α, as it had been proposed in[3] from qualitative considerations. Formula (18) reveals the essential difference of orientation dynamics of FN from that of pure nematics. In the latter, the relaxation rate is proportional to H_0^2 whereas for FN more likely are linear dependences.

According to Eqs. (17), this holds in full for magnetization-induced contributions φ, $\psi \propto \exp(-\lambda_2 t)$, but for the director-induced ones φ, $\psi \propto \exp(-\lambda_1 t)$ is valid only as long as $H_0 < H_a$. Above H_a the decrement λ_1 loses its dependence of H_0, because since then the main body couple driving the director rotations is $I_s f H_a$ and not $I_s f H_0$. This fact is in a very close analogy with the dynamic behavior of a single-domain uniaxial ferromagnetic particle—see[4,9]. One should remember, however, that since the diamagnetic term is proportional to H_0^2 (see Eq. (1), for example), at high H_0 it anyway overrides all the others. Therefore, the ultimate range of validity of Eqs. (17) is $H_0 < I_s f/\chi_a$.

In[3] for the experimental data on transient processes in a FN layer when oriented by an external filed are reported. The geometry of the cell and the fields configuration were as considered in the preceding section. The authors dealt with a lyotropic calamitic nematic mesophase doped with ionic ferrofluids. The mean size of the magnetic grains of the latter was $\sim 10\,\text{nm}$. With an optical set-up, the changes of transmittance in Oz direction due to birefringence induces by a pulse field H_1 have been measured.

According to the observations, in general the optical response consisted of three exponents with the characteristic namely

$$\tau_s \sim 1\,\text{s (short)}, \quad \tau_m \sim 10\,\text{s (medium)}, \quad \tau_l \sim 10^2\,\text{s (long)}.$$

Of the two samples (S1,S2) investigated in[3], the data on S1 looks more consistent and reads:

$$\tau_s^{-1} = \tau_{0s}^{-1} + \alpha_s H_0 , \quad \tau_m^{-1} = \tau_{0m}^{-1} + \alpha_m H_0 , \quad \tau_l = \text{const}(H_0),$$

thus confirming the dependence of the type (18) for the two more fast processes. The obtained values of the tangents $d\tau^{-1}/dH_0$ read $\alpha_s \approx 2 \cdot 10^{-4}$ and $\alpha_m \approx 10^{-5}\,\mathrm{G^{-1}s^{-1}}$.

Since the ferronematic dispersion used in the experiment was indeed rather a complex micellar system, the present model may be applied to it with a considerable caution. The method of observation adds to the indeterminacy of the data since the recorded signal $\propto \varphi^2(t)$ squares up the sum of partial birefringence contributions and produces combinational terms like $\exp[-(\lambda_1 + \lambda_2)t]$. Given all that, in the following interpretation we are rather after qualitative similarities than quantitative agreements.

Long relaxation time. Setting $K_2 \sim 10^{-6}$ dyn and $\gamma_1 \sim 10\,\mathrm{P}$, for the long time one finds $\tau_n \sim \tau_l \sim 10^2\,\mathrm{s}$. Since the parameters W and d enter Eq. (17) and the following only as a ratio, we take $W/d \sim 10^4$. It may mean, for example, $W \sim 10^{-2}$ dyn/cm and $d \sim 10^{-6}$ cm that seems reasonable by the order of magnitude. For ferrite grains $I_s \approx 300\,\mathrm{G}$, which upon substitution yields $H_a \sim 10^2\,\mathrm{G}$. The field scale of the experiment was $H_0 \geq 10^3\,\mathrm{G}$ that gives $H_a/H_0 \ll 1$ and along with formula (17) for λ_1 makes the field-independence of τ_l understandable.

We remark that the proposed in[3] interpretation of τ_l as a characteristic time of the director gliding on the layer walls may also be valid. However, discussing of this issue demands a detailed consideration which will be published elsewhere.

Medium relaxation time. From the experimental value of $\tau_{0m} \approx 30\,\mathrm{s}$ and at $W/d \sim 10^4$, one finds for the particle rotational viscosity $\gamma_{1p} \sim 10^6\,\mathrm{P}$. According to Eq. (9) it expands as $\gamma_{1p} = 6\eta\delta$, where η is the effective viscosity. Definitely, in lyotropic FN the magnetic phase does not consist of single-domain grains brought on by the ferrofluid. A reasonable guess is (see Refs.[1,10]) that a phase separation occurs producing multi-grain aggregates. Under the field H_0, the demagnetizing effect makes them to stretch and to assume a strongly elongated shape increasing formfactor δ up to 10^2 or higher. That yields the mesoscopic viscosity $\eta = \gamma_{1p}/6\delta \sim 10^3\,\mathrm{P}$. This high value should not be too surprising given the complex structure of the micellar system. Then the tangent α_m of the function $\lambda_2(H_0)$ determined from Eq. (17) is

$$\alpha_m = 1/\tau_p H_a = I_s/\gamma_{1p} \sim 10^{-4}\,\mathrm{G^{-1}s^{-1}}\,, \tag{19}$$

that is about an order of magnitude greater than the experimental value. Qualitatively, this must have been expected. Indeed, for a loose aggregate of magnetic grains the magnetization (specific magnetic moment) should be many times smaller than I_s used in Eq. (19).

Short relaxation time. The shortest time τ_s has been encountered in short-pulse experiments, different from those yielding τ_m and τ_l. Developing the hypothesis that the "magnetic particles" of a lyotropic FN are indeed aggregates, we speculate that τ_s has a mesoscopic origin and as such is not accounted for in our macroscopic model. Associating τ_s with internal motions in the aggregate, where the grains must retain some of their rotational degrees of freedom, links it directly to the Debye rotary diffusion time τ_D of a single grain. The estimation gives

$$\tau_D \sim \eta d^3/k_B T \sim 0.1 - 1\,\mathrm{s}\,,$$

for $\eta \sim 10^3$ at room temperature, thus justifying a possibility to identify τ_{0s} with τ_D.

5. Conclusions

Assuming that a ferronematic liquid crystal is a suspension of elongated magnetically rigid single-domain particles in a nematic matrix, the basic equations of motion have been derived. The estimates drawn from them show that in the continuum model the orientational response of a FN sample comprises two modes which may be conditionally called director- and magnetization-induced. Their rates are noticeably different in magnitude and have different field-strength dependences. The performed comparison with the available experimental data on lyotropic FN though displaying several points of agreement, is not to the full extent satisfactory. More work both in theory and in experiment is to be done.

6. Acknowledgments

We are grateful to J.-C. Bacri and A. M. Figueiredo Neto for an opportunity to obtain Ref. 3 yet as a manuscript and for interesting discussions.

This work was partially supported by grants: RMD 300 from the International Science Foundation and Russian Government, 95–02–03953 from the Russian Basic Research Foundation and 94–31.2–9 from the Russian State Committee for Higher Education.

7. References

[1] L. Liebert and A. Martinet, *J. Phys. Lett. (France)* **40** (1979) L363.

[2] S.-H. Chen and N. M. Amer, *Phys. Rev. Lett.* **51**, (1983) 2298.

[3] J.-C. Bacri and A. M. Figueiredo Neto, *Phys. Rev. E* **50** (1994) 3860.

[4] Yu. L. Raikher and S. V. Burylov, *Mol. Cryst. and Liquid Cryst.* **258** (1995) 107.

[5] P. G. de Gennes, *The Physics of Liquid Crystals* (Clarendon Press, Oxford, 1974) Chap. 3.

[6] J. Happel and H. Brenner, *Low Reynolds Number Hydrodynamics* (Prentice-Hall, N. Y., 1974) Chap. 5.

[7] V. N. Pokrovskii, *Statistical Mechanics of Dilute Suspensions* (Nauka, Moscow, 1978) Chap. 2.

[8] F. Brochard and P. G. de Gennes, *J. Phys. (France)* **31** (1970) 691.

[9] Yu. L. Raikher and M. I. Shliomis, *Adv. Chem. Phys.* **87** (1994) 595.

[10] A. M. Figueiredo Neto, Y. Galerne, A. M. Levelut and L. Liebert, *J. Phys. Lett. (France)* **46** (1985) L499.

Effective Viscosity of an Electrorheological Fluid

J. M. Sun and R. Tao
Department of Physics, Southern Illinois University
Carbondale, Illinois 62901, USA

ABSTRACT

The viscosity of an electrorheological fluid in an electric field is studied by computer simulations. The fluid viscosity increases with the field through three stages. In a weak field, the fluid remains Newtonian although its viscosity increases. At this stage, while drifting in the flow direction, particles diffuse in the direction perpendicular to the flow. In an intermediate electric field, the fluid has tilted and broken chains moving with the flow and the fluid becomes non-Newtonian. The viscosity η and the shear rate $\dot{\gamma}$ has the following relationship $\eta = \eta_0 e^{-\dot{\gamma}\tau}$ where τ is the relaxation time and η_0 is exponentially proportional to the dipolar interaction energy and the volume fraction. In a strong electric field, the fluid contains condensed chains which provide yield stress and hysteresis.

1. Introduction

A conventional ER fluid is made of fine dielectric particles suspended in a liquid of low dielectric constant.[1–5] The large mismatch between the particles' dielectric constant (or conductivity) and the liquid's makes the system easily polarizable in an electric field. The electric-field induced change of viscosity in ER fluids has provided opportunities for many new technological applications.[2] The crucial point of ER application is now in materials technology. Because of the density mismatch between the particles and the liquid, sedimentation is a problem in several conventional ER fluids. A recent discovery of the ER effect of liquid crystal polymer solution is thus very interesting.[6] The polymer solution is a one-component polarizable liquid which has no settling.[7] They may be suitable for applications which cannot tolerate any settling.

In this paper, we will investigate the viscosity of this class of ER fluids in Couette geometry via computer simulations. The details of simulations are presented in Section 2. In Section 3, we discuss our results.

2. Model and Simulations

Our three-dimensional system is based on the dipole model frequently used in the study of ER fluids.[8] We have N particles placed between two parallel electrodes which are planes $z = L/2$ and $z = -L/2$. The motion of each particle is determined by a classical motion equation. The electric force on each particle is the sum of dipolar forces exerted by other particles as well as by images. Experiments have shown that an electric field parallel to the flow direction has little effect on the liquid viscosity. Therefore, we examine the case in which the flow is moving along the x-direction while the applied electric field is along the z direction. In both x and y direction, a periodic boundary condition is imposed.

We introduce two wall-layers which are adjacent to each electrode respectively. The particles in the wall-layers are distinct from the particles in the flow (bulk particles). While the bulk particles are allowed to move in all three directions, the wall-layer particles only move in the x and y directions and do not exchange with the bulk particles. Thus, the wall-particles cannot leave the electrode they attach. In addition to the dipolar interactions, the wall-particles adjacent to the top electrode experience an external force in the x direction and the wall-particles adjacent to the bottom electrode experience an external force in the opposite x direction. As these shear forces act on the wall-particles, a velocity gradient in the z direction is established.

Our wall layers are based on observations made in ER fluid flow experiments which find two layers of dielectric particles cumulated on the two electrodes, distinct from the particles in the flow. Ashurst and Hoover also used similar wall layers in their simulation of dense fluid shear flow.[9]

In an electric field, each particle obtains a dipole moment, $\vec{p} = \alpha(\sigma/2)^3\vec{E}_{loc}$ where $\alpha = (\epsilon_p - 1)/(\epsilon_p + 2)$ and $\vec{E}_{loc}$ is the local field and ϵ_p is the dielectric constant of the particles. The motion of the i'th particle is described by a classical equation

$$md^2\vec{r_i}/dt^2 = \vec{F_i} \tag{2.1}$$

where $\vec{F_i}$ is the total force acting on it.

The dipolar force acting on the particle at $\vec{r_i}$ by a particle at $\vec{r_j}$ is given by

$$\vec{f_{ij}} = \frac{3p^2}{\epsilon_f r_{ij}^4}[\vec{e_r}(1 - 3\cos^2\theta_{ij}) - \vec{e_\theta}\sin(2\theta_{ij})] \tag{2.2}$$

where $\vec{r_{ij}} = \vec{r_i} - \vec{r_j}$ and $0 \le \theta_{ij} \le \frac{\pi}{2}$ is the angle between the z direction and the joint line of the two dipoles. We use $\vec{e_r}$ as a unit vector parallel to $\vec{r_{ij}}$ and $\vec{e_\theta}$ as a unit vector parallel to $\vec{e_r} \times (\vec{e_r} \times \vec{E_0})$.

A dipole $\vec{p}$ inside the capacitor at $\vec{r_i} = (x_i, y_i, z_i)$ produces an infinite number of images at $(x_i, y_i, -z_i)$ and $(x_i, y_i, 2Lk \pm z_i)$ for $k = \pm 1, \pm 2, \cdots$. The force between a dipole and an image has the same form as Eq.(2.2). The jth particle and its infinite images produce an electric force on the ith particle,[8,10]

$$
\begin{aligned}
f_{ij,x} &= \frac{p^2}{\epsilon_f L^4} \sum_{s=1}^{\infty} \frac{4s^3\pi^3(x_i - x_j)}{\rho_{ij}} K_1(\frac{s\pi\rho_{ij}}{L}) \cos(\frac{s\pi z_i}{L}) \cos(\frac{s\pi z_j}{L}) \\
f_{ij,y} &= \frac{p^2}{\epsilon_f L^4} \sum_{s=1}^{\infty} \frac{4s^3\pi^3(y_i - y_j)}{\rho_{ij}} K_1(\frac{s\pi\rho_{ij}}{L}) \cos(\frac{s\pi z_i}{L}) \cos(\frac{s\pi z_j}{L}) \\
f_{ij,z} &= \frac{p^2}{\epsilon_f L^4} \sum_{s=1}^{\infty} 4s^3\pi^3 K_0(\frac{s\pi\rho_{ij}}{L}) \sin(\frac{s\pi z_i}{L}) \cos(\frac{s\pi z_j}{L})
\end{aligned}
\tag{2.3}
$$

where $\rho_{ij} = \sqrt{(x_i - x_j)^2 + (y_i - y_j)^2}$ and K_0 and K_1 are modified Bessel functions. The force on the ith particle by its own images is in the z direction and denoted as $\vec{f}_i^{self}$,

$$
f_{i,z}^{self} = \frac{3p^2}{8\epsilon_f} \left(-\frac{1}{z_i^4} + \sum_{s=1}^{\infty} [\frac{1}{(z_i - sL)^4} - \frac{1}{(z_i + sL)^4}] \right) . \tag{2.4}
$$

In addition to the dipolar interactions, two spheres repel each other strongly whenever they are colliding each other. There have been several forms of this short-range interaction in the study of ER fluids, ranging from a hard-core interaction, soft-core, to a power rule.[8,11] In our model, we follow Melrose[12] to use a power rule, $\vec{f}_{ij}^{rep} \sim \vec{r}_{ij}/r_{ij}^{n+2}$, where $\vec{r}_{ij} = \vec{r}_i - \vec{r}_j$. The results reported here are related to $n = 23$, a value large enough to prevent the particles from collapsing.

Now for the bulk particles $\vec{F}_i$ in Eq.(2.1) is given by

$$
\vec{F}_i = \sum_{j \neq i} (\vec{f}_{ij} + \vec{f}_{ij}^{rep}) + \vec{f}_i^{self}. \tag{2.5}
$$

As mentioned earlier, the wall-particles are experiencing additional external forces in the x direction which produce the shear flow.

A viscous flow always produces heat which may cause temperature increase if the heat is not removed. In our simulations, the temperature in a layer is defined by the following equation,

$$
T = \frac{m}{3k_B(N_l - 1)} \sum_{i=1}^{N_l} (\vec{v}_i - \vec{v}_a)^2 \tag{2.6}
$$

where N_l is the number of particles in the layer and the average velocity $\vec{v}_a$ is given by $\vec{v}_a = \sum_{i=1}^{N_l} \vec{v}_i / N_l$. To maintain the temperature at a constant T_0, we apply a renormalization procedure to the wall-layers and the two layers immediately adjacent to the wall-layers.

$$\vec{v}_i^R = \vec{v}_a + (\vec{v}_i - \vec{v}_a)\sqrt{T_0/T} \tag{2.7}$$

where T is determined by Eq.(2.6). This is equivalent to removing heat in the system from the two electrodes.

In our simulations, the electrodes have $L_x = L_y = 16a$ while the distance between the electrodes is $L = 14a$. We denote a as the particle radius. Each wall-layer has 20 particles. There are other 100 particles in the bulk. The volume fraction is thus 16.4% which is quite low and makes the dipole model valid. Our theoretical calculation has already shown that when $L_z \leq 12a$, a single-chain structure is the preferred solid structure in ER fluids; only when $L_z > 12a$, ER fluids are able to have thick columns.[13] In real experiments, a has the order of micrometer, while L_z has the order of millimeter. Therefore, we take $L_z = 14a$ to satisfy the above condition. The parameter λ measures the dipolar energy to the thermal energy and also indicates the strength of the electric field,[14] $\lambda = p^2/(a^3 k_B T_0)$. In our simulations, we use $\sqrt{mk_BT/a^4}$ as the unit of viscosity, $\sqrt{ma^2/k_BT}$ as the unit of time, and $\sqrt{k_BT/ma^2}$ as the unit of shear rate.

We have applied a fourth-order Runge-Kutta method of a fixed step-size to integrate Eq.(2.1). The step-size is $0.02\sqrt{ma^2/(k_BT_0)}$. If $T_0 = 300$ K, $a = 1.0$ μm, and the particles have density $\rho = 1.2$ g/cm^3, then one step is about 2×10^{-5}s.

3. Results and Discussions

As mentioned in Section 2, at each step, we extricate heat from the two wall-layers and two layers immediately adjacent to the wall-layers. A careful examination of the velocity distribution and average temperature reveals that this approach works well in keeping the bulk nearly uniform in temperature although the middle layer still has a slightly higher temperature. This situation is normal because we assume that the heat dissipates from the two electrodes.

3.1. Newtonian flow

At a low λ, i.e. a weak field, our simulations find a good Newtonian flow. In Fig.1, we plot the effective viscosity as a function of shear rate at $\lambda = 10$ and 35. As shown there, the effective viscosity is independent on the shear rate. The parallel line is a clear indication of a Newtonian fluid. The result is similar to a neutral dilute fluid. In our simulations, we also have a low volume fraction 16.4%.

3.2. Non-Newtonian flow

At a high field, the flow shows Non-Newtonian characteristics. In Fig.2, we plot the viscosity as a function of the shear rate $\dot{\gamma}$ for $\lambda = 55$, 70 and 80. The effective viscosity is increasing as the shear rate is reducing. The changes in the rheological behaviors can be interpreted as a transition from the streaming type of transport to a potential type of transport where the interchanges of momentum between layers are mainly accomplished through potential interaction.

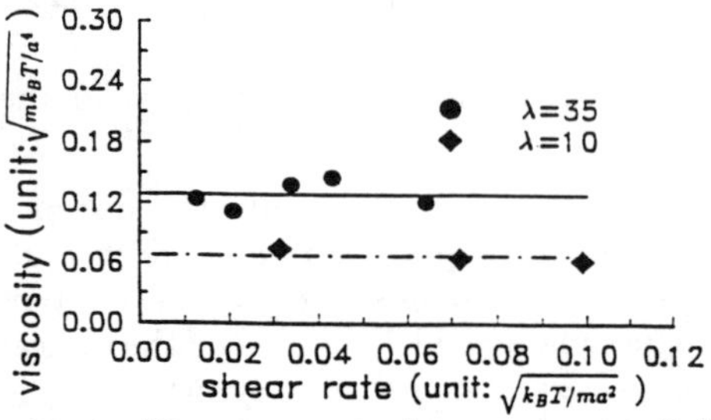

Fig.1. Viscosity versus shear rate at small λ ($\lambda = 10$ and 35).

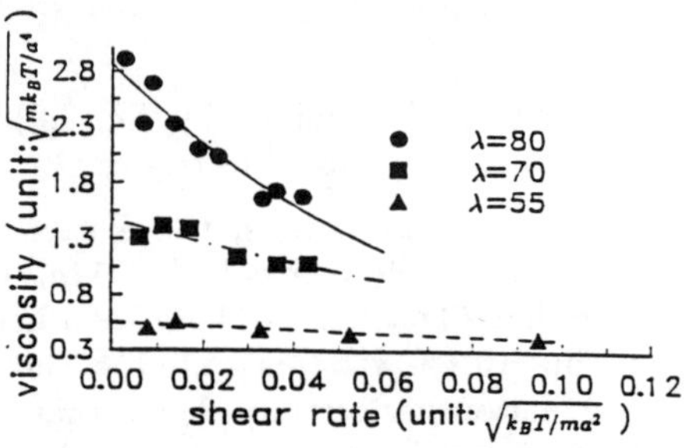

Fig.2. Viscosity versus shear rate at large λ ($\lambda = 55$, 70, and 80).

For a fluid with a strong intermolecular interaction, the dependence of viscosity on shear rate can be described by the Ree-Eyring relation,[15] $\eta = (\eta_0/\tau\dot{\gamma})\sinh^{-1}(\tau\dot{\gamma})$ where η_0 is the viscosity at zero shear rate and τ is the relaxation time of the system.

From the curve in Fig.2, we have found that the viscosity fits the following relationship,

$$\eta = \eta_0 e^{-\dot{\gamma}\tau} \tag{3.1}$$

where τ can be also defined as a relaxation time. For $\lambda = 55, 70$, and 80 we have $\tau = 3.04, 7.55$, and 14.53 (unit: $\sqrt{ma^2/k_BT}$). The relaxation time increases with the increasing of electric field. This is consistent with the following fact that the stronger is the electric field, the longer is the time for the system to reach its dynamic equilibrium state.

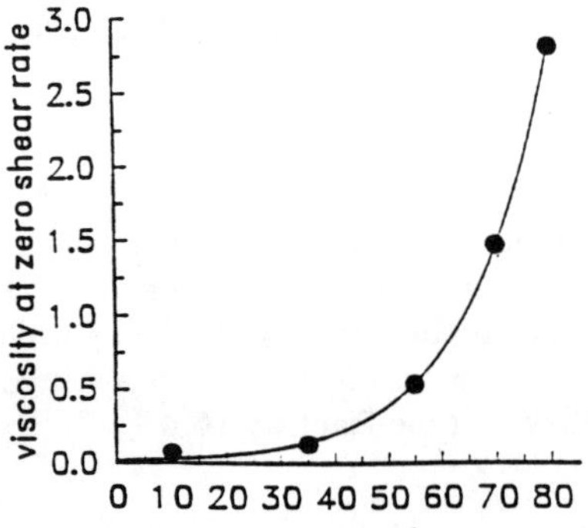

Fig.3. Viscosity at zero shear rate versus λ. The solid line is the exponential fitting.

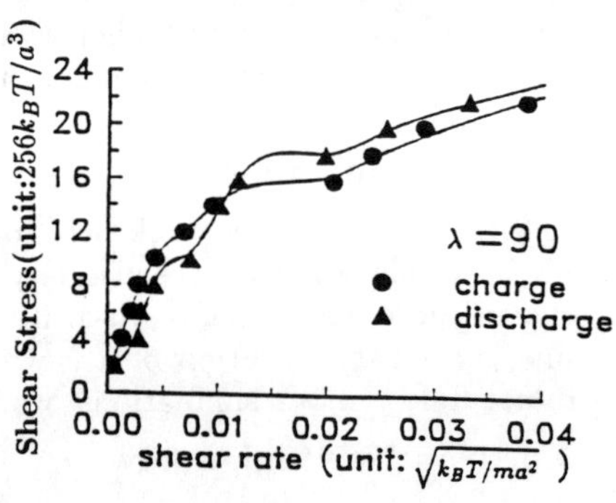

Fig.4. Shear stress versus shear rate at $\lambda = 90$. The discharge does not follow the same route of the charge curve, indicating a hysteresis.

The Eyring theory has viscosity of a normal liquid, $\eta = A\exp(\Delta E/k_BT)$, where ΔE is the energy barrier required to open up a hole for the transposed particle to move from one potential minimum to another minimum.[17] The leading term of this energy barrier has the order of the dipolar energy p^2/a^3. In addition, it is easy to open a hole when the volume fraction ϕ is low and vice verse. Therefore, we expect the viscosity at zero shear rate, η_0, has the following form

$$\eta_0 = c_0 \exp[b\phi p^2/(a^3 k_B T)] \tag{3.2}$$

where c_0 and b are constants. In Fig.3, we plot η_0 against λ and find the curve fits the above relationship very well. The two-fitting constants are $c_0 = 0.014$ and $b = 0.402$. We want to emphasize that Eq.(3.2) is only valid for the liquid state. In deriving Eq.(3.2), we only consider the contribution of dipolar interactions. Therefore, when λ is small, Eq.(3.2) may underestimate the viscosity, because the other contributions to the viscosity cannot be ignored.

3.3. Bingham Plastic

At $\lambda = 90$, we observe a new phenomenon. There is a yield stress which has a value between $2k_BT/a^3$ to $4k_BT/a^3$. The particles do not move when the applied stress is below the yield stress. Therefore, our system becomes a Bingham plastic. The relationship between shear rate and shear stress also depends on the flow history now. Similar hysteresis in experiments of ER fluids has been reported by E. Lemaire, G. Bossis, and Y. Grasselli.[18] In our simulations, we first increase the shear stress at a rate $2k_BT/a^3$ in 100,000 time steps. Then we decrease the shear stress at the same rate. The simulation result is shown in Fig.4. It is clear that the discharge does not follow the same route of the charge curve, indicating a hysteresis. The difference in the shear rate at a same shear stress can be as high as 60%.

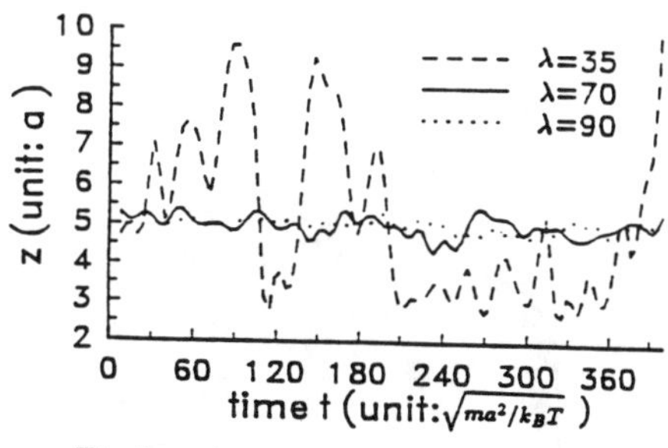

Fig.5. A typical particle motion in the z-direction at different electric fields. At a small electric field, the particles wander up and down in the z direction, but at a strong electric field, the particle moves little in the z direction.

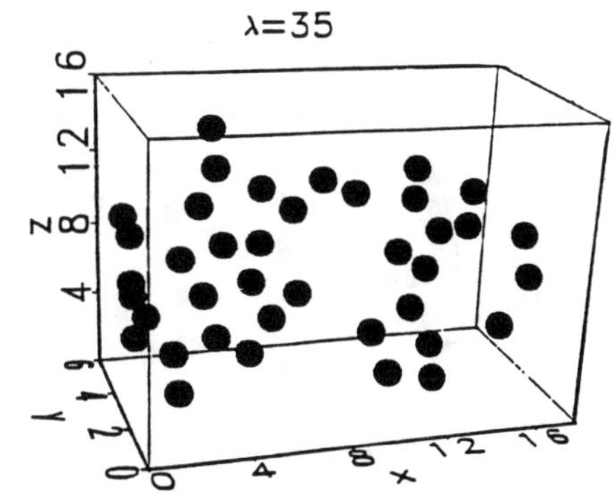

Fig.6. An instant slab of flow at $\lambda = 35$, parallel to the x-z plane. The system has no obvious ordering. Length unit: a.

3.4. Structural change

All these changes of rheology with the field can be related to the change of structure of the polarizable fluid under an electric field. Although the work reported here focuses on the effective viscosity, we have paid some attention to the movements of particles.

In Fig.5, we plot a typical particle motion in the z-direction at different electric fields. For a small electric field, such as $\lambda = 35$, the particles wander up and down in the z direction while drifting along the x direction. This characterizes a diffusive motion which provides a momentum transfer between different layers. This motion behavior explains why the liquid is a Newtonian fluid. At strong electric field, such as at $\lambda = 90$, the particle does little motion in the z direction.

In Fig.6, we plot an instant microscopic flow structure at $\lambda = 35$. In order to have a clear view, we only show a slab of the flow, parallel to the x-z plane. It is clear there that the system has no obvious ordering; therefore, it is a Newtonian liquid.

At $\lambda = 70$ the particles do not wander in the z direction, although we see some oscillations around its equilibrium position in the z direction. The instant microscopic flow structure is in Fig.7. The flow has tilted chains and broken chains. These chains are distinct and well separated. The tilting is caused by the sheared motion. This indicates that the dipolar interactions play much more important role than the diffusion now. Therefore, the fluid shows non-Newtonian behavior.

At $\lambda = 90$ the particles are tightly bound to their equilibrium position in the z direction with very small fluctuations. The dominant contribution to the viscosity is the dipolar interactions. The microscopic flow structure shows chains condensed (Fig.8). In fact, a number of chains are clinging to each other to form a thick and stable structure. This structure provide a yield stress to the system and hysteresis.

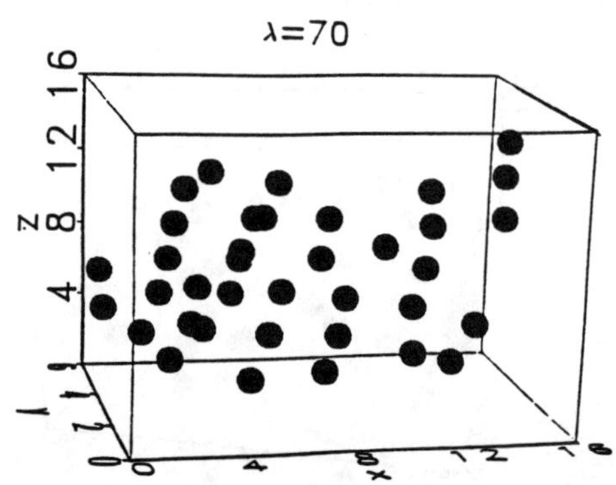

Fig.7. An instant slab of flow at $\lambda = 70$, parallel to the x-z plane. There are tilted chains and broken chains, well separated. Length unit: a.

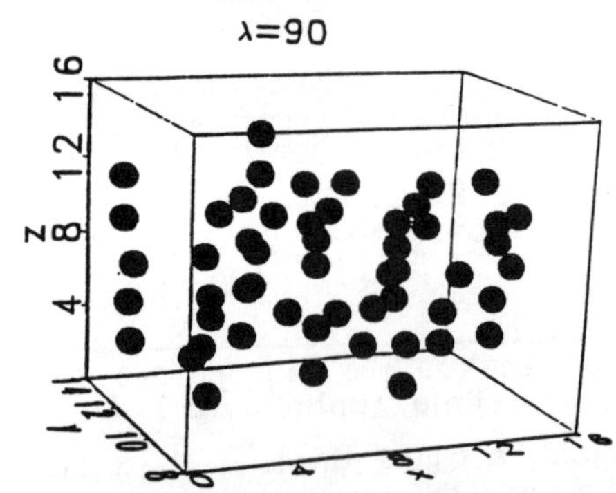

Fig.8. An instant slab of flow at $\lambda = 90$, parallel to the x-z plane. A number of chains are clinging to each other to form a thick and stable structure. Length unit: a.

In concluding our paper, we would like to compare our simulation result with experiments of Er fluids made of liquid crystal polymer solution. Under $E = 3$ kV/mm, each molecule of liquid crystal polymers has dipole moment $p = 10^{-15}$

(cgse unit) and $a \sim 5 \times 10^{-7}$ cm. At $T = 298$ K, $k_B T = 4 \times 10^{-14}$ erg. Then, $\lambda = 200$. At $\phi = 0.18$, from Eq. (3.2) we have the viscosity $\eta_0 = 2699$ Pa-S which is very close to the experimental data.[18]

The volume fraction 16.4% in our simulations is low. If the volume fraction is increases, the dipole model will be limited. However, we expect that the three stages in the viscosity change are still valid. However, it is conceivable that the fluid at a higher volume fraction will be easier to show non-Newtonian behavior and have a yield stress. The relationship between the flow structure and viscosity change should be further investigated.

Acknowledgements

This research is supported by a grant from Illinois Consortium for Advanced Radiation Sources and a grant from US Naval Surface Warfare Center.

References

1. W. M. Winslow, J. Appl. Phys. 20, 1137 (1949)
2. For example, see, *Electrorheological Fluids*, edited by R. Tao and G. D. Roy (World Scientific Publishing Comp., Singapore, 1994).
3. H. Block and J. P. Kelly, US Patent 4,687,589 (1987).
4. F. E. Filisko and W. E. Armstrong, US patent No. 4 744 914 (1988).
5. T. C. Halsey, J. E. Martin, and D. Adolf, Phys. Rev. Lett.**68**, 1519 (1992).
6. I. K.Yang and A. D. Shine, J. Rheol. **36**, 1079 (1992).
7. E. N. Andrade and C. Dodd, Proc. Roy. Soc. (London) A204, 449 (1951).
8. R. Tao and Q. Jiang, Phys. Rev. Lett **73**, 205 (1994).
9. W. T. Ashurst and W. G. Hoover, Phys. Rev. A 11, 658 (1975).
10. R. Tao, Phys. Rev. E. **47**, 423 (1993).
11. D. J. Klingenberg, F. van Swol, and C. F. Zukoski, J. Chem. Phys. **91**, 7888 (1989).
12. J. R. Melrose, Phys. Rev. A**44**, R4787, (1991).
13. R. Tao and J. M. Sun, Phys. Rev. Lett. **67**, 398 (1991).
14. P. M. Adriani and A. P. Gast, Phys. Fluids **31**, 2757 (1988).
15. F. H. Ree, T. Ree, and H. Eyring, Ind. Eng. Chem. 50, 1036 (1958).
16. H. Eyring, J. Chem. Phys. 4, 283 (1936).
17. E. Lemaire, G. Bossis, and Y. Grasselli, Langmuir 8, 2957, (1992).
18. A. D. Shine, I.-K. Yang, and K. L. Tse, in *ELectro-Rheological Flows - 1993*, edited by D. A. Siginer, J. H. Kim, and R. A. Bajura (The American Society of Mechanical Engineers, New York) p.49.

Non-linear Parameter Identification of an Electro-Rheological Fluid Damper

G. Mui, D. L. Russell and J. Y. Wong

Transport Technology Research Laboratory
Department of Mechanical and Aerospace Engineering
Carleton University, Ottawa, Canada, K1S 5B6

ABSTRACT

Parameters for a non-linear model of an electro-rheological (ER) damper were identified from experimental data. The ER fluid used in the test damper was composed of zeolite particles with diameters less than 10 microns suspended in 10 centistoke silicon oil. Motion of the damper piston forced the fluid to flow through annular gaps between three concentric cylindrical electrodes. Experiments were performed on the damper by using a fatigue testing machine made by Testline Mechanical Test Systems to provide a controlled movement of the damper. The measured data included the damper force, the piston displacement as well as the voltage applied to the damper electrodes and the resulting current. The parameters were identified using a quasi-linearization technique which has been found suitable for similar non-linear parameter model identification problems. The model used in the study included the effects of fluid inertia, as well as viscous and Coulomb friction components. Results of the identification indicate that, although the parameters generally were found to vary from one oscillation cycle to the next, average parameters were able to predict the force in the damper as a function of damper motion and the applied voltage. Some evidence suggests that the predictive capabilities of future models may be improved by adding fluid compressibility and hysteresis to the model.

1. Introduction

One promising application of electro-rheological (ER) fluids is to use them as the working fluid in automotive suspension systems. Relatively efficient and mechanically simple damping systems can be designed based on the rheological properties of ER fluids that vary as a function of the strength of an electric field applied to them. The design of control systems for ER fluid based dampers, however, generally requires a model that can predict the behaviour of the fluid under dynamic conditions. Modelling of the fluid is difficult because the mechanism behind the ER effect is not fully understood. Although abundant information is available on the behaviour of ER fluids under dynamic shearing conditions the same can not be said for dynamic flow conditions. In this paper a lumped parameter model of a fluid damper is constructed and parameters for the model are identified based on data from a set of well controlled dynamic experiments. The approach used here is similar to the work of Stanway, Sprostan and Firoozian[6], however, the work presented in this paper is based on a flow mode damper, a different identification method and identifies an additional fluid inertia parameter. The model is intended for use in the design and control of ER fluid based suspensions.

Previous attempts at modelling ER fluid behaviour have identified varying degrees of viscoelastic and plastic behaviour that depends on the strength of the electric field applied to the fluid and on the motion of the damper. Hysteresis has been identified at low flow rates by Bullough[1]. Nakano and Yonekawa[2] found transient delays in the development of flow resistance from the time of application of an electric field to an ER fluid to be as high as 0.38 seconds. These delays have been attributed to compressibility of the fluid. ER fluids under steady flow conditions and constant applied electric fields have been modelled as idealized Bingham plastics by Peel and Bullough[3].

Sinusoidal variations in applied electric field at frequencies above 50 Hz were shown by Lou et al[4] to result in constant levels of flow resistance. This indicates an upper limit in the useful frequency response of these devices and has implications for the design of controllers for ER fluid based suspension systems. Stanway et al[5] identified parameters for a telescopic ER fluid based damper. They observed that, for their damper, at field strengths greater than 1.5 kV/mm nonlinear effects became significant. They were unable to identify a power law based constitutive model for their ER fluid that would successfully describe their experimental results at high field strengths. This is perhaps an indication that difficulties may exist in controlling dampers that rely on high voltage levels.

2. Materials and Methods

2.1 Apparatus and Experimental Protocol

The ER damper examined in the experiments operates on the flow mode principle and is shown schematically in Figure 1. It consisted of a stainless steel cylindrical bucket with a cavity 200 mm deep and a diameter of 96 mm. Two concentric aluminum cylinders of mean length 178 mm together with the inner wall of the bucket formed two 3 mm thick annular gaps. Fluid was forced to flow through these gaps by a piston inside the smaller cylinder. The bucket and the smaller cylinder served as ground electrodes while the remaining middle cylinder was attached to the high voltage supply. The flow directions shown in figure 1 are for the case of the piston moving downward. Complex flow conditions may exist around the ends of the cylinders.

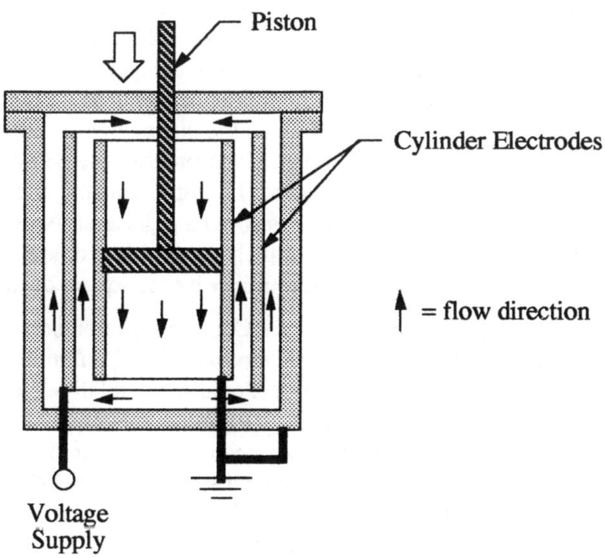

Figure 1: Schematic of ER Damper.

The damper used an ER fluid composed of zeolite particles with diameter less than 10 microns suspended in 10 centistoke silicon oil. The fluid was mixed with a mass ratio of 2 parts zeolite to 5 parts silicon oil. The resulting fluid had a density of 1145 kg/m^3. The fluid was found to be slightly dilatant with a mean viscosity of 0.1 kg/ms when tested over a range of shear rates at zero electric field strength[8].

The experimental apparatus was designed to monitor all possible interactions, including mechanical, electrical and thermal. All of the measured signals were sampled with a Computer Boards Inc. Model CIO-AD08 analog to digital converter at 1 kHz at a resolution of 12 bits and stored in a computer. Data was recorded in 4-second samples. Preliminary experiments indicated very small temperature variations during operation and as a consequence thermal effects were neglected in the results presented in this paper.

The damper was secured in a Mechanical Test Systems (MTS) fracture and fatigue testing machine. Previous experiments have shown that this apparatus can provide a low amplitude motion input to the damper with a useful range of amplitude. The force sensor used for these experiments was a 1.3 kN Lebow Tension and Compression load cell, model 3397. The motion applied to the damper was sinusoidal with an amplitude of 10mm and frequencies of 1, 2 and 5 Hz.

A Universal Voltronics supply was used to generate a step change in the field strength in the damper from 0 kV/mm to strengths of 0.5, 1.0, and 1.5 kV/mm. Both the voltage across and the current supplied to the damper were measured.

2.2 Selection of the Damper Model

The damper was modelled as a mass with both viscous and Coulomb frictional forces acting on it. The mass was an equivalent inertia that accounted for both the mass of the damper and the inertial effects of the fluid moving in the cylinder gaps. The dynamics of the system are written in state space form where the states are the momentum of the system, p_1, and the position of the damper, d. The identification technique requires that the parameters of the model be added to the state equations as extra states whose values are held constant. As the difficulties in modelling the system are primarily centered around the ER fluid and its properties the equivalent inertia was written in the fluid domain. That is, the inertia, I_1, represents the rate of change of fluid flow rate per unit of applied pressure and therefore its units are N-sec^2/m^5 or kg/m^4. While unconventional these units are a natural choice in this problem. To convert the inertia to an equivalent mass in kilograms, multiply by the square of the piston area.

The resistance to flow was modelled as a Coulomb friction component which had a constant value of R_2 and always opposed the motion. This Coulomb friction was combined with a viscous force that was proportional to the relative velocity of the damper, where the resistance parameter is R_1. Although this model is a simplified version of the Bingham plastic model that has been successful in other ER modelling application the results of this paper indicate that a significant level of predictive performance can be achieved with this simplified version. A final parameter, the load cell bias u_0, was added to the model to account for a small amount of drift observed in the signal from the load cell from one data set to the next.

The equations describing this model are:

$$\frac{d}{dt}\begin{Bmatrix} d \\ p_1 \\ I_1 \\ R_1 \\ R_2 \\ u_0 \end{Bmatrix} = \begin{bmatrix} \frac{p_1}{I_1 A} \\ \frac{u-u_0}{A} - \frac{R_1 p_1}{I_1} - R_2 sgn(p_1) \\ 0 \\ 0 \\ 0 \\ 0 \end{bmatrix} \tag{1}$$

Where u is the force in the damper in Newtons, A is the piston area in m^2, and sgn represents the sign function. Note that the only non-linearity in the model is in the Coulomb friction term.

2.3 Parameter Identification Technique.

The parameter identification method used in this study was the quasi-linearization method as described in Kalaba and Springarn[7]. This technique is based on a least squares approach and is suitable for continuous time, non-linear systems in state-space form. The technique is briefly outlined here. An approximation of the (k+1) derivative of the state vector is expressed in terms of the previous derivative of the state vector by using a Taylor series expansion and retaining only the linear terms,

$$\frac{d}{dt}\overline{X}^{k+1} = \frac{d}{dt}\overline{X}^{k} + J(\overline{X}^{k},u)\left[\overline{X}^{k+1} - \overline{X}^{k}\right] \tag{2}$$

where u is the system input, in this case the force generated in the damper, and the state vector is given in this case as

$$\overline{X} = \{d \quad p_1 \quad I_1 \quad R_1 \quad R_2 \quad u_0\}^T \tag{3}$$

As the initial condition on the first state element, the damper position, was known from the measured data, a solution to equation (2) is,

$$\overline{X}^{k+1} = \overline{P}(t) + r(t)\left[p_1(0) \quad I_1 \quad R_1 \quad R_2 \quad u_0\right] \tag{4}$$

where $\overline{P}(t)$ represents the particular solution and the remaining product represents the homogeneous solution. Constitutive equations governing $\overline{P}(t)$ and the matrix r(t) are found by substituting equation (4) into equation (2). A cost function reflecting the error between measured and predicted damper positions is written as

$$S = \sum_{i=1}^{L}\{d(t_i) - b_i\}^2 \tag{5}$$

Here L is the number of data points used in the identification algorithm, $d(t_i)$ is the predicted damper position and b_i is the measured damper position. This cost function is used in conjunction with equation (4) which predicts $d(t_i)$ and is minimized with respect to the unknown state elements. The equations that result from the minimization process are linear and can be solved for the parameters.

The algorithm was programmed in C++ and used a fourth-order Runge-Kutta integration routine. The performance of the identification was validated using known parameters to generate a data set. Errors in the inertial and resistance parameters were less than 1% while the error in the load cell bias term was less than 10%.

3. Results

Parameter values were identified as a function of the frequency of the damper motion and the strength of the electric field applied to the fluid in the damper. The results, shown in Table 1, indicate that the parameters are a function of both the strength of the electric field and the frequency of oscillation. The inertia parameter varies by a factor of four as the electric field strength is increased from 0 to 1.5 kV/mm but also changes by a factor of four as the oscillation frequency is increased from 1 Hz to 5 Hz. At the highest frequency, 5 Hz, the inertial parameter appears to be insensitive to the strength of the electric field. Similar effects can be observed in both of the resistance parameters although at 5 Hz there is an unexplained variation in the Coulomb friction parameter. The level of the load cell bias is small and does not have a significant effect on the results.

An example of the measured force data and the predicted force data is shown in Figure 2. Except at zero crossings in the force a good agreement between the predicted and measured forces is indicated. The discrepancy at zero crossings represents a lag in the response of the real system and may be due to fluid compliance that was not included in the model. For the sample shown in the figure, the electric field was turned on at approximately 0.5 sec from the start of the run and this accounts for the sudden change in the level of the force. The small impulse-like spikes in the data are attributed to a stick-slip component in the friction. Note that the construction of the damper is such that strict symmetry in the force data in the up and down directions should not be expected.

The degree of agreement between the measurements and the predictions is summarized in Table 2. An examination of the data indicated that there may be a change in parameter values from oscillation to oscillation. To examine this the identification procedure was performed on segments of the data corresponding to the period of oscillation. The identification assumes the parameters are constant over that time period but by shifting the data window used forward by the sample time an indication of any changes in the parameters could be evaluated in an approximate way. The results of these studies indicated that there may be some pattern to the variation in parameter values and that a large degree of variability may be present in the parameters at high levels of electric field strength. Two RMS errors are included in Table 2. The first indicates the error in predicting the damping force based on a model using a single parameter identified from the entire data set and the second is the prediction error based on the parameter values identified from single period segments of the data. The data indicates that there is no significant improvement in the quality of the model when parameters are identified on single period segments of the measured data and are thus allowed to vary slowly from cycle to cycle.

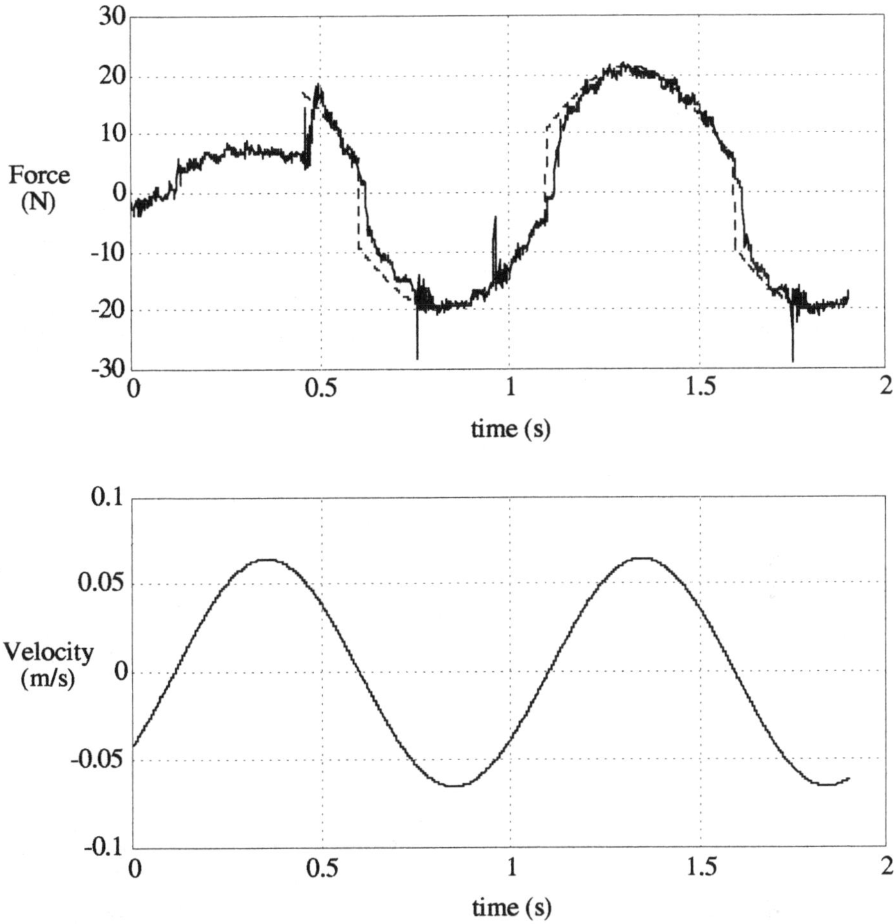

Figure 2: Measured force (solid line), predicted force (dashed line) and relative velocity for a damper at an oscillation frequency of 1 Hz and an electric field strength of 1.0 kV/mm.

4. Discussion

The parameter values identified and shown in Table 2 indicate a dependence on the frequency of oscillation. This may mean that the higher velocities in the fluid gap found at higher frequencies of oscillation may prevent the build up of ER particles on the electrode surfaces. If true, this would mean that at slower speeds more fluid would build up on the walls and thus the area for flow would be reduced. This would increase the

Table 1: Parameter values identified from experimental data.

Damper Oscillation Frequency (Hz)	Field Strength (kV/mm)	Equivalent Inertia I_1 (kN-sec^2/m^5)	Flow Resistance R_1 (MPa-sec/m^3)	Coulomb Resistance R_2 (Pa)	Load Cell Bias u_0 (N)
1	0.0	212	4.5	340	-0.03
	0.5	209	7.3	590	0.16
	1.0	382	12.1	1830	0.85
	1.5	769	19.4	3680	0.76
2	0.0	214	4.6	290	-0.07
	0.5	221	6.5	600	0.16
	1.0	273	11.5	1050	1.01
	1.5	326	15.8	1910	1.50
5	0.0	203	4.8	450	0.94
	0.5	210	5.9	930	1.41
	1.0	221	8.6	760	2.45
	1.5	227	12.3	270	3.24

Note that while the inertia and resistance parameters generally increase with electric field strength the magnitude of the effect is less significant at higher oscillation frequencies. The exception is the Coulomb friction parameter at an oscillation frequency of 5 Hz. The reason for this effect is unknown. Also note that one bit corresponds to a change in force on the load cell of 0.33N so that the load cell bias terms generally correspond to changes in force equivalent to the least significant bit.

Table 2: RMS prediction errors for the model with the identified parameters.

Damper Oscillation Frequency (Hz)	Field Strength (kV/mm)	RMS Prediction Error with Constant Parameters (N)	RMS Prediction Error with slow parameter variation (N)
1	0.0	1.1	1.1
	0.5	1.6	1.6
	1.0	3.2	3.0
	1.5	8.2	6.9
2	0.0	3.0	3.0
	0.5	3.2	3.2
	1.0	3.6	3.7
	1.5	9.2	9.0
5	0.0	10.7	10.7
	0.5	11.2	10.2
	1.0	10.6	10.0
	1.5	12.7	11.5

Note that the RMS error for the slow parameter variation represents the accuracy achievable if the parameters are identified based on data from a single period of the oscillation following the time in question. This approach was based on an observation that the damper performance may have changed from oscillation to oscillation. There is no indication that a significant improvement in predictive ability can be gained by using the parameters identified while allowing for cycle to cycle changes in parameter value.

effective inertia at low speeds since an equivalent change in pressure would act on a smaller area and thus cause a lower change in fluid flow rate.

A second possibility is that the increased agitation of the fluid at higher frequencies disrupt the formation of structures within the ER fluid. It is important that in this type of damper the fluid is continually passing through the electric field and will thus pass from areas of little or no electric field to areas of maximum electric field strength.

The variation in the effect of changes in field strength as a function of oscillation frequency means that in a control system the gains used to determine the response of a damper to excitation must be a function of the frequency of excitation. Higher frequency inputs will require different gain levels due to the decreased effect of the electric field on the ER fluid. This may be accomplished using derivative control or through more complex non-linear techniques.

5. Conclusions

The results indicate that the model used in this paper containing inertia and both viscous forces and Coulomb friction can provide a good level of predictive capability. However, the parameter values used in the model were found to have a significant dependence on the frequency of mechanical oscillation. The degree of accumulation of ER particles on the surface of the electrodes may account for this phenomenon. This is an important feature to consider when designing controllers for dampers based on these fluids. The model had some difficulty in predicting the performance of the damper near zero crossings in the damper force. The model may be improved in this respect by adding hysteresis effects or the effects of fluid compliance.

6. Acknowledgments

The authors wish to acknowledge the support of the Natural Sciences and Engineering Research Council of Canada.

7. References

1. W. A. Bullough, *Proc. of the Second Int. Conf. on ER Fluids* (World Scientific, 1989) pp. 115-123.
2. M. Nakano and T. Yonekawa, *Proc. of the Fourth Int. Conf. on ER Fluids* (World Scientific, 1994) pp. 477-489.
3. D. J. Peel and W. A. Bullough, *Proc. of the Fourth Int. Conf. on ER Fluids* (World Scientific, 1994) pp. 538-567.
4. Z. Lou, R. D. Ervin and F. E. Filisko, *Proc. of the Third Int. Conf. on ER Fluids* (World Scientific, 1992) pp. 398-423.
5. R. Stanway, J. L. Sproston and N. G. Stevens, *Journal of Electrostatics* **20** (1987) pp. 167-184.
6. R. Stanway, J. L. Sprostan and R. Firoozian, *Journal of Dynamic Systems, Measurement and Control*, **111** (1989) pp. 91-96.
7. R. Kalaba and K. Springarn, *Control Identification and Input Optimization* (Plenum Press, New York, 1982).
8. J. Catania, *M.Eng. Thesis* (unpublished), Carleton University, Ottawa, Canada (1994)

NONLINEAR OPTICAL STUDY OF ER FLUIDS

J.Y. WU, L.K. SHEN, L.W. ZHOU
T.D. Lee Physics Laboratory & Department of Physics, Fudan University, Shanghai 200433, China

ABSTRACT

A nonlinear optical method is used to study the mechanism of ER fluids. The fluid used in this study is diatomite dispersed in transformer oil. Because of the non-homogeneity of the ER fluid, second harmonic generation (SHG) powder technique is employed. Electric field induced SHG signal of the fluid has been found and the response time of the fluid has been measured. Two kinds of polarization are deduced to contribute to the ER effect of the sample at different periods. A two-stage process of ER effect is adapted to explain the experiment results.

1. Introduction

As recent improvement on polymeric and anhydrous ER materials, ER fluids have again attracted people's interest[1]. The study on its mechanisms has been being expansively taken, and understanding of ER effects is deeper than before[2]. In order to develop new ER fluids to meet practical demand, the necessary first step is to understand mechanisms of ER effects. The sophisticated knowledge of ER effects can help to synthesize and develop new practical ER fluids and plot suitable ER devices for marketplace. However, we still do not understand thoroughly the mechanisms of particle interaction, filamentation and field-induced transformation that underlies the ER effect. Better understanding the mechanism of ER fluids is a key to reach revolutionary advancement on designing, synthesizing and formulating ER fluids that have desired properties.

Many work have been done trying to disclose the underlied mechanisms of ER effect. Mason's theory[3] explained that the enhancement of effective viscosity was due to orientational lining up of polar molecules under strong electric field. Deinega[4] and other scientists thought that the strong electric fields would develop interaction of particles, form chain and column structure, which then cause the force connection between two electrodes, and enhance effective viscosity. Up to now, a lot of experiments and phenomenological theories have been established to interpret the ER effects. The most widely accepted description is that the electric field induced polarization results in ER effect. However, the yielding stress according to a simple electric polarization theory is smaller than the experimental results of some best ER fluids. The distinction between theories and experiments implies that present knowledge of ER fluids is limited, and more study and more experimental data are needed to build better theoretical models. Study of ER effect is an experiment leading science. Various experimental methods have been employed to study ER effects, such as the measurements of dielectric properties[5], nonlinear dielectric properties[6], electric conductivity[7], viscosity and so on. Recently, optical methods have also been employed to broaden the knowledge of ER phenomena. Halsey and his colleagues[8] have studied a model ER fluid with optical scattering method. They built a phenomenological theory by using "independent droplet" model of the suspension structure, and fitted their experimental results in good agreement. Tao's group[9] used laser diffraction technique to study the column structure in ER fluid of glass beads, and found that the microscopic structure of column was a body-centered tetragonal (bct) lattice structure. They also used computer simulation[10] to study the temporal evolution of three-dimensional column structure, and theoretically verified the bct lattice structure.

In this paper, we report the nonlinear optical study of an ER fluid with second harmonic generation (SHG) measurement[11]. As ER effect is the results of the cooperation between strong electric field and ER fluids, the symmetry of the ER fluids transforms when electric field is applied. The polarization field destroys the centrosymmetry of the system, causes the ER fluid polarized and makes it possess an additional second-order nonlinear optical effect, which appends to the original SHG effect of ER fluids when there is no electric field is applied. From SHG signal, one can determine the type of electric field induced polarization that contributes mainly to ER effect and analyze the evolution of ER process. This group[12] has reported the electric field induced SHG (EFISH) study of ER fluids, and found additional SHG signals as a function of electric field.

In our experiment, the ER fluid is considered to be an optically non-homogeneous system. SHG powder method used in this experiment has been widely accepted in studying nonlinear optical coefficient of polymeric materials and organic molecules[13]. The ER fluid used in this study is a water-activated diatomite suspended in transformer oil. The results of experiments show that the sample inhabits quite large SHG effect even without applying electric field. We found an SHG signal modification when electric field was applied. We also measured the temporal evolution of the SHG signal under the field. We found that the sample perceived a fast enhancement process, followed by a slow returning process in an order of 10 seconds.

2. Experiments

The SHG powder technique is a simple, quick, but powerful experimental technique, which was first systematically investigated by Kurtz[14]. With a reasonable high reliability, the method enables one to classify a new material into one of five categories according to its symmetry and nonlinear coefficients. Furthermore, the SHG effect of materials reveals electronic polarization contribution to ER effect. It is also helpful in recognizing which type of polarization is the main incentive of ER effect among the electronic polarization in microscopic molecules or the interfacial polarization in ER suspensions. In this method, a reflected laser beam is analyzed to gain SHG information. The advantage of the method is that it allows using ER suspension with high solid phase concentration

Fig. 1 is a basic configuration for the study of second-harmonic generations in non-homogeneous systems such as powders or suspensions. It consists of an active & passive mode-locked YAG laser. The fundamental wavelength of its output is 1064 nm, and each pulse is about 50 picosecond in width and 1 mJ in energy. The output pulses pass through a filter glass to cut the beam with wavelengths other than 1064 nm, and, after two focusing glass lenses, fall onto the sample. The sample box is made of glass for collecting SHG

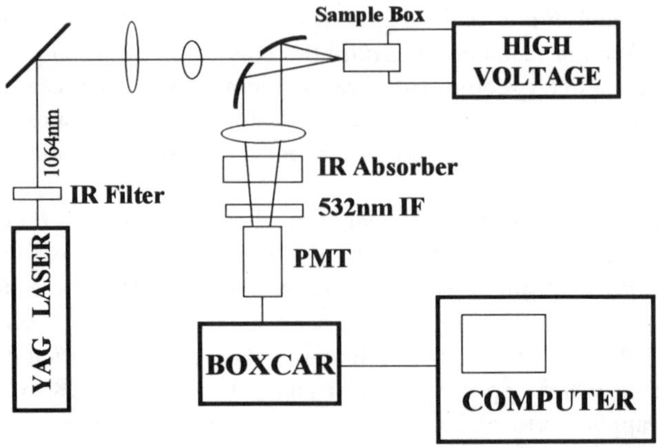

Fig. 1 Apparatus used for study of SHG in non-homogeneous systems.

signal easily. The transparent electrodes for applying high voltage are made of ITO glass. The diffracting SHG signals from the sample at every direction are collected by a concave reflector. The fundamental beam is filtered out by passing a set of absorption glass, and then the rest of the 532nm beam passes through a 532 nm interfere filter to ensure 532 nm SHG signal, which is focused and detected by a photomultiplier. Signals are averaged by a boxcar system and recorded by a computer.

The sample used in this study is an ER fluid of diatomite dispersed in a transformer oil with solid phase weight fraction less than 20%. The solid phase is made from natural diatomite using chemical methods. Its main component is about 97% of SiO_2 with some impurities such as FeO, Al_2O_3, water and so on. This fluid is a water-activated system in which the optimal water content is 7% by weight. The average diameter of particles is 10 μm. The dielectric and rheological measurements are reported elsewhere[15].

3. Results and Discussion

The sample owns effective ER effect. While the high electric field being applied, chain and column structure are clearly observed. The process of their formation can also be perceived. The evolution of structure from chains to columns takes about 10-second.

The SHG of pure transformer oil is very small. Its SHG signal is below the sensitivity of our detecting system even when 1000 V/mm of electric field is applied. The diffracting SHG signals of the sample at all directions have been measured, and a quite large SHG signal is detected even without applied electric field. The suspensions with higher concentration produce higher SHG signal at backward direction. However, the signal at the forward direction is smaller, which is believed to be the results of optical non-homogeneity and opaqueness of the ER suspensions. The backward direction is an appropriate direction of studying the SHG effect of ER fluids.

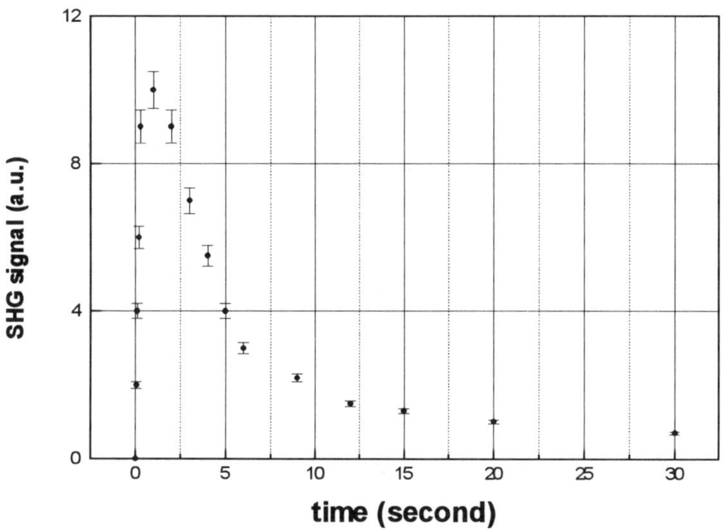

Fig. 2 Time response of additional SHG signal, where the boxcar delay has not been subtracted.. Zero point on time axis is the beginning of applying electric field. SHG signal without electric field is subtracted.

According to Kurtz's theory[14], the SHG effect of the ER fluid without electric field gives a testimony that the water-activated diatomite is a kind of crystalline powder without centrosymmetry. After a high electric field is applied, the change of SHG signal has been found. First, the signal goes up with an enhancement as small as below 10 percent of the original signal without electric field. But the additional SHG signal is not steady under electric field, and has a slow recovering process to its original zero-field level. The time constants of the processes of signal enhancement and recovery can be seen from Fig. 2. The SHG signal without electric field has been subtracted in Fig. 2; however, the boxcar delay time has not been subtracted from the time-axis. That is, the enhancement process is very fast, and its process time is shorter than the time resolution of our detecting system if the delay time of the boxcar is considered. According to the experimental arrangement, the time delay of boxcar is near 1 second, which causes the process of fast SHG enhancement cannot be distinguished. The time constant of the slow recovery process is the order of 10 seconds. For polarized samples, in which column structure is formed, the column structure remains for a quite long period of time after the electric field is turned off. If the electric field is turned back on in this period of time, the additional SHG signal becomes much smaller and even can not be detected.

After electric field is applied, more than one kind of polarization are possible to be incited. Different polarization may take different way of influencing SHG signal. Analyzing the results of nonlinear optical experiments is helpful in understanding the source of polarization in ER fluids.

If the additional SHG signal under electric field were caused by polarization of particles, the larger the particle polarization, the larger the additional SHG signal. The polarization is given as $\boldsymbol{P} = (\varepsilon_r - 1)\varepsilon_0 \boldsymbol{E}$, where $\boldsymbol{E}$ is the external static electric field. The formation of chains and columns enhances the polarization inside particles in columns as the dielectric constant of particles is larger than that of base oil and the electric field remains constant. In turn, the SHG signal should be gradually enhanced and remains at a steady value. Obviously, it contradicts with the SHG experimental results shown in Fig. 2. The SHG results confirms that SHG is not from the electric field induced polarization inside particle.

The SHG signal may related to the interfacial polarization in sample ER fluid. When the electric field is applied, polarization of particles causes net electron charges appearing at the local area of the surface and attracting charges in oil, and then forms the interfacial dipole. This kind of interfacial dipoles can induce effective SHG signal because of two different phases and no centrosymmetry[11]. When chains and columns are formed, particles are connected directly and the number of interfacial dipoles between particle and liquid becomes smaller. Thus, SHG signal becomes smaller. With all of particles being in columns, the additional SHG signal approaches to zero. It agrees with our experimental observation. It is observed that the time of the column forming process is the order of 10 seconds, which agrees also with the SHG reduction time shown in Fig. 2.

From many experiments, the response time of ER effect is the order of millisecond. However, the formation of chains and columns is connected to spatial movement of particles and takes relative longer time. This suggests that two stages with different characteristic time constants may be involved in ER effect after electric field is applied. First, the induced interfacial polarization between solid and liquid phases is formed as soon as electric field applied, which is an instantaneous process. The interaction between two phases confines the movement of particles and resists the flow of fluids in a short time. The viscosity of ER fluids grows very quickly in an order of millisecond. Second, particles aggregate because of static electric force and Brownian movement. This causes the number reduction of induced interfacial dipoles. Particles attract each other by electric dipole to form chains and columns. The movement process can also be understood from the point of view of least energy level. R. Tao has got a result from simulation[10] and experiments[9] that the column has body-centered tetragonal (bct) lattice structure, of which the energy is lowest. The column structure is then the reason of ER effect.

ER effect at different stages may have different polarization mechanisms. Our experimental observation indicates that more than one polarization mechanisms may lead to ER effect.

4. Conclusions

The SHG powder technique is a simple, quick, but powerful experimental technique, especially for an instantaneous process such as initial evolution of ER effect. From SHG signal, one can determine the type of electric field induced polarization that dominants ER effect, and analyze the evolution of ER process. The second harmonic generation (SHG) of diatomite ER fluid has been measured. The additional SHG signal and its two time processes have been observed. From the experimental phenomena, we deduce that the interfacial polarization is the source of electric field induced SHG. The ER effect of our sample have two different stages, which also have different polarization mechanisms. The mechanism of first stage of ER effect is interfacial polarization and that of the second stage is chains and columns by polarization in particles.

Acknowledgment

The sample used in this experiment is supplied by Korobko of A.V. Luikov Heat and Mass Transfer Institute, Belarussian Academy of Sciences. The work is supported by the National Natural Science Foundation of China, under Grant Nos. 19374018 and 69408001.

Reference

1. *Electrorheological(ER) Fluids: A Research Needs Assessment, Final report*, (US Dept. of Energy, 1993).
2. R.Tao and G.D.Roy, *Electrorheological Fluids: Mechanisms, Properties, Technology and Applications,* ed. R.Tao (World Scientific, Singapore, 1994).
3 A. Okagawa, *et al., J. Coll. Int. Sci,* **47** (1974) 568.
4. Y.F. Deinega, *et al., Rheo. Acta,* **23** (1984) 636.
5. H. Block, J.P. Kelly, *J. Phys. D. Appl. Phys.* **21** (1988) 1661.
6. L.W. Zhou, *et al., Polymer Preprints*, **34** (1994) 323.
7. J.C.R. Turner, *Chem.Eng.Sci,* **31** (1976) 487.
8. T.C. Halsey and J.E. Martin, *Electrorheological Fluids: Mechanisms, Properties, Technology and Applications,* ed. R.Tao, (World Scientific, Singapore, 1994) p. 115.
9. T.J. Chen, R.N. Zitter and R. Tao, *Phys. Rev. Lett.* **68** (1992) 2555.
10. R.Tao and Q. Jiang, *Electrorheological Fluids: Mechanisms, Properties, Technology and Applications,* Ed. R. Tao,(World Scientific. Singapore, 1994).p. 129.
11. Y.R. Shen, *The Principles of Nonlinear Optics* (John Wiley & Sons, New York, 1984).
12. L.W. Zhou, *et al., Inter. J. of Mod. Phys.* **B8** (1994) 2921.
13. D.S. Chemla, J. Zyss, *Nonlinear Optical Properties of Organic Molecules and Crystal* (Academic Press, New York 1987).
14. S.K. Kurtz and T.T. Perry, *J. Appl.Phys.*, **39** (1968) 3798.
15. Z.Y. Qiu, *et al.*, in *Proceedings of 5th International Conference on ER Fluids, MR Suspensions and Associated Technology*, Sheffield, UK.

MAGNETORHEOLOGICAL FLUID - BASED SEAL

W.I. KORDONSKY and S.R. GORODKIN
Academic Scientific Complex "A.V.Luikov Heat and Mass Transfer Institute" Academy of Sciences of Belarus, 15 P.Brovka str. , Minsk, 220072 Belarus and Byelocorp Scientific, Inc., 70 Pine Street, New York, NY 10270 USA

ABSTRACT

The present experimental work has been done to study the characteristics of the simplest one-step magnetofluid seal (MFS) of the rotary shaft in which the MR fluid (MRF) serves as a sealing medium.

The conducted experiments have found that the characteristics of the MRF-based seal differ very much from those of the ferrofluid (FF)-based one. In the static regime (stationary shaft), all other things being equal, the MRF holds an approximately three-fold pressure drop. The distinguishing feature of the functioning of such a seal is revealed and lies both in sharply elevating the pressure drop in the dynamic regime (rotary shaft) as against the static one and in having a maximum on the dependence of the pressure drop on the shaft rotation speed. In this regime, the MRF efficiency grows almost by the order of magnitude. However, the seal under study has an increased friction torque due to the growth of the shear stress of the MRF volume subject to shearing in the magnetic field.

1. Introduction

It is well known that the majority of the modern mechanical and hydraulic devices and mechanisms cannot operate without sealing the connecting pieces. That is why, to our mind, any attempt to improve the existing sealing facilities or to design new ones is important and urgent. Nowadays there is an extensive list of the types and designs of seals and materials, of which they are fabricated, from rubber cups in the bicycle pump to the most complicated multistep seals of vacuum and cryogenic technique. Among all this variety, the field exists in which MFSs[1,2] have found use, and their simplest design is shown in Fig.1. These are contactless slotted seals embodying the concept of a hydraulic back-pressure valve, in which the FF is confined by the magnetic field in the working gaps between mating parts. The basic good points of the MFS are as follows: provision of high sealing, which is of special importance for sealing corrosive and ecologically dangerous media, whose leaks are intolerable; low intrinsic friction torque; actual absence of wear of mating parts; ability to operate at high eccentricities and significant mechanical runs-out; simplicity in restoring the serviceability when the FF is broken.

Interest in the study of the operating characteristics of the MRF-based MFS is motivated by the following factors.

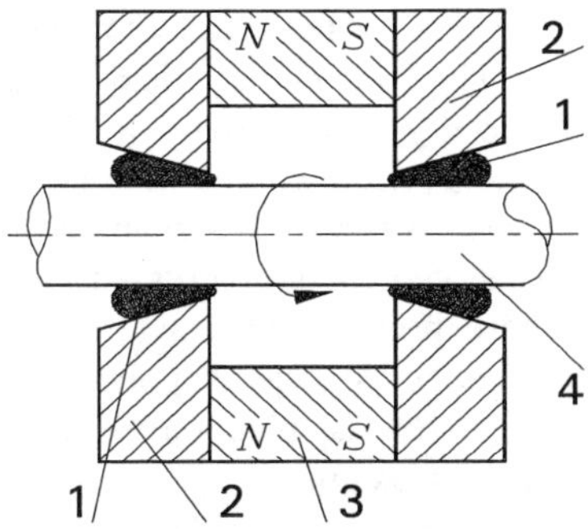

Fig. 1. Schematic of the MFS: 1, ferrofluid; 2, poles; 3, permanent magnet; 4, magnetic shaft

First, the magnetization, M, of the MRF is higher than that of the FF. So , e.g. in experiments with the compositions when acted on by the magnetic field with the intensity H=150 kA/m we have M_{FF}=50 kA/m and M_{MRF}=120 kA/m. As the critical pressure held by one step of the MFS is determined by the relation[3]

$$P*_m = \mu_0 MHh/b, \quad (1)$$

where μ_0 is the permeability of vacuum, h and b are the height and the average thickness of the sealing ring, it is obvious that the MRF has some advantage.

Second, the MRF under the magnetic field takes on the clearly defined plastic properties characterized by the yield stress $\tau_0 \sim H^2$, what the FF lacks. As a result, the sealing ring of the MRF, e.g. in the static regime, can hold the pressure P* both due to the magnetic component $P*_m$ (Eq.1) and due to the plastic component

$$P*_p = 2\,\tau_0\, h/b\,, \quad (2)$$

the resultant pressure being equal to a sum of these components

$$P* = P*_m + P*_p. \quad (3)$$

Thus, on the one-step seal with $b=10^{-4}$ m and $h=10^{-3}$ m the design pressure for the MRF-based MFS will be of the order $P*_{MRF} = 3 \times 10^5$ Pa ($\tau_0 \approx 0.5 \times 10^4$ Pa)[4] and for the FF $P*_{FF} = 10^5$ Pa, i.e. a three-fold excess of the sealing ability is seen. However, the fact should be allowed for that the growth of the plasticity and effective viscosity of the MRF when acted on by the magnetic field implies an increase in the friction torque in the dynamic regime, and hence, a loss of one of the MFS advantages.

Third, a promising use of the MRF as the working fluid, e.g. in the elements of the electric hydraulic control systems, positioning pneumatic and hydraulic drives, vibration isolation systems, etc.[5], invites evaluation of the possibility to employ it and also as the sealing agent.

2. Experimental Setup

Experiments were conducted on the setup, whose schematic is shown in Fig. 2. Rotary viscometer 1 served as the rotation drive and the friction torque meter. Working shaft 2 was mounted rigidly to the output shaft of the viscometer. The working shaft 20 mm in dia was fabricated from magnetic steel and aligned in annular magnetic circuit by bearing 8. Magnetic circuit piece 4 and shaft 2 formed working gap of the height h and the

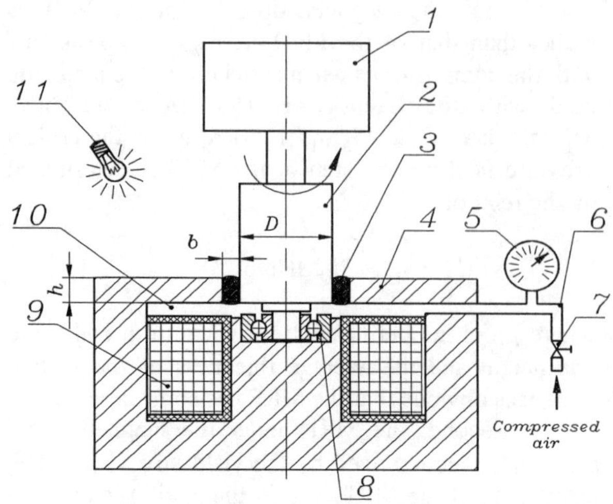

Fig. 2. Schematic of the experimental setup

thickness b filled with the MRF 3. The magnetic field was induced in the gap by a direct current passed round winding 9. Cavity 10 served as a chamber to be sealed. Compressed air, whose pressure was determined by gauge 5, was supplied for cavity 10 via pipeline 6 when valve 7 was open.

In chamber 10, the air pressure was increased progressively, and the moment of breaking the seal was determined by visualizing the fluid free surface illuminated by lamp 11. At the same time, the reading of gauge 5 fixed the critical pressure P_*, at which the seal was broken.

In experiments, the thickness and the height of the sealing ring, the shaft rotation speed n, and the winding current I were varied. The parameter range was: h = 1 to 10 mm; b = 0.075 and 0.15 mm; n = 0 to 240 r.p.m.; I = 0.25 to 2.5 A. For comparison, all experiments were duplicated with the use of ferrofluid.

3. Experimental Results and Discussion

The behavior of breaking the seal strongly depends on the type of fluid to be used. Under all conditions, the ferrofluid ring fails abruptly, the seal is ejected from the gap, and the pressure in chamber 10 (valve 7 is closed) drops dramatically. When the MRF is in the gap, the seal break is followed by the progressive accumulation of the finest air bubbles over the free surface and by the slow pressure variation.

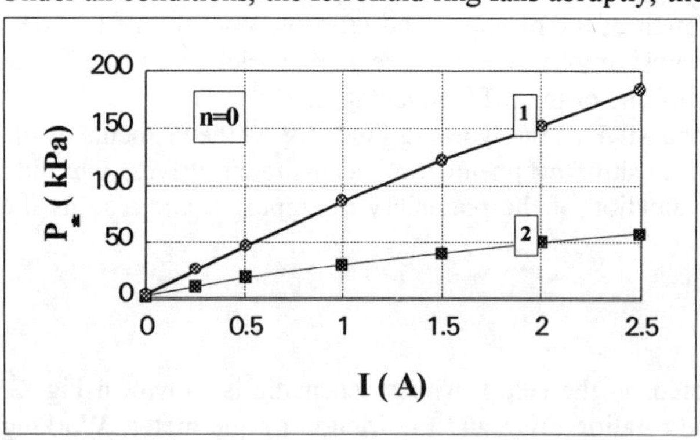

Fig. 3. Static characteristic of the MFS: 1 = MRF; 2 = FF.

In the static regime (n=0), the pressures held by the MRF-based MFS are far in excess of those held by the FF-based one. As seen from Fig. 3, for I=2.5 A an approximately three-fold pressure excess is seen, which is consistent with the above-mentioned estimates of the pressure.

In the dynamic regime, as the shaft rotation speed grows the sealing ability of the FF progressively reduces (Fig. 4, curve 5). For the MRF, as n increases , the critical pressure unexpectedly much grows, reaching its maximum (at which a seven-fold excess of $P*_{MRF}$ over $P*_{FF}$ is seen), and then it decreases, remaining nevertheless greater than the static pressure, at which the seal is broken for the winding magnetizing current I> 1A. The shaft rotation speed, at which the maximum pressure $P*_{MRF}$ is observed, has a tendency to increase as I grows. By varying the ring thickness b and, hence, the MRF deformation rate

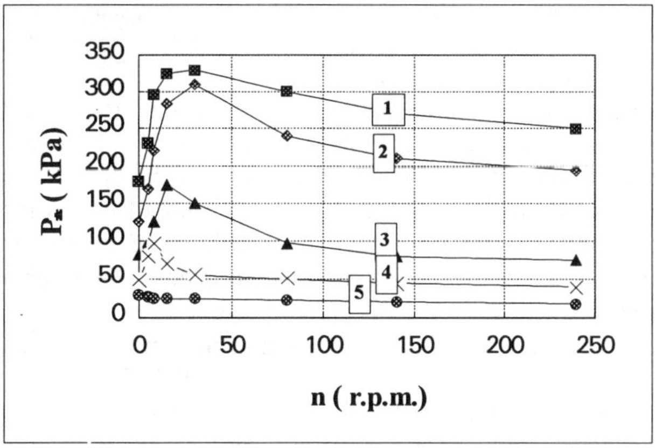

Fig. 4. Effect of the shaft rotation speed on the critical pressure. For the MRF: 1, I=2.5A; 2, 1.5A; 3, 1.0A; 4, 0.5A. For the FF: 5, I=1A.

$$\gamma = \frac{\pi n(D/2+b)^2}{15b(D+b)} \tag{4}$$

the ordinate of a maximum on the dependence $P*_{MRF} = f(n)$ is shifted, i.e. by decreasing the gap size the ordinate of a maximum is shifted down in the rotation speed.

To our mind qualitatively, the revealed specific features of the MFS can be explained as follows. Unlike the FF, the MRF is a significantly heterogeneous structurizing medium. The leak-proofness of the sealing MRF ring in the static regime is disturbed by the finest air cores which "penetrate" through the structural skeleton (formed by the anisotropic aggregates composed of interacting particles of the ferromagnetic dispersed phase of fluid) and primarily at the places of its defects. Low shear stresses originating at small n exert some random influence on the microstructure, thus increasing the averaged number of parallel connections between individual aggregates, decreasing the size and the number of defects, breaking and disintegrating the air cores. Owing to this, the critical pressure, at which the seal is broken, grows. For a value of γ exceeding some optimal one (corresponding to the ordinate of a maximum on the $P*_{MRF} = f(n)$ dependence) under the assigned conditions, the critical pressure begins to drop because of an intense progressive destruction of the aggregates and the microstructure as a whole when acted on by the shear fluid flow. In relatively small fields (e.g., for I<1A) and for sufficiently high values of γ, the continuous microstructure vanishes, and the favorable influence of the shear on the sealing ability of the MRF degenerates. The growth of the winding magnetizing current, that elevates the field intensity in the gap and the axial gradient, favors the structure sealing and the aggregate reinforcement, which is shifted up in the values of γ and in the values of n corresponding to them, at which $P*_{MRF}$ is maximum.

Experiments have revealed that from the standpoint of the highest critical pressure the optimal relation exists between the height and the thickness of the sealing ring. From Fig. 5 it is seen that $P*_{MRF}$ has a maximum for h/b=27 (h=2mm; b=0.075 mm). At a defined value of I, decreasing h and, correspondingly, the contact area of the magnetoconducting wall and the sealing ring causes the enhancement of the magnetic field intensity in the working gap, and the $P*_m$ increase, on the one hand, and the $P*_p$ decrease, on the other. As in the total value of the critical pressure the magnetic component is dominant, decreasing h causes the $P*_{MRF}$ growth. This tendency is preserved unless the magnetoconducting wall transforms to the saturation state, at which there exists the optimal value of h/b. Since a further decrease of the sealing ring thickness does not amplify the magnetic field, $P*_{MRF}$ begins to drop.

The fact that the curves for the critical pressure as a function of winding current (Fig. 6) reach the plateau is also associated with the saturation effect. This appreciably shows itself when the height of the sealing ring (curves 1 and 4) is the smallest. As already mentioned, use of the MRF in the MFS results in a substantial growth of the friction torque (Fig. 7) which quadratically depends on the value of the magnetizing current.

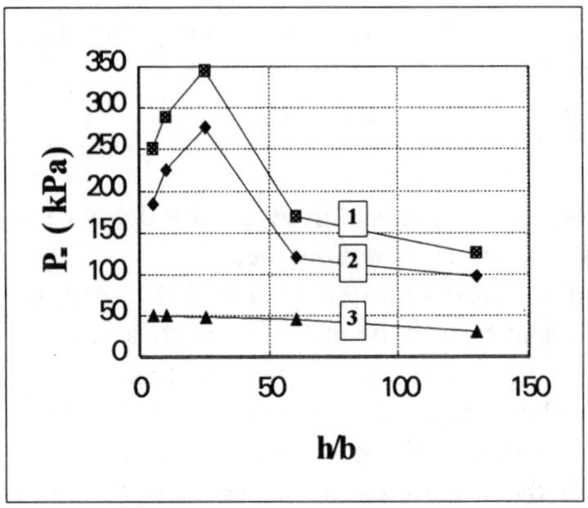

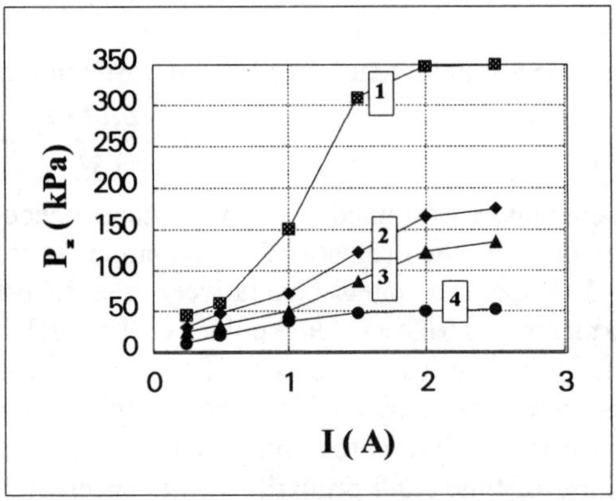

Fig. 5. Critical pressure vs. geometrical sizes of the sealing ring. I=1A. For the MRF: 1, n=27r.p.m.; 2, 81r.p.m. For the FF: 3, n=27r.p.m.

Fig. 6. Effect of the value of the winding magnetizing current on the critical MFS pressure. n=27r.p.m.; b=0.075mm. For the MRF: 1, h=2mm; 2, 4.5mm; 3, 10mm. For the FF: 4, h=2mm.

4. Conclusions

1. It is quite possible to use the MRF as the working medium of the MFS. The efficiency of such a seal increases at small rotation speeds of the shaft to be sealed or in the static regime when the friction torque is absent or is not significant. Under such operating

conditions, the advantages of the MRF are substantial over that of the FF because, under all other things being equal, the critical pressure, at which the sealing MRF ring is broken, has a three-fold pressure excess in the static regime and a seven-fold one in the dynamic regime when the rotation speed is optimum.

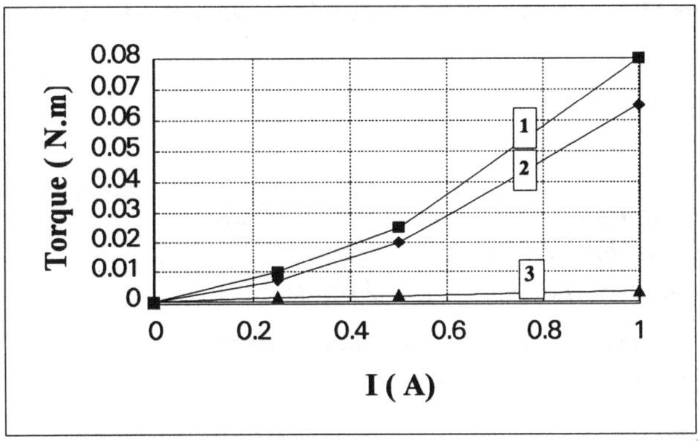

Fig. 7. Torque in the MFS. h=4.5mm; b=0.075mm. For the MRF: 1, n=81r.p.m.; 2, 27r.p.m. For the FF: 3, n=81r.p.m.

2. The specific features of the functioning of the MRF-based MFS consisting in elevating the critical pressure in the dynamic regime, as compared to the static one, and the presence of a maximum on the dependence $P*_{MRF} = f(n)$ are revealed.

3. The considered MRF- based MFS passed successful tests as a seal of the bearing unit of the shaft of the laboratory mill used for preparing the MR fluid.

5. References.

1. R. E. Rosensweig, G. Miskolczy and F. D. Ezekiel, *Machine design*, **40** (1968) 145.
2. Yu. O. Mikhalev, *Mechanization and Automation of Production*, **4** (1990) 21 (in Russian).
3. R. L. Baley, B. A. Hands and L. M. Vokins, in *Fluid seal* (Cranfield, 1975), p. A-5/85.
4. W. I. Kordonsky et al., *J. Mag. and Mag. Mat.*, **85** (1990) 114
5. W. I. Kordonsky, *J. Mag. and Mag. Mat.*, **122** (1993) 395

VARIATION WITH DISTANCE OF THE ATTRACTION FORCE BETWEEN SPHERES AND ESTIMATION OF STATIC YIELD STRESS OF ER FLUIDS

P. ATTEN†, K.Q. ZHU*, C. BOISSY† and J.-N. FOULC†

† *Laboratoire d'Electrostatique et de Matériaux Diélectriques*
CNRS et Université Joseph Fourier - Grenoble 1
B.P. 166 – 38042 Grenoble Cedex 9 (France)

* *Department of Engineering Mechanics, Tsinghua University,*
Beijing 100084 (P.R. China)

ABSTRACT

In order to estimate the static yield stress of a suspension of monodispersed spheres subjected to a DC applied field, the knowledge of the interaction force between spherical particles is required. This question of variation of the attraction force between spheres as a function of their distance s is investigated on the basis of the conduction model in the particular case of saturation régime of bulk electrical conduction of the liquid. The prediction is tested using a scaled-up two-sphere experimental model for the particles interaction. Then a rough estimate of the static yield stress is derived using a linear approximation of the force dependence on the spacing s between spheres.

1. Introduction

The particular rheological properties of ER fluids induced by an applied electric field result from the electrostatic forces between particles suspended in a carrying liquid and the subsequent structure formation. Up to now the attempts to predict the yield stress[1-4] for example, were based on the permittivity model in which the interaction between particles is ascribed to the mismatch of permittivities, ε_S (solid) and ε_L (liquid). This model only applies in the case of AC applied fields of high enough frequency. For DC and low frequency AC fields, Anderson[5] and Davis[6] showed that conduction effects play the major role in determining the field and surface charge distributions. A conduction model[7] was proposed recently to estimate the attraction force between two very slightly conducting spheres as a function of the applied field E_0 and of the ratio $\Gamma = \sigma_S/\sigma_L$ of solid (σ_S) and liquid (σ_L) conductivities. But there is no prediction of the yield stress in this case of DC fields.

Here is a first attempt to estimate the yield stress under DC applied fields. The starting point is a close examination[8] of the passage of current through the fibres of an ER fluid leading to refined expressions for the attraction force between two spheres. We focus first on the variation of this force as a function of the distance between the spheres. The conduction model is extended to this situation of a small gap between spheres in the particular case of the current in the liquid phase being limited by the dissociation of the electrolytic species. Then the predictions are tested on a scaled-up two-sphere model. Finally some very rough estimates of the yield stress are derived for a suspension of monodispersed spherical particles.

2. Attraction force between spheres at a distance s

We consider a monodisperse suspension of spherical particles. When the spheres are at a high enough distance s so that the electrical current plays but a negligible role, the attraction force is easily calculated because the spheres can be considered as equipotential. For small values of s, conversely, the passage of current plays a fundamental role and determines the force. In this case it is necessary to make extra assumptions concerning the conduction processes in the liquid. In another paper[8] we point out that for liquids of very low conductivity, the conduction current is "saturated", i.e. it involves only the ions just created by dissociation in the liquid bulk, the dissociation rate being field dependent. It is this process which is retained for the present development and we assume that there is no injection of ions at the solid/liquid interfaces[9] (the occurrence of such an injection would lead to increased currents and reduced attraction forces).

2.1 Moderate and high distances

In a dilute suspension, the electric potential of spherical particles to a first approximation is equal to the potential of their centres. For distances s greater than about the radius R, the interaction between two spheres can be described through the equivalent dipoles induced by the field $\mathbf{E}_0$. This gives for the radial component F of the force[10] :

$$F = 12\pi R^2 \varepsilon_L E_0^{\,2} \left(\frac{R}{2R+s}\right)^4 [3\cos^2\theta - 1] \tag{1}$$

ε_L being the liquid permittivity (for the other notations see Fig. 1). For $s < R$, relation 1 is no longer valid and another expression is obtained[10] by considering the field between the two spheres whose potential difference is $U = (2R+s)E_0\cos\theta$:

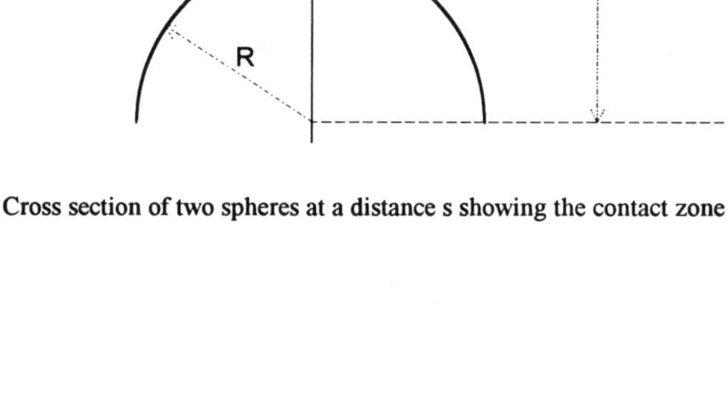

Fig. 1. Cross section of two spheres at a distance s showing the contact zone of radius δ.

$$F = 2\pi R^2 \varepsilon_L E_o^2 \cos^2\theta \left(1+\frac{s}{2R}\right)^2 \frac{R}{s} \qquad (2)$$

2.2 Small distances s

For small values of s, the field between the spheres takes high values and the force is calculated as in the conduction model by considering two zones in the region between the spheres : The outer zone for $x > \delta$ (see Fig.1) where the surfaces of the spheres are assumed to be equipotential and the contact zone ($x \leq \delta$) where the ohmic potential drop in the solid phase limits the field in the liquid gap[7]. The radius δ of the contact zone is determined through the condition of equal conductances of the solid and the liquid ($x > \delta$). For the conductance of solid spheres we have[7] $C_S = \alpha\sigma_S\delta$ (α : constant ~ 1; σ_S : conductivity of solid phase). We estimate the conductance of the liquid from the current flowing in the outer zone. In the present case of the current in the liquid phase being limited by the generation of ions in the bulk, the saturation current density j_{sat} takes the general expression[8,9] :

$$j_{sat} = \frac{\sigma_L^2 y}{2K\varepsilon_L} f\left(\frac{E}{E_c}\right) \qquad (3)$$

where σ_L is the liquid conductivity, y the local liquid thickness, K the ionic mobility, $f(E/E_c)$ the Onsager expression for the field enhancement of dissociation and E_c a characteristic field[8,9]. The elementary liquid conductance dC_L between cylinders of radii x and x+dx is :

$$dC_L = \frac{di}{U} = 2\pi x\, dx \frac{j_{sat}}{U} = \pi \frac{\sigma_L^2 y}{K\varepsilon_L U} f\left(\frac{E}{E_c}\right) x\, dx \qquad (4)$$

U being the difference of potential between the spheres centres. Approximating the field E by $E = U/y$ and taking for $f(E/E_c)$ the following asymptotic relation[8] (valid for high values of E/E_c) :

$$f\left(\frac{E}{E_c}\right) \cong \sqrt{\frac{2}{\pi}} \left(\frac{E_c}{E}\right)^{3/4} \exp\left(\sqrt{\frac{E}{E_c}}\right) \qquad (5)$$

we obtain :

$$dC_L \cong \sqrt{2\pi} \frac{\sigma_L^2}{K\varepsilon_L U} \left(\frac{E_c}{U}\right)^{3/4} \exp\left(\sqrt{\frac{U}{y E_c}}\right) y^{7/4}\, x\, dx \qquad (6)$$

where $y = s+x^2/R$ is the distance between spheres surfaces at radius x and $U = (2R+s)E_o \cos\theta$. Integrating after a change of variables and a further approximation[7,8] leads to :

$$C_L \cong \frac{\sqrt{2\pi}\,\sigma_L^2}{K\varepsilon_L E_c} \beta^{-9/2} \left(1+\frac{s}{2R}\right)^{-7/4} \frac{\delta^{13/2}}{R^{9/2}} \left(1+\frac{sR}{\delta^2}\right)^{13/4} \exp\left[\beta \frac{R}{\delta}\sqrt{\frac{1+s/2R}{1+Rs/\delta^2}}\right] \qquad (7)$$

with the notation $\beta = [2(E_o/E_c)\cos\theta]^{1/2}$. From the balance between the conductances[7,8] C_L and C_S we obtain the following implicit expression for R/δ :

$$\frac{R}{\delta} \cong \sqrt{\frac{E_c}{2E_o\cos\theta}}\sqrt{\frac{1+Rs/\delta^2}{1+s/2R}}\ \ln\left[\frac{\alpha\Gamma^2}{\sqrt{2\pi}}\frac{\varepsilon_L KE_c}{\sigma_S R}\frac{(1+s/2R)^{9/4}}{\left(1+Rs/\delta^2\right)^{13/4}}\beta^{9/2}\left(\frac{R}{\delta}\right)^{11/2}\right] \tag{8}$$

where $\Gamma = \sigma_S/\sigma_L$. By expanding 8 to first order in s/R in the case $Rs/\delta^2 \ll 1$ and defining the coefficient a from : $\frac{R}{\delta} \approx \frac{R}{\delta_o}\left[1+a\frac{s}{R}\right]$ we deduce (δ_o : value at s = 0) :

$$a = \frac{\frac{1}{2}\left(\frac{R}{\delta_o}\right)^2 - \frac{13}{4\beta}\frac{R}{\delta_o} - \frac{1}{4} + \frac{9}{8\beta}\frac{\delta_o}{R}}{1 - \frac{11}{2\beta}\frac{\delta_o}{R}} \tag{9}$$

For conditions such that $E_o \geq 1$ kV/mm and $R/\delta_o > 3$ we have $a > 0$ which corresponds to a decrease of δ versus s. This is qualitatively consistent with the fact that the "contact" disappears when s increases.

The attraction force can be estimated by making the same assumptions and approximations as usual[7], i.e. taking $E(x) = U/(s+x^2/R)$ in the outer zone and $E = Cst = E(\delta)$ in the contact zone and assuming $\delta^2 \ll R^2$. After some calculations we obtain :

$$F = 2\pi R^2\,\varepsilon_L E_o^2\,\cos^2\theta\left(\frac{R}{\delta}\right)^2\frac{\left(1+\frac{s}{2R}\right)^2}{1+\frac{Rs}{\delta^2}}\left(1 + \frac{1}{1+\frac{Rs}{\delta^2}}\right) \tag{10}$$

Defining the coefficient b from :

$$F \approx F_o\left[1-b\frac{s}{R}\right] \tag{11}$$

and expanding to first order in s/R finally leads to ($\beta = [2(E_o/E_c)\cos\theta]^{1/2}$) :

$$b = \frac{\frac{1}{2}\left(\frac{R}{\delta_o}\right)^2 - \frac{7}{4\beta}\frac{R}{\delta_o} - \frac{1}{2} + \frac{13}{4\beta}\frac{\delta_o}{R}}{1 - \frac{11}{2\beta}\frac{\delta_o}{R}} \tag{12}$$

For $E_o \geq 1$ kV/mm and R/δ high enough the coefficient b is positive and on the order of $R^2/2\delta^2$; this gives b ranging typically from 10 to 100.

3. Experimental study

In order to study the variation of the force with distance between spheres, we slightly modified the set-up consisting of two half-spheres of polyamid ($\sigma_S \cong 1.7\times10^{-8}$ S/m, R = 7mm) immersed in a liquid (the set-up is described with some detail in references 7 and 8). The upper half-sphere was fixed rigidly to the weighing pan of the electronic scales (Sartorius BA310S) which has no detectable displacement under the action of the force and the lower half-sphere was positioned using a micrometric translation table. The attraction force F and the current through the spheres I_S were measured for mineral oil Elf TF50 as the suspending liquid ($\sigma_L \cong 1.2\times10^{-12}$ S/m giving $\Gamma \cong 14000$).

As expected the measured force exhibits a decrease with distance s between spheres (see Figs. 2 and 3). Fig. 2 shows that above a gap of about 100 μm (s/R from 0.01 to 0.02) there is a well characterised decrease : $F \propto s^{-1}$. We can compare with the prediction given by relation 2 in case of the current playing a negligible role. Indeed the measured current decreases rapidly with s and for s > 100 μm it takes values which are more than one order of magnitude lower than for s = 0. Fig. 2 shows an excellent agreement between measured and predicted force values.

For very small spacings ($s/R \leq 10^{-2}$) such that the current induced potential drops in the spheres cannot be ignored (contact zone), the relevant formula is relation 10 where R/δ is given by 8. We have drawn on Fig. 3 the variations given by the first order expansion 11 where b is calculated using 12 (asymptotic behaviour for very small values of s/R). We see that the agreement is surprisingly good if we remember the various

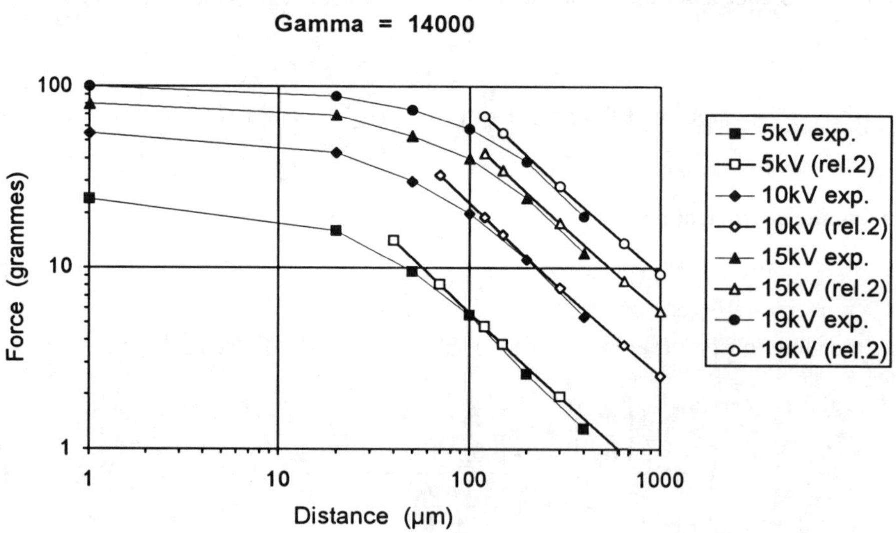

Fig. 2. Dependence on distance of the attraction force between two half-spheres immersed in mineral oil of weak conductivity for different applied voltages (points plotted at s = 1 μm are for spheres in contact). The predictions given by relation 2 are drawn for s ≥ 50 μm.

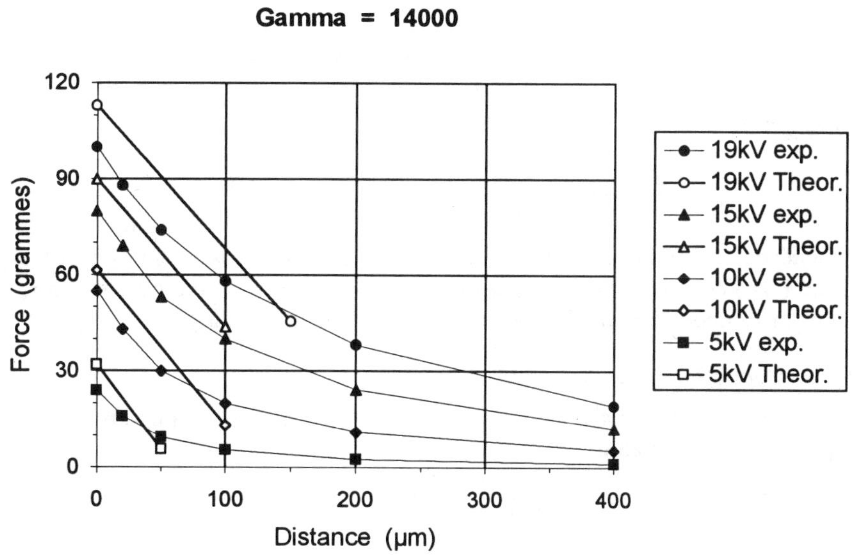

Fig. 3. Dependence on distance of the attraction force (linear scale) for different applied voltages. Comparison of measured values and of predictions given by the first order expansion 10.

assumptions made and the further approximations in the derivation of formulae 7 to 12. Such a fairly good agreement has been obtained[8] for the force F_0. We have here the result that the slope of F(s) at s = 0 is also fairly well predicted by the refined conduction model.

4. Estimation of static yield stress.

4.1. General considerations

The yield stress has been previously estimated by taking the dipolar expression for the attraction force between two spherical particles[1,3,4]. We aim here at giving other expressions of the yield stress of ER fluids subjected to DC fields, based on the conduction model. As in the previous derivations we consider a suspension of monodispersed spheres stressed by a DC electric field and initially at rest such that the chains of spheres are aligned along the field $\mathbf{E}_0$. Imposing a strain, the chains make an angle θ with the field and this gives rise to a shear stress τ :

$$\tau = N F \sin\theta \tag{13}$$

where $N = 3\phi/(2\pi R^2)$ is the number of chains per unit surface (ϕ being the volume fraction) and F the attraction force between spheres. The chains are slightly elongated and a spacing s appears at least between some of the spheres (we assume that the number of spherical particles per chain remains constant). For small values of θ and s/R, from relations 10 and 11 we deduce :

$$\tau = 6\phi\, \varepsilon_L E_0^2 \left(\frac{R}{\delta_0}\right)^2 \left(1 - b\frac{s}{R}\right) \cos^2\theta \sin\theta \tag{14}$$

or, by using 8 for s = 0 where the logarithmic term is denoted by A :

$$\tau = 3\phi\, A^2 \varepsilon_L E_0 E_c \left(1 - b\frac{s}{R}\right) \cos\theta \sin\theta \tag{15}$$

The dependence of A on θ is very weak and we neglect it; we also consider that the coefficient b is a constant. Two asymptotic expressions for the static yield stress τ_s can be derived from 15 depending on the postulated way of chain breaking.

4.2. Equal spacings between particles

Considering as in previous works[1,3,4] that the chains elongate very regularly, the equal spacings take the value :

$$\frac{s}{R} = 2\,\frac{1 - \cos\theta}{\cos\theta} \tag{16}$$

The shear stress then writes :

$$\tau = B\left[\cos\theta \sin\theta - 2b\,(1 - \cos\theta)\sin\theta\right] \tag{17}$$

with $B = 3A^2\phi\, \varepsilon_L E_0 E_c$. The static yield stress τ_s corresponds to the root of :

$$\frac{d\tau}{d\theta} = (1 + 2b)\,(2\cos^2\theta - 1) - 2b\cos\theta = 0 \tag{18}$$

It is easily seen that for b » 1 the angle θ_m at which τ is maximum takes small values. Expanding $\sin\theta_m$ and $\cos\theta_m$ finally leads to :

$$\theta_m = \frac{1}{\sqrt{3b + 2}} \tag{19}$$

which gives for the static yield stress :

$$\tau_s = 3A^2\phi\, \varepsilon_L E_0 E_c\, \frac{2 + 2/(3b)}{3 + 2/b}\, \frac{1}{\sqrt{3b + 2}} \tag{20}$$

The asymptotic expression in the case b » 1 is :

$$\tau_s \approx 2A^2\phi\, \varepsilon_L E_0 E_c \frac{1}{\sqrt{3b}} \approx 4\phi\, \varepsilon_L E_0^2 \left(\frac{R}{\delta_0}\right)^2 \frac{1}{\sqrt{3b}} \tag{21}$$

Taking the typical values $3b \cong 100$; $\varepsilon_L \cong 2\times10^{-11}$ F/m, $E_0 \cong 1$ kV/mm, $R/\delta_0 \cong 10$ and $\phi \cong 0.3$, we obtain $\tau_s \cong 240$ Pa and $\theta_m \cong 0.1$ rd.

Relation 21 gives a plausible order of magnitude of τ_s although the value appears to be lower than many of those found experimentally. Let us first remark that for volume fractions of solid phase greater than about 0.1, the chains interact and tend to gather and form clusters. Then the particles have more than the 2 contacts with adjacent particles existing in isolated chains and from the study of Gulley and Tao[4] it is clear that the shear stress is greater than the one calculated from the contribution of non interacting chains (the derivation based on straight chains as developed above is expected to give a lower bound). The assumption of equal spacings between the spheres of every chain also appears as a drastic simplification. We can nevertheless remark that the clusters subjected to a strain exhibit a certain plasticity, i.e. the elongation of chains involves spacings of numerous pairs of particles because of the multiple interactions. In this case it is not unreasonable to assume equal spacings between the spheres of a chain and this should lead to a plausible order of magnitude for τ_s.

The functional dependence of yield stress can be obtained from the expression of b as given by 12. For high enough applied DC fields, b is on the order of $R^2/(2\delta_o^2)$. This gives :

$$\tau_s \sim A\phi\, \varepsilon_L E_c^{1/2} E_o^{3/2} \tag{22}$$

This dependence of yield stress on applied field has been predicted by Davis[11]. There is no clear experimental evidence for such a variation of the yield stress on E_o.

4.3. Single spacing in the chains

For very small volume fractions, we can picture the structured fluid as a set of chains which are isolated and do not interact. The assumption of a slightly elongated chain with equal spacings between the spheres is no longer tenable because such a configuration is highly unstable. In this case, taking for granted that the chains are anchored on the electrodes, the elongation of the chains very likely occurs through the opening of only one spacing. Consistently with the Peierls-Landau instability, it is assumed that the chains under strain have one gap s only and break down when the strain reaches a critical value[4]. The relation between s and θ is :then :

$$\frac{s}{R} = \frac{d}{R}\,\frac{1 - \cos\theta}{\cos\theta} \tag{23}$$

where d is the distance between electrodes. The derivation of τ_s is identical to that of §4.2 when simply replacing 2b by bd/R. Thus the asymptotic expression is :

$$\tau_s \approx 2A^2\phi\, \varepsilon_L E_o E_c \sqrt{\frac{2R}{3bd}} \approx 4\phi\, \varepsilon_L E_o^2 \left(\frac{R}{\delta_o}\right)^2 \sqrt{\frac{2R}{3bd}} \tag{24}$$

which again gives a $E_o^{3/2}$ dependence for τ_s. The interesting point in relation 24 is the dependence of the static yield stress on the size of the particles. The $R^{1/2}$ variation law obtained here appears to agree qualitatively with the observations of Conrad et al.[12]. Unfortunately, there is no quantitative agreement : taking the same values as in section 4.2 and $R/d \cong 10^{-2}$ we obtain $\tau_s \sim 35$ Pa (extrapolation for $\phi = 0.3$), a value much lower than the observed yield stresses.

5. Conclusion

The conduction model for calculating the attraction force between two spheres in a liquid subjected to a DC electric field has been extended to the case of a small separation s between spheres in the saturation régime of current in the liquid phase. The estimate of the slope dF/ds at s = 0 is confirmed by a scaled-up two-sphere experiment for high values of the conductivity parameter Γ. This approach could be generalised to other régimes of conduction including the injection processes.

Investigating the question of static yield stress τ_s of an ER fluid under DC applied field, we derive new general expressions in terms of the rate of variation of the force with the distance s between spherical particles. For a structuration of an ER fluid in the form of isolated chains, a dependence on the size of the particles is obtained for the yield stress.

Acknowledgements

This work was performed with the joint financial support of the Centre National de la Recherche Scientifique and the company Elf-Aquitaine (under contract CNRS/SNEA No 50.8016). The authors are grateful to the company Elf-Atochem for providing samples of dielectric liquids.

References

1. H. Conrad, Y. Chen and A. F. Sprecher, *Proceed. 3rd Int. Conf. ER fluids*, ed. R. Tao (World Scientific, Singapore, 1992) 195.
2. F. E. Filisko, *Proceed. 3rd Int. Conf. ER fluids*, ed. R. Tao (World Scientific, Singapore, 1992) 116.
3. E. Lemaire, G. Bossis and Y. Grasselli, *Langmuir* **8** (1992) 2957.
4. G. L. Gulley and R. Tao, *Phys. Rev.* **E48** (1993) 2744.
5. R. A. Anderson, *Proceed. 3rd Int. Conf. ER fluids*, ed. R. Tao (World Scientific, Singapore, 1992) 81.
6. L. C. Davis, *J. Appl. Phys.*, **72** (1992) 1334.
7. P. Atten, J.-N. Foulc and N. Félici, *Intern. J. Mod. Phys. B*, **8**, (1994) 2731.
8. C. Boissy, P. Atten and J.-N. Foulc, "The conduction model of electrorheological effect revisited", *communication to this Conference*.
9. P. Atten, J.-N. Foulc and H. Benqassmi, in "*Progress in Electrorheology*", Plenum, 1995 (in press).
10. P. Atten, *J. Electrostatics* **30** (1993) 259.
11. L. C. Davis and J. M. Ginder, *Polymer Preprints* **35** (1994) 321.
12. Y.-H. Shih and H. Conrad, *Intern. J. Mod. Phys. B*, **8**, (1994) 2835.

CORRELATION BETWEEN ELECTRICAL AND RHEOLOGICAL PROPERTIES OF ELECTRORHEOLOGICAL FLUIDS

J.N. FOULC, P. ATTEN and C. BOISSY

Laboratoire d'Electrostatique et de Matériaux Diélectriques
CNRS and Université Joseph Fourier de Grenoble
B.P. 166 – 38042 Grenoble Cedex 9 (France)

ABSTRACT

We present results of simultaneous measurements of shear stress and current flowing through a cellulose/mineral oil suspension subjected to shear and DC electric field. Single curves for the variation of apparent viscosity and normalised current are obtained when taking a simplified Mason number as variable. The different regimes observed are discussed in terms of the degree of structuration of the suspension reflected by the Mason number.

1. Introduction

For a long time the mechanism proposed to explain the electrorheological (ER) effect retained only the dipolar interaction due to the polarization of particles. In addition to the fact that this approach does not satisfactorily account for the order of magnitude of the ER effect[1,2] it has been suggested recently[3,4,5] that, in DC or low frequency AC fields, the conduction properties of both liquid and solid constituents play the major role in the ER effect. In this way, the conduction model[6] shows that ER activity exists only when the electrical conductivity of the solid particles is greater than that of the carrying liquid. Practically, the ratio of solid and liquid conductivities must be greater than about 100 to obtain significant ER effects.

Then the current flowing through an ER fluid at rest or under low shear rate, and subjected to electric field, essentially flows through fibres of particles. Conversely, in the case of high shear rate, the structuration of particles in fibres disappears and very likely the current is lower because there is no longer paths of rather low electric resistance. Therefore, one can expect some correlation between the current flowing through the fluid and the stress tranferred (both depending on the degree of fibration). In this case, the monitoring of current can give some information about the structuration of ER fluid. The aim of this paper is to present and discuss preliminary experimental results.

2. Experimental procedure

2.1. ER fluid

The model ER fluid we utilized was a mixture of microcristalline cellulose (typical size ~ 40 μm, water content ≅ 6 % by weight, conductivity ≈ 10^{-9} S/m) and mineral oil Elf

TF50 (dynamic viscosity $\eta \cong 16\times10^{-3}$ Pa.s at T = 23°C, permittivity $\varepsilon_L \cong 2.2\ \varepsilon_0$, conductivity $\sigma_L \approx 10^{-12}$ S/m). Before using it, the ER fluid was mixed strongly during at least 5 minutes. The volume fraction ϕ of particles was 0.08. Such a rather small ϕ value was chosen in order to have a well marked influence of structuration on the current.

2.2. Rheological and electrical measurements

We used a modified Couette-type-controlled-rate-rheometer (Mettler 115A). The diameter and height of the bob were respectively 25 mm and 37.5 mm. The electrode gap was 1 mm. The outer cylinder was electrically insulated up to 5 kV and the inner cylinder (bob in rotation) was grounded using a special arrangement with a pin-conducting liquid contact (copper sulphate in solution).

DC high voltage was applied from a 5 kV Mesco power supply and the current was monitored on an oscilloscope and on a X(t) recorder. A laboratory built apparatus using an optical coupling was used to measure the current (range 1nA to 1mA in two scales) directly on the high voltage electrode. This arrangement allows us to ground the whole device (rheometer, test cell inner cylinder) and then to significantly reduce the noise/signal ratio.

2.3. Experimental procedure

The shear stress τ and the current I flowing through the ER fluid were measured as functions of the shear rate $\dot{\gamma}$, for different values of the electric field E. During the experiment the temperature was maintained constant (T = 23 ± 0.5°C). The imposed shear rate was controlled in the following way : $\dot{\gamma}$ was increased by steps of geometric variation on two successive ranges, the first one from 0.1 to 100 s^{-1} and the second one from 100 to 1000 s^{-1}. Each range consisted of 15 steps of 5 seconds in duration. Then, when $\dot{\gamma}$ reached 1000 s^{-1}, after 5 seconds, the shear rate was decreased down to 0.1 s^{-1} with the reverse programme. The current and shear stress were measured simultaneously. As the results (rheological and electrical responses) generally showed a more regular behaviour when the shear rate was decreasing, we present only data obtained for decreasing $\dot{\gamma}$.

3. Results

Figure 1 shows the variations of shear stress τ as a function of shear rate $\dot{\gamma}$ in double logarithmic scale. The plateau stress at low shear rate represents the dynamic yield stress τ_0. At high $\dot{\gamma}$ values, there appears a limiting curve of slope 1 which corresponds to the viscosity η_F of the unelectrified suspension ($\eta_F \cong 30\times10^{-3}$ Pa.s). Therefore the rheological behaviour of our ER suspension can be described to a first approximation by the classical Bingham plastic model:

$$\tau(E,\dot{\gamma}) = \tau_0(E) + \eta_F\,\dot{\gamma} = \eta_{app}\,\dot{\gamma} \qquad (1)$$

At intermediate values of $\dot{\gamma}$ we notice a decrease in shear stress for large applied electric fields (E ≥ 1 kV/mm). This fact also observed by several authors[7-9] on various ER fluids (particles of silica, zeolite, cellulose,... in mineral oil) reveals a more complex behaviour not accounted by relation (1). This real behaviour shows that the "viscosity" η_s of the suspension is not strictly a constant but depends on both the electric field and the shear rate. Relation (1) can be generalized by the following expression[7]:

$$\tau(E,\dot{\gamma}) = \tau_0(E) + \eta_s(E, \dot{\gamma})\,\dot{\gamma} = \eta_{app}(E, \dot{\gamma})\,\dot{\gamma} \qquad (2)$$

Some authors have proposed to replace (1) by (2). This is not satisfactory because (2) implies that $\eta_s(E, \dot{\gamma})$ should be negative on some interval of shear rate, which is not physically possible (formally eqn.(2) should be further generalized by considering that τ_0 also depends on $\dot{\gamma}$).

The variations of the current I as a function of $\dot{\gamma}$ are shown on Fig. 2. Below a $\dot{\gamma}$ value which increases with the field E, the current is practically constant. At greater shear rates, I decreases and for $E \leq 0.5$ kV/mm reaches a minimum value and then increases. The influence of $\dot{\gamma}$ and E on the current I can be partly understood in terms of the fluid structuration. At very small values of shear rate ($\dot{\gamma} < 10\ s^{-1}$) the cellulose powder is nearly wholly structured in quasi-steady fibres and the current passes mainly through these fibres because of the cellulose conductivity being greater than the liquid one. For moderate values of the field (~ 0.5 kV/mm) and of the shear rate (~100 s^{-1}) one can note a decrease of I presumably due to a noticeable alteration of the fluid structuration. Under large shear rates (~ 1000 s^{-1}) and small electric fields (~ 0.5 kV/mm) we observed an unexpected rise of the current which cannot be ascribed to the total destructuration of the fluid.

It is interesting to look for parameters which allow to obtain universal curves. This has been done for the apparent viscosity[10,11] by taking the Mason number Mn as the non dimensional variable. The choice of this variable was justified by Marshall et al.[10] who analysed the various forces governing the interaction of particles in ER suspensions. They concluded that for usual ER fluids (particle size ~ 10 μm), only viscous and electrical forces act significantly on particles. They further introduced the Mason number Mn defined as the ratio of viscous and polarisation forces : $Mn = \eta\dot{\gamma}/[2\varepsilon_L(\beta E)^2]$, ε_L being the liquid permittivity and β (≤ 1) depending on ε_L and ε_S (ε_S : permittivity of particles), and showed that when plotting all values of apparent viscosity of the suspension as a function of the Mason number, a single curve is obtained[10]. In the following, we retain a simplified Mason number defined by :

$$Mn = \eta\dot{\gamma}/(\varepsilon_L E^2) \qquad (3)$$

This number can also be viewed as the ratio of two times, $\dot{\gamma}^{-1}$ and the electroviscous time[12] $\eta/(\varepsilon_L E^2)$. The variations of apparent viscosity $\eta_{app} = \tau/\dot{\gamma}$ as a function of Mn to a first approximation define a single curve (Fig. 3). This is very similar to previous observations[10,11] and the single curve can be described by the following relation derived from the Bingham model[10] :

$$\eta_{app} = K/Mn + \eta_F \qquad (4)$$

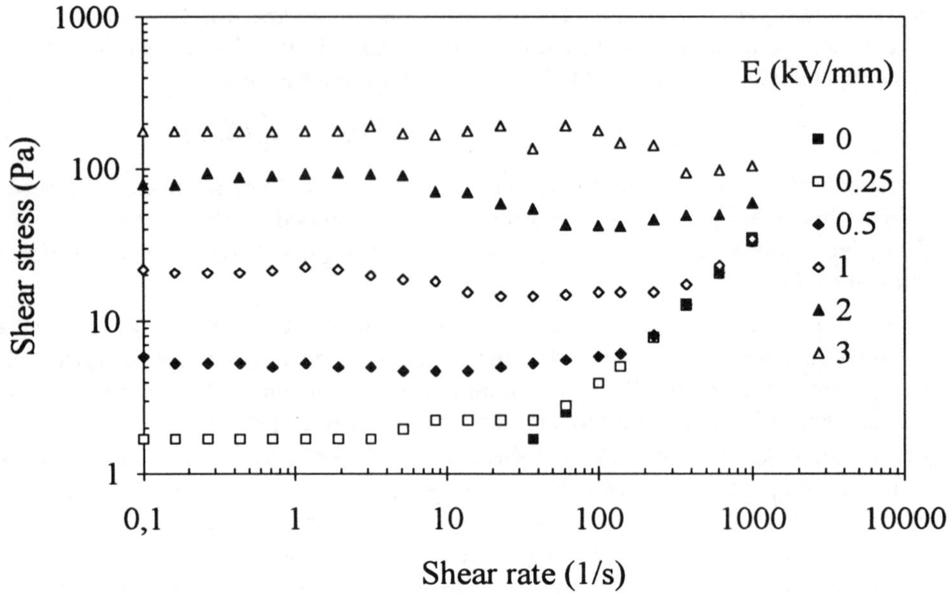

Fig. 1. Shear stress versus shear rate for a cellulose/mineral oil suspension at 0.08 volume fraction (as a model ER fluid) for different values of DC applied electric field.

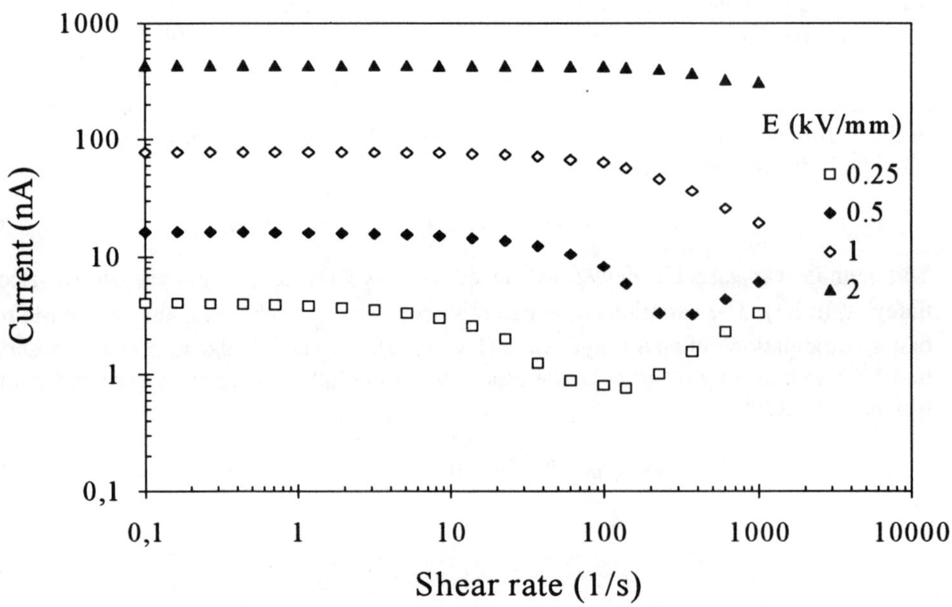

Fig. 2. Dependence of the current I flowing through the ER fluid on the shear rate $\dot{\gamma}$ for different values of the applied electric field E.

where K is a constant depending on the physical properties of materials (we see in Fig. 3 that the η_{app}(Mn) curve decreases with a slope -1 for low values of Mn and tends to the constant value η_F for high Mn). We must nevertheless point out the noticeable dispersion of data for Mn ranging between 10^{-3} and 1 with a seemingly systematic decrease of η_{app} as the electric field is increased. This dispersion arises from the non Bingham behaviour of the suspension at high enough applied electric field obtained (and discussed) on Fig. 1. Such a complex behaviour of ER suspensions for the intermediate values of Mn was previously noted by Klingenberg et al.[11]. However their explanation in terms of an extra stress does not satisfactorily account for our observations.

An interesting new result is that another single curve is obtained when we plot the normalised current I_n as a function of Mn (the normalised current is the ratio of the currents flowing through the suspension in shear ($\dot{\gamma} \neq 0$) and no shear ($\dot{\gamma} = 0$) conditions respectively). Fig. 4 shows that there are three regimes of variation of I_n i) for very low Mn values the current is practically constant ii) for small to moderate Mn values ($0.01 < Mn < 1$) I_n progressively decreases, which reveals a degradation in fluid structuration iii) for high Mn values (>1) the current increases, a behaviour which might be due to extra phenomena. We observe on Fig. 4 that the dispersion of data in the intermediate range of Mason number is much smaller than for η_{app}. The first conclusion that arises from Figs. 3 and 4 is that two universal curves are obtained by using the same dimensionless parameter and this implies the existence of some correlation between rheological and electrical properties.

4. Discussion

The Mason number is usually defined as the ratio of viscous and electrical forces and it appears as the relevant parameter to account for the variations of apparent viscosity (shear mode). In some sense it gives a qualitative measure of the degree of destructuration of the suspension subjected to both the electric field (which tends to fibrillate the fluid) and shear stress (which tends to break-up fibres). This can be viewed also by comparing the typical times τ_v and τ_e where $\tau_v \sim \dot{\gamma}^{-1}$ is the time required to have a relative displacement of two neighbouring particles on the order of their diameter (in absence of field) and $\tau_e = \eta/(\varepsilon_L E^2)$ gives the order of magnitude of the time required for the particles to come into contact (in case of no shear). Thus the Mason number can be viewed qualitatively as giving a measure of the mean relative time of absence of contact between particles in the suspension (at a given volume fraction).

With this picture we can propose a qualitative explanation for the variations of the normalised current as Mn increases. For Mn values lower than about 10^{-3}, there is a strong structuration of the fluid and we can consider that fibres which are the "natural" paths for current flow exist almost always and therefore there is no significant influence of Mn on I. As Mn increases ($10^{-3} < Mn < 1$), the mean time of existence of fibres decreases and this is reflected in a decreases of I.

For Mn > 1 (this value of 1 is fortuitous) very likely fibres no longer exist (time of existence $\cong 0$) and the viscous forces which are dominant produce a strong mixing and a

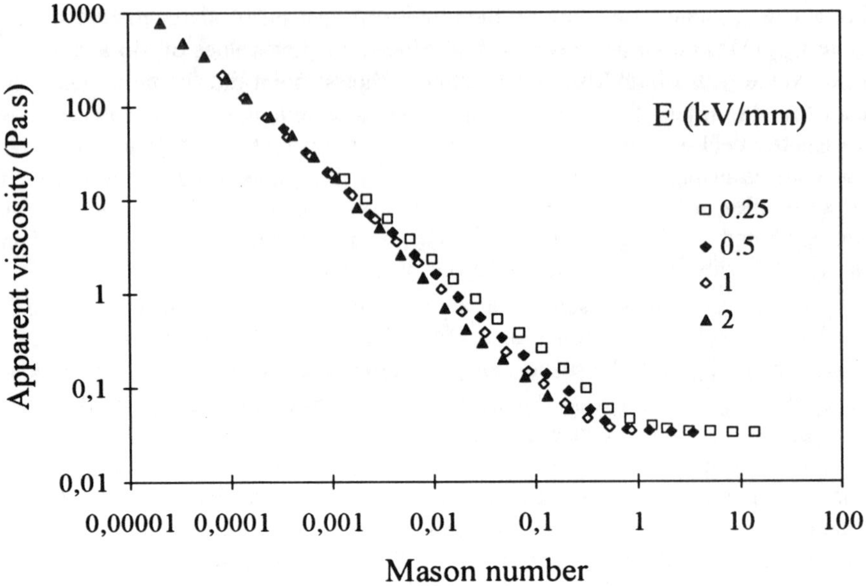

Fig. 3. The apparent viscosity (ratio of shear stress and shear rate) as a function of the Mason number $Mn = \eta\dot{\gamma}/\varepsilon_L E^2$ for different field values.

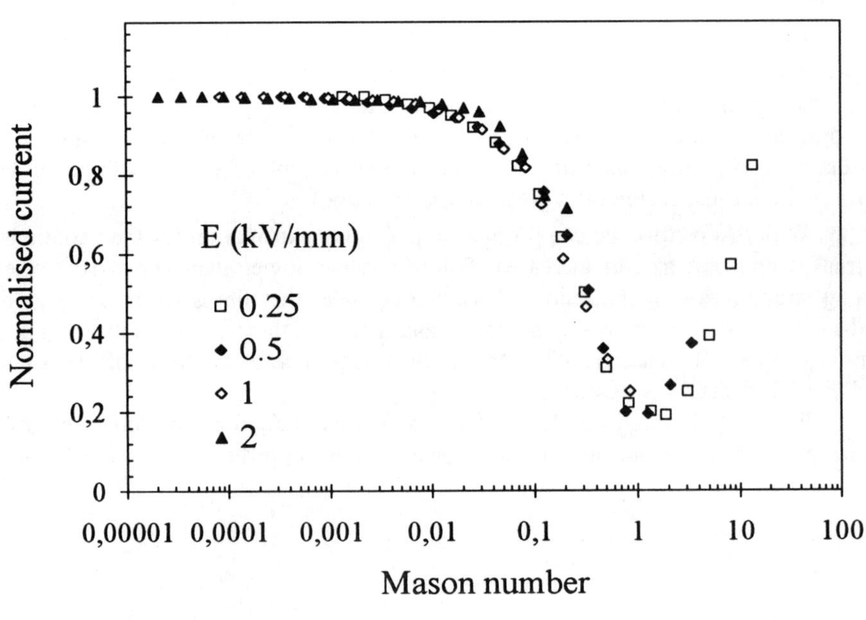

Fig. 4. The normalised current I_n as a function of the Mason number $Mn = \eta\dot{\gamma}/\varepsilon_L E^2$.

radom distribution of the particles quite analogous to the one prevailing in absence of applied field. This picture is supported by the fact that $\eta_{app} \cong \eta_F$ (Fig. 3) for Mn $\geq$ 1. In this case we would expect a constant current independent of Mn. The observed increase of I for Mn above 1 is surprising and there is no obvious mechanism which might explain this regime. A clear indication of the correlation between rheological and electrical responses of the studied ER fluid is given by the fact that the drastic changes in variations of apparent viscosity and normalised current both occur at the same Mn value ($\approx$ 1).

Another point which deserves to be discussed is relative to the dispersion of the η_{app} data for Mn ranging from a few 10^{-3} to a few 10^{-1}. The dispersion is noticeable and there appears a dependence on the applied field. A possible explanation is that the Mason number Mn is not really the relevant parameter in this case. Indeed, under DC fields, the attraction force between particles is determined by the conduction properties of the constituents and does not vary as E^2 but rather as E_cE (see Boissy et al.[13]). Thus, taking an electric force F proportionnal to $\varepsilon_L E^2$ for all applied fields results in an overestimate of F for high fields and, therefore, in an underestimate for the actual parameter comparing viscous and electrical forces. Introducing a generalised Mason number taking into account the real dependence of the attraction force on the applied field likely would reduce the observed dispersion in this interval.

5. Conclusion

The rheological response of a suspension of microcristalline cellulose in mineral oil at 0.08 volume fraction confirms that the dependence of the stress tranferred in the suspension as a function of electric field and shear rate can be grossly described by a Bingham model. As previously observed, the experimental data of apparent viscosity to a first approximation give a single curve when they are plotted as a function of a Mason number Mn (defined by $\eta\dot{\gamma}/\varepsilon_L E^2$). The main result of this study is that the normalised current flowing through the suspension also leads to a universal curve as a function of Mn. This reveals a clear correlation between rheological and electrical behaviours which are both under the dependence of the fluid structuration. From this viewpoint it is very interesting to examine the current variations which give information on the structuration.

A closer examination however suggests that the relevant parameter is not exactly the same for both properties because the shear stress is mainly due to the mechanical resistance of sets of particles in contact whereas the current which flows through the network of conducting paths spanning the electrodes is more sensitive to the length scale of instantaneous fibres. In particular the dispersal in apparent viscosity (for different applied field strengths) should be reduced by taking into account the real dependence of attraction force between particles on the field.

Acknowledgements

This work was performed with the joint financial support of the Université Joseph Fourier de Grenoble, the Centre National de la Recherche Scientifique and the company

Elf-Aquitaine (under contract CNRS/SNEA No 50.8016). The authors are grateful to the compagny Elf-Atochem for providing samples of dielectric liquids.

References

1. Y. Chen, A. F. Sprecher and H. Conrad, *J. Appl. Phys.*, **70** (1991) 6796.
2. T. C. Halsey, *Science*, **258** (1992) 761.
3. R. A. Anderson, in *Proc. 3th Int. Conf. ER Fluids*, R. Tao, ed. (World Scientific, Singapore, 1992), 81.
4. L. C. Davis, *J. Appl. Phys.*, **72** (1992) 1334.
5. J.-N. Foulc, N. Félici and P. Atten, *C. R. Acad. Sci. Paris*, **314** (1992) Série II, 1279.
6. P. Atten, J.-N. Foulc and N. Félici, *Int. J. Mod. Phys. B*, **8** (1994) 2731.
7. A. F. Sprecher, J. D. Carlson and H. Conrad, *Materials Sci. and Engineering*, **95** (1987) 187.
8. H. Conrad, Y. Chen and A. F. Sprecher, in *Proc. 3th Int. Conf. ER Fluids*, R. Tao, ed. (World Scientific, Singapore, 1992), 195.
9. J. N. Foulc and P. Atten, in *Proc. 4th Int. Conf. ER Fluids*, R. Tao, ed. (World Scientific, Singapore, 1994), 358.
10. L. Marshall, C. F. Zukoski and J. W. Godwin, *J. Chem. Soc., Faraday Trans.* 1, **85** (1989) 2785.
11. D. J. Klingenberg, D. Dierking and C. F. Zukoski, *J. Chem. Soc., Faraday Trans.*, **87** (1991) 425.
12. J. R. Melcher, "*Continuum Electromechnics*", (MIT Press, Cambridge, Mass., 1981).
13. C. Boissy, P. Atten and J.-N. Foulc, "The conduction model of electrorheological effect revisited", *communication to this Conference*.

EFFECTS OF CELL CONFINEMENT ON THE EVOLUTION OF FIELD-INDUCED STRUCTURES IN A MAGNETORHEOLOGICAL FLUID

JING LIU, TAWEI MOU, YUN ZHU, ESMAEL HADDADIAN, JACKIE POUSSET
Department of Physics and Astronomy
California State University, Long Beach, California 90840

and

SHARLENE R. LIM
St. Joseph High School, Lakewood, CA 90713

ABSTRACT

The evolution of field-induced structure in ferrofluid emulsions is studied during the "gas-solid" phase transition. The chain formation and coarsening are measured by monitoring the average separation between aggregates d using static light scattering and optical microscopy. A power-law relation, $d \sim t^{\alpha}$, is obtained for all measured cell thicknesses ranging $10 \leq L \leq 700$ μm along the field direction. The rate of chain coarsening strongly depends on the chain length, weakly depends on the particle volume fraction and is independent of the field strength at a finite cell thickness. A theoretical model based on thermal fluctuation induced coupling and nucleation theory has been developed to explain the data at large L successfully.

1. Introduction

When a magnetorheological (MR) fluid is subjected to a magnetic field, a structural transition takes place from a gas of suspended particles to a solid of aggregates of particles[1,2] due to the induced dipolar interactions. This structural transition leads to the change of the rheological properties of the system from a liquid to a solid, which has shown many potential applications [3]. A important question in the application of MR fluids as well as in the study of basic interaction mechanism of ER effect is the kinetics of the structural transition or the detailed structural evolution of MR fluid induced by the magnetic field.

The evolution of field-induced structure of an ER fluid has been studied recently [4]. The dipolar interaction aligns particles into chains, and different chains further coalesce to form columns. The proposed theory based on thermal fluctuations induced coarsening [5] predicts that the chain coarsening follows a power law of the form: $d \sim t^{5/9}$, where d is a characteristic length of the structure. The experiment from the same study for a cell thickness of 0.7 mm results $d \sim t^{0.4}$. In ER fluids, due to image dipoles formed at the conducting boundaries, the effective chain length along the field direction is essentially infinite even though the physical cell thickness is finite. Therefore, the result in ER fluid represents the extreme limit of coarsening of chains of infinite length.

The structure evolution in magnetorheological fluids is more complicated due to the additional variable of chain length. In MR fluid, applied magnetic field induces a magnetic dipole moment in each particle suspended in a non-magnetic liquid. The induced dipolar interaction also aligns the particles into chains which may coalesce to form columns or other aggregates [1]. If the magnetic field is applied without magnetic boundaries such as using a magnetic coil, the chain length is limited by the cell thickness. Since the interaction between chains depends on the chain length, the cell thickness also affects the structural evolution as it does to the equilibrium structures of MR fluids [2]. So far, studies on structural evolution in MR fluids have been limited to strong cell confinement. The work performed in polystyrene magnetic colloids shows that the characteristic length, d, does not change over time [6]. However, in our earlier work with strongly confined cells we observed d changes with time with a lower coarsening rate in ferrofluid emulsion [7,8].

In this paper, we present further experimental study of the evolution of field-induced structure in a model MR fluid, ferrofluid emulsion with extended range of cell thickness used. We show that coarsening does exist even when the chain length or the cell thickness is as short as 10 μm [9]. The coarsening process measured by separation of aggregates d as a function of time follows a power law, $d \sim t^{\alpha}$, for all measured cell thickness L over two orders of magnitude along the field direction at volume fraction $\phi =$ 3%. As L increases from 10 to 700 μm, rate of chain coarsening α increases with chain length L also as a power law, $\alpha \sim L^{0.27}$, from 0.1 to 0.27. α seems to saturate to 0.27 at L = 300 μm at the field used in our measurement which corresponding to $\lambda = 78$. λ is defined as the ratio of the interaction energy of two parallel and contact dipoles and kT [2,10]. Therefore, the behavior of chain coarsening in the limit of infinite chain in our MR fluid is observed to be quite different from that measured in ER fluids. The structure formation is complete or reaches saturation after a finite time t_s which depends almost linearly on L. In addition, the coarsening rate depends weakly on the volume fraction at $L = 25$ and 100 μm. Furthermore, when the field strength is increased, measurement done at a finite cell thickness of 200 μm shows that the rate of chain coarsening is independent of the field but the coefficient K in the relation $d = Kt^{\alpha}$ increases with field H as $K \sim H^{0.22}$. A theoretical model based on chain-fluctuation coupling and nucleation theory is proposed[11]. It predicts that the coarsening rate for infinite long chains is 1/3 which is close to the experimental data measured at large L but less than what is predicted for ER fluids. This model also predicts the dependence of structural evolution on field strength and volume fractions as $d \sim H^{1/4}$ and $d \sim \phi^{1/6}$. These results explain the experimental data very well for large L.

2. Experimental procedures

In our experiments, a nearly monodisperse and density-matched ferrofluid emulsion is used which is described else where [12,2]. Different sizes of samples are used with the oil droplets having an average diameter ranging from 0.26 to 0.5 μm with a standard deviation of less than 20% of each. D_2O is used to almost density match the droplets to reduce the effect of sedimentation. The volume fraction used in all the measurements is 3% if not otherwise indicated.

The structure evolution of the ferrofluid emulsion is studied using optical microscopy and static light scattering techniques. In both cases, the sample is loaded in either flat thin glass cells, wedge cells, or capillary cells covering the range of cell thicknesses from 10 to 700 μm. A homogeneous external field is set perpendicular to the cell surface and is produced by a coil to prevent the magnetic boundary from forming image dipoles. The applied magnetic field is ramped up in a step function from 0 to 38 - 45 mT with a rise time constant of 60 ms.

Static light scattering is performed for the cell thickness ranging 10 - 300 μm. a pair of car batteries and a H-P power supply are used to supply the current to the coil providing field from 0 to 45 mT. A He-Ne laser of wavelength 632.8 nm is directed perpendicular to the surface of the cell. Scattered light is displayed on a screen behind the cell, collected by a CCD (Dage 72) camera, and recorded by a VCR. The signal from the super VHS vidco tape is then sent one frame at a time to a Macintosh computer for processing [9]. This method with the advantage of large ensemble average is limited to thin cells and low volume fractions to avoid multiple light scattering.

Optical microscopy is used to study the induced structures for thicker cells $L \geq 300$ μm to directly measure the column separation. The sample is centered in a magnetic coil which is on the stage of an Nikon (Diaphot) inverted optical microscope to provide a magnetic field normal to the stage. The microscope is focused on the top of the sample cell. Images of the induced structures are recorded by a Super-VHS VCR for about fifteen minutes after the magnetic field is applied through a NEC (TI-324) CCD camara attached to the optical microscope. The signal from the VCR is then sent to a Macintosh computer through a Raster Ops video grabber board. The images are captured and processed using Adobe Photoshop software. In processing the images, a box of known area is defined and the "dots" inside the box are counted. Each dot represented a chain or a column at a later time of the magnetic droplets, where single droplets moves too fast to yield a clear image and therefore are discarded in the counting. Dividing the area by the number of dots inside the box, and taking the square root, the average separation of chains (or columns) d are then obtained.

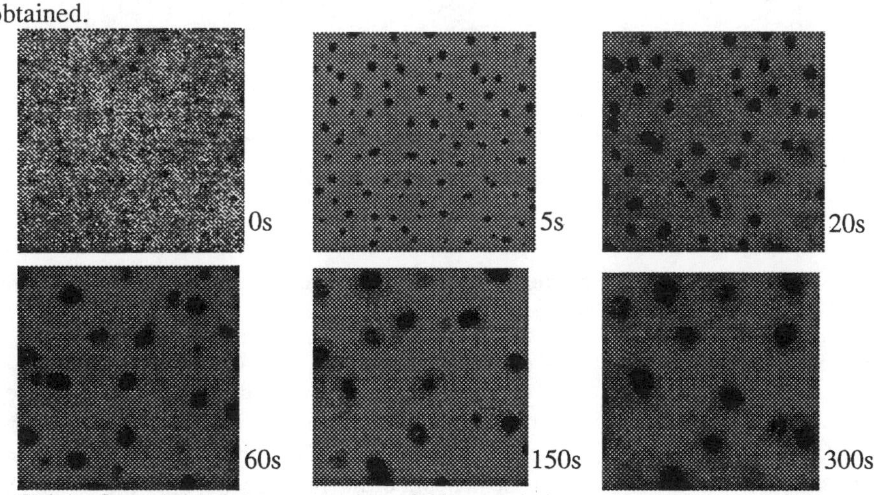

Fig. 1. Snap shots of the bottom view of microscopic images during the structural evolution process in ferrofluid emulsion taken at different time after the field is applied. Here $L = 700$ μm, $\phi = 3\%$, $2a = 0.26$ μm, and $H_{max} = 39$ mT giving $\lambda = 78$. The width of each image is 25 μm.

3. Results and Discussions

3.1. Optical Microscopy

Figure 1 shows a series of microscopic images during the evolution of a typical field-induced structure of a ferrofluid emulsion. The image is taken at cell thickness L = 700 μm, volume fraction $\phi = 3\%$, particle diameter 2a = 0.26 μm, and the maximum field used H_{max} = 39 mT corresponding to $\lambda = 78$. The imaging plane is normal to the field direction and is at the bottom of the cell so that ends of aggregates are viewed here. At t = 0 s, the randomly dispersed droplets as shown in Fig. 1 are in Brownian motion which moves too fast to image clearly. At t = 5 s after the field is applied, droplets form chains which further aggregates to thin columns. The columns have almost no order at this low volume fraction and thicker cell, but have an average separation d between them. As time increases, thin columns combine to form larger columns such that the number of columns decreases and the separation between them increases at the same time. By measuring the average separation d between columns for the images taken at different times, the structure evolution can be studied.

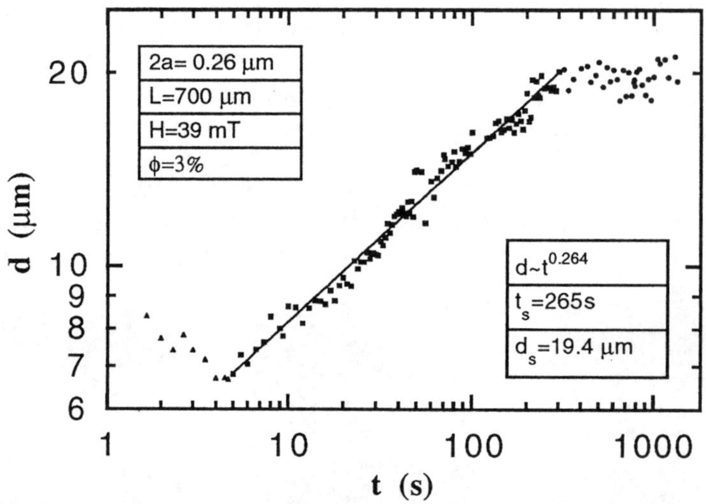

Fig. 2. The average separation d between aggregates as a function of time obtained from microscopic images in Fig. 1.

Figure 2 shows the result of the analyzed data obtained from images as those in Fig. 1. As seen from the figure, d depends on time in three stages. In the first stage where $t \leq 4$ s, chains grow from single droplets which may occur in much short time than one second. These chains are the nuclei for the formation of columns. Nucleation occurs in the entire sample, but due to the small focus depth of the microscope, only the bottom view of the capillary cell can be seen. Small chains and free droplets are evident in the sample whose motion is still too fast to measure accurately. The cause for the initial decrease in separation is not clear yet.

In the second stage where $5 < t < 272$ s, actual column growth is seen. Chains attract each other and aggregate to form columns. Smaller columns then aggregate further and form larger ones. The column growth shows a linear behavior in the log-log plot which corresponds to a power-law relation: $d \sim t^{\alpha}$ with $\alpha = 0.264$ measuring the column growth rate. In the third stage where $t > 270$ s, columns stop their growth and structures reach a saturation. The saturated time and separation in this case is $t_S = 272$ s and $d_S = 19.5$ μm. Similar behavior is also observed for cells with $L = 300$ μm where the column growth rate is slightly different, $\alpha = 0.28$ but the saturated time and separation are quite different: $t_S = 96$ s and $d_S = 16$ μm.

3.2. Static Light Scattering

To study the structure evolution further down to thin cells, static light scattering is used. It can give the same information quickly with larger ensemble average for cells where the multiple light scattering can be neglected. Fig. 3 shows a typical result of structural evolution measured by light scattering for $2a = 0.5$ μm, $\phi = 0.03$, $L = 25$ μm, and $H_{max} = 38$ mT. The scattered light intensity, I, is plotted as a function of the scattering wave vector, q, over a period of two seconds after the magnetic field is applied. Here, I is radially averaged over a constant q, defined as $4\pi\sin(\theta/2)/\lambda_o$, where θ is the angle between the incident and scattered light and λ_o is the light wavelength in the sample.

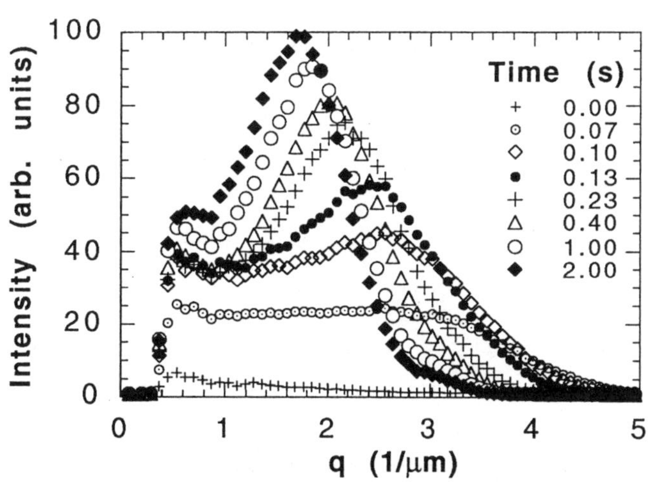

Fig. 3. Scattered light intensity as a function of scattering wave vector, q, for various times after applying a magnetic field. Here $2a = 0.5$ μm, $L = 25$ μm, $\phi = 3\%$, $H_{max} = 38$ mT and the field ramping time is 60 ms.

At $t = 0$ the scattering pattern is essentially flat, corresponding to a gas phase with droplets dispersed randomly. At $t = 0.1$ s, a ring appears in the scattering pattern. The peak intensity, I_{max}, of the ring increases with time, whereas the peak position, q_{max}, of the ring decreases with time. This indicates the occurrence of the phase transition from a gas of randomly dispersed droplets to a partially ordered solid structure. After two seconds, no further change in the ring pattern is observed; the structure "freezes". From microscopic images, the solid structure at this thin cell consists of equally spaced and locally ordered columns very similar to the structure formed at equilibrium when the field is applied slowly [13]. q_{max} is inversely proportional to the column spacing and column width. Therefore, the aggregates separation or the characteristic length can be obtained from $d = 2\pi/q_{max}$ in the scattering pattern at each time.

When we measure the ring peak position $q_{\max}$ and obtain d as a function of time, we again find a power-law behavior, $d \sim t^{\alpha}$ [7]. However, the measured chain coarsening rate or the column growth rate is $\alpha = 0.16$, and the saturation time is $t_s = 1$ s, which are all less than those obtained for $L = 700\ \mu m$ as shown in Fig. 2. Clearly the structural evolution, just like the equilibrium structural formation [2], depends strongly on the chain length.

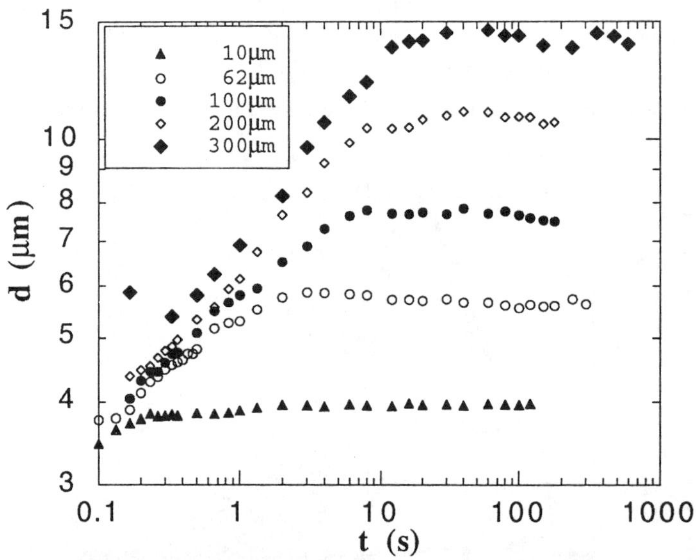

Fig. 4. Characteristic length d measured as a function of time for various cell thicknesses at $2a = 0.4\ \mu m$, $\phi = 3\%$, and $H_{\max} = 45$ mT corresponding to $\lambda = 315$.

3.2a. Cell Thickness Dependence

We then repeat the same experimental procedures for cell thickness ranging from 10 to 300 μm at a fixed volume fraction of 3%, droplet diameter of 0.4 μm and final field of 45 mT. This field corresponds to the coupling constant $\lambda = 315$. Fig. 4 shows the measured characteristic length d as a function of time for various cell thickness. As seen from the plots on a log-log scale, a power-law relation, $d \sim t^{\alpha}$, can be approximated for all the data before saturation. When the cell thickness varies from 10 to 300 μm, both the rate of coarsening before saturation α and the saturation time t_s increase with the chain length.

Figure 5 shows quantitatively the dependence of rate of coarsening on the cell thickness obtained from Fig. 4. α increases with L from 0.1 to 0.27 for L ranging from 10 to 300 μm. From fitting the data, a power law of $\alpha = 0.053\ L^{0.28}$ is obtained. In this figure, we also include the two data points with open circle for larger cell thicknesses obtained from optical microscopy. A good overlap of data at $L = 300\ \mu$m is found which gives us confidence of the data obtained from these two methods. It also shows the saturation of α on L for $L \geq 300\ \mu$m and the independence of α on field strength or λ.

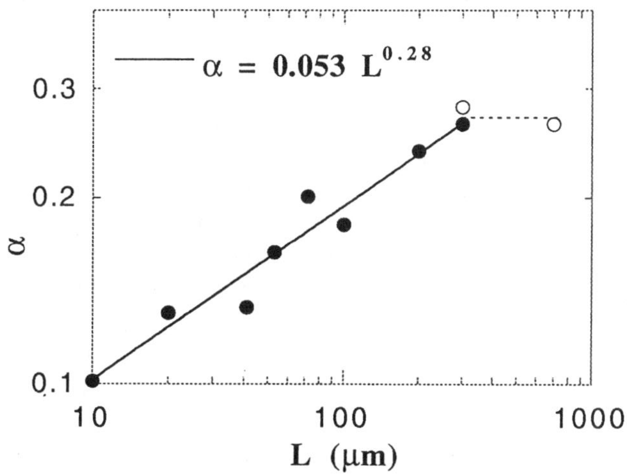

Fig. 5. Chain coarsening rate α measured as a function of cell thickness obtained from Fig. 4 at $2a = 0.4\ \mu$m, $\phi = 3\%$, and $H_{max} = 45$ mT shown as solid circles. Also shown here as open circles are data obtained from optical microscopy at $L = 300$ and $2a = 0.26\ \mu$m, and 700 μm and $2a = 0.3\ \mu$m where both corresponds to the same $H_{max} = 39$ mT.

The dependence of coarsening rate on chain length can be understood from the repulsive interaction between the ends of the chains or columns. In the limit of a single layer of dipoles, the interaction energy between any two dipoles are completely repulsive with a long interaction range as $\sim 1/r^3$ where r is the separation of the dipoles. No coarsening occurs in this case and the structure is a 2-D HCP crystal lattice with each particle occupying one site [14]. The structure formation is almost instantaneous with a time too short to measure with the VCR resolution.

In the other limit of infinite long chains such as in ER fluids, the interaction between chains are limited to a very short range $\sim \exp(-\rho\pi/a)$ [15,16], where ρ is the separation between chains and a is the particle radius. Thermal fluctuation of chains induces long-range interaction which causes the chain aggregates [4]. The effect of repulsion between ends of chains on the rest of chains are essentially zero. Therefore, the chain coarsening occurs with a higher rate which is observed as α=0.4 [4] and the coarsening time on the order of several 10 minutes.

When the chain length is finite and limited by the cell thickness such as in MR fluids without magnetic boundaries, the repulsive force between chains and columns presumably persists with its range reduced compared to that in a single layer of dipoles. This force is due to the demagnetization field between the same poles from ends of chains or columns. It limits the size of columns and determine the order of the columns [2,17]. In addition to the strong effect of repulsive force among the ends of short chains which is screened among long chains but not short chains, the thermal energy required to excite short wavelength modes is higher for short chains than for long chains. Therefore the thermally induced fluctuation in a short chain is smaller than that of long chains. As a result, the chain-chain coarsening is expected to be slower at short chain length and to increase the rate as the chain length increases.

Fig. 5 also shows that α approach asymptotically to a constant at $L = 300$ μm. We can understand this from the thermal fluctuation of chains. As chain length increases, thermal fluctuation amplitude also increases. It breaks the chain and limits the length to a finite value. Therefore, the coarsening rate reaches saturation. However, the value observed here is less than what is observed and predicted for ER fluids.

A theoretical model is built to understand the structure evolution [11] of MR fluids for infinite long chains. It is based on chain fluctuation induced coupling proposed by Halsey [4]as well as nucleation theory. Thermal fluctuation of a long chain or a thin column of width R produces a long-ranged field: $H_f \sim (kT\,R^2/a)^{1/2}/\rho^2$ at distance ρ away from the columns. Therefore the long chain or column acts as a nucleation center to attract nearby short chains or single droplets to them. Balancing the fluctuation induced attractive force and the viscous force, the separation of aggregates is predicted to follow the relations: a) $d \sim t^{1/3}$; b) $d \sim H^{1/4}$; c) $d \sim \phi^{1/6}$.

The growth rate is predicted to be $\alpha = 1/3$ instead of 5/9 in this case. This is close to the measured value of 0.27 for our large cell thicknesses used. The growth of columns is not just evenly combining of chains with equal length but a heterogeneous growth process; therefore the predictions here differ from that of Halsey. The discrepancy in the coarsening rate between the theory and experiment is probably due to the weak field used in our experiments which sets the limit of the chain length to be finite. This needs to be proved experimentally.

Figure 6 shows the saturation time obtained from Fig. 4. A linear dependence of the saturation time on the cell thickness is found, $t_s = 0.39+0.053L$. The structure formation is complete or reaches saturation in a time of 0.2 to 15 seconds for $10 \leq L \leq 300$ μm after the application of the magnetic field. The two data points measured with microscope corresponds to different droplets size and field strength. The saturation time is much longer for small dipolar interaction strength. However, the ratio of t_s to L follows the same linear relation, $t_S \sim L$. Because the reduced effect of repulsion between ends of chains to the chain coarsening process for large chain length, the columns grow bigger and therefore it takes longer time to reach saturation even though the rate of growth is higher. Our theoretical model also counts this behavior [11].

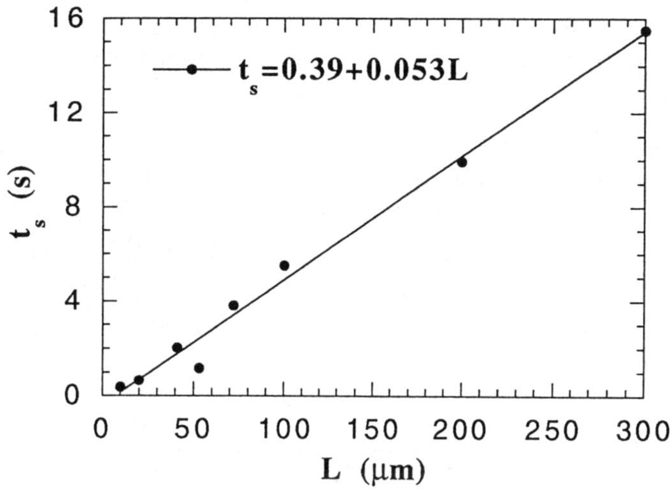

Fig. 6. Saturation time measured as a function cell thickness obtained from Fig. 4 at $2a = 0.4\ \mu m$, $\phi = 3\%$, and $H_{max} = 45$ mT. The solid line is a linear fitting.

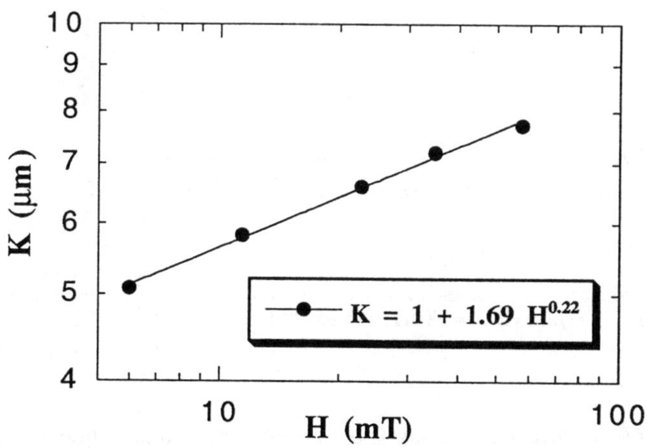

Fig. 7. Field strength dependence of the coefficient K in the relation of $d = Kt^{\alpha}$ obtained for $L = 200\ \mu m$, $\phi = 3\%$, and $2a = 0.4\ \mu m$.

3.2b. Field Strength Dependence

Figure 7 shows the influence of the field strength on the chain coarsening at $L = 200\ \mu m$, and $2a = 0.4\ \mu m$. The solid line is the fit of a power law where a constant 1 in the equation is set to account for the average distance between droplets at zero field. We found in this initial experiment that the coarsening rate is independent of the field strength within the experimental error. The aggregate separation d does depend on the field which means the coefficient K in the power-law relation, $d = Kt^{\alpha}$, depends on field. From the log-log plot, a power-law relation is obtained as $K \sim H^{0.22}$. This result agrees with our model $d \sim H^{0.25}$ predicted for infinite chains.

3.2c. Volume Fraction Dependence

Figure 8 shows droplet volume fraction dependence of the measured aggregate separation d for $2a = 0.5\ \mu m$, $\phi = 0.03$, $L = 100\ \mu m$, and $H_{max} = 38$ mT. Within the experimental error, d is found to depends weakly on ϕ. In our earlier work at $L = 25\ \mu m$, d is found basically independent of ϕ. The solid line is the fitting based the prediction of our model Our model $d \sim \phi^{1/6}$. A good agreement between the experiment and theory is seen from the fitting.

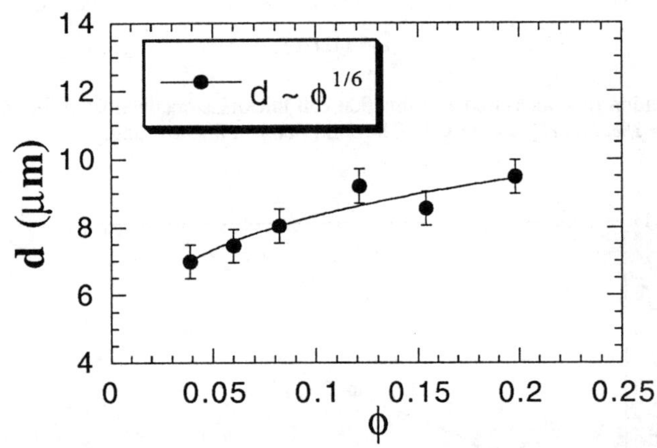

Fig. 8. Dependence of the column separation on droplet volume fraction at saturation obtained for $L = 100\ \mu m$, $H_{max} = 38$ mT, and $2a = 0.5\ \mu m$.

4. Conclusion

Our experiments, with cell thickness over two order of magnitude, demonstrate that the confinement of ferrofluid emulsions in the applied field direction strongly affects the rate of structural evolution and saturation time. The stronger the confinement is, the slower the coarsening rate is, and the quicker the saturation in structure formation is reached. This indicates that chain formation is the first step in structure evolution. The next step is the coarsening of chains to form columns. Here two driving forces compete with each other which determines the rate of chain coarsening: The repulsive interactions between ends of chains or thin columns resist the chain coarsening and controls the final width of columns; the thermal fluctuation of longer chains or thin columns provides the force and nucleation centers to attract their neighbors to them and to assist the coarsening process. The fact that coarsening rate approaches a constant at cell thickness of 300 μm and the good agreement is achieved between experiment and theory, further support the above picture of evolution mechanism.

5. Acknowledgments

We would like to thank K. Javier, G. Hu, Dr. J. Bibette, Prof. R. Sheng, and Prof. E. Wong for technical support, and Dr. Halsey for helpful discussion. This work was partially supported by an award from the Research Corporation, and by CSULB University Research and Academic Affairs Office. Acknowledgment is made also to the Donors of The Petroleum Research Fund, administered by the American Chemical Society, and NSF DMR for partial support of this research.

6. References

1. Y. Zhu, M.L. Ivey, P. Sheaffer, J. Pousset, and J. Liu, *5th Inter. Conf. on ER and MR Suspensions and associated Tech.*, Univ. of Sheffield, UK, July 10-14 1995.
2. J. Liu et. al. , *Phys. Rev. Lett.*, **74**, 2828 (1995).
3. R. Tao ed. *4th Conf. on Electrorheological Fluids*, Feldkirch, Austria, 20-24 July, 1993, (World Scientific, Singapore).
4. J.E. Martin, J. Odinek, and T.C. Halsey, *Phys. Rev. Lett.*, **69**, 1524 (1992).
5. T.C. Halsey and W. Toor, *J. Stat. Phys.*, **61**, 1257 (1990).
6. Y.H. Hwang and X.L. Wu, *Phys. Rev. E*, **49**, 3102 (1994).
7. J. Liu, T. Mou, G.A. Flores, J. Bibette, and J. Richard, *4th Int. Conf. on Electrorheological fluids*, Feldkirch, Austria, July 19-24, 1993, R. Tao (Ed.), (World Sci., Singapore).
8. J. Liu, T. Mou, and J. Pousset, *Electrorheological Symposium, ACS Meeting.*, Washington, D.C., August 1994, F. Filisko & K. Havelda (Ed.), (Plenum, 1995).
9. T. Mou (1995) The Evolution of Field-Induced Structure of Confined Ferrofluid Emulsions. MS Thesis, Calif. State Univ. Long Beach.
10. A.P. Gast and C.F. Zukoski, *Adv. Coll. Inter. Sci.*, **30**, 153 (1989).
11. Y. Zhu and J. Liu, *5th Inter. Conf. on ER and MR Suspensions and associated Tech.*, Univ. of Sheffield, UK, July 10-14 1995,
12. M. Hagenbuechle, P. Sheaffer, Y. Zhu, and J. Liu, *5th Inter. Conf. on ER and MR Suspensions and associated Tech.*, Univ. of Sheffield, UK, July 10-14 1995.
13. J. Liu et. al. , *4th Int. Conf. on Electrorheological fluids*, Feldkirch, Austria, July 19-24 1993, R. Tao (Ed.), (Singapore, World Sci., 1994).
14. A.T. Skjeltorp, *J. Appl. Phys.*, **57**, 3285 (1985).
15. R. Tao and J.M. Sun, *Phys. Rev. Lett.*, **67**, 398 (1991).
16. T.C. Halsey and W. Toor, *Phys. Rev. Lett.*, **65**, 2820 (1991).
17. A. Wu and J. Liu, *"A cylindrical Model of Field-Induced Structure of dipolar colloids", In preparation.*

RHEOLOGY OF A MAGNETORHEOLOGICAL FLUID

D. W. FELT, M. HAGENBÜCHLE, J. LIU
Department of Physics and Astronomy,
California State University, Long Beach, CA 90840.
and
J. RICHARD
Rhone-Poulenc Recherches, 52 Rue de la Hale Coq, 93308 Aubervilliers,
France

ABSTRACT

We report on experimental investigations of the rheological behavior of aqueous magnetic suspensions. The suspended particles are monodisperse colloidal polystyrene spheres which contain magnetic Fe_2O_3-grains. In the absence of a magnetic field these suspensions behave as Newtonian fluids, whereas under the influence of a magnetic field due to the formation of an ordered structure the apparent viscosity of the suspension increases up to three orders of magnitude and they clearly exhibit Non-Newtonian properties, such as shear thinning and yield stress. The apparent viscosity depends on the magnetic field according to $\eta \sim H^{2\Delta}$. Increasing the volume fraction of the particles in the range of $0.014 < \phi < 0.12$ results in a linear increase in apparent viscosity and yield stress. Both apparent viscosity and yield stress depend also on the particle size as first measurements on particles with diameters ranging from 0.5 μm to 1.0 μm clearly show. All measurements were carried out with a rotation viscometer using the cone-plate configuration so that the applied shear rate was well defined.

1. Introduction

Electrorheological (ER) fluids and magnetorheological (MR) fluids are often called intelligent or smart fluids. This is due to the fact that the rheological behavior of these fluids can be changed easily and, even more important, reversibly by applying an external electric or magnetic field. A broad range of possible applications is presumably the reason for the intense research on these fluids in recent years[1]. The underlying mechanism which is responsible for the "tunable rheological properties" is the same for both ER and MR fluids. The external field induces a dipole moment in the suspended particles. When the strength of the dipole-dipole interaction between the particles overcomes the thermal energy, they stick together and form chains along the field lines. A further increase of the external field results in the aggregation of these chains into columns. The structural changes upon application of the external field are responsible for the changes in the fluid's rheology. Apart from the common description of ER and MR fluids given above, there are, however, also some main differences[2,3,4]. For example, in MR fluids there is no counterpart to the electrical breakdown exhibited by ER fluids at high field strengths. Until now, research has mainly focused on the ER fluids whereas MR fluids have attracted much less interest, although some features of MR fluids are more promising for possible applications[5].

In this paper we present our rheology measurements on a model MR fluid. A main difference between our and previous studies[6,7,8] is that our samples are monodisperse. Therefore we are not only able to examine the influence of volume fraction, shear rate, and magnetic field strength on apparent viscosity and yield stress of the MR fluid but also the influence of particle size. Although the range and number of the different particle sizes under investigation is small, both apparent viscosity and yield stress show a strong particle size dependence.

2. Experiment

The magnetorheological fluid we used in our study is an aqueous suspension of magnetizable polystyrene spheres. The polystyrene particles are coated with the surfactant *sodium dodecyl sulfate (SDS)* which provides an effective stabilization against aggregation. The spheres are magnetizable because they contain 67.6 % by weight of magnetic Fe_2O_3 grains. The size of these grains is approximately 10 nm, whereas the average diameter of the polystyrene spheres as received from the manufacturer is 0.8 μm. However, this initial sample is extremely polydisperse. The standard deviation is roughly 100 %. In order to get well defined samples and, moreover, since we were also interested in studying the effect of the particle size on the magnetorheological behavior, we needed to purify this initial crude suspension. The purification and particle size separation was done by applying the method of fractionated crystallization. This technique, which has been reported to give good results with various suspensions and emulsions[9,10], yielded three samples with different sizes and enough sample to carry out the rheology measurements. The particle diameters of these three samples were found to be 1.00 μm with a standard deviation σ of 0.07 μm, 0.80 μm with σ = 0.10 μm, and 0.50 μm with σ = 0.08 μm. The volume fraction ϕ was determined prior to each measurement by accurately weighing the samples. Measurements at different volume fractions were carried out with the 1.0 μm spheres because those were the only ones where enough sample was available.

We used a Brookfield LVDV III rotational rheometer with a cone-plate configuration for our measurements. The main reason we used this configuration and not a plate-plate configuration is that it provides a well defined shear rate. The Brookfield rheometer is a controlled-shear-rate rheometer which means that the shear rate $\dot{\gamma}$ is given and the shear stress τ is measured. Approximately 0.5 ml of sample are placed in the gap between the cone and the plate. The minimal cone-plate separation is 0.013 mm which is still one order of magnitude larger than the diameter of the biggest particles. The angle between cone and plate is 0.8°. A magnetic field parallel to the rotation axis of the rheometer is produced by a coil which surrounds the cone-plate configuration. Since the maximum separation between the cone and the plate is 0.5 mm and the length of the coil is

15 cm, the magnetic field gradient is negligible and the field can be regarded as homogeneous. Field strengths up to 800 G could be generated with the coil. During the measurements the temperature was maintained constant at 22.5 °C by a thermal bath.

3. Results

Prior to the field measurements we measured the dependence of the shear stress τ on the shear rate $\dot{\gamma}$ with no magnetic field applied. In the concentration range $0.014 < \phi < 0.12$ the shear stress was found to be a linear function of the shear rate starting at the origin which is characteristic for Newtonian fluids. The viscosity η which is the slope of the $\tau/\dot{\gamma}$ curve is slightly increasing with the particle concentration. Within the experimental error the measured data was well fitted by the Einstein equation:

$$\eta = \eta_0(1+2.5\phi). \tag{1}$$

Here η denotes the viscosity of the suspension and ϕ is the volume fraction of the suspended particles. The viscosity of the suspending liquid (distilled water and excess SDS) η_0 is 1.01 cP. From these preliminary results we conclude that without magnetic field our samples are well described by a suspension of hard spheres, which is still dilute enough that steric interaction has no major influence on the Einstein behavior.

Figure 1 demonstrates the Non-Newtonian behavior of our MR fluids when a magnetic field is applied. It is obvious that the shear stress τ is no longer a linear function of the shear rate $\dot{\gamma}$. The slope which is equivalent to the apparent viscosity η decreases with increasing shear rate. This effect is usually called shear thinning and has also been reported for ER fluids[11]. The second and, in terms of applications, more important change caused by the magnetic field is the appearance of a yield stress τ_y. The yield stress is obtained in Fig. 1 by extrapolating $\tau(\dot{\gamma})$ to $\dot{\gamma} = 0$. For $\tau < \tau_y$ there is no flow which means the MR fluid behaves like a solid. If the applied shear stress exceeds τ_y, then the MR fluid begins to flow.

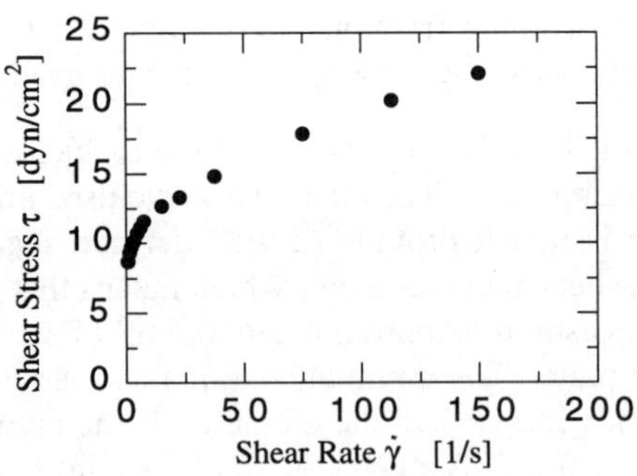

Fig. 1. Shear stress τ vs. shear rate $\dot{\gamma}$ at a field strength of 154 G for a = 1 μm and ϕ = 0.052. The yield stress τ_y is approximately 8.0 dyn/cm^2.

Figure 2 shows the dependence of the apparent viscosity on the Mason number Mn for various magnetic field strengths. $Mn \propto \eta_0 \dot{\gamma}/(\beta H)^2$ (gaussian units) is a dimensionless number, expressing essentially the ratio of the viscous to the magnetic energy. The parameter $\beta = (\mu_p - \mu_s)/(\mu_p + 2\mu_s)$ arises from the different permeability of the particles compared to that of the solvent. Since the particles have induced dipole moments, β depends on H. But for small fields up to about 200 G, β is approximately constant. We observe a dramatic increase of the apparent viscosity with decreasing Mason number. At the lowest Mason number, the difference of the measured apparent viscosity to the zero field viscosity is three orders of magnitude. The decrease of $\eta(Mn)$ demonstrates the shear thinning effect. With exception of the high Mason number region the data can be well fitted by a power law:

$$\eta \propto Mn^{-\Delta} \quad \Rightarrow \quad \eta \propto \dot{\gamma}^{-\Delta} \text{ and } \eta \propto H^{2\Delta} \tag{2}$$

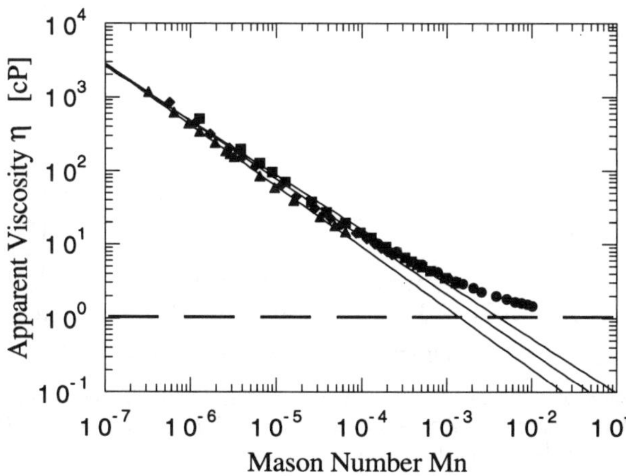

Fig. 2. Log-log plot of the apparent viscosity η vs. Mason Number Mn for various magnetic fields; a = 1 µm, $\phi = 0.052$. Circles: H = 38.5 G, squares: H = 77 G, diamonds: H = 115.5 G, triangles: H = 154 G. The solid lines are power law fits. They yield a slope of $\Delta = 0.74$ for 77 G, $\Delta = 0.78$ for 115.5 G, and Δ = 0.83 for 154 G. The broken line denotes the zero field viscosity.

Obviously, the power law exponent Δ doesn't show a pronounced dependence on the magnetic field strength. However, due to our data, there might be a slight increase of Δ with increasing field. Our Δ values are in excellent agreement with measurements on ER fluids where the same power-law behavior of shear thinning was observed[11] with $0.68 < \Delta < 0.93$. At high Mason numbers, $\eta(Mn)$ seems to approach a plateau. This is a logical consequence if we recall that the apparent viscosity of the MR fluid depends on the structure and that the field induced structure, i.e. the chains and columns, will be completely destroyed when the shear force is much bigger than the magnetic force.

Figure 3 is a plot of the apparent viscosity vs. the magnetic field strength for various concentrations of the 1 μm spheres at a constant shear rate of 3.75 1/s. The influence of the volume fraction ϕ on η is clearly demonstrated by this plot: the higher the volume fraction, the stronger the increase of the apparent viscosity with increasing magnetic field. For field strengths up to a value of about 200 G the data can be fitted with a function $\eta \sim H^{2\Delta}$. For high magnetic fields $\eta(H)$ deviates from the $H^{2\Delta}$ behavior and, as the low volume fraction curves indicate, might approach an asymptotic value. Saturation effects of the particle magnetization is a possible explanation for this deviation[12]. This is supported by a strong decrease of $\beta(H)$ in this region[13]. Due to equipment limitations we were unfortunately not able to measure the saturation of $\eta(H)$ for the higher volume fractions.

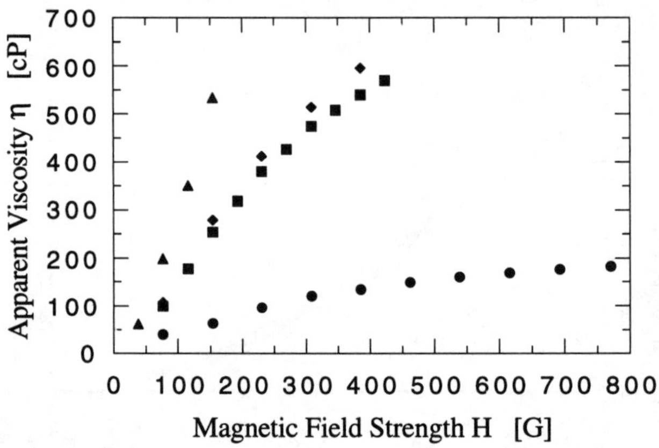

Fig. 3. The dependence of the apparent viscosity η on the magnetic field strength H for various concentrations of the 1 μm particles at a constant shear rate of 3.75 1/s. Circles: $\phi = 0.014$; Squares: $\phi = 0.052$; Diamonds: $\phi = 0.056$; Triangles: $\phi = 0.120$.

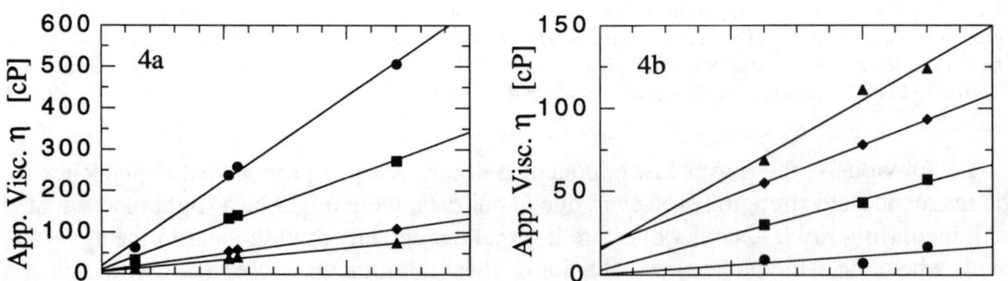

Fig. 4. Volume fraction and particle size dependence of the apparent viscosity. a) $H = 150$ G; circles: $\dot{\gamma} = 3.75$ 1/s, squares: $\dot{\gamma} = 7.5$ 1/s, diamonds: $\dot{\gamma} = 22.5$ 1/s, triangles: $\dot{\gamma} = 37.5$ 1/s. b) $\dot{\gamma} = 10$ 1/s; circles: $H = 38.5$ G, squares: $H = 77$ G, diamonds: $H = 115.5$ G, triangles: $H = 154$ G.

The measurements shown in Fig. 3 were repeated at higher shear rates. Qualitatively we found the same behavior. The absolute values changed of course because of the shear thinning effect. Figure 4a summarizes these measurements by comparing, for a given field strength of 150 G, the apparent viscosity obtained for different volume fractions and different shear rates. This plot quantifies the above mentioned volume fraction dependence of η, i.e. we find that $\eta \sim \phi$. Also, the shear thinning effect is again demonstrated by Fig. 4a.

Figure 4b presents η as a function of the particle diameter a for different magnetic fields at a constant shear rate of 10 1/s. Although we had only three different particle sizes for these measurements, our results give clear evidence for a linear relationship between η and a. Additional measurements are planed to confirm this. All particle size dependent measurements were of course done at the same volume fraction of roughly 0.05, since we know from Fig. 4a that η is also proportional to ϕ.

Up to this point we reported how the apparent viscosity η responds to changes of various parameters. The remaining graphs will deal with the question how the yield stress τ_y depends on volume fraction, particle size and magnetic field strength. However, first we emphasize that the yield stress reported here is a "dynamic yield stress" which means that it is determined by the behavior of the MR fluid in the post yield flow region. Furthermore, it is important to notice that this yield stress is not a measured quantity but obtained by extrapolation of the τ versus $\dot{\gamma}$ measurements as described in the context of Fig. 1. This method corresponds to the definition given by Klingenberg et al.[14] but differs considerably from the commonly used Bingham model yield stress[15].

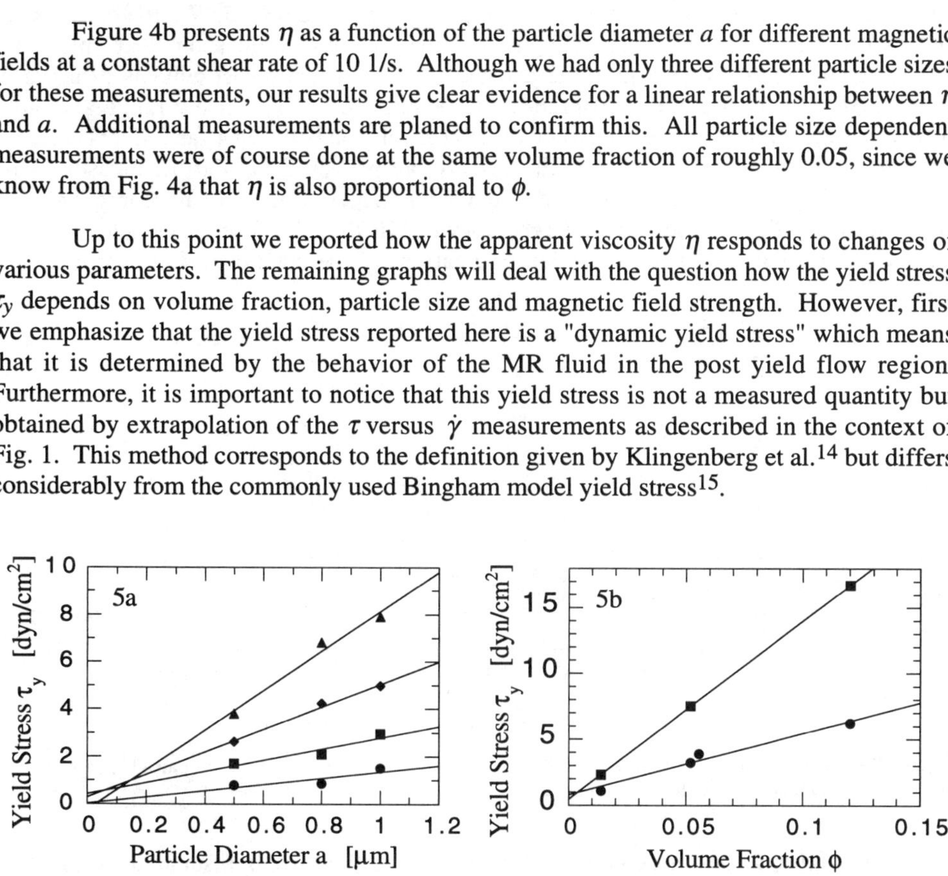

Fig. 5. The dependence of the yield stress on particle diameter and volume fraction. a) Circles: $H = 38.5$ G, squares: $H = 77$ G, diamonds: $H = 115.5$ G, triangles: $H = 154$ G; b) Circles: $H = 77$ G, squares: $H = 154$ G

Fig. 5a and 5b show the dependence of τ_y on particle size and volume fraction, respectively. The yield stress is found to be an increasing function of both the particle diameter and the volume fraction. Both graphs indicate a linear relationship. Bossis et al. argued that τ_y is mainly determined by the restoring force which counteracts the shearing of the particle chains[15] and obtained:

$$\tau_y \propto \phi \cdot H^2 . \tag{3}$$

Our data in Fig. 5b confirms this relation in terms of the volume fraction dependence. For ER fluids, Klingenberg et al. reported measurements and also simulations on the volume fraction dependence of the yield stress[16]. They found that τ_y increases roughly linearly with ϕ until it reaches a plateau at $\phi \approx 0.35$. We therefore expect that expanding our measurements to higher volume fractions would also reveal a plateau in the τ_y versus ϕ plot. Concerning the particle size dependent measurements it has to be noted that we had only three different particle sizes in a very narrow range available. Therefore, although Fig. 5a seems to favor a linear relation between τ_y and a, it is certainly too early to draw this conclusion on the basis of our present data. Further measurements are needed in order to clarify this point. In a recent paper Lemaire et al. reported also on the influence of the particle size on the rheology of MR fluids[17]. They were able to explain their experimental results by Brownian fluctuations of the positions of the particles in a chain.

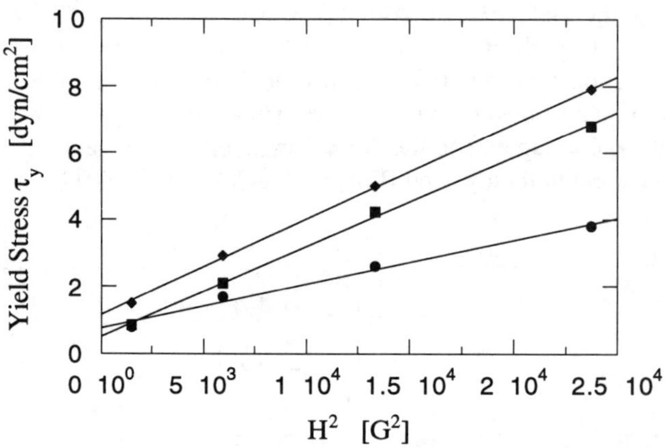

Fig. 6. Yield stress as a function of the magnetic field strength squared for different particle sizes. Circles: $a = 0.5$ μm, squares: $a = 0.8$ μm, diamonds: $a = 1.0$ μm.

In Figure 6 τ_y is presented as a function of the field strength squared for the three different particle sizes. A quadratic dependence is expected for both ER fluids and MR fluids (see Eq. 3). For ER fluids this dependence was observed by several authors in experiment and also simulation[16,18,19,20]. Our data compares well to these ER fluid measurements and confirms that $\tau_y \sim H^2$, but until now we have no explanation for the different slopes of the 0.5 μm spheres on the one side and the 0.8 μm and 1.0 μm on the other side.

4. Conclusion

In this paper we presented an extensive, but in some respect, preliminary study of the rheology of a model MR fluid. Exposure of these fluids to a magnetic field causes the formation of fibrous structures on a microscopic scale which macroscopically finally result in a Non-Newtonian rheology, i.e. the occurrence of yield stress and shear thinning. We examined the dependence of the apparent viscosity on shear rate, field strength, volume fraction, and diameter of the suspended particles. The dynamic yield stress increases proportional to both the volume fraction and the particle size. In addition, we find that τ_y is proportional to the square of the applied field. The pronounced dependence of apparent viscosity and yield stress on the particle size strongly suggests the use of monodisperse samples in the study of mechanism of rheology in MR fluids. In the whole we find that the rheological behavior of MR fluids is very similar to that of their electrical counterparts, the ER fluids. This is of course not surprising since the rheology is related to the structure which, in both cases, has its origin in the field-induced dipole-dipole interaction of the suspended particles.

For confined MR fluids, a unique phase behavior was reported which distinguishes them from ER fluids[2,3,4]. Due to the fixed geometry of the cone plate configuration used in our experiment, cell thickness is not constant across the cell and, additionally, can not be changed. This is the reason why we were not able to detect an effect of the confinement in our measurements. However, we expect that yield stress and apparent viscosity of confined MR fluids should also depend on the cell thickness. We hope that with future measurements we will not only extend the range of data concerning the volume fraction and particle size dependence, but also clarify this still open question. Finally, we agree with other authors that the commonly used Bingham model gives an insufficient rheological description since it doesn't account for the shear thinning effect[11,21].

5. Acknowledgments

We gratefully acknowledge the support of NASA (NAG 3-1634) and the Research Corporation (Cottrell College Science Award, C-3540).

6. References

1 *Proceedings of the Conference on Electrorheological Fluids*, Feldkirch, AUSTRIA, 20 - 23 July, 1993, R. Tao and G. D. Roy, Eds., (World Scientific, Singapore, 1994).

2 E. M. Lawrence, M. L. Ivey, G. A. Flores, and Jing Liu, *Int. J. Mod. Phys. B* **8** (1994) 2765.

3 T. Mou, G. A. Flores, and Jing Liu, J. Bibette, and J. Richard, *Int. J. Mod. Phys. B* **8** (1994) 2779.

4 J. Liu, E.M. Lawrence, A. Wu, M. L. Ivey, G. A. Flores, K. Javier, J. Bibette, and J. Richard, *Phys. Rev. Lett.* **74** (1995) 2828.

5 Talk of J. D. Carlson, Lord Corporation, at *Spring Meeting of the Materials Research Society* in San Francisco, 17. - 21. April.
6 E. Lemaire and G. Bossis, *J. Phys. D: Appl. Phys.* **24** (1991) 1473.
7 G. Bossis and E. Lemaire, *J. Rheol.* **35** (1991) 1345.
8 Z. P. Shulman, V. I. Kordonsky, and E. A. Zaltsgendler, *Int. J. Multiphase Flow* **12** (1986) 935.
9 J. Bibette, *J. Coll. Int. Sci.* **147** (1991) 474.
10 J. Bibette, *J. Magnetism Magn. Materials* **122** (1993) 37.
11 T. C. Halsey, J. E. Martin, and D. Adolf, *Phys. Rev. Lett.* **68** (1992) 1519.
12 J. M. Ginder and L. C. Davis, *Appl. Phys. Lett.* **65** (1994) 3410.
13 Y. Zhu, M. L. Ivey, and J. Liu submitted to the *Int. J. Mod. Phys. B*.
14 D. J. Klingenberg, F. van Swol, and C. F. Zukoski, *J. Chem. Phys.* **94** (1991) 6160.
15 G. Bossis, C. Mathis, Z. Mimouni, and C. Paparoditis, *Europhys. Lett.* **11** (1990) 133.
16 D. J. Klingenberg, F. van Swol, and C. F. Zukoski, *J. Chem. Phys.* **94** (1991) 6170.
17 E. Lemaire, A. Meunier, G. Bossis, J. Liu, D. Felt, P. Bashtovoi, and N. Matoussevitch, submitted to *J. Rheol.*.
18 L. Marshall, C. F. Zukoski, and J. W. Goodwin, *J. Chem. Soc. Faraday Trans. I* **85** (1989) 2785.
19 C. A. Randall, D. E. McCauley, C. P. Bowen, T. R. Shrout, and G. L. Messing, in *Proceedings of the Conference on Electrorheological Fluids*, Feldkirch, AUSTRIA, 20 - 23 July, 1993, R. Tao and G. D. Roy, Eds., (World Scientific, Singapore, 1994).
20 Y. D. Kim and D. J. Klingenberg, in *Proceedings of the Conference on Electrorheological Fluids*, Feldkirch, AUSTRIA, 20 - 23 July, 1993, R. Tao and G. D. Roy, Eds., (World Scientific, Singapore, 1994).
21 R. T. Bonnecaze and J.F. Brady, *J. Chem. Phys.* **96** (1992) 2183.

Nucleation Theory of Structure Evolution in Magnetorheological Fluid

Yun Zhu, Mark Gross, and Jing Liu
Department of Physics and Astronomy
California State University, Long Beach, CA 90840, USA

Abstract

A phenomenological model of nucleation-controlled structure evolution in dilute magnetorheological (MR) fluid is suggested. We find that the critical diameter of column nuclei is related to the external magnetic field H_0, the increase in permeability $\Delta\mu$ of the nuclei, and to the nucleus surface tension σ, by a relation of the form $D_c = \frac{2\sigma}{|\Delta g| + \frac{1}{8\pi}({H_0}^2 - \tilde{H_0}^2)\frac{\Delta\mu}{\bar{\mu}^2}}$. The growth rate of columnar structure, and the field and volume fraction dependencies of the separation between the columns are calculated, and found to be given respectively by $d \sim t^{2/7}$, $d \sim H^{3/14}$, and $d \sim \phi^{1/7}$. These results fit the experimental data in ferrofluid emulsion systems very well.

The interest in the evolution of structure in electrorheological (ER) and magnetorheological (MR) fluid has grown sharply in recent years. The dynamics of structure formation in ER and MR fluid is under extensive investigation. There are two focuses of the research: 1) the phase transition and the phase diagram[1–3]; 2) the growth rate of columnar structure[4–6]. The ferrofluid emulsion is a typical MR fluid under study, for which it is easy for researchers to choose the colloid particle size and to control polydispersity of the particle sizes.

The magnetic field-induced phase behavior of ferrofluid emulsion is studied by light scattering and optical microscopy[1]. It is found that when the strength of an applied magnetic field increases at a slow enough rate, ferrofluid emulsion exhibits a gas to ainduced Nematic liquid crystal phase transition at H_{c_1} followed by ainduced Nematic liquid crystal to solid phase transition at H_{c_2}. In the transition at H_{c_1}, randomly distributed emulsion droplets start to form chains at random positions and with various lengths. The result is called ainduced Nematic liquid crystal phase. In the second transition at H_{c_2} the fluctuation coupling is the dominant interaction and it drives the chains to coalesce into separated columns which form a locally ordered columnar structure. When the field just exceeds H_{c_2}, it is found that some chains are closer to one another than others. They grow more rapidly than others with increasing field, and eventually become solid columns. This strongly indicates that the nucleation plays an important role in the induced Nematic liquid crystal -

columnar solid phase transition at H_{c_2}. Because of the cylindrical symmetry of the interchain potential, this is a two-dimensional phase transition in the plane orthogonal to the field lines. It turns out to be first-order [7–9].

The growth rate is another important experimentally observable quantity which provides a sensitive test of the theories of structure evolution in MR and ER fluids. With a sudden application of a strong uniform external field, the average distance d between columns is investigated as a function of time t, field H (or E), and volume fraction ϕ. One theoretical study[4] was carried out in an ER fluid. Assuming that coarsening is due to thermal fluctuations of chains, the calculation resulted in a prediction for the coarsening of infinitely long chains: $d \sim t^{5/9}$, $d \sim E^{4/9}$, and $d \sim \phi^{2/3}$. However the experimental result for a cell of thickness 0.7 mm shows that the chain coarsening varies in time in accordance with the power law $d \sim t^{2/5}$. The experimental situation corresponds to the case of cell thickness or width along the field direction essentially infinite, due to image dipoles formed at the boundaries. Furthermore, recent experiments in ferrofluid emulsions demonstrate an even slower coarsening rate[9]. For cell thickness greater than 300 μm, the number of droplets in a chain will be in thousands so the rate should show a tendency to approach the result of infinitely long chains. But, surprisingly, there was found a very slow power law of the form: $d \sim t^{0.27}$, $d \sim H^{0.22}$, and d weakly dependent on ϕ. These exponents are only about half of the results from the previously existing theory[4].

On the other hand, spinodal decomposition (SD) mechanisms have predicted a very slow rate of nuclei growth[10]. Lifshitz et al predicted $\sim t^{1/3}$ behavior. It seems that these SD theoretical values are pretty close to the experimental result. However the mechanisms are inconsistent with the optical microscopic observation of nucleation in the ferrofluid emulsion system, which will be discussed later. In this paper, we describe a nucleation mechanism to explain the induced Nematic liquid crystal - columnar solid structure evolution which accounts for the slow coarsening rate found in ferrofluid emulsion systems.

(I) A columnar nucleation model

As is well known, in a first order phase transition there is always the possibility of metastable states and hence also of formation of nuclei[11]. Spontaneous fluctuations can form stable nuclei which exceed the critical nucleus size. This is often termed homogeneous nucleation. At large enough magnetic field, the induced Nematic liquid crystal phase in the ferrofluid emulsion system is a metastable state, because of the short range attractive interaction produced directly by dipole chains and the long-range fluctuation coupling. Within the metastable phase, nuclei will form and initiate the new phase. Since the possibility of nucleating columns of the new phase increases rapidly with increasing size of nuclei, the beginning of the phase transition is determined by the probability that nuclei of a minimum necessary size occurs. The probability that a nuclei of this minimum size will form is proportional to $exp(-R_{min}/k_B T)$, where T is the temperature and the energy barrier R_{min} is the work required to form a critical nucleus. The critical nucleus is in unstable equilibrium with the medium; its shape is such that at a given width of nucleus the free energy of the body will be a minimum, but its width corresponds to a *maximum* of the free energy. Thus a critical nucleus corresponds to a saddle point of the free energy[11]. In an isotropic system the critical nucleus is spherical. But in the ferrofluid emulsion in the presence of a sufficiently strong external magnetic field, the structure formation begins by the chaining of droplets. The nuclei of columns consist of a few overlapping chains. The shape of the nucleus is actually a long cylinder except at the ends where the width narrows. The nucleus shape has been discussed by several groups previously[12]; we do not intend to concentrate on it here. We will simply assume that each nuclei is a uniform cylinder.

R_{min} is the change in the Gibbs free energy, G, from the initial state made up of an induced Nematic liquid crystal of chains, to the final state, consisting of widely separated nuclei. These nuclei need not be stable. We make a mean field approximation to the initial and final states. The initial susceptibility will be taken to be a constant, μ_i. The nuclei in the final state will be taken to form a regular array with one nuclei of area A_n within each area A, with $A_n \ll A$. In the final state we take the susceptibility to be μ_{out} outside the nuclei and μ_{in} inside the nuclei. The ferrofluid emulsion is confined within a cell of thickness L in an applied magnetic field $\mathbf{H}_0$ in the z direction. We consider only the limit $L \to \infty$.

The Gibbs free energy density g, is given by

$$g = g^{(0)}(\rho, T) - \frac{1}{4\pi}\int_0^H \mathbf{B}\cdot d\mathbf{H}\,, \tag{1}$$

where $g^0(\rho, T)$ is the Gibb's free energy density in the absence of a magnetic field, as a function of density and temperature. ρ is the density of MR fluid (or not only the density of droplets). $\mathbf{B}$ and $\mathbf{H}$ are the magnetic induction and field respectively. Solving Maxwell's equations with the appropriate boundary conditions we obtain

$$\begin{cases} \mathrm{H}_i = H_0/\mu_i \\ \mathrm{H}_{out} = H_0/\bar{\mu}\,, \end{cases} \tag{2}$$

where $\bar{\mu}$ is the average susceptibility in the final state; $\bar{\mu} = \mu_i + \Delta\mu\, A_n/A$, $\Delta\mu \equiv \mu_{in} - \mu_{out}$. Writing $\mathbf{B}_i = \mu_i \mathbf{H}_i$, etc, we obtain for the Gibbs free energy density in the initial state,

$$g_i = g_i^{(0)}(\rho_i, T) - \frac{1}{8\pi} H_0^2/\mu_i\,. \tag{3}$$

In the final state, outside and inside the nuclei,

$$g_{out} = g_{out}^{(0)}(\rho_{out}, T) - \frac{1}{8\pi}\frac{\tilde{H_0}^2}{\mu_i} - \frac{1}{8\pi}\frac{\mu_{out}}{\bar{\mu}^2}(H_0^2 - \tilde{H_0}^2) \tag{4}$$

and

$$g_{in} = g_{in}^{(0)}(\rho_{in}, T) - \frac{1}{8\pi}\frac{\tilde{H_0}^2}{\mu_i} - \frac{1}{8\pi}\frac{\mu_{in}}{\bar{\mu}^2}({H_0}^2 - \tilde{H_0}^2) . \tag{5}$$

H_0 is the external field for which the density has a change from ρ_i to ρ_{out} outside the nuclei and ρ_{in} inside the nuclei, and the susceptibility jumps from the value μ_i to μ_{out} outside the nuclei and μ_{in} inside. Per nuclei, R_{min} is given by

$$R_{min} = [\sigma(\partial A_n) + g_{in}A_n + g_{out}(A - A_n) - g_i A]L . \tag{6}$$

Here $\sigma(\partial A_n)L$ is a surface tension term. If we assume the nuclei have circular cross section with diameter D,

$$R_{min}/L = \sigma\pi D + \frac{\pi D^2}{4}[\Delta g - \frac{1}{8\pi}({H_0}^2 - \tilde{H_0}^2)\frac{\Delta\mu}{\bar{\mu}^2}] + \dots , \tag{7}$$

where $\Delta g \equiv g_{in}^{(0)} - g_{out}^{(0)}$, and the terms left off are independent of A_n. Here, we see that R_{min} exhibits the expected maximum in D when $D = D_c$, where

$$D_c = \frac{2\sigma}{\frac{1}{8\pi}({H_0}^2 - \tilde{H_0}^2)\frac{\Delta\mu}{\bar{\mu}^2} - \Delta g} . \tag{8}$$

The interpretation of this equation is that for a given field H_0 the nuclei must have a diameter larger than D_c in order to be stable. When $D < D_c$, a decrease of the size D is energetically favorable and the nucleus will be absorbed; when $D > D_c$, an increase of D is favorable and the nucleus will grow up and initiate a new phase. D_c goes to infinity when $H_0 = H_{0c}$ where

$$H_{0c} = \sqrt{\tilde{H_0}^2 + \frac{8\pi\bar{\mu}^2}{\Delta\mu}\Delta g} . \tag{9}$$

If the external field is below H_{0c}, no stable nuclei is formed; but above H_{0c}, the critical size of nuclei is finite, and the stable nuclei occur. Hence H_{0c} is the critical external field for the structure evolution.

(II) Nucleation-controlled coarsening

Eq.(9) indicates that during the coarsening, there are two types of nucleate columns created, the stable columns with $D > D_c$ and the unstable columns with $D < D_c$. Therefore two separate processes are expected: (1) Growth of the stable nuclei. The stable nucleate columns will absorb the unstable columns and single chains. (2) Merging of the nuclei. During their growth, two nuclei may very quickly move together and merge into one larger nucleus, which then continues to grow. Microscopic observations of the structure evolution in ferrofluid emulsions have confirmed such nucleation processes [9]. The overall coarsening rates result from the compounding of the rates of these two separate processes. As observed by experiment[9], it takes shorter time for the process (2) than the process (1), but process (1) is observed to happen so much more often as to be dominant. A model was proposed for ER fluid that the fluctuating field could drive a chain to move toward another chain and then clump together into a baby column. These columns coalesce again to form the second generation of columns, so on and so forth, up to the final columnar structure[4]. All columns at each stage have an equal opportunity for coalescing and no one has preference. This model excludes the nucleation which we observed in the field induced structure evolution of MR fluid. It includes only the process (2) (or merging of the nuclei), which may be dominant in ER fluids, but not in our ferrofluid emulsions. The model presented below emphasizes that the dominant coarsening process in ferrofluid emulsions is the growth of the stable nucleate columns.

The nucleate columns with $D < D_c$ are created and destroyed rapidly. The columns with $D > D_c$ capture the individual chains from the unstable columns and recruit them for their growth. The basic coupling responsible for the growth process is the interaction between a stable column and a single chain. In the case of low volume fraction, the coupling between a chain and a column stems from a fluctuating field induced by the dipole density fluctuation[14], because of Peierls-Landau's instability. This is crucially important in the induced Nematic liquid crystal phase where each chain is a one dimensional sub-system[15]. The ordinary dipole-dipole interactions are screened, leading to an exponential decay with the distance between chains. Suppose that a chain is undergoing small longitudinal fluctuations, with droplet position $z_l = l2a + U_l$ where U_l is the displacement of the lth droplet and a is the droplet radius. The local density at long wavelengths is $n(z) = 1/(2a + U_{l+1} - U_l) \approx \frac{1}{2a}(1 - \frac{\partial U(z)}{\partial z})$. The field induced by a fluctuating chain, a distance ρ away, is[14,15]

$$H_z(\rho, z) \approx \frac{m}{\rho^2} \int d\psi F(\psi) n(\rho\psi + z), \tag{10}$$

where m is the magnetic dipole momemnt and $F(\psi) \equiv \frac{2\psi^2-1}{(\psi^2+1)^{5/2}}$. The fluctuation field induced by a column is given by

$$H_z(\rho, z) \approx \sum_{i=1}^{N} \frac{m}{|\vec{\rho} - \vec{\rho}_i|^2} \int d\psi F(\psi) n_i(|\ \vec{\rho} - \vec{\rho}_i \ |\ \psi + z), \tag{11}$$

where N is the number of chains in the column. The field-field correlation function is

$$< H_z(\rho, z) H_z^*(\rho, z') > \approx \sum_{i=1}^{N} \sum_{j=1}^{N} \frac{m^2}{|\vec{\rho} - \vec{\rho}_i|^2 |\vec{\rho} - \vec{\rho}_j|^2} \int \int d\psi d\psi' F(\psi) F(\psi') \\ < n_i(|\ \vec{\rho} - \vec{\rho}_i \ |\ \psi + z) n_j^*(|\ \vec{\rho} - \vec{\rho}_j \ |\ \psi' + z') > . \tag{12}$$

The density-density correlation is

$$< n_i(z)n_j^*(z') >\approx \frac{1}{4a^2}\{1+ < [\frac{\partial U_i(z)}{\partial z}][\frac{\partial U_j^*(z')}{\partial z'}] >\}$$
$$= \frac{1}{4a^2}[1 + \sum_k \sum_{k'} exp(ikz - ik'z')kk' < U_i(k)U_j^*(k') >]\,. \tag{13}$$

During the coarsening we assume that the stable column is not laterally tightly fused, in which case phonons excited in the different chains are not coherent with each other. Taking any reasonable inter-droplet potential and using the equipartition theorem, we obtain

$$< U_i(k)U_j^*(k') >\approx< U_i(k)U_i^*(k') > \delta_{ij} \sim \frac{k_BT}{m^2/a^5}(Nk^2a^2)^{-1}\delta_{ij}\delta_{k,k'}. \tag{14}$$

The long-wavelength approximation has used because the most significant contribution to the field will come from the long-wavelength fluctuations. Substituting Eqs. (13) and (14) into Eq. (12), the field-field correlation function can be rewritten as

$$< H_z(\rho, z)H_z^*(\rho, z') >\sim k_BT\sum_{i=1}^{N}\frac{a^2}{|\vec{\rho}-\vec{\rho}_i|^5}\int d\psi F(\psi)F(\psi + \frac{z-z'}{|\vec{\rho}-\vec{\rho}_i|}). \tag{15}$$

The mean square field induced by fluctuations of a column is

$$<| H_z(\rho) |^2>\sim k_BT\sum_{i=1}^{N}\frac{a^2}{|\vec{\rho}-\vec{\rho}_i|^5}\int d\psi[F(\psi)]^2. \tag{16}$$

Since the size of the column is typically much smaller than the separation between the stable and the unstable columns, at a transverse distance ρ from this column, the fluctuation induced field, or the square root mean field, is

$$H_f \equiv \sqrt{< H_z^2 >} \sim \frac{\sqrt{k_BTNa^2}}{\rho^{\frac{5}{2}}} \sim \frac{\sqrt{k_BTR^2}}{\rho^{\frac{5}{2}}}, \tag{17}$$

where $N \sim \frac{R^2}{a^2}$ with R the column radius. A droplet in a chain located at a distance ρ from the nucleate column has a dipole moment $\beta a^3 H$, where $\beta = \frac{\mu-\mu'}{\mu+2\mu'}$, with μ' the permeabilitity outside the droplet. It will experience a force which is given by

$$F_d \sim \frac{\sqrt{k_BT}\beta H a^3 R}{\rho^{\frac{7}{2}}}. \tag{18}$$

This force would drive the chains to move. The growth rate is determined by the rate at which these chains in the unstable nucleus are drawn to the stable column. The latter rate can be estimated by equating the force $F_d(\rho)$ to the viscous force, that is

$$C\eta a\frac{d\rho}{dt} = F_d, \tag{19}$$

where η is the solvent viscosity, $\frac{d\rho}{dt}$ is the velocity of the chain moving to the nucleate column, and C is a constant. Since $R \sim \phi^{1/2}\rho$ where ϕ is the volume fraction, we obtain a collision time

$$t_c \sim \eta(k_BT)^{-1/2}(\beta H)^{-1}a^{-2}\phi^{-1/2}\rho^{7/2}. \tag{20}$$

We now can find the qualitative dependencies of the characteristic length d, here taken to be the mean separation between columns, on the growth rate, the field, and the volume fraction as follows:

$$\begin{cases} d \sim t^{2/7}, \\ d \sim (\beta H)^{2/7} \sim H^{3/14}, \\ d \sim \phi^{1/7}, \\ d \sim \eta^{-2/7}. \end{cases} \tag{21}$$

We carried out a numerical, statistical mechanical calculation of $\beta(H)$. The result indicated $\beta \sim H^{-1/4}$ for $50G < H < 400G$ which was used in the second of Eqs. (15).

(III) Comparison with experimental results

The structural evolution of the ferrofluid emulsion has been studied using optical microscopy and static light scattering techniques, described in detail elsewhere[9]. Comparison of our theoretical model with the experimental results is listed in the following table and thereafter explained.

Power Law	Theoretical Values	Experimental Values
$d \sim t^{\alpha}$	$\alpha = 2/7 \approx 0.286$	$\alpha \approx 0.27$
$d \sim H^{\beta}$	$\beta = 3/14 \approx 0.21$	$\beta \approx 0.22$
$d \sim \phi^{\gamma}$	$\gamma = 1/7 \approx 0.14$	$\gamma \approx 0.17$
$d \sim (\sqrt{T}/\eta(T))^{\zeta}$	$\zeta = 2/7$	$\zeta \approx$?
$t_c \sim H^{-\nu}$	$\nu = 3/4 = 0.75$	$\nu \approx 0.81$

a) There are three sets of experimental data for volume fraction $\phi = .03$. (i) The light scattering measurement shows a power law of the form $d \sim t^{0.27}$, where the droplet diameter is 0.4μm, the sample thickness is 300μm, and the field is 450G. (ii) For the sample with droplet diameter 0.3μm and sample thickness 300 μm in a magnetic field of 380G, the optical microscopy measurement of the coarsening rate shows $d \sim t^{0.28}$. (iii) For the sample with diameter 0.26μm and thickness 700 μm in a magnetic field of 390G, it was found that $d \sim t^{0.263}$ by microscopy measurements. All measurements are very close our theoretical result $d \sim t^{0.286}$. It was also found that the chain length is proportional to $\lambda^{\frac{2}{3}}$, where $\lambda \approx \frac{\beta^2 H^2 a^3}{k_BT}$; for a sample with droplet diameter 0.3μm in a field $H = 400G$, the chain length is about $300\mu m$[15]. If the length of chain is increased by increasing λ, , α may tend toward the value $\frac{2}{7} \approx 0.286$, which was theoretically obtained assuming infinite chain length.

b) The field dependence of the separation between the columns was measured by light scattering, resulting in $d \sim H^{0.22}$. This is in good agreement with our theoretical result $d \sim H^{0.21}$.

c) The same measurement was performed for various volume fractions with the same cell thickness, $100\mu m$. It was found that all follow a similar power-law relation, $d \sim t^{\alpha}$. The volume fraction dependence of separation d fits our theoretical result $d \sim \phi^{1/7}$ quite well.

d) Although the coarsening rates, or α are the same in both microscopic measurements of (ii) and (iii), different saturation times and saturation separations are found, which are given by $t_{s2} = 87$s with $d_{s2} = 16.4\mu$m, and $t_{s3} = 265$ with $d_{s3} = 19.5\mu$m. The saturation time t_{s3} is more than three times t_{s2}. At first glance, it looks like a puzzle. But from Eq.(20) we have

$$\frac{t_{s2}}{t_{s3}} \approx (\frac{T_2}{T_3})^{-1}\frac{\eta(T_2)}{\eta(T_3)}(\frac{a_3\lambda_3}{a_2\lambda_2})^{1/2}(\frac{\phi_2}{\phi_3})^{-1/2}(\frac{d_{s2}}{d_{s3}})^{7/2}. \tag{22}$$

Substituting all parameters into the right hand side of Eq.(22), and setting $T_3 = T_2$, we obtain $(\frac{a_3\lambda_3}{a_2\lambda_2})^{1/2}(\frac{d_{s2}}{d_{s3}})^{7/2} = 0.42$, which is close to the experimental data given by $\frac{t_{s2}}{t_{s3}} \approx \frac{87s}{265s} \approx 0.33$.

e) In the growth rate measurement for ER fluid, λ is much larger than in MR fluid, and the existence of imaging charges in ER fluid enhances the symmetry along the chain direction. Hence the chains in the stable column would likely be coherent with each other. In that case, the fluctuation field H_f in Eq.(17) becomes $H_f \equiv \sqrt{< H_z^2 >} \sim \frac{\sqrt{k_BTN^2a^2}}{\rho^{\frac{5}{2}}} \sim \frac{\sqrt{k_BTR^4/a^2}}{\rho^{\frac{5}{2}}}$, which is proportional to R^2, instead of R. Thus we find $d \sim t^{2/5}$. This result fits the experimental data much better than the theory of Refs.4 which includes a somewhat paradoxical assumption.

f) The dependence of the growth rate on magnetic field in ferrofluid emulsion is found to be[9] $\frac{1}{\tau} \sim H^{0.81}$, which is the same as what was measured in ER fluid[4]. From Eq.(20) and $\beta \sim H^{-1/4}$ we have $\frac{1}{\tau} \sim H^{0.75}$. This good agreement between calculation and measurement rules out the direct dipole-dipole interaction mechanism responsible for the coarsening process, and provides significant evidence for fluctuation theory as suggested by Halsey and Toor[14].

Although our nucleation model and spinoidal decomposition both give low growth rates, there are essential differences between the two mechanisms. SD requires no thermal activation energy; that is, it occurs in the unstable rather than in the metastable region of a phase diagram. Secondly, in the nucleation mechanism the new phase starts from small regions, the nuclei, which then proceed to grow in extent. In SD mechanism the composition changes gradually in both directions from the average;

the growth is not in extent but in amplitude of density with a characteristic wavelength. Thirdly, in nucleation, a surface energy barrier is necessary, but for SD no surface energy barrier is needed; hence a very high degree of connectivity is a good indication of the SD mechanism. Our experimental results show the lack of the characteristics of the SD mechanism. We believe that the mechanism of coarsening is nucleation and its growth, and that the dominant interaction for coarsening of the columns is the fluctuation coupling caused by Peierls-Landau's instability.

ACKNOWLEDGMENTS. We would like to thank T. Mou, E. Haddadian, K. Javier, E. Lawrence, J. Pousset and S. Lim for providing us information about their measurements. We appreciate support from Dr. M. Hagenbuechle, Dr. R. Tao, and Dr. P. Pincus in the form of helpful discussions. This work is supported by the National Science Foundation DMR No. 9321201.

References

1. Yun Zhu, M. L. Ivey, P. Sheaffer, J. Pousset, and J. Liu, to be published.
2. Y. H. Hwang and X. I. Wu, Phys.Rev.E49, (1994) 3102.
3. E. Lemaire, Y. Grasselli and G. Bossis, J. Phys. II France 2, (1992) 359.
4. J. E. Martin, J. Odinek, and T. C. Halsey, Phys. Rev. Lett. 69, (1992) 1524.
5. J. Liu et al, Phys. Rev. Lett. 74, 2828 (1995).
6. M. Fermigier and A. P. Gast, J.of Colloid and Interface Science 54, (1992) 522.
7. R. Tao, Phys. Rev. B47, 423 (1993); R.Tao and Qi Jiang, Phys. Rev. Lett.73, (1994) 205.
8. R. D. Kamien and D. R. Nelson, J. of Statistical Physica, **71**, (1993) 23.
9. J. Liu, Tawei Mou, Yun Zhu, E. Haddadian, Xiao Lin and J. Pousset, to be submitted.
10. I. M. Lifshitz and V. V. Slyozov, J. Phys. Chem. Solids, **19**, (1961) 35.
11. L.D.Landau and E.M.Lifshitz, *Statistical Physics*, 2nd ed. (Addison-Wesley), 1969.
12. Hao Wang, Yun Zhu, C. Boyd, Wei-li Luo, A. Cebers, and R. E. Rosensweig, Phys. Rev Lett., **72**, (1994) 1006.
13. I. A. Privorotskii, Soviet Physics JEPT Lett., **5**, (1972) 429; JEPT, **32**, (1972) 625; L.D.Landau and E.M.Lifshitz, *Electrodynamics of Continuous Media*, 2nd ed. (Addison-Wesley), 1969.
14. T. C. Halsey and W. Toor, J. Stat. Phys. 61, (1990) 1257.
15. Yun Zhu and Jing Liu, et al, in preparation for publication.

ON THE ROLE OF CONDUCTIVITIES AND FREQUENCY IN THE ELECTRORHEOLOGICAL EFFECT

C. BOISSY, P. ATTEN and J.-N. FOULC
Laboratoire d'Electrostatique et de Matériaux Diélectriques
CNRS et Université Joseph Fourier - Grenoble 1
B.P. 166 - 38042 Grenoble Cedex 9 (France)

ABSTRACT

The aim of this investigation is to establish in the case of DC field or low frequency field the role of conductivities of both suspending liquid and particles in the electrorheological (ER) effect. We approach the problem in considering the case $\sigma_S<\sigma_L$ and the case $\sigma_S>\sigma_L$. For $\sigma_S>\sigma_L$ we estimate an upper limit for σ_S to obtain stable chains. We also investigate the question of the transition between a behaviour determined by the conductivities of the constituents and the behaviour determined by their permittivities.

1. Introduction

The aggregation of particles in electrorheological (ER) fluids was long ascribed to polarisation of materials due to the dielectric constant mismatch between suspending liquid and solid particles. However this explanation is unsatisfactory, especially in the case of DC or low frequency applied field where the electric field distribution is controlled by the conduction properties of both liquid and solid constituents. Although emphasised by many authors, the role of conduction has not been frequently detailed. The works of Anderson[1] and Davis[2] give very good indications on the role of conductivities and recently Atten *et al.*[3] addressed the question of estimating the attraction force between two spheres in contact.

It now clearly appears that the electrorheological effect is promoted by a conductivity mismatch in the case of DC and low frequency AC fields and by a dielectric constant mismatch for high frequency fields. This is the reason why ER fluids seldom have a good behaviour in the whole frequency range. In the case of DC fields it is not necessary that the particle have a bulk conductivity, surface conductivity playing the same role.

In this paper we examine, in the case of DC fields, the various behaviours of particles in ER fluids depending on the values of the conductivities σ_S and σ_L of the solid and the liquid respectively. Qualitative considerations[4] suggest that an aggregation of particles (and hence an ER effect) is possible only if $\sigma_S > \sigma_L$. This is confirmed by experiments which reveal, in some cases, a "negative" effect. When σ_S takes very high values, the ER effect should disappear because two particles coming in contact can exchange charges quickly and separate, therefore preventing the formation of chains. Thus we can expect an optimum value of σ_S for which the ER effect is maximum. The question of transition between a behaviour determined by the conductivities in DC field and a behaviour determined by the permittivities in AC field is also briefly considered.

2. Observation of a "negative" ER effect ($\sigma_S<\sigma_L$)

When the solid particles are more insulating than the liquid, a qualitative argument shows that the particles cannot form stable chains[4], which implies that the ER effect is impossible. Experiments on suspensions such that $\sigma_S < \sigma_L$ show that the fluids, as expected, do not present an ER effect but sometimes even exhibit a "negative" effect : the apparent viscosity of the suspension decreases. Let us briefly recall recent observations on two fluids[5]. In the first one, the solid phase was a powder of PMMA ($\phi \sim 15$ μm, $\sigma_S \sim 10^{-14}$ S/m) dehydrated during 40 hours in vacuum, dispersed in a more conducting ($\sigma_L \sim 3\ 10^{-10}$ S/m) mixture of two liquids, Ugilec T and mineral oil TF50 (both from Elf-Atochem Company), in proportion such as to have a liquid density equal to that of the solid phase. The second fluid was a suspension of dehydrated high density polyethylene particles ($\phi \sim 300$ μm, $\sigma_S \sim 2\ 10^{-16}$ S/m) in a mixture of silicone oils again of density equal to that of the solid phase ($\sigma_L \sim 10^{-13}$ S/m). The rheograms of Fig.1, obtained with PMMA powder, illustrate the "negative" effect, i.e. the decrease of the shear stress with applied field. The rheograms with PE exhibit the same behaviour.

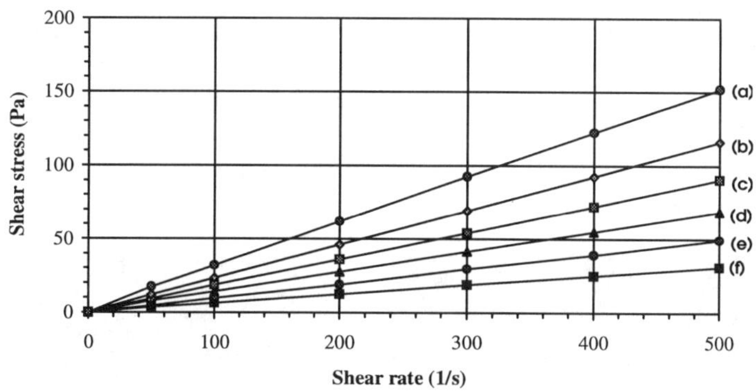

Fig. 1 Variation of the shear stress as a function of shear rate for an ER fluid consisting of 30% PMMA powder in Ugilec T- Elf TF50 oil mixture : No field (a); E = 0.5 kV/mm (b); E = 0.7 kV/mm (c); E = 1 kV/mm (d); E = 1.5 kV/mm (e); E = 3 kV/mm (f).

A scaled-up experiment helped us to understand the origin of this decrease in apparent viscosity. Observations on millimetre size particles ($\phi \sim 2$ mm; polyethylene granules in silicone oil) between two coaxial cylindrical electrodes (gap $\cong$ 20 mm) showed that electrophoretic phenomena occur for these particles less conducting than the liquid (Fig. 2). Electrophoresis is often observed in DC field with electrically insulating ceramic particles[6] and, as pointed out by Stangroom[7], the layer of pure liquid left by particles act as a lubricant and leads to a decrease in apparent viscosity.

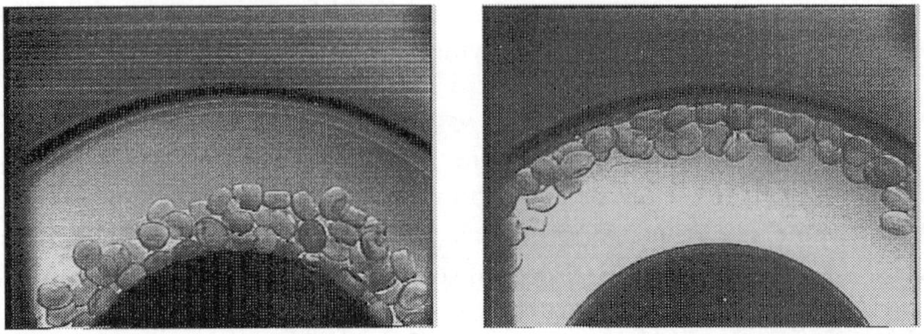

Fig. 2 Behaviour of polyethylene granules ($\sigma_S \sim 2\ 10^{-16}$ S/m) in silicone oil ($\sigma_L \sim 10^{-13}$ S/m) in DC field and for the two polarities.

The decisive role of the conductivity parameter $\Gamma = \sigma_S/\sigma_L$ is illustrated further by another experiment in the same cell using polyamide granules ($\sigma_S \cong 1.7\ 10^{-8}$ S/m) in a mixture of two liquids (TF50 and Ugilec T). Initially the liquid had a conductivity lower than that of the solid, $\sigma_L \cong 10^{-10}$ S/m ($\Gamma \cong 200$) and we observed the formation of rather rigid chains (Fig. 3-a) (and an increase of the shear stress). After adding a salt which resulted in a higher conductivity value $\sigma_L \cong 2\ 10^{-7}$ S/m ($\Gamma \cong 0.2$), the granules migrated very slowly towards the outer cylindrical electrode (Fig. 3-b) whatever the polarity of the field. In this case, electrophoresis does not play a role and the very slow motion is due to the interfacial polarisation which draws the granules in the regions of low field (dielectrophoretic-like effect). Clearly there is no significant interaction between granules and the only effect in this case might be a very slight decrease of the shear stress.

We can conclude that the condition $\sigma_S > \sigma_L$ is a necessary condition to obtain an ER effect in DC field.

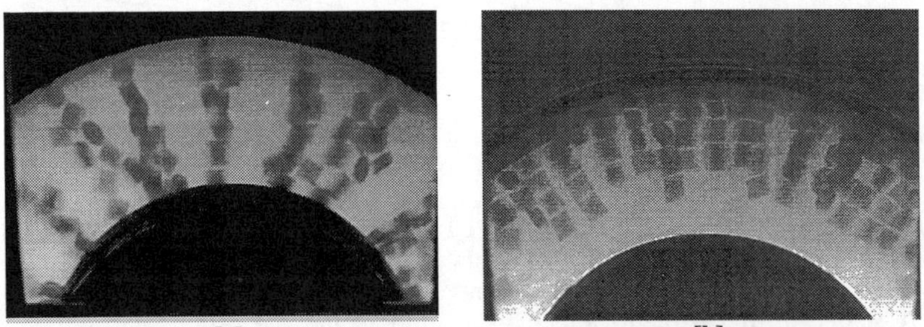

[a] [b]

Fig.3. Behaviour of polyamide granules ($\sigma_S \cong 1.7\ 10^{-8}$ S/m) in TF50/UgilecT mixture: a) $\sigma_L \sim 10^{-10}$ S/m; b) with added AOT : $\sigma_L \cong 2\ 10^{-7}$ S/m (the behaviour is the same for the two polarities).

3. Formation of chains in the case $\sigma_S > \sigma_L$

Anderson[1] pointed out the fact that if two conducting particles are subjected to an electric field they are drawn together by their induced moment and, when they are in contact, there is a redistribution of charge that leads to their separation (Fig. 4). But he also pointed out that if the particles have enough time to form chains between the electrodes, a steady current passes through the chains and provide an electrostatic force that stabilise them. It is this idea which is developed in the conduction model[3].

Thus we can give a schematic view of the two different behaviours of the particles when a field is applied (Fig. 5). Without electric field the particles are randomly dispersed [a]; when the electric field is applied [b] the particles attract each other and many pairs of close particles tend to come into contact (see pairs 2-3 and 4-5 for example) and some particles drift towards the electrodes (1). After this step there are two possibilities : in the first one [c], the particles have time long enough to build up longer and longer segments which finally span the gap and the current stabilises the chain; in the second one [d], the particles separate before the complete formation of the chain and we then expect a weak electrorheological effect.

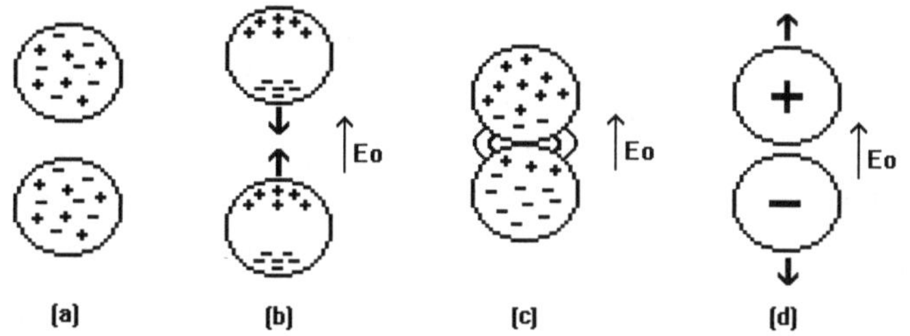

Fig. 4 Behaviour of two close isolated slightly conducting spheres subjected to an electric field.

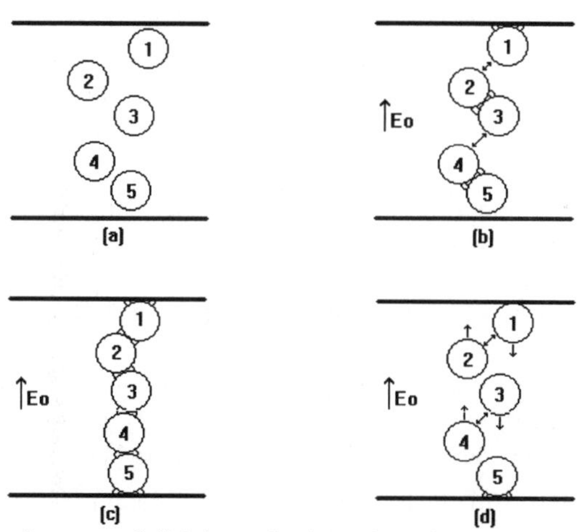

Fig. 5 Behaviour of a group of slightly conducting spheres between two electrodes subjected to an electric field. [a] randomly dispersed without electric field. [b] they attract each other or with an electrode when the field is applied. [c] they have enough time to create a chain before separating. [d] they separate before a chain can be formed.

The characteristic time for charge transfer between two particles once in contact is[8] $\tau \sim 10$ to $20\ \varepsilon_L/\sigma_S$ (the dependence on ε_L comes from the charge density at the solid/liquid interface being determined by the electric field in the liquid). We have to compare τ with the time t_{ch} required to pass from pairs of particles in contact to the completed chain. If we consider that the latter time is on the order of the mean time of initial formation of particles pairs, we have $t_{ch} \sim \eta/\varepsilon_L E_o^2 \sim 1$ ms with the typical values $\eta = 2\ 10^{-2}$ Pa.s, $\varepsilon_L = 2.2\ \varepsilon_o$ and $E_o = 1$ kV/mm. We then obtain the following condition for the formation of stable chains[8] : $\sigma_S \leq \sigma_{max}$ with $\sigma_{max} \sim 10^{-7}$ S/m. From the expression of t_{ch} we see that the condition $\tau > t_{ch}$ involves the electric field. Therefore with particles of greater conductivity σ_S, the ER effect would require a larger applied field E_o to manifest. This might explain why with some fluids the ER effect manifests above a certain field threshold.

Gow and Zukoski[9] report that Kelly and Block[10] examining the influence of solid conductivity observed a maximum in yield stress occurring for σ_S about 10^{-5} S/m. Uejima[11] reports that with cellulose, the ER effect decreases when the amount of adsorbed water is above 12% (in weight). From some measurements of conductivity of wet cellulose we performed, we found that this amount of water corresponds to the same value $\sigma_S \sim 10^{-5}$ S/m. We examined qualitatively the role of solid conductivity on a mixture of cellulose powder and mineral oil. By using a cylindrical plexiglass cell with circular electrodes at a distance d = 3 cm and working with very low volume fractions it was possible to observe different behaviours of the fluid. In Fig. 6-a the dried cellulose ($\sigma_S < \sigma_L$) is such that the particles are attracted by the cathode (electrophoresis). Fig. 6-b is relative to particles containing about 5% of water giving $\sigma_S \sim 10^{-9}$ S/m $> \sigma_L$: the chains are stable. In the case of Fig. 6-c, the cellulose particles are much more conducting and exhibit a tendency to form chains; but the partial chains do not span the gap and go to and fro between the electrodes. Increasing further the solid conductivity ($\sigma_S \sim 10^{-4}$ S/m corresponding to 14% in weight of water) results in particles not forming chains except very unstable ones on the plexiglass surface (Fig. 6-d). The same behaviours have been obtained with polyaniline powder of different conductivities. Rheological measurements carried out with these two powders confirm the observations : a well characterised positive ER effect could be obtained only in case [b] (in the other cases the effect was insignificant or negative).

As a conclusion, from the experimental observations on ER fluids, a value on the order of 10^{-5} S/m is an upper bound for the conductivity of solid particles to obtain a significant ER effect under DC field. The difference between this value of σ_{max} and our rough estimate ($\sigma_{max} \sim 10^{-7}$ S/m) is not very surprising. Many parameters can influence the charge transfer as the geometric details of the contact zone between particles, the strength of the electric field, the conduction processes in the liquid, etc . . . The formation of the chains and the redistribution of charge between particles in contact, which both depend on the volume fraction of solid particles, must be analysed more closely in order to derive a more realistic prediction for σ_{max}.

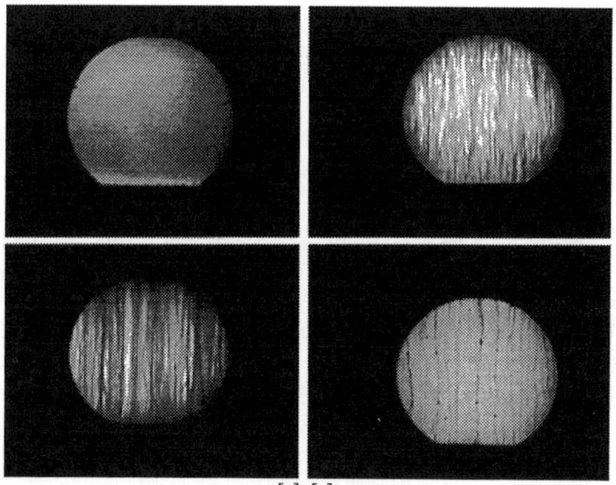

Fig. 6 Behaviour of cellulose in oil TF50 ($\sigma_L \sim 10^{-12}$ S/m) : [a] $\sigma_S < \sigma_L$ particles are attracted by the upper electrode, [b] $\sigma_S \sim 10^{-9}$ S/m the particles form stable chains [c] the particles are wet and the chains go to and fro between the electrodes [d] the amount of water is greater and there is no chains.

4. Influence of frequency

As we recalled, there are two basic mechanisms which promote the polarisation of particles. The first one (interfacial polarisation) operates only in the case of DC or low frequency fields and depends on the conductivities of solid and liquid materials. The second one works only at high frequency and depends on the dielectric constant mismatch between particles and suspending liquid. This is the reason why insulating ceramic particles show greater ER effect under AC fields than under DC fields[6,12] and why water promoted fluids are more effective in DC fields (see for example a work by Ginder and Ceccio[13] who present the dependence of the shear stress on field frequency for a fluid from AFS – 12-15% water – and another one with barium titanate). Thus there is a crossover frequency[1,12], f_c, between the two regimes which can be estimated by using a Maxwell-Wagner approach[2].

On Fig. 7 we present the variations, versus field frequency, of the interaction force between two half-spheres of radius 0.7 cm (the experimental apparatus and procedure are described in Atten *et al.*[3]). The polyamide half-spheres of conductivity $\sigma_S \sim 1.7\ 10^{-8}$ S/m are immersed in mineral oil Elf-TF50. The conductivity of the oil and, therefore, the conductivity ratio $\Gamma = \sigma_S/\sigma_L$ can be changed by adding an ionic surfactant (AOT). We can estimate[1] that $f_c \sim 8$ Hz. For technical reasons it is not possible to realise good measurements below 20 Hz and we can compare only the measurements performed for frequencies above 20 Hz and for DC field (values plotted at f = 1 Hz on Fig. 7). We see that, to a first approximation, the force between the half-spheres is the same whatever Γ for f > 20 Hz; thus in the regime dominated by permittivity effects, the force is roughly independent of the conductivities. The slight decrease of the force with frequency is mainly due to the decrease with frequency of the permittivity of the solid material and of the subsequent dielectric mismatch. One interesting fact to note from Fig.7 is that, depending on Γ, the force under DC field can be greater (Γ ≥ 300) or lower (Γ = 30) than that under AC field (f = 20Hz).

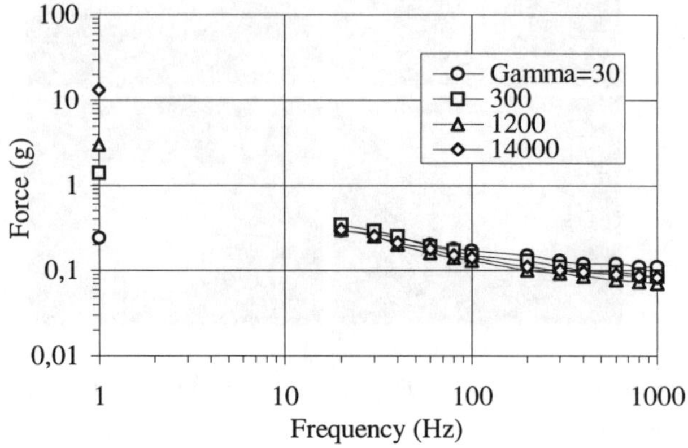

Fig. 7 Dependence of the force between two half-spheres on field frequency for different gamma and for a tension U=4 kV.

5. Discussion and conclusion

In the light of the different behaviours described in this paper we can conclude that there are two mechanisms which lead to the ER effect. The first one, widely described in litterature, involves the permittivity effects which are dominant at high enough frequencies. In the other mechanism at work under DC and low frequency AC fields, the conductivities of both constituting materials play the dominant role. In the latter case, we have to distinguish between the conductivity of the solid particles being greater ($\sigma_S > \sigma_L$) or smaller ($\sigma_S < \sigma_L$) than that of the suspending liquid.

Results presented in this paper recall that for $\Gamma = \sigma_S/\sigma_L < 1$, the ER effect under DC field or low frequency field is impossible[4] and, in some cases, appears as "negative"[5]. Measurements on two half-spheres[3] show that, in the case $\sigma_S > \sigma_L$, the conductivity of the solid particles has to be at least three orders of magnitude greater than the liquid conductivity in order to obtain an important ER effect. As the attraction force appears to increase smoothly[14] with Γ, it can be infered that it is desirable to work with the highest possible value of σ_S. But there is an intrinsic limitation in the conductivity of the solid. The general argument presented here suggests that for $\sigma_S \geq \sigma_{max}$ the formation of chains under DC field or low frequency field should be impossible because of the rapid redistribution of charge between two particles coming into contact and their subsequent repulsion. The practical limiting value ($\sigma_{max} \sim 10^{-5}$ S/m) is much greater than the very rough estimate and this question requires further investigation. In practice a stronger limitation for σ_S arises from the constraint relative to the current flowing through an ER fluid. The condition on current density $j \leq 10\ \mu A/cm^2$ leads to $\sigma_S \leq \sigma_m$ with $\sigma_m \sim 10^{-7}$ S/m under a field on the order of 3 kV/mm. We then conclude that the practical range of values of σ_S is :

$$10^3\ \sigma_L \leq \sigma_S \leq 10^{-7}\ \text{S/m}$$

This also gives a condition for σ_L : $\sigma_L < 10^{-10}$ S/m which is fulfilled by numerous liquids.

In an ER fluid under shear the chains which tend to form are continuously broken and do not span the gap between the electrodes. Here the current is expected to be shear dependent and to take lower values than in the case of no shear[15]. Now it is not easy to give an idea of the "structuration" in this case. If we take the picture of pieces of chains anchored on the electrodes, we are facing contradictions because after some time these pieces will tend to separate from the electrodes; moreover in the absence of a current passing along the chains, there is no mechanism of stabilisation of these pieces of chains. It is clear that the behaviour of particles and the structuration of an ER fluid under shear raise very complicated questions which require acute observations and refined analysis.

6. Acknowledgements

This work was performed with the financial support of the company Elf-Aquitaine (under contract CNRS/Elf-Aquitaine No 50.8016). The authors wish to express their thanks to Pr. N. Felici for fruitful discussions.

7. References

1. R.A. Anderson, *Proc. 3rd Int. Conf. on ER Fluids*, R. Tao, ed. (World Scientific, Singapore, 1992) 81.
2. L.C. Davis, *J. Appl. Phys.*, **72** (4), (1992) 1334.
3. P. Atten, J.-N. Foulc and N. Félici, *Int. J. Mod. Phys. B*, **8**, (1994) 2731.
4. J.N. Foulc, N. Félici, and P. Atten, *C. R. Acad. Sci. Paris*, **t. 314 Série II** (1992) 1279.
5. C. Boissy, P. Atten and J.N. Foulc, *J. Electrostatics*, (1995) in press.
6. C.A. Randall *et al.*, *Proc. 4th Int. Conf. on ER Fluids*, R. Tao & G.D. Roy, eds. (World Scientific, Singapore, 1994) 60.
7. J. E. Stangroom, *J. Stat. Phys.*, **64**, (1991) 1059.
8. N. Felici (private communication).
9. C.J. Gow and C.F. Zukoski, *J. Coll. Int. Sci.*, **136** (1), (1989) 175.
10. J.P. Kelly and H. Block, *Proc. 3rd Annual Chem. Congres of North America, Toronto, Canada,* 1988.
11. H. Uejima, *Jap. J. Appl. Phys.*, **11** (3), (1972) 319.
12. T. Garrino, D. Adolf and B. Hance, *Proc. 3rd Int. Conf. on ER Fluids*, R. Tao, ed. (World Scientific, Singapore, 1992) 167.
13. J. M. Ginder and S.L. Ceccio, *J. Rheol.*, **39** (1), (1995) 211.
14. P. Atten, C. Boissy and J.-N. Foulc, "The conduction model of electrorheological effect revisited", (in this conference).
15. J.-N. Foulc and P. Atten, *Proc. 4th Int. Conf. on ER Fluids*, R. Tao & G.D. Roy, eds. (World Scientific, Singapore, 1994) 358.

The Numerical Analysis on Channel Flows of ER Fluid

Y. J. Jang, M. S. Suh, M. S. Yeo
Agency for Defense Development,
P.O. Box 35 Yoosung, Taejon 305-600 KOREA(R.O.K)
and
S. B. Choi
Smart Structures and Systems Laboratory
Department of Mechanical Engineering, Inha University
Incheon 402-751, KOREA(R.O.K)

ABSTRACT

A numerical analysis is performed on the flow characteristics of an electro rheological (ER) fluid when fully developed ER fluid flow in a channel is instantly subjected to electric fields. The Bingham plastic model is used to describe the ER fluid behavior. The flow field is assumed to be incompressible and isotropic. The SIMPLER algorithm with staggered grid system is employed to solve the momentum and continuity equations governing the steady state, laminar flow. The effects of electric field intensity and upper wall velocity changes on velocity profiles are investigated.

1. Introduction

Some applications of Electrorheological(ER) fluids to industrial devices include a damper[1], an engine mount[2], a valve, a clutch, and so on. In the automotive industry, the major items of these devices are the damper and engine mount, and these are related to the damping of mechanical vibrations.

The mechanism of damping in these ER devices consists of many channels filled with ER fluids. Electrical fields on the channels change the characteristic of ER fluids and restrict ER fluid from flowing through the channel and finally generate the large pressure drop which results in increasing the damping force. In most of these ER devices, the ER fluid flow passing through a channel forms a Poiseuille-flow configuration. The velocity profiles in a channel dramatically change with the electric field strength(E). That is, as E is increased, a plug flow region(non-sheared region) appears around a channel center and expands to electrode walls. These changes enable us to control the damping force of the damper or the engine mount.

In the analysis of the ER fluid flow, the specification of the viscous term in the Navier-Stokes equation is the most important step. In general, Non-Newtonian fluid has a non-linear viscosity between the shear stress and the shear rate. Phillips[3] investigated theoretically the velocity profiles, the pressure gradients and the flow rate in a channel with the Bingham plastic model. Wang[4] et al. developed the mathematical formulation for solution of coupled Non-Newtonian flow with the Bingham plastic model and the electric field using the finite element method.

In the present study Numerical simulation is performed on the flow characteristics of the ER fluid when fully developed ER fluid flow in a channel is instantly subjected to electric fields. The momentum and continuity equations are solved by the finite difference method(F.D.M). The SIMPLER algorithm with staggered grid system is employed to improve the convergence of solution and the Bingham plastic model is used to describe the ER fluid behavior.

2. ANALYSIS

2.1 Bingham Plastic Model

The ER fluid model to be pursued in this research is that of Bingham plastic whose constitutive equation is represented by Eq. 1.

$$\tau = \tau_y(E) + \eta \cdot \dot{\gamma} \tag{1}$$

The ER fluid possesses a yield stress, τ_y. When stressed beyond the yield point, the ER fluid is assumed to flow with constant plastic viscosity, η. The η is taken to be independent of the electric field strength. The $\dot{\gamma}$ is the shear rate. The τ_y is assumed to behave as following equation.

$$\tau_y(E) = \alpha \cdot E^{\beta} \tag{2}$$

The α and β are constants which are determined according to the characteristics of the ER fluid.

To analyze the ER fluid flow, we introduce a variable viscosity model[3] in which the viscosity of ER fluid is varied locally. To do this, consider Fig. 1. Here Bingham plastic behavior is contrasted to Newtonian fluid. A Newtonian fluid has the following relation between the shear stress and the shear rate.

$$\tau = \mu \frac{du}{dy} \quad (\mu = \frac{\tau}{du/dy}) \tag{3}$$

where μ is the Newtonian viscosity. For Bingham plastic flow, the shear stress τ_A at the point A can be described by

$$\tau_A = \mu_A (\frac{du}{dy}) \tag{4}$$

Using Eq. 1 for shear stress

$$\mu_A = \frac{\tau_y}{(du/dy)} + \eta \tag{5}$$

Eq. 5 is graphically described in Fig. 2. Note that as the shear rate approaches zero, the viscosity becomes very large, and that as the shear rate increases, the apparent viscosity approaches the plastic viscosity.

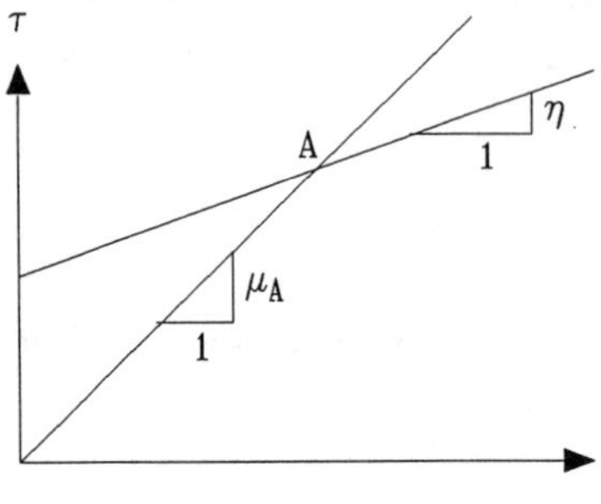

Fig. 1. Comparison of Bingham plastics and Newtonian fluids

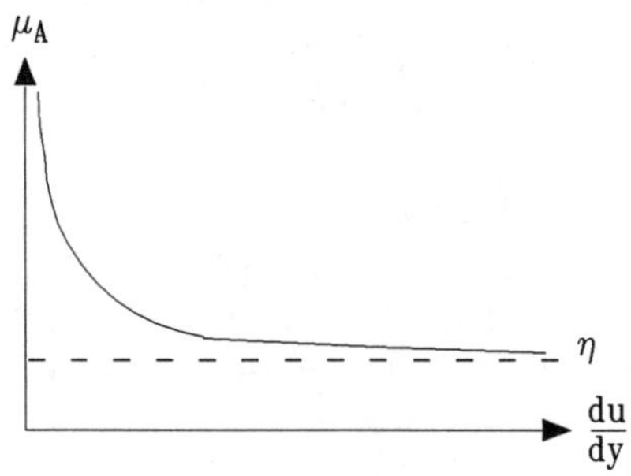

Fig. 2. Relationship of the apparent viscosity and the plastic viscosity

2.2 Governing Equations

The governing equations describing the ER fluid behavior are simplified by adopting following assumptions.

(1) Steady, two-dimensional, laminar flow.

(2) Incompressible flow which satisfies the continuity equation.

(3) Isotropic flow which has the same physical characteristic in all directions.

(4) The body force can be ignored.

The continuity equation and the momentum equations can be written as follows.

Continuity Equation :

$$\frac{\partial u}{\partial x}+\frac{\partial v}{\partial y}=0 \tag{6}$$

Momentum Equation : (ρ is the density and p is the pressure)

$$\frac{\partial}{\partial t}(\rho u)+\frac{\partial}{\partial x}(\rho u\cdot u)+\frac{\partial}{\partial y}(\rho v\cdot u)=\frac{\partial}{\partial x}\left(\mu\frac{\partial u}{\partial x}\right)+\frac{\partial}{\partial y}\left(\mu\frac{\partial u}{\partial y}\right)-\frac{\partial p}{\partial x} \tag{7}$$

$$\frac{\partial}{\partial t}(\rho v)+\frac{\partial}{\partial x}(\rho u\cdot v)+\frac{\partial}{\partial y}(\rho v\cdot v)=\frac{\partial}{\partial x}\left(\mu\frac{\partial v}{\partial x}\right)+\frac{\partial}{\partial y}\left(\mu\frac{\partial v}{\partial y}\right)-\frac{\partial p}{\partial y} \tag{8}$$

The difficulty in the calculation of the velocity field lies in the unknown pressure field because there is no obvious equation for obtaining pressure. In case of calculating the velocity and pressure for the same grid points, we have the unrealistic solutions. To avoid such troublesome solutions, we adopt the staggered grid in which the velocity and pressure components are calculated for the each different grid points as sketched in Fig.3.

The discretization is accomplished by finite differences on a staggered grid with power-law scheme using the method provided by Patankar[5]. The Algorithm employed for numerical analysis is the SIMPLER which is a revised version of SIMPLE algorithm which corrects the velocities and pressures to satisfy the continuity equation after solving the momentum equations with the guessed pressure field.

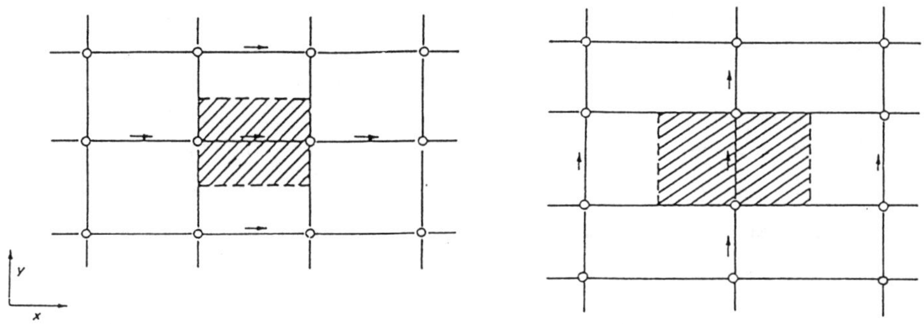

(a) Control volume for u (b) Control volume for v

Fig. 3. Staggered locations for u and v

(→ : stands for u, ↑ : stands for v, o : stands for pressure)

2.3 Geometry and Grid

The geometry of the ER channel is presented in Fig. 4. The ratio of the height to the length of the channel is 1:100. The region which is subjected to electric fields is behind fully developed velocity profiles. This region is about 14% of the channel length. In the computational plane we use uniform grids of 60×20 as sketched in Fig. 5.

2.4 Boundary Conditions

All velocity components at the wall are zero based on the assumption of no-slip condition at the solid wall boundary.

Inlet velocities :

$$u = u_{in}, \quad v=0 \tag{9}$$

At the outlet, the Neumann conditions are assumed as follows.

$$\frac{\partial u}{\partial x} = 0 , \quad \frac{\partial v}{\partial x} = 0 \tag{10}$$

The bottom wall is fixed and the upper wall is fixed or given a constant velocity in the x-direction.

2.5 Numerical analysis

One of the major problems encountered in using the Bingham plastic model is the theoretically infinite viscosity in regions of plug flow[6]. That is, a singularity will happen when the shear rate(du/dy) becomes identically zero(Eq. 5). In this research, this phenomenon happens at the point y=0.5mm between the point 10 and point 11 in Fig. 6 ($\because\ u_{10}=u_{11}$). To eliminate this problem the viscosity μ_A which is due to the velocity difference between the point 1 and point 2 is used for the viscosity of the point 2, the viscosity between the point 2 and 3 for the point 3, · · ·, the viscosity between the point 9 and 10 for the point 10, and the viscosity between the point 11 and 12 for the point 11. In the end we don't use the viscosity of the singularity point. Near the walls the shear rate is very high, so the plastic viscosity is used for the point 1 and 20.

We use the under-relaxation factors[5] for the velocities u,v and the pressure p to improve the convergence of solution. As for the convergence conditions, the difference between the flow rates to get in and get out of the control volume should be under 10^{-6} at each iterating calculation and the four significant figures of the physical quantity must not be changed at each iteration.

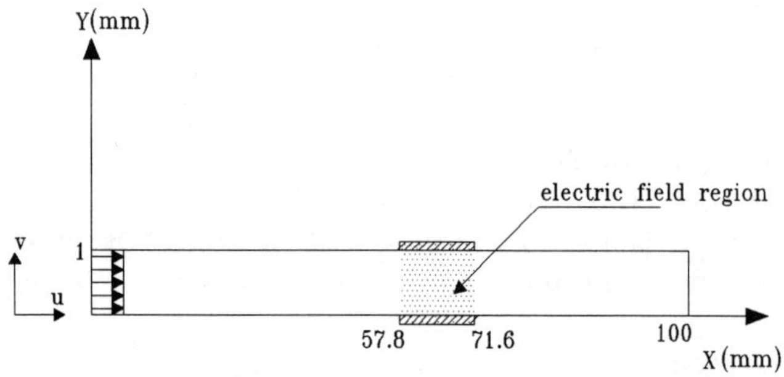

Fig. 4. Flow geometry

Fig. 5. Grid generation(60×20)

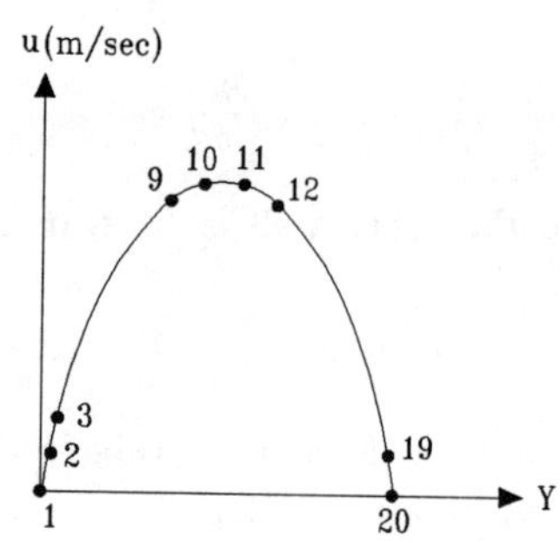

Fig. 6. Fully developed velocity profile

3. Results and Discussion

The behavior of the ER fluid is significantly changed according to the characteristic of its particles and base liquid. In this research, we chose experimentally the values of the main factors in the ER fluid and investigated the ER fluid behavior case by case. For the constants α and β in Eq. 2., we chose 71.8, 1.31, respectively from the many experiment data. The plastic viscosity η is fixed to $\eta = 0.1\ Pa \cdot \sec$. The inlet velocity of the channel is taken to be $u_{in} = 0.05$m/sec and the density $\rho = 1{,}010\ kg/m^3$. The variations of the centerline velocity along axial distance are shown in Fig. 7. As seen, when the electric field is not applied, the velocity profile in a channel is fully developed around the point x=5mm. In the region subjected to the electric field E=3kV/mm, the centerline velocities are abruptly reduced.

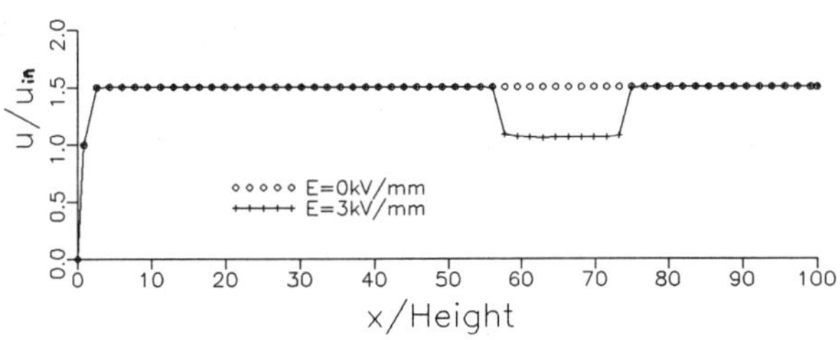

Fig. 7. Variation of the centerline velocity with axial distance

3.1 when the upper wall is fixed

The velocity vectors along the channel (49.1mm < x < 80.2mm) are shown in Fig. 8, when the electric field of 3kV/mm is applied. For the behavior of ER fluid at the entrance(x=56.0mm) of the region subjected to the electric field, the flow at the center(y=0.5mm) is retarded due to the electric field region and the rest of the flow rate in this region is moving to the both walls(that is, in this region v velocities exist). In the electric field region, we can see that the plug behavior(non-sheared region) is developed along the channel. Just after the electric field region(x=73.3mm), the ER fluid is moving to the center region from the both walls. We can conclude that this phenomenon happens because the resistance caused by no-slip condition at the walls is greater than that at the center region. However, the plug behavior region still exists.

In Fig. 9 the above velocity vectors are compared with each other on the graphs. As seen from the Fig. 9-(a) the plug flow is fully developed at the point x=61.2mm. The region from the point x=56.0mm to the point x=61.2mm is the transient one. The recovery from the plug behavior to Newtonian fluid as in the upstream is completed at the point x=76.7mm as shown in Fig. 9-(b).

In Fig.10 we changed the electric field strength from 0kV/mm to 5kV/mm and investigated the velocity profiles. The comparison point is x=64.7mm in which the plug behavior is fully developed. As the electric field strength is increased, the plug flow regions expand to the both walls.

3.2 when the upper wall is moving in the downstream

The velocity profiles at E=1kV/mm in the presence of a moving upper wall are shown in Fig.11. The upper wall velocity is varied from u_p=0.0m/sec to u_p= 0.1m/sec. It is observed that as the upper wall velocity is increased, the core(plug region) of ER flow is close to the moving wall. The core becomes attached to the moving wall when the upper wall velocity is u_p=0.05m/sec.

4. Conclusion

To investigate the flow characteristics of the ER fluid, the Bingham plastic model has been employed in conjunction with the assumption of isotropic fluid flow. The finite difference scheme with the SIMPLER algorithm which has been developed, comparatively well depicted the ER fluid behavior. The FDM which is the most widely used in fluid flow simulations for Newtonian fluids can be adopted for the solution of ER flows which is complicated, and non-linear. It produces physically reasonable results. It is important to note that the staggered grid system is adequate to improve the convergence of the solutions.

To the authors' knowledge, the results about the realistic pressure drop in the ER fluid flow which is numerically simulated have not been yet published. To learn the damping force, the investigation of the pressure drop in a channel is needed and it is anticipated that the pressure drop can be predicted by using this algorithm and appropriately altering boundary conditions.

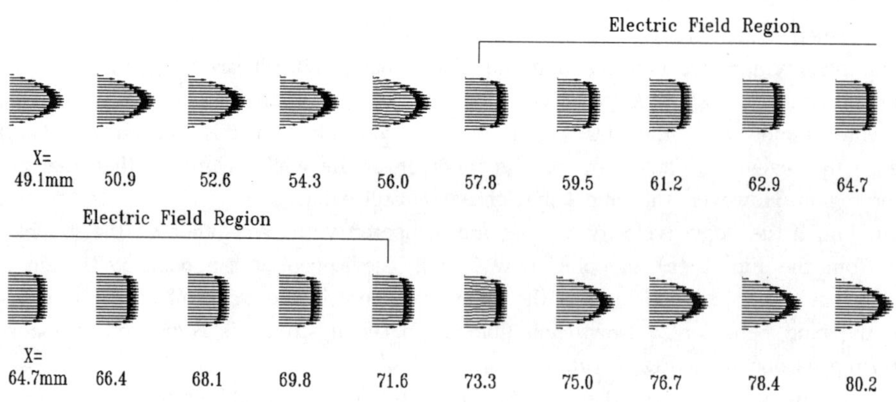

Fig. 8. Velocity vector profiles in a channel(E=3kV/mm)

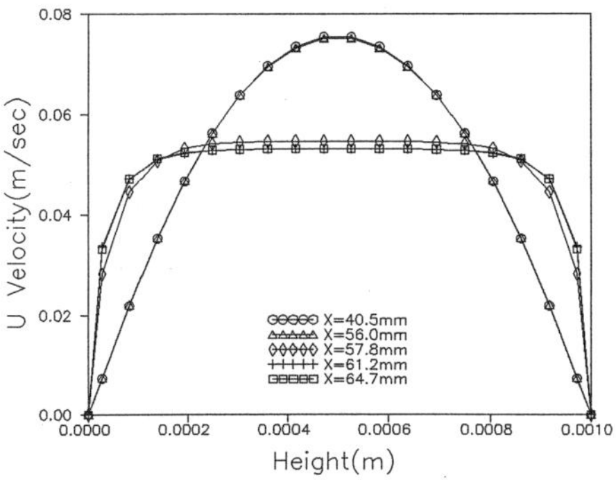

Fig. 9.(a) Streamwise velocity profiles (56.0mm<x<64.7mm)

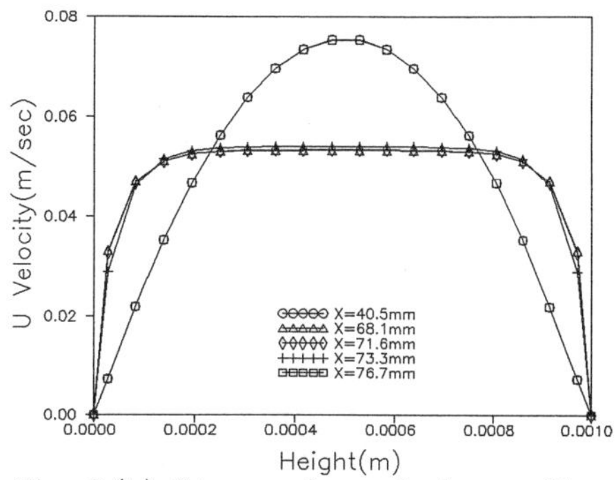

Fig. 9.(b) Streamwise velocity profiles (68.1mm<x<76.7mm)

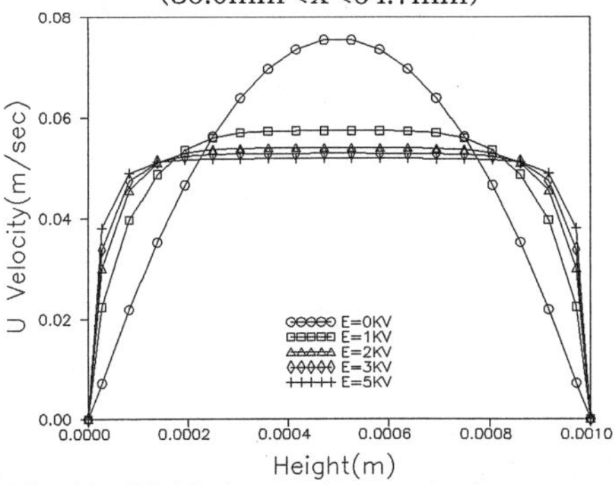

Fig.10. Field depentdent velocity profiles at X=64.7mm

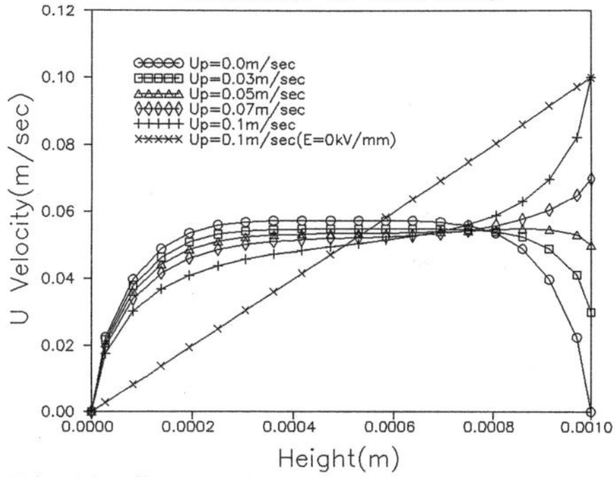

Fig.11. Comparison of velocity profiles at X=64.7mm when the upper wall is moving (Up=upper wall velocity)

References

1. Nicholas K.Petek, *An Electronically Controlled Shock Absorber Using Electrorheological Fluid.*, February 24-28, 1992, SAE TECHNICAL PAPER SERIES 920275
2. Theodove G. Duclos, *An Externally Tunable Hydraulic Mount which uses Electro-Rheological Fluid.*, SAE Noise and Vibration Conference and Display, April 28-30, 1987.
3. Robert William Phillips, *Engineering Applications of Fluids with a Variable Yield Stress.*, Ph.D. Thesis, University of California, 1969.
4. K.C.Wang, R.McLay and G.F.Carey, *ER Fluid Modeling.*, Proceedings of the second international conference on ER Fluids, August 7-9,1989.
5. Suhas V. Patankar, *Numerical Heat Transfer and Fluid Flow.*, McGRAW-HILL BOOK COMPANY. 1980.
6. N.Patel and D.B.Ingham, *Mixed Convection Flow of a Bingham Plastic in an Eccentric Annulus.*, Int. J. Heat and Fluid Flow, Vol. 15, No.2, April 1994.